Monograph No. 5
Museum of Natural History, The University of Kansas
ix + 640 pp., November, 1976

Editor: Richard F. Johnston

Lawrence · Kansas

Cover illustration by Frances Stiles

ISBN: 0-89338-001-6

PRINTED BY
THE UNIVERSITY OF KANSAS PRINTING SERVICE

Selected Readings in Mammalogy

Selected from the original literature
and introduced with comments by

J. Knox Jones, Jr.
Texas Tech University, Lubbock

Sydney Anderson
The American Museum of Natural History, New York City

AND

Robert S. Hoffmann
The University of Kansas, Lawrence

Monograph No. 5
Museum of Natural History
The University of Kansas
1976

PREFACE

This anthology is intended as an introduction to the study of mammals, principally for those who already have some biological background and who want to know the general scope of the field of mammalogy. The subdisciplines or specialties of mammalogy, its relationship to other biological fields, and specific examples of the type of work done by mammalogists are here introduced by means of a selection of complete papers in their original form. In addition to serving those already committed to the study of mammals, we hope this volume will help college students looking forward to graduate work in biology to obtain a realistic general view of mammalogy as a possible specialty. Also, beginning graduate students in related disciplines such as ornithology, mammalian physiology, or ecology, or undergraduate majors in wildlife management, may find their perspectives broadened by perusal of the present selection of papers and the introductory commentaries.

The published literature on the scientific study of mammals, which, broadly speaking, comprises the field of mammalogy, includes about 115,000 separate papers, and new papers are being published at the rate of 5000 to 6000 each year, the actual number depending on where one draws the borders of the discipline. Precise borders do not exist. Mammalogy, like other scientific fields, draws from, and contributes to, various areas of human knowledge. Our selection of the 65 papers here reproduced was influenced by: (1) our concept of the scope of mammalogy and of a reasonable and representative balance of its parts at this time; (2) our desire to illustrate various ways in which a wide variety of information can be presented in published form; and (3) our awareness that most of our readers will be English-speaking Americans, which led us to use articles published in English and selected predominantly from American sources. Nevertheless, we judge that the broad sweep of concepts and methods portrayed is relevant to students of mammalogy in all parts of the world. We have selected short papers, in general less than 20 pages in length, in preference to either longer papers or excerpts therefrom—chiefly because of space, but also because we want the serious student, who may later contribute to the literature himself, to see each published work in its entirety as one tangible contribution to knowledge. He can then grasp its concept, its methodology, its organization, its presentation, its conclusions, and perhaps even its limitations. On the latter score we would note that, although we think the papers selected are worthy contributions and make the points we wish to emphasize, we do not pretend to have selected the finest papers ever published. We could have used a somewhat different selection to serve much the same purpose, and we are sure others would use a different selection to represent their views of mammalogy. Limitations imposed by the format of this volume made it impossible to include selections from sources with a considerably larger page size, for example from recent volumes of SCIENCE.

Although many college and university libraries have some or most of the journals and other sources from which papers were selected, we decided an anthology was warranted for those who want an overview of the field, who may not know where to find the relevant literature, or who want the convenience of a collection of separate papers. Many undergraduate students are largely unaware of the existence of, or the nature of, the technical literature of science, although their textbooks are replete with terminal citations. Hopefully, many of our readers will be provoked to go to the library to seek out additional information on mammals, once they learn the interest and value of subjects treated in the pages of the JOURNAL OF MAMMALOGY and other scientific sources.

In the six years since publication of our original anthology, READINGS IN MAMMALOGY (1970), which is now out of print, the total literature in the discipline has increased by approximately 30 per cent. Moreover, many new concepts and techniques have been developed. Rather than a simple reprinting, therefore, we felt a revised and expanded version of the anthology was timely. Some of the papers originally reproduced have been retained and 29 new selections have been added. We are grateful for suggestions received from many persons and also acknowledge editors, publishers, and living authors for permission to include works in SELECTED READINGS IN MAMMALOGY.

J. KNOX JONES, JR.
SYDNEY ANDERSON
ROBERT S. HOFFMANN

CONTENTS

Introduction 3

Section 1—Systematics

Grinnell, J.
The museum conscience. Museum Work, 4:62-63, 1922 9

Thomas, O.
Suggestions for the nomenclature of the cranial length
measurements and of the cheek-teeth of mammals. Proc.
Biol. Soc. Washington, 18:191-196, 1905 11

Lidicker, W. Z., Jr.
The nature of subspecies boundaries in a desert rodent and its
implications for subspecies taxonomy. Syst. Zool., 11:160-171, 1962 ... 17

Glass, B. P., and R. J. Baker
Vespertilio subulatus Say, 1823: proposed suppression under the
plenary powers (Mammalia, Chiroptera). . . . Bull. Zool.
Nomenclature, 22:204-205, 1965 29

Allen, J. A.
Two important papers on North-American mammals. Amer.
Nat., 35:221-224, 1901 31

Merriam, C. H.
Descriptions of two new species and one new subspecies of
grasshopper mouse, with a diagnosis of the genus Onychomys,
and a synopsis of the species and subspecies. N. Amer.
Fauna, 2:1-5, pl. 1, 1889 35

Handley, C. O., Jr.
Descriptions of new bats (*Choeroniscus* and *Rhinophylla*) from
Colombia. Proc. Biol. Soc. Washington, 79:83-88, 1966 42

Benson, S. B.
The status of Reithrodontomys montanus (Baird). J.
Mamm., 16:139-142, 1935 48

Genoways, H. H., and J. K. Jones, Jr.
Systematics of southern banner-tailed kangaroo rats of the
Dipodomys phillipsii group. J. Mamm., 52:265-287, 1971 52

Forman, G. L., R. J. Baker, and J. D. Gerber
Comments on the systematic status of vampire bats (family
Desmodontidae). Syst. Zool., 17:417-425, 1968 75

Bowers, J. H., R. J. Baker, and M. H. Smith
Chromosomal, electrophoretic, and breeding studies of selected
populations of deer mice (*Peromyscus maniculatus*) and black-
eared mice (*P. melanotis*). Evolution, 27:378-386, 1973 84

Genoways, H. H., and J. R. Choate
A multivariate analysis of systematic relationships among
populations of the short-tailed shrew (genus *Blarina*) in
Nebraska. Syst. Zool., 21:106-116, 1972 93

Section 2—Anatomy and Physiology

Hooper, E. T.
The glans penis in *Sigmodon, Sigmomys,* and *Reithrodon*
(Rodentia, Cricetinae). Occas. Papers Mus. Zool.,
Univ. Michigan, 625:1-11, 1962 107

Hughes, R. L.
Comparative morphology of spermatozoa from five marsupial
families. Australian J. Zool., 13:533-543, pl. 1, 1965 118

Evans, W. E., and P. F. A. Maderson
Mechanisms of sound production in delphinid cetaceans: a
review and some anatomical considerations. Amer.
Zool., 13:1205-1213, 1973 130

Noback, C. R.
Morphology and phylogeny of hair. Ann. New York Acad. Sci.,
53:476-492, 1951 139

Hildebrand, M.
Motions of the running cheetah and horse. J. Mamm.,
40:481-495, 1959 156

Rabb, G. B.
Toxic salivary glands in the primitive insectivore *Solenodon.*
Nat. Hist. Misc., Chicago Acad. Sci., 170:1-3, 1959 171

Heller, H. C., and T. Poulson
Altitudinal zonation of chipmunks (Eutamias): adaptations to
aridity and high temperature. Amer. Midland Nat.,
87:296-313, 1972 174

Pearson, O. P.
The oxygen consumption and bioenergetics of harvest
mice. Physiol. Zoöl., 33:152-160, 1960 192

Bartholomew, G. A., and R. E. MacMillen
Oxygen consumption, estivation, and hibernation in the kangaroo
mouse, Microdipodops pallidus. Physiol. Zoöl., 34:177-183, 1961 201

Scholander, P. F., and W. E. Schevill
Counter-current vascular heat exchange in the fins of
whales. J. Applied Physiol., 8:279-282, 1955 208

Section 3—Reproduction and Development

Millar, J. S.
Evolution of litter-size in the pika, *Ochotona princeps*
(Richardson). Evolution, 27:134-143, 1973 215

Sharman, G. B.
The effects of suckling on normal and delayed cycles of reproduction in the red kangaroo. Z. Säugetierk., 30:10-20, 1965 225

Bronson, F. H.
Rodent pheromones. Biol. Reproduction, 4:344-357, 1971 236

Wright, P. L., and M. W. Coulter
Reproduction and growth in Maine fishers. J. Wildlife
Mgt., 31:70-87, 1967 250

CONAWAY, C. H.
Ecological adaptation and mammalian reproduction.
Biol. Reproduction, 4:239-247, 1971 _____ 268

CLARK, T. W.
Early growth, development, and behavior of the Richardson
ground squirrel (Spermophilus richardsoni elegans). Amer.
Midland Nat., 83:197-205, 1970 _____ 277

ALLEN, J. A.
Cranial variations in Neotoma micropus due to growth and
individual differentiation. Bull. Amer. Mus. Nat. Hist.,
6:233-246, pl. 4, 1894 _____ 286

LINZEY, D. W., and A. V. LINZEY
Maturational and seasonal molts in the golden mouse,
Ochrotomys nuttalli. J. Mamm., 48:236-241, 1967 _____ 301

SECTION 4—ECOLOGY AND BEHAVIOR

CAUGHLEY, G.
Mortality patterns in mammals. Ecology, 47:906-918, 1966 _____ 309

TAST, J., and O. KALELA
Comparisons between rodent cycles and plant production in Finnish
Lapland. Suomalaisen Tiedeakatemian Toimituksia (Ann. Acad.
Sci. Fenn.). ser. A, IV Biol., 186:1-14, 1971 _____ 322

BURT, W. H.
Territoriality and home range concepts as applied to
mammals. J. Mamm., 24:346-352, 1943 _____ 336

WILLIAMS, T. C., L. C. IRELAND, and J. M. WILLIAMS
High altitude flights of the free-tailed bat, Tadarida brasiliensis,
observed with radar. J. Mamm., 54:807-821, 1973 _____ 343

CONGDON, J.
Effect of habitat quality on distribution of three sympatric species
of desert rodents. J. Mamm., 55:659-662, 1974 _____ 358

DAVIS, D. E., and J. J. CHRISTIAN
Changes in Norway rat populations induced by introduction
of rats. J. Wildlife Mgt., 20:378-383, 1956 _____ 362

PEARSON, O. P.
A traffic survey of Microtus-Reithrodontomys runways.
J. Mamm., 40:169-180, 1959 _____ 368

ESTES, R. D., and J. GODDARD
Prey selection and hunting behavior of the African wild
dog. J. Wildlife Mgt., 31:52-70, 1967 _____ 380

LAYNE, J. N.
Homing behavior of chipmunks in central New York.
J. Mamm., 38:519-520, 1957 _____ 399

SUTHERS, R. A.
Comparative echolocation by fishing bats. J. Mamm.,
48:79-87, 1967 _____ 401

ROBERTS, M. W., and J. L. WOLFE
   Social influences on susceptibility to predation in cotton rats.
   J. Mamm., 55:869-872, 1974 _____ 410

McCARLEY, H.
   Ethological isolation in the cenospecies *Peromyscus*
   leucopus. Evolution, 18:331-332, 1964 _____ 414

GRODZINSKI, W.
   Influence of food upon the diurnal activity of small rodents.
   Pp. 134-140, *in* Symposium Theriologicum, Czech. Acad.
   Sci., Prague, 1962 _____ 416

BARLOW, G. W.
   Galapagos sea lions are paternal. Evolution, 28:476-478, 1974 _____ 422

SAMARAS, W. F.
   Reproductive behavior of the gray whale *Eschrichtius robustus*,
   in Baja California. Bull. Southern California Acad. Sci.,
   73:57-64, 1974 _____ 425

RONGSTAD, O. J., and J. R. TESTER
   Behavior and maternal relations of young snowshoe hares.
   J. Wildlife Mgt., 35:338-346, 1971 _____ 433

SECTION 5—PALEONTOLOGY AND EVOLUTION

REED, C. A.
   The generic allocation of the hominid species *habilis* as a problem in
   systematics. South African J. Sci., 63:3-5, 1967 _____ 445

ZAKRZEWSKI, R. J.
   Fossil Ondatrini from western North America. J. Mamm.,
   55:284-292, 1974 _____ 448

WILSON, R. W.
   Type localities of Cope's Cretaceous mammals. Proc. South
   Dakota Acad. Sci., 44:88-90, 1965 _____ 457

RADINSKY, L. B.
   The adaptive radiation of the phenacodontid condylarths and the
   origin of the Perissodactyla. Evolution, 20:408-417, 1966 _____ 460

WOOD, A. E.
   Grades and clades among rodents. Evolution, 19:115-130, 1965 _____ 470

NADLER, C. F., R. S. HOFFMANN, and K. R. GREER
   Chromosomal divergence during evolution of ground squirrel
   populations (Rodentia: *Spermophilus*). Syst., Zool.,
   20:298-305, 1971 _____ 486

GUTHRIE, R. D.
   Variability in characters undergoing rapid evolution, an analysis
   of *Microtus* molars. Evolution, 19:214-233, 1965 _____ 494

JANSKY, L.
   Evolutionary adaptations of temperature regulation in
   mammals. Z. Säugetierk., 32:167-172, 1967 _____ 514

Heithaus, E. R., P. A. Opler, and H. G. Baker
  Bat activity and pollination of *Bauhinia pauletia:* plant-pollinator
  coevolution. Ecology, 55:412-419, 1974 ⟶ 520

Smith, M. H., R. W. Blessing, J. L. Carmon, and J. B. Gentry
  Coat color and survival of displaced wild and laboratory reared
  old-field mice. Acta Theriol., 14:1-9, 1969 ⟶ 528

Section 6—Zoogeography and Faunal Studies

Kurtén, B.
  Notes on some Pleistocene mammal migrations from the
  Palaearctic to the Nearctic. Eiszeitalter u. Gegenwart,
  14:96-103, 1963 ⟶ 539

Guilday, J. E.
  Pleistocene zoogeography of the lemming, *Dicrostonyx.*
  Evolution, 17:194-197, 1963 ⟶ 547

Koopman, K. F., and P. S. Martin
  Subfossil mammals from the Gómez Farías region and the tropical
  gradient of eastern Mexico. J. Mamm., 40:1-12, 1959 ⟶ 551

Brown, J. H.
  Mammals on mountaintops: nonequilibrium insular biogeography.
  Amer. Nat., 105:467-478, 1971 ⟶ 563

Findley, J. S., and S. Anderson
  Zoogeography of the montane mammals of Colorado.
  J. Mamm., 37:80-82, 1956 ⟶ 575

Hansen, R. M., and G. D. Bear
  Comparison of pocket gophers from alpine, subalpine and shrub-
  grassland habitats. J. Mamm., 45:638-640, 1964 ⟶ 578

Wilson, D. E.
  Bat faunas: a trophic comparison. Syst. Zool., 22:14-29, 1973 ⟶ 581

Hagmeier, E. M.
  A numerical analysis of the distributional patterns of North
  American mammals. II. Re-evaluation of the provinces.
  Syst. Zool., 15:279-299, 1966 ⟶ 597

Wilson, J. W., III
  Analytical zoogeography of North American mammals.
  Evolution, 28:124-140, 1974 ⟶ 618

Literature Cited ⟶ 635

SELECTED READINGS IN MAMMALOGY

INTRODUCTION

The overall unity of the different fields of science and of other aspects of human experience, or at least their interdependence, is evident in both theory and practice. Nevertheless, this is an age of specialization. The sheer volume of information, the current rate of increase in knowledge, the changing and often elaborate techniques that must be learned, and human limitations all have contributed to the production of specialties.

A definition of the specialty of mammalogy as "all scientific study of mammals" is too broad, for that definition encompasses, for example, all parts of animal physiology in which any mammal, such as a white rat in the laboratory, may happen to be used. It also would include much of medical practice, because humans are mammals. Generally speaking, those scientists who call themselves mammalogists are interested in the mammal as an animal—as an organism—not just as a specific case of some more general phenomenon, be it the nature of life or the nature of the nerve impulse. For example, a physiologist who is interested in comparative studies between different mammals or in the function of a physiological process as an adaptive mechanism may regard himself as a mammalologist. A physiologist who studies one kind of laboratory animal and is interested in explaining a process in terms of progressively simpler mechanisms rarely will regard himself as a mammalogist. Both types of study, of course, contribute to biological knowledge.

Life is best comprehended in terms of four basic concepts: first, that biology, as all of science, is monistic, assuming one universe in which the same natural laws apply to living and non-living things; second, that life is a dynamic and self-perpetuating process; third, that the patterns of life change with time; finally, that these factors together have resulted in a diversity of living forms.

Different branches of biology tend to focus or concentrate on different concepts. Thus, the above four concepts are focal points, respectively, of (1) physiology and biochemistry, (2) ecology, (3) evolutionary biology, and (4) systematic biology. Mammalogy is the study of one systematic group or taxon, the Class Mammalia. Studies emphasizing different aspects of mammalian biology are evident in our section headings and in the specific papers reproduced. Biology as a whole and mammalogy specifically may be best likened to a woven fabric rather than to a series of compartments.

We feel that a unifying conceptual scheme for "mammalogy" lies in the realm of "systematic mammalogy." This scheme is unifying because it includes the basis for subsequent study and the only meaningful framework for the synthesis of existing knowledge of mammals. On this point, George Gaylord Simpson, in introducing his classic *The Principles of Classification and a Classification of Mammals* (1945) wrote (Simpson at that time used the term "taxonomy" as we use "systematics"):

> "Taxonomy is at the same time the most elementary and the most inclusive part of zoology, most elementary because animals cannot be discussed or treated in a scientific way until some taxonomy has been achieved, and most inclusive because taxonomy in its various guises and

3

branches eventually gathers together, utilizes, summarizes, and implements everything that is known about animals, whether morphological, physiological, psychological, or ecological."

Knowledge of the identity of any animal studied is essential so that the results may be compared with other knowledge about the same kind of animal and with the same kind of knowledge about different animals.

We originally had hoped to develop the history of mammalogy along with our other objectives, but when the hard fact of page limitation was faced, some selections whose chief justification was historical were sacrificed. In the comments beginning each section, some historical information helps place the selections in an understandable framework. To attain variety we have included papers both of restricted and of general scope; for example, papers pertaining to local faunas and continental faunas, to higher classification and infraspecific variation, and to contemporary serum proteins and millions of years of evolution are reproduced here.

Every serious student of mammalogy, whether amateur or professional, researcher or compiler, writer or reviewer, artist or teacher, must learn to use the literature. One does not learn all about mammals because that is impossible. One learns what one can, where to look for further information, and, more important, how to evaluate what one finds.

Most of the literature on mammals is in technical journals, a few of which are devoted exclusively to mammalogy: JOURNAL OF MAMMALOGY (USA), MAMMALIA (France), ZEITSCHRIFT FÜR SÄUGETIERKUNDE (Germany), SÄUGE-TIERKUNDLICHE MITTEILUNGEN (Germany), LUTRA (Benelux countries), LYNX (Czechoslovakia), ACTA THERIOLOGICA (Poland), MAMMAL REVIEW (Great Britain), THERIOLOGY (USSR), THE JOURNAL OF THE MAMMALOGICAL SOCIETY OF JAPAN, and AUSTRALIAN MAMMALOGY. Also there is the specialized FOLIA PRIMATOLOGICA, an international journal of primatology, founded in 1963, and a number of serial publications such as JOURNAL OF WILDLIFE MANAGEMENT, EAST AFRICAN WILDLIFE JOURNAL, BULLETIN OF THE WILDLIFE DISEASE ASSOCIATION, SOUTHWESTERN NATURALIST, and journals issued by various game departments and conservation agencies that may deal in large part, but not exclusively, with mammals. However, much of the published information on mammals, as on most aspects of biology, is widely scattered. About 40 journals include 50 per cent of the current literature, but to cover 70 per cent, at least 150 journals must be consulted. Articles in the JOURNAL OF MAMMALOGY (now approximately 900 pages each year) comprise only about three per cent of all current titles on mammals, for example.

Some categories of literature other than journals are books, symposia, transactions of various meetings or groups such as the Transactions of the North American Wildlife and Natural Resources Conference (the 40th was issued in 1975), yearbooks such as the International Zoo Yearbook (the 15th was published in 1975), newsletters such as the Laboratory Primate Newsletter, Carnivore Genetics Newsletter, Mammalian Chromosome Newsletter, or Bat Research News, major revisions or compilations of special subjects, bibliographies, and abstracts. The chief bibliographic sources for mammalogists are the JOURNAL OF MAMMALOGY, through its lists of Recent Literature,

Säugetierkundliche Mitteilungen, through its "Schriftenschau" section, the Zoological Record, published by the Zoological Society of London, Biological Abstracts, and the quarterly Wildlife Review that is issued by the U.S. Fish and Wildlife Service (along with the three collections of Wildlife Abstracts—a misnomer because only citations are included—compiled therefrom and published in 1954, 1957, and 1964); one especially useful bibliography to older papers on North American mammals is that compiled by Gill and Coues (*in* Coues and Allen, 1877). Some individuals and institutions maintain records in the form of card files, or collections of separates, or both, over many years for special subjects, special geographic areas, or other more general purposes. It is important for the student to remember that large-scale faunal reports, catalogues, revisionary works, and the like often are valuable as bibliographic sources as well as sources of other information. Some of these reports are mentioned in the introductory remarks to several sections.

An individual who delves into the literature on a particular subject usually begins with one or more pertinent recent works and proceeds backward in time by consulting publications cited in the later works or found in other bibliographic sources.

An amazing amount of published information on a given subject frequently is available to the person willing to look for it. However, paradoxically, there is often no published record for what one might suppose to be nearly common knowledge. The questioning mind must return to nature when the literature holds no answer, exactly what the authors of papers reproduced in this anthology have done.

A few decades ago only a small number of American colleges and universities offered a formal course in mammalogy, and only since about 1950 have such courses been widely offered. It is not surprising, therefore, that only three textbooks, Cockrum's *Introduction to Mammalogy* (1962), *Principles in Mammalogy* by Davis and Golley (1963), and *Mammalogy* by Vaughan (1972) have been published in English. Some instructors use general works such as *Recent Mammals of the World, A Synopsis of Families* (edited by Anderson and Jones, 1967) or *Mammals of the World*, a three-volume work by Walker *et al.* (1964 and subsequent editions) as texts or as references along with other suggested readings. *A Manual of Mammalogy* by DeBlase and Martin (1974) is another useful reference. Accounts of the mammals of certain states or regions also may be used as texts by persons in those places. Other general works of reference value are Bourlière's *Natural History of Mammals* (1954), Young's *The Life of Mammals* (1957), Crandall's *Management of Wild Mammals in Captivity* (1964), the fascicles on mammals in the *Traité de Zoologie* (edited by Grassé, 1955 and later), and four volumes on mammals in the *Encyclopedia of Animal Life* (edited by Grzimek, 1972 and later). Two classic general works less readily available are *An Introduction to the Study of Mammals, Living and Extinct* by Flower and Lydekker (1891) and *Mammalia* by Beddard (1902) in the Cambridge Natural History series.

Compact field guides to the mammals of a few parts of the world are available, such as those of Burt and Grossenheider (1964), Palmer (1954), and Anthony (1928) for parts of North America, Van den Brink (1967) for Europe,

Prater (1965) for India, Flint *et al.* (1965) for the Soviet Union, and Dorst and Dandelot (1970) for the larger mammals of Africa. Mörzer-Bruyns' (1970) field guide to cetaceans is world-wide in scope.

The dates in the two preceding paragraphs suggest the recent expansion in the volume of work in mammalogy. Another such measure is membership in The American Society of Mammalogists, which grew from 252 in 1919 to almost 4000 in 1976. Interested persons are invited to apply for membership in this society, members of which receive the JOURNAL OF MAMMALOGY.

Human medicine, veterinary medicine, animal husbandry, and animal physiology (including much work with a comparatively few species of mammals in the laboratory), all preceded mammalogy as separate disciplines dealing with mammals. Many of the first mammalogists (as defined here) trained themselves in one of these disciplines and some also practiced in fields other than mammalogy. C. Hart Merriam, who founded the U.S. Biological Survey, studied medicine, as did E. A. Mearns, who wrote on mammals of the Mexican boundary (1907). Harrison Allen wrote much of his first review of North American bats (1864) while on furloughs from duty as a surgeon in the Union army during the Civil War. Mammalogists continue to interact with specialists in the above-mentioned fields to their mutual benefit.

Another largely separate but partly overlapping field that flowered slightly later than mammalogy is genetics. We have included no papers on mammalian genetics as such, although the relevance of genetics is evident in some of our selections. The book on *Comparative Genetics of Coat Color in Mammals* by Searle (1968) contains about 800 references, including some fascinating works on species other than the oft-studied mouse (*Mus musculus*).

Our six groupings of papers are somewhat arbitrary. Ecology is as closely allied to physiology or zoogeography as to behavior, and anatomy could as well have been placed with development as with physiology. The present arrangement as to the sequence of sections and their contents seems to be about as convenient as any other, and that is the extent of our expectations. We imply no hierarchy of subdisciplines.

In selecting works to be included here, we have, in addition to the considerations already noted, sought papers in which different kinds of mammals were compared, and in which different approaches, styles, and methods of presentation were used. Individual papers often pertain to more than one area of study. In fact, we favored papers that illustrated the relevance of different disciplines and methods of study to each other. Perhaps the reader will be able to appreciate our moments of anguish as the final selections were made for this anthology.

Our introduction for each section is brief. We hope that our comments aid the reader in considering (1) some historical aspects that make the papers more meaningful, (2) the major areas of study and some major concepts that the papers illustrate, (3) the existence of related literature, to which we can only call attention by citing a few examples, and (4) the continuous transfer of ideas, methods, and results from one worker to another, from one field of science to another, and between science and other fields of human endeavor.

SECTION 1—SYSTEMATICS

A sound classification provides the necessary framework upon which other information about mammals can be ordered. In order to classify organisms, it first is necessary to assess their similarities and differences; in other words, structures and their functions need to be observed, described, and compared, and taxa need to be recognized and named before a useful and meaningful classification can be constructed. The field of study relating to classification frequently is called "taxonomy," although the broader term "systematics" is also widely used and is preferred by us.

The goal of scientific nomenclature, one aspect of systematics, is to assure that each kind of organism has a unique name, and only one name. The International Code of Zoological Nomenclature (latest edition, Stoll, 1964) forms the accepted framework for dealing with nomenclature, both past and present. The presently reprinted paper by Glass and Baker points up one kind of nomenclatorial problem faced by the systematist (see also Bull. Zool. Nomenclature, 22:339-340, 1966, and Glass and Baker, 1968, for further commentary on this same problem). The Code is administered by the International Commission on Zoological Nomenclature but, as Blair (1968) pointed out, the Commission "has no way of enforcing its decisions, and the burden of holding names to conformity with the [Code] falls on the individual worker and on editors of scientific publications."

Prior to the first decade or so of the 20th century, mammalian systematics generally was based on a "hit-or-miss" typological approach, which, although it fostered considerable advancement in cataloguing the faunas of the world, was limited in perspective and potential. The development of evolutionary thought and the spectacular growth of genetics have led to the "new systematics," the biological species concept of today, as discussed in detail in such syntheses as Huxley (1943), Mayr et al. (1953), Simpson (1961), and Mayr (1963 and 1969), among others. Blackwelder's (1967) and Ross's (1974) texts on systematics also are deserving of mention here, as are treatises on special taxonomic methods such as Sneath and Sokal (1973) on phenetics and numerical taxonomy, and Hennig (1966) on cladistics.

However, to infer that all early taxonomic treatments of mammals were either inconsequential or poorly conceived would be a gross error. Pallas' (1778) revision of rodents, for example, was a monumental work far advanced for its day, as were many other outstanding contributions by 18th and 19th century mammalogists. Nevertheless, one has only to compare the descriptions and accounts of Pallas with those found in papers reprinted here by Merriam, Handley, and Genoways and Jones to appreciate the tremendous revolution in systematic practice. The short contribution by J. A. Allen not only provides an example of a review, but deals in some detail with two substantial revisionary works published at the turn of the century. Among the larger modern revisionary studies that might be recommended to the student are those of Osgood (1909), Jackson (1928), Hooper (1952), Pearson (1958), Lidicker (1960), Packard (1960), Musser (1968), Choate (1970), Smith (1972), Birney (1973), and Genoways (1973); the last of these is especially noteworthy for its completeness in coverage of a mammalian genus. Eller-

7

man's (1940, 1941, 1948) well organized review of living rodents and Hill's (1953 and subsequent volumes) somewhat more rambling and less critical coverage of the primates also are noteworthy attempts to summarize selected bodies of knowledge of important groups of mammals.

Development of technologically advanced means of collecting, preparing, and storing specimens has resulted in the accumulation of series of individuals of the same species (the invention of the break-back mouse trap and Sherman live trap might be mentioned here along with the relatively recent widespread use of mist nets and specialized traps for capturing bats). This in turn allowed for assessment of variations within and between populations. Sophisticated studies of intergradation, hybridization, and speciation are examples resulting from technological and conceptual advances in this area, and have been greatly enhanced by development of multivariate statistics and computerized programs for data analysis.

The emphasis in this section is mostly at the level of species and subspecies (for example, the papers by Benson and by Lidicker). Higher categories are dealt with primarily in Section 5, although the contribution included here by Forman *et al.* deals with systematics at the familial level. Attempts over the years by systematists to standardize techniques and definitions are illustrated in the paper by Thomas. The short essay by Grinnell also bears on this point.

Our final four selections illustrate the application of these techniques and concepts to specific systematic problems, all of which were clarified in ways that might have been impossible otherwise. Those by Genoways and Jones and by Forman *et al.* have been mentioned previously. The treatment by Bowers *et al.* of the problem of the relationship between two species of *Peromyscus* employs data from karyological and electrophoretic studies as well as breeding experiments. Genoways and Choate have used multivariate analysis to elucidate relationships among populations of the short-tailed shrew, including identification of hybrid individuals.

Many papers in other sections of this anthology touch on systematics in one way or another, and the usefulness to the taxonomist of information from a variety of sources will be immediately evident to the reader. Two journals devoted to the concepts and practices of systematics and in which contributions to mammalogy regularly appear are SYSTEMATIC ZOOLOGY and ZEITSCHRIFT FÜR ZOOLOGISCHE SYSTEMATIK.

The Museum Conscience

THE scientific museum, the kind of museum with which my remarks here have chiefly to do, is a storehouse of facts, arranged accessibly and supported by the written records and labeled specimens to which they pertain. The purpose of a scientific museum is realized whenever some group of its contained facts is drawn upon for studies leading to publication. The investment of human energy in the formation and maintenance of a research museum is justified only in proportion to the amount of real knowledge which is derived from its materials and given to the world.

All this may seem to be innocuous platitude—but it is genuine gospel, never-the-less, worth pondering from time to time by each and every museum administrator. It serves now as a background for my further comments.

For worthy investigation based upon museum materials it is absolutely essential that such materials have been handled with careful regard for accuracy and order. To secure accuracy and order must, then, once the mere safe preservation of the collections of which he is in charge have been attended to, be the immediate aim of the curator.

Order is the key both to the accessibility of materials and to the appreciation of such facts and inferences as these materials afford. An arrange-
ment according to some definite plan of grouping has to do with whole collections, with categories of specimens within each collection, with specimens within each general category, with the card indexes, and even with the placement of the data on the label attached to each specimen. Simplicity and clearness are fundamental to any scheme of arrangement adopted. Nothing can be more disheartening to a research student, except absolute chaos, than a complicated "system," in the invidious sense of the word, carried out to the absolute limits recommended by some so-called "efficiency expert." However, error in this direction is rare compared with the opposite extreme, namely, little or no order at all.

To secure a really practicable scheme of arrangement takes the best thought and much experimentation on the part of the keenest museum curator. Once he has selected or devised his scheme, his work is not *done*, moreover, until this scheme is in operation throughout all the materials in his charge. Any fact, specimen, or record left out of order is lost. It had, perhaps, better not exist, for it is taking space somewhere; and space is the chief cost initially and currently in any museum.

The second essential in the care of scientific materials is *accuracy*. Every item on the label of each specimen, every item of the general record in the accession catalog, must be precise as to fact. Many errors in published literature, now practically impossible to "head off," are traceable to mistakes on labels. Label-writing having to do with scientific materials is not a chore to be handed over casually to a "25-cent-an-hour" girl, or even to the ordinary clerk. To do this essential work correctly requires an exceptional genius plus training. The important habit of reading every item back to copy is a thing that has to be acquired through diligent attention to this very point. By no means *any* person that happens to be around is capable of doing such work with reliable results

Now it happens that there is scarcely an institution in the country bearing the name museum, even though its main purpose be the quite distinct function of exhibition and popular education, that does not lay more or less claim to housing "scientific collections." Yet such a claim is false, *unless* an adequate effort has been expended both to label accurately and to arrange systematically all of the collections housed. Only when this has been done can the collections be called *in truth scientific.*

My appeal is, then, to every museum director and to every curator responsible for the proper use as well as the safe preservation of natural history specimens. Many species of vertebrate animals are disappearing; some are gone already. All that the investigator of the future will have, to indicate the nature of such then extinct species, will be the remains of these species preserved more or less faithfully, along with the data accompanying them, in the museums of the country.

I have definite grounds for presenting this appeal at this time and in this place. My visits to the various larger museums have left me with the unpleasant and very distinct conviction that a large portion of the vertebrate collections in this country, perhaps 90 per cent of them, are in far from satisfactory condition with respect to the matters here emphasized. It is admittedly somewhat difficult for the older museums to modify systems of installation adopted at an early period. But this is no valid argument against necessary modification, which should begin at once with all the means available—the need for which should, indeed, be emphasized above the making of new collections or the undertaking of new expeditions. The older materials are immensely valuable historically, often irreplaceable. Scientific interests at large demand special attention to these materials.

The urgent need, right now, in every museum, is for that special type of curator who has ingrained within him the instinct to devise and put into operation the best arrangement of his materials—who will be alert to see and to hunt out errors and instantly make corrections—who has the *museum conscience.*

March 29, 1921.

VOL. XVIII, PP. 191-196   SEPTEMBER 2, 1905

PROCEEDINGS

OF THE

BIOLOGICAL SOCIETY OF WASHINGTON

SUGGESTIONS FOR THE NOMENCLATURE OF THE
CRANIAL LENGTH MEASUREMENTS AND OF
THE CHEEK-TEETH OF MAMMALS.

BY OLDFIELD THOMAS.

Although various reasons prevent the general success of such
a wholesale revolution in scientific terms as is described in
Wilder and Gage's Anatomical Technology (1882), where the
many arguments in favor of accurate nomenclature are admirably put forth, yet in various corners of science improvements
can be suggested which, if the workers are willing and in touch
with each other, may be a real help in reducing the inconvenience
of the loose or clumsy terminology commonly in vogue.

Two such suggestions, due largely to the instigation of Mr.
Gerrit S. Miller, Jr., form the subject of the present paper.

I. LENGTH MEASUREMENTS OF THE SKULL AND PALATE.

In giving the length measurement of the skull, not only do
different authors at present use different measurements in describing the skulls of similar or related animals, but in doing so
they designate these measurements by terms of which it is often
difficult or impossible to make out the exact meaning. Such a
name as " basal length " has I believe been used by one person
or another for almost every one of the measurements to be here-

after defined, and readers are expected to know by heart everything that the user has ever written on the subject, footnotes
and all, in order to understand what is meant by the particular
term employed. Such a state of things has many inconveniences,
and it is hoped the present communication, if it meets with the
approval of other workers on the subject, may do a little toward
putting an end to the existing confusion.

As long ago as 1894,* by agreeing with Dr. Nehring for the
definition of the terms basal and basilar in our own future writ-

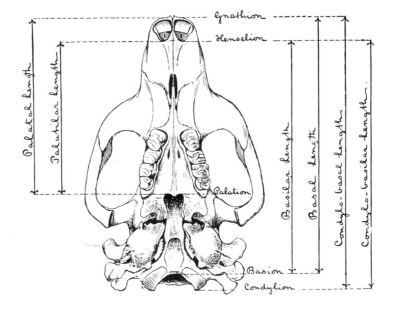

ings, I made a first step in this direction, and the present is
an amplification of the principle then adopted.

All the difficulty has arisen from the fact that both at the
anterior and the posterior ends of the skull there are two measurement points, so that there are four different ways in which
the basal length of the skull may be taken, and under that
name some authors have adopted nearly every one of them.

It is clear that if a definite name be given to each one of the
four measurements, authors, by using these names, will be en-
abled to give the measurements they fancy without causing confusion in the minds of their readers as to their exact meaning.

*Ann. & Mag. Nat. Hist., Ser. 6, XIII, p. 203.

The different points are:

Anteriorly: 1. THE GNATHION, the most anterior point of the premaxillæ, on or near the middle line.

           2. THE HENSELION, the back of the alveolus of either of the median incisors, the point used and defined by Prof. Hensel in his craniological work.

Posteriorly: 3. THE BASION, a point in the middle line of the hinder edge of the basioccipital margin of the foramen magnum.

           4. THE CONDYLION, the most posterior point of the articular surface of either condyle.

A fifth measuring point to be referred to below is the PALATION, the most anterior point of the hinder edge of the bony palate, whether in the middle line or on either side of a median spine.

Now using these words for.the purposes of definition, I would propose, as shown in the diagram, the following names for the four measurements that may be taken between the points above defined:—

1. BASAL LENGTH, the distance from Basion to Gnathion.
2. BASILAR LENGTH, the distance from Basion to Henselion.
3. CONDYLO-BASAL LENGTH, the distance from Condylion to Gnathion.
4. CONDYLO-BASILAR LENGTH, the distance from Condylion to Henselion.

     In addition there may be:

5. GREATEST LENGTH, to be taken not further divergent from the middle line than either condylion. A long diagonal to a projecting bulla or paroccipital process would thus be barred. If however the words "between uprights" be added the measurement would be between two vertical planes pressed respectively against the anterior and posterior ends of the skull at right angles to its middle line.
6. UPPER LENGTH, from tip of nasals to hinder edge of occipital ridge in middle line.

The difference between the words basal and basilar, which at first seemed trivial and indistinctive, is founded on the use of the English word basal by the older writers, such as Flower and others, who used the measurement from the gnathion; while basilar is an adaptation of the German of Hensel and his school, who used the "*basilar-länge*" from the henselion. These names again, combined with condylo-, readily express the points which are used by those who like to adopt the condylion as a posterior measuring point.

But further, the association of the ending "al" with a measurement from the gnathion, and "ilar" with one from the henselion, if once defined and fixed, may be utilized in a second case of similar character.

The length of the bony palate is a measurement given by all careful describers, but the anterior measuring point used is again either the gnathion or henselion, doubt as to which is being used often nullifying the value of the measurement altogether.*
To avoid this doubt I would suggest, exactly as in the other case, that the name of the measurement from the gnathion should end in "al" and that from the henselion in "ilar." We should then have:

PALATAL LENGTH, the distance from gnathion to palation.

PALATILAR LENGTH, the distance from henselion to palation.

The indeterminate "palate length" would then be dropped altogether.

II. The Names of the Cheek-teeth of Mammals.

Although the cheek-teeth of mammals, the molars and premolars, have been studied and written about ever since the birth of zoology, no uniform system of naming them has been evolved and there is the greatest divergence between the usage of different workers on the subject. In old days all were called molars or grinders; then the premolars were distinguished from the true molars (although French zoologists, Winge in Denmark, and Ameghino in Argentina, continued to use a continuous notation for the two sets of teeth combined) and the usual habit among zoologists in general was to speak of them individually as "second premolar," "third molar," and so on. Even here, however, an important difference cropped up owing to Hensel

---

* I may explain that in my own descriptions the palate of any given animal has always been measured from the same anterior point, gnathion or henselion, as the skull itself, this latter being indicated by the use of the words basal or basilar.

and his school in Germany numbering the premolars from behind forwards, while naturalists of other nations counted from
before backwards, as with the incisors and molars, a difference
often productive of fatal confusion.

Of late years, however, partly owing to an increasing concensus
of opinion that the seven cheek-teeth of Placentals, four premolars and three molars, are serially and individually homologous with the seven of Marsupials, formerly reckoned as three
premolars and four molars, many naturalists have again begun
to think that a continuous numeration might be the best one.

But the difficulties in the way of its adoption are very great,
largely owing to the absence of any convenient and suitable word
in English less clumsy than "cheek-tooth," to express a tooth
of the combined premolar and molar series. To speak of the
"first cheek-tooth" or of the "predecessor to the fourth cheek-
tooth" would be so retrogressive a step that I am sure no
one would adopt it. But if instead of trying to find a word
for the series combined with a numeral to show the position,
we were to have a name for each tooth, we should get something of the immense convenience we have all realized in having
definite names for the canine and the carnassial teeth, the latter
name being found of value in spite of the fact that the upper
and lower carnassials are not homologous with each other. Such
names might be made from the positions of the teeth if their
meanings were not so obtrusive as to confuse the minds of persons who do not readily understand how a tooth should be called
"the second" or "secundus" when it is actually the most anterior of the series.

Now it fortunately happens that while the Latin terms "primus," "secundus," etc., express the serial positions too clearly
for the convenience of weak minds, Latinized Greek terms have
just about the right amount of unfamiliarity which would enable
them to be used as names without their serial origin being too
much insisted on. Moreover, their construction is similar to
the process we all use in making generic names, and so far as I
know they have never been previously utilized in zoology.

Then, after Latinizing the Greek ordinal terms πρωτος, etc.
for the cheek-teeth of the upper jaw, the same modification as
is already used in cusp nomenclature might be adopted for those
of the mandible.

We should thus have, counting from before backwards:

		UPPER JAW.	LOWER JAW.
Cheek-tooth	1	Protus	Protid
"	2	Deuterus	Deuterid
"	3	Tritus	Tritid
"	4	Tetartus	Tetartid
"	5	Pemptus	Pemptid
"	6	Hectus	Hectid
"	7	Hebdomus	Hebdomid

To avoid any doubt, I would expressly allocate these names to the permanent teeth of placentals, leaving the names of the marsupial teeth to be settled in accordance with their placental homologies.

For the milk teeth a further modification would be available by prefixing the syllable Pro- to the names of the respective permanent teeth. We could thus for example in the case of a third lower milk premolar call it the protritid, and so use one word instead of four.

Of course I have no supposition that this system would ever be frequently or generally used, but I am convinced that in many special cases, and particularly in such descriptions and catalogues of isolated teeth as paleontologists often have to give, it might result in considerable convenience and saving of space.

The Nature of Subspecies Boundaries in a Desert Rodent and its Implications for Subspecies Taxonomy

WILLIAM Z. LIDICKER, JR.

IT SEEMS to me that the wide diversity of opinion which exists concerning the usefulness of trinomial nomenclature revolves in large measure on the more basic issue of whether or not it is possible to recognize infraspecific categories which reflect genetic relationships. As recently pointed out by Sneath (1961), taxonomic categories which are not based on relationships are thereby rankless and cannot logically be included in a taxonomic hierarchy. Thus if the subspecies category is used merely as an instrument for describing geographic variation in a few characters or as a device for cataloging geographic variants (as is done by many taxonomists), artificial classifications of convenience are characteristically produced. Such convenience classifications usually contain rankless groups (the "false taxa" of Sneath) which cannot be allocated in the taxonomic hierarchy. This is simply because categories based on a few arbitrary characters are themselves arbitrary, and lead to the objection of Brown and Wilson (1954) and others that trinomials based on one group of characters need not bear any relation to those based on different traits. Many of the same philosophical difficulties apply to systems such as that recently proposed by Edwards (1954) and Pimentel (1959) in which the subspecies would become a measure of isolation, by restricting its use to completely isolated and "obviously different" populations.

If on the other hand studies of infraspecific populations are focused on discovering evolutionary diversity or degrees of relationship between the various populations, I see no philosophical objection to the use of the trinomen. The question then reduces to one of the feasibility and/or desirability of searching for such relationships, and of deciding what level of dissimilarity if any justifies use of the formal trinomen. It is primarily these two subsidiary questions which are examined in this paper, with the frank hope that the subspecies can be rescued from the rankless limbo of the morph, ecotype, and form. If this rescue operation should prove successful, attention can then be directed to other problems of greater biological interest, such as whether or not determinations of genetic relationships within a species, which are based on phenotypes, can serve as a basis for speculations on phylogeny. Obviously geographic relations would have to be considered at this level, but, assuming that such information is taken into account, it would be highly informative to contrast phenetic and phylogenetic subspecies classifications. In any case, analyses of infraspecific relationships would very likely provide valuable clues concerning the environmental forces which have influenced the development of the existing evolutionary diversity.

In a previous paper (1960) I attempted to determine the genetic relationships among populations within a species of kangaroo rat (*Dipodomys merriami* Mearns, 1890) by a careful analysis of 20 morphological features. I concluded at that time that at least in well-known terrestrial species an attempt to recognize relative relationships within a species was at least possible. And, at the same time it was apparent that (besides the philosophical objections already pointed out) the subspecies category by itself was completely inadequate for describing the complex geographic variation occurring in that species. It is the raw data from this former investigation that I have used here to test further the reliability of those tentative conclusions.

The search for relationships among populations of the same species implies a search for total genetic differentiation (or at least its phenotypic manifestations), and hence of lineages with partially independent evolutionary origins such that they have some internal homogeneity and their own adaptive tendencies. To detect this kind of differentiation it seems important to analyze, among other things, the populations occurring at the boundaries between differentiating groups, just as in the analysis of species relationships it is the boundaries between them, or areas of sympatry, where the most significant information on relationships is to be found. This is not to say that information concerning the regions of greatest divergence or adaptive peaks (in this case peaklets) of infraspecific populations is not important, but only that such data should not be the only source material for taxonomic judgments. Thus it is the intent of this paper to focus attention on the previously all but ignored subspecies boundaries, and to examine the nature of these areas in *Dipodomys merriami* as I had previously and without the benefit of this analysis defined them (Lidicker, 1960). Because the determination of these intraspecific units was guided in this case by a desire to find populations of comparable evolutionary relationship, careful scrutiny of the intergrading zones between them and surrounding areas should be of particular interest. Comparisons will also be made with levels of differentiation in areas in which no subspecies boundary was recognized, as well as with one region in which species level differentiation was postulated to have been reached by an island isolate.

The second and related purpose of the paper is to describe a method which helps to accomplish the first objective by measuring total differentiation, or lack of similarity, in many diverse characters, and hence is proposed as a criterion of relationship. But at the same time the technique does not require the hard working taxonomist to have either access to a digital computer or facility with matrix algebra.

The Method

Most quantitative techniques available to the systematist, which concern themselves with determining relationships, and hence with similarities as well as differences, either involve the analysis of qualitative or discontinuous characters and thus are most useful at the species or genus level (e.g., Michener and Sokal, 1957; Lysenko and Sneath, 1959), or involve calculations sufficiently complex (e.g., Williams and Lance, 1958) that they are avoided by most practicing systematists. What seems to be needed is an additional technique which is sufficiently adaptable to handle continuously variable, as well as discontinuous, characters of diverse types (and so is useful in infraspecific studies), and which at the same time is sufficiently practical that it will be widely useful. To this end the following proposed method is dedicated. It is not intended as a substitute for discriminant function analysis (Fisher, 1936; Jolicoeur, 1959) and related methods which attempt to discriminate between previously conceptualized populations by using combinations of variables.

An analysis of relationship should ideally compare relative similarities and not differences. However, since the number of similarities between populations within a species is very large, it is much easier to measure their differences and consider that the reciprocal of the amount of difference represents a measure of similarity. Thus as the amount of difference approaches zero, the reciprocal approaches infinity. The problem then becomes one of summing the amount of difference in many diverse characters. To do this we must be able to express the differentiation for each trait by a pure number (no units). Cain and Harrison (1958), for example, accomplished this by dividing the differences which they observed between means by the maximum value recorded for each character. The resulting ratios, which they called "reduced values," express the observed differences in terms of a fraction of the maximum size of each character. Although this permits the comparison of diversity among traits of different absolute size, it does not take into account either the possibility that various characters may have different variabilities, or the statistical significance of the observed mean differences. Furthermore, maximum size would seem to be a statistic of dubious biological importance in continuously varying characters. On the other hand, all of these important variables, the variance of each trait, character magnitude, as well as a consideration of whether or not mean differences have a high probability of representing real differences, are taken into account by expressing differentiation as a proportion between the observed differences between samples and the maximum amount of difference expected on the basis of chance sampling variation. Only mean differences greater than that amount which may be due to chance would then be considered as real differences. For our purposes the maximum chance variation expected in any comparison can be equated to the minimum difference required for statistical significance (at any given confidence level). This minimum significant difference (msd) can be calculated in a number of ways. One possibility is to determine the standard error of the mean for each character for each sample. Then in comparing two samples for this character, $2\,SE_{\bar{x}_1} + 2SE_{\bar{x}_2} = msd$. This provides a conservative estimate of msd with confidence limits usually well in excess of

95%. For large studies, however, these calculations would be extremely laborious, as well as perhaps overly conservative, and a short-cut is suggested.

If we can assume that each quantitative character in a given species exhibits a characteristic variability throughout its range, then calculations would be tremendously reduced if we were able to determine the expected or pooled standard deviation (s_p) and standard error ($s_{p_{\bar{x}}}$) of samples for which say $n \geqslant 20$. Very small samples would have to be grouped with adjacent samples whenever possible, or if necessary either ignored or have separate msd-values calculated for them. Under these conditions $4s_{p_{\bar{x}}}$ represents our best estimate of msd. Unfortunately confidence limits cannot be calculated for its reliability, although again it is generally a conservative estimate. The statistic s_p can be conveniently determined by averaging the weighted variances for several samples of adequate size (Hald, 1952: 395). Note that as the estimate of s_p improves it approaches the population standard deviation (σ), and hence is applicable to a wider range of sample sizes. Better estimates of s_p require knowledge of the total number of individuals in each of the populations sampled (see Cochran, 1959: 72), an obvious impossibility in this type of problem. In the examples given in this paper $4s_{p_{\bar{x}}}$ was estimated by using the standard deviation of one large sample collected near the center of the species' range, and by assuming $n = 20$. This expediency seemed justified because of the close similarity in values of s calculated for a given trait among several samples, and because of the likelihood that s approaches σ under these circumstances.

Still another method of deriving the statistic msd, but one not used in this report, involves a more laborious, but statistically more precise, procedure. The confidence limits for the difference between any pair of means can be calculated (see Dixon and Massey, 1957:128) whether or not we assume that we know the variance characteristic of each trait (s_p^2) or use only the pertinent sample variances (s_1^2 and s_2^2). For a large study, the calculations are very much reduced if one can estimate s_p (see above), and perhaps even use only samples in which $n \gtrless 20$. If these simplifications are possible, a pair of confidence limits can be computed which will be characteristic for each trait studied. In either case, one confidence limit gives us our *msd,* since mean differences greater than this can with a known probability be considered real. We need not be concerned with the possibility that the mean differences are even larger than those observed.

Consider then only those characters in which the differences in the mean values ($\overline{x}_1 - \overline{x}_2$) for a given pair of locations (samples) are greater than the minimum significant differences. Now, divide these significant differences in mean values by the minimum significant difference characteristic for that trait (or for that pair of samples). This procedure gives us our pure number which can be designated as $d_1, d_2, \ldots \ldots d_n$ for successive characters, each representing a measure of differentiation in one character between one pair of samples. Having defined the amount of differentiation in each character in terms of a pure number, we can now add these to arrive at an estimate of total differentiation in the characters studied (Σd_i). In interpreting this statistic in any real situation, however, it seems apparent that the distance between the two samples compared should be taken into account. Obviously an amount of total differentiation exhibited between two samples which are close together geographically would be more significant than the same amount of differentiation between samples geographically distant. To compensate for this effect of distance, I have divided the total differentiation by the distance (in miles) between the two samples. The resulting figure, which I have called D or the Index of Differen-

tiation,[1] represents the proportion of significant change that occurs between the two locations per mile. Then the reciprocal of D easily gives us our measure of similarity between populations. D need now only be further divided by the total number of characters studied, including those of course in which no differentiation occurred, to arrive at the mean character differentiation per mile ($MCD/mi.$).

The importance of considering distance between samples will depend in large measure on the specific problem under investigation. Obviously air-line distance between samples does not always accurately reflect the real magnitude of the distance or barriers between them. I feel that this is not a serious difficulty, however, since we are concerned with the abruptness of differentiation between adjacent populations and not with barriers *per se.* Moreover, in some ways D acts as a measure of restriction on gene flow, because, if distance is kept constant, D will tend to increase as gene flow is reduced. Another potential difficulty with the distance calculation is that it carries the assumption that if the two localities being compared were actually closer together, the amount of total differentiation shown would be less. This is not always true because not only are there sometimes large areas which exhibit very little geographic variation, but also there exist unavoidable gaps in specimen collections. For these reasons I felt that in the present analysis of *D. merriami* it was necessary to consider both Σd_i and D in assessing differentiation.

One further complication seems worth considering. This concerns variation in the direction of change between different characters. It seemed to me more significant if one or two characters were found to change significantly in a direction opposite to that of the other characters, than if they all changed in the same di-

[1] Note that this is in no way similar to the "differentiation index" of Kurtén (1958) which compares growth gradients.

rection. Thus for each such direction change, I arbitrarily added one half the d-value for that specific character to Σd. This also serves to oppose any tendency to give too much weight to characters which may not be entirely independent in their variation, or to those varying allometrically. Otherwise no allowance has been made for differentially weighting characters which might be considered to have greater phylogenetic importance. Presumably this could be readily done, if there were some sound basis for making such judgments. Sneath (1961), however, points out some of the dangers inherent in attempts to do this, and argues for considering each character equally.

Table 1 summarizes the calculations for Σd, D, and $MCD/mi.$ for one pair of localities in southeastern Arizona. Note that a value of 3.476 has been included in Σd for color changes. Ordinarily color characters should be quantified so that they can easily be added into this scheme. Unfortunately in my study, I did not quantify in numerical terms the six color features analyzed. This necessitated my determining when significant changes had occurred by reference to the color descriptions of each sample. Whenever important color changes were found be-

tween samples, I included in Σd for each such change a figure which represented the average d-value for all pairs of localities in the boundary region under study which exhibited the same number of color changes as the sample pair being calculated. For example, if two samples differed in three color features and if the average d for all pairs of samples in that region which also differed by three color features was 1.50, then 4.5 (3 x 1.50) would represent the combined value of d for the three color traits. Although this represents an unfortunate complication, it should not detract from the validity of the overall method being proposed.

Figures 1, 2, and 3 show the differentiation observed in 20 characters in *D. merriami* in selected portions of its large range. It is important to emphasize that these 20 characters were chosen in the original investigation (Lidicker, 1960) independently of the conclusions of other authors concerning what they considered important characters in distinguishing subspecies. The list thus includes not only most of the "taxonomically important" characters of other authors but numerous additional features as well. I chose for illustration regions which demonstrate various levels of differentiation

TABLE 1—CALCULATIONS FOR TOTAL DIFFERENTIATION, THE INDEX OF DIFFERENTIATION, AND THE MEAN CHARACTER DIFFERENTIATION PER MILE IN *Dipodomys merriami* FOR A PAIR OF LOCALITIES IN SOUTHEASTERN ARIZONA (VICINITIES OF THE HUACHUCA AND SANTA RITA MOUNTAINS 54 MILES APART).

CHARACTER *	$\bar{x}_1-\bar{x}_2$ (mm.)	msd	d_i
hind foot length	1.40	0.68	2.059
ear length	0.62	0.52	1.192
basal length of the skull	0.54	0.52	1.039
cranial length	0.78	0.48	1.625
rostral width	0.19	0.06	3.167
1 direction change (ear)	($\frac{1}{2}d_2$)		0.596
2 color changes	($2\bar{x}_d$ for those pairs of localities with two color changes)		3.476

$$\Sigma d_i = 13.154$$
$$D = 0.244$$
$$MCD/mi. = 0.012 **$$

* See Lidicker (1960) for a description of these characters.
** Total of 20 characters studied.

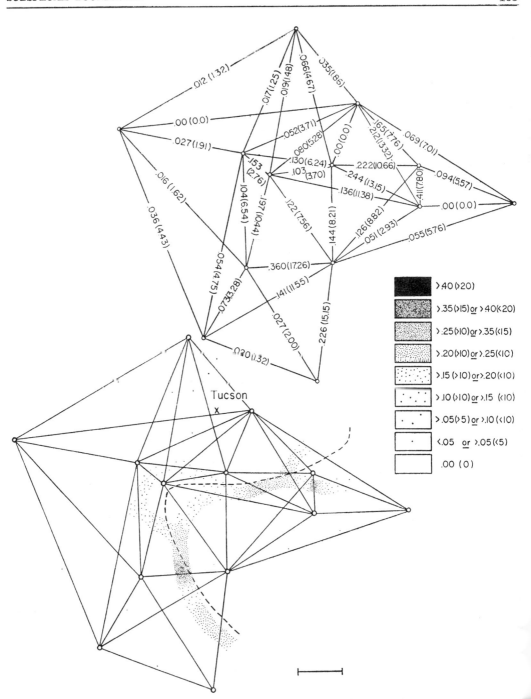

FIG. 1. Observed differentiation of *Dipodomys merriami* in southeastern Arizona and adjacent Mexico and New Mexico. Numbers on the lines connecting the various sample localities represent the calculated values for *D* and in parenthesis Σd_i. The key to the intensity of stippling is based on these same statistics. The scale associated with each map represents a distance of 25 miles. See also the text for a more complete explanation of the figures.

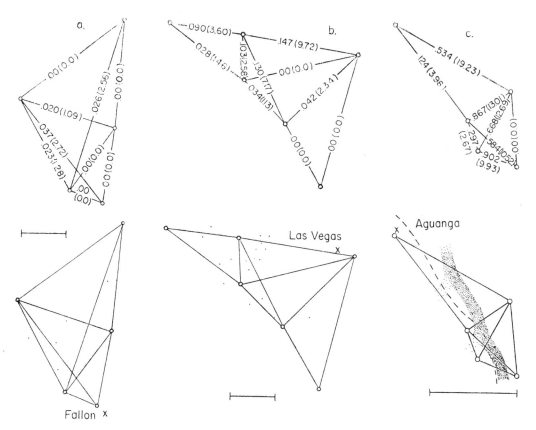

Fig. 2. Observed differentiation of *D. merriami* in a) northern Nevada, b) southern Nevada and adjacent Mojave desert of California, and c) small area in extreme southern California. For a more complete explanation of the figures and a key to stippling intensity, see the text and Figure 1.

ranging from essentially none to that judged to be at the species level. Figure 1 shows the boundary region between *D. merriami merriami* and *D. m. olivaceus* (nomenclature based on Lidicker, 1960) in southeastern Arizona and adjacent Mexico and New Mexico. Figure 2 illustrates areas in northern Nevada (a), southern Nevada and the adjacent eastern Mojave desert of California (b), and finally a small area in southern California at the boundary of *D. m. collinus* and *D. m. arenivagus* (c). Figure 3 represents the southern tip of the Baja California peninsula (a), and southern Sonora where the boundary between *D. m. merriami* and *D. m. mayensis* is found (b).

The first of these (3a) is of particular interest as it shows the entire range of *D. m. melanurus* and the adjacent island populations of *D. m. margaritae* and the presumed allopatric species *D. insularis*. Notice that the key takes into account both *D* and Σd (but gives greatest weight to *D*) and is arranged so that increased intensity of stippling represents increased differentiation. Heavy dashed lines represent the locations of previously established subspecies boundaries, and double dashed lines previously established species boundaries (see Lidicker, 1960). Each drawing also indicates the location of one prominent town so that each chart can be placed geographically; all are oriented with north upward.

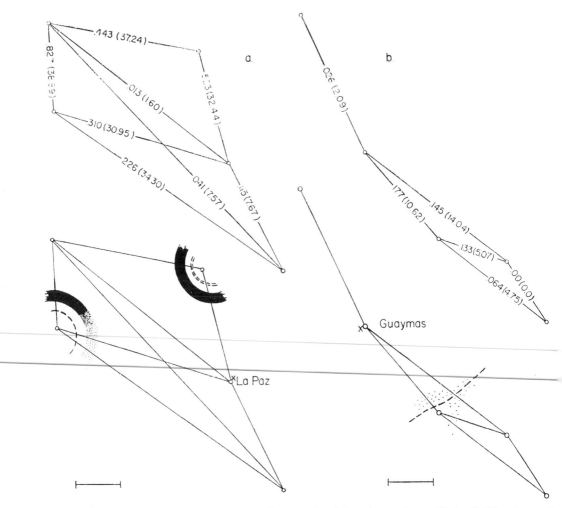

FIG. 3. Observed differentiation of *D. merriami* in a) southern Baja California, and b) southern Sonora. For a more complete explanation of the figures and a key to stippling intensity, see the text and Figure 1.

Discussion of the Method

The method described and its pictorial representation as shown in the figures gives us a geographically oriented summary of statistically significant differentiation in the characters studied. Its most important feature is that it takes into account the variability of each character as well as its magnitude, and concerns itself only with diversity which has a high probability of being real. Clearly, the more characters examined by the investigator, the greater will be his chance of discovering all of the existing differences between populations, and the better will be his estimate of genetic diversity. In the present case there is a remarkable correlation between the subspecies and species boundaries as previously described by the author and the bands of rapid character changes as defined by the Index of Differentiation.[2] It is clear that in this case subspecies boundaries uni-

[2] No particular correlation is evident, however, with many of the taxonomic conclusions of previous authors.

formly appear as relatively narrow zones of high levels of differentiation or low levels of similarity, and, although it cannot be determined from the figures, these are usually, but not always, in areas of partial or complete isolation between populations. If the Index method truly describes genetic diversity, then our confidence is bolstered in the possibility of using the subspecies category for characterizing infraspecific lineages.

Besides the degree of differentiation, other suggested criteria for the recognition of such lineages include the following considerations: 1) the continuity of the zone of differentiation; 2) diversity of the two postulated adaptive peaks; 3) differences in the environments to which the adjacent populations are adapted, or consideration of the possibility that the two populations are adapted to the same environment in a different way; 4) geologic or paleontologic evidence of separate evolution. Moreover, it would seem to be a simple matter to devise modifications of the Index of Differentiation so as to incorporate discontinuous and qualitative characters. This would extend the usefulness of the method, not only to infraspecific populations which differ by such characters, but also to the species level. However, above the infraspecific level the problems of convergence, giving different weight to different characters (that is identifying primitive or generalized characters), and correlated characters (see especially discussions by Cain and Harrison, 1960) are aggravated. It might be added in passing, however, that these sources of error are not so great a problem as might be expected, because the proposed method emphasizes large numbers of characters and overall similarities. Under these circumstances a few convergent or pleiotropic traits would alter the results very little. Moreover, the problem of differentially weighting characters often leads into circular arguments as pointed out by Sneath (1961).

Although the proposed method incorporates a number of compromises with mathematical sophistication, I think that it is sufficiently accurate to be of considerable utility to the practicing taxonomist. Furthermore, several modifications are suggested for improving precision if this seems appropriate. The method will not of course make any decisions for the investigator, as it should not, but it will give him additional objective criteria on which to base his decisions. The fact that the conclusions suggested by the calculations and analysis of D's are similar if not identical to those proposed without the benefit of the method suggests that the method does not produce unreasonable results, and therefore must not suffer unduly from its lack of statistical elegance.

Discussion of Results

An obvious, but important, conclusion derived from a study of the figures is that statistically significant differences can be found between the vast majority of the population pairs. This serves to emphasize what is really intuitively obvious, namely that the ability to prove that two populations are statistically different in one or several characters is only a measure of the persistence and patience of the systematist. To base formal subspecific descriptions on this kind of evidence seems to me to be almost meaningless as well as a contribution to the degradation of the subspecies category to the extent of losing it as a legitimate member of the taxonomic hierarchy. Furthermore, this is precisely the philosophy which usually seems to nurture the widespread emphasis on naming with its often accompanying neglect of relationships, which has stimulated so much critical comment (see for example Wilson and Brown, 1953, and Gosline, 1954).

The description of differentiation provided in the figures carries the further implication that all levels of differentiation are found in *D. merriami*, and no obvious dividing line between subspecies and non-subspecies, and species for that matter, is thereby indicated. The method thus gives us information regarding how different (or similar) populations are, but does not tell us which ones we should call subspecies. This finding is consistent with current concepts of intraspecific variation, and permits the systematist to decide what degree of relationship has phylogenetic significance for the particular organism involved, and finally what level, if any, he wants to recognize with formal subspecies descriptions. In the present example subspecific boundaries are found to be usually associated with continuous bands of differentiation characterized by D-values greater than 0.15.

The fact that this study has failed to reveal some biologically meaningful division marking the subspecies level does not mean of course that some such division will not be possible in the future. However, such a line of demarcation is obviously not a prerequisite to the success of the proposed method, which only concerns recognition of degrees of evolution. Nevertheless, one possible criterion for such a division which occurs to me is the relationship between the observed gene flow between two adjacent populations and that amount expected on the basis of the extent of physical contact existing between them. If the observed gene flow turned out to be less than that expected, or discriminating in terms of what genes were allowed to flow, this would serve as an indication that partially independent lineages were involved. This idea would not diminish in any way the obvious importance of geographical barriers in inhibiting gene flow, but merely suggests that some day it may be possible to ask the question—would a high level of differentiation persist between two geographically partially isolated populations if the barrier were reduced or eliminated? Or to put it in another way, how much reduction in the physical barrier between them can these two populations resist before gene flow becomes free flowing? This genetic concept of a subspecies argues that there are numerous infraspecific populations which by virtue of their past isolation (not necessarily complete) show some inhibition of gene flow between them and their neighbors, which would tend to slow down the dedifferentiation process. If the geographic isolation is current, the argument must be stated that such a reduction in gene flow would occur if they were not so isolated. This reasoning is merely a corollary of the fact that not all attempts by a species for isolation and differentiation result in species formation. There are a number of reasons why gene flow might be inhibited in such cases, and one of these is interdeme genetic homeostasis (Lerner, 1954). Other factors might be partial ecological or behavioral barriers to free interbreeding.

Although this suggestion for a biologically meaningful subspecies criterion is mainly speculative, it seems to me to be one possible direction that future developments in intraspecific analysis might take. The following definition of a subspecies is thus perhaps premature, but is offered because it is only a slight modification of widely used current definitions, but yet incorporates the concept outlined above; at the same time it does not commit one to any specific criteria for the recognition of subspecies. A subspecies is a relatively homogeneous and genetically distinct portion of a species which represents a separately evolving, or recently evolved, lineage with its own evolutionary tendencies, inhabits a definite geographical area, is usually at least partially isolated, and may intergrade gradually, although over a fairly narrow zone, with adjacent subspecies. This does not say that subspecies are "incipient species." It does say that subspecies are populations which have made initial steps in the direction of species formation, such that they might form species if suitable isolating conditions should develop, or they may be populations which have not reached the species level and are dedifferentiating. Obviously most subspecies will not become species, and likewise the process of dedifferentiation may become relatively stabilized through diverse selective pressures on either side of the intergrade zone.

It seems to me then that the Index of Differentiation or some similar device can give us an often needed additional criterion for judging relationship between populations. And it is these relative relationships that are of primary interest; and if used as guide lines to the recognition of subspecies will permit the legitimate retention of this category in the taxonomic hierarchy. Such an evolutionary philosophy applied to infraspecific analysis has a number of important advantages, not least of which is that it focuses attention on the speciation process and not on geographic variation *per se,* and thus emphasizes that the steps which can lead to species divergence must be initiated long before the process is actually completed. Other advantages not already alluded to include a consistency in applying the concept of relationship to all taxa and hence justifying to some extent the nomenclatorial equivalence of species and subspecies, provision of a more uniform goal for infraspecific systematists, and greater usefulness of subspecies to non-taxonomists because of the greater nomenclatorial stability and more reliable predictability of genetic differences in unstudied traits that would result.

There is little doubt that this approach will be considered impractical in some groups of organisms, but this seems of relatively little importance to the present discussion. Whereas a technique must be usable, no limits should be placed on the conceptualization of direction and significance of inquiry. I have confidence that systematists are not so unimaginative that appropriate procedures will not rapidly follow perception of important and necessary goals, as they have already done to some extent. Present day taxonomy is fraught with practicality, but is nevertheless shaken by criticism as to where it is all leading.

Summary

A growing dissatisfaction with much of what is now subspecies taxonomy and the associated indiscriminant use of the trinomen has caused many taxonomists to re-examine the basic tenets of intraspecific analysis. This "soul searching" has raised the important questions of whether or not it is possible or even desirable to use the subspecies category as a rankable taxon below the species level in the taxonomic hierarchy and at what level of dissimilarity, if any, formal trinomial nomenclature becomes appropriate. It is argued here that if the subspecies is to be preserved from degradation to the level of the rankless morphs, ecotypes, and forms, it must be based on degrees of relationship or evolutionary divergence. Moreover, the determinations of relative genetic relationships implies an emphasis on similarities between the various subpopulations comprising a species, as well as careful scrutiny of events occurring in the boundary regions between them. This paper is therefore concerned with characterizing some of these postulated boundary areas, as well as some areas of lesser and greater amounts of differentiation, in the kangaroo rat *Dipodomys merriami.*

To accomplish this, a method is outlined which serves to sum the observed statistically significant differentiation in many diverse characters between adjacent populations. In doing this, the method takes into account the variability and magnitude of each character. The estimate of total differentiation thus obtained can then be divided by the distance between the samples being compared to give the Index of Differentiation (D). The reciprocal of this statistic can also be taken as a measure of similarity. The Index of Differentiation can be further divided by the number of characters studied to give the mean character differentiation per mile ($MCD/mi.$). The system involves no complicated mathematical procedures, and yet contains only minor compromises with statistical sophistication. Furthermore it is readily adapted to visual portrayal and analysis.

The results of this analysis demonstrate a very close agreement between levels of differentiation as determined by the Index of Differentiation and the taxonomic conclusions previously arrived at, when an attempt was made to base subspecies on the relative relationships among infraspecific populations. Under these conditions subspecies boundaries are uniformly characterized by a high level of differentiation which occurs over a relatively narrow zone, and is usually but not always associated with partial or complete isolation between populations. Moreover the analysis has emphasized the nearly ubiquitous occurrence of statistically significant differences between populations, and hence of the futility of basing formal subspecies on this kind of evidence. And finally a continuum of levels of differentiation was found, ranging from none at all to the species level.

It is concluded from this evidence that it is indeed possible to gather evidence on the relative relationships of the various portions of a species, and it is suggested that data of this sort should form the foundation for subspecific diagnosis. This approach tends to focus attention on the speciation process itself instead of on geographic variation *per se*. Various other advantages of this system are pointed out, and speculation is presented concerning the possible determination of a biologically meaningful division between subspecies and lesser categories.

Acknowledgments

I am greatly indebted to the following individuals who have critically read this manuscript, but who do not necessarily share the views which I have expressed: S. B. Benson, N. K. Johnson, O. P. Pearson, F. J. Sonleitner, and C. S. Thaeler. The figures were prepared by G. M. Christman of the Museum of Vertebrate Zoology.

REFERENCES

BROWN, W. L., JR., and E. O. WILSON. 1954. The case against the trinomen. System. Zool., 3:174–176.

CAIN, A. J., and G. A. HARRISON. 1958. An analysis of the taxonomist's judgment of affinity. Proc. Zool. Soc. London, 131:85–98.

——— 1960. Phyletic weighting. Proc. Zool. Soc. London, 135:1–31.

COCHRAN, W. G. 1959. Sampling techniques. John Wiley, New York.

DIXON, W. J., and F. J. MASSEY, JR. 1957. Introduction to statistical analysis. McGraw-Hill, New York.

EDWARDS, J. G. 1954. A new approach to infraspecific categories. System. Zool., 3:1–20.

FISHER, R. A. 1936. The use of multiple measurements in taxonomic problems. Ann. Eugenics, 7:179–188.

GOSLINE, W. A. 1954. Further thoughts on subspecies and trinomials. System. Zool., 3:92–94.

HALD, A. 1952. Statistical theory with engineering applications. John Wiley, New York.

JOLICOEUR, P. 1959. Multivariate geographical variation in the wolf, *Canis lupus* L. Evolution, 13:283–299.

KURTÉN, B. 1958. A differentiation index, and a new measure of evolutionary rates. Evolution, 12:146–157.

LERNER, I. M. 1954. Genetic homeostasis. Oliver and Boyd, London.

LIDICKER, W. Z., JR. 1960. An analysis of intraspecific variation in the kangaroo rat *Dipodomys merriami*. Univ. California Publs. Zool., 67:125–218.

LYSENKO, O., and P. H. A. SNEATH. 1959. The use of models in bacterial classification. Jour. Gen. Microbiol., 20:284–290.

MEARNS, E. A. 1890. Description of supposed new species and subspecies of mammals, from Arizona. Bull. Amer. Mus. Natur. Hist., 2:277–307.

MICHENER, C. D., and R. R. SOKAL. 1957. A quantitative approach to a problem in classification. Evolution, 11:130–162.

PIMENTEL, R. A. 1959. Mendelian infraspecific divergence levels and their analysis. System. Zool., 8:139–159.

SNEATH, P. H. A. 1961. Recent developments in theoretical and quantitative taxonomy. System. Zool., 10:118–139.

WILLIAMS, W. T., and G. N. LANCE. 1958. Automatic subdivision of associated populations. Nature, 182:1755.

WILSON, E. O., and W. L. BROWN, JR. 1953. The subspecies concept and its taxonomic application. System. Zool., 2:97–111.

WILLIAM Z. LIDICKER, JR. is Assistant Curator of Mammals at the Museum of Vertebrate Zoology and Assistant Professor in the Department of Zoology at the University of California, Berkeley.

VESPERTILIO SUBULATUS SAY, 1823: PROPOSED SUPPRESSION
UNDER THE PLENARY POWERS (MAMMALIA, CHIROPTERA).
Z.N.(S.) 1701

By Bryan P. Glass and Robert J. Baker (*Department of Zoology,
Oklahoma State University, Stillwater, Oklahoma, U.S.A.*)

1. The purpose of this application is to request the International Commission on Zoological Nomenclature to use its plenary powers to suppress the specific name *subulatus* Say, 1823, as published in the combination *Vespertilio subulatus* (James' account of Long's Expedition from Pittsburgh to the Rocky Mountains, **2** : 65), and thus to ensure that the specific name *Myotis yumanensis* H. Allen, 1864 (*Smithsonian Musc. Coll.* **7** (Publ. 165) : 58) is conserved.

2. In 1823 Say collected a specimen of a species of *Myotis* near the 104th meridan on the Arkansas River, and described it in his notes, using the species name *Vespertilio subulatus*. His description was published verbatim as a footnote in James' account of the expedition. Say did not state that the specimen was preserved; however, as far as is known, all of his natural history collections were deposited in the Philadelphia Museum (Peale's Museum) which was later destroyed by fire.

Pertinent parts of Say's description read as follows:
". . . flew rapidly in various directions, over the surface of the creek. . . .
Ears longer than broad, nearly as long as the head, hairy on the basal half, a little ventricose on the anterior edge, and extending near the eye; tragus elongated, subulate; the hair above blackish at base, tip dull cinereous; the interfemoral membrane hairy at base, the hairs unicolored, and a few also scattered over its surface, and along its edge, as well as that of the brachial membrane; hair beneath black, the tip yellowish white; hind feet rather long, a few setae extending over the nails; only a minute portion of the tail protrudes beyond the membrane. Total length 2 9-10 inches. tail 1 1-5."

3. The description of Say fits *M. yumanensis*, not *M. subulatus* (of Miller and G. M. Allen, *USMN Bull.* 144, 1928, and of later authors):

M. yumanensis	*M. subulatus*
Dorsum dull cinereous	Dorsum bright chestnut
Uropatagium hairy at base	Uropatagium naked at base
Hind feet long	Hind feet short
Setae over nails	No setae over nails
Flies close to water	Flies high

4. *Myotis yumanensis* is at present the only species of *Myotis* (other than the species currently referred to as *subulatus*) known from the vicinity of Say's type locality, but the recognition of *yumanensis* in this region dates only from 1957. Other western *Myotis* possibly occurring in the vicinity may be excluded on the basis of one or more characters listed by Say: *Myotis velifer*—ears not hairy on basal half, hairs not blackish at base, size much too large; *Myotis thysanodes*—ears too long and not hairy on basal half, fringed interfemoral

Bull. zool. Nomencl., Vol. 22, Part 3. August 1965.

29

membrane, size much too large; *Myotis volans*—color brown, not dull cinereous, interfemoral membrane naked, size too large; *Myotis lucifugus*—color brown with burnished tips to hairs, not dull cinereous; *Myotis californicus*—color not dull cinereous, foot too small.

5. In 1855 Le Conte (*Proc. Acad. Nat. Sci. Philad.* : 435) applied the name *Vespertilio subulatus* Say to bats from his plantation in the tidewater country near Riceboro, Liberty County, Georgia. Miller and G. M. Allen (*USNM Bull.* 144 : 42, 1928) have indicated that Le Conte presumed that he had two species, to one of which he applied the name *M. subulatus* Say, but they presumed that all the specimens were actually *M. lucifugus*. Whatever the species actually was, it certainly was not the saxicolous species currently bearing the name *M. subulatus*, which is absent from the south eastern United States.

6. In 1864 Harrison Allen (*Smith Miscl. Coll.* No. 165 : 51) applied Say's name to the eastern form of the long-eared bat, which usage was accepted until the revision of the genus by Miller and G. M. Allen (*USNM Bull.* 144, 1928) wherein they correctly rejected *M. subulatus* for the eastern long-eared *Myotis* in favor of the name *M. keeni* Merriam 1895, which is currently accepted, but erroneously applied the name *Myotis subulatus* Say to the form currently bearing the name. Miller and Allen (*op. cit.* p. 28) based this change, in part, on their imperfect knowledge of the bats known to occur in south eastern Colorado.

7. Identification of the species that Say had in hand when he wrote his description places in jeopardy the species name *yumanensis* which has stood unchallenged for 101 years. Such a change is not in keeping with the intent of the rules to promote stability.

8. The oldest species name available for the bat currently carrying the name *M. subulatus* Say is *leibi* published as *Vespertilio leibii* Audubon and Bachman, *Jour. Acad. Nat. Sci. Philad.* (1) **8** : 124, 1842. Suppression of the name *subulatus* requires that the subspecies of this taxon be as follows:

Myotis leibi leibi Audubon and Bachman 1842, Type locality Erie County, Ohio.

Myotis leibi ciliolabrum H. Allen, 1893, Type locality Near Banner, Trego County, Kansas.

Myotis leibi melanorhinus Merriam 1890. Type locality Little Spring, North base of San Francisco Mountain, Coconino County, Arizona, Altitude 8,250 feet.

8. For the reasons listed above we now request the International Commission on Nomenclature:

(1) to use its plenary powers to suppress the specific name *subulatus* Say, 1823, used originally in the combination *Vespertilio subulatus*, for the purposes of the Law of Priority but not for those of the Law of Homonymy;

(2) to place the name *yumanensis* H. Allen, 1864, as published in the binomen *Vespertilio yumanensis* on the Official List of Specific Names in Zoology; and

(3) to place the specific name *subulatus* Say, 1823, as published in the binomen *Vespertilio subulatus*, on the Official List of Rejected and Invalid Specific Names in Zoology.

REVIEWS OF RECENT LITERATURE.

ZOÖLOGY.

Two Important Papers on North-American Mammals. — The literature relating to recent work on North-American mammals is so scattered, and the results have been the outcome of investigations by such a number of different workers, and based on such varying amounts of material, that it is a great gain when a competent authority on any given group can go over it and coördinate the efforts of his predecessors in the light of, practically, all of their material, combined with a vast amount in addition. In other words, the monographic revision of any of the larger genera of North-American mammals by an expert is a distinct advance, for which all mammalogists may well feel grateful. It is with pleasure, therefore, that we call attention to two recent contributions of this character — Mr. Vernon Bailey's "Revision of American Voles of the Genus Microtus," and Mr. W. H. Osgood's "Revision of the Pocket Mice of the Genus Perognathus."

Mr. Bailey's revision[1] of the American voles, or meadow mice, is "based on a study of between five thousand and six thousand specimens from more than eight hundred localities, including types or topotypes of every recognized species with a known type locality, and also types or topotypes of most of the species placed in synonymy." With such material at command, and with a wide experience with the animals in life, and personal knowledge of the actual conditions of environment over a large part of the range of the group, Mr. Bailey has had peculiar advantages for his work, and his results are subject to revision only at points where material is still deficient, or from some other point of view. This revision, while obviously not final, presents a new starting point for future workers, and is likely to be a standard for many long years to come.

The little animals here treated are the short-tailed field mice,

---

[1] Revision of American Voles of the Genus Microtus. By Vernon Bailey, Chief Field Naturalist, Division of Biological Survey, U. S. Department of Agriculture. Prepared under the direction of Dr. C. Hart Merriam, Chief of the Division. *North American Fauna*, No. 17, pp. 1–88, with 5 plates and 17 text-figures. Issued June 6, 1900.

familiarly typified by our common "meadow mice" of the Eastern
States.   The group is divisible into several well-marked subgenera,
formerly generally known under the generic term "Arvicola," which
has had to give way to the less known but older term "Microtus."
The group is especially distinctive of the northern hemisphere north
of the tropics, and is found throughout North America from the
mountains of Guatemala and southern Mexico northward, increasing
numerically, both in species and individuals, from the south northward till it reaches its greatest abundance in the middle and colder
temperate zones, again declining thence northward to the Arctic
coast.   They are vegetable feeders, and often do considerable damage to trees and crops ; they are active in the winter, forming long
burrows or tunnels under the snow ; they are also very prolific, breeding several times a year, young being found throughout the warmer
months.

The seventy species and subspecies recognized by Mr. Bailey are
arranged in nine subgenera ; between the extreme forms the differences are strongly marked, but the intermediate forms present gradual stages of intergradation.   The subgenus Neofiber, of Florida,
embracing the round-tailed muskrat, and the subgenus Lagurus, of
the semi-arid districts of the northwestern United States, present the
most striking contrast, not only in size but in many other features.
The former is perhaps the largest known vole, while the latter group
includes the smallest.

Mr. Bailey's paper, being a synopsis rather than a monograph,
leaves much to be desired in point of detail, but is admirable in its
way, and covers the ground with as much fullness as his prescribed
limits would permit.   Of the twenty-six synonyms cited, it is noticeable that thirteen relate to our common eastern meadow mouse, and
date from the early authors, while two other eastern species furnish
three others, also of early date.   Only six of the remaining ten are
of recent date, showing that of some fifty-five forms described within
the last ten years, by nine different authors, forty-eight meet with
Mr. Bailey's approval.   Four of the remaining seven are identified
with earlier names which for many years have been considered
indeterminable, but which Mr. Bailey claims to have established on
the basis of topotypes.

While he may be correct in these determinations, it would have
been of interest to his fellow-specialists if he had stated the basis of
his determination of certain type localities, notably those of Richardson's species, described as from the "Rocky Mountains," or similarly

vague localities. If he has some "inside history" to fall back upon, it is only fair that the secret should be made public.

It may be said further, in the way of gentle criticism, that it is hardly fair wholly to ignore such knotty points as the allocation of a few names which he omits, since they form part of the literature of the subject, as, for example, *Hypudæus ochrogaster* Wagner, *Arvicola noveboracensis* Richardson, and some of Rafinesque's names. Mr. Bailey describes as new two species and one subspecies.

Mr. Osgood's "Revision of the Pocket Mice"[1] is an equally welcome contribution, and has been prepared upon much the same lines, with equal advantages in the way of material and field experience. The pocket mice of the genus Perognathus are confined to a limited portion of North America, being found only west of the Mississippi, and ranging from the southern border of British Columbia south to the valley of Mexico. They are strictly nocturnal and live in burrows, are partial to arid regions and seem to thrive even in the most barren deserts. Their habits are hence not well known, as they are very shy and even difficult to trap. They are mouse-like in form, but only distantly related to the true rats and mice. Their most obvious character is the possession of cheek pouches which open externally.

The pocket mice vary greatly in size, form, and in the nature of their pelage, which may be either soft or hispid ; but between the wide extremes there are so many closely connecting links that it is difficult to find any sharp lines of division, although two subgenera are fairly recognizable. The whole number of forms here recognized is 52 — 31 species and 21 additional subspecies, about equally divided between the subgenera Perognathus and Chætodipus. Of these, thirteen are here for the first time described. Out of a total of 61 specific and subspecific names applied to forms of this group, 9 are relegated to synonymy. Of these 61 names, it is interesting to note that 52 date from 1889 or later, and that of these, eight prove to be synonyms, three of them having become so through the identification of older names thought ten years ago to be indeterminable, but since reëstablished on the basis of topotypes.

A previous revision of this group was made in 1889 by Dr. C. Hart Merriam, on the basis of less than two hundred specimens —

---

[1] Revision of the Pocket Mice of the Genus Perognathus. By Wilfred H. Osgood, Assistant Biologist, Biological Survey, U. S. Department of Agriculture. Prepared under the direction of Dr. C. Hart Merriam, Chief of Division of Biological Survey. *North American Fauna*, No. 18, pp. 1–72, Pls. I–IV, and 15 text-cuts. Issued Sept. 20, 1900.

all of the material then available — when the number of currently recognized forms was raised from six to twenty-one.   Dr. Merriam's work, however, cleared the way for a better conception of the group, rectifying important errors of nomenclature and making known many new forms.   Mr. Osgood, with fifteen times this amount of material, seems to have settled all of the remaining doubts regarding the application of certain early names, and, besides coördinating the work of his predecessors, has immensely extended our knowledge of the group.   The paper is admirable from every point of view and does great credit to its author.   J. A. A.

DESCRIPTIONS OF TWO NEW SPECIES AND ONE NEW SUBSPECIES OF GRASSHOPPER MOUSE,

WITH A DIAGNOSIS OF THE GENUS ONYCHOMYS, AND A SYNOPSIS OF THE SPECIES AND SUBSPECIES.

By C. HART MERRIAM, M. D.

A. DESCRIPTIONS OF NEW SPECIES AND SUBSPECIES.

ONYCHOMYS LONGIPES sp. nov.

(TEXAS GRASSHOPPER MOUSE.)

Type $\frac{3307}{3859}$ ♀ ad. Merriam Collection. Concho County, Texas, March 11, 1887. Collected by William Lloyd.

Measurements (taken in the flesh by collector).— Total length, 190^{mm}; tail, 48 [this measurement seems to be too short]; hind foot, 25; ear from crown, 13 (measured from dry skin).

General characters.—Size larger than that of the other known representatives of the genus, with larger and broader ears, and much longer hind feet. Ears less hairy than in *O. leucogaster*, with the lanuginous tuft at base less apparent; tail longer and more slender.

Color.—Above, mouse gray, sparingly mixed with black-tipped hairs, and with a narrow fulvous stripe along each side between the gray of the back and white of the belly, extending from the fore-legs to the root of the tail; under parts white.

Cranial characters.—Skull longer and narrower than that of *O. leucogaster* (particularly the rostral portion), with much longer nasals, and a distinct supraorbital " bead" running the full length of the frontals and there terminating abruptly. The nasals overreach the nasal branch of the premaxillaries about as far as in *leucogaster*. The incisive foramina, as in *O. leucogaster*, barely reach the anterior cusp of the first molar. The roof of the palate extends further behind the last molar than in *leucogaster*, and gives off a median blunt spine projecting into the pterygoid fossa. The palatal bones end anteriorly exactly on a line

1

with the interspace between the first and second molars. The presphenoid is excavated laterally to such a degree that the middle portion is reduced to a narrow bar less than one-third the width of its base. The condylar ramus is lower and more nearly horizontal than in *leucogaster*, and the angular notch is deeper. The coronoid process resembles that of *leucogaster*.

ONYCHOMYS LONGICAUDUS sp. nov.

(LONG-TAILED GRASSHOPPER MOUSE.)

Type ♂♀♂♀ ♂ ad. St. George, Utah, January 4, 1889. Collected by Vernon Bailey.

Measurements (taken in the flesh by the collector).—Total length, 145; tail, 55; hind foot, 20; ear from crown, 10 (measured from dry skin).

General characters.—Similar to *O. leucogaster*, but smaller, with longer and slenderer tail. Pelage longer, but not so dense. General color above, cinnamon-fawn, well mixed with black-tipped hairs.

Cranial characters.—Skull smaller and narrower than that of *O. leucogaster*; zygomatic arches less spreading; nasals less projecting behind nasal branch of premaxillaries. The coronoid and condylar processes of the mandible are shorter, and the coronoid notch is not so deep as in *leucogaster*. The presphenoid shows little or no lateral excavation. The incisive foramina do not quite reach the plane of the anterior cusp of the first molar. The shelf of the palate projects posteriorly considerably beyond the molars, and terminates in a nearly straight line without trace of a median spine.

ONYCHOMYS LEUCOGASTER MELANOPHRYS subsp. nov.

(BLACK-EYED GRASSHOPPER MOUSE.)

Type, ♂♀♂♀ ♂ ad. Kanab, Utah, December 22, 1888. Collected by Vernon Bailey.

Measurements (taken in the flesh by collector).—Total length, 154; tail, 41; hind foot, 21. Ear from crown 10 (measured from the dry skin).

Size of *O. leucogaster*. Ear a little smaller. Hind foot densely furred to base of toes. Color above, rich tawny cinnamon, well mixed with black-tipped hairs on the back, and brightest on the sides; a distinct black ring round the eye, broadest above. This ring is considerably broader and more conspicuous than the very narrow ring of *leucogaster*.

Cranial characters.—Skull large and broad; very similar to *O. leucogaster* in size and proportions, but with zygomatic arches less spreading posteriorly, interparietal narrower, nasals not reaching quite so far beyond the nasal branch of premaxillaries, and antorbital slit narrower. Presphenoid moderately excavated, as in *leucogaster*. The incisive foramina reach past the plane of the first cusp of the anterior molar. The condylar ramus is longer and directed more obliquely upward than in *leucogaster*, with the coronoid and infra-condylar notches deeper.

NOTE.—In order to render the preceding diagnoses of new forms more useful, the following brief descriptions of the skulls of the two revious ly known species are appended for comparison, together with figures of the skull of the type of the genus (*O. leucogaster*):

Onychomys leucogaster Max.—Skull large and broad, with zygomatic arches spreading posteriorly. Antorbital slit larger than in the other known species. Palate hort, ending posteriorly in a short median spine (see figure).

Onychomys torridus Coues.—Skull small, narrow, with zygomatic arches not spreading, and vault of cranium more rounded than in any other member of the genus. Interparietal relatively large. Nasals projecting far beyond nasal branch of premaxillary. Incisive foramina very long, extending back to second cusp of first molar. Shelf of palate produced posteriorly nearly as far as in *longicaudus*, and truncated. Presphenoid slightly excavated laterally. Mandible much as in *longicaudus*, but with coronoid process more depressed and condylar ramus more slender.

B. DIAGNOSIS OF THE GENUS ONYCHOMYS.

The striking external differences which distinguish the Missouri Grasshopper Mouse from the other White-footed Mice of America (*Hesperomys* auct.) led its discoverer, Maximilian, to place it in the genus *Hypudæus* (=*Evotomys*, Coues), and led Baird to erect for its reception a separate section or subgenus, which he named *Onychomys*. Coues, the only recent monographer of the American Mice, treats *Onychomys* as a subgenus, and gives a lengthy description of its characters. Since, however, some of the statements contained in this description are erroneous, and the conclusions absurd,* and since the most important taxonomic characters are overlooked, it becomes necessary to redefine the type. A somewhat critical study of the cranial and dental characters of *Onychomys* in comparison with the other North American White-footed Mice has compelled me to raise it to full generic rank. It may be known by the following diagnosis :

Genus ONYCHOMYS Baird, 1857.

Baird, Mammals of North America, 1857, p. 457 (*subgenus*).
Type, Hypudæus leucogaster, Max. Wied, Reise in das innere Nord Amerika, II, 1841, 99–101 (from Fort Clark, Dakota).
Hesperomys auct.

First and second upper molars large and broad; third less than half the size of the second. First upper molar with two internal and three external cusps, the anterior cusp a trefoil when young, narrow, and on a line with the outside of the tooth, leaving a distinct step on the inside. Second upper molar with two internal and two external cusps, and a narrow antero-external fold. Last upper molar subcircular in outline, smaller than in *Hesperomys*, and less indented by the lateral notches.

---

* Coues says: "Although unmistakably a true Murine, as shown by the cranial and other fundamental characters, it nevertheless deviates much from *Mus* and *Hesperomys*, and approaches the Arvicolines. Its affinities with *Evotomys* are really close." (Monographs of North American Rodentia, 1877, p. 106.) As a matter of fact, *Onychomys* has no affinities whatever with *Evotomys*, or any other member of the Arvicoline series, its departure from *Hesperomys* being in a widely different direction.

Lower molar series much broader than in *Hesperomys*. First lower molar with an anterior, two internal, and two external cusps, and a postero-internal loop. In *Hesperomys* the anterior cusp is divided, so that there are three distinct cusps on each side. Second lower molar with two internal and two external cusps, an antero-external and a postero-internal fold. Third lower molar scarcely longer than broad, sub-circular in outline, with the large posterior lobe of *Hesperomys* reduced to a slight fold of enamel, which disappears with wear.

Coronoid process of mandible well developed, rising high above the condylar ramus and directed backward in the form of a large hook (see accompanying cut). Nasals wedge-shaped, terminating posteriorly considerably behind the end of the nasal branch of the premaxillaries.

FIG. 1.   FIG. 2.

1. Lower jaw of *Onychomys leucogaster*.   2. Lower jaw of *Hesperomys leucopus*.

Body much stouter and heavier than in *Hesperomys*. Tail short, thick, and tapering to an obtuse point.

Fore feet larger than in *Hesperomys*; five-tuberculate, as usual in the Murine series. Hind feet four-tuberculate, and densely furred from heel to tubercles. Tubercles phalangeal, corresponding to the four anterior tubercles of *Hesperomys*, that is to say, the first is situated at the base of the first digit, the second at the base of the second digit, the third over the bases of the third and fourth digits together, the fourth at the base of the fifth digit. The fifth and sixth (or metatarsal) tubercles of *Hesperomys* are altogether wanting.

C. SYNOPSIS OF SPECIES AND SUBSPECIES.

(1) By External Characters.

Length, about 150^{mm}; tail, about 40; hind foot, about 21; ear from crown, 10. Color above, mouse-gray; black ring around eye inconspicuous .... *O. leucogaster*.

Size of *O. leucogaster*. Color above, rich tawny cinnamon, brightest on the sides; black ring round eye conspicuous .... *O. leucogaster melanophrys*.

Length, about 145^{mm}; tail, about 55; hind foot, 20; ear from crown, 10. Color above, cinnamon fawn .... *O. longicaudus*.

Length, about 190^{mm}; tail, about 50; hind foot, 25; ear from crown, 13. Color above, mouse-gray, with a narrow fulvous stripe along the sides ...... *O. longipes*.

Length, about 135^{mm}; tail, about 45; hind foot, 20; ear from crown, 10. Color above, uniform dull tawny cinnamon; no black ring around the eye. Tail thick with a dark stripe above reaching three-fourths its length; rest of tail white.

O. torridus.

(2) By Cranial Characters.

Palate ending posteriorly

with a blunt median spine
- a distinct supraorbital bead..............*longipes.*
- no distinct supraorbital bead...........*leucogaster.*

with straight or slightly convex edge
- skull large and broad.................*melanophrys.*
- skull smaller and narrower
  - incisive foramina barely reach plane of first molar........*longicaudus.*
  - incisive foramina reach second cusp of first molar............*torridus.*

Cranial measurements of the known forms of the genus Onychomys.

	O. leucogaster, Fort Buford, Dakota.		Melanophrys, Kanab, Utah.		Longipes, Concho County, Texas,
	4418 ♀	4419 ♂	5893 ♂	5894 ♂	3839 ♀
Basilar length of Hensel (from foramen magnum to incisor).	22	22	22.3	21.6	23.3
Zygomatic breadth	15	15.2	15.4	15.5	15.5
Greatest parietal breadth	12.9	12.7	12.8	12.5	12.2
Interorbital constriction	4.5	4.5	5.2	4.8	4.4
Length of nasals	10.8	11.6	10.7	10.7	12.5
Incisor to post-palatal notch	12	12	11.7	11.5	12.4
Foramen magnum to incisive foramina	14.7	14.6	15	14.5	15.7
Foramen magnum to palate	9.7	10	10.2	9.9	10.6
Length of upper molar series (on alveolæ)	4.5	4.2	4.6	4.8	4.4
Length of incisive foramina	5	5.7	5	5	5.3
Length of mandible	15.5	15.8	15.7	15.3	16
Height of coronoid process from angle	6.5	7.3	6.8	6.8	7.2
Ratios to basilar length:					
Zygomatic breadth	68.1	69	69	71.7	66.6
Parietal breadth	58.9	57.7	57.3	57	52
Nasals	49	52.7	47.9	49.5	52.3
Molar series (on alveolæ)	20.4	19	20.6	22	20
Incisive foramina	22.7	25.9	22.4	23.1	22.7
Foramen magnum to incisive foramen	66	66.3	67.3	67	67.3
Foramen magnum to palate	44	45.4	45.7	45.8	45.4

	Longicaudus, St. George, Utah.			Torridus, Grant County, N. Mex.
	5895 ♀	5896 ♂	5897 ♂	2839 ♂
Basilar length of Hensel (from foramen magnum to incisor)	19.3	19.3	19.4	18.5
Zygomatic breadth	13	13	13.1	12.5
Greatest parietal breadth	11.2	11.5	11.2	11.4
Interorbital constriction	4.7	4.7	4.8	4.2
Length of nasals	10	9.5	9.7	9.6
Incisor to post-palatal notch	10.5	10.5	10.4	10
Foramen magnum to incisive foramina	13.5	13.4	13.3	12.5
Foramen magnum to palate	8.8	8.7	8.7	8.5
Length of upper molar series (on alveolæ)	3.8	3.8	3.8	3.5
Length of incisive foramina	4.3	4.3	4 4	5
Length of mandible	13.4	13.5	13.2	13.2
Height of coronoid process from angle	6.2	6.3	6.2	5.8
Ratios to basilar length:				
Zygomatic breadth	67.3	67.3	68	67.5
Parietal breadth	58	59.5	57.7	61.6
Nasals	51.8	49.2	50	51.8
Molar series (on alveolæ)	19.6	19.6	19.5	18.9
Incisive foramina	22.2	22.2	22.6	27
Foramen magnum to incisive foramen	68.3	69.4	68.5	67.5
Foramen magnum to palate	45.5	45	44.8	45.8

PLATE I.

Figs. 1, 2, 3, 4, and 5, *Onychomys leucogaster*, ♂ young.   (Skull No. 4422.)   Fort Buford, Dakota.

    1. Skull from above, and left under jaw from outside (× 2).

    2. Crowns of left upper molars from below (× 10).

    3. Crowns of left lower molars from above (× 10).

    4. Crowns of right upper molars from the side (× 10).

    5. Crowns of right lower molars from the side (× 10).

Figs. 6 and 7, *Onychomys leucogaster*, ♀ ad.   (No. 5012).   Valentine, Nebraska.

    6. Crowns of left upper molars from below (× 10).

    7. Crowns of left lower molars from above (× 10).

Figs. 8 and 9, *Onychomys longicaudus*, ♂ ad.   (No. 5896).   St. George, Utah.

    8. Crowns of left upper molars from below (× 10).

    9. Crowns of left lower molars from above (× 10).

38

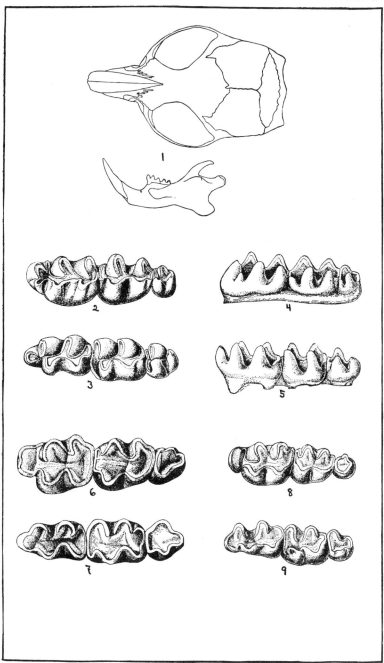

1-5. *Onychomys leucogaster,* ♂ young.
6,7. *Onychomys leucogaster,* ♀ adult.
8,9. *Onychomys longicaudus,* ♂ adult.

Vol. 79, pp. 83–88   23 May 1966

PROCEEDINGS

OF THE

BIOLOGICAL SOCIETY OF WASHINGTON

DESCRIPTIONS OF NEW BATS (*CHOERONISCUS* AND *RHINOPHYLLA*) FROM COLOMBIA

By Charles O. Handley, Jr.

U. S. National Museum, Washington, D. C.

An imperfectly known endemic mammalian fauna is found on the Pacific coast and Andean foothills of northwestern Ecuador and Colombia and northward into Panama, where it crosses to the Caribbean slope and continues into Costa Rica and Nicaragua and in some instances even into Mexico. The relatives of its endemic species are mostly South American, but some are Mexican. Species characteristic of this fauna, such as *Carollia castanea*, *Vampyressa nymphaea*, *Heteromys australis*, *Oryzomys bombycinus*, and *Hoplomys gymnurus*, were among the mammals collected in the course of virological studies of the Rockefeller Foundation on the Pacific coast of Colombia in 1962 and 1963. In addition there were striking new species of *Choeroniscus* and *Rhinophylla*.

I am grateful to Wilmot A. Thornton, Center for Zoonoses Research, University of Illinois, Urbana (formerly at Universidad del Valle, Cali, Colombia) for the opportunity to study the Colombian material here reported. Richard G. Van Gelder, American Museum of Natural History (AMNH); Philip Hershkovitz and J. C. Moore, Chicago Natural History Museum (CNHM); Bernardo Villa-R, Instituto de Biología, Mexico (IB); Barbara Lawrence, Museum of Comparative Zoology, Harvard University (MCZ); J. Knox Jones, Jr., Museum of Natural History, University of Kansas (KU); William H. Burt, Museum of Zoology, University of Michigan (UMMZ); A. Musso, Sociedad de Ciencias Naturales La Salle (LS); and Juhani Ojasti, Universidad Central de Venezuela (UCV) kindly permitted me to study comparative material. Specimens in the U. S. National Museum are designated by the abbreviation (USNM). Studies which led to the following descriptions were supported in part by National Science Foundation Grant G–19415.

All measurements are in millimeters. For definition of cranial measurements see Handley (1959: 98–99). Capitalized color terms are From Ridgway (1912).

CHOERONISCUS

There are few specimens of the poorly known glossophagine genus *Choeroniscus* in collections. The limits of variation in the genus are incompletely known (Sanborn, 1954), and until now its separation from *Choeronycteris* has been questionable. A specimen of a new species of *Choeroniscus* from the west coast of Colombia greatly extends knowledge of the genus and strengthens its stature as a genus distinct from *Choeronycteris*.

Choeroniscus periosus, new species

Holotype: USNM no. 344918, adult female, alcoholic and skull, collected 1 February 1963, by Wilmot A. Thornton, at the Río Raposo, near sea level, 27 km south of Buenaventura, Departamento de Valle, Colombia, original number 592.

Etymology: Greek *periosus*, immense.

Distribution: Known only from the type-locality.

Description: Body size large (forearm 41.2; greatest length of skull 30.3). Dorsal mass effect coloration (after three month's submersion in formalin) rich blackish-brown; basal three-fourths orange-brown in dorsal hairs; underparts but slightly paler than dorsum. Vibrissae abundant and conspicuous on snout and chin. Ears, chin, noseleaf, lips, membranes, legs, feet, and fingers blackish. Lancet of noseleaf relatively narrow, with three notches on each side near tip, and with prominent vertical median ridge on anterior face. Membranous "tongue-channel" on chin unusually well developed, protruding 1.5 mm forward and 2.0 mm up from lower lip; dorsal and anterior edges scalloped. Ear short, tip rounded, antitragus well defined; tragus spatulate, 3.8 mm long, with margins entire (except for prominent posterior notch opposite anterior base), and with anterior edge and posterior basal lobe thickened. Interfemoral membrane broad, naked. Hind legs naked. Calcar shorter than foot, not lobed.

Rostrum longer than braincase; cranium little elevated from basicranial plane; profiles of rostrum and cranium evenly tapered, without sharp angle in between; no orbital ridges or processes; zygoma absent; lambdoidal crest low; sagittal crest absent; maxillary toothrows subparallel; palate relatively broad anteriorly and narrow posteriorly; posterolateral margin of palate not notched; postpalatal extension parallel-sided, tubular, reaching posterior to level of mandibular fossae; mesopterygoid fossa reduced to a straight-sided, V-shaped notch; hamular processes greatly inflated and approaching, but not quite touching, auditory bullae; basial pits prominent, separated by broad median ridge.

Dentition weak. Dental formula $\frac{2}{0}, \frac{1}{1}, \frac{2}{3}, \frac{3}{3} = 30$. Upper incisors small, unicuspid; inner upper incisors (I^1) separated by a space three to four times the width of the teeth; larger, outer upper incisor (I^2) separated by somewhat less than its own width from I^1 and from canine.

Upper canine with small posterobasal cusp. Upper premolars very narrow; median cusp, particularly of anterior premolar, very little higher than well-defined anterior and posterior cusps. Upper molars with cusps greatly reduced; M^1 and M^2 similar in size and shape, M^3 slightly shorter and broader. Upper premolars widely spaced; molars closer together, but not touching. Lower premolars narrow, with well-defined, subequal anterior, posterior, and median cusps. Metaconid cusps of lower molars enlarged and protoconid cusps reduced; paraconid cusps in line with protoconids, not inflected. Anterior lower premolar close behind, but not touching, canine; spaces between premolars great, but spaces between P$_4$ and M$_1$ and between other molars, much less.

Measurements (All external dimensions taken from specimen in alcohol): Total length 62, tail vertebrae 10, hind foot 12, ear from notch 15, forearm 41.2, tibia 13.3, calcar 7.9.

Greatest length of skull 30.3, zygomatic breadth 11.0, postorbital breadth 4.7, braincase breadth 9.9, braincase depth 7.4, maxillary tooth row length 10.8, postpalatal length 7.0, palatal breadth at M^3 5.2, palatal breadth at canines 4.6.

Comparisons: *C. periosus* can be distinguished from all other species of *Choeroniscus* by its longer (longer than braincase), more robust rostrum; more inflated hamular process; and larger size (*e.g.*, forearm 41.2 *vs.* 32.4–36.9; greatest length of skull 30.3 *vs.* 19.3–24.4; maxillary tooth row 10.8 *vs.* 0.5–0.9). It is allied with the Amazonian species *C. minor*, *C. intermedius*, and *C. inca*, and distinguished from the Central American and northern South American *C. godmani*, in having the posterolateral margin of the palate unnotched and the cranium not so markedly elevated from the basicranial plane.

Remarks: With the addition of *C. periosus*, the genus *Choeroniscus* includes five nominal species. *C. periosus* is much the largest species; *C. inca* Thomas, *C. intermedius* Allen and Chapman, and *C. minor* Peters are intermediate in size; and *C. godmani* Thomas is smallest.

Choeroniscus is the most specialized of a group of nominal glossophagine genera which may be characterized briefly as follows:

Teeth nearly normal	pterygoids normal	*Lichonycteris*
Teeth slightly reduced	pterygoids?	*Scleronycteris*
Teeth reduced; PM high	pterygoids slightly specialized	*Hylonycteris*
Teeth reduced; PM high	pterygoids specialized	*Choeronycteris*
Teeth greatly reduced; PM low	pterygoids greatly specialized	*Choeroniscus*

Lichonycteris has 26 teeth and the other genera have 30.

As here understood, the genus *Choeronycteris* includes *Musonycteris harrisoni* Schaldach and McLaughlin, which is distinguished from *Choeronycteris mexicana* Tschudi principally by its strikingly elongated rostrum and associated modifications in proportions. The disparity in rostral proportions is much greater, however, between *Choeroniscus godmani* and *Choeroniscus periosus* than between *Choeronycteris mexicana* and *Choeronycteris harrisoni*. Thus, to distinguish *C. harrisoni* as representative of a separate genus tends to obscure relationships in this segment of the Glossophaginae. *Musonycteris* should be regarded as a synonym of *Choeronycteris*.

Specimens examined: *Choeroniscus godmani*. COLOMBIA: Meta: Restrepo, 1 (MCZ). COSTA RICA: Vicinity of San José, 3000 ft, 5 (AMNH). HONDURAS: Cantoral, 1 (AMNH); La Flor Archaga, 2 (AMNH). MEXICO: Chiapas: Pijijiapan, 50 m, 1 (UMMZ); Guerrero: 1 mi. SE San Andrés de la Cruz, 700 m, 1 (UMMZ); Oaxaca: 16 km ENE Piedra Blanca, 1 (IB); Sinaloa: San Ignacio, 700 ft, 1 (KU). NICARAGUA: El Realejo, 1 (KU), 2 (USNM). VENEZUELA: Bolivar: 38 km S El Dorado, 1 (UCV); Distrito Federal: Caracas (Santa Monica), 900 m, 1 (LS); Chichiriviche, 1 (UCV). *Choeroniscus inca*. BRITISH GUIANA: Kamakusa, 1 (AMNH); Kartabo, 1 (AMNH). ECUADOR: Los Pozos, 2 (AMNH). VENEZUELA: Bolivar: Chimantá-tepuí, 1300 ft, 9 (CNHM). *Choeroniscus intermedius*: TRINIDAD: Irois Forest, 1 (AMNH); Maracas, 1 (AMNH), Princestown, 1 (holotype of *C. intermedius*, AMNH); Sangre Grande, 1 (AMNH). *Choeroniscus minor*. BRAZIL: Pará: Belém, 3 (USNM). PERU: Pasco, San Juan, 900 ft, 1 (USNM); Puerto Melendez, above Marañon, 1 (AMNH). *Choeroniscus periosus*. COLOMBIA: Valle: Río Raposo, 1 (holotype of *C. periosus*, USNM). Also, numerous specimens of *Lichonycteris*, *Hylonycteris*, and *Choeronycteris* (including *C. harrisoni*).

RHINOPHYLLA

The carolliinine genus *Rhinophylla* has until now been known only from the basin of the Rio Amazonas and the lowlands of northeastern South America (Husson, 1962: 152–153). The sole representative of the genus, *R. pumilio* Peters, has been regarded as closely related to, but more specialized than, the species of the abundant and widespread genus *Carollia* (Miller, 1907: 147). It is thus rather surprising to find in the collection of W. A. Thornton from the west coast of Colombia a number of specimens of a striking new species of *Rhinophylla* that is even more strongly differentiated from *Carollia* than is *R. pumilio*.

Rhinophylla alethina, new species

Holotype: USNM no. 324988, adult male, skin and skull, collected 13 July 1962, by Wilmot A. Thornton, at the Río Raposo, near sea level, 27 km south of Buenaventura, Departamento de Valle, Colombia, original number 172.

Etymology: Greek, *alethinos*, genuine.

Distribution: Known only from the type-locality.

Description: Size large for genus (forearm 34.9–37.2 mm). Coloration blackish, darkest anteriorly, paler posteriorly. In holotype, head and nape black, shading to Fuscous-Black on rump; underparts varying from black on chin to Fuscous-Black on chest and to Natal Brown on abdomen. Another specimen (Univ. del Valle 220) slightly paler: Fuscous-Black anteriorly and Natal Brown posteriorly on dorsum, and correspondingly paler on underparts. Hairs of dorsum and abdomen sharply tricolor: at mid-dorsum Slate-Black basally, with broad Benzo Brown median band; on sides, neck, and shoulders median band pales almost to Ecru-Drab and shows through to surface rather prominently. Noseleaf, lips, ears, tragis, fingers, forearms, legs, feet, and all membranes blackish. Fur soft, woolly; legs, feet, interfemoral membranes, and basal two-thirds of forearm hairy; interfemoral membrane fringed. Interfemoral membrane narrow (about 5 mm at base); calcar short (less than length of metatarsals); tibia and forearm stout; pinna with anterior margin convex, posterior margin concave, tip blunt, antitragus triangular; tragus usually blunt, with upper posterior margin entire or notched; lancet of noseleaf longer than broad, upper margins slightly concave; horseshoe of noseleaf with median half of base bound to lip; chin ornament composed of four parts—a central triangular element (apex down), a pair of narrow, elongated lateral elements converging ventrally but not meeting (their outer margin more or less scalloped), and a small, circular median ventral element.

Skull like that of *Rhinophylla pumilio* but rostrum slightly heavier (broader and deeper anteriorly), and a distinct low sagittal crest present.

Dentition, with the exception of inner incisors, extremely weak and reduced; formula $\frac{2}{2} - \frac{1}{1} - \frac{2}{2} - \frac{3}{3} = 32$. Inner upper incisor ($I^1$) large, adz-shaped, with cutting edge entire; outer upper incisor (I^2) small, featureless. Canine simple, without cingulum or subsidiary cusps. Anterior upper premolar (P^1) small and featureless; posterior upper premolar (P^4) almost rectangular, longer than broad, with large median cusp and tiny posterior cusp. M^1 short and M^2 shorter, almost triangular in occlusal shape, each with a single prominent internal cusp (the metacone); protocone obliterated; paracone barely indicated in M^1, obliterated in M^2; parastyle and metastyle, particularly the latter, low and weakly developed; M^3 reduced to a tiny featureless spicule.

Inner lower incisors (I_1) large, trilobed (occasionally bilobed); I_2 small, unicuspid. Canine simple, without accessory cusps. Premolars simple, unicuspid; anterior premolar wider than any succeeding tooth. Molars very narrow, tricuspid; anterior and posterior cusps low on M_1 and M_2 and more or less obliterated on M_3.

Measurements (Extremes in parentheses, preceded by means and followed by number of individuals (only adults included). Measurements of the total length, ear, and weight were made by the collector in the field. All other measurements were made by me in the laboratory.): Total length ♂ 55, 58; hind foot ♀ 11 (11–11) 4, ♂ 11 (10–11) 6; ear from notch ♂ 15, 16; forearm ♀ 36.4 (35.5–37.2) 4, ♂ 35.7 (34.9–36.6) 4; tibia ♀ 12.3 (11.2–12.9) 4, ♂ 12.0 (11.5–12.5) 4; calcar ♀ 3.1 (3.0–3.5) 4, ♂ 3.4 (3.3–3.5) 5. Weight ♂ 12 gm, 16 gm.

Cranial measurements of male holotype: Greatest length of skull 19.5, zygomatic breadth 10.7+; postorbital breadth 5.3, braincase breadth 8.9, braincase depth 7.5, maxillary tooth row length 4.9, postpalatal length 7.2, palatal breadth at M^2 6.4, palatal breadth at canines 5.1.

Comparisons: Specimens of *R. alethina* are slightly larger than specimens of *R. pumilio* from the valley of the Rio Amazonas; have the interfemoral membrane narrower; calcar shorter; hind legs stouter; legs, feet, and interfemoral membrane (including posterior margin) more hairy; fur more woolly in texture; and coloration, including that of lips, ears, and membranes, darker, more blackish. As noted in the description, the skulls of the two species are very similar. However, except for the inner incisors, the teeth of *R. alethina* are smaller and weaker, and the tooth rows are shorter than in *R. pumilio*. *R. alethina* has cutting edges of I^1 and I_2 entire rather than notched; P^4 shorter; cusps of upper molars more reduced; and I_2, P^1, M^3, and lower molars notably smaller.

Aside from its relative *R. pumilio*, *R. alethina* is likely to be confused only with the Glossophaginae and with *Carollia castanea*. Its non-extensible tongue and lack of rostral elongation are sufficient to distinguish it from the Glossophaginae. From *Carollia castanea* it can be distinguished easily by its blacker coloration, narrow, fringed interfemoral membrane, hairy legs, simple chin ornament, and smaller, simplified teeth. In most of these characteristics *R. alethina* differs more from the species of *Carollia* than *R. pumilio* does.

Specimens examined: *Rhinophylla alethina.* COLOMBIA: Valle Río Raposo, 11 (including the holotype, USNM), 1 (Univ. del Valle). *Rhinophylla pumilio.* BRAZIL: Pará: Belém, 52 (USNM). ECUADOR: Boca de Río Curaray, 2 (USNM). PERU: Pasco: San Juan, 900 ft, 4 (USNM).

Literature Cited

Handley, C. O., Jr. 1959. A revision of American bats of the genera Euderma and Plecotus. Proc. U.S. Nat. Mus., 110: 95–246, 27 figs.

Husson, A. M. 1962. The bats of Suriname. Zoologische Vernhandelingen, Rijksmus. Nat. Hist., Leiden, 58: 1–282, 30 pls., 39 figs.

Miller, G. S., Jr. 1907. The families and genera of bats. U.S. Nat. Mus. Bull. 57, xvii + 282 pp., 14 pls., 49 figs.

Ridgway, R. 1912. Color standards and color nomenclature. iv + 44 pp., 53 pls.

Sanborn, C. C. 1954. Bats from Chimantá-tepuí, Venezuela, with remarks on *Choeroniscus.* Fieldiana–Zoology, 34: 289–293.

THE STATUS OF REITHRODONTOMYS MONTANUS (BAIRD)

By Seth B. Benson

The status and relationships of *Reithrodontomys montanus* have been uncertain ever since this harvest mouse was named and described by Baird in 1855. Study of the type specimen, and of specimens collected in the type locality of *R. montanus*, has revealed that all the specimens, except the type itself, are examples of the species *Reithrodontomys megalotis* (Baird). Confusion has arisen because these specimens have been mistakenly referred to *R. montanus*.

The nomenclatural history is as follows. Baird (1855, p. 355) described *Reithrodon montanus* on the basis of a single specimen collected by a Mr. Kreutzfeldt [= J. Creutzfeldt, botanist of Gunnison's expedition] at "Rocky Mountains, Lat. 38°." Later, Baird (1857, p. 450) gave the locality as "Rocky Mountains, 39°." Coues (1874, p. 186) listed *Ochetodon montanus* as a questionable species. This he also did later (1877, p. 130), stating "The single specimen is too imperfect to permit of final characterization, or to enable us to come to any positive conclusion; but if the size and coloration it presents are really permanent, we should judge it entitled to recognition as a valid species. At present, however, we regard it with suspicion and are unwilling to endorse its validity."

This remained the status of the name until Allen (1893, p. 80), after examining the type specimen, stated "I have therefore no hesitation in recognizing *Reithrodontomys montanus* (Baird) as a well-marked, valid species, which will probably be found to range from the eastern base of the Rocky Mountains eastward to middle Kansas."

When Allen (1895) revised the harvest mice the type of *montanus* was still unique. In his treatment of the species (pp. 123–125) he determined the type locality to be the upper part of the San Luis Valley in Colorado. He stated that "Until this region has been thoroughly explored for 'topotypes' of *R. montanus*, it would be obviously improper to reject this species as unidentifiable or to give the name precedence over *R. megalotis* for the form here recognized under that name."

At this time the species currently recognized as *albescens* was not known, although Allen actually had specimens which he confused with the form now known as *R. megalotis dychei* (see Howell, 1914, p. 31). Subsequently Cary (1903, p. 53) described *Reithrodontomys albescens* from Nebraska, stating that the species required "no close comparison with any described *Reithrodontomys*." Bailey (1905, p. 106) described *Reithrodontomys griseus* from Texas, and remarked that it probably graded into *albescens*.

In 1907 Cary visited Medano Springs Ranch in search of topotypes of *R. montanus*. He collected twenty specimens, most of them immature, which he identified as *montanus*. Cary (1911, pp. 108–110), following a manuscript of A. H. Howell, regarded *R. montanus* as a species related to *R. albescens* and *R. griseus*. He placed *albescens* as a subspecies of *montanus*.

When Howell (1914) revised the harvest mice, the specimens from the type locality of *montanus* consisted of the type specimen and the specimens collected by Cary at Medano Springs Ranch. In this revision Howell altered his earlier opinions concerning the relationships of *montanus*. He wrote (p. 26) "The species, although combining in a remarkable degree the characters of the *megalotis* and *albescens* groups, seems not to be directly connected with either of them. It is perhaps best placed in the *megalotis* group, but seems not to intergrade with any member of it." He pointed out that the relationships of the species were yet not clear, since the type specimen did not agree with any of the "topotypes" collected by Cary, but instead resembled specimens of *R. a. griseus* from Texas. Because the color of the "topotypes" agreed with the original description of *montanus*, he decided to "consider the type skull aberrant, and to continue to use the name for the form represented

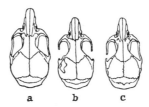

a   b   c

Fig. 1. Drawings Made from Photographs of Skulls of Harvest Mice. Natural Size

    a. Reithrodontomys megalotis subsp., no. 61120, Mus. Vert. Zool., from Medano Ranch, 15 miles northeast of Mosca, Alamosa County, Colorado.
    b. Reithrodontomys montanus montanus, type specimen, no. 1036/441, U. S. Nat. Mus., from upper end of San Luis Valley, Colorado.
    c. Reithrodontomys montanus griseus, no. 58737, Mus. Vert. Zool., from 3 miles north of Socorro, Socorro County, New Mexico.

by the modern series." It might be mentioned here that no specimens of the *albescens* group have as yet been taken near the type locality of *montanus*.

    In 1933, Miss Annie M. Alexander and Miss Louise Kellogg collected harvest mice from several localities in Colorado and New Mexico including the type localities of *R. megalotis aztecus* and *R. montanus*. Two of the three specimens from the Medano Ranch, 15 miles northeast of Mosca, Alamosa County, Colorado, were adults similar to adult topotypes of *aztecus*. The third, a young individual, was so much smaller that at first I judged it was of a different species. I suspected then that Howell's treatment of *montanus* was the result of confusing two distinct species, one a small form like *albescens*, the other a larger one like *megalotis*. The occurrence of two species at this locality seemed possible, since *albescens* occurs with *dychei* in Nebraska, *griseus* is known to occur with *dychei* in Kansas, and in 1933 I collected *megalotis* and *griseus* together in the bottom-land of the Rio Grande, three miles north of Socorro, New Mexico, which is in the same drainage system as San Luis Valley.

At my request the Bureau of Biological Survey loaned me 16 of the specimens collected by Cary at Medano Ranch. Only one of these was fully adult (the one whose skull is figured as *montanus* in Howell's revision). Among the younger specimens were some which matched the smallest of the three specimens collected by Miss Alexander and Miss Kellogg, and the rest formed a series approaching the largest specimens. The adult specimen collected by Cary is smaller than the other two adults, yet is similar to them in most characters. It was obvious that all belonged to a single species. I concluded that all the Medano Ranch specimens I had examined were of the species currently known as *megalotis*. It was also obvious that if the type of *montanus* were conspecific with the other San Luis Valley specimens, *megalotis* would become a synonym of *montanus*, since *montanus* has priority.

Through the courtesy of Dr. Remington Kellogg and others in charge of the collection of mammals in the United States National Museum, I was granted the loan of the type specimens of *R. megalotis* and *R. montanus*. After studying these specimens I reached the following conclusions: (1) The Medano Ranch specimens are conspecific with the type of *megalotis*; (2) the type specimen of *montanus* is specifically distinct from *megalotis*, and is conspecific with *albescens* and *griseus*. Some characters in which the type of *montanus* and specimens of *griseus* (MVZ no. 41192, from Hemphill Co., Texas; no. 56220, from 44 miles northwest of Roswell, N. M.; and no. 58737, from 3 miles north of Socorro, N. M.) differ from *megalotis* are: smaller size; shorter, more depressed rostrum; narrower interorbital space; relatively shorter brain case.

As a result, *megalotis* is not a synonym of *montanus*, and *montanus* becomes the specific name for the species currently known as *albescens*. Until additional specimens of *montanus* from San Luis Valley are available to allow a more thorough appraisal of its characters, it seems best to regard *albescens* and *griseus* as valid races of *montanus*, although it is quite likely that *griseus* may become a synonym of *montanus*. The three races here recognized are:

    Reithrodontomys montanus montanus (Baird)
    Reithrodontomys montanus albescens Cary
    Reithrodontomys montanus griseus Bailey.

It may be well to remark here that all the available information indicates that the species *R. montanus* is rarely abundant and that it prefers more arid, sandier ground than does its relative *R. megalotis*, although both species may be found together.

The racial identity of the San Luis Valley *megalotis* has also presented some problems. At first I referred them to the race *aztecus* because some of them fell within the range of variation present in specimens from within the distributional area assigned to *aztecus* in Howell's revision. In addition, there was so much variation in size in the few adults available to me that I felt it was possible they did not truly represent the population, and so could not serve as a satisfactory basis for the description of a new race.   However, Mr. Howell, who has restudied the problem with the aid of a greater amount of material than was available to me, has concluded that the San Luis Valley *megalotis* represent an unnamed race.   He will describe this race in another article.

LITERATURE CITED

ALLEN, J. A. 1893. List of mammals collected by Mr. Charles P. Rowley in the San Juan Region of Colorado, New Mexico and Utah, with descriptions of new species. Bull. Amer. Mus. Nat. Hist., vol. 5, pp. 69–84. April 28, 1893.

———— 1895. On the species of the genus *Reithrodontomys*. Bull. Amer. Mus. Nat. Hist., vol. 7, pp. 107–143. May 21, 1895.

BAILEY, V. 1905. Biological Survey of Texas. U. S. Dept. Agric., North Amer. Fauna no. 25, 222 pp., 16 pls., 24 figs. October 24, 1905.

BAIRD, S. F. 1855. Characteristics of some new species of North American Mammalia, collected chiefly in connection with U. S. Surveys of a Railroad route to the Pacific. Proc. Acad. Nat. Sci. Philadelphia, vol. 7, April, 1855, pp. 333–336.

———— 1857. General report upon the mammals of the several Pacific Railroad routes. U. S. Pac. R. R. Expl. and Surv., vol. 8, pt. 1, xxxiv + 764 pp., 60 pls.

CARY, M. 1903. A new Reithrodontomys from western Nebraska. Proc. Biol. Soc. Washington, vol. 16, pp. 53–54. May 6, 1903.

———— 1911. A biological survey of Colorado. U. S. Dept. Agric., North Amer. Fauna no. 33, 256 pp., 12 pls., 39 figs. August 17, 1911.

COUES, E. 1874. Synopsis of the Muridae of North America. Proc. Acad. Nat. Sci. Philadelphia, 1874, pp. 173–196.

———— 1877. No. I.—Muridae, *in* Coues and Allen, Monog. North Amer. Rodentia (= U. S. Geol. Surv. Terr. [Hayden], vol. 9), pp. 481–542.

HOWELL, A. H. 1914. Revision of the American harvest mice. U. S. Dept. Agric., North Amer. Fauna no. 36, 97 pp., 7 pls., 6 figs. June 5, 1914.

Museum of Vertebrate Zoology, University of California, Berkeley, California.

SYSTEMATICS OF SOUTHERN BANNER-TAILED KANGAROO RATS OF THE *DIPODOMYS PHILLIPSII* GROUP

Hugh H. Genoways and J. Knox Jones, Jr.

Abstract.—Both nongeographic and geographic variation was assessed in southern banner-tailed kangaroo rats of the nominal species *Dipodomys phillipsii* and *D. ornatus*. Univariate and multivariate analyses were employed in consideration of geographic variation. *D. ornatus* is arranged as a subspecies of *D. phillipsii*, in which four races (*phillipsii, ornatus, perotensis,* and *oaxacae*) are recognized. Some observations on natural history also are included.

The southern banner-tailed kangaroo rat, *Dipodomys phillipsii*, originally was described by Gray (1841:522), based on a specimen from "near Real del Monte," Hidalgo, and is the type species of the genus *Dipodomys*. The holotype remained the only known specimen of *phillipsii* until Merriam (1893) reported on material and field observations obtained by E. W. Nelson in the vicinity of Mexico City, and in the states of Tlaxcala, Puebla, and Veracruz. In the following year, Merriam (1894:110–111) named *Dipodomys ornatus* and *Dipodomys perotensis*, based on specimens from Berriozábal, Zacatecas, and Perote, Veracruz, respectively, as species that resembled *D. phillipsii*. Davis (1944:391) reduced *perotensis* to subspecific status under *phillipsii*, but *ornatus* has stood until now in the literature as a distinct species. Finally, Hooper (1947) described *Dipodomys phillipsii oaxacae* from Teotitlán, Oaxaca, distinguishing it from other known races on the basis of small size and pale coloration. For the currently recognized groups of kangaroo rats, see Lidicker (1960a:134).

Although southern banner-tailed kangaroo rats have been mentioned in various publications dealing with the mammalian faunas of northern and central México, no previous attempt has been made to assess systematically the relationships within this group. Our study is based on 251 specimens, many of them obtained in the last two decades, and includes analysis of both nongeographic and geographic variation in these rats. We have also summarized the available information on natural history.

Methods and Acknowledgments

All measurements recorded beyond are in millimeters; those recorded for crania were taken by means of dial calipers, whereas external dimensions are those recorded on specimen labels by field collectors. The measurement depth of cranium was taken using a glass microscope slide as described by Hooper (1952:10). Bacula were measured under a binocular microscope fitted with an ocular micrometer and were drawn with the aid of an ocular grid. Variation in color was assessed using a Photovolt Photoelectric Reflection Meter, Model 610, which gives reflectance readings as a percentage of pure white (see Lawlor, 1965, and Dunnigan, 1967). Readings were taken with red, green, and blue filters in the middorsal region on skins with unworn pelage.

All statistical analyses were performed on the GE 635 computer at The University of Kansas. Univariate analyses were carried out using a program (UNIVAR) written by Power (1970). This program yields standard statistics (mean, range, standard deviation, standard error of the mean, variance, and coefficient of variation) and, when two or more groups are being compared, employs a single-classification analysis of variance (F-test, significance level .05) to test for significant differences between or among the means of the groups (Sokal and Rohlf, 1969). When means were found to be significantly different, the Sums of Squares Simultaneous Test Procedure (SS-STP) was used to determine the maximally nonsignificant subsets.

Multivariate analyses were performed using the NT-SYS programs developed at The University of Kansas by F. J. Rohlf, R. Bartcher, and J. Kishpaugh. Matrices of Pearson's product-moment correlation were computed, and phenetic distance coefficients were derived from standardized character values. Cluster analyses were conducted using UPGMA (unweighted pair-group method using arithmetic averages) on the correlation and distance matrices. A matrix of correlation among characters then was computed and the first three principal components extracted. A three-dimensional stereogram was not prepared from these data because two-dimensional plots were sufficient to depict relationships among the samples studied of the southern banner-tailed kangaroo rat. Discussions of the theory underlying these tests are given by Schnell (1970:42–44) and Atchley (1970:206–212); Choate (1970) and Rising (1970) have used these techniques in studies similar to ours.

In order to obtain samples of a sufficient number of specimens for statistical analysis, it was necessary in many cases to group specimens from several geographic localities. In so doing, we attempted to keep the area concerned as small as possible, and we did not include specimens from more than one major physiographic region nor cross any previously recognized taxonomic boundaries. Specimens labeled with reference to the following geographic places comprised the samples used in our analysis (see Fig. 2): sample 1—*Durango* (Durango, Laguna de Santiaguillo, Morcillo, Vicente Guerrero); sample 2—*Durango* (La Pila); sample 3—*Zacatecas* (Hda. San Juan Capistrano, Valparaíso); sample 4—*Jalisco* (La Mesa María de León); sample 5—*Zacatecas* (Berriozábal, Fresnillo, Plateado, Trancoso, Villanueva, Zacatecas); sample 6—*Aguascalientes* (Aguascalientes, Rincón de Romos), and *Jalisco* (Encarnación de Díaz, Villa Hidalgo); sample 7—*Jalisco* (Guadalupe de Victoria, Lagos, Matanzas) and *San Luis Potosí* (Arenal, Bledos); sample 8—*Guanajuato* (León); sample 9—*Querétaro* (Tequisquiapam); sample 10—*Distrito Federal* (Ajusco, Mexico City, Tlalpam) and *México* (Amecameca, Texcoco); sample 11—*Tlaxcala* (Huamantla); sample 12—*Puebla* (Lago Salido) and *Veracruz* (Guadalupe Victoria, Limón, Perote); sample 13—*Puebla* (Chalchicomula, Puebla); sample 14—*Oaxaca* (Teotitlán); sample 15—*Puebla* (Tehuitzingo).

We gratefully acknowledge the following curators who made material in their institutions available for study (abbreviations of institutional names used in the accounts are given in parentheses): Gorden B. Corbet, British Museum (Natural History) (BM); Robert T. Orr, California Academy of Sciences (CAS); Ticul Alvarez, Escuela Nacional de Ciencias Biológicas, México (ENCB); George H. Lowrey, Jr., Museum of Natural Science, Louisiana State University (LSU); Rollin H. Baker, The Museum, Michigan State University (MSU); Dilford C. Carter, Texas Cooperative Wildlife Collection, Texas A & M University (TCWC); Emmet T. Hooper, Museum of Zoology, The University of Michigan (UMMZ); Bernardo

Villa-R., Instituto de Biología, Universidad Nacional Autonóma de México (UNAM); James S. Findley, Museum of Southwestern Biology, University of New Mexico (UNM); Clyde Jones and Charles O. Handley, Jr., United States National Museum (USNM). Specimens in the collections of the Museum of Natural History of The University of Kansas (KU) also were used.

Part of the research on this project was done under the aegis of a contract (DA-49-193-MD-2215) with the U. S. Army Medical Research and Development Command. A. Alberto Cadena translated the summary into Spanish.

NONGEOGRAPHIC VARIATION

Variation with age.—All specimens examined were assigned to one of three age categories and these were studied in order to determine which should be used in taxonomic comparisons. Age categories, modified after Hall and Dale (1939:49–50) and Lidicker (1960a:128), were as follows:

juvenile—deciduous premolars present;
young—permanent premolars present, but slightly worn, and re-entrant enamel angle still present on lingual edge of upper premolars;
adult—re-entrant angle no longer present on upper premolars, occlusal surface of premolars oval in outline.

No "old adults" were found among the available specimens. Juveniles and young were present only in limited numbers and variation with age, therefore, was not tested statistically; it was obvious, however (Table 1), in comparing age classes in a sample from the vicinity of Perote, Veracruz, that individuals classed as juveniles or young were much smaller than adults. In only one measurement, depth of cranium, did young individuals have the same mean as adults, and juveniles averaged smaller than young in all measurements tested. We used only those individuals classed as adults in our analysis of geographic variation.

Juvenile pelage is grayer and darker dorsally than that of adults; furthermore, the individual hairs of juveniles seem to be finer than those of adults and juveniles are not so densely haired. We detected no pelage intermediate between those we have termed "juvenile" and "adult." There may be, however, an earlier pelage, as suggested by Eisenberg (1963:71), not represented in our specimens.

Individual variation.—The three external measurements used in this study had coefficients of variation in a series of adults from Chalchicomula, Puebla (see Table 2), that ranged from 1.8 (length of hind foot for females) to 4.1 (length of tail for males), whereas the six cranial measurements had a range of 1.4 (depth of cranium for males) to 4.1 (length of maxillary toothrow for males and interorbital breadth for females). These values are well within the range of those for rodents of similar size as cited by Long (1968:210–213, 1969:300–301). Males had higher coefficients of variation in four measurements (total length, length of tail, length of hind foot, and greatest length of skull), and females were more variable in the other five.

TABLE 1.—*Variation with age in a sample of* Dipodomys phillipsii *from vicinities of Perote and Limón, Veracruz.*

Measurement	Juvenile		Young		Adult	
	N	Mean(Range)	N	Mean(Range)	N	Mean(Range)
Total length	4	234.3(215.0–244.0)	4	266.2(261.0–270.0)	15	270.7(254.0–304.0)
Length of tail	4	144.0(137.0–151.0)	4	159.0(154.0–166.0)	15	164.8(149.0–190.0)
Length of hind foot	4	38.0(34.0–40.0)	4	39.8(39.0–40.0)	16	40.5(38.0–44.0)
Greatest length of skull	2	33.4(32.0–34.8)	4	36.4(35.9–36.9)	13	37.4(35.3–38.7)
Length of max. toothrow		———	4	4.8(4.5–5.1)	17	4.9(4.2–6.0)
Depth of cranium	3	13.2(12.8–13.4)	4	13.4(13.0–13.8)	12	13.4(12.9–13.7)
Mastoid breadth	4	21.6(21.0–22.5)	4	22.8(22.2–23.1)	16	23.4(22.7–24.5)
Maxillary breadth	2	18.3(17.7–18.8)	4	19.8(19.5–20.1)	15	20.9(20.0–22.1)
Interorbital constriction	3	10.5(9.5–12.0)	4	12.6(12.3–12.7)	12	13.0(12.3–14.1)

Color is geographically variable in southern banner-tailed kangaroo rats, but varies to a greater degree among rats from a single locality than do external or cranial measurements. For example, among 14 geographic samples analyzed by us, reflectance of red ranged from 5.5 to 13.1 in coefficient of variation.

Secondary sexual variation.—Analysis of variance was used to test each of nine measurements in a sample from Puebla (Table 2) to determine if the means were significantly different between sexes. Males were found to be significantly longer than females at the .01 level for total length and at the .05 level for length of tail. No significant differences were found between the sexes in the remaining seven measurements, although males averaged slightly larger in all except depth of cranium. We combined values for the two sexes in geographic analysis, but attempted to keep males and females in similar proportions in the samples analyzed.

Seasonal variation.—Molt from one adult pelage to another appears to occur semiannually. The first molt probably begins in late March or early April (9 April earliest date recorded), with all individuals completing the molt by late July (23 July latest date recorded). The second molt begins in mid-September (14 September earliest date recorded) and is completed by mid- or late December (14 December latest date recorded). The pattern of molt is similar to that recorded for *Perognathus parvus* by Speth (1969). Molt begins dorsally just behind the ears and progresses anteriorly and posteriorly as well as ventrally. We detected no seasonal differences in color of pelage in the material at hand.

TABLE 2.—*Secondary sexual variation in adult* Dipodomys phillipsii *from Chalchicomula, Puebla.*

Sex	N	Mean ± 2 sᴇ (Range)	CV	Fₛ/F
		Total length		
Male	17	278.4 ± 4.62(252.0–291.0)	3.4	9.84**
Female	13	268.2 ± 4.29(258.0–283.0)	2.9	4.20
		Length of tail		
Male	17	172.2 ± 3.46(158.0–182.0)	4.1	4.48*
Female	13	167.0 ± 3.29(159.0–178.0)	3.6	4.20
		Length of hind foot		
Male	19	41.3 ± 0.43(40.0–43.0)	2.3	3.33ns
Female	13	40.8 ± 0.42(40.0–42.0)	1.8	4.17
		Greatest length of skull		
Male	18	37.5 ± 0.44(35.7–38.8)	2.5	2.23ns
Female	13	37.0 ± 0.44(35.7–38.3)	2.1	4.18
		Length of maxillary toothrow		
Male	19	4.9 ± 0.09(4.5–5.3)	4.1	3.70ns
Female	13	4.8 ± 0.11(4.4–5.2)	4.3	4.17
		Depth of cranium		
Male	18	13.3 ± 0.09(12.7–13.5)	1.4	1.49ns
Female	13	13.4 ± 0.16(12.9–14.0)	2.2	4.18
		Mastoid breadth		
Male	19	23.1 ± 0.21(22.3–24.0)	2.0	0.12ns
Female	13	23.0 ± 0.34(22.0–23.9)	2.6	4.17
		Maxillary breadth		
Male	17	20.9 ± 0.23(20.2–21.9)	2.3	0.65ns
Female	13	20.7 ± 0.46(19.5–22.2)	4.0	4.20
		Interorbital constriction		
Male	15	12.7 ± 0.15(12.3–13.3)	2.2	0.95ns
Female	10	12.5 ± 0.32(11.8–13.4)	4.1	4.28

GEOGRAPHIC VARIATION

Specimens of the nominal species *Dipodomys ornatus* and *D. phillipsii* were grouped into 14 samples for univariate analysis of geographic variation; a fifteenth sample consisting of a single individual was added for the multivariate analysis (see section on methods and Fig. 2). Samples 1 to 8 are from the geographic range of *Dipodomys ornatus* (as understood at the outset of this study), whereas samples 10 to 15 are from within the known range of *D. phillipsii*. Sample 9 is from an area intermediate between the previously known ranges of the two taxa. Table 3 gives standard statistics for external and cranial measurements and for color reflectance of rats from the 15 samples.

Univariate Analysis

External measurements.—Total length and length of the tail exhibit little significant geographic variation. The SS-STP yielded only two nonsignificant subsets for both measurements. For total length the first subset is composed of all localities except number 14 and the second subset contained samples 1, 2, 8, 12, and 14. Examination of means for total length revealed that only two samples, 2 (264.0) and 14 (244.3), yielded values that did not fall between 270 and 280. The first subset for length of tail again included all samples excepting 14, whereas the second subset included all save 9; the mean for specimens comprising sample 14 (157.0) is much smaller than the others, which have values ranging from 162.3 (sample 2) to 176.7 (sample 9). Length of the hind foot exhibited slightly more geographic variation than did other external measurements, being divided into three nonsignificant subsets (Table 4): the first contained all samples excepting 5 (central Zacatecas), 2 (La Pila, Durango), and 14 (Teotitlán, Oaxaca); sample 3 (western Zacatecas) with the largest mean (42.0) and sample 14 with the smallest (36.3) were the only samples not included in the second subset, whereas the last subset consisted of samples 2, 5, and 14. Inspection of means for length of hind foot revealed no noteworthy breaks in the continuum of variation, except for that between sample 14 and the next nearest mean value, 40.0 (sample 2).

Cranial measurements.—Means for greatest length of skull were arranged in five broadly overlapping nonsignificant subsets (Table 4). Specimens from Tlaxcala (sample 11), Veracruz (12), central Puebla (13), and Querétaro (9) have, on the average, the longest skulls (mean values for greatest length, 37.2 to 37.5). Samples from Zacatecas, Jalisco, Aguascalientes, San Luis Potosí, and Guanajuato (3 to 8) as well as the sample (10) from Estado de México and the Distrito Federal have means ranging from 36.2 to 36.5, and the two samples from Durango (1 and 2) are only slightly smaller (35.5 and 35.8). Specimens from Teotitlán, Oaxaca (sample 14), averaged smallest, with a mean length of skull of 34.1.

Length of the maxillary toothrow exhibited little geographic variation, having only two broadly overlapping subsets. The first contained all samples except 13 (central Puebla), 7 (northeastern Jalisco and western San Luis Potosí), and 14 (Teotitlán, Oaxaca), whereas the second contained all samples except the largest individuals, represented by sample 9 (Querétaro). All samples have a mean length of maxillary toothrow in the range of 4.8 to 5.0 with the exception of sample 9 (mean 5.3) and sample 14 (4.4).

Means for depth of cranium were arranged into three nonsignificant subsets. The first contained all samples with the exception of 1, 2, and 14, which have the shallowest crania. Omitted from the second subset are samples 6 (with the deepest cranium) and sample 14 (with the shallowest). The third subset is made up of samples 1, 2, 4, 9, 10, and 14. Visually, the means for this measurement form an unbroken series from 13.0 (value of samples 1 and 2) to 13.6 (value of sample 6), except for sample 14, in which the depth of cranium averaged only 12.5.

Means for mastoid breadth fell into five broadly overlapping nonsignificant subsets. Little can be discerned from examination of these subsets, but the size-order of means is of interest. Samples from Veracruz, Puebla, and Tlaxcala (11 to 13), along with those from Querétaro and Guanajuato (8 and 9), averaged largest in mastoid breadth with values of 22.9 to 23.4. Next in order of decreasing size are samples from Jalisco, Aguascalientes, Zacatecas, and San Luis Potosí (3 to 7) with means of 22.6 to 22.8. The sample from México and Distrito Federal (10) averaged 22.5, followed by two samples from Durango (1 and 2) with values of 22.4 and 22.2. Specimens from northern Oaxaca (14) averaged narrowest in mastoid breadth (21.3) among the samples studied.

Means for maxillary breadth formed five broadly overlapping subsets and are of considerable interest when compared with means of mastoid breadth. The average maxillary breadth in samples 8 and 9 (22.5 and 22.0) is large, as were the means for mastoid breadth in these samples. Next in size in maxillary breadth are the samples from Jalisco, Aguascalientes, and San Luis Potosí (means 21.4 to 21.7), and these also grouped together in mastoid breadth. Near the middle of the range of variation for the species is the sample from the vicinity of Mexico City (10) and the two Durangan samples (1 and 2), with values of 21.0 to 21.3, these three groups had the smallest means, with the exception of sample 14, for mastoid breadth. The three samples from Veracruz, Puebla, and Tlaxcala (11 to 13) averaged relatively narrow (mean values 20.8 and 20.9), but it is noteworthy that specimens from these samples averaged among the broadest in mastoid breadth. Specimens from Oaxaca (14) averaged narrowest in maxillary breadth (18.8) as they did also in mastoid breadth.

Means for interorbital constriction fell into six broadly overlapping subsets (Table 4). For this measurement, sample 10 had the largest mean although for maxillary breadth it was near the middle of the range of variation and for mastoid breadth it was among the smallest. Samples 8 and 9 again are among the broadest, as they were for the two previous breadth measurements. Specimens from Jalisco, Aguascalientes, Zacatecas, San Luis Potosí, and Durango (1 to 7) fell in the middle part of the range of variation, as they did for other breadth measurements. Specimens from Veracruz, Puebla, and Tlaxcala (11 to 13) have, on the average, narrow interorbital regions compared with those of other samples, and specimens from northern Oaxaca (14) averaged narrowest in interorbital constriction.

Color reflectance.—The six nonsignificant subsets into which means for red reflectance fell are shown in Table 4. No overall trend in geographic variation is apparent. Specimens from Jalisco and San Luis Potosí (sample 7) and those from northern Oaxaca (14) had the highest reflectance readings. Most of the northern populations had high readings with the exception of samples

TABLE 3.—Geographic variation in external dimensions, cranial dimensions, and color among 15 samples (see text and Fig. 2) of southern banner-tailed kangaroo rats.

Statistics	Sample 1	Sample 2	Sample 3	Sample 4	Sample 5	Sample 6	Sample 7	Sample 8	Sample 9	Sample 10	Sample 11	Sample 12	Sample 13	Sample 14	Sample 15
Total length															
N	12	7	7	8	13	9	19	4	11	16	3	12	30	3	1
Mean	271.5	264.0	275.4	275.8	274.1	277.8	277.6	273.0	279.0	275.3	279.7	270.7	274.0	244.3	252.0
Minimum	255.0	253.0	252.0	260.0	259.0	264.0	257.0	263.0	264.0	259.0	268.0	254.0	252.0	230.0	
Maximum	288.0	282.0	289.0	302.0	295.0	302.0	297.0	282.0	290.0	285.0	295.0	304.0	291.0	255.0	
2 SE	5.82	8.32	9.08	8.92	5.98	8.65	4.75	7.87	4.26	4.71	16.01	6.66	3.67	14.89	
Length of tail															
N	12	7	7	8	13	9	19	4	11	16	3	15	30	3	1
Mean	169.3	164.3	173.3	168.5	172.7	171.6	172.2	169.5	176.7	172.9	175.0	164.8	169.9	157.0	155.0
Minimum	151.0	155.0	164.0	160.0	162.0	162.0	156.0	167.0	167.0	160.0	165.0	149.0	158.0	150.0	
Maximum	188.0	175.0	183.0	192.0	188.0	190.0	190.0	174.0	185.0	182.0	186.0	190.0	182.0	161.0	
2 SE	6.12	5.32	5.84	8.25	4.17	6.76	4.30	3.32	2.66	3.66	12.17	5.51	2.57	7.02	
Length of hind foot															
N	14	7	10	9	16	9	18	6	11	17	3	16	32	3	1
Mean	40.3	40.0	42.0	41.1	40.2	40.4	40.5	41.2	41.8	41.2	41.0	40.5	41.1	36.3	34.0
Minimum	38.0	38.0	39.0	39.5	38.0	39.0	38.0	40.0	41.0	39.0	40.0	38.0	40.0	36.0	
Maximum	42.0	42.0	45.0	44.0	42.5	44.5	42.5	42.0	43.0	43.5	42.0	44.0	43.0	37.0	
2 SE	0.57	1.06	1.39	0.84	0.60	1.24	0.54	0.61	0.47	0.57	1.15	0.91	0.32	0.67	
Greatest length of skull															
N	11	8	10	9	14	8	18	6	9	16	2	13	31	3	1
Mean	35.5	35.8	36.5	36.2	36.5	36.3	36.4	36.4	37.2	36.5	37.5	37.4	37.3	34.1	34.4
Minimum	34.4	34.4	34.3	35.0	35.4	34.7	34.7	35.5	36.2	35.0	37.0	35.3	35.7	34.0	
Maximum	36.9	36.8	37.6	37.2	37.7	38.1	37.7	37.2	38.0	37.5	38.0	38.7	38.8	34.3	
2 SE	0.58	0.60	0.76	0.53	0.35	0.78	0.35	0.56	0.41	0.36	1.00	0.51	0.32	0.18	
Length of maxillary toothrow															
N	14	8	9	9	17	9	19	6	11	16	3	17	32	3	1
Mean	5.0	4.9	5.0	4.9	4.9	5.0	4.8	4.9	5.3	5.0	5.0	4.9	4.8	4.4	4.4
Minimum	4.2	4.5	4.6	4.4	4.6	4.7	4.3	4.5	5.0	4.7	4.9	4.2	4.4	4.4	
Maximum	5.4	5.3	5.3	5.1	5.4	5.3	5.2	5.1	5.5	5.3	5.0	6.0	5.3	4.5	
2 SE	0.17	0.24	0.18	0.16	0.11	0.15	0.09	0.20	0.12	0.09	0.07	0.21	0.07	0.07	
Depth of cranium															
N	12	7	10	8	14	8	13	6	7	14	3	12	31	3	1
Mean	13.0	13.0	13.3	13.2	13.3	13.6	13.2	13.3	13.2	13.2	13.2	13.4	13.3	12.5	12.6
Minimum	12.6	12.5	12.8	13.0	13.0	12.8	12.9	13.1	12.8	12.8	13.1	12.9	12.7	12.4	
Maximum	13.6	13.3	13.6	13.6	13.6	14.4	13.6	13.5	13.6	13.6	13.3	13.7	14.0	12.5	
2 SE	0.15	0.20	0.20	0.14	0.13	0.36	0.13	0.14	0.20	0.13	0.13	0.13	0.09	0.07	

TABLE 3. *Continued*

Statistics	Sample 1	Sample 2	Sample 3	Sample 4	Sample 5	Sample 6	Sample 7	Sample 8	Sample 9	Sample 10	Sample 11	Sample 12	Sample 13	Sample 14	Sample 15
Mastoid breadth															
N	13	8	10	9	17	8	19	6	11	15	3	16	32	3	1
Mean	22.4	22.2	22.6	22.7	22.8	22.8	22.8	23.3	23.0	22.5	22.9	23.4	23.0	21.3	21.4
Minimum	21.7	21.4	20.8	22.0	21.9	21.5	21.7	22.9	22.3	22.0	22.2	22.7	22.0	21.1	
Maximum	23.9	22.5	24.0	23.6	23.4	23.3	23.3	23.8	23.5	23.0	23.3	24.5	24.0	21.6	
2 SE	0.36	0.28	0.68	0.34	0.20	0.55	0.22	0.30	0.22	0.16	0.73	0.28	0.18	0.31	
Maxillary breadth															
N	15	7	9	9	15	9	16	5	11	11	3	15	30	3	1
Mean	21.1	21.2	21.4	21.4	21.7	21.6	21.6	22.5	22.0	21.3	20.8	20.9	20.8	18.8	18.9
Minimum	19.6	20.2	19.7	20.6	19.4	20.8	20.9	21.8	21.3	20.4	20.0	20.0	19.5	18.4	
Maximum	21.9	21.9	22.3	21.9	22.9	23.1	22.1	22.9	22.6	22.1	21.2	22.1	22.2	19.5	
2 SE	0.32	0.43	0.62	0.29	0.48	0.47	0.20	0.44	0.24	0.39	0.77	0.31	0.24	0.70	
Interorbital constriction															
N	15	6	10	9	17	9	19	6	10	13	2	12	25	3	1
Mean	13.2	13.5	13.1	13.2	13.5	13.4	13.2	14.0	14.0	14.1	13.1	13.0	12.6	12.1	11.2
Minimum	12.6	12.9	12.7	12.5	12.5	13.1	12.5	13.4	13.3	13.2	13.0	12.3	11.8	11.7	
Maximum	13.5	14.0	13.7	13.9	14.3	13.8	14.0	14.2	14.6	14.5	13.1	14.1	13.4	12.5	
2 SE	0.13	0.28	0.21	0.29	0.25	0.15	0.19	0.26	0.32	0.20	0.10	0.31	0.15	0.47	
Reflected red															
N	7	6	4	8	7	9	8	6	8	11	2	13	12	2	1
Mean	15.6	15.5	14.6	14.4	16.9	17.2	18.2	15.2	16.3	12.1	17.3	15.4	14.1	18.0	15.0
Minimum	14.5	14.0	13.0	11.5	15.0	15.0	17.0	14.0	15.0	10.5	16.5	12.5	13.0	17.0	
Maximum	17.0	17.0	17.0	17.5	21.0	19.5	20.0	17.0	18.5	13.0	18.0	17.5	16.0	19.0	
2 SE	0.75	0.97	1.70	1.33	1.50	0.91	0.80	0.92	0.78	0.40	1.50	0.88	0.69	2.00	
Reflected green															
N	7	6	4	8	7	9	8	6	8	11	2	13	12	2	1
Mean	9.4	8.3	8.0	8.1	9.2	9.3	9.3	8.3	8.1	6.6	9.5	8.7	8.1	10.3	8.0
Minimum	8.0	7.0	7.0	6.5	7.5	8.5	8.0	7.0	7.0	6.0	8.5	7.0	7.0	9.5	
Maximum	12.0	9.5	9.0	10.5	12.0	11.0	10.5	10.5	9.0	7.0	10.5	10.5	10.5	11.0	
2 SE	0.97	0.72	0.82	1.04	1.21	0.60	0.60	1.17	0.45	0.24	2.00	0.62	0.66	1.50	
Reflected blue															
N	7	6	4	8	7	9	6	6	8	11	2	13	12	2	1
Mean	7.5	7.3	6.4	6.3	7.6	7.9	7.6	6.9	6.8	5.9	8.0	7.5	6.9	8.8	7.5
Minimum	6.5	6.0	6.0	5.0	6.5	7.5	6.5	6.5	5.5	4.5	7.5	6.5	6.0	8.5	
Maximum	8.5	9.5	7.0	7.5	9.5	9.0	8.5	7.5	7.5	6.5	8.5	8.5	10.5	9.0	
2 SE	0.49	0.99	0.48	0.63	0.89	0.39	0.43	0.40	0.50	0.38	1.00	0.44	0.72	0.50	

TABLE 4.—*Results of four typical SS-STP analyses of geographic variation in southern banner-tailed kangaroo rats. Geographic origin of samples is shown in Fig. 2. Horizontal lines connect means of maximally nonsignificant subsets at the .05 level.*

Length of hind foot

Sample	3	9	10	8	13	4	11	12	7	6	1	5	2	14
Mean	42.0	41.8	41.2	41.2	41.1	41.1	41.0	40.5	40.5	40.4	40.3	40.2	40.0	36.3

Greatest length of skull

Sample	11	12	13	9	5	10	3	7	8	6	4	2	1	14
Mean	37.5	37.4	37.3	37.2	36.5	36.5	36.5	36.4	36.4	36.3	36.2	35.8	35.5	34.1

Interorbital constriction

Sample	10	9	8	2	5	6	4	7	1	3	11	12	13	14
Mean	14.1	14.0	14.0	13.5	13.5	13.4	13.2	13.2	13.2	13.1	13.1	13.0	12.6	12.1

Color (red reflectance)

Sample	7	14	11	6	5	9	1	2	12	8	3	4	13	10
Mean	18.2	18.0	17.3	17.2	16.9	16.3	15.6	15.5	15.4	15.2	14.6	14.4	14.1	12.1

3 and 4, which are from relatively high elevations as compared with other samples from Jalisco and Zacatecas. Specimens from Tlaxcala (11) also had high values of reflected red. The two samples (12 and 13) that often grouped with sample 11 in external and cranial measurements fell among those with the low mean values for red reflectance. The sample from the vicinity of Mexico City (sample 10) had the lowest mean reading for red reflectance. The average value for this sample is a full 2 per cent less in reflectance than is the next lowest mean value (sample 13).

Samples fell into three broadly overlapping, nonsignificant subsets for reflectance of green and four for reflectance of blue. They revealed approximately the same relationship as for reflected red. Sample 10 had a much lower mean than the other samples for green and blue, as it did for red.

Multivariate Analysis

Means for each sample for the three external and six cranial measurements and three color reflectance values were used in a NTSYS-multivariate analysis. Phenograms diagramming the phenetic relationships of southern banner-tailed kangaroo rats were computed by cluster analysis from both distance and cor-

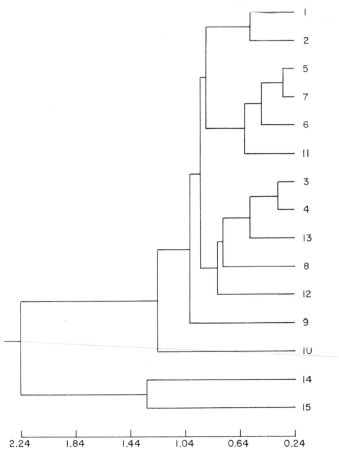

Fig. 1.—Distance phenogram resulting from cluster analysis of 15 geographic samples (see Fig. 2) of southern banner-tailed kangaroo rats.

relation matrices; the phenogram based upon the distance matrix is presented in Fig. 1. The samples in this phenogram are divided into two major clusters, one consisting of samples 14 and 15 and the second containing all others. Samples 14 and 15 are distantly separated in the first cluster. In the second cluster there are at least four major subclusters. Two of these consist of single samples 9 and 10, but the other two contain five (3, 4, 8, 12, and 13) and six samples (1, 2, 5, 6, 7, and 11). The coefficient of cophenetic correlation for the distance phenogram was 0.931.

Fig. 2 indicates the approximate areas from which the samples were drawn and the distance coefficient between the connected samples. In most cases, for ease of diagrammatic presentation, we have connected only adjacent samples. The largest distance coefficients were found between samples 13 and 14 (2.568) and between 13 and 15 (1.815). Distance coefficients of more than

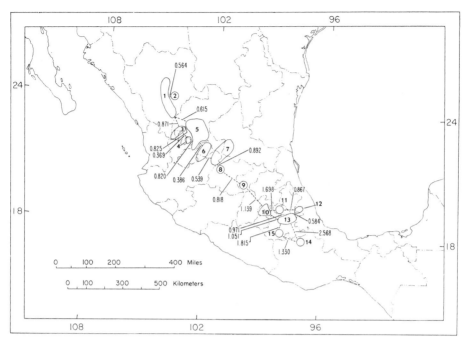

Fig. 2.—Map showing geographic location of 15 samples of southern banner-tailed kangaroo rats used for numerical analysis in this study, and distance coefficients between adjacent samples.

1.00 were generated between samples 9 and 10, 10 and 11, 10 and 13, and 14 and 15, and a coefficient of 0.971 was found between samples 11 and 13. Distance coefficients of 0.8 to 0.9 were common, being found between six sets of localities.

The first three principal components were computed from the matrix of correlation among the 12 characters. The first principal component expresses 64.16 per cent of the phenetic variation, the second 19.09, and the third 6.30. Two-dimensional plots of the three principal components are shown in Fig. 3. From the results of the factor analysis (Table 5), it appears that both external and cranial size had a strong influence on the first component. With respect to positioning of samples along component I, the sample containing specimens that were smallest overall (14) is located on the far right; from that point, samples are arranged in ascending order relative to size, with the sample consisting of the largest individuals (sample 9) on the far left of the plot. Major factors in the second component were the color reflectance ratings, although depth of cranium and mastoid breadth had some influence. The positioning of samples along component II reveals that sample 10, which contained the darkest individuals, is near the top of the plot, and that samples 3 and 4, which also had low reflectance readings, occupy a somewhat lower

63

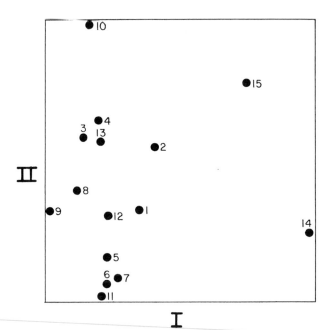

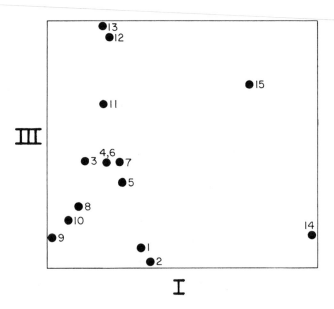

FIG. 3.—Two-dimensional projections of the first three principal components, illustrating the phenetic position of 15 samples of southern banner-tailed kangaroo rats. Top, component I plotted against component II; bottom, component I plotted against component III.

TABLE 5.—*Factor matrix from correlation among the 12 characters studied.*

Measurement	Factor Component I	Factor Component II	Factor Component III
Total length	−0.922	−0.135	0.150
Length of tail	−0.867	−0.159	−0.043
Length of hind foot	−0.947	−0.072	−0.077
Greatest length of skull	−0.848	−0.163	0.423
Length of maxillary toothrow	−0.891	−0.109	−0.224
Depth of cranium	−0.867	−0.246	0.195
Mastoid breadth	−0.854	−0.230	0.330
Maxillary breadth	−0.891	−0.137	−0.277
Interorbital constriction	−0.805	−0.093	−0.505
Reflected red	0.332	−0.904	−0.106
Reflected green	0.467	−0.852	−0.050
Reflected blue	0.659	−0.720	0.065

position on the component. Near the bottom of component II are samples that averaged high in reflectance readings. Factor analysis (Table 5) indicates that loading in the third component had high positive values for greatest length of skull and mastoid breadth and high negative values for maxillary breadth and interorbital constriction. The three samples (11, 12, and 13) that separate from the others in the third principal component are those that were shown in the univariate analysis to have among the largest means for mastoid breadth and among the smallest for maxillary breadth and inter-orbital constriction. The single specimen in sample 15 also appears to fall in the third component with samples 11, 12, and 13, but it is widely separated from these samples in component I, which indicates that in overall size this specimen is much smaller than those in samples 11, 12, and 13.

Bacular Morphology

The bacula of *Dipodomys ornatus* (Lidicker, 1960*b*:496) and *Dipodomys phillipsii* (Burt, 1960:45) have been figured previously, but no comparison between them has been made. Burt (*op. cit.*) stated that the morphology of the single baculum from a Oaxacan specimen that he examined differed from all others in the genus in that the tip is upturned at a sharp angle from the shaft. This is not the case, however, in bacula of two specimens that we examined from México and Veracruz. These do have an upturned tip (Fig. 4), but the angle between the tip and shaft appears similar to that figured by Burt (1960:pl. 12) for other *Dipodomys*. The bacula of four specimens from within the previously understood range of *ornatus* (two from Jalisco, one each from Zacatecas and Guanajuato) agree morphologically with those figured by Lidicker (1960*b*:496) from Aguascalientes.

Little difference is evident in construction and morphology of the bacula of *ornatus* and *phillipsii*. That of *ornatus* is somewhat the larger, but it should be noted that the bacula of *phillipsii* we examined were from young adults,

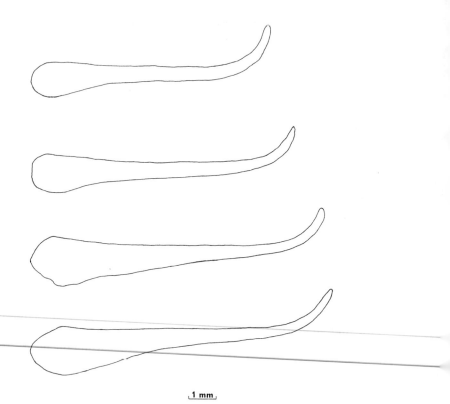

Fig. 4.—Bacula of four southern banner-tailed kangaroo rats. Specimens represented (from top to bottom) are as follows: KU 19384, 2 km E Perote, Veracruz; KU 48987, 5 mi. S, 1 mi. W Texcoco, México; KU 48975, 8 mi. SE Zacatecas, Zacatecas; KU 48986, 4 mi. N, 5 mi. W León, Guanajuato.

because no other material was available. Bacular measurements are as follows (specimens from, respectively, México, Veracruz, Burt's rat from Oaxaca, Zacatecas, two from Jalisco, Guanajuato, mean and range of five bacula from Aguascalientes examined by Lidicker): length of baculum, 10.0, 9.1, 10.5, 11.3, 11.3, 11.1, 11.5, 12.2 (11.7–12.7); height of base, 1.3, 1.4, 1.3, 1.6, 2.0, 2.0, 1.4, 1.8 (1.4–2.2); width of base, 1.0, 0.9, 1.2, 1.6, 1.5, 1.6, 1.8, 1.6 (1.5–1.9).

Taxonomic Conclusions

We interpret the univariate and multivariate analyses as revealing that southern banner-tailed kangaroo rats represent one geographically variable species. The relatively minor cranial variations that Merriam (1894:110–111) used to characterize *D. ornatus* (such as a flatter cranium) are the result of individual and geographic variation within a population of this species. Therefore, in the following accounts all southern banner-tailed kangaroo rats are treated as a single species, *Dipodomys phillipsii* Gray.

Within *D. phillipsii*, we recognize four subspecies. In the north, from Querétaro to central Durango, is *Dipodomys phillipsii ornatus*, which is characterized by medium to large size, relatively pale coloration, and medium to broad cranium. The nominate race, *Dipodomys phillipsii phillipsii*, is confined to the Valle de México and immediate vicinity; it is characterized by medium size, dark coloration, and broad interorbital region. *Dipodomys phillipsii perotensis*, which occurs in Tlaxcala, Puebla, and Veracruz, can be distinguished by large size, coloration intermediate between that of *ornatus* and *phillipsii*, a broad mastoid region, and narrow interorbital and maxillary regions. The fourth subspecies, *Dipodomys phillipsii oaxacae*, known from northern Oaxaca and southern Puebla, is much smaller than the others and pale in color.

Synopsis of Subspecies

The four recognized subspecies of *D. phillipsii* are briefly described in the following accounts, and pertinent commentary is included on distribution and infrasubspecific variation. In the lists of specimens examined, localities in italic type are not plotted on the accompanying distribution map (Fig. 5) because crowded symbols would have resulted.

Dipodomys phillipsii phillipsii Gray

1841. *Dipodomys phillipii* [*sic*] Gray, Ann. Mag. Nat. Hist., ser. 1, 7:522 (see Coues, 1875:325, and Coues and Allen, 1877:540, for emendation of spelling). Type locality—"near Real del Monte," Hidalgo.

Distribution.—Confined to Valle de México and immediately adjacent areas in Hidalgo, México, and the Distrito Federal (see Fig. 5).

Remarks.—The nominal subspecies is characterized by dark dorsal coloration, broad maxillary and interorbital regions relative to mastoid breadth, and in being medium for the species in general size. For comparison of *D. p. phillipsii* with other subspecies of the species, see accounts of those taxa.

According to Merriam (1893:84–86), after collecting a large series of *D. phillipsii* near Mexico City, E. W. Nelson attempted to obtain specimens in the vicinity of the type locality, Real del Monte, Hidalgo, at the extreme northern edge of the Valley of Mexico. His search in the vicinities of Real del Monte, Pachuca, Tula, San Agustín, and Irolo, all in Hidalgo, proved unsuccessful, although en route from Pachuca to Irolo, Nelson noted an area south of Pachuca that he believed might be suitable habitat for these kangaroo rats. Based on his failure to obtain specimens near the type locality, Nelson concluded that the locality recorded by Gray was erroneous and that the holotype most likely had originated from somewhere near Tlalpam, which was one of the important cities in the Valley of Mexico in the mid-1800's, and a place where *D. phillipsii* was abundant. Later, however, a specimen was obtained on 22 August 1942 (Davis, 1944:391) at a place 85 km N Mexico City (approximately 9 km S Pachuca, Hidalgo) casting considerable doubt on Nelson's conclusion. We believe it best to consider that the holotype of *Dipodomys phillipsii* came from the vicinity of Real del Monte, Hidalgo, at least until more convincing evidence to the contrary is available.

The one specimen examined from south of Pachuca (TCWC 3028) is a male with deciduous premolars and still in juvenile pelage. The pelage of this specimen is darker than in juveniles of *ornatus*, but paler than in juveniles of typical *phillipsii*. The possibility exists that specimens from this area represent intergrades between *ornatus* and *phillipsii*, but

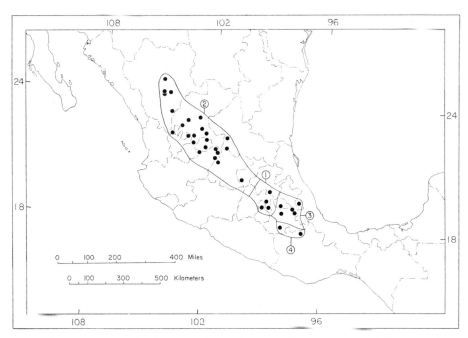

Fig. 5. Distribution of subspecies of *Dipodomys phillipsii*: 1, *D. p. phillipsii*; 2, *D. p. ornatus*; 3, *D. p. perotensis*; 4, *D. p. oaxacae*.

adults are needed before any definitive statement can be made. One of us (Jones) did, however, examine the holotype (a badly preserved skin unaccompanied by skull) in the British Museum; its dorsal coloration is relatively dark, more or less typical of that found in specimens herein assigned to *phillipsii*.

Merriam (1893:86), quoting Nelson's field notes, stated that "they [*D. phillipsii*] were noted close to the peak of Huitzilac, near the Cruz del Marquez, at an altitude of 9000 feet." We have not seen specimens from this locality and it is not clear whether Nelson collected the rats there.

Specimens examined (38).—HIDALGO: near Real del Monte, 1 (BM—the holotype); 85 km N Mexico City, 8200 ft, 1 (TCWC). MEXICO: 5 mi. S, 1 mi. W Texcoco, 7350 ft, 1 (KU); *2 km S Huatongo, 2700 m,* 1 (UNAM); Amecameca, 3 (USNM). DISTRITO FEDERAL: *17 km ESE Mexico City,* 1 (TCWC) [labeled as in México, but reported as in the Distrito Federal by Villa-R., 1953:404, and Alvarez, 1961:409]; *Cerro de la Caldera, 2300 m,* 1 (ENCB); Tlalpam, 26 (USNM); *km 20 de la cerretera México-Tláhuac,* 1 (UNAM); *Ajusco,* 2 (USNM).

Dipodomys phillipsii ornatus Merriam

1894. *Dipodomys ornatus* Merriam, Proc. Biol. Soc. Washington, 9:110, 21 June. Type locality—Berriozábal, Zacatecas.

Distribution.—Recorded from the Mexican states of Durango, Zacatecas, Jalisco, Aguascalientes, San Luis Potosí, Guanajuato, and Querétaro. The northernmost record of occurrence is in the vicinity of Santa Cruz, Durango, and the southernmost is at Tequisquiapam, Querétaro (see Fig. 5).

Remarks.—This subspecies occupies the northern segment of the geographic range of the southern banner-tailed kangaroo rat. It is characterized by medium to large size and pale coloration. *D. p. ornatus* can be distinguished from *D. p. phillipsii* by its much paler color and relatively narrow interorbital region (57.9 to 60.9 per cent of mastoid breadth in nine samples of *ornatus* as compared with an average of 62.7 per cent in *phillipsii*). From *D. p. perotensis*, the subspecies *ornatus* differs in having a somewhat shorter skull on the average (greatest length in nine samples of *ornatus* ranged from 35.5 to 37.2 and in three of *perotensis* from 37.3 to 37.5), relatively broad maxillary and interorbital regions (maxillary breadth 94.2 to 96.6 per cent of mastoid breadth in *ornatus* and 89.3 to 90.8 in *perotensis*, interorbital breadth 57.9 to 60.9 per cent of mastoid breadth in *ornatus* and 54.8 to 57.2 in *perotensis*), and somewhat paler color. *D. p. ornatus* is easily distinguished from *D. p. oaxacae* by its much larger size.

A general increase in size in several cranial measurements was noted from north to south within the geographic range of *ornatus*. Samples from Durango generally had the smallest mean values, whereas those from Zacatecas, Aguascalientes, Jalisco, and San Luis Potosí were intermediate in size, and samples from Guanajuato and Querétaro had the largest mean values. Specimens from relatively high elevations (especially samples 3 and 4) average slightly darker in color than do those from lower areas.

Baker (1960:315–316) reported that individuals from the Guadiana lava fields (sample 2) were somewhat darker than typical specimens of *ornatus*. We find they are darker than rats from the vicinity of the type locality (sample 5), but that they are only slightly darker (revealed only in reflected green) than other specimens from Durango (sample 1), and that specimens from samples 3 and 4 are darker than those from the lava field in all three color readings taken (see Table 3).

Specimens from Tequisquiapam, Querétaro (9), show some tendencies toward *D. p. phillipsii*, but clearly are assignable to *ornatus*. They have the broadest interorbital region relative to mastoid breadth (60.9 per cent) of any sample of *ornatus*, and are somewhat darker than adjacent populations (as seen in reflectance readings of green and blue). Nevertheless, in all of these characters, the specimens from Querétaro resemble *ornatus* to a greater degree than *phillipsii*.

Alvarez (1961:409) cited from Dugés a record of *Dipodomys phillipsii* from San Diego de la Unión, Guanajuato. This record is of interest because it fills an otherwise rather broad gap in the known distribution of the species.

Specimens examined (141).—DURANGO: SE end Laguna de Santiaguillo, Santa Cruz, 4 (KU); 9 mi. N Durango, 6200 ft, 1 (KU); 6 mi. NW La Pila, 6150 ft, 10 (MSU); *4 mi. S Morcillo, 6450 ft*, 1 (MSU); Durango, 4 (USNM); 16 mi. S, 20 mi. W Vicente Guerrero, 6675 ft, 6 (MSU). ZACATECAS: 12 mi. N, 7 mi. E Fresnillo, 4 (UNM); Laguna Valderama, 40 mi. W Fresnillo, 7800 ft, 6 (CAS); Valparaíso, 16 (USNM); Zacatecas, 4 (USNM); *2 mi. S, 5 mi. E Zacatecas, 7700 ft*, 1 (MSU); *8 mi. SE Zacatecas, 7225 ft*, 4 (KU); *2 mi. ESE Trancoso, 7000 ft*, 1 (KU); Hda. San Juan Capistrano, 3 (USNM); Berriozábal, 2 (USNM); 2 mi. N Villanueva, 6500 ft, 1 (KU); Plateado, 5 (USNM). SAN LUIS POTOSI: 1 km N Arenal, 1 (LSU); *1 mi. W. Bledos*, 1 (LSU); Bledos, 1 (LSU). JALISCO: La Mesa María de León, 7400 ft, 14 (KU); 10 mi. NW Matanzas, 7550 ft, 5 (KU); 1 mi. NE Villa Hidalgo, 6550 ft, 5 (KU); *5½ mi. N, 2 mi. W Guadalupe de Victoria, 7700 ft*, 1 (MSU); *8 mi. W Encarnación de Díaz, 6000 ft*, 2 (KU); 2 mi. SW Matanzas, 7550 ft, 13 (KU); Lagos, 1 (USNM). AGUASCALIENTES: *7 mi. N Rincón de Romos*, 1 (UNAM); 5 mi. NNE Rincón de Romos, 2 (KU); 3 mi. SW Aguascalientes, 6100 ft, 1 (KU). GUANAJUATO: 4 mi. N, 5 mi. W León, 7000 ft, 8 (KU). QUERETARO: Tequisquiapam, 12 (USNM).

Dipodomys phillipsii perotensis Merriam

1894. *Dipodomys perotensis* Merriam, Proc. Biol. Soc. Washington, 9:111, 21 June. Type locality—Perote, Veracruz.
1944. *Dipodomys phillipsii perotensis*, Davis, J. Mamm., 25:391, 21 December.

Distribution.—Known from Tlaxcala, a limited area in west-central Veracruz in the vicinity of the type locality, and from eastern Puebla (see Fig. 5).

Remarks.—From *D. p. phillipsii*, the subspecies *perotensis* is distinguishable by its somewhat longer cranium (see Table 3), narrower maxillary breadth and interorbital constriction, and paler dorsal coloration. Comparisons of *perotensis* with other subspecies are in the accounts of those taxa. Specimens from Tlaxcala (sample 11) are paler than specimens in the other two samples of *perotensis* studied, but in other respects the three samples are fairly homogeneous.

Merriam (1893:86–88), quoting from the field notes of E. W. Nelson, stated that southern banner-tailed kangaroo rats were known from the northern and eastern base of Cerro de Malinche and from San Marcos, both places in Tlaxcala, and several localities in Puebla including Cañada Morelos, Esperanza, San Juan de los Llanos, and Ojo de Agua. We have not seen specimens from any of these localities and it is unclear (except for the last-mentioned place) from the account whether Nelson had specimens in hand or simply based his notes on field observations. Nelson did see a specimen from Ojo de Agua, Puebla, in a small collection at a college in the city of Puebla.

Specimens examined (67).—TLAXCALA: Huamantla, 3 (USNM). VERACRUZ: 2 km N Perote, 8000 ft, 1 (KU); *2 km W Perote, 8000 ft, 1* (KU); *Perote, 7* (USNM); *2 km E Perote, 8300 ft, 7* (KU); *Guadalupe Victoria (6 km SW Perote), 8300 ft, 5* (TCWC); *3 km W Limón, 7500 ft, 3* (KU); *2 km W Limón, 7500 ft, 4* (KU). PUEBLA: *Laguna Salada (near Alchichia), 8000 ft, 2* (TCWC); 2 km W Atenco de Aljojuca, 1 (UNAM); *10 km W Chalchicomula, 8300 ft, 1* (TCWC); Chalchicomula, 31 (USNM); 7 mi. S, 3 mi. E Puebla, 6850 ft, 1 (KU).

Dipodomys phillipsii oaxacae Hooper

1947. *Dipodomys phillipsii oaxacae* Hooper, J. Mamm., 28:48, 17 February. Type locality—Teotitlán, 950 m, Oaxaca.

Distribution.—Known only from the type locality and one place in southern Puebla (see Fig. 5).

Remarks.—This subspecies is easily distinguished from all others of the species by its small size. Also, the color of *oaxacae* is much paler than that found in populations in adjacent areas of Puebla and Veracruz. Specimens we have examined exhibit the narrow maxillary (88.3 per cent) and interorbital (56.8 per cent) breadths relative to mastoid breadth that is characteristic also of *D. p. perotensis.*

D. p. oaxacae originally was described by Hooper (1947:48) on the basis of four specimens from Teotitlán, Oaxaca, until now the only known representatives of this distinctive subspecies. We have examined a specimen obtained by R. W. Dickerman at a place 1½ mi. W Tehuitzingo, Puebla, on 15 August 1954 that also appears assignable to *oaxacae* (this is the single individual in sample 15). This specimen is a young adult, but its appearance does not suggest that it ever will attain the size of the larger *perotensis,* which occurs to the northeast. We regard this specimen as only tentatively assigned to *oaxacae* until additional material becomes available from southern Puebla. It extends the known range of the subspecies approximately 130 kilometers to the west-northwest.

Specimens examined (5).—PUEBLA: 1½ mi. W Tehuitzingo, 3570 ft, 1 (KU). OAXACA: Teotitlán, 950 m, 4 (UMMZ).

Natural History

Habitat

Although there has been no extensive ecological study of the southern banner-tailed kangaroo rat, notes on natural history of the species have appeared in several publications (Merriam, 1893:88–89; Davis, 1944:391; Villa-R., 1953:404; Dalquest, 1953:117; Baker and Greer, 1962:103; Hall and Dalquest, 1963:282–283). The accounts of Merriam and of Hall and Dalquest are especially noteworthy and both contain descriptions of the burrows of *D. phillipsii*. Most of the accounts record these kangaroo rats as commonest on sandy soils in areas of short grass where large clumps of prickly pear or nopal cactus and low thornbrush are found. It is interesting to compare the record given by Merriam (*op. cit.*) of E. W. Nelson's field accounts, written in 1892 and 1893, in which it was noted that the species was abundant in the vicinity of Tlalpam in the Valley of Mexico, and that given by Villa-R. (*op. cit.*), written in the early 1950's, in which it was stated that the species was scarce in the vicinity of Tlalpam; in fact, Villa was unable to obtain specimens from that area. Authors agree that this kangaroo rat is extremely difficult to trap, possibly accounting for Villa's inability to obtain specimens, but an alternative is that the species may have been displaced from the vicinity of Tlalpam by urbanization.

In the following paragraphs, we have given brief descriptions of seven representative localities at which *D. phillipsii* was obtained by field parties from the Museum of Natural History and for which field notes are available. These portray the situations in which this species may be found and list other species of mammals that may be expected to be found in association with the southern banner-tailed kangaroo rat.

8 mi. SE Zacatecas, 7225 ft, Zacatecas.—R. H. Baker and a group of students visited this locality on 12–13 July 1952. Soils of the area are of volcanic origin and volcanic rocks were evident on the hills west of their campsite. Much of the land was under cultivation and many traps were placed along the edges of cornfields. Others were placed around clumps of grass and nopal cactus in a ravine near the camp. More than 190 mice were taken in 350 traps on the one night of trapping. Surprisingly, seven other species of heteromyid rodents were taken along with *Dipodomys phillipsii*—*Perognathus flavus, P. hispidus, P. nelsoni, Dipodomys merriami, D. ordii, D. spectabilis,* and *Liomys irroratus*. Other small mammals collected in this area included *Thomomys umbrinus, Reithrodontomys fulvescens, R. megalotis, Peromyscus maniculatus, P. melanophrys,* and *Neotoma albigula*.

La Mesa María de León, 7400 ft, Jalisco.—This locality, situated on a mesa approximately 1000 feet above the country immediately to the east, was visited from 21 to 24 June 1966 by P. L. Clifton and Genoways. The top of the mesa was a grassland supporting scattered oak trees, thus giving those areas not under cultivation a park-like appearance; the eastern edge of the mesa was steep and rock outcroppings were common there. Stands of trees and brush were much denser on the escarpment, with oak and manzanita being most abundant. Other species of mammals obtained at this place included *Didelphis marsupialis, Sylvilagus floridanus, Lepus callotis, Spermophilus mexicanus, S. variegatus, Perognathus flavus, Peromyscus boylii, P. maniculatus, P. melanophrys, Sigmodon hispidus, Neotoma albigula, Urocyon cinereoargenteus, Spilogale putorius,* and *Mephitis macroura*.

10 mi. NW Matanzas, 7550 ft, Jalisco.—P. L. Clifton described the area northwest of Matanzas in mid-May 1966 as consisting of thousands of acres of unbroken grassland, with scattered patches of nopal cactus. Low stands of oaks grew on small hills scattered through the area. Other mammals collected were *Sylvilagus floridanus, Lepus callotis, Spermophilus spilosoma, Thomomys umbrinus, Peromyscus difficilis, P. maniculatus, P. melanophrys, P. truei, Neotoma albigula, Canis latrans,* and *Spilogale putorius*.

TABLE 6.—*Distribution by month of capture of 220 southern banner-tailed kangaroo rats of three age classes.*

Month	Juvenile	Young	Adult	Total
January	—	—	—	—
February	1	1	1	3
March	—	—	—	—
April	0	1	30	31
May	2	5	11	18
June	2	4	33	39
July	1	2	28	31
August	0	1	6	7
September	3	2	11	16
October	1	1	16	18
November	0	2	6	8
December	15	10	24	49

2 mi. SW Matanzas, 7550 ft, Jalisco.—As the above locality, this place was essentially an unbroken prairie with scattered clumps of nopal cactus and thornbrush. Many traps were placed under clumps of nopal, which were surrounded by grass and weeds. The following species were obtained along with southern banner-tailed kangaroo rats on 13–14 October 1965. *Spermophilus spilosoma, Thomomys umbrinus, Perognathus flavus, P. hispidus, Peromyscus difficilis, P. maniculatus, Onychomys torridus, Sigmodon fulviventer, and Mephitis macroura.*

8 mi. W Encarnación de Díaz, 6000 ft, Jalisco.—Vegetation in this part of Jalisco was primarily grassland, scattered with mesquite and other thorny bushes. Some cultivation (mostly corn) also prevailed. Traps were set along the edge of a cornfield and among weeds along a rock fence. Mammals obtained in the period 6 to 10 October 1965 included *Sylvilagus audubonii, Spermophilus mexicanus, S. spilosoma, Perognathus flavus, P. hispidus, Dipodomys ordii, D. phillipsii, Reithrodontomys fulvescens,* and *Mus musculus.*

4 mi. N, 5 mi. W León, 7000 ft, Guanajuato.—This locality was visited by R. H. Baker and his field party shortly after they visited the locality in Zacatecas discussed above. The party camped on a grassy hillside where thornbrush and nopal cactus were abundant. There were numerous rock fences in the area along which many traps were set. Other rodents trapped at this locality included *Perognathus flavus, P. hispidus, Reithrodontomys fulvescens, Peromyscus maniculatus, P. melanophrys, P. truei,* and *Baiomys taylori.*

1½ mi. W Tehuitzingo, 3570 ft, Puebla.—Vegetation west of Tehuitzingo was dominated by mesquite; some areas were under cultivation with the fields planted mainly to corn. R. W. Dickerman, a field representative of the Museum of Natural History, trapped on 15 August 1954 along a sandy river bed and among brush on the dry slope above the river. Aside from a single specimen of *D. phillipsii,* the only other small mammal obtained there was *Liomys irroratus.*

Reproduction

Of 27 adult female southern banner-tailed kangaroo rats that were examined for reproductive data, only three were found to contain embryos. A female taken on 1 June 1954 at the southeast end of Laguna de Santiaguillo, Durango, carried two embryos that measured 25 in crown-rump length, and two females obtained on 25 October 1950 at a place 1 mi. NE Villa Hidalgo, Jalisco, each contained three embryos that measured 16 (KU 40014) and 24

(KU 40015) in crown-rump length. The remaining females, mostly collected in the months of June, July, and August, evinced no gross reproductive activity, and Hall and Dalquest (1963:283) reported that no females taken in Veracruz from late September to mid-November were pregnant. Length of testes for three adult males were 9 (3 June), 12 (23 June), and 11 (15 August).

Table 6 records 220 of the specimens we have examined by month of collection and by age. Only two months (January and March) are unrepresented by specimens. In only three months—April, August, and November—of the remaining 10 were no juveniles present in the sample and young individuals were present in all months. These data would seem to indicate a more prolonged reproductive period than can be deduced from the meager data on known reproduction in females.

Resumen

En un estudio de las ratas canguros del grupo *Dipodomys phillipsii*, se calcularon diversas variaciones geográficas y no geográficas. En la muestra proveniente de la vecindad de Perote, Veracruz, se reconocieron tres clases de edades (juveniles, subadultos, y adultos). En la muestra de Chalchicomula, Puebla, la variación estadística de los caracteres sexuales secundarios medidos fué poco significativa. Las variaciones geográficas de las medidas externas, craneales y del color del pelage fueron estudiados por medio de análisis de varianza y multivarianza para 15 muestras geográficas, las cuales revelaron que *Dipodomys ornatus* debe ser colocado como subespecie de *D. phillipsii*. Otras razas válidas son *phillipsii*, *perotensis*, y *oaxacae*.

Se incluye también algunas notas sobre la morfología del baculum, los lugares de vida y la reproducción de este grupo.

Literature Cited

Alvarez, T. 1961. Sinopsis de las especies Mexicanas del genero Dipodomys. Revista Soc. Mexicana Hist. Nat., 21:391–424.
Atchley, W. R. 1970. A biosystematic study of the subgenus *Selfia* of *Culicoides* (Diptera: Ceratopogonidae). Univ. Kansas Sci. Bull., 49:181–336.
Baker, R. H. 1960. Mammals of the Guadiana lava field Durango, Mexico. Publ. Mus., Michigan State Univ., Biol. Ser., 1:303–327.
Baker, R. H., and J. K. Greer. 1962. Mammals of the Mexican state of Durango. Publ. Mus., Michigan State Univ., Biol. Ser., 2:25–154.
Burt, W. H. 1960. Bacula of North American mammals. Misc. Publ. Mus. Zool., Univ. Michigan, 113:1–76.
Choate, J. R. 1970. Systematics and zoogeography of Middle American shrews of the genus Cryptotis. Univ. Kansas Publ., Mus. Nat. Hist., 19:195–317.
Coues, E. 1875. A critical review of the North American Saccomyidae. Proc. Acad. Nat. Sci. Philadelphia, pp. 272–327.
Coues, E., and J. A. Allen. 1877. Monographs of North American Rodentia. Bull. U. S. Geol. Surv. Territories, 11:xii + x + 1–1091.
Dalquest, W. W. 1953. Mammals of the Mexican state of San Luis Potosi. Louisiana State Univ. Studies, Biol. Sci. Ser., 1:1–229.
Davis, W. B. 1944. Notes on Mexican mammals. J. Mamm., 25:370–403.

DUNNIGAN, P. B.   1967.   Pocket gophers of the genus Thomomys of the Mexican state of Sinaloa. Radford Rev., 21:139–168.

EISENBERG, J. F.   1963.   The behavior of heteromyid rodents. Univ. California Publ. Zool., 69:iv + 1–100.

GRAY, J. E.   1841.   A new genus of Mexican glirine Mammalia. Ann. Mag. Nat. Hist., ser. 1, 7:521–522.

HALL, E. R., AND F. H. DALE.   1939.   Geographic races of the kangaroo rat, Dipodomys microps. Occas. Papers Mus. Zool., Louisiana State Univ., 4:47–62.

HALL, E. R., AND W. W. DALQUEST.   1963.   The mammals of Veracruz. Univ. Kansas Publ., Mus. Nat. Hist., 14:165–362.

HOOPER, E. T.   1947.   Notes on Mexican mammals. J. Mamm., 28:40–57.

———.   1952.   A systematic review of the harvest mice (genus *Reithrodontomys*) of Latin America. Misc. Publ. Mus. Zool., Univ. Michigan, 77:1–255.

LAWLOR, T. E.   1965.   The Yucatan deer mouse, Peromyscus yucatanicus. Univ. Kansas Publ., Mus. Nat. Hist., 16:421–438.

LIDICKER, W. Z., JR.   1960a.   An analysis of intraspecific variation in the kangaroo rat Dipodomys merriami. Univ. California Publ. Zool., 67:125–218.

———.   1960b.   The baculum of *Dipodomys ornatus* and its implication for superspecific groupings of kangaroo rats. J. Mamm., 41:495–499.

LONG, C. A.   1968.   An analysis of patterns of variation in some representative Mammalia. Part I. A review of estimates of variability in selected measurements. Trans. Kansas Acad. Sci., 71:201–227.

———.   1969.   An analysis of patterns of variation in some representative Mammalia. Part II. Studies on the nature and correlation of measures of variation. Pp. 289–302, *in* Contributions in mammalogy (J. K. Jones, Jr., ed.), Misc. Publ., Univ. Kansas Mus. Nat. Hist., 51:1–428.

MERRIAM, C. H.   1893.   Rediscovery of the Mexican kangaroo rat, *Dipodomys phillipsi* Gray. Proc. Biol. Soc. Washington, 8:83–96.

———.   1894.   Preliminary descriptions of eleven new kangaroo rats of the genera Dipodomys and Perodipus. Proc. Biol. Soc. Washington, 9:109–115.

POWER, D. M.   1970.   Geographic variation of Red-winged Blackbirds in central North America. Univ. Kansas Publ., Mus. Nat. Hist., 19:1–83.

RISING, J. D.   1970.   Morphological variation and evolution in some North American orioles. Syst. Zool., 19:315–351.

SCHNELL, G. D.   1970.   A phenetic study of the suborder Lari (Aves). I. Methods and results of principal components analyses. Syst. Zool., 19:35–57.

SOKAL, R. R., AND F. J. ROHLF.   1969.   Biometry: the principles and practice of statistics in biological research. W. H. Freeman and Co., San Francisco, xiii + 776 pp.

SPETH, R. L.   1969.   Patterns and sequences of molts in the Great Basin pocket mouse, *Perognathus parvus*. J. Mamm., 50:284–290.

VILLA-R., B.   1953.   Mamíferos silvestres del Valle de México. An. Inst. Biol. México, 23:269–492.

Museum of Natural History, The University of Kansas, Lawrence, 66044. Accepted 20 February 1971.

COMMENTS ON THE SYSTEMATIC STATUS OF VAMPIRE BATS (FAMILY DESMODONTIDAE)

G. Lawrence Forman, Robert J. Baker and Jay D. Gerber

Abstract

Immunologic analyses of serum proteins, studies of karyotypes, and morphology of spermatozoa reveal that vampire bats (family Desmodontidae) are more closely related to members of the family Phyllostomatidae than is suggested by conventional morphological characters. Immunologic tests show *Desmodus* to be related to the Phyllostomatidae through the subfamilies Phyllostomatinae and Glossophaginae. When fundamental and diploid numbers of chromosomes are plotted, two monotypic desmodontid genera (*Desmodus* and *Diaemus*) have karyotypic values that fall in the area of highest concentration of phyllostomatids. Spermatozoa of *Desmodus* and the third monotypic desmodontid genus, *Diphylla*, are indistinguishable in general morphology from those of representatives of five subfamilies of phyllostomatids. It is suggested that the vampires may represent only a subfamily of the Phyllostomatidae.

Adaptations of bats to specialized feeding habits are reflected by changes in dental, cranial, and other gross morphological features. In some cases, the resultant modifications have been extensive, making it necessary to employ other than conventional morphological criteria to determine phylogenetic relationships. The three monotypic genera (*Desmodus, Diaemus,* and *Diphylla*) of vampire bats comprising the New World family Desmodontidae have undergone extreme modifications associated with their sanguineous food habits, and these structural adaptations have been the basis for assignment of vampire bats to a distinct family. However, recently acquired evidence suggests that vampires are much more closely related to members of the New World family Phyllostomatidae than implied by the current classification of bats.

Machado-Allison (1967) noted host-ectoparasite relationships that closely allied *Desmodus rotundus* with the Phyllostomatidae and suggested that a re-evaluation of the status of desmodontids might be in order. Immunologic and electrophoretic analysis, karyotype studies, and comparison of sperm morphology also indicate that desmodontids are closely allied with the Phyllostomatidae. The findings recorded below further suggest that a reappraisal of the familial status of Desmodontidae is necessary. Of the studies here reported, Gerber is responsible for the serological work, Baker for the karyotypic analysis, and Forman for the comments on sperm morphology. Specimens mentioned by Gerber and Forman are deposited in the Museum of Natural History at The University of Kansas; those recorded by Baker are housed in the collections at Texas Technological College or the University of Arizona.

IMMUNOLOGIC AND ELECTROPHORETIC COMPARISONS

Serum samples from 25 species of New World bats, representing six families, were compared by immunoelectrophoresis and two-dimensional, micro-Ouchterlony immunodiffusion tests. The 25 species were studied using antiserum prepared against the sera of 7 species representing 4 families. In a second series of tests, sera from 16 species, including 15 of Phyllostomatidae and 1 of Desmodontidae, were studied using antisera prepared against the sera of 10 species of Phyllostomatidae and antiserum against the serum of *Desmodus rotundus*.

Bats were bled by cardiac puncture, the sera separated by centrifugation, and preserved by freezing to $-15°C$ or by adding "Merthiolate" to inhibit bacterial growth. Antisera against whole bat sera were prepared in rabbits.

417

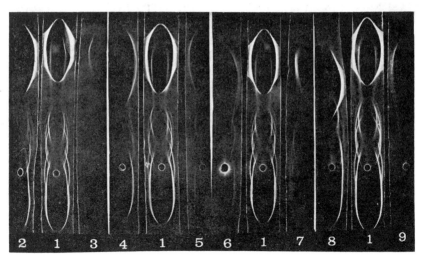

Fig. 1.—Immunoelectropherograms of reactions of anti-*Desmodus rotundus* antiserum with sera of five families of bats: phyllostomatid sera—*Sturnia lilium* (no. 2), *Glossophaga soricina* (no. 4), *Phyllostomus discolor* (no. 8); emballonurid serum—*Saccopteryx bilineata* (no. 3); vespertilionid sera—*Plecotus townsendii* (no. 5), *Myotis velifer* (no. 6); noctilionid serum—*Noctilio leporinus* (no. 7); molossid serum—*Tadarida brasiliensis* (no. 9). The reference antigen, *Desmodus* serum (no. 1), is in the center of each slide, and the anode is toward the top. Note that all families, except the Phyllostomatidae, give a weak cross-reaction with anti-*Desmodus rotundus* antiserum.

Protein-nitrogen concentrations for immunoelectrophoresis tests were determined by using the Aloe-Hitachi hand protein-refractometer. Protein-nitrogen determinations for immunodiffusion tests, where concentration is critical, were made using the Dittebrandt modification of the Biuret method and the Beckman 151 Spectro-Colorimeter, which is specially designed for micro-samples (0.1 ml).

Micro-immunoelectrophoresis was carried out in Michalis buffer, pH 8.7, for 50 minutes at 40 volts (Fig. 1). Equal amounts of standardized serum (1% protein per ml) were added to each well. For micro-Ouchterlony immunodiffusion tests, all serum samples were standardized to 400 μg protein-nitrogen/ml. To the antigen well, 8 lambda of serum were added; to the antiserum well, 16 lambda of antiserum were added in two equal portions. Diffusion of the reactants in both the immunoelectrophoresis and immunodiffusion tests was allowed to proceed for 20 to 24 hours after which the slides were washed in three changes of borate-buffered saline for three

days, rinsed in distilled water for one day, and allowed to dry. Precipitin arcs were stained with Amido-Schwartz dye. The unbound stain was removed from agar on the slides by two washes in acid-alcohol (70% ethanol, 1% acetic acid) and one wash in distilled water. To prevent cracking and peeling of the agar, slides were soaked 30 minutes in a 2% glycerol solution.

Immunoelectrophoretic relationships were demonstrated in two ways: (1) by comparing the number of arcs in the cross-reaction with those in the reference reaction, and (2) by considering the differences in density of the arcs, assigning a value of five if the arcs in the reference and cross-reaction were the same, and a value of four if the corresponding cross-reacting arc was less dense than the reference arc. Summated values for the cross-reactions were divided by the summated values for the reference-reactions and multiplied by 100 to give a per cent immunological correspondence.

The zones of precipitate of immunodiffusion tests were scanned by using transmitted light in a Joyce-Loebl densitometer

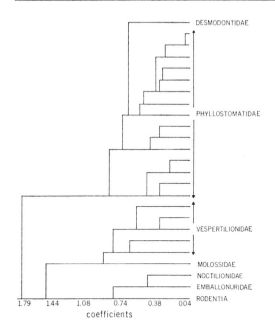

FIG. 2.—Phenogram representing distance coefficients of similarity of sera from six families of Chiroptera and a rodent, *Neotoma floridana*. The lowest distance coefficient represents the closest relationship. Note that *Desmodus* has a high affinity for species of Phyllostomatidae.

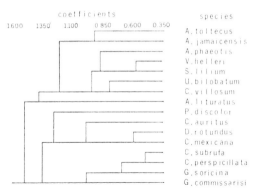

FIG. 3.—Phenogram representing distance coefficients of immunological affinities of sera from 15 phyllostomatids and *Desmodus rotundus*. The lowest distance coefficient represents the closest relationship.

(40° cam, sensitivity 2.5 to 3.0) to obtain a quantitative evaluation of the relative similarities of the cross-reacting sera to the reference serum.

Both immunoelectrophoretic and immunodiffusion data were analyzed on a GE 625 computer using programs for multivariate analysis devised by Rohlf, Kishpaugh, and Bartcher at The University of Kansas. The results of the "classical numerical taxonomy" program were expressed as distance coefficients, and a cluster analysis was performed. Phenograms were constructed from the distance coefficients, illustrating immunological relationships among bats (Figs. 2 and 3). A factor analysis was performed on the data, and a three-dimensional configuration of relationships was constructed (Fig. 4).

Fig. 1 shows immunoelectrophoresis tests illustrative of the 175 analyzed. The reference reaction is *Desmodus rotundus* and the cross-reacting sera represent five additional families of Chiroptera. The three phyllostomatids had a high degree of cross-reactivity with *Desmodus rotundus*, whereas the sera from species of Noctilionidae, Molossidae, Emballonuridae, and Vespertilionidae had little cross-reactivity. Distance coefficients, computed using all of the immunoelectrophoresis data suggest to us that *Desmodus rotundus* is more closely related to some members of the Phyllostomatidae than are certain phyllostomatids to each other.

Interfamilial affinities based on immunodiffusion tests can be seen in Figs. 2 and 4. Except for *Desmodus rotundus*, all the families studied showed a marked immunologic separation from the phyllostomatids. To analyze more precisely the affinities of *Desmodus* and the phyllostomatids, sera from 15 species of Phyllostomatidae and *Desmodus* were compared using antisera prepared against the sera of 10 species representing five subfamilies of Phyllostomatidae and *Desmodus*. Again *Desmodus rotundus* showed a high immunological affinity for the phyllostomatids, being more similar to some species of phyllostomatids than other phyllostomatids are to each other (Fig. 3). Distance coefficients within the Phyllostomatidae ranged from 0.500 to 1.520. *Desmodus* showed a distance coefficient of 0.600 to *Choeronycteris mexicana*,

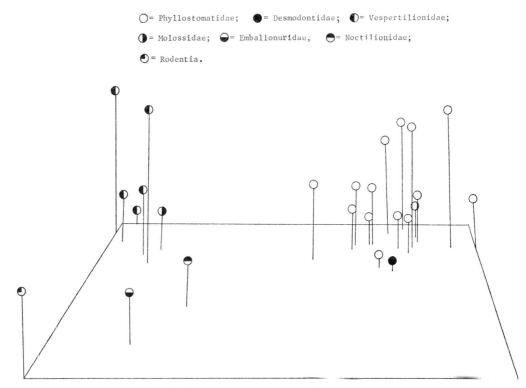

○= Phyllostomatidae;   ●= Desmodontidae;   ◑= Vespertilionidae;

◑= Molossidae;   ◔= Emballonuridae,   ◓= Noctilionidae;

◕= Rodentia.

Fig. 4.—Three-dimensional plot illustrating immunological affinities of 25 species of six families of bats studied.

1.00 to *Chrotopterus auritus,* and 1.275 to *Phyllostomus discolor.* Distance coefficients for the immunoelectrophoresis data ranged from 0.475 to 1.575. *Desmodus* was definitely within this range having a distance coefficient of 0.825.

ANALYSIS OF KARYOTYPES

The rate of chromosomal change in bats, based on the amount of variation found within genera and between closely related genera, seems to be low (Baker, 1967; Baker and Patton, 1967; Hsu, Baker, and Utakoji, 1968). For this reason, similarities in chromosome morphology deserve serious consideration as possibly being the result of close phylogenetic relationships. The chromosomes of specimens of *Desmodus rotundus* (2N = 28) from Veracruz, Mexico, were illustrated by Hsu and Benirschke (1967). In this study, additional specimens

of *Desmodus* have been examined from Oaxtepec, Morelos (two males, two females), Ojo de Agua del Rio Atayac, Veracruz (two females), 42 km west of Cintalapa, Chiapas (three males), and Guayaguayare, Trinidad (one male, three females). The chromosomes of these individuals were indistinguishable from those illustrated by Hsu and Benirschke. The chromosomes of a female from Guayaguayare, Trinidad (Fig. 5) show only one autosome has an obvious nucleolus organizer (see arrow), the same thing being found in some specimens from all localities studied.

The fundamental number (FN) of *Desmodus* is 52 (see Baker and Patton, 1967). All autosomes are biarmed and, except for one medium-sized pair of subtelocentrics, are metacentrics or submetacentrics. The X chromosome is the largest submetacentric.

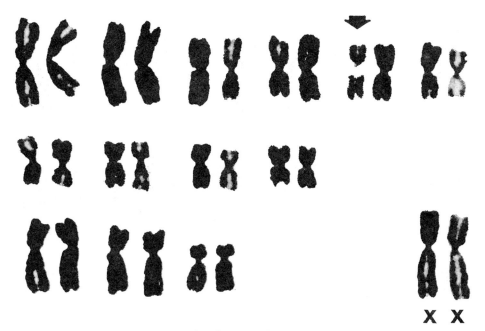

X X

Fig. 5.—Representative karyotype of a female *Desmodus rotundus* from Guayaguayare, Trinidad (TT 6502).

The Y chromosome usually appears as a minute acrocentric; however, in one Trinidadian specimen it had a biarmed appearance.

Four specimens of *Diaemus youngi* (2N = 32, FN = 60, Fig. 6) were examined. Chromosomes were biarmed and, except for one pair of medium-sized subtelocentrics, all were metacentric or submetacentric in nature. The X chromosome was a large submetacentric and the Y chromosome a minute acrocentric. The smallest pair of autosomes had a nucleolus organizer on the longest arm near the centromere.

Karyotypic data are available for two of the three known species of desmodontids. The diploid number varies between the two species by four, and the fundamental number varies by eight. An analysis of karyotypic variation at various taxonomic levels has been made for two families of bats, Phyllostomatidae (Baker, 1967) and Vespertilionidae (Baker and Patton, 1967). In both these families karyotypic variation between closely related genera was found to be both greater and less than that illustrated by the two desmodontids, although the degree of variation in fundamental number usually was less.

In the family Vespertilionidae (33 species examined) the diploid number varies from 20 to 50, and the fundamental number varies from 28 to 56 (Baker and Patton, 1967; Capanna and Civitelli, 1964). In the family Phyllostomatidae (22 species examined) the diploid number varies from 16 to 46 and the fundamental number from 26 to 68 (Baker, 1967). Although there is broad overlap between the two families in both diploid number and fundamental number, when the two are plotted against each other little overlap occurs (see Fig. 7). Because diploid number limits the fundamental numbers possible, the degree of separation of the plotted values for the two families is significant.

The values of the phyllostomatid *Uroderma bilobatum* (2N = 44, FN = 50) fall in the area occupied by most members of the Vespertilionidae; the values of the

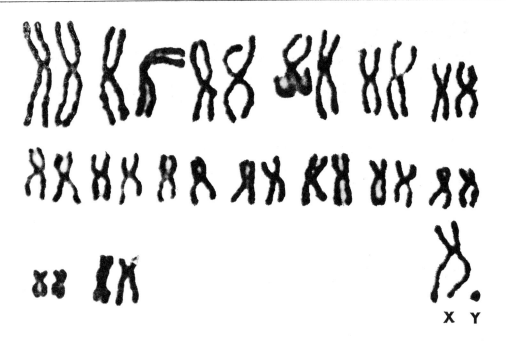

X Y

F<small>IG</small>. 6.—Representative karyotype of a male *Diaemus youngi* from Las Cueveus, Trinidad (TT 5411).

vespertilionid *Pipistrellus subflavus* (2N = 30, FN = 56) fall in the area of most members of Phyllostomatidae. Vespertilionidae is characterized by relatively lower fundamental numbers and higher diploid numbers than Phyllostomatidae. Values plotted for the few members of Rhinolophidae that have been studied (Dulic, 1967; Capanna and Civitelli, 1964) do not fall in the area of either family because of their high fundamental and diploid numbers.

Desmodus and especially *Diaemus* have karyotypic values that fall in the area of the highest phyllostomatid concentration. These data imply that the desmodontids are more closely related to the Phyllostomatidae than to the other two families for which information is available. In fact, the values for *Desmodus* and *Diaemus* are well within the variation already known for the Phyllostomatidae, and vampires share with species of the latter group the chromosomal characteristics of relatively high fundamental number and lower diploid number.

MORPHOLOGY OF SPERMATOZOA

Morphology of spermatozoa has proved useful in systematic studies of the Chiroptera (Forman, 1968). The following account compares gross morphology of spermatozoa of *Desmodus rotundus* and *Diphylla ecaudata* to that of selected representatives of three other families of North American bats.

Testes were taken from freshly killed males and preserved in a propio-alcohol fixative. Smears were prepared by smashing a short section of tubule on a slide and staining the sperm with a lacto-phenol cotton blue stain. Measurements (in microns) were taken from photographs on which 1.082 mm equaled 1 micron. Two specimens of *Desmodus rotundus murinu* and two specimens of *Diphylla ecaudata* all from Nicaragua, were examined and are described below.

Desmodus rotundus.—**Head** (measure ments based on 20 spermatozoa) with ape: narrowly rounded but blunt, generally egg shaped or ovate; tapering abruptly in latera

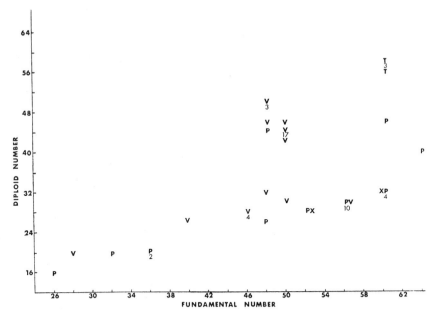

FIG. 7.—Chromosomal values (diploid number plotted against fundamental number) for species of four families of bats: Vespertilionidae (V), Phyllostomatidae (P), Desmodontidae (X), and Rhinolophidae (T). Numbers below symbols indicate number of species at that coordinate.

view to a fine point at apex; base symmetrical and concave; length 4.66 (range 4.44–4.81), width 3.27 (3.19–3.38), depth 0.97 (0.92–1.02). **Neck** region not observed, indistinct or absent. **Midpiece** (measurements based on 15 spermatozoa) tapering posteriorly; junction with head indistinct but recognizable neck region probably absent; demarcation with tail distinct in dorsal and ventral view; length 11.71 (11.23–12.29), width 0.84 (0.83–0.88). **Tail** tapers gradually to narrow filament; length about 78 in four measurements.

Diphylla ecaudata.—**Head** (measurements based on 20 spermatozoa) rounded with blunt but rounded apex, generally more circular than *Desmodus rotundus*; broad in basal half; tapering abruptly to fine point at apex in lateral view, sometimes with slight curvature; base concave, appearing symmetrical; length 5.10 (4.90–5.27), width 4.06 (3.88–4.15), depth 0.53 (0.51–0.54). **Neck** region not observed; midpiece appears continuous with base of head. **Midpiece** (measurements based on

10 spermatozoa) tapering posteriorly although sides nearly parallel in anterior two-thirds; not centrally attached to head; helical configuration observed throughout length; length 9.77 (9.18–10.64), width 0.79 (0.75–0.83). **Tail** not observed as complete.

The sperms of *Desmodus rotundus* (Fig. 8A) and *Diphylla ecaudata* (Fig. 8B) are similar in general morphology and dimensions of head and midpiece to spermatozoa of representative species of five subfamilies of the Phyllostomatidae previously examined (Forman, 1968). The head of *Desmodus rotundus* appears slightly more compressed laterally, and the midpiece somewhat longer than in sperms of *Phyllostomus, Glossophaga, Anoura, Carollia, Sturnira,* and *Artibeus* (Fig. 8C–G), but is otherwise indistinguishable from them in gross structure.

The sperm of *Diphylla ecaudata* is similar to phyllostomatids previously examined, differing only in point of attachment of the midpiece to the head. In *Diphylla,* the midpiece is not centrally attached. Point

81

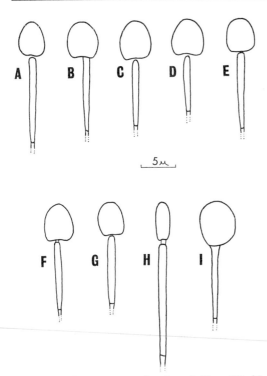

FIG. 8.—Spermatozoa of selected New World bats. **A**, *Desmodus rotundus* (KU 106263); **B**, *Diphylla ecaudata* (KU 115129); (C-G, Phyllostomatidae) **C**, *Phyllostomus discolor*; **D**, *Anoura cultrata*; **E**, *Sturnira ludovici*; **F**, *Carollia castanea*; **G**, *Artibeus jamaicensis*; **H**, *Myotis volans* (Vespertilionidae); **I**, *Molossus molossus* (Molossidae). Note that the midpieces of some spermatozoa are shown without connection to the heads. Although the anterior limits of the midpieces were demonstrable, the exact nature of the midpiece-head connection could not be resolved in these species.

of attachment of head to midpiece of spermatozoa is variable within families of mammals as illustrated by Friend (1936) and Hughes (1965), although this condition has not previously been demonstrated in Chiroptera.

Examination of sperms of other families of North American bats (e.g., Vespertilionidae and Molossidae) clearly reveals familial distinctness in general morphology (Fig. 8H–I) far exceeding the differences that separate *Desmodus rotundus* and *Diphylla ecaudata* from species of the Phyllostomatidae.

SUMMARY

Immunologic and electrophoretic tests show *Desmodus* to be related to the Phyllostomatidae through the subfamilies Phyllostomatinae and Glossophaginae. The glossophagine serum for which *Desmodus* serum shows the highest affinity is that of *Choeronycteris mexicana*. The phyllostomatine serum for which *Desmodus* serum shows the highest affinity is that of *Chrotopterus auritus*. The remaining four families studied show a distinct immunological separation from the Phyllostomatidae.

Karyotypic data are yet so meager for Chiroptera that much additional work needs to be done before relationships can be properly analyzed. Nonetheless, it can be stated that the available karyotypic data generally support, and in no way refute, the assignment of the vampire bats as a subfamily of the Phyllostomatidae. Three subfamilies of phyllostomatids (Phyllostomatinae, Glossophaginae, Stenoderminae) have members with karyotypic values similar to those of *Desmodus* and *Diphylla*. From chromosome studies, a relationship of vampires to one of these three subfamilies appears likely.

In addition, spermatozoan morphology shows *Desmodus rotundus* and *Diphylla ecaudata* to be notably similar in general structure to the phyllostomatids examined, when compared to the morphological distinctness of spermatozoa of North American representatives of two other families of bats. The sperms of *Desmodus* and *Diphylla* resemble those of *Artibeus*, *Phyllostomus*, *Anoura* and other genera representing five subfamilies of the Phyllostomatidae to such a degree that the familial distinctness of the Desmodontidae appears questionable. Evidence from immunologic and karyotypic comparisons and studies of morphology of spermatozoa suggests that vampire bats should be classified as a subfamily within the Phyllostomatidae.

ACKNOWLEDGMENTS

The authors are indebted to a number of individuals for assistance during the course of the separate studies presented herein, but especially wish to thank Drs. J. Knox Jones, Jr., and Charles A. Leone for comments and suggestions concerning preparation of the manuscript. Many specimens used in this study were collected under the aegis of a contract (DA-49-193-MD-2215) from the Medical Research and Development Command, U. S. Army. Serological studies were conducted with support from a USPHS Traineeship (grant no. 5 T01 00256) to Gerber. Baker's work was supported by NIH Fellowship no. 1-F1-GM-28, 182-02 and NSF grant no. GB-1867.

REFERENCES

BAKER, R. J. 1967. Karyotypes of phyllostomid bats and their taxonomic implications. Southwestern Natur., 12:407–428.

BAKER, R. J., AND J. L. PATTON. 1967. Karyotypes and karyotypic variation of North American vespertilionid bats. J. Mamm., 48:270–286.

CAPANNA, E., AND M. V. CIVITELLI. 1964. I cromosomi di alcune specie di microchirotteri italiani. Boll. di Zool., 31:533–540.

DULIC, B. 1967. Comparative study of the chromosomes of the spleen of some European Rhinolophidae (Mammalia, Chiroptera). Bull. Sci., Conseil Acad. RSF Yougoslavie, 12:63–65.

FORMAN, G. L. 1968. Comparative gross morphology of spermatozoa of two families of North American bats. Univ. Kansas Sci. Bull., 47:901–928.

FRIEND, G. F. 1936. The sperms of British Muridae. Quart. J. Micros. Sci., 78:419–443.

HSU, T. C., R. J. BAKER, AND T. UTAKOJI. 1968. The multiple sex chromosome system of American leaf-nosed bats (Chiroptera, Phyllostomidae). Cytogenetics, 7:27–38.

HSU, T. C., AND K. BENIRSCHKE. 1967. An atlas of mammalian chromosomes. Vol. 1, Springer-Verlag, New York, 50 folios.

HUGHES, R. L. 1965. Comparative morphology of spermatozoa from five marsupial families. Australian J. Zool., 14:533–543.

MACHADO-ALLISON, C. E. 1967. The systematic position of the bats Desmodus and Chilonycteris, based on host-parasite relationships (Mammalia: Chiroptera). Proc. Biol. Soc. Washington, 80:223–226.

Museum of Natural History, The University of Kansas, Lawrence, Kansas 66044; Department of Biology, Texas Technological College, Lubbock, Texas 79409; and Department of Zoology, The University of Kansas, Lawrence, Kansas 66044. Present address (third author): Department of Physiology, Stanford University School of Medicine, Stanford, California 94305.

CHROMOSOMAL, ELECTROPHORETIC, AND BREEDING STUDIES OF SELECTED POPULATIONS OF DEER MICE (*PEROMYSCUS MANICULATUS*) AND BLACK-EARED MICE (*P. MELANOTIS*)

J. Hoyt Bowers,[1] Robert J. Baker,[2] and Michael H. Smith[3]

Received October 17, 1972

The deer mouse, *Peromyscus maniculatus*, has the widest geographic distribution of the native species of rodents in North America. This is correlated with a high degree of genetic diversity as reflected by (1) the amount of morphological divergence and the large number of subspecies (Hall and Kelson, 1959; King, 1968), (2) the unusual amount of geographic chromosomal polymorphism (Arakaki and Sparkes, 1967; Duffey, 1972; Kreizinger and Shaw, 1970; Ohno et al., 1966; Singh and McMillan, 1966; Sparkes and Arakaki, 1966; Bradshaw and George, 1969) and (3) the degree of biochemical variation (Rasmussen, 1964, 1968, and 1970; Rasmussen and Koehn, 1966; Rasmussen et al., 1968; Brown and Welser, 1968; and Jensen and Rasmussen, 1971). The complexity of this species is further compounded because there are several peripheral populations, subspecies and closely related "species" which are distinguishable from "typical" *P. maniculatus* by varying degrees of morphological divergence (Blair, 1950).

Although this situation poses a nightmare to the taxonomist, an analysis of genetic relationship of such populations provides a valuable approach to the study of mammalian evolution. Rasmussen (1970) has already reported on one such study involving the degree of genetic diversity as reflected by electrophoretic patterns. He chose certain populations of deer mice from the higher elevations in southern Arizona because of their degree of monomorphism and their divergence from other mountain populations of *P. maniculatus*. Rasmussen assumed that this divergence and monomorphism became established as a result of drift after these populations became isolated on the respective mountain tops which now function as terrestrial islands. Divergence of these mountain populations from those of more typical *P. maniculatus* from southern Arizona has been pointed out on the basis of karyotype (Kreizinger and Shaw, 1970). Our data show that some of these populations are not *P. maniculatus*, but members of another species, *P. melanotis*, and that this monomorphism and divergence was characteristic of *P. melanotis* before the isolation of these populations on the mountain tops. There are no published karyotypic data for *P. melanotis*, and the electrophoretic data for this species consist of results from a single specimen (Brown and Welser, 1968).

METHODS AND MATERIALS

Only live-trapped mice (see specimens examined) or their F_1 offspring were used for karyotypic and electrophoretic studies. Our karyotypic techniques using bone marrow are described by Baker (1970). At least 10 spreads were counted per specimen. Voucher specimens are deposited in the collection of mammals, The Museum, Texas Tech University. Representative spreads were photographed and idiograms prepared. We follow the chromosomal nomenclature of Patton (1967), although in most cases we refer only to the number of acrocentric versus biarmed elements

[1] Department of Biology and The Museum, Texas Tech University, Lubbock, Texas 79409. Present address: Department of Biology, Wayland Baptist College, Plainview, Texas 79072.

[2] Department of Biology and The Museum, Texas Tech University, Lubbock, Texas 79409.

[3] Savannah River Ecology Laboratory, Drawer E, Aiken, South Carolina 29801.

EVOLUTION 27:378–386. September 1973

378

TABLE 1. *Chromosomal characteristics of* P. maniculatus *and* P. melanotis.

Subspecies	Number of Specimens ♂	Number of Specimens ♀	Number of Acrocentric Chromosomes	Fundamental Number	Authority
P. maniculatus austerus	1	0	18	74	Hsu and Arrighi
P. maniculatus bairdii	9	8	18	74	Singh and McMillan
P. maniculatus bairdii	4	4	8, 9, 10, 11	81, 82, 83, 84	Ohno et al.
P. maniculatus blandus	17	3	6, 8, 10	82, 84, 86	This paper
P. maniculatus fulvus	1	2	8	84	This paper
P. maniculatus gambelii	1	0	17	75	Kreizinger and Shaw
P. maniculatus gracilis	6	9	15	77	Singh and McMillan
P. maniculatus hollesteri	3	0	18	74	Arakaki and Sparkes
P. maniculatus hollesteri	3	5	12, 14, 18, 19	73, 74, 78, 80	Ohno et al.
P. maniculatus luteus	27	34	8, 10, 11, 12	80, 81, 82, 84	This paper
P. maniculatus nubiterrae			9	83	Bradshaw and George
P. maniculatus oreas	1	0	9	83	Kreizinger and Shaw
P. maniculatus ozarkiarum	1	1	10	82	This paper
P. maniculatus rufinus[1]	1	0	11	81	Kreizinger and Shaw
P. maniculatus rufinus[1]	1	0	10	82	Hsu and Arrighi
P. maniculatus rufinus[1]	37	10	7, 8, 9, 10, 11, 12, 13	79, 80, 81, 82, 83, 84, 85, 86	This paper
P. maniculatus sonoriensis	1	0	12	80	Kreizinger and Shaw
P. melanotis[2]	2	2	30	62	Kreizinger and Shaw
P. melanotis[2]	1	0	30	62	Hsu and Arrighi
P. melanotis[2]	8	1	30	62	This paper
P. melanotis	24	7	30	62	This paper

[1] Specimens from: Colorado: Larimer and Weld counties; New Mexico: Lincoln, Otero, Taos and Torrance counties; Arizona: Coconino County.
[2] Specimens from: Arizona: Cochise, Graham and Pima counties (currently considered *P. maniculatus rufinus*).

(metacentric, submetacentric and subtelocentric of Patton). Fundamental number is the number of arms of the autosomal complement. Morphology of the sex elements cannot be determined accurately with our techniques (Kreizinger and Shaw, 1970), and we considered the X and Y to be biarmed. Processing of mice and subsequent electrophoretic techniques were the same as those by Selander et al. (1971) utilizing hemolysate and plasma samples. Identification numbers of specimens used for electrophoretic analysis refer to accession numbers in catalogs of the laboratory of R. K. Selander.

Crosses were made in cages 38 cm × 30 cm × 20 cm under 14 hours of light and 10 hours of darkness. At least 6 weeks were allowed before an attempted cross was considered negative.

RESULTS AND DISCUSSION

The diploid number of all 133 specimens of *P. maniculatus* was 48 with individual variation in the number of acrocentric ranging from 6–19 (Figs. 4–6). Several workers have referred to the unique amount of morphological chromosomal variation within *P. maniculatus* (Ohno et al., 1966; Singh and McMillan, 1966; Hsu and Arrighi, 1968; Arakaki et al., 1970). A compilation of our results and literature reports reveal individuals of *P. maniculatus* with 6, 7, 8, 9, 10, 11, 12, 13, 14, 17, 18 and 19 acrocentrics respectively (Fig. 7). Forty specimens, including those from the Chiricahua, Pinaleno, and Santa Catalina mountains currently considered *P. maniculatus* had a diploid number of 48 with 30 acrocentrics, resulting in a fundamental number of 62 (Table 1 and Figs. 1–3). The geographic distribution of the number of acrocentrics found in the karyotypes of *P. maniculatus* and *P. melanotis* is shown in Fig. 7.

The degree of electrophoretic pattern variation in *P. maniculatus* is also striking (Birdsall et al., 1970; Canham et al., 1970;

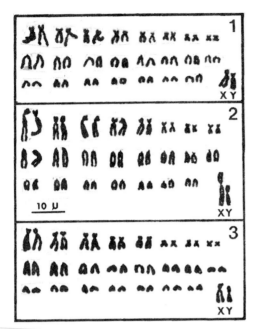

FIGS. 1–3. Representative karyotypes of male *P. melanotis.* 1) Number 13684 collected from 4 mi. E. Perote, Veracruz, Mexico, 2) Number 13677 collected from 29.1 mi. W. El Salto, Durango, Mexico, and 3) Number 13661 collected from the Santa Catalina Mountains, Pima County, Arizona.

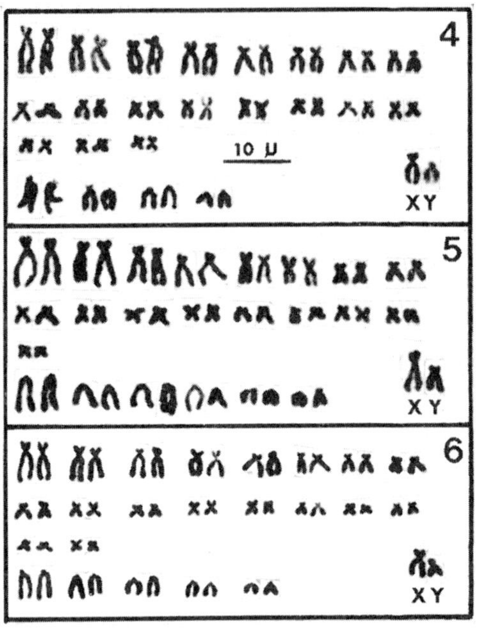

FIGS. 4–6. Representative karyotypes of male *P. maniculatus.* 4) *P. m. fulvus,* Number 10343 from 7 mi. S.S.E. of Perote, Veracruz, 5) *P. m. rufinus,* Number 13622 from Coconino County, Arizona, and 6) *P. m. blandus,* Number 13577 from 2.2 mi. S.E. Portal Cochise County, Arizona.

Rasmussen, 1970; Savage and Cameron, 1971; Jensen and Rasmussen, 1971). The mountain top populations that have 30 acrocentric chromosomes differed from the other populations studied (Rasmussen, 1970; Jensen and Rasmussen, 1971) by having different gene frequencies and by being monomorphic for several loci where *P. maniculatus* is usually polymorphic. Our electrophoretic data are from a limited number of specimens (Table 2); however, most (80–90%)of the genetic variation in a species of *Peromyscus* can be detected in mice from a single population (Selander et al., 1971). On the assumption that the alleles in the Chiricahua populations were identical in both studies, our data are consistent with those of Rasmussen (1970) with one exception (Table 2). We did not detect the rare hemoglobin variant in the Chiricahua Mountain population. Albumin and transferrin are monomorphic and

hemoglobin essentially monomorphic in the populations from the Chiricahua, Pinaleno (sometimes referred to as the Graham Mountains) and Santa Catalina mountains, Arizona, and El Salto, Perote, and Districto Federal, Mexico. The various populations of *P. melanotis* (the samples from Mexico) are identical to those of supposed *P. maniculatus* from southern Arizona in their albumin, transferrin and hemoglobin banding patterns.

We also processed two individuals of *P. maniculatus* from near Portal, Arizona and three from the mountains near Flagstaff, Arizona. Exact allele designation in accordance with Rasmussen's designation is impossible for each allele without running comparison gels using his specimens. However, it is probable that Trf-B, Trf-C and Alb-B and Alb-C were present at both locations as at least one mouse from each site was heterozygous for each system, and

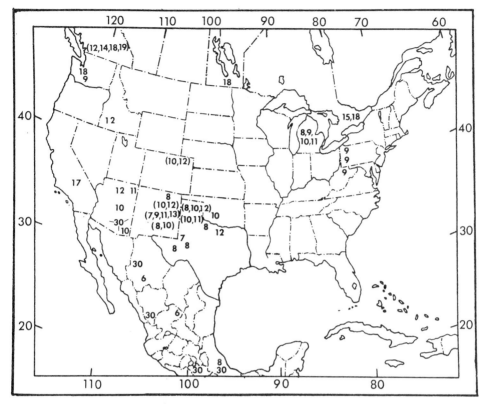

Fig. 7. Variation in the number of acrocentric chromosomes of *P. maniculatus* and *P. melanotis*. Values enclosed in parentheses represent polymorphism in a single population. Data are from this study and available literature.

neither band corresponded to Alb-D and Trf-C of *P. melanotis*. All five *P. maniculatus* were homozygous for the common hemoglobin allele in *P. maniculatus*. Neither albumen band corresponded to Alb-C and neither transferrin band corresponded to Trf-C of *P. melanotis*.

The probable explanation of this large amount of karyotypic and electrophoretic variation in presumed *P. maniculatus* is that more than one species is involved. We believe this to be the case concerning the forms from the Chiricahua, Pinaleno and Santa Catalina mountains of southern Arizona. A comparison of the karyotypes of *P. melanotis* from Perote, Veracruz (Fig. 1), and El Salto, Durango (Fig. 2), with that of *P. maniculatus rufinus* (Fig. 3) from Pima County, Arizona, reveals the three

karyotypes to be indistinguishable from each other. All populations of *P. maniculatus* that have thirty acrocentrics are restricted to higher life zones (coniferous forest) of the isolated mountain ranges (Fig. 7). The lowland populations from southern Arizona have 10 acrocentrics. The ecological range of *P. melanotis* is restricted to the high coniferous forest of central and northern Mexico and its associated grasslands.

That the similarity of habitat, karyotype and electrophoretic mobility patterns reflect a true genetic relationship is also supported by the results of breeding studies. Forty-two crosses have been attempted between animals from the southern Arizona mountain populations having 30 acrocentric chromosomes with specimens of *P. melanotis* from Perote, Veracruz, Popo-

TABLE 2. *Estimated allelic frequencies in montane populations of* Peromyscus.

		Albumins					Transferrins			Hemoglobins	
	N	A	B	C	D	E	A	B	C	A	O
Kaibab Plateau[1]	85	0.01	0.04	0.95	0.01		0.01	0.81	0.18	0.61	0.39
Flagstaff[1]	47	0.01	0.03	0.95	0.01			0.76	0.24	0.62	0.38
Mingus Mtn. I[1]	24	0.02	0.17	0.81				0.83	0.17	0.69	0.31
Mingus Mtn. II[1]	34		0.04	0.96				0.63	0.37	0.54	0.46
White Mtns., 1966[1]	37	0.03	0.11	0.86				0.85	0.15	0.36	0.64
White Mtns., 1968[1]	18		0.14	0.83		0.03		0.72	0.28	0.42	0.58
White Mtns., 1969[1]	18	0.06	0.17	0.78			0.03	0.69	0.28	0.61	0.39
Pinaleno Mtns.[1]	56				1.00				1.00	1.00	
Pinaleno Mtns.	2				1.00				1.00	1.00	
Chiricahua Mtns.[1]	62				1.00				1.00	1.00	0.06
Chiricahua Mtns.	9				1.00				1.00	1.00	
Santa Catalina Mtns.[1]	38				1.00				1.00	1.00	
El Salto, Durango	3				1.00				1.00	1.00	
Districto Federal	3				1.00				1.00	1.00	
Perote	3				1.00				1.00	1.00	

Protein Alleles

[1] Data from Rasmussen (1970) and Jensen and Rasmussen (1971).

cateptel, Districto Federal, and El Salto, Durango. Twenty-four of these attempted crosses were successful (57%). Successful crosses between individuals from each of three southern Arizona mountains (Chiricahua, Pinaleno and Santa Catalina) have been made with specimens of *P. melanotis* from each of the above mentioned localities. Attempted crosses of specimens from the southern Arizona mountain top populations with 30 acrocentrics times *P. maniculatus* with 10 acrocentric elements (30 crosses attempted) collected from the lowlands of Arizona southeast of Portal and times individuals from the coniferous forest of the mountains of Arizona and New Mexico (58 crosses attempted) have failed to produce a single litter. Crosses made between individuals of *P. melanotis* from the same locality resulted in 33 of 77 attempted crosses being successful (43%).

Thus, the populations in southern Arizona isolated on the tops of the Chiricahua, Pinaleno and Santa Catalina mountains have 30 acrocentrics, are monomorphic for Alb-D and Trf-C and do not readily interbreed with other populations of *P. maniculatus*. These characteristics are the same as those of populations of *P. melanotis* from within its currently recognized distribution. Cross-fertility of the populations of *P. melanotis* with those of southern Arizona further argues their conspecific nature.

Rasmussen (1970) argued that the relatively high degree of genetic monomorphism was due to drift in these isolated mountain populations. Our study suggests this is an artifact produced by combining data from two species, *P. melanotis* being monomorphic and *P. maniculatus* polymorphic for the loci that happened to be studied. However, *P. melanotis* shows variability in other genetic systems that were not presented in his paper (e.g., phosphoglucomutase-3). There is no evidence for the occurrence of genetic drift in these isolated mountain populations, and the danger of using only a limited number of loci and improper species identification should be apparent. Jensen and Rasmussen (1971) and Brown and Welser (1968) concluded that albumin migration patterns would be of little or no taxonomic value. Our study and those of Smith et al. (1973) indicate this conclusion was premature and that albumin patterns may actually be quite useful.

We have studied only a small part of the *P. maniculatus* complex, but certain karyotypic patterns have become evident and breeding studies have supported the genetic reality of these patterns. The agreement between karyotypic and electrophoretic mobilities characteristic of these populations and the taxonomic lines suggested by the karyological and breeding data is impressive. Our study reveals that when preliminary data from a character obviously do not fit present taxonomic lines, it should not be concluded that the character will be of no apparent taxonomic or phylogenetic value. We believe that more intensive study of the karyotypes of *P. maniculatus*, (especially with heterochromatin and Giemsa banding techniques) taken in conjunction with breeding and electrophoretic studies, will result in a much better understanding of the evolution and systematics of this species. Certainly no character can be unconditionally accepted to imply a close phylogenetic relationship. For instance, the karyotype of *Peromyscus floridanus* as published by Hsu and Arrighi (1968), is very similar to that shown for *P. melanotis* (Figs. 1–3). These two species are presently placed in separate subgenera, and we do not know if this similarity of karyotypes has resulted from convergent evolution or from both remaining unchanged since divergence from a common ancestor. The point is that this is only one character.

As far as the relationship of *P. maniculatus* to *P. melanotis* is concerned, we have no data to suggest natural hybridization. Further, none of the animals with 30 acrocentrics have been successfully crossed with any *P. maniculatus* from a southwest desert population or from the adjacent mountain populations of New Mexico or

Arizona. The one successful cross reported in the literature involved specimens from the Ann Arbor, Michigan, area times *P. melanotis* from El Salto, Mexico (Clark, 1966). One of our laboratory stocks is from El Salto, Durango, and, as indicated above, all crosses of this stock (16 attempted) times various *P. maniculatus* were negative. In Clark's study (1966), only one of four attempted matings was successful, and this mating produced all males (Clark does not mention the number of young produced). It may be that in the zone where the two species have been in contact for some time, isolating mechanisms have evolved; however, where the two forms are more distantly associated from a geographical point, their isolating mechanisms may not have been established. Also, this may represent an isolated instance of compatibility between the two species. If this cross is valid, it certainly suggests some interesting implications; however, additional work is in order before we place too much importance on the data.

Blair (1950:270) discussed the evolution and speciation of *P. maniculatus* and related species. He concluded that species on the margins of the range of *P. maniculatus* have resulted from "peripheral isolates" which have undergone "both adaptive and nonadaptive differentiation in morphological, physiological and psychological characters." Blair concluded that *P. maniculatus* was the parental stock that gave rise to *melanotis, polionotus, sitkensis, slevini* and *sejugis*. We agree with this interpretation. In addition, we feel that the preliminary data indicate that the *P. maniculatus* complex may closely fit the "centrifugal speciation" model proposed by Brown (1957) which involves the classical concept of adaptation of these groups to local conditions. A critical component of the "centrifugal speciation" hypothesis is that the center is the principal source of evolutionary change leading to "potential new species." In our particular case, we interpret "the center" to be the species *Peromyscus maniculatus*. In Brown's model, the peripheral populations will be more primitive than the central stock, and it is possible that the peripheral populations may be more genetically compatible with each other than with the more rapidly evolving central stock.

The following discussion will serve to show the relationship of our data to Brown's model.

1) Hsu and Arrighi (1968) and Baker and Mascarello (1969) hypothesized that the primitive karyotype for *Peromyscus* was 48 with a large number of acrocentrics. In the *P. maniculatus* complex the populations with higher numbers of acrocentrics are found in the peripheral isolates (Fig. 7) and in the more peripheral populations. *Peromyscus melanotis* has 30 acrocentrics; *P. polionotus* has 24–26 acrocentrics. In these two species, which are composed of numerous populations that are completely isolated from each other, only one minor chromosomal polymorphism and no chromosomal races have been described (Te and Dawson, 1971). When the chromosomal data for these two species are contrasted with the many races and polymorphisms of *P. maniculatus*, one is forced to conclude that *P. maniculatus* is in a more dynamic evolutionary state.

2) *Peromyscus melanotis* is not genetically compatible with the nearest populations (from the Mogollon rim, Arizona, or the Sacramento Mountains of New Mexico) or with its ecological equivalent, *P. maniculatus rufinus* or with the geographically adjacent grassland population *P. maniculatus blandus*. *Peromyscus melanotis* will hybridize with *P. maniculatus bairdii* from Michigan (Clark, 1966). Bowen (1968) suggested that *P. maniculatus bairdii* gave rise to *P. polionotis*. Clearly additional data are needed. However, should the peripheral isolates and other peripheral populations be more genetically compatible with each other than with the central stock of *P. maniculatus* from the central U.S.

and Mexico, then this complex will closely fit the model for centrifugal speciation.

SUMMARY

An unusual amount of chromosomal polymorphism and geographic variation in chromosomes and electrophoretic pattern have been reported for *Peromyscus maniculatus*. Karyological, electrophoretic and breeding data indicate that populations of *Peromyscus* from the Chiricahua, Pinaleno and Santa Catalina Mountains in southern Arizona are conspecific with *P. melanotis*, not with *P. maniculatus*. These data also argue against the assumptions that monomorphism in the isolated populations arose by drift. Rather they support a model of centrifugal speciation.

ACKNOWLEDGMENTS

We gratefully acknowledge Dr. Robert K. Selander for his criticisms and for usage of his laboratory. John Avise, Dale Berry, William Bleier, Joanne Bowers, Mark Bowers, Brent Davis, Genaro Lopez, Bob Martin, Kenneth Matoka, Rick McDaniel, Paul Ramsay, Jim Reichman, Sherry Shackelford, Gordon Smith, Melinda Smith and Bradley Wray assisted in collecting specimens and Ruth Barnard and Mike Jackson gave technical assistance. Supported in part by a NSF Science Faculty Fellowship awarded to Bowers and NSF Grant GB 8120, NIH Grant GM-15769 and AEC Contract At(38-1)-310.

APPENDIX

Specimens Examined

Peromyscus maniculatus: *P. m. blandus* Osgood. TEXAS: Jeff Davis County, 9.3 mi. W. Balmorhea on Texas 17, one female, 13580. ARIZONA: Cochise County, 6.5 mi. S.E. Portal, 3 males, 13567, 13573, 13575; 2.5 mi. S.E. Portal, eight males, 13564–13566, 13569, 13570, 13572, 13576, 13579, one female, 13574; 2.2 mi. S.E. Portal, 4 males, 13568, 13571, 13577, 13578. MEXICO: CHIHUAHUA: 47.2 mi. S. Jiminez, 2 males, not yet catalogued. ZACATECAS: 7 mi. E. Mazapil, one female, not yet catalogued. *P. m. fulvus* Osgood. MEXICO: VERACRUZ: 7 mi. S.S.W. Perote, one male, 10345, two females,

10343, 10344. *P. m. luteus* Osgood. TEXAS: Andrews County, 18 mi. E. and 2 mi. N. Andrews, one male, 10393. Ector County, 10 mi. E. Odessa, one male, 10387. Hale County, 1.5 mi. W. Plainview, 3 females, 13582–13584. 3.2 mi. N. Plainview, 3 males, 10352, 10395, 13858, 8 females, 10359, 10360, 10362, 10363, 10386, 13587, 13588, 13586. Hardeman County, 2.5 mi. N.E. Quanah, 2 males, 10372, 13592, 6 females, 10370, 10371, 10385, 13589, 13590, 13591. Hockley County, 8.5 mi. N.W. of Levelland, two males, 10378, 10379, three females, 10377, 13593, 13594. Lamb County, 7.2 mi. S. Olton on F. M. 168, four males, 10366, 10382–10384, four females, 10364, 10365, 10380, 10381. Lubbock County, 0.5 mi. N. of Lubbock Lake Site, six males, 10368, 13597–13601, nine females, 10367, 10369, 13596, 13602–13607. McCulloch County, 5 mi. S.E. Brady, two males, 10391, 10392, one female, 13595. OKLAHOMA: Texas County, 10 mi. E. Hardesty, one male, 10361. Washita County, 2 mi. W. Burns Flat, five males, 10373–10376, 13581. *P. m. ozarkiarum* Black. TEXAS: Wichita County, Spillway of Lake Wichita, Wichita Falls, one male and one female, uncatalogued. *P. m. rufinus* (Merriam). COLORADO: Larimer County, 13 mi. W. Ft. Collins, one male, 13623. Weld County, 25 mi. N.E. Ft. Collins, one male, 13624. NEW MEXICO: Taos County, Taos, two females, 13649, 13650. Torrance County, Red Canyon, Cibola National Forest, three males, 10357, 10358, 13651, two females, 10355, 10356. Lincoln County, 2 mi. W. Bonita Lake, Lincoln National Forest, five males, 10353, 13625, 10354, 13626, 13628, one female, 10350. Otero County, Cloudcroft, thirteen males, 13629–13632, 13635–13637, 13642–13646, 13648, five females, 13633, 13634, 13639, 13640, 13641. ARIZONA: Coconino County, 2 mi. E. U.S. Hwy. 89 on the Sunset Crater Road, fourteen males, 13609–13622. *Peromyscus melanotis*: Allen and Chapman. ARIZONA: Pima County, Santa Catalina Mountains, Bear Wallow Campground, one male, 13661. Graham County; Coronado National Forest, Arcadia Campground, two males, 13657, 13658, one female, 13660. 35.0 mi. W. junction of U.S. Hwy. 666 and Arizona 366, one male, 13659. Cochise County, 14 mi. W. Portal, four males, 13653–13656. MEXICO: CHIHUAHUA: 8.2 mi. S. San Juanita, one male, 13662, one female, 13663. DURANGO: 25.2 mi. W. El Salto, six males, 13664–13666, 13673, 13675, 13676, one female, 13674. 29.1 mi. W. El Salto, four males, 13677–13680. 31 mi. W. El Salto, four males, 13667, 13668, 13670, 13671, two females, 13669, 13672. VERACRUZ: 4 mi. E. Perote, three males, 13682–13684, one female, 13681. DISTRICTO FEDERAL, 17.3 mi. S.E. Amecameca on the road to Popocatepetl, six males, 13686–13688, 13691–13693, two females, 13689, 13690.

The following specimens were processed electrophoretically and are preserved in the laboratory of Dr. Robert K. Selander, Department of Zoology, The University of Texas at Austin. The specimens are identified by Dr. Selander's laboratory catalogue numbers:

Peromyscus melanotis — ARIZONA: Graham County, Coronado National Forest, Arcadia Campground, 2(3615–16); Cochise County, 14 mi. W. Portal, 8(2404, 2410, 2415, 2420, 2422, 2426, 2430, 2432); MEXICO: DURANGO, 52 km W. El Salto, 3(3602–04); VERACRUZ, 7 km E. Perote, 3(3605–7); DISTRICTO FEDERAL, 29 km S.E. Amecameca on the road to Popocatepetl, 3(3608–10); F₁ hybrids, ARIZONA; Cochise County, 14.0 mi. W. Portal times MEXICO: Durango, 4(3611–14).

LITERATURE CITED

ARAKAKI, D. T., AND R. S. SPARKES. 1967. The chromosomes of *Peromyscus maniculatus hollesteri* (deer mouse). Cytologic 32:180–183.

ARAKAKI, D. T., I. VEOMETTI, AND R. S. SPARKES. 1970. Chromosome polymorphism in deer mouse siblings (*Peromyscus maniculatus*). Experientia 26:425–426.

BAKER, R. J. 1970. Karyotypic trends in bats, p. 65–96. *In* W. A. Wimsatt (ed.), Biology of Bats, Academic Press, London/New York.

BAKER, R. J., AND J. T. MASCARELLO. 1969. Karyotypic analyses of the genus *Neotoma* (Cricetidae, Rodentia). Cytogenetics 8:187–198.

BIRDSALL, D. A., J. A. REDFIELD AND D. G. CAMERON. 1970. White bands on starch gels stained for esterase activity: A new polymorphism. Biochemical Genetics 4:655–658.

BLAIR, W. F. 1950. Ecological factors in the speciation of *Peromyscus*. Evolution 4:253–275.

BOWEN, W. W. 1968. Variation and evolution of gulf coast populations of beach mice, *Peromyscus polionotus*. Bull. Fla. State Mus. 12:1–91.

BRADSHAW, W. N., AND W. W. GEORGE. 1969. The karyotype in *Peromyscus maniculatus nubiterrae*. J. Mamm. 50:822–824.

BROWN, J. H., AND C. F. WELSER. 1968. Serum albumin polymorphisms in natural and laboratory populations of *Peromyscus*. J. Mamm. 49:420–426.

BROWN, W. L., JR. 1957. Centrifugal speciation. Quart. Rev. Biol. 32:247–277.

CANHAM, R. P., D. A. BIRDSALL AND D. G. CAMERON. 1970. Disturbed segregation at the transferrin locus of the deer mouse. Genet. Res. 16:355–357.

CLARK, D. L. 1966. Fertility of a *Peromyscus maniculatus* × *Peromyscus melanotis* cross. J. Mamm. 47:340.

DUFFEY, P. A. 1972. Chromosome variation in *Peromyscus*: A new mechanism. Science 176: 1333–1334.

HALL, E. R., AND K. R. KELSON. 1959. The mammals of North America. The Ronald Press Co., New York, Vols. 1 and 2.

HSU, T. C., AND F. E. ARRIGHI. 1968. Chromosomes of *Peromyscus* (Rodentia, Cricetidae). I. Evolutionary trends in 20 species. Cytogenetics 7:417–446.

JENSEN, J. N., AND D. I. RASMUSSEN. 1971. Serum albumins in natural populations of *Peromyscus*. J. Mamm. 52:508–514.

KING, J. A. (ed.). 1968. Biology of *Peromyscus* (Rodentia). American Society of Mammalogists, Special Publication No. 2.

KREIZINGER, J. D., AND M. W. SHAW. 1970. Chromosomes of *Peromyscus* (Rodentia, Cricetidae). II. The Y chromosome of *Peromyscus maniculatus*. Cytogenetics 9:52–70.

OHNO, S., D. WEILER, J. POOLE, L. CHRISTIAN AND C. STENIUS. 1966. Autosomal polymorphism due to pericentric inversions in the deer mouse (*Peromyscus maniculatus*) and some evidence of somatic segregation. Chromosoma 18:177–187.

PATTON, J. L. 1967. Chromosome studies of certain pocket mice, genus *Perognathus* (Rodentia—Heteromyidae). J. Mamm. 48:27–37.

RASMUSSEN, D. I. 1964. Blood group polymorphism and interbreeding in natural populations of the deer mouse, *Peromyscus maniculatus*. Evolution 18:219–229.

——. 1968. Genetics in biology of *Peromyscus* (Rodentia), p. 340–372. *In* King, J. A. (ed.), Publ. 2, American Society of Mammalogists.

——. 1970. Biochemical polymorphisms and genetic structure in populations of *Peromyscus*. Symp. Zool. Soc. Lond. 26:335–349.

RASMUSSEN, D. I., J. N. JENSEN AND R. K. KOEHN. 1968. Hemoglobin polymorphism in the deer mouse, *Peromyscus maniculatus*. Biochemical Genetics 2:87–92.

RASMUSSEN, D. I., AND R. K. KOEHN. 1966. Serum transferrin in polymorphism in the deer mouse. Genetics 54:1353–1357.

SAVAGE, E., AND D. G. CAMERON. 1971. Blood group complexity: the Pm locus in *Peromyscus maniculatus*. Anim. Blood Groups Biochem. Genet. 2:23–29.

SELANDER, R. K., M. H. SMITH, S. Y. YANG, W. E. JOHNSON AND J. B. GENTRY. 1971. Biochemical polymorphisms and systematics in the genus *Peromyscus*. I. Variation in the oldfield mouse (*Peromyscus polionotus*). Studies in Genetics VI. Univ. Texas Publ. 7103:49–90.

SINGH, R. P., AND D. B. McMILLAN. 1966. Karyotypes of three subspecies of *Peromyscus*. J. Mamm. 47:261–266.

SMITH, M. H., R. K. SELANDER, W. E. JOHNSON AND Y. J. KIM. 1973. Biochemical polymorphism and systematics in the genus *Peromyscus*. III. Variation in the Florida deer mouse (*Peromyscus floridanus*), a Pleistocene relict. J. Mamm. 54:(*in press*).

SPARKES, R. S., AND R. S. ARAKAKI. 1966. Intrasubspecific and intersubspecific chromosomal polymorphism in *Peromyscus maniculatus* (deer mouse). Cytogenetics 5:411–418.

TE, G. A., AND W. D. DAWSON. 1971. Chromosomal polymorphism in *Peromyscus polionotus*. Cytogenetics 10:225–234.

A MULTIVARIATE ANALYSIS OF SYSTEMATIC RELATIONSHIPS AMONG POPULATIONS OF THE SHORT-TAILED SHREW (GENUS *BLARINA*) IN NEBRASKA

Hugh H. Genoways and Jerry R. Choate

Abstract

Genoways, H. H., and J. R. Choate (*Museum of Natural History, The Univ. Kansas, Lawrence, Kansas 66044. Present addresses: The Museum, Texas Tech University, Lubbock, Texas 79409 and Division of Biological Sciences and Agriculture, Fort Hays Kansas State College, Hays, Kansas 67601). 1972. A multivariate analysis of systematic relationships among populations of the short-tailed shrew (genus* Blarina) *in Nebraska. Syst. Zool., 21:106–116.*— The genus *Blarina* (Mammalia: Soricidae) is represented in Nebraska by two well-differentiated, geographically exclusive phena that generally have been regarded as subspecies. Field studies conducted along their zone of contact resulted in the collection of representatives of both phena at each of five localities. Cluster analysis of distance matrix readily separated reference samples of the phena as well as test samples from near the zone of contact. A three-dimensional projection of the specimens onto their first three principal components, together with a discriminant function analysis, served further to elucidate the degree of differentiation among the phena and to confirm that their characteristic differences are maintained even where they occur sympatrically. The latter technique also indicated that one specimen not singled out by other analyses might be a natural hybrid, but none of the analyses provided even the slightest evidence for panmictic intergradation. The possibility that the phena represent the ends of a circularly intergrading species is considered, as is the possibility that the phena are distinct, biological species. Two means of speciation, one "classical" and the other involving formation of "stasipatric species," are discussed. [Multivariate analysis; Systematics; *Blarina*; Populations; Hybridization.]

Shrews of the Nearctic genus *Blarina* (Mammalia: Soricidae) historically have been classified as representatives of two species—one, *B. brevicauda* (Say), wide-ranging and geographically variable, and the other, *B. telmalestes* Merriam, restricted to the Dismal Swamp region of coastal Virginia and North Carolina and of uncertain taxonomic status (see Hall and Kelson, 1959:53, 55). This arrangement stems primarily from Merriam's (1895) revision of *Blarina* and Bole and Moulthrop's (1942) later synopsis of the genus. As presently understood, the ranges of four nominal subspecies—*B. b. brevicauda, B. b. kirtlandi* Bole and Moulthrop, *B. b. churchi* Bole and Moulthrop, and *B. b. talpoides* (Gapper)—all characterized by large external and cranial dimensions, geographically abut the range of *B. b. carolinensis* (Bachman), which is characterized by much smaller size. This zone of contact extends from Nebraska to Mary-

land and effectively divides the range of the species into two parts—a northern segment occupied by comparatively large shrews and a southern segment occupied by smaller shrews (Hall and Kelson, 1959: 53).

Jones and Findley (1954), and subsequently Jones and Glass (1960), studied the geographic relationships of taxa of *Blarina* west of the Mississippi River. They demonstrated the presence of a clinal increase in size from the Gulf coastal region to northern Nebraska. The cline exhibited a significant "step" in southern Nebraska, which was considered to constitute the line of demarcation between *B. b. brevicauda* and *B. b. carolinensis*. The magnitude of the "step" is such that Nebraskan specimens of *B. brevicauda* invariably can be assigned to subspecies without regard for location of capture; external and cranial dimensions in *B. b. brevicauda* are substantially greater than (and seldom overlap) those in *B. b. carolinensis* (Jones, 1964:67, 69, 72). Jones (1964:67) found no specimens from Nebraska that could be described as exactly intermediate between *brevicauda* and *carolinensis*, and (1964:28) regarded the two phena as "markedly different subspecies . . . that now meet along a fairly well-defined line in Nebraska with little intergradation between them."

The geographic relationship of large northern taxa to small southern taxa of *Blarina* apparently has remained unchanged, except for latitudinal shifts in position, for a long period of time. Parmalee (1967:135–136) reported two distinctive phena of *Blarina* in a Recent bone deposit in Illinois; Oesch (1967:171) found the two phena in a Pleistocene (late Wisconsin) deposit in Missouri; and Guilday et al. (1964:147–151) described large and small phena of *Blarina* from Pleistocene (Wisconsin) deposits in Pennsylvania and Virginia. These findings generally have been interpreted to demonstrate that climatic fluctuations have effected sequential geographic replacement of one subspecies by another although, as Parmalee (1967:136) admitted, "it is problematical as to whether these races were contemporaneous or occupied the . . . [areas] during different periods." Hibbard (1970:423) treated the two phena as distinct species in earlier (Illinoian) Pleistocene deposits, and preliminary cytogenetic studies (Elmer C. Birney, personal communication; Meylan, 1967; Lee and Zimmerman, 1969:337; Hoffman and Jones, 1970:389) have indicated that, indeed, more than one species might be involved (*brevicauda* has 48–50 chromosomes and *carolinensis* 46).

The initial purpose of this study, therefore, was to search the zone of contact between the nominal subspecies *B. b. brevicauda* and *B. b. carolinensis* in Nebraska (see Jones, 1964:66) for evidence of intergradation or hybridization, and thereby to shed light on the systematic relationships of these taxa.

METHODS AND MATERIALS

Intermittent field studies were conducted in the period 1965 to 1969 in three areas of Nebraska where the ranges of *B. b. brevicauda* and *B. b. carolinensis* were thought to be contiguous (see also Choate and Genoways, 1967; Genoways and Choate, 1970). One area in northeastern Adams and northern Clay counties was selected because a specimen identified as *carolinensis* had been obtained previously by Genoways at a place 1½ mi. N and 6 mi. E Hastings, Adams County, and Jones (1964:68) had reported one specimen of *brevicauda* from just 18 miles to the east at Saronville, Clay County. Additional collecting indicated that the zone of contact was between Harvard and Saronville in northern Clay County, and specimens tentatively identified as *brevicauda* were caught together with specimens of *carolinensis* at each of three localities (1 mi. N and 3 mi. W Saronville; 1 mi. N and 2 mi. W Saronville; 1 mi. N and 1 mi. W Saronville). At the first two localities representatives of both taxa were taken together in the same traplines on 20 December 1965, whereas at the last locality a specimen identified in the field as *brevicauda* was caught on 20 November 1965 and a specimen identified as *carolinensis* was taken on 2 April 1966.

In eastern Saline County the two taxa also were known from only 18 miles apart (*brevicauda* from 4 mi. NE Crete and *carolinensis* from 1½ mi. W De Witt). We were unable to define the exact area of contact in Saline County, but the known distance between the taxa was reduced to seven miles with the capture of a specimen tentatively identified as *carolinensis* at a place 5 mi. S and 3 mi. E Crete. The third area in which field studies were conducted was in Cass County, which was selected because a specimen definitely identified as *brevicauda* was known from just north of the county line at a place 1 mi. W Meadow, Sarpy County, and two undoubted specimens of *carolinensis* had been reported previously (Jones, 1964:70).

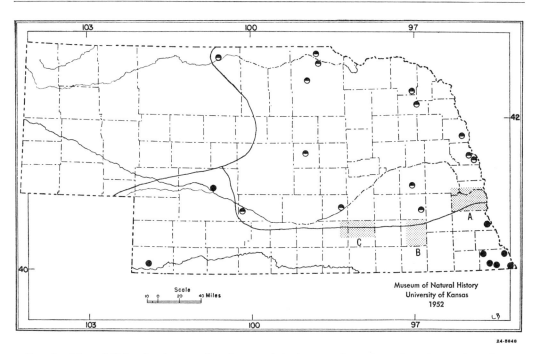

Fig. 1.—Map of Nebraska showing distribution of the *brevicauda* (half solid circles) and *carolinensis* (solid circles) phena of *Blarina* (modified from Jones, 1964:66). Localities plotted are those from which specimens were drawn for reference samples (see text). The shaded areas are enlarged in Fig. 2.

from the county to the south (1 mi. SE Nebraska City, Otoe County); furthermore, Jones (1964:67) assigned five specimens from Louisville (in extreme northern Cass County) to *B. b. brevicauda*, but remarked that they "are smaller externally and average slightly smaller cranially than topotypes of that subspecies. . . ." By trapping along a transect extending southward from Louisville to a place just south of Weeping Water, we were able to locate the zone of contact between the two taxa. On 24 November 1968, three specimens (one tentatively identified as *brevicauda* and two as *carolinensis*) were caught together in the same trapline at a place 1 mi. S and 1½ mi. W Weeping Water. In addition, two specimens identified in the field as *brevicauda* were caught on the same morning at a place 2 mi. N and 2 mi. W Weeping Water, and four specimens of *brevicauda* and 27 of *carolinensis* were caught at other localities in the same area.

Specimens thus collected (together with a few reported from near the zone of contact by Jones, 1964) were tested against reference samples from Nebraska of *brevicauda* and *carolinensis*. Localities of reference samples in the following lists are arranged from north to south and correspond to localities plotted in Figure 1; localities or counties at about the same latitude are listed from west to east. Numbers in parentheses indicate how many specimens from each locality were included in analyses. Numbers (bold face) of test samples refer to localities numbered in Figure 2.

brevicauda reference sample

CHERRY CO.: 3 mi. SSE Valentine (1). KEYA PAHA CO.: 12 mi. NNW Springview (1). BOYD CO.: 5 mi. WNW Spencer (1). HOLT CO.: 1 mi. S Atkinson (1); 6 mi. N Midway (1). KNOX CO.: 3 mi.

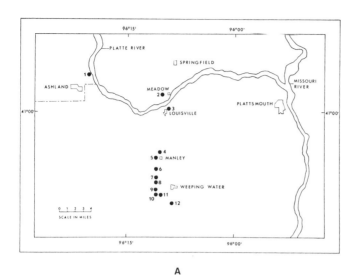

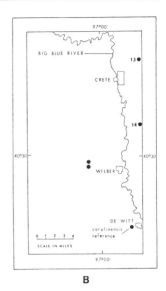

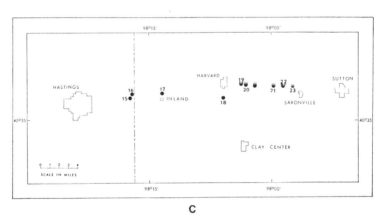

FIG. 2.—Areas in eastern Nebraska where two phena of *Blarina* were found to be contiguous or sympatric. A, Cass County and adjacent parts of southern Saunders and Sarpy counties; B, eastern Saline County; C, northeastern Adams County and northern Clay County. Numbered localities refer to test samples identified in text. Unnumbered localities indicate places where specimens of *Blarina* have been taken that could not be used in statistical analyses because we were unable to obtain some measurements from them.

W Niobrara (1). CEDAR CO.: 4 mi. SE Laurel (6). WAYNE CO.: ½ mi. W Wayne (2); Wayne (7). BURT CO.: 1 mi. E Tekamah (2). VALLEY CO.: 2½ mi. N Ord (2). WASHINGTON CO.: 6 mi. SE Blair (4). BUTLER CO.: 4 mi. E Rising City (1); 5 mi. E Rising City (2). DAWSON CO.: 5 mi. S Gothenburg (1). HALL CO.: 6 mi. S Grand Island (1). SEWARD CO.: 1 mi. N Pleasant Dale (3).

carolinensis reference sample

LINCOLN CO.: 2 mi. N North Platte (3). OTOE CO.: 3 mi. S, 2 mi. E Nebraska City (6). SALINE CO.: ½ mi. W De Witt (1). DUNDY CO.: 5 mi. N, 2 mi. W Park (18). RICHARDSON CO.: 4 mi. E Barada (4); 5 mi. N, 2 mi. W Humboldt (2); 3½ mi. S, 1 mi. W Dawson (3); 8 mi. S, 1 mi. E Dawson (1); 6 mi. W Fall City (1); 1 mi. S, 1½ mi. W Rulo (2).

Test samples

1—2 mi. NE Ashland, Saunders Co. (3).
2—1 mi. W Meadow, Sarpy Co. (1). 3—
Louisville, Cass Co. (4). 4—½ mi. N Manley, Cass Co. (1). 5—½ mi. W Manley,
Cass Co. (2). 6—2 mi. N, 2 mi. W Weeping
Water, Cass Co. (2). 7—1 mi. N, 2 mi. W
Weeping Water, Cass Co. (6). 8—⁴⁄₁₀ mi.
N, 2 mi. W Weeping Water, Cass Co. (1).
9—³⁄₁₀ mi. S, 2 mi. W Weeping Water, Cass
Co. (10). 10—1 mi. S, 2 mi. W Weeping
Water, Cass Co. (3). 11—1 mi. S, 1½ mi.
W Weeping Water, Cass Co. (3). 12—2 mi.
S Weeping Water, Cass Co. (8). 13—2 mi.
NE Crete, Saline Co. (3). 14—5 mi. S, 3
mi. E Crete, Saline Co. (1). 15—1½ mi.
N, 6 mi. E Hastings, Adams Co. (2). 16
—1²⁄₁₀ mi. N, 5⁶⁄₁₀ mi. E Hastings, Adams
Co. (2). 17—½ mi. N Inland, Clay Co. (1).
18—1 mi. S Harvard, Clay Co. (3). 19—
1½ mi. E Harvard, Clay Co. (3). 20—1 mi.
N, 6 mi. W Saronville, Clay Co. (1). 21—
1 mi. N, 3 mi. W Saronville, Clay Co. (3).
22—1 mi. N, 2 mi. W Saronville, Clay Co.
(2). 23—1 mi. N, 1 mi. W Saronville, Clay
Co. (2).

Specimens listed above are housed in
The University of Kansas Museum of
Natural History. The age of each individual selected for analysis was estimated
(Choate, 1968:253; 1970:214), and no specimen judged to be less than adult size was
included in reference or test samples. Nine
cranial measurements (Choate, 1972) were
taken from each specimen by Choate (by
means of dial calipers) as follows: occipitopremaxillary length; length of P4-M3; cranial breadth; breadth of zygomatic plate;
maxillary breadth; interorbital breadth;
length of mandible; height of mandible;
articular breadth.

Computations were performed using a
system of multivariate statistical computer
programs (NT-SYS) developed by F. J.
Rohlf, R. Bartcher, and J. Kishpaugh for
the GE 635 computer at The University of
Kansas (see Schnell, 1970:42). Matrices
of Pearson's product-moment correlations
were computed, and taxonomic distance
coefficients were derived from standardized character values. Cluster analyses
were conducted using UPGMA (unweighted pair group method using arithmetic averages) on the correlation and
distance matrices, and a phenogram was
generated for each. Phenograms were compared with their respective matrices, and
a coefficient of cophenetic correlation was
computed for each comparison. A matrix
of correlation among characters then was
computed, and the first three principal
components extracted. A three-dimensional
projection of the OTUs onto the first three
principal components was made; this projection then was drawn using a Benson-Lehner incremental plotter. Rising (1968,
1970) used principal component analysis
to assess interbreeding between species of
chickadees (genus *Parus*) and orioles
(genus *Icterus*), respectively.

Discriminant function analysis was performed using the MULDIS subroutine of
the NT-SYS system. This program uses
variance-covariance mathematics to differentially weight characters relative to
their within- and between-groups variation.
For the discriminant analysis in this paper,
two reference samples from areas geographically removed from zones of
suspected hybridization were used. A discriminant multiplier was calculated for
each character, and this was multiplied by
the value of its respective character; all
such values were summed for each individual to yield its discriminant score.
The discriminant scores were plotted on
a frequency histogram to compare individuals of the two reference samples and
to compare the test sample from the intermediate geographical areas where hybridization was suspected. A good discussion
of discriminant functions is given by
Jolicoeur (1959); Lawrence and Bossert
(1969) used this test to identify hybrids
in their study of members of the genus
Canis as did Birney (1970) in a study of
woodrats of the genus *Neotoma*.

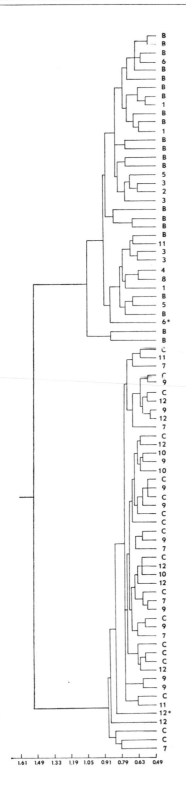

A distance phenogram (Fig. 3) was prepared using 21 reference specimens of *B. b. brevicauda* and 18 of *B. b. carolinensis*, together with 44 test specimens from localities at or near the zone of contact of those taxa in Saunders, Sarpy, and Cass counties (Fig. 2A). The phenogram is divided into two major clusters separated by an appreciable phenetic distance (1.82). The upper cluster contains all the reference specimens of *brevicauda*, whereas the lower cluster contains all the reference specimens of *carolinensis*; specimens from test samples appear in both clusters. All specimens from as far south in Cass County as sample 6 are in the upper part with the *brevicauda* reference sample. Specimens denoted by an asterisk (6* and 12*) are discussed below. The six specimens from sample 7 and the 10 from sample 9 fall in the lower cluster with the reference specimens of *carolinensis*; however, a specimen from a geographically intermediate locality (sample 8) fell with the *brevicauda* specimens. Of the three specimens from sample 11, two are in the lower part with *carolinensis*, whereas the third is in the upper part with *brevicauda*. All eight specimens from sample 12 and the three from sample 10 are grouped with *carolinensis*. The two reference specimens of *brevicauda* at the lower end of the upper cluster of the phenogram, and at a substantial "distance" from other specimens in that group, are young adults with relatively small dimensions.

A three-dimensional projection of the specimens onto the first three principal

←

FIG. 3.—Phenogram computed from distance matrix based on standardized characters and clustered by the unweighted pair-group method using arithmetic averages (UPGMA). Numbers refer to individuals from test samples identified in text and in Fig. 2. Specimens labelled "B" are from reference samples of *brevicauda*, whereas specimens labelled "C" are from reference samples of *carolinensis*. An asterisk indicates that special reference is made to the specimen in text.

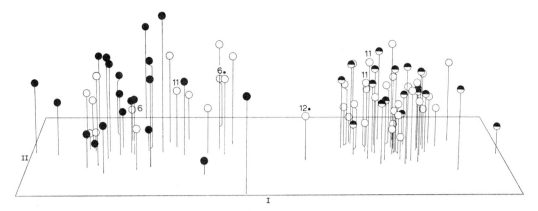

Fɪɢ. 4.—Three-dimensional projection of 83 specimens onto the first three principal components based on a matrix of correlations among 12 external and cranial measurements. I and II are indicated in the figure and III is represented by height. The first three components include approximately 92 per cent of the total variance, with component I accounting for 83.92, II for 4.88, and III for 3.15 per cent, respectively. Numbers refer to individuals from test samples identified in text and in Fig. 2. An asterisk indicates that special reference is made to the specimen in text.

components (Fig. 4) likewise shows two groups, one (on the right in the figure) containing all the reference specimens of *carolinensis* and the other (on the left) all the reference specimens of *brevicauda*. One specimen from sample 12 (designated 12*) is situated between the groups, as would be expected of a hybrid or intergrade, although it was grouped with the *carolinensis* reference specimens in the distance phenogram. However, a note made by us when that specimen was measured suggests that it may have abnormal proportions ("skull unusually long, rostrum barrel-shaped, cranium narrow relative to length"). The remainder of the specimens are clustered as would be expected on the basis of the distance phenogram. Note that the three specimens from locality 11 still are divided with two in the *carolinensis* cluster and one in the *brevicauda* cluster. Also, note the position of specimen denoted as 6* toward the lower limit of the *brevicauda* group; this specimen is discussed below.

Discriminant function analysis (Fig. 5) was conducted using reference specimens totaling 37 for *brevicauda* and 40 for *carolinensis*. From the table of discriminant multipliers (Table 1), it can be seen that

all the cranial measurements excepting length of the mandible were weighted heavily, whereas the three external measurements were weighted comparatively lightly. The discriminant scores for the *brevicauda* reference sample ranged from 38.158 to 43.330 and those for the *carolinensis* reference sample ranged from 31.199 to 34.827, thus yielding a separation between the taxa of 3.331. The specimen (12*) that fell in the intermediate area of the three-dimensional plot (Fig. 4) has a

TABLE 1. Dɪsᴄʀɪᴍɪɴᴀɴᴛ ᴍᴜʟᴛɪᴘʟɪᴇʀs ʀᴇsᴜʟᴛɪɴɢ FROM A DISCRIMINANT FUNCTION ANALYSIS COMPARING *Blarina brevicauda brevicauda* WITH *B. b. carolinensis* ɪɴ Nᴇʙʀᴀsᴋᴀ.

Character	Discriminant Multiplier
Total Length	0.045
Length of Tail Vertebrae	−0.239
Length of Hind Foot	0.274
Occipito-premaxillary Length	−0.482
Length of P4-M3	3.814
Cranial Breadth	1.023
Breadth of Zygomatic Plate	−0.358
Maxillary Breadth	−2.709
Interorbital Breadth	0.529
Length of Mandible	0.068
Height of Mandible	2.302
Articular Breadth	3.941

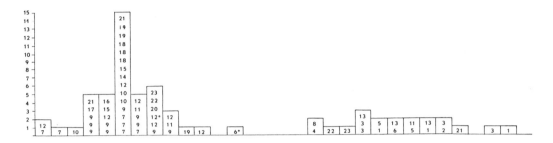

CAROLINENSIS REFERENCE   BREVICAUDA REFERENCE

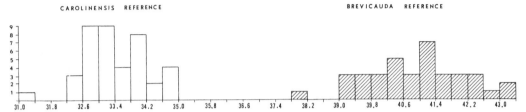

FIG. 5.—Histogram of linear discriminant scores for short-tailed shrews from Nebraska. Discriminant scores are indicated along the bottom of the histogram and frequency of individuals is indicated on the left-hand side. Individuals arranged below are from reference samples of *brevicauda*, at right, and *carolinensis*, at left. Individuals arranged above are numbered according to test samples, which are identified in text and in Fig. 2. The specimens denoted by an asterisk are discussed in text.

value of 34.027 and clearly pertains to *carolinensis*. However, two test specimens from Cass County have values between those of the reference samples, as would be expected of hybrids or intergrades (see especially Lawrence and Bossert, 1969, and Birney, 1970). One of those specimens (from sample 12) has a value of 35.034 and probably is best considered a representative of *carolinensis*. The other specimen (6*), with a discriminant score of 36.070, might actually be a hybrid between the taxa. It is of special interest to note that the discriminant score of this specimen was nearer the upper limit for *carolinensis* than the lower limit for *brevicauda*, although in both the distance phenogram and three-dimensional plot (Figs. 3 and 4) the specimen was grouped with *brevicauda*. Other specimens from Saunders, Sarpy, and Cass counties were arranged within the same phena as they were in the distance phenogram and three-dimensional plot; this includes those from sample 11, where two specimens fell with the *carolinensis* reference sample and one fell with *brevicauda*.

From Saline County (Fig. 2D), the one specimen comprising sample 14 had a discriminant score that fell within the range of the *carolinensis* reference sample, whereas the three specimens from sample 13 all fell within the range for *brevicauda*. Among Clay County (Fig. 2C) specimens, three samples (21, 22, and 23) include representatives of both taxa, thus confirming tentative field identifications. However, at none of those localities or any other locality from which we have examined specimens is there any indication of intergradation between the taxa.

DISCUSSION

Data presented herein yield no indication that the nominal subspecies *B. b. brevicauda* and *B. b. carolinensis* intergrade in Nebraska. Only one specimen (6*, Fig. 5) was found to be intermediate between the phena using discriminant function analysis; a second possibly intermediate specimen (12*) was identified using the principal components analysis. Probably only the specimen from sample

6 is a "hybrid" or "intergrade" among the 66 specimens tested from the zones of contact between the two taxa. In other words, *brevicauda* and *carolinensis* behave as good biological species where their ranges are contiguous in southern Nebraska. Unpublished data (John B. Bowles, personal communication) indicate that a similar relationship between large and small phena of *Blarina* probably exists across southern Iowa. Panmictic intergradation resulting in numerous viable hybrids between *brevicauda* and *carolinensis* almost certainly does not occur west of the Mississippi River at the present time, and we know of no conclusive published evidence for intergradation east of the Mississippi River. That an occasional viable hybrid might be produced in the zone of contact between the phena is entirely consistent with their behavior as "species."

We recognize two possible explanations for the evolutionary relationship between these phena: (1) that they are an example of circular overlap (as defined by Mayr, 1963:507–512, 664) within the same species —this has been reported for several species of mammals, including *Sorex vagrans* (Findley, 1955:14), *Thomomys talpoides* (Long, 1965:603), and *Peromyscus maniculatus* (Dice, 1931; Hooper, 1942; King, 1948; Harris, 1954)—or (2) that the phena represent distinct species as suggested by Hibbard (1970:423).

Jones (1964:28–31) provided an explanation for the circumstances that might have resulted in Nebraskan populations of *B. brevicauda* becoming the ends of a circularly intergrading species. He hypothesized that *B. brevicauda*, which is a common inhabitant of the eastern deciduous forest, became widespread on the plains during the warm, wet segment of the Hypsithermal Period during post-Wisconsin times. The species probably varied clinally in size, ranging from small in the south to large in the north in typical Bergmannian fashion. During the subsequent Xerothermic Period, a general drying occurred and the distribution of *B.*

brevicauda was divided into two segments as far east as the eastern limit of the so-called "prairie peninsula." Jones postulated that during reinvasion of the plains those populations to the northeast and southeast reached Nebraska sooner than those directly to the east; as a result, the middle portion of the cline was obliterated and two distinctly divergent phena achieved secondary contact.

One notable characteristic of the examples given for circular overlap that is lacking in *Blarina* is a high degree of ecological separation between the overlapping subspecies (defined as "microallopatry" by Smith, 1965:57). No ecological separation of the taxa is evident in *Blarina* in that all specimens from the zone of contact were trapped in grassy roadside ditches in otherwise highly agricultural areas; disruption of the original habitat, however, may have altered some original ecological differences. Another problem with this interpretation has to do with the fossil record; if available paleontological evidence is correct, the secondary zone of contact between the phena has fluctuated with regard both to latitude and longitude at least since the middle Pleistocene, long before the period of time suggested by Jones (1964) for elimination of the central part of the cline. Considering the element of time and the durability of the geographic relationship, the two taxa seem to us to be behaving more nearly like closely related species than like subspecies.

If, indeed, the phena represent distinct species, speciation classically would be interpreted as having resulted from geographic isolation of the phena during or before the Kansas glaciation, with the resultant taxa having maintained a parapatric distribution (in the sense used for mammals by Vaughan, 1967, although possibly without ecological divergence) at least since Illinoian times. Accordingly, the two species might have displaced one another north and south (and probably also east and west) across the plains in response to fluctuations in environmental factors during the Pleistocene, with one species competitively excluding the other depending on the direction of the climatic shift.

Another possible interpretation is that the large and small phena of *Blarina* represent "stasipatric species" (Key, 1968; White, 1968; White et al., 1967). With development of a "tension zone," possibly as the result of the chromosomal differences between the two emergent phena, speciation might have occurred gradually without actual geographic isolation of the main body of the parental stock, although small peripheral populations undoubtedly must have undergone isolation and differentiation in the classical sense. The tension zone could have shifted position geographically, as described by Key (1968), in response to changing environmental conditions in the Pleistocene. Hybridization would have occurred regularly across the tension zone, especially early in the evolution of this complex. However, divergence now might have progressed to the level (at least in Nebraska) at which the tension zone of intergradation has ceased to function; the presence of only one probable hybrid in our combined samples of 66 specimens from at or near the zone of contact between the two phena strongly suggests that isolating mechanisms are actively preventing, or at least restricting, hybridization, and that introgression is negligible.

ACKNOWLEDGMENTS

Field studies conducted by the authors were supported in part by a grant from the Kansas Academy of Science. Multivariate analyses were performed on the GE 635 computer at The University of Kansas Computation Center. Joyce E. Genoways provided clerical assistance and prepared the illustrations. Finally, helpful discussions were held with numerous persons at The University of Kansas Museum of Natural History, but special thanks is due Drs. Elmer C. Birney, John B. Bowles, Robert S. Hoffman, J. Knox Jones, Jr., and Carleton J. Phillips.

REFERENCES

BIRNEY, E. C. 1970. Systematics of three species of woodrats (genus *Neotoma*) in central North America. Ph.D. dissertation, The University of Kansas, Lawrence.

BOLE, B. P., JR., AND P. N. MOULTHROP. 1942. The Ohio Recent mammal collection in the Cleveland Museum of Natural History. Sci. Publ. Cleveland Mus. Nat. Hist., 5:83–181.

CHOATE, J. R. 1968. Dental abnormalities in the short-tailed shrew, *Blarina brevicauda*. J. Mamm., 49:251–258.

CHOATE, J. R. 1970. Systematics and zoogeography of Middle American shrews of the genus *Cryptotis*. Univ. Kansas Publ., Mus. Nat. Hist., 19:195–317.

CHOATE, J. R. 1972. Variation within and among Connecticut populations of the short-tailed shrew. J. Mamm., 53:116–128.

CHOATE, J. R., AND H. H. GENOWAYS. 1967. Notes on some mammals from Nebraska. Trans. Kansas Acad. Sci., 69:238–241.

DICE, L. R. 1931. The occurrence of two subspecies of the same species in the same area. J. Mamm., 12:210–213.

FINDLEY, J. S. 1955. Speciation of the wandering shrew. Univ. Kansas Publ., Mus. Nat. Hist., 9:1–68.

GENOWAYS, H. H., AND J. R. CHOATE. 1970. Additional notes on some mammals from eastern Nebraska. Trans. Kansas Acad. Sci., 73:120–122.

GUILDAY, J. E., P. S. MARTIN, AND A. D. McCRADY. 1964. New Paris No. 4: A Pleistocene cave deposit in Bedford County, Pennsylvania. Bull. Natl. Speleol. Soc., 26:121–194.

HALL, E. R., AND K. R. KELSON. 1959. The mammals of North America. Ronald Press, New York, 1:xxx+1–546+79.

HARRIS, V. T. 1954. Experimental evidence of reproductive isolation between two subspecies of *Peromyscus maniculatus*. Contrib. Lab. Vert. Biol., Univ. Michigan, 70:1–13.

HIBBARD, C. W. 1970. Pleistocene mammalian local faunas from the Great Plains lowland provinces of the United States. Pp. 395–433, *in* Pleistocene and Recent environments of the central Great Plains (W. Dort, Jr., and J. K. Jones, Jr., eds.), Spec. Publ. 3, Dept. Geol., Univ. Kansas, xii+433 pp.

HOFFMANN, R. S., AND J. K. JONES, JR. 1970. Influence of late-glacial and post-glacial events on the distribution of Recent mammals on the northern Great Plains. Pp. 356–394, *in* Pleistocene and Recent environments of the central Great Plains (W. Dort, Jr., and J. K. Jones, Jr., eds.), Spec. Publ. 3, Dept. Geol., Univ. Kansas, xii+433 pp.

HOOPER, E. T. 1942. An effect on the *Peromyscus maniculatus* rassenkreis of land utilization in Michigan. J. Mamm., 23:193–196.

JOLICOEUR, P. 1959. Multivariate geographical variation in the wolf *Canis lupus* L. Evolution, 13:283–299.

JONES, J. K., JR. 1964. Distribution and taxonomy of mammals of Nebraska. Univ. Kansas Publ., Mus. Nat. Hist., 16:1–356.

JONES, J. K., JR., AND J. S. FINDLEY. 1954. Geographic distribution of the short-tailed shrew, *Blarina brevicauda*, in the Great Plains. Trans. Kansas Acad. Sci., 57:208–211.

JONES, J. K., JR., AND B. P. GLASS. 1960. The short-tailed shrew, *Blarina brevicauda*, in Oklahoma. Southwestern Nat., 5:136–142.

KEY, K. H. L. 1968. The concept of stasipatric speciation. Syst. Zool., 17:14–22.

KING, J. A. 1948. Maternal behavior and behavioral development in two subspecies of *Peromyscus maniculatus*. J. Mamm., 39:177–190.

LAWRENCE, B., AND W. H. BOSSERT. 1969. The cranial evidence for hybridization in New England *Canis*. Breviora, Mus. Comp. Zool., 330:1–13.

LEE, M. R., AND E. G. ZIMMERMAN. 1969. Robertsonian polymorphism in the cotton rat, *Sigmodon fulviventer*. J. Mamm., 50:333–339.

LONG, C. A. 1965. The mammals of Wyoming. Univ. Kansas Publ., Mus. Nat. Hist., 14:493–758.

MAYR, E. 1963. Animal species and evolution. Harvard Univ. Press, Cambridge, xiv+797 pp.

MERRIAM, C. H. 1895. Revision of the shrews of the American genera *Blarina* and *Notiosorex*. N. Amer. Fauna, 10:5–34, 102–107.

MEYLAN, A. 1967. Formules chromosomique et polymorphisme Robertsonian chez *Blarina brevicauda* (Say) (Mammalia: Insectivora). Canadian J. Zool., 45:1119–1127.

OESCH, R. D. 1967. A preliminary investigation of a Pleistocene vertebrate fauna from Crankshaft Pit, Jefferson County, Missouri. Bull. Natl. Speleol. Soc., 29:163–185.

PARMALEE, P. W. 1967. A Recent cave bone deposit in southwestern Illinois. Bull. Natl. Speleol. Soc., 29:119–147.

RISING, J. D. 1968. A multivariate assessment of interbreeding between the chickadees, *Parus atricapillus* and *P. carolinensis*. Syst. Zool., 17:160–169.

RISING, J. D. 1970. Morphological variation and evolution in some North American orioles. Syst. Zool., 19:315–351.

SCHNELL, G. D. 1970. A phenetic study of the suborder Lari (Aves) I. Methods and results of principal components analyses. Syst. Zool., 19:35–57.

SMITH, H. M. 1965. More evolutionary terms. Syst. Zool., 14:57–58.

VAUGHAN, T. A. 1967. Two parapatric species of pocket gophers. Evolution, 21:148–158.

WHITE, M. J. D. 1968. Models of speciation. Science, 159:1065–1070.

WHITE, M. J. D., R. E. BLACKWITH, R. M. BLACKWITH, AND J. CHENEY. 1967. Cytogenetics of the *viatica* group of morabine grasshoppers. I. The "coastal" species. Australian J. Zool., 15:263–302.

(Received March 10, 1971)

103

SECTION 2—ANATOMY AND PHYSIOLOGY

Form and function are intimately related. It is difficult to consider one at all thoroughly without considering the other.

In taxonomy, classification begins with individuals and proceeds through local aggregates or populations, geographic variants, subspecies, and species, and on to groupings at the level of higher categories. In ecology, the individual organism is the basic unit, and progressively more inclusive and more complex levels are local species populations, local communities and ecosystems of many species, and finally the entire biosphere of life-supporting parts of the surface of the Earth. Similarly, in anatomy and physiology there are organizational levels. However, in these fields the individual is the largest unit instead of the smallest, except as we may speak of the anatomical characters of a species or other taxon. Form or function may be studied at the biochemical or molecular level, or at progressively higher levels through more complex molecules, organelles, cells, tissues, organs, systems, and finally to the organism in its entirety.

The study of anatomy began at the gross level and only after the invention of the microscope and development of special techniques of preparing materials did histological and cytological studies become possible. Physiology developed later than gross anatomy and in many ways paralleled chemistry and physics.

Our selection of examples is a modest one, drawn from a rich field, and while none has electron photomicrographs or histochemical analyses, they do, nevertheless, serve to illustrate some fundamental biological concepts.

The concept of homeostasis was conceived and broadly applied in physiology. We judge that homeostasis or the tendency of an organism to maintain internal conditions at a dynamic equilibrium is the most general concept of physiology, and that homology is the most general concept of anatomy. This concept is implicit in every comparative study of anatomy, and hence in some of the papers reproduced here. For an explicit treatment of the concept see Bock (1969).

Most anatomical and physiological studies are not comparative, and deal with one species only, focusing on the description or mechanisms of form and function. The contributions by Hooper and Hughes are comparative studies within one family (Cricetidae) and one order (Marsupialia), respectively. Each author studied a different part of the animals concerned, in this case reproductive structures, and attempted to relate his observations to existing knowledge within the systematic framework. Many comparative studies deal with other organ systems.

Techniques are also extremely important. Just as the light microscope opened new vistas for anatomy of cells and tissues, so the recent development of electron microscopy and the scanning electron microscope have greatly increased magnifications. The use of radioisotopes is another important recent development in technique.

The next paper, by Noback, treats hair, one of the unique features of the Class Mammalia, and theorizes about its adaptive and phylogenetic implica-

105

tious. This article is from a symposium that contains other interesting papers on hair.

The three reprinted papers of Hildebrand, Evans and Maderson, and Rabb treat form and function together, of entire animals during high speed locomotion, of part of the respiratory system as it relates to vocalization, and of the poisonous salivary glands of one species, respectively.

Among the classic works in mammalian anatomy is Weber's *Die Säugetiere* (1927, 1928). English mammalogists dating back to Richard Owen and earlier have published many comparative papers on mammalian anatomy (see for example Pocock's *The External Characters of the Pangolins*, 1924). One of the most productive American mammalian anatomists was A. B. Howell, whose *Anatomy of the Wood Rat* (1926) and *Aquatic Mammals* (1930) both have much to offer. Four good recent works of a comparative nature are Rinker's (1954) study of four cricetine genera, Vaughan's (1966) paper on flight of bats, Klingener's (1964) treatment of dipodoid rodents, and D. Dwight Davis' major work (1964) on the giant panda. The ANATOMICAL RECORD and JOURNAL OF MORPHOLOGY are two of the more important serial publications containing papers on anatomy.

Among the environmental influences that are important to organisms, and the effects of which within the organism must be mitigated, are water, oxygen and other gases, energy sources (food), ions, temperature, and radiation. Most of these factors are touched upon in one or more of the last three papers in this selection in ways that help clarify the adaptive nature of internal, behavioral, and ecological responses. In addition to these aspects of physiology, some areas of special mammalogical interest are hibernation (Kayser, 1961), estivation, thermoregulation, and sensory physiology.

A paper by Brown (1968), too long to include among our selections, is an excellent example of how physiological adaptations, related in this case to environmental temperature, can be studied comparatively. Other important contributions in mammalian physiology can be found in such journals as COMPARATIVE BIOCHEMISTRY AND PHYSIOLOGY, JOURNAL OF APPLIED PHYSIOLOGY, JOURNAL OF CELL AND COMPARATIVE PHYSIOLOGY, and PHYSIOLOGICAL ZOÖLOGY.

Number 625   May 10, 1962

OCCASIONAL PAPERS OF THE MUSEUM OF ZOOLOGY
UNIVERSITY OF MICHIGAN
Ann Arbor, Michigan

THE GLANS PENIS IN *SIGMODON, SIGMOMYS,* AND *REITHRODON* (RODENTIA, CRICETINAE)

By Emmet T. Hooper

Cotton rats (*Sigmodon* and *Sigmomys*), marsh rats, (*Holochilus*), coney rats (*Reithrodon*), and red-nosed rats (*Neotomys*) compose an assemblage which Hershkovitz (1955) considers to be natural and which he designates as the "sigmodont group." This group contrasts with oryzomyine, ichthyomyine, phyllotine, akodont, and other supraspecific assemblages which various authors (e.g., Thomas, 1917; Gyldenstolpe, 1932; Hershkovitz, 1944, 1948, 1955, 1960; and Vorontsov, 1959) have recognized in analyzing the large cricetine fauna of South America. While all of these groups are tentative, at least in regard to total complement of species in each, nevertheless some are strongly characterized and probably natural; and all, whether natural or not, are useful in that they constitute conveniently assessable segments of an unwieldly large South American cricetine fauna, now disposed in approximately 40 nominal genera. New information regarding three of those genera is provided below. It is derived from fluid-preserved and partially cleared glandes (procedures described by Hooper, 1959) as follows:

Reithrodon cuniculoides: Argentina, Tierra del Fuego, 1 adult. *Sigmodon alleni*: Michoacán, Dos Aguas, 3 adults. *S. hispidus*: Arizona, Pima Co., 1 subadult. Florida, Alachua and Osceola counties, 3 adults. Michoacán, Lombardia, 2 adults. *S. minimus*: New Mexico, Hidalgo Co., 1 juvenile. *S. ochrognathus*: Texas, Brewster Co., 1 subadult. *Sigmomys alstoni*: Venezuela, Aragua, 1 subadult.

I am indebted to Elio Massoia for the specimen of *Reithrodon* and to Charles O. Handley, Jr., for the example of *Sigmomys*. Figures 1 and 2 were rendered by Suzanne Runyan, staff artist of the Museum of Zoology. The National Science Foundation provided financial aid.

Listed below in sequence are representative measurements (in mm.)

of *Sigmodon hispidus* (averages of five adults), *Sigmomys alstoni* (one subadult), and *Reithrodon cuniculoides* (one adult). Length of hind foot: 34, 30, 33; greatest lengths of glans, 7.6, 6.6, 7.8; greatest diameter of glans, 6.2, 4.0, 5.0; length of main bone of baculum, 5.5, 4.9, 4.1; length of medial distal segment of baculum, 2.8, 2.0, 2.7; total length of baculum, 8.3, 6.9, 6.8.

DESCRIPTION OF GLANDES

Sigmodon hispidus.—In *Sigmodon hispidus* the glans is a spinous, stubby, contorted cylinder (Fig. 1), its length one-fourth to one-fifth that of the hind foot and its greatest diameter approximately three-fourths its length (see measurements). The spines which densely stud almost all of the epidermis, except that of the terminal crater, are short and thick-set; each is recessed in a rhombic or hexagonal pit. The glans is somewhat swayback and potbellied, yet in its basal one-half or two-thirds it is essentially plain and cylindrical, without lobes or folds other than a short midventral frenum which, as an indistinct raphe, continues distad to the rim of the crater. The distal third or half of the glans is conspicuously hexalobate, the six lobes separated from each other by longitudinal troughs or grooves which increase in depth distad. The lobes are unequal in size and shape; the ventral pair is largest and the least convex, the lateral pair smallest, and the dorsal pair the most convex; the latter is a key item in the swayback appearance of the glans. These lobes converge distally, and their crescentic lips form the scalloped, overhanging rim of the terminal crater.

The largest structure in the crater is the mound which houses the medial distal segment of the baculum. Nestled between the lips of the ventral lobes, it projects outside the crater approximately to the limits of the dorsal lobes. The two smaller lateral mounds, housing the lateral processes of the baculum, are closely appressed to the medial mound, and the tip of each is distinctly pointed, rather than gently rounded like the medial mound. Immediately ventral to the medial mound is the meatus urinarius which is guarded ventrally by a urethral process. This process consists of a pair of rather thick arms each of which is out-curved and tapers to an obtuse tip (Fig. 1); in one specimen the ventral face of the process is studded with spines. Dorsal to the medial mound is the dorsal papilla, which is a single distensible cone of soft tissue dotted with spines both dorsally and laterally. Two additional pairs of crater conules, here termed "dorsolateral and lateral papillae," are particularly noteworthy because, insofar as known

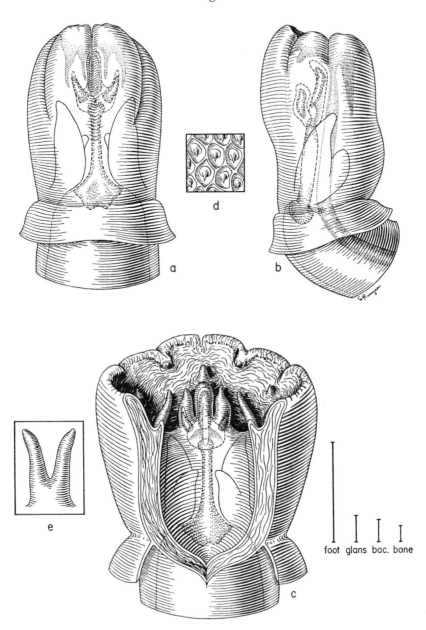

Fig. 1. Views of glans penis of *Sigmodon hispidus*: *a*, dorsal; *b*, lateral; *c*, incised midventrally exposing urethra; *d*, epidermal spines, enlarged; *e*, urethral process, enlarged, ventral aspect; UMMZ 97270, Florida.

in the New World cricetids studied to date, they are peculiar to *Sigmodon* and *Sigmomys*. All four of these are spine-studded, stubby, and smoothly rounded terminally. Each dorsolateral papilla is situated just below the crater rim at the junction of the dorsal and lateral lobes. Each lateral papilla is partly recessed in a pocket on the lower flank of the crater wall alongside a lateral bacular mound.

There is no ventral shield (a large mass of tissue between the urethral process and the ventral lip of the crater) as seen in most microtines, and the bacular mounds are relatively free within the crater, there being no partitions connecting the lateral mounds with the crater walls; the urethra empties onto the crater floor, not into a partition-encircled secondary crater within the larger crater, an arrangement seen in some rodent species.

Below the crater floor is a right and left pair of bilobed sacs (Fig. 1), each ovoid ventral lobe about 1.5 mm. in length, and each attenuate dorsal lobe approximately a millimeter longer, its tip extending distad almost to the limits of the main bone of the baculum. These sacs or sinuses emerge from tissues situated beside the corpora cavernosa penis and they extend alongside the baculum and the corpus cavernosum urethra, but they apparently are not parts of either of those structures. Composed entirely of soft tissues and engorged with blood in some specimens, they appear to be continuous with the deep dorsal vein and, thus, they seem to be part of the vascular system. Similar sacs, as illustrated in *Phyllotis* by Pearson (1958:424) for example, occur in all of those New World cricetids studied to date that have a four-part baculum; they have not been observed in *Peromyscus*, *Neotoma*, or other cricetid groups which are characterized by a simple baculum and glans.

The four-part baculum is at least as long as the glans and is one-fourth the hind foot in length (see measurements). The main bone, one-sixth the length of the hind foot, is angular and gross. The dorsal face of its wide and angular base is deeply concave between prominent lateral and proximal condyles to which the corpora cavernosa attach, while the ventral surface is almost flat except for a midventral keel of either cartilage or bone which, spanning approximately four-fifths the length of the bone, terminates at the cartilage of the digital junction. The shaft is oval in cross-section, the dorsoventral diameter exceeding the transverse one; as viewed laterally it is slightly bent and is constricted terminally, while in ventral view it is gently tapered distad before expanding to form a distinct terminal head.

The three distal segments of the baculum are subequal in length, the lateral pair slightly shorter than the medial one. They differ considerably in shape and amount of ossification. In one breeding adult they are entirely cartilaginous, while in four other adults they contain various amounts of osseous tissue in addition to cartilage; probably in very old animals they are entirely osseous. The medial segment, attached to the ventral sector of the main bone, projects distad and slightly ventrad, then it bends abruptly dorsad before terminating in a rounded tip. It is approximately oval in cross section in its distal three-fourths, but in its proximal fourth it is much wider than deep and is keeled ventrally; moreover, at the digital junction it bears a pair of lateral processes and a medial flange, the continuation of the midventral keel, which extends over the ventral face of the head of the main bone. In all specimens at hand these three processes are cartilaginous; furthermore, the osseous tissue of the three distal segments is restricted to, or concentrated in, the distal parts of each segment, indicating that ossification apparently proceeds from the tip proximad in *S. hispidus.*

The lateral segments, situated dorsolateral to the medial unit, attach onto the dorsal and lateral parts of the head of the main bone— dorsal to the flanges of the medial segment. Each is pointed and blade-shaped, the dorsoventral diameter exceeding the transverse one; and as viewed ventrally each curves gently distad and slightly laterad. Whether cartilaginous or osseous, they are situated in the lateral parts of each bacular mound, while the medial and distalmost parts of each mound consist entirely of soft tissue, a large part of which is vascular and appears to be instrumental in distention of the mounds. In some examples, the basal parts of the three distal segments of the baculum are more or less coalesced; this is particularly true of the two lateral units, and the two have been interpreted as a single horn-shaped structure (Hamilton, 1946). However, as indicated by Burt (1960) they are separate units (Fig. 1); their individual limits are clear in specimens at hand.

Sigmodon minimus, S. ochrognathus, and *S. alleni.*—I recognize no interspecific differences in the specimens of *minimus* and *ochrognathus,* both examples of which are young and rather unsatisfactory. Each closely resembles specimens of *hispidus* of like age in external size and shape, and in conformation of the six exterior lobes, dorsal papilla, dorsolateral papillae, lateral papillae, urethral process, crater mounds, and baculum. If there are interspecific differences, they are not clearly evident in the materal at hand.

The three adults from Dos Aguas, Michoacán, which are labeled *S.*

alleni, are also like adults of *hispidus.* The two series differ slightly in regard to size of glans and shape of baculum, but these are small differences and doubtfully interspecific.

A few remarks regarding the identification of the specimens from Dos Aguas are needed. Until variation in *Sigmodon* is better understood, *S. alleni* seems to be the most appropriate name to apply to these specimens and, as well, to others like them from the vicinity of Autlán, Jalisco, and Angahuan and Uruapan, Michoacán. Cranially and externally distinguishable from specimens of *S. hispidus* and *S. melanotis* from nearby localities in the same states, they appear to represent a species other than either *hispidus* or *melanotis.* They agree well with the description of *alleni,* but they have not been compared directly with the type specimen of that form.

Sigmomys alstoni.—The specimen of *Sigmomys alstoni* resembles examples of *Sigmodon* of comparable age in length (relative to hind foot), in external configuration (hexalobate, swaybacked and pot-bellied in lateral view, and covered with proximally directed, thickset, sharp, entrenched spines), shape of dorsal papilla (single, spine-studded cone), appearance of urethral process (two outcurved arms with a longitudinal row of spines on the ventral face of each), shape of the bacular mounds (the medial one large and rounded, each lateral one smaller and rounded laterally but acute medially), position of digits of baculum with respect to the main bone, presence of ventral keel and lateral arms on the medial digit, and occurrence of a midventral keel on the main bone. The specimen differs from examples of *Sigmodon* in characters as follows: glans smaller in diameter (diameter-length ratio approximately 60 per cent, compared with 70–88 per cent in *Sigmodon*); the six external lobes, particularly the dorsal pair, less prominent; dorsolateral papillae smaller, scarcely more than the spine-studded infolding of the dorsal and lateral lobes; crater more extensively spinous (spines studding most of inner wall of each lateral lobe); medial digit of baculum projecting principally distad, its tip not sharply flexed dorsad; and the osseous proximal segment flatter and wider for a larger fraction of its length.

The lateral papillae and baculum warrant additional comment. It is uncertain whether lateral papillae are present in the specimen. Two papillose vascular cores occur at sites where papillae are to be expected, but in the present damaged specimen the overlying crater floor is not correspondingly papillose, although it is strongly spinous; the spiny area occupies most of the inner face of the lateral lobe and of the adjoining crater floor. On the left side of the specimen this roughly circular spiny area is plate-like, while on the right side it is buckled distad and, thus, resembles a large papilla. If, in undamaged specimens, these areas are papillose, then the lateral papillae in *S. alstoni* are relatively larger than any yet seen in *Sigmodon*.

In ventral view, the main bone of the baculum is shaped roughly like an isosceles triangle—wide basally and tapered rather evenly distad (without pronounced incurve) almost to the slight constriction which subtends the small, round, terminal head. Its wide basal part is concave dorsally (between low lateral condyles) and almost flat ventrally; but farther distad the bone is deeper than wide and, somewhat triangular in cross section, it bears a slight midventral ridge to which a cartilaginous keel is attached. The distal segments are entirely cartilaginous. The medial one is deeper than wide in its distal half and blunt terminally; basally it bears a medial process and two lateral flanges. Each lateral segment, also deeper than wide and blunt terminally, is situated dorsolateral to the medial unit.

Reithrodon cuniculoides.—The glans of *R. cuniculoides* (Fig. 2) is stubby (diameter-length ratio 64 per cent), subcylindrical, and indistinctly lobate, the lobes defined by four, shallow, longitudinal troughs. Two of these depressions, one situated middorsally and the other midventrally, extend approximately the full length of the glans and thereby divide the surface of the glans into right and left halves; the distal limit of each is a notch in the crater rim. The shorter third pair of troughs is situated dorsolaterally in the distal half of the glans, but each terminates short of the rim. All of the epidermis as far distad as the crenate, membranous, overhanging rim of the crater is densely studded with small, conical, recessed tubercles.

The three bacular mounds, together with the underlying baculum, resemble a fleur-de-lis in ventral aspect (Fig. 2); the erect medial part extends beyond the crater, while each of the truncate lateral pair sends off an attenuate lateral segment which curves laterad and then distad before terminating in an acute tip. These lateral processes contain no cartilage or bone; they consist entirely of soft tissues, a large part of which is vascular and apparently erectile. The spine-tipped dorsal papilla is unusually small and slender; it is a single cone, but a slight cleft near its tip suggests that the papilla may consist of two conules in other specimens. The urethral process is a bilobed flap with two attenuate and erect (not outcurved) arms; it bears two longitudinal rows, each of eight tubercles, on its ventral face. There are no lateral or dorsolateral papillae, and the crater walls and floor are smooth and non-spinous.

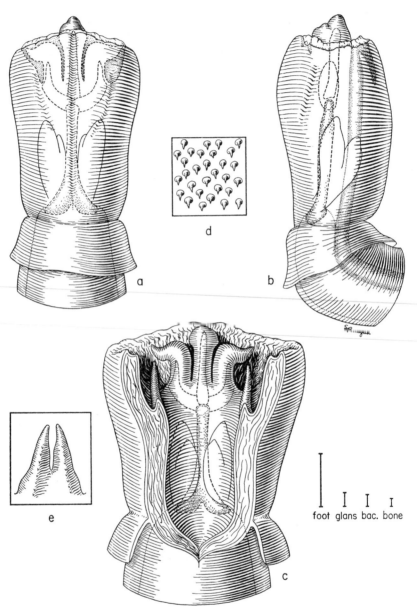

FIG. 2. Views of glans penis of *Reithrodon cuniculoides*; UMMZ 109233, Argentina. For explanation see Fig. 1 and text.

The baculum is shorter than the glans (see measurements). Its proximal, osseous segment consists of a wide basal part and a slender shaft. The basal part, which bears large, proximally directed condyles (these separated medially by a deep notch), is broadly concave ventrally and narrowly and shallowly concave dorsally. The relatively straight shaft is slightly deeper (dorsoventrally) than wide and it bears a slight ventral keel; its terminal portion is slightly expanded laterad and slightly constricted dorsoventrally (Fig. 2). The three distal segments are cartilaginous. The long medial one (its length two-thirds that of the bone) is rod-like for much of its length, but it is enlarged basally and is tapered distally to a pointed tip. The lateral units are disc-shaped in cross section, the dorsoventral diameter of each much greater than the transverse one. From its attachment on the head of the bone (the attachment dorsal and lateral to that of the medial unit) each lateral segment curves gently laterad and distad before it terminates at the base of the laterally projecting process of its lateral mound.

DISCUSSION

To judge from specimens at hand, the glandes of *Sigmodon alleni,* *S. hispidus,* *S. minimus,* and *S. ochrognathus* are fundamentally alike, although they may differ interspecifically in details which can not be appraised in present samples. In each species the stubby, swayback, tubercle-invested glans bears six prominent exterior lobes which surround the terminal crater and divide its rim into six corresponding parts. Within the crater there are five spine-studded papillae consisting of dorsolateral and lateral pairs in addition to a single cone mid-dorsally. The urethral process bears two attenuate, outcurved arms. The bacular mounds are truncate except for a small, acute medial crest on each lateral mound, and the medial distal segment of the four-part baculum bears a medial keel and a pair of lateral processes on its base, while its tip is flexed sharply dorsad. These characters, together with others, distinguish *Sigmodon* from the other New World cricetid genera which have been studied to date, with the possible exception of *Sigmomys. Sigmomys alstoni,* the only species of *Sigmomys* about which there is information on the glans, appears to be closely similar to species of *Sigmodon,* but its characters are not yet adequately known.

In contrast to the phalli of *Sigmodon* and *Sigmomys,* the glans of *Reithrodon cuniculoides* is comparatively slim and simple. There are only four exterior lobes, and these are less prominent than the lobes of *Sigmodon* or *Sigmomys.* The membranous, crenate, and non-spiny crater rim is not divided into six distinct lobes. The crater, also smooth and spineless, has no dorsolateral or lateral papillae. The slender dorsal papilla bears spines only at its tip. Each lateral mound has an attenuate lateral process, and the entire configuration of the three crater mounds as well as of the underlying baculum is distinctive. The three, long, erect distal segments of the baculum, all cartilaginous insofar as known, are essentially rod-like in form, without prominent keels or processes. These and other contrasting characters indicate that the glans of *R. cuniculoides* is morphologically quite different from that seen in *Sigmodon* and *Sigmomys*. Preliminary comparisons suggest that it may be more similar to glandes of phyllotine or other species which are not now included in the sigmodont group of rodents.

LITERATURE CITED

BURT, WILLIAM H.
    1960   Bacula of North American mammals. Miscl. Publ. Mus. Zool. Univ. Mich.,
           113:1–76, 25 pls.

GYLDENSTOLPE, NILS
    1932   A manual of Neotropical sigmodont rodents. Kungl. Svenska Veten.
           Hand., Ser. 3, no. 3: 1–164, 18 pls.

HAMILTON, WILLIAM J., JR.
    1946   A study of the baculum in some North American Microtinae. Jour.
           Mamm., 27:378–87, 1 pl., 3 figs.

HERSHKOVITZ, PHILIP
    1944   A systematic review of the neotropical water rats of the genus *Nectomys*
           (Cricetinae). Miscl. Publ. Mus. Zool. Univ. Mich., 58:1–88, 4 pls., 5 figs.
    1948   Mammals of northern Colombia, preliminary report No. 3: water rats
           (genus *Nectomys*), with supplemental notes on related forms. Proc. U.S.
           Natl. Mus., 98:49–56.
    1955   South American marsh rats, genus *Holochilus*, with a summary of sig-
           modont rodents. Fieldiana: Zoology, 37:639–73, 13 pls., 6 figs.
    1960   Mammals of northern Colombia, preliminary report No. 8: arboreal rice
           rats, a systematic revision of the subgenus Oecomys, genus Oryzomys.
           Proc. U.S. Natl. Mus., 110:513–68, 12 pls., 6 figs.

HOOPER, EMMET T.
    1959   The glans penis in five genera of cricetid rodents. Occ. Pap. Mus. Zool.
           Univ. Mich., 613:1–10, 5 pls.

PEARSON, OLIVER P.
    1958   A taxonomic revision of the rodent genus Phyllotis. Univ. Calif. Publ.
           Zool., 56:391–496, 8 pls., 21 figs.

THOMAS, OLDFIELD
    1917   On the arrangement of the South American rats allied to *Oryzomys* and
           Rhipidomys. Ann. Mag. Nat. Hist., ser. 8, 20:192–8.

VORONTSOV, N. N.
    1959   The system of hamster (Cricetinae) in the sphere of the world fauna
           and their phylogenetic relations. Bull. Mosk. Obsh. Ispyt. Prirody, Biol.
           Sec. (Bull. Moscow Soc. Naturalists), 64:134–7.

Accepted for publication February 5, 1962

COMPARATIVE MORPHOLOGY OF SPERMATOZOA FROM FIVE MARSUPIAL FAMILIES

By R. L. Hughes*

[Manuscript received April 8, 1965]

Summary

The spermatozoa of 18 marsupial species derived from five families have been examined and of these only the spermatozoon of the bandicoot *Perameles nasuta* has previously been described adequately.

The spermatozoon morphology within the families Macropodidae, Dasyuridae, Phascolarctidae, and Peramelidae was relatively homogeneous. A distinctive morphology occured between these families. Within the family Phalangeridae spermatozoa were morphologically diverse, however, as a group they were relatively separate from those of the other families studied.

The spermatozoa of the Phascolarctidae (koala, *Phascolarctos cinereus*, and wombat, *Phascolomis mitchelli*) have a unique, somewhat rat-like morphology which clearly separates them from those of the other marsupial sperm studied. This finding is of considerable taxonomic interest as most authorities consider the koala to be more closely related to the phalangerid marsupials than to the wombat.

I. Introduction

Previous descriptions of marsupial spermatozoon morphology cover six of the major marsupial groups. A considerable proportion of these accounts is devoted to a study of the spermatozoon morphology of three species, each belonging to separate marsupial families. (1) Family Didelphidae: *Didelphis* [Selenka (1887), Fürst (1887), Waldeyer (1902), Korff (1902), Retzius (1909), Jordan (1911), Duesberg (1920), Wilson (1928), McCrady (1938), Biggers and Creed (1962)]; (2) family Phalangeridae: *Phalangista vulpina* (= *Trichosurus vulpecula*) [Korff (1902), Benda (1897, 1906), Retzius (1906), Bishop and Walton (1960)]; (3) family Peramelidae: *Perameles nasuta* [Benda (1906), Cleland (1955, 1956, 1964), Cleland and Rothschild (1959), Bishop and Austin (1957), Bishop and Walton (1960)].

The spermatozoon morphology of two Dasyuridae, *Phascogale albipes* (= *Sminthopsis murina*) and *Dasyurops maculatus*, was studied by Fürst (1887), Bishop and Austin (1957), and Bishop and Walton (1960).

Benda's (1906) description of an epididymal sperm from the koala, *Phascolarctos* (family Phascolarctidae), is, as he admits, inadequate.

Spermatozoon morphology studies on members of the family Macropodidae include those of an unknown *Macropus* sp. (Benda 1906), *Macropus billardierii* (= *Thylogale billardierii*), *Petrogale penicillata*, *Onychogale lunata* (= *Onychogalea lunata*), *Bettongia cuniculus* (Retzius 1906), *Macropus giganteus* (= *Macropus canguru*) (Binder 1927), and *Potorous tridactylus* (Hughes 1964).

* Division of Wildlife Research, CSIRO, Canberra.

Aust. J. Zool., 1965, **13**, 533–43

The spermatozoa examined in the present study were obtained from members of the five Australasian marsupial families: Phalangeridae, Peramelidae, Dasyuridae, Phascolarctidae, and Macropodidae. The present series of observations has been viewed with reference to those of earlier workers and this has permitted at least an elementary discussion of the comparative aspects of spermatozoon morphology between the marsupial families examined.

II. MATERIAL AND METHODS

The testes together with the attached epididymis were removed from the scrotum soon after death and fixed in 10% neutral formalin or, more rarely, Bouin's fluid or Carnoy fixative.

(i) *Method for Adhering Spermatozoa to Microscope Slides*

The slides were labelled at one end with a diamond pencil and a 15-mm square was marked out at the opposite end. The entire surface of the slide was liberally smeared with Mayer's albumen. A small piece of epididymal tissue was placed in a drop of 10% neutral buffered formalin within the marked square and extensively teased with dissecting needles. Filter paper circles of 5·5 cm diam. were saturated with 10% formalin, drained, and placed over the specimen by a rolling action. Air bubbles were punctured with a needle. The filter paper was kept moist with 10% formalin for at least 30 min and then permitted to dry until free fluid between the slide and the filter paper had disappeared. The filter paper was then removed by a rolling action, excess tissue was removed with fine forceps, and the preparations rinsed and stored in water for staining.

(ii) *Staining of Spermatozoa*

(1) *Heidenhain's Iron Haematoxylin.*—Slides containing adhering spermatozoa were transferred from water to a 5% solution of iron alum and kept in a warm place for 2–3 days. They were then stained with Heidenhain's haematoxylin for a similar period. The area not containing the specimen was thoroughly cleaned with paper tissues during a 10–15 min rinsing period in running tap water. The preparations were then differentiated in 5% iron alum under a staining microscope at 30 sec intervals. The preparation was washed in water and re-examined after each differentiation interval. Differentiation times of between 30 sec and 5 min proved satisfactory to show the desired range of structures. The preparations were upgraded to absolute ethyl alcohol, placed in two changes of xylol, and mounted in euparal.

(2) *Periodic Acid–Schiff (with saliva controls).*—Slides containing the mounted spermatozoa were removed from water and placed horizontally in two groups on a flat tray. One group was flooded with water and the other with saliva for 1 hr at a temperature of 37°C. The slides were then thoroughly rinsed in distilled water and stained by a method described by Carleton and Drury (1957, p. 143). The Schiff's reagent used was de Tomasi (for preparation see Pearse 1961, p. 822). The preparations were mounted in euparal.

(3) *Feulgen (with and without fast green counterstain).*—Slides containing the adhering spermatozoa were removed from water and stained by a method described

by Pearse (1961, p. 823). The Schiff's solution used was de Tomasi. Half the Feulgen preparations were stained with fast green counterstain (0·5% solution in 70% ethyl alcohol) for 15–20 min. Both Feulgen and Feulgen–fast green preparations were quickly passed through three changes of 90% alcohol (dips only) to absolute ethyl alcohol and then cleared in xylol and mounted in euparal.

Slides were stored until dry in an oven at a temperature of 37°C after mounting in euparal. Preparations were not permitted to dry out during any of the earlier stages in preparation.

The drawings of spermatozoa shown in Figure 1 are based on camera lucida outlines using a ×12 eyepiece in conjunction with a ×100 oil-immersion objective.

The spermatozoon dimensions shown in Table 1 are means of 25 observations and were obtained with a special Leitz ×12·5 screw micrometer eyepiece and a ×100 oil-immersion objective. The preparations used were fixed in 10% neutral formalin or, more rarely, Bouin's fluid or Carnoy and were stained with Heidenhain's iron haematoxylin.

During the course of the observations on sperm it became apparent that the efferent ducts connecting the testis and epididymis were either multiple or single within each marsupial family. This was investigated further from frozen transverse sections stained with haematoxylin and eosin. The sections were prepared from the efferent duct or ducts at the point of their emergence from the testis and also approximately midway between the testis and epididymis.

The author follows Cleland and Rothschild (1959) in considering for the purpose of description that the flagellum is inserted into the ventral surface of the sperm head and the opposite surface is taken as dorsal.

III. RESULTS

The mature epididymal spermatozoa of 18 marsupial species have been examined. The dimensions of 13 of these spermatozoa are shown in Table 1. The gross morphology of 14 of the spermatozoa is shown in Figure 1.

Spermatozoa of each of the five marsupial families studied (Macropodidae, Phalangeridae, Dasyuridae, Peramelidae, Phascolarctidae*) exhibited sufficient homogeneity in morphology and dimensions of the head, flagellum, and fine structure to be of taxonomic value.

The heads of all marsupial spermatozoa examined showed some dorsoventral flattening. This was most marked in the Dasyuridae and Peramelidae. It was least evident in the Phascolarctidae and the genus *Pseudocheirus* of the Phalangeridae. Macropod and the other phalangerid spermatozoa exhibited an intermediate condition. The distal extremity of the head of all species when viewed dorsally was relatively rounded while the shape of the lateral margins and proximal tip varied considerably. In the Dasyuridae the spermatozoon heads of up to 12·7 μ in length in *Dasyuroides byrnei* are among the longest known for mammals (Table 1). The

* The author follows Sonntag (1923) in grouping the wombat and koala in the family Phascolarctidae.

lateral head margins of dasyurid sperm are slightly convex in dorsal view and taper gradually to a proximal point. Macropod sperm heads are considerably shorter than those of the Dasyuridae and in dorsal outline are elongated ovoids bluntly pointed proximally. The sperm head of the macropod *Megaleia rufa* (Figs. 1g and 1h) is rapidly tapering, a condition typically found in the Phalangeridae. Phalangerid sperm, when viewed dorsally, exhibit considerable variability in the convexity of the lateral head margins. The proximal region of the head is typically semicircular, although sometimes bluntly pointed as in *Pseudocheirus cupreus* (Figs. 1n and 1o).

TABLE 1

MARSUPIAL SPERMATOZOON DIMENSIONS

Family and Species	Mean ±SD (μ)				
	Head		Middle-piece		Flagellum
	Length	Width	Length	Diameter	Length
Macropodidae					
Macropus canguru	7·3±0·16	2·2±0·11	10·7±0·24	1·5±0·14	111·6± 3·60
*Megaleia rufa**	5·1±0·21	2·4±0·09	7·9±0·25	1·4±0·12	104·0± 4·74
Protemnodon rufogrisea	8·5±0·22	2·3±0·18	11·7±0·34	1·6±0·14	115·4± 8·85
Protemnodon agilis†	7·1±0·38	1·8±0·12	11·0±0·28	1·4±0·13	— —
*Thylogale stigmatica**	7·2±0·09	2·2±0·11	10·9±0·22	1·5±0·12	103·1± 4·43
Dasyuridae					
Dasyuroides byrnei	12·7±0·41	2·5±0·15	40·7±1·26	3·1±0·19	242·1± 6·77
Sarcophilus harrisii	11·1±0·45	2·2±0·17	34·4±0·84	2·6±0·13	207·4±12·02
Phalangeridae					
Petaurus breviceps‡	5·9±0·19	2·5±0·18	8·3±0·27	1·4±0·11	101·3± 4·96
Pseudocheirus cupreus‡	5·4±0·16	2·6±0·11	6·2±0·16	1·5±0·17	84·7± 2·47
Pseudocheirus peregrinus	5·9±0·38	3·8±0·18	6·9±0·21	2·1±0·22	106·9± 5·31
Phascolarctidae					
Phascolomis mitchelli	5·7±0·33	1·7±0·09	18·0±1·56	0·9±0·10	87·9± 8·23
Peramelidae					
Perameles nasuta	5·7±0·15	3·0±0·13	14·0±0·32	2·0±0·11	194·1± 5·25
Isoodon macrourus	6·0±0·13	3·3±0·18	10·7±0·19	1·8±0·14	165·1± 3·64

* Fixed in Bouin's fluid; † Carnoy fixative; ‡ from New Guinea.

Peramelid spermatozoon heads have concave lateral margins when seen in dorsal view and are relatively square proximally with a median cap. In phascolarctid sperm the proximal portion of the spermatozoon head of both the wombat *Phascolomis mitchelli*, and the koala, *Phascolarctos cinereus*, bears a strongly recurved hook.

In all sperm, a positive Feulgen reaction for nuclear material (DNA) was given by almost the entire head mass. The DNA-negative areas that took up a fast green counterstain in Feulgen preparations were the acrosome (Fig. 1; *AC*) and basal granule complex which is located at the proximal tip of the flagellum. The acrosome

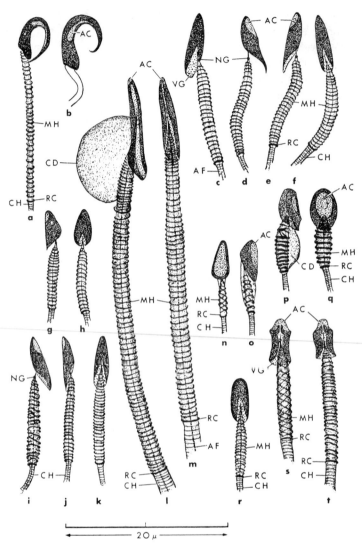

Fig. 1.—Marsupial epididymal spermatozoa: the drawings are all at the same scale and are based on camera lucida outlines of formalin-fixed Heidenhain's iron haematoxylin preparations. A ×12 eyepiece was used in conjunction with a ×100 oil-immersion lens. Fam. Phascolarctidae: *Phascolomis mitchelli*, (*a*) lateral view; *Phascolarctos cinereus*, (*b*) lateral view of spermatozoon head with flagellum outline. Fam. Macropodidae: *Protemnodon rufogrisea*, (*c*) ventral view, (*d*) lateral view; *Protemnodon agilis**, (*e*) lateral view, (*f*) ventral view; *Megaleia rufa*†, (*g*) lateral view, (*h*) ventral view; *Macropus canguru*, (*i*) lateral view; *Thylogale stigmatica*†, (*j*) lateral view; (*k*) ventral view. Fam. Dasyurinae: *Dasyuroides byrnei*, (*l*) dorsolateral view; *Sarcophilus harrisii*, (*m*) ventral view. Fam. Phalangeridae: *Pseudocheirus cupreus*, (*n*) dorsal view, (*o*) lateral view; *Pseudocheirus peregrinus*, (*p*) lateral view, (*q*) dorsal view; *Petaurus breviceps*, (*r*) ventral view. Fam. Peramelidae: *Isoodon macrourus*, (*s*) ventral view; *Perameles nasuta*, (*t*) ventral view. Key: *AC*, acrosome; *AF*, axial filament; *CD*, cytoplasmic droplet (middle-piece bead); *CH*, cortical helix of main-piece sheath; *MH*, mitochondrial helix of middle-piece; *NG*, neck granule; *RC*, ring centriole; *VG*, ventral groove.

* Fixed in Carnoy fixative. † Fixed in Bouin's fluid.

was also variably positive to periodic acid–Schiff (P.A.S.) between species and the basal granule complex was invariably strongly P.A.S.-positive. Neither acrosome nor basal granule complex exhibited any reduction in P.A.S. activity in saliva controls. A faint tinge of green over the entire head surface in Feulgen–fast green preparations presumably represents a limiting membrane.

A "nuclear rarefaction" of vacuole-like appearance results from a minute superficial nuclear indentation. The nuclear rarefaction was most conspicuous in the Dasyuridae and Peramelidae and least evident in the Macropodidae. This structure is located on the mid-median aspect of the ventral nuclear surface of all sperm with the exception of those of the Phascolarctidae, where its occurrence is also ventral and median but distal.

In most of the marsupial sperm examined acrosomal material (Fig. 1; *AC*) was apparently confined to a relatively small surface area of the head. In the Macropodidae the acrosome is relatively small and is a discrete ovoid structure embedded superficially in the extreme proximal portion of the dorsal head surface. In some of the Phalangeridae it has a definite structure as in *Pseudocheirus* (Figs. 1*n*–1*q*) where it occupies all but a marginal annular zone of the dorsal head surface and is rather deeply embedded. In other phalangerids, such as *Petaurus breviceps* (Fig. 1*r*), the dorsal head surface in Feulgen–fast green preparations gives a diffuse acrosomal reaction and bears a shallow depression which extends to all but the margins. A similar diffuse acrosomal reaction of at least the proximal half of the dorsal head surface occurred in the Dasyuridae. The proximal dorsal tip of the dasyurid sperm has a concentration of acrosomal material situated in a minute groove. The acrosomal material in the Peramelidae was found in a small distally flanged proximal cap which covered a minute nuclear protuberance. In the Phascolarctidae the acrosome is a small "comma-shaped" structure. The body of the acrosomal "comma" is embedded superficially in about the middle of the dorsal head surface and the tail of the comma extends throughout the greater portion of the inner curvature of the head hook.

In marsupial sperm the ventral surface of the head (by convention that bearing the flagellum) is typically grooved (Fig. 1; *VG*) or bears a shallow distal notch as in the case of the Phascolarctidae. At the distal extremity of the head the groove is broad and deep so that the head is here relatively broad and has the form of an extremely thin curved plate. The groove becomes shallow and narrow towards its proximal extremity; in the Macropodidae and Phalangeridae it terminates at about the mid-median portion of the ventral head surface. The groove is most extensive in the Peramelidae involving the whole of the ventral aspect of the nucleus, only the proximal acrosomal head cap is excepted. In the Dasyuridae it extends throughout the distal four-fifths of the head.

Spermatozoa are immature when they enter the head of the epididymis and were characterized by the orientation of the long axis of the head at 90° to the flagellum which was directed towards the nuclear rarefaction. The ventral surface of the spermatozoon head was supported by a somewhat cone-shaped cytoplasmic droplet (Fig. 1; *CD*) of characteristic morphology for each species. Phascolarctid sperm from the head region of the epididymis differed from the other marsupial

123

species examined in that the flagellum was most frequently observed not to meet the head at right angles and cytoplasmic droplets were small and often absent. On entering the epididymis the head hook of the phascolarctid spermatozoa were only slightly recurved or of an irregular spiral configuration. During the passage of spermatozoa through the epididymis the head hook became simple (without spiral) and more tightly recurved.

Maturation of spermatozoa is completed during their passage through the epididymis and is accompanied by shedding of the cytoplasmic droplet and rotation of the long axis of the head parallel to that of the flagellum. The neck of the flagellum of mature epididymal sperm in all species was inserted in the vicinity of the nuclear rarefaction. In the Dasyuridae the neck was inserted rather deeply into the proximal margin of the nuclear rarefaction. In the Peramelidae the proximal tip of the flagellum was also deeply inserted and extended from the proximal margin of the nuclear rarefaction to a point about midway between the anterior rim of the nuclear rarefaction and the most proximal extremity of the nucleus.

The flagellum is traversed throughout its entirety by an axial filament (Fig. 3; *AF*). The size of the flagellum varies from species to species. The smallest flagellum was that of *Phascolomis mitchelli* with a maximum diameter of $0·9\,\mu$ and a minimum length of $87·9\,\mu$ (Table 1). The giant flagella of dasyurid sperm are among the largest known for mammals. *Dasyuroides byrnei* has a minimum flagellum length of $242·1\,\mu$ and a maximum flagellum thickness of $3·1\,\mu$. In an old museum specimen of the testes of the now possibly extinct dasyurid *Thylacinus cynocephalus* (Tasmanian wolf or tiger) the flagellum of epididymal sperm in wax sections had a maximum diameter of $3·0\,\mu$ and comparable morphology to that of other dasyurids; the sperm heads, although degenerate, were in the form of a long narrow plate, dorso-ventrally flattened and with the flagellum inserted at about the mid-median ventral aspect. Peramelid sperm flagellae were also relatively large, having a maximum diameter of as much as $2\,\mu$ and a minimum length of up to about $200\,\mu$ (Table 1). Macropod and phalangerid sperm flagellae were of intermediate dimensions rarely varying from a maximum diameter of $1·5\,\mu$ and a minimum length of a little over $100\,\mu$.

The basal granule complex located at the proximal end of the flagellum consists of at least fused proximal and distal components in the Dasyuridae and Peramelidae.

The neck region of the flagellum is a slender proximally pointed cone with a smooth contour, and a small neck granule (Fig. 1; *NG*) is situated at approximately half its length. It was only possible to identify the neck granule with certainty in the Peramelidae, Dasyuridae, and Macropodidae. In the Peramelidae and Dasyuridae it seemed to be a more deeply stained, modified portion of the ground substance of the neck rather than the discrete granule found in the Macropodidae. The sperm of the dasyurids *Dasyuroides byrnei*, *Sarcophilus harrisii*, and *Thylacinus cynocephalus* had a neck length of about $3·5\,\mu$ in comparison with $2·7\,\mu$ for the peramelids *Isoodon macrourus* and *Perameles nasuta*. Macropod sperm necks ranged in length from $1·8\,\mu$ in *Thylogale stigmatica* to $2·6\,\mu$ in *Protemnodon rufogrisea*. The neck lengths of the Phalangeridae and Phascolarctidae were somewhat reduced in comparison to those of other marsupial families.

The proximal portion of the middle-piece in all species examined tapered gradually to the diameter of the neck and was particularly firmly clasped by the lateral margins of the sperm head in the Peramelidae and Dasyuridae. The remainder of the middle-piece was relatively cylindrical. A mitochondrial helix (Fig. 1; *MH*) of spiral configuration gave the entire surface of the middle-piece sheath a slightly uneven contour. The mitochondrial helix is a relatively fine structure in the Dasyuridae and Peramelidae, of moderate thickness in the Macropodidae and Phascolarctidae and *Petaurus breviceps* of the Phalangeridae. It was quite thick and granular with relatively few gyres in the genus *Pseudocheirus* of the Phalangeridae. The middle-piece is terminated distally by a ring centriole (Fig. 1; *RC*).

The flagellum undergoes an abrupt reduction in diameter on the main-piece side of the ring centriole in both the *Pseudocheirus* species and to a moderate degree in *Petaurus breviceps* and the Macropodidae, but not to any appreciable extent in the Dasyuridae, Peramelidae, and Phascolarctidae.

The main-piece of the flagellum tapers distally and in twisted specimens appears not to be circular in cross section in *Pseudocheirus peregrinus*, Peramelidae, Dasyuridae, and Macropodidae. Striations of the sheath of the main piece in all sperm indicate the presence of a fine spiral cortical helix (Fig. 1; *CH*). The tail sheath also gave a strong impression of two lateral thickenings in transverse axis in *Macropus canguru*, *Protemnodon rufogrisea*, *Pseudocheirus peregrinus*, *Perameles nasuta*, and *Isoodon macrourus*.

The axial filament protruded beyond the terminal portion of the sheath of the main-piece in apparently complete sperm of all species but this cannot be positively taken to represent a true end-piece for in all preparations terminal breakage of the main-piece was prevalent.

IV. DISCUSSION

Spermatozoa from three other Peramelidae, *Perameles gunnii*, *Isoodon obesulus*, and *Echymipera rufescens* have also been examined superficially and it can be stated that they are comparable in morphology to other peramelid sperm. The spermatozoa of marsupial mice, *Antechinus flavipes flavipes*, *A. f. leucogaster*, *A. swainsonii*, *A. stuartii*, and *Sminthopsis crassicaudata*, have a morphology typical of other dasyurids (Woolley, personal communication). This similarity in morphology also extends to two other dasyurids, *Phascogale albipes* (= *Sminthopsis murina*) (Fürst 1887) and *Dasyurops maculatus* (Bishop and Austin 1957; Bishop and Walton 1960). The spermatozoon morphology of the macropod species examined in the present study varies in only minor details from that of six other macropods previously described by Benda (1906), Retzius (1906), and Hughes (1964).

The phenomenon of conjugate spermatozoa (pairing of relatively numerous epididymal spermatozoa) redescribed and reviewed by Biggers and Creed (1962) for the American opossum, *Didelphis*, has not been observed in any of the sperm preparations from Australian marsupials; however, fresh unfixed material has been examined only for *Potorous tridactylus* (Hughes 1964) and *Phascolomis mitchelli*.

Another feature worth mentioning is that the head of the epididymis was not fused with the testis in any marsupial examined, including *Thylacinus cynocephalus* and *Dendrolagus lumholtzi*. In the Dasyuridae and Peramelidae a relatively long single efferent duct together with associated blood vessels links the epididymis to one pole of the testis long axis by way of an extensive membrane, the mesorchium. A tract of relatively long multiple efferent ducts serves the same function in the Phalangeridae, Phascolarctidae, and Macropodidae. A ligament was inserted by way of the mesorchium into the opposite pole of the testis.

In both the wombat and the koala the morphology of the sperm, particularly the head, differs strikingly from that of any marsupial sperm previously described. In both species the proximal portion of the spermatozoon head bears a strongly recurved hook not described for other marsupial sperm, and the flagellum is inserted into a notch on one side of the distal portion of the head (Plate 1, Fig. 1; and Figs. 1a and 1b). These features, although somewhat resembling those of certain murid sperm, are not strictly comparable (Plate 1, Fig. 2) (Friend 1936). The hook in the wombat sperm resembles that of *Microtus hirtus*, *Lemmus lemmus*, and several other members of the murid subfamily Microtinae in that the hook contains no supporting "rod" and its tip like that of *Lemmus lemmus* is typically extremely reflected so that it lies against the distal portion of the head (Friend 1936). The position of the hook in *Phascolomis* is not an artefact of fixation for it was observed in living spermatozoa from the epididymis of several specimens. In sperm from the head of the epididymis the curvature of the hook frequently approximated to that of rats and mice. It can be seen from Plate 1, Figure 1, and Figures 1a and 1b that the insertion notch of the flagellum of the wombat and koala sperm is located on the opposite side of the head to the hook, whereas in the hooked types of murid sperm both structures occur on the same side of the head (Plate 1, Fig. 2). A head hook is absent in at least the murine, *Micromys minutus* and in the microtine *Ondatra zibethica* (Friend 1936). In Heidenhain's iron haematoxylin preparations the head length of the wombat sperm measured from the distal extremity to the most proximal point of the curvature of the hook (i.e. excluding the recurved portion of the hook) is about $5 \cdot 7\,\mu$ in contrast to $8 \cdot 0\,\mu$ and $11 \cdot 7\,\mu$ for mouse and rat, respectively (Friend 1936). Feulgen preparations (with or without fast green counterstain) of wombat and koala sperm have shown that nuclear material (DNA) extends to the tip of the hook and occupies all but a small comma-shaped acrosomal portion of the head. Herein lies the greatest departure of wombat sperm from the hooked varieties of murid sperm. In several microtine species the hook is formed entirely from a proximal extension of the nuclear cap (acrosome). In murine sperm a hooked portion of the nucleus bearing a rod extends into the hooked nuclear cap and follows its contour almost to its proximal extremity (Friend 1936).

On the basis of skeletal and dental structure most workers consider the koala to be more closely related to the ringtail possums of the genus *Pseudocheirus* than the wombat (Wood Jones 1924; Simpson 1945). Comparisons of sperm morphology on which selection pressure would presumably be lower than that for external characters of an animal such as skeletal or dental characters, is therefore of considerable interest as a possible basis for taxonomic classification.

It can be seen from the previous descriptions that the spermatozoon of *Pseudocheirus peregrinus* is not intermediate in structure between the more typical marsupial types (Macropodidae and Dasyuridae) and those of the highly divergent wombat and koala. On the contrary, it deviates in quite a different manner from the typical marsupial patterns. The head is broad $(3 \cdot 8 \mu)$ in comparison to its length $(5 \cdot 9 \mu)$, the anterior end lacks a hook and is semi-circular in dorsal view (Plate 1, Figs. 3 and 4). Other distinguishing features are the shape and position of the acrosome previously mentioned and a relatively short middle-piece $(6 \cdot 9 \mu)$. The view that the koala is more closely related to the ringtail possum than the wombat is not supported by comparisons of sperm morphology. On the contrary, the findings reported here support the observations of Sonntag (1923) and Troughton (1957) who considered that the koala shares sufficient characters with the wombat for its classification along with the phalangers to be rejected.

V. ACKNOWLEDGMENTS

The author wishes to express his sincere thanks to Dr. E. H. Hipsley, Director, Institute of Anatomy, Canberra; to J. T. Woods, Queensland Museum; to J. A. Thomson, Zoology Department, University of Melbourne; to Dr. M. E. Griffiths, W. E. Poole, K. Keith, M. G. Ridpath, Division of Wildlife Research, CSIRO, for material; to J. Sangiau and L. S. Hall for technical assistance; to E. Slater for photography; and to Professor K. W. Cleland and Dr. A. W. H. Braden who offered helpful criticism.

VI. REFERENCES

BENDA, C. (1897).—Neuere Mittheilungen über die Histiogenese der Säugethierspermatozoen. *Verh. berl. physiol. Ges.* 1897. [In *Arch. Anat. Physiol. (Physiol. Abt.)* **1897**: 406–14.]
BENDA, C. (1906).—Die Spermiogenese der Marsupialier. *Denkschr. med.-naturw. Ges. Jena* 6: 441–58.
BIGGERS, J. D., and CREED, R. F. S. (1962).—Conjugate spermatozoa of the North American opossum. *Nature, Lond.* **196**: 1112–3.
BINDER, S. (1927).—Spermatogenese von *Macropus giganteus*. *Z. Zellforsch.* **5**: 293–346.
BISHOP, M. W. H., and AUSTIN, C. R. (1957).—Mammalian spermatozoa. *Endeavour* **16**: 137–50.
BISHOP, M. W. H., and WALTON, A. (1960).—Spermatogenesis and the structure of mammalian spermatozoa. In "Marshall's Physiology of Reproduction". (Ed. A. S. Parkes.) 3rd Ed. Vol. 1, Pt. 2, pp. 1–129. (Longmans, Green and Co.: London.)
CARLETON, H., and DRURY, R. A. B. (1957).—"Histological Technique." (Oxford University Press.)
CLELAND, K. W. (1955).—Structure of bandicoot sperm tail. *Aust. J. Sci.* **18**: 96–7.
CLELAND, K. W. (1956).—Acrosome formation in bandicoot spermiogenesis. *Nature, Lond.* **177**: 387–8.
CLELAND, K. W. (1964).—History of the centrioles in bandicoot (*Perameles*) spermiogenesis. *J. Anat.* **98**: 487.
CLELAND, K. W., and LORD ROTHSCHILD (1959).—The bandicoot spermatozoon: an electron microscope study of the tail. *Proc. R. Soc.* B **150**: 24–42.
DUESBERG, J. (1920).—Cytoplasmic structures in the seminal epithelium of the opossum. Carnegie Institute Contributions to Embryology. Vol. **9**, pp. 47–84.
FRIEND, G. F. (1936).—The sperms of the British Muridae. *Quart. J. Micr. Sci.* **78**: 419–43.
FÜRST, C. M. (1887).—Ueber die Entwicklung der Samenkörperchen bei den Beutelthieren. *Arch. mikrosk. Anat. EntwMech.* **30**: 336–65.
HUGHES, R. L. (1964).—Sexual development and spermatozoon morphology in the male macropod marsupial *Potorous tridactylus* (Kerr). *Aust. J. Zool.* **12**: 42–51.

JORDAN, H. E. (1911).—The spermatogenesis of the opossum (*Didelphis virginiana*) with special reference to the accessory chromosome and the chondriosomes. *Arch. Zellforsch.* **7**: 41–86.

KORFF, K. VON (1902).—Zur Histogenese der Spermien von *Phalangista vulpina. Arch. mikrosk. Anat. EntwMech.* **60**: 233–60.

MCCRADY, E. (1938).—The embryology of the opossum. *Am. Anat. Mem.* **16**: 1–233.

PEARSE, A. G. E. (1961).—"Histochemistry." (J. & A. Churchill: London.)

RETZIUS, G. (1906).—Die Spermien der Marsupialier. *Biol. Unters.* (N. F.) **13**: 77–86.

RETZIUS, G. (1909).—Die Spermien von *Didelphis. Biol. Unters.* (N. F.) **14**: 123–6.

SELENKA, E. (1887).—Studien über Entwickelungsgeschichte der Thiere. Das Opossum (*Didelphis virginiana*). Wiesbaden **1887**, pp. 101–72.

SIMPSON, G. G. (1945).—Principles of classification and a classification of mammals. *Bull. Am. Mus. Nat. Hist.* **85**: 1–350.

SONNTAG, C. F. (1923).—On the myology and classification of the wombat, koala and phalangers. *Proc. Zool. Soc. Lond.* **1922**: 683–895.

TROUGHTON, E. (1957).—"Furred Animals of Australia." 6th Ed. (Angus and Robertson: Sydney.)

WALDEYER, W. (1902).—Die Geschlechtszellen. In "Handbuch der vergleichend und experimentellen Entwickelungsgeschichte der Wirbelthiere". Vol. 1, Pt. 1, pp. 86–476.

WILSON, E. B. (1928).—"The Cell in Development and Heredity." (Macmillan: New York.)

WOOD JONES, F. (1924).—"The Mammals of South Australia." Pt. II. (Govt. Printer: Adelaide.)

EXPLANATION OF PLATE 1

Figures 1 and 2 are photographs of Heidenhain's iron-haematoxylin preparations from formalin-fixed epididymal material

Fig. 1.—*Phascolomis mitchelli*, mature epididymal spermatozoon, lateral view.

Fig. 2.—*Rattus norvegicus*, mature epididymal spermatozoon, lateral view.

Fig. 3.—*Pseudocheirus peregrinus*, spermatozoon head, showing centrally placed acrosomal pit, dorsal view.

Fig. 4.—*Pseudocheirus peregrinus*, epididymal spermatozoon, lateral view.

MORPHOLOGY OF SPERMATOZOA

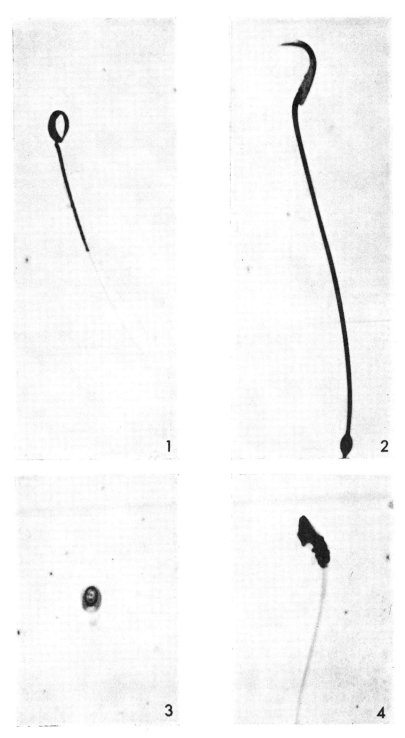

Aust. J. Zool., 1965, **13**, 533–43

Amer. Zool., 13:1205-1213 (1973).

Mechanisms of Sound Production in Delphinid Cetaceans: A Review and some Anatomical Considerations

William E. Evans

Naval Undersea Center, San Diego, California 92132

AND

Paul F. A. Maderson

Department of Biology, Brooklyn College, Brooklyn, New York 11210

synopsis. The past literature describing the possible sites of the sound-producing mechanisms in delphinid cetaceans is reviewed. The morphology of the nasal sac system of delphinids which has been implicated in the production of sounds, by most investigations, is discussed with special emphasis placed on the physical characteristics of these sounds. New data on the histological structure of the epithelia throughout the nasal region of a delphinid are presented with some suggestions as to its function. The presence and structure of glandular tissues are described along with a discussion of their potential role in the production of sound. It is concluded that the theories implicating the nasal sac systems of odontocete cetaceans in the production of sound are additionally supported by certain anatomical specializations adjacent to the tissues of this system.

All the theories to date concerning the mechanism of delphinid sound production have implicated the larynx (arytenoepiglottic tube), the complicated diverticuli associated with the blowhole mechanism, the large muscular plugs that seal off the internal nares, or various combinations of these. The driving mechanism has been thought to be pneumatic, mechanical (muscle-driven), or both. Various combinations of internal sound transmission paths have been considered: air—muscle/fat—water; air—bone—water; tissue—water (Norris, 1969; Evans, 1973).

Attempts have been made to construct conceptual models of the delphinid sound source so that the models could be compared with existing anatomical structures as an aid in localizing the sound source. Unfortunately most of the current ideas, with the exceptions of Norris and Harvey (1972) and Evans (1973), have not considered the acoustical parameters of the signals being produced by the "theoretical"

sound source. Other aspects of these various theoretical mechanisms have been discussed in detail elsewhere (Evans, 1973) and will not, therefore, be reviewed in this paper.

Evans and Prescott (1962) postulated a dual sound source involving the laryngeal mechanism (arytenoepiglottic tube) for whistles, the nasal plugs with associated sacs for pulses. Norris (1964, 1969) seems to favor the nasal sac system for pulse production and, in general, a tissue-water transmission path in which the sound is projected through the "melon" into water. In addition he endorses the frequently suggested idea that the melon acts as an acoustic lens and functions in the formation of the beam of sound. This theory has received additional support from the recent study by Norris and Harvey (1973) on sound transmission in the porpoise head and from the work of Vasanasi and Malius (1972). Recent measurements made using contact transducers on both the melon and the rostrum in two species indicate that echoranging pulses are projected equally efficiently from both the rostrum and the melon. Analysis of data recorded with a multiple array of attached

Maderson's work is supported in part by grants AM-15515 and CA-10844 from the National Institute of Health and C.U.N.Y. Doctoral Faculty Award 1574.

sensors places the source in the vicinity of the nasal plugs at a depth of 1.5-2.0 cm (Direcks et al., 1971). These data were used by Evans (1973) to propose a sound transmission path with both bone and adipose tissue components. A mechanical source without dependence on air flow, and thus independent of depth effects, could have definite advantages. Movements of the external parts of the blowhole mechanism have been observed and discussed by several authors, e.g., Norris (1968) and Evans and Prescott (1962). Even though the paired muscular nasal plugs fit tightly into the external bony nares, they are capable of considerable movement. It is suggested that as these plugs are moved mechanically or pneumatically against the hard edge of the external bony nares, "relaxation oscillations," with resultant acoustic pulses, are generated by alternate resistance and release of the plugs' movements. The sound produced would thus follow the paths described by Norris and Harvey (1973); first, from the muscular plug through the melon and into the water; or second, from the plug through the melon, along the premaxillary bones, and radiate into the water from the tip of the rostrum. This is in contrast to Purves' (1967) contention that all the sound is radiated from the rostrum.

The "relaxation-oscillation" mechanisms are appealing because of their efficient energy conversion capabilities, especially when one considers the sound levels measured. However, such mechanisms would place certain demands on the tissues of the nasal region, notably with respect to tolerance of shearing forces produced by high velocity air currents and the possible need for lubrication. We will now present new anatomical and histological data concerning the nasal pasages and diverticula of *Tursiops truncatus*, which we believe reinforce the theory of sound production derived from acoustic studies.

THE ANATOMY AND FUNCTION OF THE
NASAL SAC SYSTEM

The various theories previously discussed cast the nasal sac system in two possible extreme roles: that of the only active sound-producing system, or that of a reservoir for storing "recycled" air during the underwater vocalizations produced by another structure, e.g., the larynx. In fact, all theories imply a partial "storage role" for some or all of the nasal system components. Consideration of the two extreme roles permits certain predictions to be made regarding aspects of sac anatomy which can be investigated directly.

If the sounds were produced solely in the larynx, with the sac system serving only as a reservoir for recycling, one would predict a relatively simple system, of homogeneous gross and microscopic structure lacking noteworthy localized specializations. This prediction is not borne out by such studies as those of Lawrence and Schevill (1956) or Mead (1972). If, on the other hand, the nasal sac system were assumed to be the source of the sounds, one would predict anatomical and histological diversity, with specializations appropriate to the various roles of the component elements.

The odontocete nasal system is one which defies satisfactory verbal description and certainly reduction to two-dimensional graphic representation, although several excellent attempts to overcome the inherent complexities are available (Lawrence and Schevill, 1956; Evans and Prescott, 1962; Schenkkan, 1971; Mead, 1972). The horizontal section through the system of an adult *T. truncatus* shown in Figure 1 gives some indication of the problem.

Fundamentally, the system consists of a single nasal passage, formed by the fusion of paired passages exiting the bony skull, running toward the dorsally situated "blowhole," with several pairs (the actual number varying according to the species and the criteria of the investigator) of laterally arising diverticula. The entire system is bounded anteriorly by the nasal plug, a massive muscular organ which effects closure of the nasal passages (Lawrence and Schevill, 1956; Mead, 1972) and posteriorly by the assymetrical, concave, anterior sur-

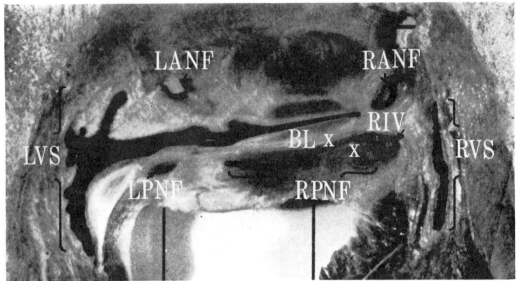

FIG. 1. A horizontal section across the nasal region of *Tursiops truncatus* approximately 1.5 cm below the blowhole, viewed from the dorsal surface. Note that the section reveals the full length of the right posterior nasofrontal (tubular) sac (RPNF) and its junction with the right inferior vestibule (RIV), while the homologous left elements lie approximately 1.0 cm ventral to this section. The grid lines are 3.8 cm apart. Other abbreviations: BL—blowhole ligament; LANF—left anterior nasofrontal sac; LVS—left vestibular sac; LPNF—left posterior nasofrontal sac; RANF—right anterior nasofrontal sac; RVS—right vestibular sac; x—location of glandular tissue (Figs. 7, 8). The anterior nasofrontal sacs have flags inserted in their lumina.

face of the cranium. The simplest possible representation of this system is shown in Figure 2, which also attempts to indicate the approximate inclinations of the various diverticula after their origin from the single nasal passage. The "distortion" of this fundamental plan can be seen if Figure 2 is compared with Figures 1 and 3, which indicate that all the sacs and tubes are flattened to a greater or lesser degree, so that: (i) the epithelial surface area is greatly increased, and (ii) there is a high probability of actual temporary physical juxtapositioning of opposing epithelial surfaces. If we add to these data the fact that fresh dissection material suggests a flexibility and mobility of the entire system comparable to that of the lips of a small boy making obnoxious noises, we can begin to appreciate how the sac system might produce sounds.

Either theory of nasal sac function suggests that a column of air passes through the system at high speed. Since air passing over an epithelial surface will exert a lateral shearing force on the tissue, *a priori* one would predict that an anatomical arrangement minimizing epithelial surface area would be present. Furthermore, if the functions of the sac system were simply that of a reservoir for the recycling of a given volume of air, not only would simple gross anatomy be predictable, but one would expect epithelial homogeneity throughout the system. Neither of these predictions is fulfilled.

In all genera, although variable from species to species, there is a pair of dorsalmost "vestibular sacs," ventralmost "premaxillary sacs" (lying beneath the posteroventral margin of the nasal plug), and between them the "tubular sacs" (nasofrontal sacs) with anterior and posterior extensions (Fig. 2). The presence of distinct "accessory sacs" (Schenkkan, 1971; Mead, 1972), "connecting sacs" (Lawrence and Schevill 1956), and paired "inferior vestibules" (Mead, 1972) seems to be somewhat variable and/or dependent on the interpretation of a particular investigator, but both are

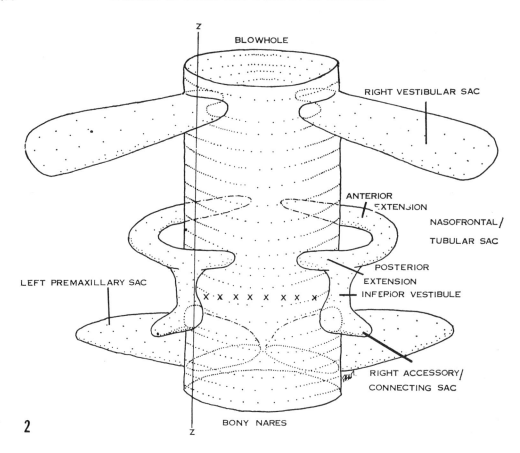

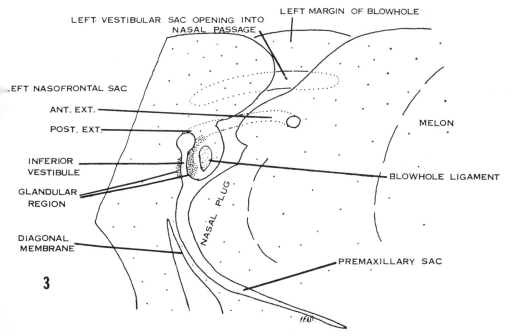

sufficiently distinct to be identified in *T. truncatus*. It is important to emphasize that not only is there variation in form and/or presence of all of the above-mention elements between genera and species, but that within at least one species—*Delphinus delphis* L.—there may be noticeable quantitative variation in size between individuals (Evans, unpublished). In all animals thus far studied, the right and left components not only show a relative asymmetry of quantitative development which is characteristic of many aspects of odontocete head anatomy, but also show a spatial asymmetry with respect to the head axes in a manner which is not predicated by the bony elements.

The entire nasal region is lined throughout by a stratified squamous parakeratotic epithelium which basically resembles that of the cetacean body epidermis (Spearman, 1972), although in the nasal region, the "corneous" layer is somewhat thinner. There are no indications of mucus-secreting or specialized sensory regions of any description. Lack of mucus-producing cells presumably reflects the fact that in an aquatic environment, the relative humidity of the air within the nasal system is always sufficiently high so that dessication is never a serious problem (Coulombe et al., 1965). The lack of sensory elements confirms the long-assumed anosmatic nature of the odontocete nose. The epithelia throughout have a general similarity to those of the human lips and buccal regions and this serves to reinforce the analogy of the potential for noise production by those parts of the human body.

The histological structure of the epithelia throughout the nasal region can be divided into three broad categories. In those regions where the gross appearance is relatively smooth and unwrinkled and where the relationships of the epithelium to the underlying tissues suggest relative immobility, i.e., the ventralmost nasal passages and the ventral aspects of the nasal plug leading into the premaxillary sacs, the epithelium is smooth, the cells are aligned parallel to the surface, and the dermal papillae are not very well developed (Fig. 4). In the vestibular sacs, which are conspicuously wrinkled, and which have been demonstrated to be capable of inflation (Evans and Prescott, 1962), the epithelial histology is quite different (Fig. 5). The dermal papillae are well developed and are oriented perpendicular to the epithelial surface. The surface of the epithelium is crenate, and above each papillar apex, the corneal cells are arranged in regularly wavy rows. All features of the epithelium in the vestibular sacs suggest a functional adaptation directed primarily towards stretching, rather than to resistance to surface shearing forces seen in the first category (Fig. 4). The epithelial structure throughout the accessory sacs, inferior vestibules, and nasofrontal sacs varies between these two extremes (Fig. 6). The farther away from the actual nasal passage, the less determinate is the epithelial structure. This is particularly true as one approaches the blind ends of the anterior and posterior extensions of the nasofrontal sacs.

Around the inferior vestibular surfaces just extending into the posterior nasofrontal sacs (Fig. 2), there are distinct modifications of the epithelial structure (Maderson, 1968). On the right side there are about 20, and on the left perhaps only half as many, crescentic pores approximate

FIG. 2. A highly schematic representation of the nasal sac system of *Tursiops truncatus* viewed as a transparent object from the posterior aspect. Note that all axes are greatly distorted, and no attempt has been made to indicate the right/left asymmetry in terms of either size of homologous units, or with respect to spatial orientation. The position X X X X X indicates the location of the blowhole ligament. Nomenclature after Lawrence and Scheville (1956) and Mead (1972).
FIG. 3. A diagrammatic drawing of a sagittal section through the nasal region of *Tursiops truncatus* taken at the axis Z-Z in Figure 2. An attempt has been made to show the general relationships of the elements further to the left of the head.

134

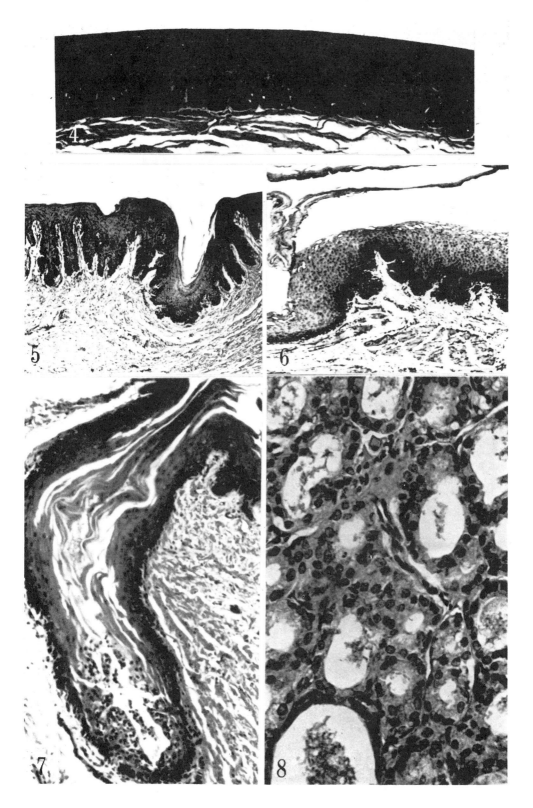

ly 1.0 mm long (Fig. 7). These lead into compound acinar, exocrine glands, running into the sub-epithelial tissues, which are larger on the right than on the left. Material in the acinar lumina probably derives by apocrine secretion from the simple cuboidal epithelium. The secreted material does not have mucinous properties and may contain some lipid (Fig. 8). Similar structures have also been seen in *Kogia,* and Mead (1972) refers to "glandular epithelia or tissues" in *Inia geoffrensis, Phocoena phocoena,* and *Phocoenoides dalli,* but offers no histological descriptions. In all cases the location includes the inferior vestibular, nasofrontal sac tissues.

DISCUSSION

The acoustic data currently available suggest most forcibly that delphinid phonations are produced in the nasal region. The figure of 1.5-2.0 cm beneath the blowhole margin provided by Diercks et al. (1971) as the site of origin would seem to correspond to the region where the lip of the nasal plug abuts the blowhole ligament (Fig. 3). While this location is only one of many which anatomical study shows close abutment of opposing epithelia, it is immediately adjacent to the glandular structures, so that it is appropriate to consider the possible function of the latter. Their small total bulk suggests that they could not be salt glands, and their anatomy is so different from such organs (Waterman, 1971, p. 587) that they cannot be interpreted as vestigial structures. They are similar in some ways to Steno's glands found in other mammals (Moe and Bojsen-Moller, 1971), but it is unlikely that they serve the humidifying function proposed by these authors. If we assume that

sound production is effected by a relaxation-oscillation mechanism involving rapid intermittent juxtapositioning of the opposing epithelia of the posterior nasal plug, and those of the ventral blowhole ligament, and circumnarial region in a rapidly moving airstream, then the glandular secretions could lubricate the tissues involved, and thus minimize mechanical damage.

Other available anatomical data seem to strengthen this postulate. Gross and microscopic analysis suggest that only the vestibular sacs regularly expand and deflate, and Norris (1964) has indicated that not only could they serve as reservoirs for recycled air, but that this activity would also permit them to act as sound reflectors. Mead (1972) suggests, therefore, that the different sizes and shapes of the vestibular sacs in different species might permit differences in the shape of sound fields. Mead (1972) states that the premaxillary sacs "are probably the best situated for storage and recycling of air for sound production." We suggest that their gross and microscopic structure does not reflect a distensible unit comparable to that of the vestibular sacs, but rather a mechanism permitting antero-postero movement of the nasal plug, ensuring a tight fit of the ventral aspect of the latter against the floor of the "sac" during intermittent contact of the posterior face with the blowhole ligament and the circumnarial tissues. Lawrence and Schevill (1956) suggested that the nasofrontal sacs (their "tubular sacs") functioned as pneumatic seals around the nasal passage. We agree with Mead's (1972) contention that this function could be better served by large muscle masses in the same area, and note his comment that in *Grampus* the left anterior nasofrontal sac is entirely

---

FIG. 4. Photomicrograph of the epithelial structure at the base of external nares in *Tursiops truncatus.*

FIG. 5. Photomicrograph of the epithelial structure of the left vestibular sac of *Tursiops truncatus.*

FIG. 6. Photomicrograph of the epithelial structure of the posterior nasofrontal sac of *Tursiops truncatus.*

FIG. 7. Photomicrograph of the opening of a glandular duct in the right inferior vestibule of *Tursiops truncatus.*

FIG. 8. Photomicrograph of the distal acini of the gland shown in Figure 7.

absent. While the intra-specific variation which we have commented upon has not yet been quantified, it seems most unlikely that any significant variation would be "permitted" by natural selection if the nasofrontal sacs played any active major role in such a sophisticated function as delphinid phonation. Schenkkan (1971) and Mead (1972) both draw attention to the considerable diversity in structure and degree of development of the accessory sacs and vestibular regions between genera. Mead (1972) makes frequent reference to the *possibility* that any and all diverticula *could* function as storage areas for recycled air. However, in the light of Norris' (1964) suggestion that an inflated vestibular sac could act as a sound reflector, it is important to note that if the nasofrontal-vestibular accessory region did become wholly or partially filled with air, then we would assume that they would play a similar reflecting role. According to this premise there should be a relationship between the details of the presence and degree of development of these various components and the shape of the sound field produced by particular genera. Since there are insufficient data available to establish such a relationship, it is simpler to assume that air storage is not the primary function of these diverticula.

Our histological data, derived from studies on *Tursiops* and *Kogia,* and comments by Mead (1972) on the other genera, suggest that glandular secretion is an important function of the vestibular-nasofrontal regions. We have suggested that lubrication of the posterior nasal plug tissues might be the function of these secretions. We have also indicated that it may be the vestibular sacs alone which serve as recycling storage areas. Therefore, the high-speed current of air which passes up the paired narial openings towards the vestibular sacs might tend to blow the glandular secretions back up the nasofrontal system, thus deflecting them away from the surfaces which should be lubricated. However, Mead (1972) suggests that the intrinsic musculature of the nasofrontal sacs

would permit their emptying and filling. Therefore, if we assume that the functions of the nasofrontal-vestibular accessory regions are secretory, and also storage of the secreted materials, we can propose the following answers to some of the problems which have been raised. During periods of non-sound production, the opening of the vestibular-nasofrontal accessory system into the nasal passage could be occluded by the nasal plug being drawn back tightly against the blowhole ligament (Fig. 2). During this time, glandular secretions could accumulate within the lumina, especially of the nasofrontal sacs due to relaxation of the intrinsic musculature. During sound production, anterior movement of the nasal plug would permit expression of the glandular secretion which, aided by active expulsion by contraction of the intrinsic musculature, would not only lubricate the opposing epithelial surfaces, but possibly also prevent the entry of air into this system of diverticula. This explanation satisfies the problem of the possibility of air-filling creating an acoustic reflector, *but,* should it be proven that there is indeed a correlation between vestibular-nasofrontal accessory anatomy and the shape of the sound fields in different genera, the model can be modified without altering the fundamental functions proposed here.

CONCLUSIONS

Following an extensive review of the nasal anatomy of a variety of odontocete genera, Mead (1972) states: "In summary, it appears that the structures most likely to be involved in sound production are those in the vicinity of the nasal plugs." We have found that this premise is supported by recent acoustic studies (Diercks et al., 1971; Evans, 1973) and by certain anatomical specializations adjacent to these tissues described in the present paper. Further substantiation of the model presented here must await demonstration of specialized glandular structures in similar regions in all other sound-producing genera, histochemical identification of the secretions produced by analysis of fresh material, and finally, correlations between the anatomical diversity now known to exist between genera and species and the acoustic properties of the sounds produced by them.

REFERENCES

Coulombe, H. N., S. H. Ridgway, and W. E. Evans. 1965. Respiratory water exchange in two species of porpoise. Science 149:86-88.

Diercks, J. J., R. T. Trochta, R. L. Greenlaw, and W. E. Evans. 1971. Recording and analysis of dolphin echolocation signals. J. Acoust. Soc. Amer. 49:1729-1732.

Evans, W. E. 1973. A discussion of echolocation by cetaceans based on experiments with marine delphinids and one species of freshwater dolphin. J. Acoust. Soc. Amer. 54:191-199.

Evans, W. E., and J. H. Prescott. 1962. Observations of the sound production capabilities of the bottlenose porpoise: a study of whistles and clicks. Zoologica 47:121-128.

Lawrence, B., and W. E. Schevill. 1956. The functional anatomy of the delphinid nose. Bull. Mus. Comp. Zool. (Harvard) 114:103-151.

Maderson, P. F. A. 1968. The histology of the nasal epithelia of *Tursiops truncatus* (Cetacea) with preliminary observations on a series of glandular structures. Amer. Zool. 8:810. (Abstr.)

Mead, J. G. 1972. On the anatomy of the external nasal passages and facial complex in the family Delphinidae of the order Cetacea. Doctoral Thesis, Univ. of Chicago.

Moe, H., and F. Bojsen-Moller. 1971. The fine structure of the lateral nasal gland (Steno's gland) of the rat. J. Ultrastruct. Res. 36:127-148.

Norris, K. S. 1964. Some problems of echolocation in cetaceans, p. 317-336. *In* W. N. Tavolga [ed.], Marine bioacoustics. Pergamon Press, New York.

Norris, K. S. 1968. The evolution of acoustic mechanisms in odontocete cetaceans, p. 297-324. *In* E. T. Drake [ed.], Evolution and environment. Yale Univ. Press, New Haven.

Norris, K. S. 1969. The echolocation of marine mammals, p. 391-423. *In* H. T. Anderson [ed.], The biology of marine mammals. Academic Press, New York.

Norris, K. S., and G. W. Harvey. 1972. A theory for the function of the spermaceti organ of the sperm whale (*Physeter catodon*, L.), p. 397-417. *In* Animal orientation and navigation, NASA SP-262, Sci. and Tech. Office NASA, Washington, D.C.

Norris, K. S., and G. W. Harvey. 1973. Sound transmission in the porpoise head. Science (In press)

Purves, P. E. 1967. Anatomical and experimental observations on the cetacean sonar system, p. 197-270. *In* R. G. Busnel [ed.], Animal sonar systems, biology and bionics. Imprimerie Louis-Jean, GAP, Haute-Alpes.

Schenkkan, E. J. 1971. The occurrence and position of the "connecting sac" in the nasal tract complex of small odontocetes (Mammalia, Cetacea). Beaufortia 19:37-43.

Spearman, R. I. C. 1972. The epidermal stratum corneum of the whale. J. Anat. 113:373-381.

Vasanasi, V., and D. C. Malius. 1972. Triacylglycerols characteristic of porpoise acoustic tissues: molecular structures of dilsovaleroylglycerides. Science 176:4037, p. 926-928.

Waterman, A. J. 1971. Chordate structure and function. Macmillan, New York.

MORPHOLOGY AND PHYLOGENY OF HAIR

By Charles R. Noback*

Department of Anatomy, College of Physicians and Surgeons, Columbia University, New York

Hair is a structure found exclusively in mammals. With this in mind, Oken named the Mammalia, Trichozoa (hair animals), and Bonnet (1892) named them Pilifera (hair bearers).

Of the many aspects of morphology and phylogeny of hair, only four will be discussed. These include (1) the principle of the arrangement of hairs in group patterns, (2) the types of hair and their relation to the principle of the group pattern, (3) a brief analysis of the structural elements of hair and their relation to the types of hair, and (4) the phylogeny of hair, with some remarks on (*a*) the relation of hair to the epidermal derivatives of other vertebrate classes and (*b*) aspects of the phylogeny of the hair and wool of sheep to illustrate that marked differences in hair coats exist between closely related animals.

Hair is the subject of a voluminous literature. Toldt (1910, 1912, 1914, and 1935), Danforth (1925a), Pinkus (1927), Pax and Arndt (1929-1938), Trotter (1932), Lochte (1938), Smith and Glaister (1939), and Stoves (1943a) discuss the problem of mammalian hair in general. Wildman (1940), von Bergen and Krause (1942), and the American Society for Testing Materials (1948) discuss the problem of fiber identification as applied to textiles.

Principle of the Group Pattern of Hairs

In the only extensive survey of the grouping of hair in mammals, DeMeijere (1894) documented the concept of the group pattern of hair (FIGURES 1–6). Unfortunately, the few studies on this phase of the problem since that time have not fully exploited the implications of this concept. DeMeijere concluded that hairs are mainly arranged in groups with the pattern of 3 hairs—with the largest hair in the middle—as the basic pattern. The concept of the basic trio as the primitive condition is accepted as an adequate working hypothesis by Wildman (1932), Galpin (1935), Höfer (1914), Gibbs (1938), Hardy (1946), and others. DeMeijere described 8 patterns: (1) 3 or less hairs behind each scale of the tail (as in the opossum, *Didelphis marsupialis*), (2) more than 3 hairs behind each scale of the tail (as in the rodent, *Loncheres* [Echimys] *cristata*), (3) 3 hairs (as in the back of the marmoset, *Midas rosalia*), (4) more than 3 hairs arranged in a regular pattern with some of greater diameter than others (as in the back hairs of *Loncheres* [Echimys] *cristata* in FIGURE 3), (5) several hairs composed of a number of fine hairs and one coarse hair (as in the back of the dog, *Canis familiaris*, in FIGURE 5D), (6) several hairs composed of a number of fine hairs and one isolated coarse hair (as in the back hairs of the mouse, *Mus decumanus*, in FIGURE 6D), (7) scatterings of fine hairs with no apparent

* The author wishes to thank Dr. Margaret Hardy, Division of Animal Health and Production, Sydney, Australia, for her valuable suggestions.

476

arrangement and a few intermingled coarse hairs (as in the back hairs of the cat, *Felis domesticus* in FIGURE 4D), and (8) hairs in irregularly scattered groups (as in the back hair of the raccoon, *Procyon cancrivorus*).

Dawson (1930) does not completely agree with DeMeijere's pattern in the guinea pig. She found variations in the pattern and no correlation between the size of hair and the arrangement of the hairs in each group. Histological study frequently shows follicle grouping which was not apparent to DeMeijere when he was examining only the skin surface, *e.g.*, in *Felis domesticus* (see Höfer, 1914). This indicates that analyses of the group pattern of hairs are needed in both common laboratory mammals and mammals in general.

In addition, DeMeijere analyzed the formation of the patterns by examining the skins of animals during their development (FIGURES 4–6). This phase of the problem has been extended to include a study of the ontogeny of the arrangement of hair follicles in sheep (Wildman, 1932, Galpin, 1935, and Duerden, 1939), in the cat (Höfer, 1914), in marsupials (Gibbs, 1938, Stoves, 1944b, and Hardy, 1946), in the mouse (Calef, 1900, Dry, 1926, and Gibbs, 1941) in the rat (Frazer, 1928), and in a number of mammals (Duerden, 1939). The terminology used by these authors in this problem is summarized in TABLE 1 (adapted from Wildman and Carter, 1939 and Carter, 1943).

Utilizing the terminology of Wildman and Carter, 1939, the following is a brief statement of the relation of the fiber generations. The first follicles to differentiate are the central trio follicles (FIGURE 7). If these follicles appear at two different times as in the opossum (Gibbs, 1938), then the follicles are called "primary X" and "primary Y." The essential point is that each of these primary follicles will be the central follicle of different hair groups. Later in development, other follicles of the hair group differentiate in relation to these central trio follicles. The trio is formed when two follicles are differentiated lateral to the primary follicles (FIGURE 8). The lateral follicles associated with primary X and primary Y are called respectively "primary x" and "primary y." If only one lateral follicle is formed adjacent to a primary follicle (X or Y), then a couplet follicle is formed. If no lateral follicles differentiate, a primary follicle (X or Y) is called a "solitary follicle." Later, another generation of follicles is differentiated—the secondary follicles. In the opossum (FIGURE 9), these secondary follicles are located between the central trio follicle and the lateral trio follicles. The ontogenetic studies of follicle arrangement have added confirmatory evidence to DeMeijere's basic concept that in mammals there is a universal and regular grouping of hair follicles (Hardy, 1946).

In general, the early differentiating follicles (central trio follicles) form the coarse overhair, while the late differentiating follicles (lateral trio follicles and secondary follicles) form the fine underhair. Lateral trio follicles sometimes at least produce overhair like that of the central follicles (*e.g.* in sheep) or intermediate types such as awns, which are classified by Danforth (1925a) as overhair. In *Ornithorhynchus anatinus* (Spencer and Sweet, 1899) and many marsupials (Gibbs, 1938, Bolliger and Hardy,

1945, Hardy, 1946), however, the lateral trio fibers are indistinguishable from those of secondary follicles, so it is difficult to place them in either the "overhair" or the "underhair" category.

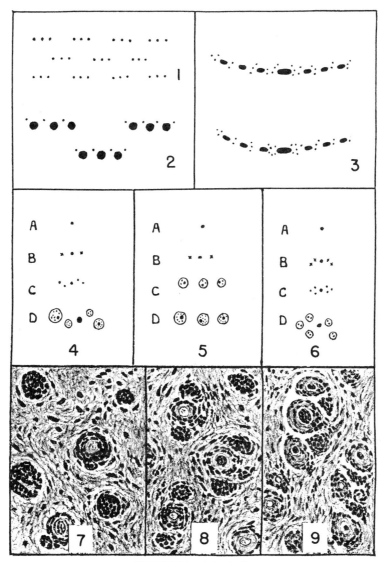

FIGURES 1-9 (see facing page).

Spencer and Sweet (1899) claimed that, in monotremes, each group of follicles was differentiated by budding from the central follicle. This has not been described in marsupials or in eutherians, in which the follicles arise independently as epidermal downgrowths. Monotremes and marsupials have in common the fact that a follicle group typically contains a large central follicle with a sudoriferous gland, and two or more clusters of smaller lateral follicles (Spencer and Sweet, 1899, Gibbs, 1938, Hardy, 1946). This arrangement is also found in some eutherians, such as the cat (Höfer, 1914) and dog (Claushen, 1933). In the cat and a few other eutherians, the first-formed lateral follicles (primary x and y of the classification of Wildman and Carter, 1939) produce hairs intermediate in type between those of the central and the other lateral follicles. There are other eutherians in which the lateral primary x and y fibers are still more like the central primary X and Y fibers, as in the pig (Höfliger, 1931) and the sheep (Carter, 1943). Except in the rodents, there is always a sudoriferous gland opening into the central primary X or Y follicle (Hardy, unpublished data). Many animals, such as the pig and sheep, also have a sudoriferous gland opening into each primary x and y follicle, but others do not (Duerden, 1939). Some of the eutherians have only primary follicles in their skin, each with a sudoriferous gland. Findlay and Yang (1948) showed that this is the arrangement in cattle, and the same is probably true in horses and in human head hair (Hardy, unpublished observations).

Types of Hair

DeMeijere's analysis leads to the classification of hair types by Toldt (1910 and 1935) and by Danforth (1925a). Many details of the hair types in many species of animals and the variations of the structure of these types are described, illustrated, and bibliographically annotated by Toldt (1935) and Lochte (1938).

TYPES OF MAMMALIAN HAIR
(after Danforth, 1925a)

1. Hairs with specialized follicles containing erectile tissue. Large, stiff hairs that are preeminently sensory. They have been variously designated as feelers, whiskers,

---

FIGURES 1–9 (*see opposite page*).

FIGURE 1. The trio hair group pattern on the back and tail of the marmoset, *Midas rosalia* (after DeMeijere, 1896). All hairs have similar diameters.

FIGURE 2. The hair group pattern of more than 3 hairs with some fibers of greater diameter than other fibers on the back of the paca, *Coelogenys paca* (after DeMeijere, 1896).

FIGURE 3. The hair group pattern of more than 3 hairs with some fibers of greater diameter than other fibers on the back of the rodent, *Loncheres* (Echimys) *cristata* (after DeMeijere, 1896).

FIGURE 4. Ontogeny of a hair group on the back of the cat, *Felis domesticus*. A, from a newborn animal; B and C, from an older animal; and D, from an adult animal (after DeMeijere, 1896).

FIGURE 5. Ontogeny of a hair group on the back of the dog, *Canis familiaris*. A, from an embryo dog; B, from a newborn animal; C, from a young dog; and D, from an adult animal (after DeMeijere, 1896).

FIGURE 6. Ontogeny of a hair group on the back of the mouse, *Mus decumanus*. A, from a 7 cm. long animal; B, from a 9 cm. long animal; C, from a 12.5 cm. long animal; and D, from an adult animal.

(FIGURES 4, 5, and 6 illustrate that the follicle of the first hair to erupt (A) will be the follicle of the coarsest hair of the hair group in the adult. The type of hair group pattern in the adult (D) in each figure is noted in the text. The X in the diagrams marks the location of erupting follicles.

FIGURE 7. The primary follicles X (the more differentiated follicles) and the primary follicles Y (the less differentiated follicles) in the transverse section of skin of a 12.5 cm. Australian opossum embryo (*Trichosurus vulpecula*). Follicles are scattered irregularly. (After Gibbs, 1938.)

FIGURE 8. Two new follicles (primary x or primary y) have become grouped with each previously differentiated follicle (primary X or primary Y) to form the typical trio arrangement. The trio would be either primary x, primary X, primary x or primary y, primary Y, primary y. Transverse section of skin of a 15.0 cm. Australian opossum embryo (*Trichosurus vulpecula*). (After Gibbs, 1938.)

FIGURE 9. Two secondary follicles have added to each trio group to form a 5 follicle group. The secondary follicles differentiate between the primary X (or Y) follicle and the primary x (or y) follicles. The five group would be either primary x, secondary follicle, primary X, secondary follicle, primary x or primary y, secondary follicle, primary Y, secondary follicle, primary y. Transverse section of skin from 20.0 cm. Australian opossum embryo (*Trichosurus vulpecula*). Note presence of a dermal capsule surrounding each 5 follicle group. (After Gibbs, 1938.)

(In FIGURES 7, 8, and 9, the terminology of Wildman and Carter (1939), noted in the text, is used.)

TABLE 1

FIBER-FOLLICLE TERMINOLOGY IN THE MAMMALIA*

Wildman and Carter (1939)		Duerden (1939) (sheep)	Hardy (1946) (marsupials)	Gibbs (1938) (opossum)	Galpin (1935) (sheep)	Wildman (1932) (sheep)	Toldt (1911) (various animals) and Höfer (1914) (cat)
Primary follicles							
(a) Central trio follicles	primary X primary Y	trio follicles central trio follicles	central follicle	primary follicles secondary follicles	X follicle Y follicle	primary follicle secondary follicle	Mittelhaar follicle
(b) Lateral trio follicles	primary x primary y	lateral trio follicles	large lateral follicle	tertiary follicles	x follicle y follicle	lateral trio follicle	Stammhaar follicles
(c) Couplet follicles	primary X + x primary Y + y						
(d) Solitary follicles	primary X or primary Y						
Secondary follicles		post-trio follicles	lateral follicles	quaternary follicles			Beihaar follicles

* Adopted from Wildman and Carter (1939).

sensory hairs, sinus hairs, tactile hairs, vibrissae, *etc.* They occur in all mammals except man, and are grouped by Botezat (1914) (Pocock, 1914) essentially as follows:

    (1). Active tactile hairs—under voluntary control.

    (2) Passive tactile hairs—not under voluntary control.

        (*a*) Follicles characterized by a circular sinus.

        (*b*) Follicles without a circular sinus.

2. Hairs with follicles not containing erectile tissue. The remaining types of hair, most of which are more or less defensive or protective in function. In many cases, the follicles have a good nerve supply, endowing the hair with a passive sensory function as well. These hairs are grouped here according to their size and rigidity.

    (1). Coarser, more or less stiffened "overhair," guard hair, top hair.

        (*a*) Spines. Greatly enlarged and often modified defensive hairs, quills.

        (*b*) Bristles. Firm, usually subulate, deeply pigmented, and generally scattered hairs. "Transitional hairs" (Botezat, 1914), "Leithaare" (Toldt, 1910), "protective hair," "primary hair," "overhair." This group also includes mane hairs.

        (*c*) Awns. Hairs with a firm, generally mucronate tip but weaker and softer near the base. "Grannenhaare" (Toldt, 1910), "overhair," "protective hair."

    (2). Fine, uniformly soft "underhair," "ground hair," "underwool."

        (*a*) Wool. Long, soft, usually curly hair.

        (*b*) Fur. Thick, fine, relatively short hair—"underhair," "wool hair."

        (*c*) Vellus. Finest and shortest hair—"down," "wool," "fuzz," "lanugo." (Danforth, 1939).

The following comments supplement the above classification. The guard hairs are listed in a series from greater to lesser rigidity (in order: spines, bristles, and awns). There are many intergrade hairs between the typical bristle and the typical awn and between the typical awn and the typical fur hair (FIGURES 10, 11, and 12).

The tactile hairs have a rich nerve supply, while the roots of some are encircled by large circular sinuses containing erectile tissue. When the pressure in the circular sinus is increased the hair becomes a more efficient pressure receptor. The overhairs have a definite nerve supply, while the underhairs have no direct nerve supply. As a general but not absolute rule, the coarser hairs appear ontogenetically earlier than the finer hairs (Gibbs, 1938, Danforth, 1925a, Duerden, 1937 (reported by Wildman, 1937), Höfer, 1914, and Spencer and Sweet, 1899).

The contour, diameter, and shape of a hair fiber changes from its root to its tip (Note awns, FIGURES 16–18). The cross-sectional outline of hairs may vary from the thick rounded porcupine quill to the eccentric flattened hairs of seals. The former serves a protective function, while the latter is adapted to hug to the skin so as not to hinder aquatic locomotion. Many details of the anatomy of hair form are noted by Stoves (1942 and 1944a), Toldt (1935), and Lochte (1938).

It is possible for a hair follicle to differentiate one type of hair at one stage and another type at another stage. The follicle of a bristle (kemp) of the Merino lamb may become the follicle of wool in the adult sheep (Duerden, 1937, reported by Wildman, 1937). A fine lanugo hair of the human fetus is associated with a follicle which will later be the follicle of a coarser hair.

The theories of hair curling are reviewed by Herre and Wigger (1939). The curling of hair in primitive sheep is independent of the arrangement of hair, existence of hair whorls, or the cross section of the hair (Pfeifer, 1929).

Wildman (1932) suggests that the shape of the follicle, especially the curve in its basal portion, is a possible factor in hair curling. Reversal of the spiral in some wool fibers may be explained according to Wildman as due to a shift in the growing point of the follicle and inner root sheath. Spiral reversal occurs in human hair (Danforth, 1926). Pfeifer (1929) doubts that curling is determined by a curve of the follicle alone and suggests that Tänzer's (1926) contention that the follicle must be saber-shaped is im-

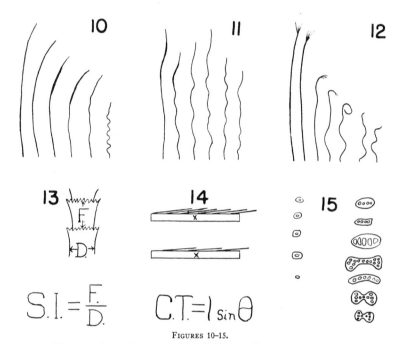

FIGURES 10–15.

FIGURE 10. The hair of the fox, *Canis vulpes* (after Toldt, 1935), illustrating intergrade hairs. From the left to the right, Toldt named the fibers Leithaar (bristles), Leit-Grannenhaar, thick Grannenhaar (awns), thin Grannenhaar, Grannen-Wollhaar, and Wollhaar (fur).

FIGURE 11. The hair of the chinchilla, *Chinchilla laniger* (after Toldt, 1935) illustrating an animal hair coat with hairs of approximately the same length. The 2 hairs on the left are awns, and the rest, either intergrade hairs or fur hairs.

FIGURE 12. The hair of the wild pig, *Sus scrofa* (after Toldt, 1935) illustrating bristles on the left and underhair on the right with some intergrade hairs between them. Note the brushlike distal ends of the bristles.

FIGURE 13. The scale index (S. I.), according to Hausman (1930), is equal to the ratio of the free proximo-distal length of a scale (F) to the diameter of the hair shaft (D).

FIGURE 14. The thickness of the cuticle (C. T.), according to Rudall (1941), is equal to the length of a cuticular scale (1) times the sine of angle (sin Θ) the scale makes with the cortex (X).

FIGURE 15. Cross sections of several regions of a fur hair (left) and an awn (right) of the rabbit (after Toldt, 1935). The sections, at the top of the figure, are from the base of the hair and, at the bottom of the figure, from the tip of the hair. Illustrates general uniformity of the diameters of the fur hair and differences in diameters and contour of awn hairs throughout their lengths.

portant. Waving of all compact wools is due at least in part to the flattening of the primary spiral and to the unequal lateral growth of the fiber (Duerden, 1927). The curling of hair in karakul sheep fetuses may be associated with the differences in the rates of growth in the various skin layers (Herre and Wigger, 1939).

The factors responsible for curling and crimping of hair are as yet not completely known.

Structural Components of Hair

The cuticle, cortex, and medulla are the three structural components in hair. They will be discussed in order.

Cuticle. The cuticle consists of thin, unpigmented, transparent overlapping scales, whose free margins are oriented toward the tip of the hair

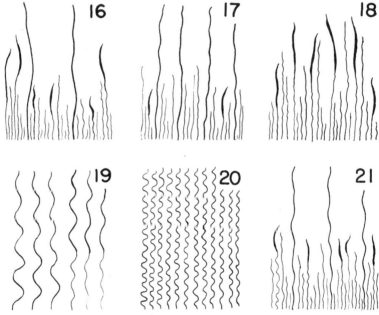

FIGURES 16–21.

FIGURE 16. Diagram of the fiber components of coat of a generalized non-wooled animal (after Duerden, 1929). Note presence of bristles (coarse fibers), awns (fibers with fine basal segments and coarse distal segments) and fur fibers (fine fibers).

FIGURE 17. Diagram of the fibers of the wild sheep (after Duerden, 1929). Note the presence of bristles (kemp), awns (heterotypes), and wool.

FIGURE 18. Diagram of the fibers of British mountain breeds (after Duerden, 1929). The fibers are mainly awns and wool. Few bristles are present.

FIGURE 19. Diagram of the fibers of the British luster breeds (after Duerden, 1929). Fibers on the left are wool fibers which are coarser than the wool fibers of wild sheep. The fibers on the right are modified awns with fine proximal segments and slightly coarse distal segments. All fibers are elongated and spiraled.

FIGURE 20. Diagram of fibers of adult Merino sheep (after Duerden, 1929). All fibers are wool. Note uniformity of all fibers as to size, length, and waviness. These wool fibers are coarser than wool fibers from wild sheep. Unlike the fibers of other breeds, the fibers of the adult Merino sheep grow from persistent germs and do not shed.

FIGURE 21. Diagram of the fibers of the Merino lamb. Note the presence of bristles (kemp), awns (heterotypes), and wool. During later development, the bristles are shed and the distal coarse segments of the awns are lost. The adult coat is formed by the persistent growth of the wool fibers of the lamb, by the replacement of wool in the follicles of the shed kemp, and by the persistence of the growth of the proximal segments of the awns.

(FIGURE 22). Within the follicle, the free margins of the hair cuticular scales interlock with the inner root sheath cuticular scales, which are oriented in the opposite direction toward the papilla. This interlocking of scales helps to secure the hair in place (Danforth, 1925a). The cuticle functions as a capsule containing the longitudinally splitable cortex (Rudall, 1941). This explains why the cortex of a hair frays at its severed end. In addition, the cuticle, with its oily layer, prevents the transfer of water (Rudall, 1941).

The cuticular scales vary in thickness from 0.5 to 3 micra (Frölich, Spötel, and Tänzer, 1929). Since the scales overlap, the number of overlapping scales at any point on the hair surface determines the thickness of the cuticle. The cuticular thickness may be expressed as being equal to the length of the scales times the sine of the angle the scale makes with the cortical surface (FIGURE 14, Rudall, 1941).

The cuticular scales may be classified into two types: coronal scales and imbricate scales (Hausman, 1930). A coronal scale completely encircles the hair shaft. They are subdivided according to the contour of the free margins as: simple, serrate, or dentate (FIGURE 23). Müller (1939) contends that a coronal scale is in reality several scales whose lateral edges are

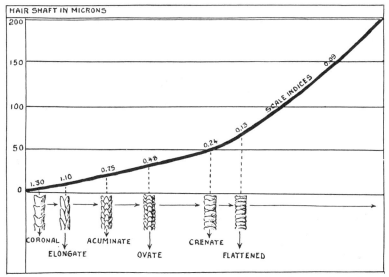

FIGURE 22. Graph illustrating the relation of the diameter of the hair to the types of cuticular scales. The finest hairs (with small diameters) have a high-scale index and coronal scales. The coarsest hairs (with large diameters) have a low-scale index and flattened scales. Diameters of hair shafts are plotted on the ordinate. General regions of the occurrence of scale forms are shown along the abscissa, the average scale indices along the curve. The figures of the scale types beneath the graph are not drawn to scale. (After Hausman, 1930.)

fused. For example, a dentate coronal scale with 5 processes in its free border is the fused product of 5 elongated pointed scales.

An imbricate scale does not completely surround the hair shaft. They are classified as ovate, acuminate, elongate, crenate, and flattened (FIGURE 22, Hausman, 1930).

Hausman (1930) devised a scale index to express the relation between the diameter of the hair shaft and the free proximo-distal dimension of the scales (FIGURE 13). The free proximo-distal dimension is actually a means of expressing the type of scale. For example, coronal scales have a large proximo-distal dimension, while crenate scales have a small dimension (FIGURE 22). An analysis of the scale indices indicates that a relation exists between the types of scales and the shaft diameters. In general, the finest hairs have large scale indices and coronal scales, while the coarsest hairs have small scale indices and crenate or flattened scales. On the basis of the above, it is concluded that the types of cuticular scales present on hair are related not to the taxonomic status of the animal possessing the hair but rather to the diameter of the hair shaft (Hausman, 1930). In hairs with both thick and thin segments, the thick segments have the scale types of large diameter hairs while the thin segments have the scale types of small diameter hairs.

A coarse guard hair has scales with free lips that are closely applied to the cortex and are scarcely raised. As a result, these hairs have a high luster (due to unbroken reflection of light from the hair surface) and do not interlock with other hairs. A fine underhair has scales with lips that have raised margins. As a result, these hairs are dull (due to broken reflection of light) and interlock with other fine hairs. Thus, mohair has a high luster but makes poor felt, while wool is dull but makes good textiles.

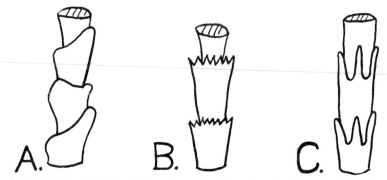

FIGURE 23. Figures illustrating the types of coronal cuticular scales. A. simple scales, B. serrate scales, C. dentate scales, (after Hausman, 1930). Note raised margins on the free lips of scales.

Many details of the cuticle in many species of animals are presented and illustrated by Lochte (1938).

Cortex. The cortex usually forms the main bulk of a hair. It is a column of fusiform keratinized cells which are coalesced into a rigid, almost homogeneous, hyaline mass (Hausman, 1932). Damaged hairs tend to split lengthwise because the elongated cortical cells are oriented longitudinally. The cortex has such a low refractive index—due to the degree of cornification—that, in the absence of pigment, it is translucent. Since cortical scales have not been analyzed in such detail as cuticular scales, no statement can be made of a relation between cortical scale morphology and hair size. The form and distribution of the pigment in the cortex and the medulla is noted by Lochte (1938), Toldt (1935), and Hausman (1930).

Hausman (1932 and 1944) analyzed the cortical air spaces known as cortical fusi—cortical in location and fusiform in shape—air vacuoles, air chambers, air vesicles, or vacuoles. As the irregular-shaped cortical cells located in the bulb rise to the follicular mouth, they carry between them cavities filled with tissue fluid. As the hair shaft dries out, the cavities lose the fluid, and air may fill the resulting spaces—the fusi. The shape of the fusi vary. They are largest, most numerous, and most prominent near the base of the hair, and they are filiform and thin or lost in the distal segments of the hair. Seldom do they persist to the tip of a hair. As a rule, they are visible only under a microscope. Hausman implies that there is a relation between fusi and hair size. Presumably, the coarser a hair segment is, the more numerous the fusi.

Ringed hair results when the fusi appear in masses at regular intervals in the shaft. Fractured fusi result when hairs are damaged sufficiently to separate the cortical cells enough to allow air to collect between them. Fusi can be distinguished from pigment granules, for they are fusiform, whereas pigment granules have blunt ends.

The presence of a thin membrane located between the cuticle and the cortex has been assumed by Lehmann (1944). Observations of pigment granules, cell nuclei, and submicroscopic fibrils are presented by Mercer (1942), Hausman (1930), and others.

Medulla. The medulla (pith), when present, is composed of shrunken and variably shaped cornified remnants of epithelial cells connected by a filamentous network. In contrast to the cortex, the medulla is less dense and has fewer and larger cells, which are more loosely held together. In the medulla are air cells or chambers, which are filled by a gas, probably air. These air cells may be intracellular (deer) or intercellular (dog, weasel, and rat) (Lochte, 1938). The intercellular air cells are classifiable according to their coarseness and arrangement (Lochte, 1934 and 1938).

Medullas are classified by Hausman (1930) as follows: absence of medulla, discontinuous medulla (air cells separate), intermediate medulla (several separate air cells of the discontinuous type arranged into regular groups), continuous medulla (air cells arranged to form a column), and fragmental medulla (air cells arranged into irregular groups). These types are illustrated in FIGURE 24 and are arranged in the order of the sizes of hairs in which they are located. In the finest hairs (underfur), the medulla is either absent or of the discontinuous type. In the coarsest hairs, the medulla is either of the continuous or the fragmental type (FIGURE 24). If a hair varies in thickness, its medulla will vary. For example, in the awns of sheep, the distal thickened segment has a medulla, while the fine proximal segment may have no medulla. The arrangement of the medullary air cells is related not to the taxonomic group of the animal possessing the hair nor the age of the hair, but rather to the diameter of the hair shaft (FIGURE 24) (Hausman, 1930; Wynkoop, 1929; and Smith, 1933). The sheens and colors of hairs are largely determined by the light reflected from the medulla (Hausman, 1944).

Although the cortex forms the bulk of the shaft in most hairs, the medulla assumes large proportions in some hairs. In rabbit hair (FIGURE 15), the medulla is composed of large air cells separated by little more than a framework of cortex (Stoves, 1944a).

The significance of the cuticle, cortex, and medulla in the commercial aspects of fur is presented by Bachrach (1946). Although many of the details of the structural elements of hair cannot be definitely utilized to identify an animal species (Hausman, 1944), it is possible that some morphological features of hair can be used (Williams, 1938).

Some chemical and physical aspects of the morphological elements of hair have been analyzed. Not only do the cuticle, cortex, and medulla exhibit different chemical and physical properties, but various segments of these structural elements may have different chemical and physical proper-

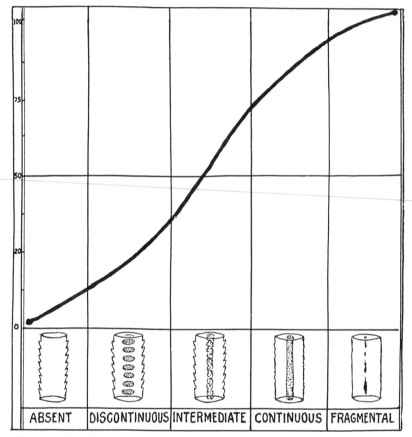

ABSENT | DISCONTINUOUS | INTERMEDIATE | CONTINUOUS | FRAGMENTAL

FIGURE 24. Graph illustrating the relation of the diameter of the hair to the types of medullas. The finest hairs have no medulla, and the coarsest hairs have a fragmental medulla. Diameters in micra of the hair shafts are plotted on the ordinate. The figures of the types of medullas beneath the graph are not drawn to scale. (After Hausman, 1930.)

ties (Rudall, 1944; Stoves, 1943b, 1945; Lustig, Kondritzer, and Moore, 1945; Leblond, 1951; and Giroud and Leblond, 1951).

Some Phylogenetic Aspects of Hair

The relation of hair to the epidermal structures in non-mammalian animals has been discussed by many authors and has been summarized by

Botezat (1913 and 1914), Danforth (1925b), and Matkeiev (1932). No direct relation between hair and non-mammalian epidermal elements has been established. Hair is most probably an analog to these structures. Danforth (1925b) and others conclude that hair is probably a *de novo* morphological entity in mammals.

Broili (1927) reports that he identified hair and hair follicles in a fossil aquatic reptile, *Rhamphorynchus*. This animal is a specialized reptile, removed from those reptiles in the evolutionary line to mammals. If established, this observation would alter the concept that only mammals produce hair.

The phylogeny of hair in related groups of animals has not been analyzed extensively. Because of the economic importance of wool, several studies of the hair types in the coat of a number of breeds of sheep have been made. One significant aspect of these studies is that they illustrate how the hair coat may vary in closely related forms.

This statement is adapted primarily from Duerden (1927 and 1929). The generalized wild animal hair coat consists of an overcoat of bristles and awns and an undercoat of fur (FIGURE 16). In the wild sheep and the black-headed Persian sheep, the hair coat is similar to that of the wild animal. These sheep have an overcoat of bristles (called kemp) and awns (called heterotypes) and an undercoat of wool (FIGURE 17). The British mountain breeds have a hair coat consisting of awns and wool (FIGURE 18). In these breeds, kemp formation is negligible. The coat of the British luster breeds have evolved in another direction. The awns retain their fine proximal segments. Their distal segments are still thicker than the proximal segments, but are thinner than the distal segments of the awns of primitive sheep. The wool undercoat fibers have thickened. Both fiber types are elongated and spiraled (FIGURE 19). The adult Merino sheep, the most efficient wool-producing sheep, has a coat consisting of elongated, regularly crimped fibers of uniform diameters and lengths (FIGURE 20). An analysis of the coat of the Merino lamb is essential for the identification of the types of fibers that form the coat of the adult sheep. The Merino lamb coat has bristle, awn, and wool fibers (FIGURE 21). During ontogeny, the bristles are shed and then replaced by wool fibers. The awn fibers lose their distal thickened segments, but the thin proximal segments persist. The wool fibers are retained, but are coarser than the wool of primitive sheep. Hence, the adult Merino sheep coat consists of wool fibers differentiating from follicles which produced kemp in the lamb, of awns deprived of their distal segments, and of wool fibers differentiating from follicles which produced wool in the lamb. A major difference between the Merino sheep and other sheep is in the nature of the hair follicles. Whereas the coat of other breeds is shed periodically and then new hairs differentiate from the follicles, the fibers of the adult Merino sheep grow from persistent germs and are not shed.

In the evolution of the sheep coat from primitive wild sheep to the various domestic breeds, several changes have occurred. As summarized by Duerden (1927), the domestic wooled sheep has evolved in the direction of the loss of the protective coat both of bristles (kemp) and awns (heterotypes), the increase in length, density, and uniformity of the fibers, and the tendency of the retained bristles to become finer but still capable of being shed. In addition, the Merino sheep has developed persistently growing hair follicles.

Important implications of the evolution of the sheep coat are that the types of hair in the hair coat may differ (1) in closely related animals and (2) at various stages of ontogeny with the same animal. Hence, data derived from a study of the coat of one animal species may not always apply to another animal species.

Bibliography

American Society for Testing Materials. 1948. Standards on textile materials. By the Society for Testing Materials. 560 pp. Philadelphia.

BACHRACH, M. 1946. Fur. 672 pp. Prentice-Hall. New York.

BOLLIGER, A. & M. H. HARDY. 1945. The sternal integument of *Trichosurus vulpecula*. Proc. Roy. Soc. New South Wales **78**: 122–133.

BONNET, R. 1892. Ueber Hypotrichosis congenita universalis. Anat. Hefte **1**: 233–273.

BOTEZAT, E. 1913. Ueber die Phylogenie der Säugetierhaare. Verhandl. d. Gesellsch. deutsch. Naturf. u. Aerzte. **85**: 696–698.

——— 1914. Phylogenese des Haares der Säugetiere. Anat. Anz. **47**: 1–44.

BROILI, F. 1927. Ein Rhamphorhynchus mit Spuren von Haarbedeckung. Sitzungsber. Math.-Naturw. Abt. Bayerische Akad. Wiss. München 49–68.

CALEF, A. 1900. Studio isologico e morfologica di un' appendice epitaliale del pelod nella pelle del Mus dectmanus var. albina e del Sus scrofa. Anat. Anz 17; 509–517.

CARTER, H. B. 1943. Studies in the biology of the skin and fleece of sheep. Council for Scientific and Industrial Research, Commonwealth of Australia. Bulletin No. 164: 1–59.

CLAUSHEN, A. 1933. Mikroskopische Untersuchungen über die Epidermatgebilde am Rumpfe des Hundes mit besonder Berücksichtigung der Schweissdrüsen. Anat. Anz. **77**: 81–97.

DANFORTH, C. 1925a. Hair, with special reference to hypertrichosis. American Medical Association. Chicago, Illinois; Arch. of Derm. and Syph. **11**: 494–508, 637–653, 804–821.

DANFORTH, C. 1925b. Hair in its relation to questions of homology and phylogeny. Am. J. Anat. **36**: 47–68.

DANFORTH, C. 1926. The hair. Nat. Hist. **26**: 75–79.

DANFORTH, C. 1939. Physiology of human hair. Physiol. Rev. **19**: 94–111.

DAWSON, H. L. 1930. A study of hair growth in the guinea pig (*Cavia cobaya*). Amer. J. Anat. **45**: 461–484.

DeMEIJERE, J. C. H. 1894. Über die Haare der Säugethiere besonders über ihre Anordung. Gegenbaur's Morphol. Jahrb. **21**: 312–424.

DRY, F. W. 1926. The coat of the mouse (*Mus musculus*). J. Genetics **16**: 287–340.

DUERDEN, J. E. 1927. Evolution in the fleece of sheep. S. Afri. J. Sci. **24**: 388–415.

DUERDEN, J. E. 1929. The zoology of the fleece of sheep. S. Afri. J. Sci. **26**: 459–469.

DUERDEN, J. E. 1939. The arrangement of fibre follicles in some mammals with special reference to the Ovidae. Trans. Roy. Soc. Edinburgh **59**: 763–771.

FINDLAY, J. D. & S. H. YANG. 1948. Capillary distribution in cow skin. Nature, Lond. **161**: 1012–1013.

FRAZER, D. A. 1928. The development of the skin of the back of the albino rat until the eruption of the first hairs. Anat. Rec. **38**: 203–224.

FRÖLICH, G., W. SPÖTTEL, & E. TÄNZER. 1929. Wollkunde; bildung und eigenschaften der wolle. 419 pp. J. Springer. Berlin.

GALPIN, N. 1935. The prenatal development of the coat of the New Zealand Romney lamb. J. Agri. Sc. **25**: 344–360.

GIBBS, H. F. 1938. A study of the development of the skin and hair of the australian opossum Trichosurus vulpecula. Proc. Zool. Soc. London **108**: 611–648.

GIBBS, H. F. 1941. A study of the post-natal development of the skin and hair of the mouse. Anat. Rec. **80**: 61–82.

GIROUD, A. & C. P. LEBLOND. 1951. The keratinization of epidermis and its derivatives, especially the hair, as shown by X-ray diffraction and histochemical studies. Ann. N. Y. Acad. Sci. **53** (3): 613.

HARDY, M. 1946. The group arrangement of hair follicles in the mammalian skin. Notes on follicle group arrangement in 13 australian marsupials. Proc. Roy. Soc. Queensland **58:** 125–148.

HAUSMAN, L. A. 1930. Recent studies of hair structure relationships. Sc. Month. **30:** 258–277.

HAUSMAN, L. A. 1932. The cortical fusi in mammalian hair shafts. Amer. Nat. **66:** 461–470.

HAUSMAN, L. A. 1934. Histological variability of human hair. Amer. J. Phys. Anthrop. **18:** 415–429.

HAUSMAN, L. A. 1944. Applied microscopy of hair. Sc. Month. **59:** 195–202.

HERRE, W. & H. WIGGER. 1939. Die Lockenbildung der Säugetiere. Kühn-Archiv. **52:** 233–254.

HÖFER, H. 1914. Das Haar der Katze, seine Gruppenstellung und die Entwicklung der Beihaare. Arch. f. Mik. Anat. **85:** 220–278.

HÖFLIGER, H. 1931. Haarkleid und Haut des Wildschweines. VII. Beitrag zur Anatomie von Sus scrofa L. und zum Domestikations-problem. Z. Ges. Anat. 1. Z. Anat. Entw. Gesch. **96:** 551–623.

LEBLOND, C. P. 1951. Histological structure of hair, with a brief comparison to other epidermal appendages and epidermis itself. Ann. N. Y. Acad. Sci. **53** (3): 464.

LEHMANN, E. 1944. Neue Reactionen der Tierischen Faser. Kolloid-Zeitschrift. **108:** 6–10.

LOCHTE, T. 1934. Untersuchungen über die Unterscheidungsmerkmale der Deckhaare der Haustiere. Deutche Zeitschr. Ges. Genchtl. Med. **23:** 267–280.

LOCHTE, T. 1938. Atlas der· Menschlichen und Tierschen Haare. 306 pp. Paul Schops. Leipsig.

LUSTIG, B., A. KONDRITZER, & D. MOORE. 1945. Fractionation of hair, chemical and physical properties of hair fractions. Arch. Biochem. **8:** 57–66.

MATKEIEV, B. S. 1932. Zur theorie der Rekapitulation. Über die Evolution der Schuppen, Federn und Haare auf dem Wege embryonaler Veränderungen. Zool. Jahrb. Abt. Anat. u. Ontog. **55:** 555–602.

MERCER, F. H. 1942. Some experiments on the structure and behavior of the cortical cells of wool fibers. J. Council Sc. Ind. Research **15:** 221–227.

MÜLLER, C. 1939. Ueber den Bau der koronalen Schüppchen des Säugetierhaares. Zool. Anz. **126:** 97–107.

PAX, F. & W. ARNDT. 1929–1938. Die Rohstoffe des Tierreichs. 2235 pp. Gebrüder Borntraeger, Berlin. Sections in this reference are: FRÖLICH, G., W. SPÖTTEL, & E. TÄNZER. 1932. Haare und Borsten der Haussäugetiere. **1:** 995–1221; MEISE, W. 1933. Menschenhaar. **1:** 1312–1363; SCHLOTT, M., & E. BRASS, W. STICHEL, & E. KLUMPP. 1930. Pelze. **1:** 405–567; SCHLOTT, M. 1933. Haare und Borsten von Wildsäugern. **1:** 1222–1311.

PFEIFER, E. 1929. Untersuchungen über das histologisch bedingte Zustandekommen der Lockung, mit besonderer Berücksichtigung des Karakullammes. Biol. Generalis. **5:** 239–264.

PINKUS, F. 1927. Die normale Anatomy der Haut. In: Handbuch der Haut- und Geschlechtskrankheiten. **1:** 1–378. J. Springer, Berlin.

POCOCK, R. 1914. On the facial vibrissae of mammalia. Proc. Zool. Soc. Lond. **40:** 889–912.

RUDALL, K. M. 1941. The structure of the hair cuticle. Proc. Leeds Philos. Soc. **4:** 13–18.

SMITH, H. H. 1933. The relationships of the medulla are cuticular scales of the hair shafts of the Soricidae. J. Morph. **55:** 137–149.

SMITH, S. & J. GLAISTER. 1939. Recent Advances in Forensic Medicine. 264 pp. P. Blakiston's Sons and Co. Phila.

SPENCER, B. & G. SWEET. 1899. The structure and development of the hairs of monotremes and marsupials. Part 1. Monotremes. Quart. J. Micro. Sci. (N. S.) **41:** 549–588.

STOVES, J. L. 1942. The histology of mammalian hair. Analyst **67:** 385–387.

STOVES, J. L. 1943a. The biological significance of mammalian hair. Proc. Leeds Phil. and Literary Soc. **4:** 84–86.

STOVES, J. L. 1943b. Structure of keratin fibres. Nature **151:** 304–305.

STOVES, J. L. 1944a. The appearance in the cross sections of the hairs of some carnivores and rodents. Proc. Roy. Soc. Edinburgh **62B:** 99–104.

STOVES, J. L. 1944b. A note on the hair of the South American opossum (*Didelphis caranophaga*). Proc. Leeds Philos. Soc. **4**: 182–183.

STOVES, J. L. 1945. Histochemical studies of keratin fibres. Proc. Roy. Soc. Edinburgh. **62**: 132–136.

TÄNZER, E. 1926. Haut und Haar beim Karakul im rassenanalytischem Vergleich. Halle a.d.S. (cited by: Pfeifer, 1929).

TOLDT, K., JR. 1910. Über eine beachtenswerte Haarsorte und über das Haarformensystem der Säugetiere. Annalen d. K. K. Naturhistorischen Hofmuseums in Wien. **24**: 195–265.

TOLDT, K., JR. 1912. Beträge zur Kenntnis der Behaarung der Säugetiere. Zool. Jahrb. Abt. Syst. **33**: 9–86.

TOLDT, K., JR. 1935. Aufbau und natürliche Färbung des Haarkleides der Wildsäugetiere. 291 pp. Deutsche Gesellschaft fur Kleintier- und Pelztierzucht. Leipsig.

TROTTER, M. 1932. The hair. Special Cytology (Cowdry, E. ed.) **1**: 39–65. P. Hoeber. New York.

VONBERGEN, W. & W. KRAUSS. 1942. Textile fiber atlas. Amer. Wool Handbook Co. New York.

WILDMAN, A. B. 1932. Coat and fibre development of some british sheep. Proc. Zool. Soc. London **1**: 257–285.

WILDMAN, A. B. 1937. Non-specificity of the trio follicles in the Merino. Nature, Lond. **140**: 891–2.

WILDMAN, A. B. 1940. Animal fibres of industrial importance: their origin and identification. 28 pp. Wool Industries Research Assoc. Torridon, Headingley, Leeds.

WILDMAN, A. B. & H. CARTER. 1939. Fibre-follicle terminology in the mammalia. Nature **144**: 783–784.

WILLIAMS, C. 1938. Aids to the identification of mole and shrew hairs with general comments on hair structure and hair determination. J. Wildlife Management **2**: 239–250.

WYNKOOP, E. M. 1929. A study of the age correlations of the cuticular scales, medullae and shaft diameters of human head-hair. Am. J. Phys. Anthrop. **13**: 177–188.

Discussion of the Paper

DOCTOR M. H. HARDY (*McMaster Laboratory, Glebe, N. S. W., Australia*): I am glad Dr. Noback mentioned sheep, because the study of the arrangement of follicles in groups on these animals has disclosed some important principles. Terentjeva,[1] Duerden,[2] and Carter[3] showed that de Meijere's trio group is the basic unit in the follicle population of sheep. The trio (primary) follicles develop first and have accessory structures (sudoriferous gland, arrector pili muscle) which are absent from the later developing (secondary) follicles of the group.[3]   In the young lamb, it is the primary follicles which produce the coarse, and frequently medullated, kemp fibers and the secondary follicles which produce the fine and usually non-medullated wool fibers. These correspond respectively to the 'overhair' and 'underhair' in Danforth's classification. The primary follicles may produce kemp in the lamb and wool in the adult sheep, as Dr. Noback has mentioned.

The size of the follicle groups, *i.e.*, the number of secondary follicles to each trio of primary follicles, varies greatly between breeds[4] and individuals[5] and also between body regions.[6] The breeds and, to some extent, individuals with the largest group size have also the greatest number of fibers to the square inch and the greatest uniformity of fiber thickness and length.[5] In the midside region, at least, the potential group size (including secondary follicle rudiments in the young lamb) is strongly inherited, but the actual group size (number of active follicles) in the mature animal varies according to the food intake in the first year of life.[7]   Thus, it is possible to alter the group size experimentally.   Varying the food intake in the second and third year of life had no marked effect on group size.[8,9]

It seems that many properties of the coat of the sheep depend on the inherited follicle group pattern and the modifications of this superimposed by the environment.   Perhaps the same principles apply to other mammals.

1. TERENTJEVA, A. A.   1939.   Pre-natal development of the coat of some fine-wooled breeds of sheep. C. R. Acad. Sci. U.R.S.S. (N.S.) **25**: 557.
2. DUERDEN, J. E.   1939.   The arrangement of fibre follicles in some mammals, with special reference to the Ovidae. Trans. Roy. Soc. Edin. **59**: 763.
3. CARTER, H. B.   1943.   Studies in the biology of the skin and fleece of sheep. 1. The development and general histology of the follicle group in the skin of the Merino. Coun. Sci. Ind. Res. (Aust.) Bull. 164: 7.
4. CARTER, H. B. & P. DAVIDSON.   Unpublished data.
5. CARTER, H. B.   1942.   "Density" and some related characters of the fleece in the Australian Merino. J. Coun. Sci. Ind. Res. (Aust.) **15**: 217.
6. CARTER, H. B. & M. H. HARDY.   1947.   Studies in the biology of the skin and fleece of sheep. 4. The hair follicle group and its topographical variations in the skin of the Merino foetus. Coun. Sci. Ind. Res. (Aust.) Bull. **215**: 5.
7. CARTER, H. B., H. R. MARSTON, & A. W. PEIRCE.   Unpublished data.
8. FERGUSON, K. A., H. B. CARTER & M. H. HARDY.   1949.   Studies of comparative fleece growth in sheep. Aust. J. Sci. Res. B **2**: 42.
9. FERGUSON, K. A., H. B. CARTER, M. H. HARDY, & H. N. TURNER.   Unpublished data.

MOTIONS OF THE RUNNING CHEETAH AND HORSE

By Milton Hildebrand

The horse is perhaps the most efficient running machine ever evolved; probably no other vertebrate has so many structural adaptations for rapid and untiring progress on the ground. The cheetah is conceded to be the fastest of all animals for a short dash, but lacks the endurance of the horse. This paper will analyze and contrast the running motions of these champions, and will reveal some of the secrets of the cheetah's superlative speed.

Several authors have noted cursorial adaptations of the cheetah (e.g., Pocock, 1927; Hopwood, 1947) but to my knowledge none has contrasted its mode of running with that of other cursorial quadrupeds. Morphological adaptations of the horse have been described by Howell (1944), Eaton (1944), Smith and Savage (1956) and many others. Those references emphasized structure; this paper stresses function. The classical study by Muybridge (1899) has remained the most important analysis of the motion of the horse. A paper by Grogan (1951) provides a concise review of the sequence of footfalls and combinations of supporting members.

MATERIALS AND METHODS

This study was inspired by the excellent film sequence of a running cheetah in the Walt Disney *True Life Adventure* picture "African Lion." I am grateful to Walt Disney Productions for furnishing film strips for analysis.

The photographer, Alfred Malotte, filmed the animal with a telephoto lens, so perspective changes slowly during the run. The chase presented is actually a combination of two dashes: a slower, shorter one, filmed at regular speed, and one taken in slow motion. Sequences of three and seven consecutive strides show the cheetah about side-on to the camera. With a Recordak Film Reader, 155 successive frames were traced. Registration points permitted these to be redrawn as a composite picture, with the images in proper spatial relationship to one another. The film outlines are not sharp, and low vegetation usually obscured the feet when they were on the ground, but there are enough nearly identical frames to establish a ground line and a reasonably accurate depiction of motions.

The analysis for the horse was made from photographs in Muybridge (1899:

481

171–179) and from the film "Horse Gaits," produced by the Horse Association of America, Inc. In the latter, action was filmed with an electric camera at 128 frames per second; the sequence analyzed shows the horse "Citation" winning the mile-and-one-quarter American Derby in 1948. The method of analysis was the same as with the cheetah.

FINDINGS

Speed.—Figure 1 shows speed records of the horse and, for comparison, of man, for distances up to 900 yards, expressed as rate of travel and lapsed time. Approximate speed of the cheetah is also indicated.

The maximum measured speed for man is 22.28 mph, over 220 yds.; the plotted curve shows that he could average 22.6 mph for 155 yds.

The horse has run ¼ mi. at 43.27 mph; it could probably average 44 mph for 300 yds.

The speed of the cheetah is legendary, yet scantily documented. Authors quote each other and the estimates of lay observers. However, there is both direct and indirect evidence of great speed. Because many artiodactyls will run parallel to a moving vehicle, accurate data are available on the speed of some of them. Einarsen (1948) reported that the pronghorn normally can run at 50 mph, and under favorable conditions can attain 60 mph. On a California desert a pet cheetah overtook a young buck pronghorn (Mannix, 1949). Craighead (1942) stated that the cheetah often runs down its quarry within 150 yds., and that it is not unusual for a cheetah to overtake an antelope that has had a head start of 100 yds. or more.

At Ocala, Florida, John Hamlet includes a cheetah in an animal show featuring species employed in hunting. The cheetah is trained to run in a long enclosure. A popular article (Severin, 1957) reported the results of a speed test stating that "from a deep crouch Okala spurted to the end of the 80 yard course in 2¼ seconds, for an average speed of about 71 miles an hour." Unfortunately, this record must be disregarded because the enclosure is, in fact, about 65 yards long, the method of timing was inexact, and there is an arithmetical error.

It is a general consensus that this remarkable cat can run at least 70 mph.

The speed of the cheetah in the film strips analyzed in this paper could be computed if the film speeds and the animal's body length were exactly known, but these can only be approximated. The studio reported film speeds of about 24 and 48 frames per second; efforts to check these figures with the photographer were unsuccessful. The animal shown is a male. Male cheetahs average about 7 ft. in length (records taken from Hollister, 1918; Shortridge, 1934; Bryden, 1936; Roberts, 1951); the largest of record measured 7 ft. 9 in. Separate calculations based on assumed animal lengths of 6½ and 7¾ ft., and (for the slow-motion sequence) on film speeds of 46 and 50 frames per second give a range of possible speeds between 37½ and 49 mph. Since the animal had to find its footing among scattered shrubs that were 6 to 24 in. high, these less-than-optimum speeds seem expectable.

Endurance.—Records of animal endurance that are accurate and comparable are difficult to secure. Figure 2 presents some relatively reliable data. The human distance records were taken from various editions of *The World Almanac.* Records for the horse are from the same source and from Howell (1944), who also cited a record (dating from 1853) of 100 mi. at 11.2 mph. If accurate, this is truly remarkable: on the basis of curves plotted from other records one would expect no more than a 9–10 mph rate for this great distance.

Andrews (1933) reported following another perissodactyl, the Mongolian wild ass (*Equus hemionus*), over open country with an automobile. One particular animal ran 16 mi. at an average speed of 30 mph "as well as could be estimated"; the next 4 mi. were covered at about 20 mph. It ran 29 mi. before it stopped from exhaustion. Since it repeatedly changed direction and speed, these figures must be taken as approximate, but it is unlikely that any other

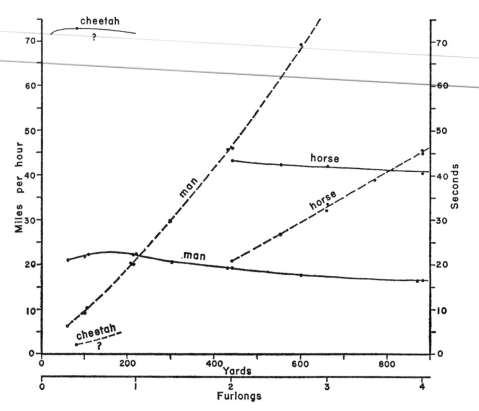

Fig. 1.—Speed records of the cheetah (approximate), horse and man, expressed as rate of travel (solid lines and left ordinate scale) and lapsed time (dashed lines and right ordinate scale). Source for man and horse: several editions of *The World Almanac.*

animal could equal this feat over distances greater than 3 mi. (The pronghorn is faster for short distances, according to Einarsen, 1948.)

In sharp contrast to the equids noted, the cheetah seldom runs more than ¼ mi. Pocock (1927) claimed that 600 yds. is the maximum distance for a chase at speed, and Bryden (1936) stated that two mongrel dogs brought one to bay in 2½ mi. Prey species are almost invariably overtaken by the cheetah, and usually knocked to the ground. However, if they can scramble to their feet and run again, the cheetah often abandons further pursuit.

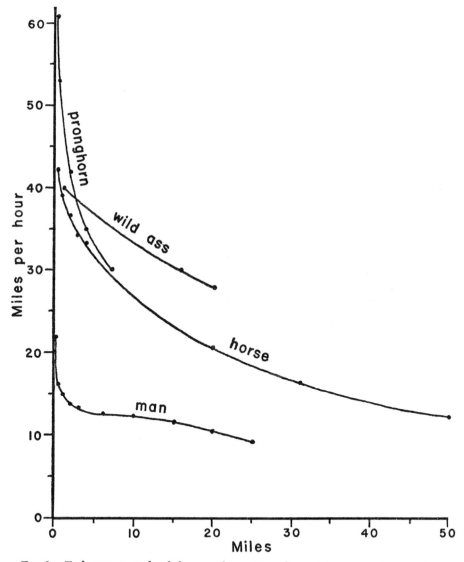

FIG. 2.—Endurance records of the pronghorn, Mongolian wild ass, race horse and man, expressed as average rate of travel for different distances. Sources cited in text.

The longer of the dashes on the film analyzed in this paper was about 325 yds.

Sequence of footfalls.—The leading front or hind foot is second of the pair to touch and leave the ground in each stride or cycle of movement. An unqualified reference to lead applies to the front feet: an animal is said to be running with a left lead if the left forefoot is placed in front of its opposite. I will call the other member of each pair the trailing limb.

In the extreme flexed position the galloping horse passes one hind foot forward of one forefoot (Fig. 4*e*). Since the legs have little lateral motion and nearly equal straddle, the animal can avoid interference only by a sequence in which the leading forefoot is followed with the hind foot on the other side of the body. Thus the front and back legs must use the same lead. This sequence of footfalls, diagrammed in Fig. 3, is termed the transverse gallop.

In the extreme flexed position the cheetah passes both hind feet forward of both forefeet (Fig. 5*h*). To avoid interference it must therefore straddle the forelimbs with the hind limbs. It would seem that the lead of the fore- and hind limbs could be independent, but in practice the leading forefoot is followed by the hind foot on the same side—a sequence of footfalls called the rotary (or lateral) gallop. If the legs on one side of the body were extended as those on the other side were gathered together, and if the spine were flexed to right and left, then the rotary sequence of footfalls would increase the reach of the limbs slightly (about 2 inches per stride for a 7° swing of shoulders and pelvis with a straddle of 8 inches), but this is not the case. Perhaps the rotary sequence provides subtle benefits to balance or muscle function.

The domestic cat commonly places the hind feet nearly opposite one another when running (a gait termed the half bound) but, curiously, it may on occasion follow the horse rather than the cheetah, using the transverse gallop (Muybridge, 1899).

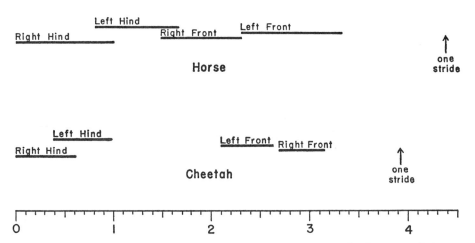

FIG. 3.—Sequence of footfalls and phases of one representative stride, shown in relation to time in tenths of seconds. The period that each foot is on the ground is shown by the length of its respective line.

Phases of the stride and their duration.—The galloping animal has all feet off the ground one or more times in each stride, and during periods of support the legs are used in different combinations. Each suspended period and each combination of supporting members is called a phase. There is much individual variation in the phases of gaits. Indeed, Howell (1944: 222) reported 16 different phase formulas for galloping horses. However, a usual phase formula can be selected for analysis. The nature and duration of the phases of such a formula of the galloping horse and cheetah are shown in Fig. 3.

The horse has all feet off the ground once in each stride—in the flexed position (see Fig. 4e). Howell (1944: 240) depicted a light horse that had a second, brief suspended phase, just before the trailing forefoot struck the ground, but this is unusual.

The cheetah is suspended when flexed, and again when extended. I believe there is sometimes a third, though fleeting, instant of suspension—between falls of the front feet (Fig. 3 and positions *d, f* and *h*, Fig. 5). Muybridge (1899: 157) anticipated this circumstance when he wrote, "It is probable that future research will discover—with the horse and some other animals—during extreme speed, an unsupported transit from one anterior foot to the other."

Analysis of Fig. 3 shows that the galloping horse characteristically has one suspended and seven supported phases (the supported transit from one forefoot to the other being almost instantaneous when galloping at good speed). The cheetah has three suspended and five supported phases.

The duration of each phase varies not only with speed but also with the

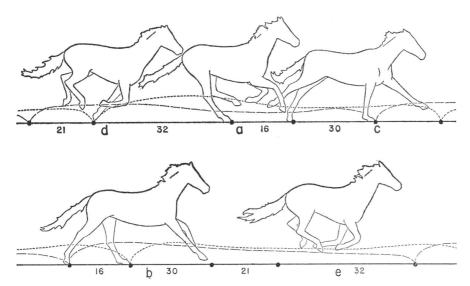

FIG. 4.—Five positions of a galloping horse shown in correct spatial relationship. Trajectories followed by the front feet are indicated above, those by the hind feet below, long dashes for right feet and short dashes for left feet. Positions of footfalls are shown by spots on the ground line. Figures below ground line give for each interval its percentage of total stride distance.

161

individual and for the same individual at different times. The following statements are as representative as the material available permits, but are only approximations of any particular performance.

When galloping at 35 mph the horse completes one stride in about .44 second, or 2¼ strides per sec.; at about 45 mph the cheetah completes one stride in about .39 second, or 2½ strides per sec. The horse is supported during ¾ of its stride and the cheetah during only half of its stride. Each animal is supported by two legs for 11 to 12 per cent of its total support period.

The trailing hind foot of the horse is on the ground about 85 per cent as long as the leading hind foot, whereas the two hind feet of the cheetah are on the ground about the same amount. The disparity betwen the animals is greater for the forefeet: the trailing forefoot of the horse is on the ground 80 per cent as long as the leading foot, whereas with the cheetah the foot that has the shorter contact is the leading foot (about 95 per cent as long as that of the trailing foot).

Change of lead.—These differences in duration of support and the asymmetry in resulting stresses require a change in lead from time to time to postpone fatigue. Further, a galloping animal can turn more sharply by leading with the inside forefoot.

Unless rider or terrain demand frequent turning, a horse changes its lead most often to compensate for the relatively great discrepancy in the duration of support provided by leading and trailing legs. Actual lead reversal is usually accomplished first by the forelimbs, but the motion of the hind limbs must be coordinated to avoid the interference that would otherwise result. Probably the spacing of the footfalls must also be altered, and it is likely that average speed will be reduced slightly if the lead is changed frequently.

The cheetah's leading and trailing legs share the exertion of running more evenly, but sharp and frequent changes of direction are usually dictated by the evasive quarry. The cheetah in the film strip changed lead three times in a sequence of 33 strides, and nine times in a sequence of 34 strides. Only once was the same lead used consecutively more than seven times, and five times it was changed after three or fewer strides.

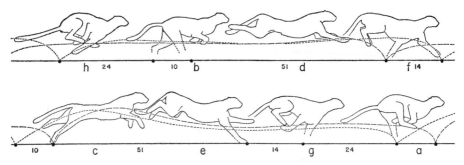

Fig. 5.—Eight positions of a galloping cheetah, shown in correct spatial relationship. Symbols and figures as for Fig. 4.

Several factors contribute to the facility with which the cheetah changes lead, and it is unlikely that speed is sacrificed. In contrast to the horse, there is a time in the stride of the cheetah (just following position *d*, Fig. 5) when the two front and two hind feet are opposite one another in both the horizontal and vertical planes. At this instant the lead can be changed as quickly and smoothly as not, and since this position immediately precedes the placement of the first (trailing) forefoot, the animal need not long anticipate the change of lead required by a turn, and cannot easily be thrown off balance by the dodging of its prey.

Length of stride.—The spacing of footfalls, and hence total length of stride, varies considerably with speed and individual performance. The data presented here are indicative of usual distances. They are based on five strides of three horses and on ten strides of a cheetah.

The strides of the galloping horse recorded by Muybridge (1899) varied from nearly 19 ft. to nearly 23 ft., and averaged 22.8 ft. Exceptional horses are reputed to cover 25 ft. at a stride (Howell, 1944: 241). Assuming the cheetah of the film strip to be 7 ft. long, the shortest of the seven strides traced was 21 ft., the longest 26 ft., and the average 23 ft.

Thus the cheetah covers at least as much ground per stride as does the horse in spite of the great disparity in body sizes: the stride of the cheetah is 8½ to 11½ times its shoulder height (with supporting forelegs vertical), compared with 4½ to 5 for the horse; or 5¼ to 6¼ times its chest-rump length (in position of maximum extension), compared with 3½ to 4 for the horse. The cursorial skill of the cheetah results in large measure from its ability to achieve so long a stride. The number and duration of the suspended phases of its gait contribute; other factors are considered further in following sections of this paper.

In Figs. 4 and 5 the footfalls are marked by dark spots on the ground lines. The per cent of total stride involved in each interval is indicated by the numbers below the ground lines. The most evident difference between the two animals in spacing of footfalls is the greater percentage of stride (51 against 30) that the cheetah achieves between the strike of the leading hind foot and that of the trailing forefoot. At this time it is bounding forward with all feet off the ground; at a corresponding time the horse is supported (compare Figs. 4*b*, and 5*d*). If we arbitrarily eliminate the difference by reducing this particular interval of the cheetah's stride to 30 per cent of total stride (as with the horse) and adjust the remaining three percentages accordingly (making the sum of the intervals again 100 per cent), the horse still has a slightly longer reach between the two hind feet and covers less ground in its suspended transit from leading front foot to trailing hind foot.

Support role of the forelegs.—It has been said that the front legs of a galloping horse do nothing that a wheel would not do better. To be strictly true, the wheel would need to be versatile at banking and at shifting track to maintain the balance of its load, yet support is certainly the principal function of the equid forelimbs. The hind quarters are closest to the ground when the hind feet are on the ground (see croup-to-ground curve, Fig. 7), but the withers, in contrast, start to rise when the first (trailing) front foot strikes the ground and continue to rise until the leading front foot is lifted. The cushioning of body impact by the digital ligaments (Camp and Smith, 1942) and the muscles that suspend the thorax between the shoulder blades does not prevent the forequarters from rising as they pass over the stiff front legs which are pivoting on the supporting feet. The variation in withers-to-ground height is only 1½ to 2 inches, about one-third of the variation in croup-to-ground height.

It is not possible to determine the deceleration of forward motion that results from the lift given the body by the front legs, but, making some reasonable assumptions, we can learn its order of magnitude. If a 1150-lb. horse galloping 40 mph lifts half of its weight 2 inches as the stiff forelegs pivot forward over the supporting feet, the resulting deceleration will be .034 mph. Conclusion: in regard to speed, a wheel would do nothing for a horse that its front legs don't do just about as well.

Figure 7 shows that the shoulders of the cheetah are falling when the trailing forefoot strikes, continue to fall all the time the front feet are on the ground, and start to rise again only as the first hind foot strikes the ground. Evidently the front legs provide little support and no deceleration, yet, before concluding that the cheetah could run without wheel *or* forelegs, we must consider other functions of its front legs.

Role of the back.—Like other carnivores the cheetah sharply flexes and extends the spine when running. For reasons considered in the next section, the heavy-bodied horse must hold its back nearly rigid, although there is some motion at the sacrum. The amounts of flexion and extension for the two animals, approximated from photographs, are shown in Fig. 6.

The angle that the pelvis makes with the scapula changes about 60° in the running horse, and about 130° in the running cheetah. The rotation of the scapula on the spine is about the same (roughly 20°) in each animal, so the 70° difference between them is attributable to the spine. In both animals the motion of the spine in the vertical plane is greater at the pelvis than at the shoulder.

Of what advantage is a supple spine to a cursorial animal?

One would expect flexion and extension of the spine to increase the swing of the limbs, thus increasing the distances covered during the suspension phases of the stride and extending the duration of the support phases. If this is true, the angles between ground line and limbs as they strike and leave the ground should be more acute for the cheetah than for the horse. The instant of impact of the feet is difficult to determine from the somewhat blurred images of the available moving-picture frames, so I cannot offer quantitative data, but it appears that these angles are indeed more acute for the cheetah.

Swing of the limbs is accomplished for the horse almost exclusively by muscles inserted on the limbs, while muscles of the back also contribute for the cheetah. This is of significance. If two muscles move one bone on another, the force of rotation is equal to the sum of the individual forces whereas the velocity is limited to that of one muscle acting alone (assuming comparable and adequate leverages and intrinsic rates of contraction). However, if a muscle moves one bone on a second while another muscle moves the second bone in the same direction on a third bone, then there is summation of both force and velocity. Thus, on the recovery stroke, the swing of a limb can be hastened by flexing several of its joints. (Shortening the limb also decreases the load on the muscles.) But when a limb is supporting the body, only a limited amount of motion is possible between the limb joints. Therefore, by swinging its limbs with two independent sets of muscles (of the limbs and of the back) the cheetah increases the speed of its stride.

Although the forward extension of the limbs when the feet strike the ground is only a little greater for the cheetah than for the horse, the more supple spine of the former contributes to substantially greater maximum forward extension before the feet start their backward acceleration preliminary to striking the ground (Fig. 6). Further, comparing the trajectories traced by the feet, as shown in Figs. 4 and 5, it is clear that, in the position of maximum forward extension, the limbs of the cheetah are held higher than are those of the horse. Indeed, they are not only higher relative to body size, but actually higher by about one-third for the front feet and trailing hind foot. It follows that the feet of the cheetah travel farther in moving to the ground. It may be inferred that they have greater backward acceleration when they strike the

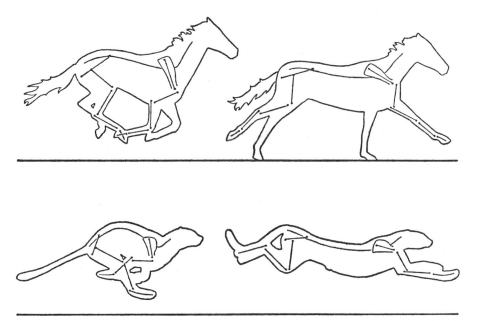

Fig. 6.—The galloping horse and cheetah, shown in positions of maximum flexion and extension of the spine and maximum rotation of the scapula on the spine.

ground and that they probably develop enough traction to prevent any deceleration from factors discussed below and in the next section.

In the flexed position the chest-buttock length of the horse is 80–90 per cent of its length in the extended position (87 in my analysis; 81 in an instance reported by Howell, 1944: 240). The flexed length of the cheetah is only about 67 per cent of its extended length. The actual shortening of the body accomplished by flexion is about 16 in. for the cheetah and 9 in. for the horse. In Fig. 7, changes in chest-buttock length are synchronized with duration of contact of each foot with the ground. For the cheetah, flexion from the position of maximum body length (high points on upper curve) is initiated when the body is unsupported. This helps impart backward acceleration to the front foot that is about to strike the ground. However, any considerable body flexion at this time would tip the shoulders forward and reduce the reach of the leading front leg, so sharp flexion is postponed to the instant the leading foot strikes. Flexion is then rapid, and is nearly completed while that foot is on the ground; only a little more body shortening is accomplished as the leading front foot follows through. Thus the fore- and hindquarters are not significantly drawn toward one another by flexion of the spine: the hindquarters alone move toward the forequarters as the latter are fixed by the forelegs (with reference to the ground, their deceleration is prevented)

In similar manner, extension of the body starts as the trailing hind foot initiates its down stroke. Again this action must help give that foot acceleration to the rear. Some extension also accompanies the unsupported follow-through of the hind legs, but most of the body extension occurs when the hind feet are on the ground. Since backward motion (deceleration) of the hindquarters is thus prevented by the hind legs, nearly all of the increase in body length resulting from extension is added to the length of the stride.

We see that the body of the cheetah moves forward like that of the measuring worm. The added distance is nearly 15 in. per stride, giving an increment in speed of 2 to 2¼ mph at a rate of about 40 mph. What the increment might be at greater speeds will depend on the relative roles played by increased length of stride and increased rate of stride as the animal moves faster. It seems probable that at 60 mph the animal adds in this manner at least 3 mph to its rate of travel.

A limber spine contributes to speed in still another way. As the cheetah's trailing foreleg strikes the ground, its forequarters and hindquarters are moving with equal velocity. But while the front feet are on the ground, the body is flexed on the forelimbs so that, at the instant the leading foot leaves the ground, the hindquarters have greater forward velocity than the forequarters. (The energy necessary to bring this about is here considered to be exerted by muscles of the back and forelimbs, against the ground as traction.) The difference between the velocity of the shoulders and of the center of mass of the entire body is nearly 3½ ft. per sec. when the animal is running at 45 mph. (The figure was derived by estimating the positions of the respective points on tracings of the animal plotted from successive moving-picture frames and then measuring their relative motion in a known time interval.) In other words, when the forelimbs are on the ground, the portion of the body to which they are joined is moving forward nearly 2½ mph. slower than the body as a whole. Similarly, when the hind feet are on the ground the pelvis is also moving slower than the body as a whole. It is reasonable to surmise that speed is benefited by this circumstance, for it reduces the backward velocity (though not the force) required of the legs in order to propel the body forward.

Body size, speed and endurance.—The speed at which an animal can run is a function of length and duration of stride. Each of these factors is related to body size.

If it were possible to disregard mass, then animals of like form would run at the same speed regardless of body size, because length of stride varies in direct proportion to linear measure whereas intrinsic rate of muscle contraction, and hence rate of stride, varies inversely with linear measure (Hill, 1950).

It is true that the red fox can run as fast as a horse although it is one-tenth as long, but mass cannot be neglected: the horse weighs 100 times as much as the fox, and with like form could scarcely run at all. The force of contraction of a muscle is proportional to the cross-sectional area of its fibers, therefore varying as the square of linear measure. The mass of the body varies as the cube of linear measure, so largeness places the muscles at a disadvantage even when the body is at rest. In motion the disadvantage is greater (Hill, *op. cit.*) because as body size increases the power the muscles can deliver does not quite keep up with the demands placed on them to control the kinetic energy developed in oscillating parts of the body.

To avoid impossible stresses, a large animal must therefore modify the form and function of its body to reduce the load placed on its muscles and supportive tissues. Since momentum is the product of mass and velocity, this can be done by minimizing the motion of one part of the body relative to another, by causing its center of mass to move in as nearly a rectilinear fashion as possible, and by reducing the mass of such structures as must change their velocities. These principles, and related structural adaptations, are noted in publications cited above and in the introduction to this paper.

To run at all, the horse must have a degree of efficiency that assures both speed and endurance. The fox has both speed and endurance for a different reason: its mass is so small that inertia does not increase sufficiently with speed to cause distress. What of the cheetah?

At 125 lbs. the cheetah is only about one-ninth as heavy as the horse, but it is about 14 times as heavy as the fox. *Miohippus,* some litopterns, and many artiodactyls are (or were) of comparable size, but have cursorial mechanisms that conserve energy more effectively than does that of the cheetah. Why is not this cat either smaller or more like the horse in the form of its body and the way that it runs?

The answer is that the cheetah does not need to be efficient; it needs to be fast, and its size is about optimum for maximum speed. Its muscles can stand the strain long enough for the animal to run the necessary 400 to 600 yds., so greater efficiency is not needed. However, if its body were heavier, then even for such short distances it could not employ every mechanism for gaining speed while disregarding those that improve efficiency. Its speed, then, imposes an upper limit on its body size. There are probably several reasons why the cheetah is not smaller: its size gives it wide vision, independence of irregularities in the terrain, and enough weight to bring down its prey.

SUMMARY

The cheetah is the fastest of animals for a short dash, and the horse has superlative endurance. These animals differ greatly in body size, so it is instructive to compare their ways of running.

Analysis was made from slow-motion moving-picture sequences by tracing images of successive frames and arranging them in correct spatial relation to one another.

The cheetah can sprint at 70 to 75 mph; the horse can attain 44 mph for 300 yds. The cheetah seldom runs more than ¼ mi., the horse can run at 20.5 mph for 20 mi., and its rate

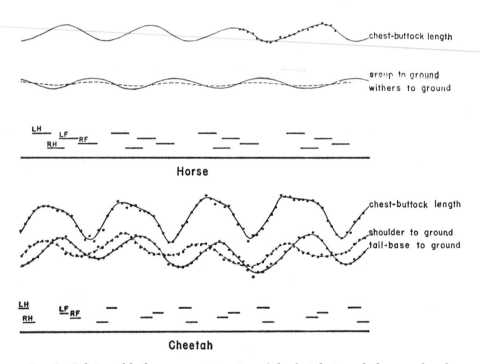

Fig. 7.—Relation of body movement to action of the feet during a little more than four strides. Motion is from left to right. Lower broken lines show, in the manner of Fig. 3, the periods of contact of the feet with the ground; letters *R*, *L*, *H* and *F* mean right, left, hind and front, respectively. Upper curves indicate, by distance above the base lines, variation in chest–buttock length. Middle curves depict height of shoulders (withers) and tail base (or croup) above the ground. All distances above the base line are in proportion to maximum chest–buttock length, which is equated for the two animals.

of travel declines only slowly as distances increase over 30 mi. The endurance of the Mongolian wild ass is apparently superior to that of the horse.

The horse uses the transverse gallop, usually covers 19 to 25 ft. per stride and completes about 2¼ strides per sec. at 35 mph. Its body is suspended once in each stride, during one-quarter of the stride interval. The leading front and trailing hind limbs support the body longer than their opposites. A change of lead usually occurs first for the front feet, but must be anticipated well before the trailing front foot strikes the ground. The forward motion of the front limbs as they pivot on the supporting feet raises the forequarters, but the resulting deceleration of the body is negligible. Its mass and inertia require that the horse minimize the motion of one part of the body relative to another and move its center of mass in a nearly rectilinear fashion: the feet are not lifted high, there is little up-and-down motion of withers and croup, and the back is relatively rigid.

The cheetah uses the rotary gallop, covers as much ground per stride as the horse, and at 45 mph completes about 2½ strides per sec. The body has two long periods of suspension (and probably a short one) in each stride, adding up to half of the stride. The trailing front foot is on the ground a little longer than the leading foot; the two hind feet have about equal periods of support. Changes of lead are smoothly accompilshed, and can be initiated an instant before the trailing front foot strikes the ground. The front limbs do not raise the forequarters. Body size is about optimum for maximum speed: it is small enough so body form and motion can be adapted for speed with little regard for efficiency, yet large enough to gain a long and rapid stride, as noted below. The feet are lifted high. There is pronounced up-and-down motion of shoulders and pelvis, and marked flexion and extension of the spine.

Flexion and extension of the back contribute to speed by: (1) increasing the swing of the limbs, thus increasing the distance covered during suspended phases of the stride and increasing the duration of the supported phases; (2) advancing the limbs more rapidly, since two independent groups of muscles (spine muscles and intrinsic limb muscles) acting simultaneously can move the limbs faster than one group acting alone; (3) contributing to increased maximum forward extension of the limbs, which permits their greater backward acceleration before they strike the ground; (4) moving the body forward in measuring-worm fashion; and (5) reducing the relative forward velocity of the girdles when their respective limbs are propelling the body.

Speed is the product of stride rate times length. Relative to shoulder height, the length of the cheetah's stride is about twice that of the horse. Factors contributing to its longer stride are: (1) two principal suspension periods per stride instead of one; (2) greater proportion of suspension in total stride; (3) greater swing of limbs, so they strike and leave the ground at more acute angles; and (4) flexion and extension of the spine synchronized with action of the limbs so as to produce progression by a measuring-worm motion of the body.

The rate of the cheetah's stride is faster than that of the horse because: (1) its smaller muscles have faster inherent rates of contraction; (2) its limbs are moved simultaneously by independent groups of muscles; (3) its feet move farther after starting their down strokes before striking the ground, thus developing greater backward acceleration; (4) the forelimbs have a negligible support role and probably actively draw the body forward; (5) the limbs are flexed more during their recovery strokes; and (6) the shoulders and pelvis move forward slower than other parts of the body at the times that their respective limbs are propelling the body.

LITERATURE CITED

ANDREWS, R. C. 1933. The Mongolian wild ass. Nat. Hist., 33: 3–16.

BRYDEN, H. A. 1936. Wild life in South Africa. Geo. G. Harrap & Co., Ltd., London. pp. 282.

CAMP, C. L. AND NATASHA SMITH. 1942. Phylogeny and functions of the digital ligaments of the horse. Memoirs Univ. Calif., 13 (2): 69–124.

CRAIGHEAD, JOHN AND FRANK. 1942. Life with an Indian prince. Nat. Geogr. Mag., 81: 235–272.

EATON, T. H., JR. 1944. Modifications of the shoulder girdle related to reach and stride in mammals. Jour. Morph., 75 (1): 167–171.

EINARSEN, A. S. 1948. The pronghorn antelope. Wildlife Management Inst., Wash. D. C. pp. 238.

GROGAN, J. W. 1951. The gaits of horses. Jour. Amer. Vet. Med. Assn., 119 (893): 112–117.

HILL, A. V. 1950. The dimensions of animals and their muscular dynamics. Science Progress, 38 (150): 209–230.

HOLLISTER, NED. 1918. East African mammals in the United States National Museum. Bull. U.S. Nat. Mus., 99, Pt. 1.

HOPWOOD, A. T. 1947. Contribution to the study of some African mammals. III, Adaptations in the bones of the fore-limb of the lion, leopard, and cheetah. Jour. Linn. Soc. (Zool.), 41: 259–271.

HOWELL, A. B. 1944. Speed in animals. Univ. Chicago Press. pp. 270.

MANNIX, JULE. 1949. We live with a cheetah. Sat. Evening Post, 221 (37): 24 ff.

MUYBRIDGE, EADWEARD. 1899. Animals in motion. Chapman & Hall, Ltd., London, pp. 264. [Republished with minor changes, 1957. Ed. by L. S. Brown. Dover Publ., Inc. pp. 74, 183 pls. References in present paper are to 1899 ed.]

POCOCK, R. I. 1927. Description of a new species of cheetah (*Acinonyx*). Proc. Zool. Soc. London, 1927: 245–252.

ROBERTS, AUSTIN. 1951. The mammals of South Africa. Contr. News Agency So. Africa, pp. 700.

SEVERIN, KURT. 1957. Speed demon. Outdoor Life, 119 (4): 54 ff.

SHORTRIDGE, G. C. 1934. The mammals of Southwest Africa, vol. I, W. Heinemann, Ltd., London. pp. 437.

SMITH, J. M. AND R. J. G. SAVAGE. 1956. Some locomotory adaptations in mammals. Jour. Linn. Soc. (Zool.), 42 (288): 603–622.

Dept. of Zoology, Univ. of California, Davis. Received March 28, 1958.

ADDENDA

The last paragraph on p. 482 notes that estimates of speed for the cheetah studied were based on information from Disney Studios that the film analyzed was made at 48 frames per second. The photographer, Alfred Malotte, could not be reached when the paper went to press, but now informs me that the film speed was 64 frames per second. This increases my estimate of the animal's speed to 56 mph and the rate of the stride (top page 487 and Fig. 3) to one stride in .28 seconds or about 3½ strides per second.

On p. 491 it is stated that the cheetah adds about 15 inches to each stride by a measuring worm motion of the back. This should have been 15 inches two times per stride (when the back is flexed and again when it is extended), and, with the revised estimate of stride rate, the benefit to speed becomes about 6 mph.

… # NATURAL HISTORY MISCELLANEA

Published by

The Chicago Academy of Sciences

Lincoln Park - 2001 N. Clark St., Chicago 14, Illinois

No. 170 October 30, 1959

Toxic Salivary Glands in the Primitive Insectivore *Solenodon*

GEORGE B. RABB*

In 1942 O. P. Pearson demonstrated the toxic property of the saliva of *Blarina brevicauda,* a common shrew of the eastern United States, and identified its principal source as the submaxillary gland. Comparative studies at that time and subsequently revealed that similar poisonous factors were not present in the salivary glands of other soricid and talpid insectivores (Pearson, 1942, 1950, 1956). I had an unexpected opportunity to make a crude check on the salivary glands of *Solenodon paradoxus,* a remote relative of the shrews, when three of these animals died at the Chicago Zoological Park within two months after their arrival in 1958 from the Dominican Republic.

Parts of the submaxillary and parotid glands of one animal that had died one to two hours beforehand were ground separately with sand, diluted to 10 per cent by weight solutions with 0.9 per cent NaCl solution, and filtered, following the procedure of Pearson (1942). These solutions were injected into a small series of male white mice that ranged in weight from 29 to 44 grams.

All of the mice injected with extract from submaxillary gland showed some reaction — at least urination and irregular or rapid breathing for several minutes. Five that received intravenous doses of extract of .09 to .38 mg. submaxillary gland per gram of body weight did little more than this and recovered within 30 minutes. Five that received intravenous doses of .38 to .55 mg. per gram additionally exhibited protruding eyes, gasping, and convulsions before dying within two to six minutes. Two animals that had intraperitoneal injections of extract of .56 and .66 mg. per gram died in about 12 hours, and one injected at the level of 1.02 mg. per gram died in 13 minutes. Urination, cyanosis, and depression were observed in these animals. Three "control" mice injected intravenously with extract of 1.02, 1.68, and 1.87

*Chicago Zoological Park, Brookfield, Illinois

mg. of parotid gland per gram of body weight showed no distress except for initially very rapid breathing in the last case.

In general these results are very like those described for *Blarina* extracts. It may be noted that the twentyfold lesser potency evident here of *Solenodon* extract as compared to that of *Blarina* may be due to postmortem inactivation of the toxic principle as reported by Eil.s and Krayer (1955) for fresh *Blarina* material. Further tests with the refined techniques of these authors using acetone treated glands will be necessary for a fairer assessment of the potency of *Solenodon* toxin.

Sections were made of the submaxillary glands and stained w.th hematoxylin and eosin and also with a modification of Mallory's triple stain. These sections showed some large cells with coarse acidophilic granules and small nuclei in the secretory ducts. Pearson (1950) suspected that such cells in *Blarina* might be concerned in the production of the saliva's toxic principle, although somewhat similar cells are found in other soricids.

The submaxillary glands of *Solenodon* are rather enormous and conspicuous structures (see fig. 47 in Mohr, 1938). Each gland weighs three to four grams in adult animals. According to Allen (1910), the duct of the submaxillary gland ends at the base of the large deeply channeled second incisor tooth of the lower jaw (see fig. 19D in Mc Dowell, 1958). Presumably toxic saliva would be conducted thereby into a wound. I could not induce *Solenodon* to bite live mice and therefore have no direct evidence on this point. However, in 1877 Gundlach reported inflammatory effects of bites by Cuban *Solenodon* to himself and a mountaineer (although he dismissed the possibility of venomous action on the basis of authority!). Of his hand bite he said: ". . . I was bitten by the tame individual, which gave me four wounds corresponding to the [large] incisors: those from the two upper incisors healed well, but those from the lower ones inflamed."

Moreover, there are indications that *Solenodon* is not immune to its own venom. Autopsy of the third animal disclosed multiple bite wounds on the feet and no obvious internal evidence of other causes of death. Sections of the liver show considerable congestion in that organ. The snout, lips, limbs, and tail were very pale the afternoon preceding death. Mohr (1937, 1938) gave accounts of several cases in which death was the outcome of fighting with cage mates although only slight foot wounds were inflicted. Pearson (1950) reported that *Blarina* was relatively immune to its own venom, although the single test animal died and the interpretation was problematical. The utility of the venom for

2

Solenodon in its natural environment is unknown and is certainly not indicated by its insectivorous habits. The explanation may be phylogenetic and historical rather than one of present-day function.

I wish to acknowledge the help of the park's veterinarian, W. M. Williamson, and medical technician, Ruth M. Getty.

Literature Cited

Allen, Glover M.
    1910.   *Solenodon paradoxus.* Mem. Mus. Comp. Zool., 40: 1-54.
Ellis, Sydney and Otto Krayer
    1955.   Properties of a toxin from the salivary gland of the shrew, *Blarina brevicauda.* Jour. Pharmacol. and Exptl. Therap., 114: 127-37.

Gundlach, Juan
    1877.   Contribucion a la mamalogia Cubana. Havana, G. Monteil, 53 pp.
McDowell, Samuel B., Jr.
    1958.   The Greater Antillean insectivores. Bull. American Mus. Nat. Hist., 115(3): 113-214.

Mohr, Erna
    1937.   Biologische beobachtungen an *Solenodon paradoxus* Brandt in Gefangenschaft. III. Zool. Anz., 117: 233-41.
    1938.   Biologische beobachtungen an *Solenodon paradoxus* Brandt in Gefangenschaft. IV. Ibid., 122: 132-43.

Pearson, Oliver P.
    1942.   On the cause and nature of a poisonous action produced by the bite of a shrew (*Blarina brevicauda*). Jour. Mamm., 23: 159-66.
    1950.   The submaxillary glands of shrews. Anat. Record, 107: 161-69.
    1956.   A toxic substance from the salivary glands of a mammal (short-tailed shrew). pp. 55-58 in Venoms, ed. E. E. Buckley and N. Porges, American Assoc. Adv. Science Publ. No. 44, xii + 467 pp.

Altitudinal Zonation of Chipmunks (Eutamias): Adaptations to Aridity and High Temperature[1]

H. CRAIG HELLER and THOMAS POULSON

*Department of Biology, Stanford University, Stanford, California 94305 and
Department of Biology, University of Notre Dame, Notre Dame, Indiana 46556*

ABSTRACT: Fecal, urinary and evaporative water losses were measured at 15 C, 50-75% relative humidity for four species of western chipmunks (*Eutamias*) which are contiguously allopatric and altitudinally zoned on the eastern slope of the Sierra Nevada, California. Evaporative loss and hyperthermia were also studied for acute exposures to 25, 35 and 40 C. Differences in total water budgets, calculated for 35 C and 11% relative humidity are not important in determining the lines of contact, starting from the alpine and descending toward the desert, between *E. alpinus* and *E. speciosus* or between *E. speciosus* and *E. amoenus*. But they may play a role in preventing *E. amoenus* from colonizing the desert sagebrush habitat occupied by *E. minimus*. *E. minimus* can be active in the open areas of the hot, arid sagebrush desert by minimizing evaporative water loss and tolerating increased body heat content; this species frequently retreats to its burrows to unload excess body heat. When large patches of shade are available from piñon pines the aggressively dominant *E. amoenus* can occupy the sagebrush habitat. Hence, in the field area of this study the line of contact between *E. amoenus* and *E. minimus* coincides with the lower limits of the piñon pine.

INTRODUCTION

Four species of western chipmunks (genus *Eutamias*) are zoned altitudinally on the eastern fault scarp of the Sierra Nevada, Calif. (Heller, 1971). These species do not have overlapping ranges where populations meet, but rather they maintain lines of contact and can be described as contiguously allopatric (Fig. 1).

A study was undertaken to determine the relative importance of physiological adaptations and behavioral adaptations in limiting the distributions of these species (Heller, 1970). Aridity and high temperatures are extremely important sources of stress in the habitats of the lower two species, *E. amoenus* and *E. minimus*. The ambient temperatures are never very high in the alpine habitat of *E. alpinus*, but levels of incident radiation are high and in late summer this habitat becomes quite arid. It was suspected that differential physiological adaptations of the species to these sources of stress might play a role in limiting their distributions. A comparative study of the water balance and tolerance to temperature stress of the four species was therefore undertaken and the results are reported herein.

MATERIALS AND METHODS

Animals.—Eutamias alpinus Merriam, *E. speciosus frater* J. A. Allen, *E. amoenus monensis* Grinnell and Storer and *E. minimus scru-*

---

[1] Part of a dissertation submitted by H. C. Heller to the faculty of Yale University in partial fulfillment of requirements for the Ph.D. degree.

296

tator Hall and Hatfield (Hall and Kelson, 1959) were live-trapped on an E-W transect of the Sierra Nevada, Calif., through Yosemite National Park (38° N lat). The taxonomy and specific habitats of the chipmunks of the Sierra Nevada have been described by Johnson (1943).

Laboratory conditions.—The animals were brought into the laboratory in September 1966, 1967 and 1968, caged individually as previously described (Heller and Poulson, 1970), and maintained under constant conditions of 15 ± 2 C and a photoperiod of 12L:12D. Food in the form of sunflower seeds (27% protein) and Purina rat chow (23% protein, both 8 to 11% water) was available ad lib. except where indicated otherwise. Water was available ad lib. (except during water deprivation) from inverted 100-ml graduated cylinders fitted with L-shaped drinking tubes.

Ad libitum water consumption. — Daily measurements of water consumption were made by recording water level in the drinking tubes every day at lights-on. *Eutamias* spp. show a depressed water consumption during their inactive season due to lower activity and lower routine metabolism (Heller and Poulson, 1970). Ad lib. consumptions reported are from active season records, those most ecologically relevant to the subject of this paper. The average ad lib. water consumption for each species was determined from records of at least nine individuals for 45 to 60 days during their first full summer in captivity. The relative humidity in the laboratory during summer varied between 50 and 75% at a temperature of 15 ± 0.2 C.

Water deprivation.—To determine the minimum water requirements for weight maintenance and the maximum abilities of the

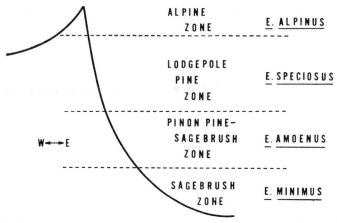

Fig. 1.—The "Yosemite" transect of the Sierra Nevada, Calif., at 38° N lat showing the dominant vegetation of the altitudinal life zones and the altitudinal ranges of the four *Eutamias* species on the eastern slope from approximately 2500 to 4000 m

animals to minimize fecal and urinary water loss, the animals were stressed by a high protein diet and limited access to water. Four water deprivation experiments were performed:

A. In early summer 1967, three *E. alpinus,* six *E. speciosus* and three *E. amoenus* were acutely stressed by progressively reducing their water rations by 25 to 50% every 2-4 days until each could just maintain body weight.

B. In the early summer and also in the autumn of 1968 at least five individuals of all four species were chronically stressed by progressively reducing their daily water rations by 10% or less of their ad lib. values until the ration was reached for each on which it could just maintain body weight.

C. In the summer of 1969 four individuals of each species were stressed by a diet of ground Purina rat chow mixed with 5% urea by weight. Their daily water rations were progressively reduced, as under B, until each could just maintain body weight. Water was then completely withdrawn and the urea content of the food was increased to 10% for 1 day before the urine and fecal samples were taken.

Urine concentration. — Urine was collected when an animal reached the minimum daily ration of water on which it could maintain body weight. It was deprived of all water for 1 day and was then placed in a cylindrical cage (10 x 15 cm) with a wire-mesh bottom. One hour after an animal was put into such a collection cage, a dish of mineral oil was placed underneath. Samples were collected from most urine drops under the oil with micropipettes. The melting points of the samples were measured immediately or else the micropipettes were sealed at both ends with sealing wax and frozen. Melting points were measured to ± 0.01 C with a Kalber Direct Reading Biological Cryostat (Clifton Technical Physics, New York, N. Y.).

Fecal water content. — Defecation was elicited by handling the animals when they were taken out of the urine collection cages after water deprivation "C." The fresh weight of the feces ±0.1 mg was determined, they were dried at 90 C, and then cooled in a desiccator until constant weight was reached.

Relative medullary thickness.—The relative medullary thickness of the kidneys of the animals was measured to determine their theoretical maximum potential to concentrate urine. Kidneys were preserved in 5% formalin. All excess fat was removed and the surface was quickly dried by rolling the kidney on a paper towel. The kidney was then weighed to the nearest milligram. Each kidney was cut with a razor blade along the frontal axis through the longest part of the renal papilla. The longest possible loop of Henle was measured to the nearest 0.01 mm by measuring, under a dissecting microscope with an ocular micrometer, the longest path parallel to the loops of Henle extending from the cortex to the tip of the renal papilla. The relative medullary thickness (RMT) was calculated according to the formula from Sperber (1944):

$$RMT = \frac{\text{Medullary Thickness x 10}}{\sqrt[3]{\text{Kidney Size}}}$$

Kidney size is generally measured as length x width x thickness, but in this study the weight of the kidney was used as the index of size. For comparative purposes the RMT's of at least four animals of each species were calculated using both weight and the product of the linear dimensions as indices of kidney size. Linear dimensions were measured with vernier dial calipers to ± 0.01 mm.

Evaporative water loss.—Evaporative water loss was measured by direct weighing (Lasiewski *et al.,* 1966) in a room with temperature controlled to ± 0.5 C and relative humidity to ± 3.0% (Environmental Growth Chambers, Chagrin Falls, Ohio). Throughout the experiments the water content of the air was maintained at 4.3 g/kg dry air, which is a relative humidity of 20% at 25 C, 11% at 35 C, and 8% at 40 C. An animal was placed in a wire-mesh cage (10 x 10 x 12 cm) suspended under a Mettler balance accurate to 0.01 g. Urine and feces were collected in a weighed pan of mineral oil under the cage. The cage was shielded from any direct air currents, and mass air flow in the room was very slow. Before and after each run at 40 C the rectal temperature of the animal was measured (±1 C) with a 36-gauge iron-constantan thermocouple. The measurements were made during the diurnal active period of the animals, but the lights were switched off to minimize activity. The data from a run were discarded if activity was observed during that run. Control animals handled in the same way as experimental animals, *i.e.,* placed in a duplicate small cage for the same period of time but kept at 16 C rather than 40 C, did not show a change in body temperature.

The per cent of metabolic heat dissipated by evaporative water loss was calculated using data on the resting, postabsorptive metabolic rates of the four species (Heller and Gates, 1971). It was assumed that the metabolic consumption of 1.0 ml·O_2 results in the production of 4.8 cal and that the heat of evaporation of 1.0 ml H_2O at slightly less than core body temperature is 580 cal.

Results

Ad lib. water consumption.—The daily ad lib. water consumption of *E. alpinus, E. speciosus* and *E. amoenus* in the laboratory during their active season averages 16% of body weight; the average value for *E. minimus* is 11% of body weight (Table 1). The ad lib. water consumption of *E. minimus* is significantly lower than that of *E. amoenus* (p < .05) and *E. speciosus* (p <.05) but not that of *E. alpinus* (p < .10, t-test).

Minimum water requirement for weight maintenance.—The weight specific minimum daily consumption (Table 1) of *E. amoenus* was significantly lower than that of *E. speciosus* (p < .05) and tended to be lower than that of *E. alpinus* (p < .10). This indicates that *E. amoenus* has a higher capacity for water conservation under water deprivation stress than do the other two species. *E. minimus* does not show a minimum consumption significantly different from any of the other three species.

The average extent to which each species can decrease its water

TABLE 1.—Mean water consumption, fecal water content, RMT, and urine concentration

Parameter	*Eutamias*	
	alpinus	*speciosus*
Body weight	39. 1±0. 4	70. 7±2. 4
(g) ±1 S.E.	n=16	n=22
Ad lib. water consumption	16. 1±2. 6	15. 9±0. 9
(per cent Wt_B) ±1 S.E.	n= 9	n=11
Minimum water consumption	7. 2±0. 7	7. 9±0. 7
(per cent Wt_B) ±1 S.E.	n= 8	n=12
Fecal water content	49. 9±2. 3	47. 0±1. 0
(per cent wet wt.) ±1 S.E.	n= 5	n= 7
Relative medullary thickness	11.15±0.15	9.32±0.14
±1 S.E.	n= 5	n=14
Urine concentration		
(Melting-pt. depression, C)	....	....
Mean maximum	7.20	6.72
	n=10	n=13
Five maximum values	>9.53	8.87
	7.73	8.32
	7.50	7.85
	7.50	7.50
	7.16	7.12

TABLE 1.—(continued)

Parameter	*Eutamias*	
	amoenus	*minimus*
Body weight	42. 4±0. 9	35. 2±0. 8
(g) ±1 S.E.	n=18	n=18
Ad lib. water consumption	16. 1±0. 8	10. 8±0. 8
(per cent Wt_B) ±1 S.E.	n=11	n= 9
Minimum water consumption	5. 6±0. 4	6. 3±0. 6
(per cent Wt_B) ±1 S.E.	n= 9	n= 6
Fecal water content	47. 2±1. 0	46. 1±2. 0
(per cent wet wt.) ±1 S.E.	n= 6	n= 3
Relative medullary thickness	10.39±0.16	12.08±0.35
±1 S.E.	n= 9	n=15
Urine concentration		
(Melting-pt. depression, C)	....	....
Mean maximum	7.93	8.02
	n= 9	n= 9
Five maximum values	>9.53	>9.53
	>9.53	>9.53
	>9.53	>9.53
	>9.53	8.45
	7.17	8.27

requirement when stressed is 65% for *E. amoenus,* 55% for *E. alpinus,* 50% for *E. speciosus* and 43% for *E. minimus.* The already low ad lib. consumption of *E. minimus* may allow it less scope for additional water conservation than is available to *E. amoenus.*

Fecal water content.—No differences were evident in the fecal water content of the four species after water deprivation and urea loading. The means were between 46 and 50% of wet fecal weight (Table 1).

Maximum urine concentration.—All four species produced very concentrated urines when deprived of water or deprived of water and urea-loaded (Table 1). *E. minimus* and *E. amoenus* produced the most and *E. speciosus* the least concentrated urines, but there is a great deal of overlap in the maximal concentrations of the urine produced by the four species. All freshly dropped urine samples contained numerous unidentified small crystals.

If urine concentration is at all important regarding competition, then the maximum urine concentration that individuals of each species can achieve will determine whether any species can occupy an arid habitat that is physiologically inaccessible to other species. Melting-point depressions showed a great deal of variability; for example, one animal during a single 12-hr collecting period can drop samples differing in their melting points by 3-4 C. "Fright diuresis" resulting from placing the animal in the urine collection cage may have contributed to this variation in spite of the precautionary measure of not collecting urine samples during the 1st hr an animal was in the cage. Also, there may be species differences in fright diuresis, and, therefore, maximum values were taken from the data. The mean maximum melting points (Table 1) were determined from the maximal value for each individual even though that individual may have been used in more than one experiment. The five maximum values given are the five highest values for each species. The ranges and means for *E. minimus* and *E. amoenus* are in reality lower, but the cryostat used did not measure below −9.53 C.

Relative medullary thickness.—RMT's calculated from the weight of the kidney agree well with RMT's calculated from the linear measurements, but in the range of measurements for the *Eutamias* species RMT by weight averages 1 RMT unit higher than RMT by linear dimensions.

The RMT's of all four species are significantly different from each other (Table 1), with that of *E. minimus* > *E. alpinus* > *E. amoenus* > *E. speciosus.*

Evaporative water loss.—There are no significant differences between the weight specific rates of evaporative water loss of *E. alpinus, E. speciosus* and *E. amoenus* at 25 C, but *E. minimus* shows a significantly higher value than any of the other three species (Table 2).

The values for evaporative water loss/metabolic heat production at 25 C are in Table 2. Evaporative water loss accounts for 21-32% of metabolic heat production in all four species at this temperature.

The rate of evaporative water loss of *E. amoenus* at 35 C showed great variability and overlapped that of all the other species. The value for *E. minimus* at 35 C is significantly lower than that for *E. alpinus* and *E. speciosus*. This relative shift in the ranges of values for the four species results from roughly parallel, significant increases in the evaporative water losses of *E. alpinus*, *E. speciosus* and *E. amoenus* between 25 and 35 C, whereas the evaporative water loss rate of *E. minimus* remained the same at 35 as it was at 25 C. The evaporative water loss rates of *E. alpinus*, *E. speciosus* and *E. amoenus* increased between 25 C and 35 C while their metabolic rates decreased slightly; hence, the proportion of heat production dissipated by water loss (E.W.L./M.H.P.) increased. The E.W.L./M.H.P. ratio of *E. minimus* remained constant, however, over this same temperature range. The maintenance of a low E.W.L./M.H.P. ratio conserves water and is adaptive in the arid habitat of *E. minimus*.

The rates of evaporative water loss of all four species increased three- to fourfold between 35 C and 40 C (Table 2). *E. minimus* still had the lowest mean value, significantly lower than *E. alpinus* and *E. amoenus* (p < .05, t-test).

TABLE 2.—Mean evaporative water and heat loss rates, metabolic rates and heat storage rates

Parameters and conditions	Eutamias	
	alpinus	*speciosus*
Evaporative water loss rate (ml x 10^3/g/hr) ±1 S.E.		
T_A = 25 C	3. 2± 0.2	2.8 ± 0.3
	n= 4	n= 4
T_A = 35 C	6. 2± 0.4	5.1 ± 0.2
	n= 4	n= 4
T_A = 40 C	24. 8± 1.7	18.0 ± 1.8
	n=17	n=22
Evaporative heat loss rate at 40 C (cal/45 min) ±1 S.E.	393. 5±16. 9	525. 7±31. 4
Metabolic rates (Heller and Gates, 1971) (ml O_2 $g^{-1}hr^{-1}$) ±1 S.E.		
T_A = 25 C	1.65± .03	1.51± .03
	n= 4	n= 4
T_A = 35 C	1.43± .04	1.47± .01
	n= 3	n= 3
E.W.L. x 100/M.H.P. (per cent)		
T_A = 25 C	23.5	30.4
T_A = 35 C	50.7	41.1
Rate of rise in body-heat content at T_A = 40 C		
(cal/g/45 min) ±1 S.E.	2.29± 0.68	2.20± 0.65
Mean body size (g) ±1 S.E.	39. 1± 0. 9	70. 7± 4.3
	n=16	n=22
(Total cal/45 min) ±1 S.E.	93. 2± 9. 1	156. 6±15.7

All of the animals showed hyperthermia at 40 C. Therefore, uniform 45-min exposures were used to measure evaporative water loss at this temperature. Body-temperature changes were recorded to allow calculation of the rates of heat storage. The results are presented in Table 2 and Figure 2 on a weight specific basis and in Figure 3 on an absolute basis. The specific heat of animal tissue was taken as 0.8 cal/g. The areas shown in Figures 2 and 3 for each species enclose all of the data points obtained for that species.

There are no significant differences in rate of weight specific increase in body heat content at 40 C. Significant differences in calories lost by evaporation exist, however, between *E. minimus* and *E. alpinus*, *E. minimus* and *E. amoenus*, and *E. alpinus* and *E. speciosus*. *E. speciosus* is significantly higher in both parameters than the other three species when the data are plotted as total calories lost through evaporative water loss vs. total calories gained through hyperthermia for a 45-min exposure (Fig. 3). The other three species are significantly different from each other in total calories lost through evaporative water loss but not in total calories gained through hyperthermia.

Table 2.—(continued)

Parameters and conditions	*Eutamias*	
	amoenus	*minimus*
Evaporative water loss rate (ml x 10^3/g/hr) ± 1 S.E.		
$T_A = 25$ C	3. 4$\pm$ 0.4	4.9 $\pm$ 0.2
	n= 3	n= 4
$T_A = 35$ C	6. 0$\pm$ 0.7	4.5 $\pm$ 0.2
	n= 4	n= 8
$T_A = 40$ C	20. 7$\pm$ 1.4	14.6 $\pm$ 1.3
	n=18	n=18
Evaporative heat loss rate at 40 C (cal/45 min) ± 1 S.E.	328. 8$\pm$17. 0	217. 8$\pm$17. 0
Metabolic rates (Heller and Gates, 1971) (ml 0_2 $g^{-1}hr^{-1}$) ± 1 S.E.		
$T_A = 25$ C	1.82$\pm$ .06	1.81$\pm$ .02
	n= 3	n= 4
$T_A = 35$ C	1.70$\pm$ .09	1.61$\pm$ .15
	n= 3	n= 5
E.W.L. x 100/M.H.P. (per cent)		
$T_A = 25$ C	21.6	32.0
$T_A = 35$ C	41.2	33.8
Rate of rise in body-heat content at $T_A = 40$ C (cal/g/45 min) ± 1 S.E.	2.01$\pm$ 0.31	2.92$\pm$ 0.49
Mean body size (g) ± 1 S.E.	42. 4$\pm$ 1. 8	35. 2$\pm$ 1. 5
	n=18	n=18
(Total cal/45 min) ± 1 S.E.	84. 5$\pm$ 4. 9	101. 9$\pm$ 5. 4

TOTAL WATER BUDGETS

Any adaptations to a hot, arid habitat that are due to interspecific differences in water balance can best be judged and compared by constructing total water budgets for a given set of conditions (Table 3). The conditions used in the present study are relatively hot and arid, but not extreme: 35 C ambient temperature, 11% relative humidity, and a diet of air-dried sunflower seeds. All calculations were made for 4-hr periods because the aboveground activity of *E. minimus* in nature is mostly limited to a 4-hr period before noon (Heller, 1970).

The expected total water requirement is calculated by summing fecal, urinary and evaporative water loss. The water from food, preformed water and water derived from oxidative metabolism is sub-

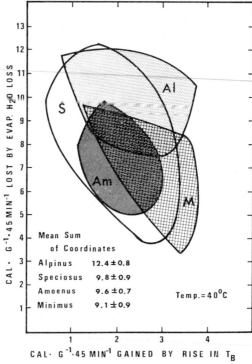

Fig. 2.—Calories per gram gained through hyperthermia and calories per gram lost through evaporation when resting animals are exposed to 40 C, 8% R.H. for 45 min. The areas enclose all of the data points for each species. S = *E. speciosus*, Al = *E. alpinus*, M = *E. minimus*, and the remaining area is for *E. amoenus*. The sum of the coordinates of each data point is an index of the degree to which the conditions stressed the animals (*see* Discussion). The mean sum of the coordinates for each species is given at the lower left. Data in Table 2

tracted from the total water requirement to get the free water requirement. The weight specific free water requirement is multiplied by the species' mean body weight to arrive at the absolute amount of free water an animal of that species must have to remain in neutral energy balance under the hypothetical conditions for 4 hr without suffering a net water loss. These calculations make no allowance for activity, nor are the conditions unusually stringent in comparison to the natural habitat. A T_A of 35 C, which is below the upper critical temperature of all four species, is realistic for the habitat of *E. minimus* near noon on a clear summer day, but the soil surface temperatures and temperatures of the air strata that this species is exposed to are considerably higher.

Fecal water loss.—The dry weight of feces produced per gram of animal (unpublished data) during measurements of routine metabolism at 5.0 C (Heller and Poulson, 1970) was corrected for the lower metabolic rates at 35 C (Heller and Gates, 1971). The fecal water loss per gram of animal per 4 hr was calculated using the data on fecal water content after water deprivation (Table 1).

Urinary water loss.—The expected urinary water loss was calculated using data on metabolic rate, protein content and caloric value of food, and urine concentration. Sunflower seeds have a caloric

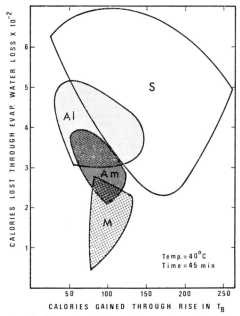

Fig. 3.—Total calories gained through hyperthermia and total calories lost through evaporation when resting animals are exposed to 40 C, 8% R.H. for 45 min. Notation is the same as in Figure 2. Data in Table 2

value of 6.1 kcal/g and a protein content of 28%. Assuming the standard N content of protein, 16%, 7.52×10^{-3}g N must be excreted for every kcal produced from the metabolism of sunflower seeds. Grams of N/kcal were multiplied by the metabolic rate of the animal in kcal/g animal/4 hr to arrive at g N excreted/g animal/4 hr. Assuming all N is excreted as urea and each mole of urea contains two moles of N, dividing g N/g animal/4 hr by 0.028 gives mM urea/g animal/4 hr; and dividing that figure by the maximal urine concentration in mM urea/ml gives the minimal weight specific urine volume or urinary water loss for a 4-hr period. The maximum urine concentration was inferred from the RMT's. Electrolytes from the food also contribute to the total osmolarity of the urine. We have not taken this into account, so our calculated urinary water loss will be slightly too low.

Evaporative water loss.—Evaporative water loss in ml/hr/g animal was taken from Table 2 and multiplied by four.

Water gain.—The water the organism acquires from ingested food depends on the preformed water content of the total food ingested and the water formed from the oxidation of the ingested food which is assimilated. The amount of seeds assimilated was calculated from the metabolic rate at 35 C and the caloric value of sunflower seeds. The water derived from the sunflower seeds assimilated was calculated by dividing grams of seeds into grams of protein, grams of fat and grams of carbohydrate, and each was multiplied by the water of oxidation it yields per gram. The average composition of sunflower seeds (Winton and Winton, 1932) is 8% water, 28% protein, 46% fat and 12% carbohydrate. Metabolized protein yields 0.40 g H_2O/g; fat yields 1.07 g H_2O/g, and carbohydrate yields 0.56 g H_2O/g (Schmidt-Nielsen and Schmidt-Nielsen, 1951).

TABLE 3.—Expected water budgets of resting animals exposed to 35 C, 11% R.H. for 4 hr in nature

| | *Eutamias* | | | |
	alpinus	*speciosus*	*amoenus*	*minimus*
Water loss (ml g^{-1} hr^{-1} x 4)				
Fecal water loss	.0004	.0005	.0006	.0004
Urinary water loss	.0013	.0016	.0017	.0014
Evaporative water loss	.0248	.0204	.0232	0.180
Total water loss	.0265	.0225	.0255	.0198
Water gain (ml g^{-1} hr^{-1} x 4)				
Water of oxidation	.0031	.0031	.0036	.0034
Preformed water	.0005	.0006	.0006	.0006
Total water gain	.0036	.0037	.0042	.0040
Free water requirement (= total loss − total gain) (ml g^{-1} hr^{-1} x 4)	.0229	.0188	.0213	.0158
Free water requirement per animal (ml hr^{-1} x 4)	.894	1.335	.895	.554

The preformed water in all of the food ingested contributes to the water budget even if all of that food is not assimilated. So the total amount of sunflower seeds ingested times their water content gives the amount of preformed water in the food. The total amount of sunflower seeds ingested can be determined from the amount assimilated if the efficiency of assimilation is known. The efficiency of assimilation was calculated by comparing the caloric value of sunflower seeds ingested at 5 C (Heller and Poulson, 1970) with metabolic rate at 5 C as measured by O_2 consumption (Heller and Gates, 1971). The values for all species were between 64 and 68%. It was assumed that the efficiency of assimilation would be the same at 35 C. Two studies on steers and one on rabbits at unspecified temperatures show calculated assimilation efficiencies of 68, 69 and 66%, respectively (Brody, 1945, p. 80). The assimilation efficiencies of English sparrows varied only 9% between 0 and 34 C (Davis, 1955). The maximum was at 18 C, and the assimilation efficiency at 4 C was 77% and at 34 C, 78% for birds on a 10-hr photoperiod.

Free water requirement.—The amount of free water required per gram animal per 4 hr is obtained by subtracting total water acquired through food from total water loss. This figure for each species is then multiplied by the mean weight of that species to arrive at total requirement of free water per animal per 4 hr.

Calculated water budgets.—Over 90% of the water loss of all species is attributable to evaporation at 35 C and 11% relative humidity (Table 3). *E. minimus* has the lowest weight specific water requirement due to its low evaporative loss, and its total water requirement is considerably less than that of the other three species. The total water requirement of *E. speciosus* is considerably greater than that of the other three species.

DISCUSSION

Differences exist in the water balance of the four species of *Eutamias* studied, but are any of these differences of sufficient magnitude to physiologically exclude any of the species from the habitat of any other? It has been shown that *E. speciosus* is excluded from the more arid habitats of *E. alpinus* and *E. amoenus* by the aggressive dominance of those two species (Heller, 1971), so physiological limitations on *E. speciosus* are probably not primary in restricting its local distribution. The interesting case is that of *E. minimus* and *E. amoenus,* both of which are found in hot, arid habitats. *E. amoenus* aggressively excludes *E. minimus* from *E. amoenus* habitat (Heller, 1971), but is *E. amoenus* physiologically excluded from *E. minimus* habitat? To answer this question with respect to water balance it is necessary to compare the species' adaptations to all sources of water loss and to calculate their total water budgets. Total water loss is the sum of fecal, urinary and evaporative water losses. No differences were found in fecal water loss, so only urinary and evaporative water losses will be discussed in detail.

Urinary water loss.—Relative Medullary Thickness is an index

directly related to aridity of habitat (Sperber, 1943) and the maximum possible urine-concentrating ability (B. Schmidt-Nielsen and R.
O'Dell, 1961; Heisinger and Breitenbach, 1969), but it is also related
to body size (Blake, 1967). If one determines the differences in the
RMT's of the *Eutamias* spp. expected solely on the basis of body size,
then the residual differences can be implicated as adaptation to aridity. RMT is plotted as a function of body size in Figure 4B. The lower
line was fitted by eye to the mean RMT's of samples of *E. minimus,
E. amoenus, E. quadrivittatus, E. umbrinus, E. townsendii, Tamias
striatus* and *Spermophilus lateralis* (Blake, 1967). Blake's RMT's are
consistently lower than ours. One RMT unit of this difference is
attributable to our use of kidney weight as the index of kidney size,
and at least 0.6 RMT units difference are due to Blake's selection of
data for lowest RMT's. The balance of the discrepancy is unexplained.
A line drawn parallel to Blake's line and passing through the RMT
values for *E. speciosus* also passes through the mean value for *E.
amoenus,* but the values for *E. minimus* and *E. alpinus* are considerably above this line. This indicates that the RMT differences between

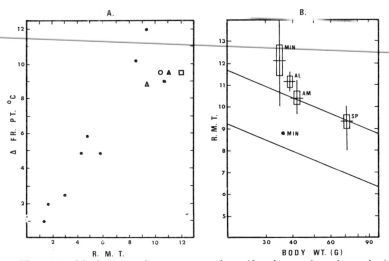

Fig. 4A.—Maximum urine concentration (freezing point depression)
plotted vs. mean R.M.T. ▲ = *E. alpinus,* △ = *E. speciosus,* ○ = *E. amoenus,* □ = *E. minimus.* Solid dots are from B. Schmidt-Nielsen and O'Dell
(1961) and left to right represent beaver, pig, man, dog, cat, rat, kangaroo
rat, jerboa and sand rat. 4B.— Relative Medullary Thickness vs. body weight.
The lower line (Blake, 1967) represents five *Eutamias* species, *Tamias striatus*
and *Spermophilus lateralis,* all from relatively mesic habitats. The dot (Blake,
1967) represents a population of *E. minimus* from an arid environment. The
upper line was drawn parallel to the lower line and made to pass through the
mean value for *E. speciosus* (this study). Vertical lines are ranges, horizontal
lines are means, and bars represent two standard errors on each side of the
mean. Data in Table 1

E. speciosus and *E. amoenus* could be expected solely as adaptations related to differences in body size, but that *E. minimus* and *E. alpinus* show a relatively greater adaptation to aridity in their kidney morphologies.

The high maximal urine concentrations of all four *Eutamias* species are in the range of other small desert rodents (Schmidt-Nielsen *et al.,* 1948; Schmidt-Nielsen and O'Dell, 1961; Hudson, 1962; Hudson and Rummel, 1966; Carpenter, 1966; MacMillen and Lee, 1969). An accurate comparison of RMT and maximal urine concentration in chipmunks is not possible because the maximal melting-point depressions of *E. minimus* and *E. amoenus* samples were beyond the range of the cryostat. The maximum values found in this study plotted as a function of RMT correlate well with data for other animals (Fig. 4A). The values from this study would fall even closer to Schmidt-Nielsen and O'Dell's curve if the actual maximum urine concentrations for *E. minimus* and *E. amoenus* were known and if the correction of −1 RMT unit were made so that our values were equivalent to those based on linear measurements of the kidney.

Evaporative water loss.—Evaporative water loss is an effective heat-dissipating mechanism in a hot, arid environment (Hudson, 1962, 1964), but also the main source of water loss for a small animal (Schmidt-Nielsen and Schmidt-Nielsen, 1951). If an animal adapts to the aridity of the environment by reducing evaporative water loss, concomitantly it must increase its efficiency of dealing with heat stress via alternative mechanisms. *E. alpinus, E. speciosus* and *E. amoenus* dissipated a greater per cent of metabolic heat through the evaporation of water at 35 C than they did at 25 C, but *E. minimus* maintained a constant E.W.L./M.H.P. ratio over this temperature range (Table 2). Therefore, *E. minimus* must be able to increase the efficiency of its evaporative cooling (*e.g.,* by decreasing cutaneous relative to pulmonary loss), or to enhance the effectiveness of convection, conduction or radiation as T_A approaches T_B, or to permit a slight increase in body-heat content. T_B was not measured during the 35 C evaporative-water-loss experiments; however, the possibility that *E. minimus* used heat storage to maintain a low E.W.L./M.H.P. ratio at 35 C is given credence by the fact that at 40 C *E. minimus* showed hyperthermia more than the other three species. An increase in T_B of *E. minimus* of only 0.12 C/hr would be equivalent in calories to the increase in evaporative water loss shown by the other species between 25 and 35 C.

At a T_A of 40 C heat must be lost through the evaporation of water, or alternatively, the heat content of the body must increase if the T_B starts out below 40 C. The data in Figure 2 indicate that the strategy of *E. minimus* to cope with dry heat is to depend more on hyperthermia and less on evaporative water loss during short exposures than do the other three species. The sum of increase in heat content through hyperthermia and heat loss by evaporation (Fig. 2, insert) is an indication of the degree to which the animals were stressed by dry heat.

Although *E. minimus* was stressed the least, the differences between it and *E. speciosus* and *E. amoenus* are not significant (p > .05, t-test). *E. alpinus* was stressed by the dry heat significantly more than the other species.

The low E.W.L./M.H.P. ratio of *E. minimus* (Table 2) and its tendency to show hyperthermia in dry heat (Fig. 2) indicate that in nature it may rely chiefly on hyperthermia to cope with its hot, arid environment. Like the antelope ground squirrel which is found in the same type of habitat (Hudson, 1962), *E. minimus* probably alternates periods of aboveground activity, during which it allows its body-heat content to increase, with periods of rest in a burrow where it can dissipate the incurred heat load. *E. minimus* was observed to spend less and less time aboveground as the T_A rose and reached its maximum in early afternoon (Heller, 1970).

Body size.—Body size is also an adaptive feature in a hot, arid environment. Regardless of the weight specific values, it is the total amount of water evaporated which must be replenished and the total increase in heat content which will have to be dissipated. It is interesting to note the relative positions of the data "areas" in Figure 3 where total calories gained through hyperthermia and total calories lost through evaporation are plotted. *E. speciosus* experiences the greatest total flux of energy and *E. minimus* the least. There is almost complete separation of *E. minimus* and *E. amoenus,* both of which are of similar body sizes and are from hot, arid habitats.

Water budgets.—The calculation of the total water budgets for the four species (Table 3) reveals that evaporation accounts for over 90% of the total water loss under the experimental conditions. This high evaporative water loss relative to urinary loss probably holds in nature even though the evaporative water loss was not measured on animals acclimated to high T_A. The antelope ground squirrel, which weighs about 90 g, does not show significant differences in evaporative water loss between individuals that are and are not acclimated to 35 C (Hudson, 1962). The evaporative loss of this species at 35 C is similar to the values for the *Eutamias* species reported here.

Water budgets calculated for aboveground summer environmental conditions indicate that significant improvements in the water conserving ability of small diurnal mammals can only occur through the reduction of evaporative water loss. Even relatively large changes in the abilities of the closely related *Eutamias* species to decrease fecal and urinary water loss would have a rather small effect on the total water loss incurred while active aboveground. When the animals are in the atmosphere of a burrow, the urinary and fecal water losses would be a larger per cent of the total loss, and, therefore, a larger per cent of the 24-hr water budget than of the 4-hr budget for above-ground. To be sure, there have been strong selective pressures to achieve the high renal efficiency observed in these animals, but these renal adaptations to aridity are not of primary importance in physiologically limiting the time intervals over which these diurnal animals can be active aboveground in the desert environment.

Does the fundamental niche of E. amoenus *include the habitat of* E. minimus *in the study area?*—*E. minimus* is excluded from the habitat of *E. amoenus* by the aggressive dominance of *E. amoenus;* therefore, *E. minimus* has a realized niche smaller than its fundamental niche in the study area (Heller, 1970, 1971). What prevents *E. amoenus* or any of the other species from colonizing the habitat of *E. minimus?* The habitat of *E. amoenus* is also hot and arid, but it contains numerous piñon pine trees. The piñon pines provide large patches of shade in comparison to the small dispersed patches of shade available in the sagebrush habitat of *E. minimus.* Both *E. amoenus* and *E. minimus* experience similar T_A's and relative humidities, but the large patches of shade enable *E. amoenus* to escape from direct incident radiation and to avoid high soil-surface temperatures. *E. amoenus* frequently climbs in the piñon pines and sits on well-shaded branches where heat loss by convection and radiation is maximized. On the edge of Diamond Valley near Woodfords, Calif., the range of *E. amoenus* extends far beyond the limit of piñon pines and into the sagebrush-shrub habitat (personal observation). The desert shrubs in that area, however, are more dense and generally twice as tall as they are on the Yosemite transect; hence, large patches of shade were available at all hours of the day. No *E. minimus* were trapped in this habitat.

In spite of similar body sizes, *E. minimus* is better adapted to coping with dry heat than is *E. amoenus* and may use a different strategy for doing so. The activity pattern of *E. minimus* in nature indicates that after 11 AM, P.D.S.T., this species, in contrast to the others, spends most of its time in its burrow (Heller, 1970). Energy budget calculations (Heller and Gates, 1971) indicate that *E. minimus* is physiologically prevented from remaining active aboveground for the greater portion of the day in spite of its adaptation to dry, hot sagebrush desert. If *E. minimus* is marginally adapted to the sagebrush desert habitat in the study area, then it is certain that *E. amoenus* and the other two species are physiologically incapable of colonizing this habitat. Hence, the realized niche of *E. minimus* in the study area is outside of the fundamental niches of the other species.

This and other studies (Heller and Gates, 1971) have shown considerable overlap in the physiological adaptations of the four species of chipmunks with the most marked differences between *Eutamias minimus* and the other three species. The most important present-day factor that determines their altitudinal zonation is interspecific aggression with some reinforcement by habitat selection both by the aggressive and subordinate species (Heller, 1971). Thus *E. alpinus* is aggressively dominant to *E. speciosus* and *E. amoenus* is dominant to both *E. speciosus* and *E. minimus.* The only case where physiology helps to explain the contiguously allopatric zonation is with the *E. amoenus/E. minimus* contact. Our conclusion from the present report is that the fundamental niche of *E. amoenus* does not include the hot, arid sagebrush habitat occupied by *E. minimus*. Thus, *E. amoenus* is the only one of the four species which is physiologically excluded from another species habitat.

Acknowledgments.—This research was supported by NSF Grant GB-6212 to T. L. Poulson, an N.D.E.A. Title IV Fellowship to H. C. Heller, and funds from the Department of Biology, Yale University. We are grateful to Yosemite National Park for permission to do field work and to Mr. and Mrs. Gary Colliver for untiring assistance with field work. Also we are greatly indebted to Mr. Vincent Salerno for assistance in the laboratory.

REFERENCES

BLAKE, B. H. 1967. A comparative study of energy and water conservation throughout the annual cycle in ground-dwelling Sciuridae. Ph.D. Dissertation, Yale University. 188 p. University Microfilms, Ann Arbor, Michigan.

BRODY, S. 1945. Bioenergetics and growth, with special reference to the efficiency complex in domestic animals. Reinhold Publishing Co., New York. 1023 p.

CARPENTER, R. E. 1966. A comparison of the thermoregulation and water metabolism in the kangaroo rats *Dipodomys agilis* and *Dipodomys merriami. Univ. Calif. Publ. Zool.*, **78**:1-36.

DAVIS, E. A. 1955. Seasonal changes in the energy balance of the English sparrow. *Auk*, **72**:385-416.

HALL, E. R. AND K. R. KELSON. 1959. The mammals of North America. Vol. I. The Ronald Press Co., New York. 546 p.

HEISINGER, J. F. AND R. P. BREITENBACH. 1969. Renal structural characteristics as indexes of renal adaptation for water conservation in the genus *Sylvilagus. Physiol. Zool.*, **42**:160-172.

HELLER, H. C. 1970. Altitudinal zonation of chipmunks (genus *Eutamias*): Interspecific aggression, water balance, and energy budgets. Ph.D. Dissertation, Yale University. University Microfilms, Ann Arbor, Mich.

————. 1971. Altitudinal zonation of chipmunks (genus *Eutamias*): Interspecific aggression. *Ecology*, **52**:312-319.

———— AND D. M. GATES. 1971. Altitudinal zonation of chipmunks (genus *Eutamias*): Energy budgets. *Ibid.*, **52**:424-433.

———— AND T. L. POULSON. 1970. Circannian rhythms: II. Endogenous and exogenous factors controlling reproduction and hibernation in chipmunks (*Eutamias*) and ground squirrels (*Spermophilus*). *Comp. Biochem. Physiol.*, **33**:357-383.

HUDSON, J. W. 1962. The role of water in the biology of the antelope ground squirrel *Citellus leucurus. Univ. Calif. Publ. Zool.*, **64**:1-56.

————. 1964. Temperature regulation in the round-tailed ground squirrel *Citellus tereticaudus. Ann. Acad. Sci. Fenn. Ser. A, IV, Biol.*, **71**:217-233.

———— AND J. A. RUMMEL. 1966. Water metabolism and temperature regulation of the primitive Heteromyids *Liomys salvani* and *Liomys irroratus. Ecology*, **47**:345-354.

JOHNSON, D. H. 1943. Systematic review of the chipmunks (genus *Eutamias*) of California. *Univ. Calif. Publ. Zool.*, **48**:63-148.

Lasiewski, R. C., A. L. Acosta and M. H. Bernstein. 1966. Evaporative water loss in birds — II. A modified method for determination by direct weighing. *Comp. Biochem. Physiol.,* **19**:459-470.

MacMillen, R. E. and A. K. Lee. 1969. Water metabolism of Australian hopping mice. *Ibid.,* **28**:493-514.

Schmidt-Nielsen, B. and K. Schmidt-Nielsen. 1950. Evaporative water loss in desert rodents in their natural habitat. *Ecology,* **31**:75-85.

———— and ————. 1951. A complete account of the water metabolism in kangaroo rats and an experimental verification. *J. Cell. Comp. Physiol.,* **38**:165-181.

———— and R. O'Dell. 1961. Structure and concentrating mechanism in the mammalian kidney. *Amer. J. Physiol.,* **200**:1119-1124.

————, K. Schmidt-Nielsen, A. Brokaw and H. Schneiderman. 1948. Water conservation in desert rodents. *J. Cell. Comp. Physiol.,* **32**: 331-360.

Sperber, A. 1944. Studies on the mammalian kidney. *Zool. Bidrag Uppsala,* **22**:249-432.

Winton, A. L. and K. B. Winton. 1932. The structure and composition of foods. Vol. I. John Wiley and Sons, Inc., New York. 613 p.

Submitted 5 March 1971   Accepted 15 April 1971

THE OXYGEN CONSUMPTION AND BIOENER-
GETICS OF HARVEST MICE

OLIVER P. PEARSON

Museum of Vertebrate Zoölogy, University of California, Berkeley

RATES of metabolism or of oxygen consumption have been reported for many species of small mammals, but little effort has been made to relate such measurements to the energy economy of small mammals in the wild. Such effort has been avoided because the rate of metabolism varies so much with changes of the ambient temperature and with activity of the animal. I believe, however, that these variables can be handled with sufficient accuracy so that one can make meaningful estimates of the 24-hour metabolic budget of free-living mice in the wild. In this study I have measured the oxygen consumption of captive harvest mice under different conditions, and from these measurements I have estimated the daily metabolic exchange of wild harvest mice living in Orinda, Contra Costa County, California.

The harvest mice used in the study (*Reithrodontomys megalotis*) are nocturnal, seed-eating rodents living in grassy, weedy, and brushy places in the western half of the United States and in Mexico. In Orinda they encounter cool wet winters (nighttime temperatures frequently slightly below 0° C.) and warm dry summers (daytime temperatures sometimes above 35° C., but nights always cool). They do not hibernate.

MATERIAL AND METHODS

Five adult harvest mice were caught on January 29 and 30, 1959, and were kept in two cages in an unheated room with open windows so that the air temperature would remain close to that outside the building. They were fed a mixture of seeds known as "wild bird seed." Metabolic rates were tested between January 29 and April 1 in a closed-circuit oxygen consumption apparatus similar to the one described by Morrison (1947) but without the automatic recording and refilling features. All tests except the 24-hour runs were made during the daytime and without food. Since harvest mice are strongly nocturnal, several hours had usually elapsed between their last meal and the measuring of their oxygen consumption. When placed in the apparatus, the mice usually explored the metabolic chamber and groomed their fur for about half an hour and then went to sleep on the wire mesh floor of the chamber. One hour or more was allowed for the animals to become quiet and for the system to come to temperature equilibrium. The animals usually were left in the chamber until from five to ten determinations of oxygen consumption had been made, during which they had remained asleep or at least had made no gross movements. Each determination lasted between 9 and 24 minutes. The mice were weighed when they were removed from the apparatus. Oxygen consumptions are reported as volume of dry gas at 0° C. per gram of mouse.

RESULTS

SIZE × RATE OF METABOLISM

Adult harvest mice weigh between 7 and 14 grams. Larger individuals consume oxygen at a lower rate per gram of body weight (Fig. 1). For example, at 12° C. a 12-gram mouse would use only 1.17 times as much oxygen per hour as an 8-gram mouse, although it is 1.5 times as heavy. The various points in the regression of body weight against rate of oxygen consumption can be fitted adequately with a straight line, and from the slopes of such lines illustrating the regression at different ambient temperatures it may be seen (Fig. 1) that at cold temperatures a variation of 1 gram in body weight causes a greater change in metabolic rate than at 30° C. At 1°, 12°, and 24° a change of 1 gram in weight is associated with a change in oxygen consumption of 0.98, 0.48, and 0.35 cc/g/hr, respectively.

At warm and moderate temperatures there was little variation in the measurements of each mouse during any one run (Fig. 1), but at 1° C. the variation was sometimes enormous. Since each measurement was made over a period while the mouse was inactive, the variation must stem from a real difference in the resting metabolism of each mouse at different times. I believe that lability of body temperature is the cause. Harvest mice exposed to cold and hunger in box traps sometimes are found to be torpid and with a cold body temperature. If they are tagged and released, they can be recaptured in good health at subsequent trappings, demonstrating that harvest mice have a labile body temperature and can recover from profound hypothermia. During the metabolic tests at 1° C., especially those with the mouse in a nest, there was a tendency for most of the measurements to lie at one level; but there would be a few very low readings and a few intermediate readings, presumably as the animal entered and emerged from the low-metabolic condition (best shown by the 11½-gram mouse in Fig. 1). In response to cold coupled

with restful surroundings, as in a nest, the animals probably relaxed their temperature control temporarily. This explanation seems plausible in view of the known lability of the body temperature of some rodents such as *Peromyscus* (Morrison and Ryser, 1959), *Dipodomys* (Dawson, 1955), and *Perognathus* (Bartholomew and Cade, 1957) under similar circumstances. Birds permit their body temperature to drop about 2° C. when they sleep at night, and this is accompanied by a drop of as much as 27 per cent in rate of metabolism (De Bont, 1945). The 40 per cent drop shown by some of the mice may have been accompanied by a drop in body temperature of several degrees.

RESTING METABOLISM AT DIFFERENT
TEMPERATURES

Since the weights of adult harvest mice vary so much, it is desirable to eliminate the size variable by adjusting all rates of metabolism to a single average size (9 grams). This has been done by using the series of regression lines in Figure 1. Where each of these lines crosses the 9-gram ordinate, that value is taken as the appropriate rate for a "standard" 9-gram harvest mouse and is used in Figure 2.

The middle curve in Figure 2 shows that the minimum rate of oxygen consumption of harvest mice (2.5 cc/g/hr for a 9-gram mouse) is reached at the relatively high ambient temperature of 33° or 34° C. and that there is almost no zone of thermal neutrality. Rate of metabolism almost certainly begins to increase before 36° C. is reached so that the zone of minimum metabolism could not include more than 3°. The critical temperature (33–34°) is remarkably close to the upper lethal temperature. The single animal tested at 37° died after two hours at this temperature but provided several good measurements before entering the

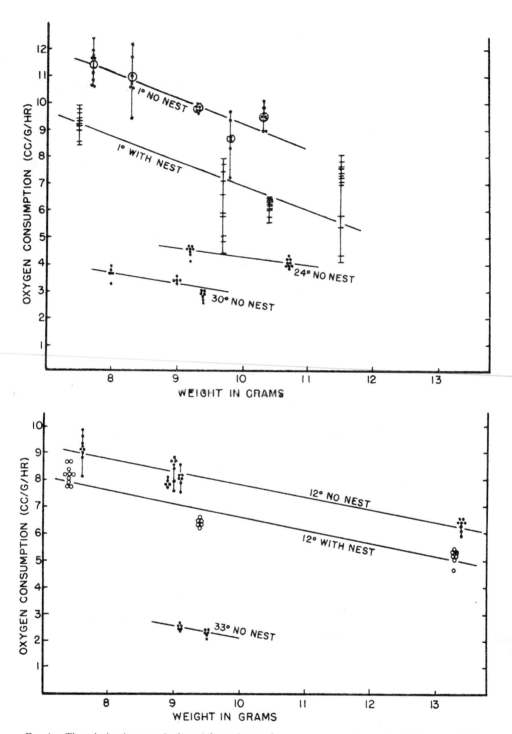

Fig. 1.—The relation between body weight and rate of oxygen consumption under different conditions, showing also the variation in individual measurements. Each cluster or vertical array of points represents a series of values obtained from a single individual.

194

final coma. Because of the large exposed surface of calcium chloride and soda lime in the metabolic chamber, relative humidity was probably low; heat death would probably occur at an even lower temperature under humid conditions in which cooling by evaporation would be limited.

ered body temperature. Inclusion of these low values causes the apparent decrease of the slope of the two curves between 12° and 1°. No body temperatures, however, dropped to the torpid level. *Reithrodontomys megalotis* is able to maintain its temperature well above the torpid level even when sleeping in cold sur-

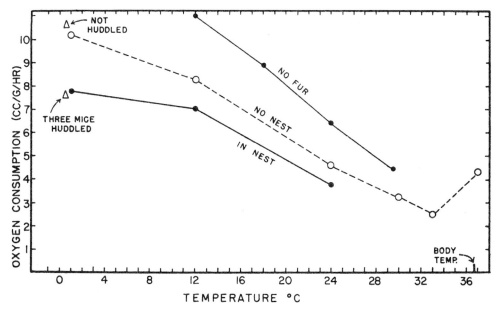

FIG. 2.—The rate of oxygen consumption of resting harvest mice at different temperatures in a nest, without nest, and without fur. All three curves have been adjusted, on the basis of the regression lines shown in Fig. 1, to represent a 9-gram mouse. Triangles indicate rate of oxygen consumption of three mice huddled together without a nest compared with the expected rate for the same three mice singly (average weight 8.5 grams). I am grateful to Martin Murie for supplying the value for deep body temperature, which was the average of many determinations made during the day and night at ambient temperatures between 14° and 27° C.

The increase in rate of metabolism at cool temperatures is almost linear between 33° and 12°; each drop of 1° C. causes an increase in the rate of oxygen consumption of 0.27 cc/g/hr. This rate of change, possibly because of the small size of harvest mice, is greater than that of any of the rodents listed by Morrison and Ryser (1951) and by Dawson (1955). The averages used for the two points at 1° C. include several low values obtained while the animals probably had a slightly low-

roundings. In this respect it differs from the pocket mouse (*Perognathus longimembris*), a mouse with which it should be compared because of its similarly small size. When pocket mice are caged at cold temperatures with adequate food, they either drop into torpor or are continually awake and active. They may even be *unable* to maintain a high body temperature during a prolonged period of sleep at cool temperatures (Bartholomew and Cade, 1957).

The only other report on the rate of oxygen consumption of harvest mice lists a rate of 3.8 cc/g/hr at 24° C. for mice with an average weight of 9.6 grams (Pearson, 1948a). This rate is almost 10 per cent lower than the comparable rate obtained from Figure 1 and is below the range of variation obtained at this temperature. The difference may be accounted for by the fact that the mice used in the earlier study were acclimated to a warmer temperature (for discussion of the effect of acclimatization on metabolism see Hart, 1957).

INSULATING EFFECTIVENESS OF FUR

Figure 2 shows also the metabolic effect of removing all the fur (277 mg. in

cent at intermediate temperatures and 24 per cent at 1° C. (lowest curve in Fig. 2). To obtain these measurements, individual mice placed in the metabolic chamber were provided with a harvest mouse nest collected from the wild (shredded grass and down from Compositae), and this the mouse quickly rebuilt into an almost-complete hollow sphere about three inches in diameter. Metabolic rates were counted only when a mouse was resting quietly deep in the nest.

THERMAL ECONOMY OF HUDDLING

The metabolic economy of huddling was measured on one occasion with three mice at an environmental temperature of 1° C. without nesting material. The rate

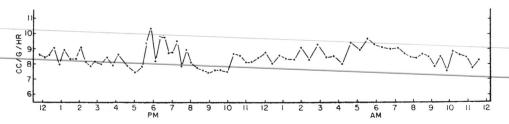

FIG. 3.—Rate of oxygen consumption of a 9-gram harvest mouse for 24 hours at 12° C.

the single 8.8-gram specimen used) with an electric clipper. When calculating the points for the curve in Figure 2, 0.28 grams was added to the naked weight and then this rate was adjusted to that for a 9-gram animal on the basis of the regression lines shown in Figure 1. The rate of metabolism of the naked mouse was about 35 per cent higher at each of the temperatures used, and the rate increased 0.38 cc/g/hr for each 1° C. drop in air temperature.

INSULATING EFFECTIVENESS OF NESTS

When normal, fully furred mice were given an opportunity to increase their insulation by constructing nests, their metabolic rates were lowered about 17 per

of metabolism per gram of huddled mice was 28 per cent less than it would have been for a single one of the mice (Fig. 2). The metabolic saving would probably be greater when more mice were huddled together and less when only two mice were huddled, as is true for feral *Mus* (Pearson, 1947) and laboratory mice (Prychodko, 1958).

24-HOUR OXYGEN CONSUMPTION IN CAPTIVITY

Figure 3 illustrates the rate of oxygen consumption of a mouse kept in the apparatus at 12° C. without nesting material but with food and water for 24 hours. The mouse consumed 1,831 cc. of oxygen to give an average rate of 8.48 cc/g/hr. This is equal to a heat production of about 8.8 Calories per day. In agreement with the fact that activity of harvest mice in the wild is greatest shortly after dusk (Pearson, 1960), the oxygen consumption was greatest at that time. The prolonged low period lasting from about 8:30 to 10:00 P.M. was unexpected in this nocturnal animal.

Pearson (1947) used as an indicator of the nocturnality of different species the ratio of the total amount of oxygen consumed at night (6:00 P.M. to 6:00 A.M.) to that consumed in the daytime. For the harvest mouse described in Figure 3, the ratio is low—1.02—but it should be pointed out that the record was made at 12° C., which is colder than the temperature used for the species in the earlier report. Temperature affects the night/day ratio of oxygen consumption because the difference in amount of oxygen con-

sumed during rest and during activity is proportionately great at warm temperatures and small at cold temperatures.

EFFECT OF ACTIVITY ON METABOLISM

An athlete is able, for short periods, to raise his rate of metabolism to a level 15 to 20 times his basal rate, but small mammals do not match this effort. The peak metabolic effort of mice running in a wheel is only 6 to 8 times their basal rate (Hart, 1950). At 0° C. lemmings running in a wheel at a speed of 15 cm/sec increase their oxygen consumption less than 35 per cent above the level of resting lemmings (Hart and Heroux, 1955). At cool ambient temperatures, such as this, small mammals expend so much energy at rest that a considerable amount of activity causes only a proportionately small increase in oxygen consumption;

TABLE 1

THE 24-HOUR OXYGEN CONSUMPTION (IN CC.) OF A 9-GRAM HARVEST MOUSE DURING DECEMBER AND JUNE AT ORINDA, CALIFORNIA

| | | DECEMBER | | | JUNE | |
		Without Nest	With Underground Nest		Without Nest	With Underground Nest
Nocturnal habit	4 hr. above ground at 1° C.*	367	367	4 hr. above ground at 12° C.†	297	297
	20 hr. under ground at 10° C.‡	1,548	1,296	20 hr. under ground at 18° C.§	1,152	954
	Activity correction‖	+119	+119	Activity correction‖	+119	+119
		2,034	1,782 cc. (8.55 Cal.)#		1,568	1,370 cc. (6.58 Cal.)#
Diurnal habit	20 hr. under ground at 10° C.‡	1,548	1,296	20 hr. under ground at 18° C.§	1,152	954
	4 hr. above ground at 6° C.**	333	333	4 hr. above ground at 25° C.††	155	155
	Activity correction‖	+119	+119	Activity correction‖	+119	+119
		2,000	1,748 cc. (8.39 Cal.)#		1,426	1,228 cc. (5.89 Cal.)#

* Mean temperature in runways at time of passage of harvest mice in December.
† Mean temperature in runways at time of passage of harvest mice in June.
‡ Underground temperature in December.
§ Underground temperature in June.
‖ Add 40 per cent of the oxygen consumption on the surface at a temperature of 12° C.
Assumed 4.8 Cal. per liter of oxygen.
** Mean half-hourly temperature in runways between 6 A.M. and 6 P.M. in December.
†† Mean half-hourly temperature in runways between 6 A.M. and 6 P.M. in June.

and at cold temperatures the metabolic cost of keeping warm may be so high as to leave little or no capacity for exercise (Hart, 1953). During measurement of the resting metabolism of harvest mice, numerous measuring periods had to be discarded because the mouse was moving around in the metabolism chamber. Such activity rarely raised the oxygen consumption more than 40 per cent above the level of a resting animal at the same temperature. During the 24-hour run at 12° C., the highest metabolic rate occurred during an 11-minute period when the average oxygen consumption was 10.36 cc/g/hr. This is only 40 per cent greater than the lowest rate recorded for that mouse during any one measuring period. The maximum metabolic effort recorded for any harvest mouse was that of an 8.6-gram mouse at 1° C. This animal persisted in gnawing, exploring, and trying to escape from the chamber for more than two hours. During one 10-minute period its oxygen consumption averaged 15.8 cc/g/hr, which is 50 per cent higher than the rate of a resting mouse at the same air temperature and six times the minimum value for the species at thermal neutrality. This is probably not far from the peak metabolic effort of the species.

On several occasions I have watched undisturbed harvest mice carrying on their normal activities in the wild, and I have been impressed by their leisurely approach to life. Hard physical labor and strenuous exercise must occur quite infrequently. Most normal activities of harvest mice are probably accomplished without a rise in metabolic rate more than 50 per cent above what it would be in a resting animal at the same air temperature.

24-HOUR METABOLISM IN THE FIELD

The preceding observations indicate that ambient temperature is a much more important variable than activity in the 24-hour energy budget of harvest mice in the wild. By use of automatic devices that record the temperature in mouse runways whenever a mouse passes by, the temperature encountered by harvest mice during their nightly periods of activity are known (Pearson, 1960). I have also recorded throughout the year the temperature five inches below the surface of the ground. This gives an approximation of the temperature encountered by the mice while they are in their retreats during the daytime. Some of these surface and underground temperature measurements have been used in the calculations summarized in Table 1.

To complete the calculations in Table 1, it has been necessary to estimate how many hours of each 24 the mouse spends on the surface of the ground and how many below the surface. No good data exist, so I have made an estimate based on the behavior of captive animals and on automatic recordings made at the exit of an underground nest box being used by wild harvest mice. Admittedly this estimate (4 hours on the surface each night) could be wrong by 50 per cent or more, but it should be noted that an error of two hours in this estimate would only alter the answer (the total 24-hour metabolism) by about 25 per cent. Assuming that the rate of oxygen consumption during above-ground activity is 40 per cent higher than the rate of a mouse resting at 12° C. (see above), the activity correction used in Table 1 can be calculated.

In 24 hours in December a harvest mouse uses 8.55 Calories, and in June, 6.58 Calories (Table 1), assuming that the mouse has the benefit of a nest. A nest reduces his daily energy budget by about 12 per cent. These estimates of daily metabolic demands seem reasonable when compared with the values actually obtained by measuring the 24-hour oxygen consumption of captive animals, as reported above. The average metabolic impact, or daily degradation of energy, by a single harvest mouse should be somewhere between that in December and that in June, perhaps 7.6 Calories. This is about the same as that of a hummingbird in the wild (Pearson, 1954)—less than half that of a much heavier English sparrow (Davis, 1955).

BIOENERGETICS

In seasons when harvest mice are abundant, there may be twelve of them per acre (Brant, 1953). At that population density, the species would be dissipating at the rate of 91 Calories per acre per day the solar energy captured by photosynthesis, or something like $\frac{1}{2}$ of 1 per cent of the energy stored each day by the plants in good harvest-mouse habitat in the Orinda area. This percentage was calculated using a net productivity of 20,000 Cal/acre/day, which was estimated by assuming 4 Calories per gram of dry vegetation (based on data in Brody, 1945, pp. 35, 788) and an annual crop of 1,800 kg. of dry vegetation per acre (based on Bentley and Talbot, 1951). The harvest mice on this hypothetical acre are causing about the same caloric drain on the environment as all the small mammals in the acre of forest described by Pearson (1948b).

By using caloric units, direct comparison can be made of the metabolic impact of different species, as in the example above. Similarly, the metabolic cost of different activities and different habits can be compared (Pearson, 1954). For example, harvest mice are strongly nocturnal (Pearson, 1960), in spite of the fact that air temperatures are much colder at night and force mice to consume more oxygen and more food than if they were diurnal. Since evolution has permitted nocturnality to persist, it seems logical to assume that the value of nocturnality to harvest mice is greater than the metabolic cost. I estimate that during a 24-hour period in December a 9-gram harvest mouse uses 0.16 more Calories by being nocturnal than it would if it were diurnal (Table 1). In summer, the difference is even greater, 0.69 Calories. The average is 0.42 Calories, or about $3\frac{1}{2}$ grains of wheat. This is a rough estimate of the price each harvest mouse pays for nocturnality. Some environmental pressure makes harvest mice remain nocturnal, and this pressure must be more than 0.42 Calories per mouse per day. If harvest-mouse nocturnality evolved for one reason only—to avoid predation by hawks—then we would have discovered a minimum estimate of the predation pressure of hawks on harvest mice. Surely the situation is not this simple; nevertheless, it is interesting to measure the pressure that makes harvest mice nocturnal even if the cause of the pressure is not known.

SUMMARY

Oxygen consumption of harvest mice reaches a minimum of 2.5 cc/g/hr at an ambient temperature of 33° C., and the zone of thermal neutrality is not more than 3°. Each drop of 1° in ambient temperature causes an increase in the rate of metabolism of 0.27 cc/g/hr. Removing the fur raises the rate of metabolism about 35 per cent, and use of a nest lowers it 17 to 24 per cent. Huddling by three mice at 1° reduces the rate 28 per cent.

Exercise at cool temperatures causes a relatively small increase in the rate of metabolism, whereas change of ambient temperature has a great effect. Making use of the temperatures that harvest mice are known to encounter in the wild, the 24-hour oxygen consumption of a wild harvest mouse was calculated to be 1,782 cc. in December and 1,370 cc. in June. The average (1,576 cc.) is equivalent to about 7.6 Calories per day. A dense population of harvest mice would dissipate about 91 Calories per day per acre, which is about $\frac{1}{2}$ of 1 per cent of the energy stored by the plants each day.

By being nocturnal, harvest mice encounter cooler temperatures, and this habit increases the daily energy budget of each mouse by 0.42 Calories, or about $3\frac{1}{2}$ grains of wheat.

LITERATURE CITED

BARTHOLOMEW, G. A., and CADE, T. J. 1957. Temperature regulation, hibernation, and aestivation in the little pocket mouse, *Perognathus longimembris*. Jour. Mammal., 38:60–72.

BENTLEY, J. R., and TALBOT, M. W. 1951. Efficient use of annual plants on cattle ranges in the California foothills. U.S. Dept. Agriculture, Circular No. 870, 52 pp.

BRANT, D. H. 1953. Small mammal populations near Berkeley, California: *Reithrodontomys, Peromyscus, Microtus*. Doctoral thesis, University of California, Berkeley.

BRODY, S. B. 1945. Bioenergetics and growth. New York: Reinhold Publishing Corp.

DAVIS, E. A., JR. 1955. Seasonal changes in the energy balance of the English Sparrow. Auk, 72:385–411.

DAWSON, W. R. 1955. The relation of oxygen consumption to temperature in desert rodents. Jour. Mammal., 36:543–53.

DE BONT, A.-F. 1945. Métabolisme de repos de quelques espèces d'oiseaux. Ann. Soc. Roy. Zoöl. Belgique, 75 (1944):75–80.

HART, J. S. 1950. Interrelations of daily metabolic cycle, activity, and environmental temperature of mice. Canadian Jour. Research, D, 28:293–307.

———. 1953. The influence of thermal acclimation on limitation of running activity by cold in deer mice. Canadian Jour. Zoöl., 31:117–20.

———. 1957. Climatic and temperature induced changes in the energetics of homeotherms. Revue Canadienne de biol., 16:133–74.

HART, J. S., and HEROUX, O. 1955. Exercise and temperature regulation in lemmings and rabbits. Canadian Jour. Biochem. & Physiol., 33:428–35.

MORRISON, P. R. 1947. An automatic apparatus for the determination of oxygen consumption. Jour. Biol. Chem., 169:667–79.

MORRISON, P. R., and RYSER, F. A. 1951. Temperature and metabolism in some Wisconsin mammals, Federation Proc., 10:93–94.

———. 1959. Body temperature in the white-footed mouse, *Peromyscus leucopus noveboracensis*. Physiol. Zoöl., 32.90 103.

PEARSON, O. P. 1947. The rate of metabolism of some small mammals. Ecology, 28:127–45.

———. 1948a. Metabolism of small mammals, with remarks on the lower limit of mammalian size. Science, 108:44.

———. 1948b. Metabolism and bioenergetics. Scientific Monthly, 66:131–34.

———. 1954. The daily energy requirements of a wild Anna Hummingbird. Condor, 56:317–22.

———. 1960. Habits of harvest mice revealed by automatic photographic recorders. Jour. Mammal. (in press).

PRYCHODKO, W. 1958. Effect of aggregation of laboratory mice (*Mus musculus*) on food intake at different temperatures. Ecology, 39:500–503.

Reprinted for private circulation from
PHYSIOLOGICAL ZOÖLOGY
Vol. XXXIII, No. 2, April 1960
Copyright 1960 by the University of Chicago

OXYGEN CONSUMPTION, ESTIVATION, AND HIBERNATION IN THE KANGAROO MOUSE, MICRODIPODOPS PALLIDUS[1]

GEORGE A. BARTHOLOMEW AND RICHARD E. MacMILLEN

Departments of Zoölogy, University of California, Los Angeles,
and Pomona College, Claremont, California

THE pallid kangaroo mouse occurs only in the desert parts of western Nevada and extreme eastern California. Its habitat is restricted to areas of fine sand which support some plant growth. Like its relatives, the kangaroo rats (*Dipodomys*) and the pocket mice (*Perognathus*), it is nocturnal, fossorial, and gramnivorous and can under some circumstances live indefinitely on a dry diet without drinking water. The general life history (Hall and Linsdale, 1929) of this kangaroo mouse and the details of its distribution (Hall, 1946) are known, but virtually no quantitative data on its physiology are available.

The present study was undertaken to compare the thermoregulation of *Microdipodops* with that of the better-known genera, *Dipodomys* and *Perognathus*. These three genera belong to the family Heteromyidae, which has been more successful in occupying the arid parts of western North America than any other group of mammals.

[1] This study was aided in part by a contract between the Office of Naval Research, Department of the Navy, and the University of California (Nonr 266[31]).

MATERIAL AND METHODS

Experimental animals.—The twenty-three kangaroo mice used were trapped in sand dunes four miles south of Arlemont Ranch, Esmeralda County, Nevada, in April, 1959, and May, 1960. They were housed individually in small terraria partly filled with fine sand, kept in a windowless room on a photoperiod of 12 hours, and fed on a diet of mixed bird seed supplemented occasionally with small pieces of cabbage. Survival was excellent, and some of the animals were kept for over ten months.

Body temperatures.—All temperatures were measured with 30-gauge copper-constantan thermocouples connected to a recording potentiometer. All body temperatures were taken orally by inserting a thermocouple to a depth of at least 2 cm.

Ambient temperatures.—The ambient temperatures were monitored with thermocouples and controlled by insulated chambers equipped with automatic heating and cooling units, blowers, and lights.

Oxygen consumption.—Oxygen con-

177

sumption was measured by placing a mouse in an air-tight 500-cc. glass container equipped with a thermocouple and ports for the introduction and removal of air. The bottom of the container was covered to a depth of about 1 cm. with fine dry sand. The glass container with animal inside was placed in a temperature-control chamber, and dry air was metered through the container at a rate

the response of body temperature (T_B) to moderately low ambient temperatures (T_A), kangaroo mice were placed at T_A of 7°–9° C. for five days starting May 11, 1959, with food available in excess; measurements of T_B were made at 24-hour intervals. There were no apparent changes in T_B during the test period, nor was the mean T_B significantly different from that of animals maintained at room

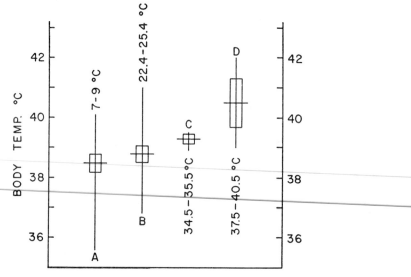

Fig. 1.—Body temperatures of *M. pallidus* at various ambient temperatures. *A*, 47 measurements on twelve animals; *B*, 38 measurements on thirteen animals; *C*, 7 measurements on four animals; *D*, 9 measurements on four animals (three other animals tested at this temperature died). The horizontal lines indicate the means (*M*). The rectangles inclose $M \pm \sigma_M$. The vertical lines indicate the range.

of 250 cc/min and then delivered to a Beckman paramagnetic oxygen analyzer which, used in conjunction with a recording potentiometer, gave a continuous record of oxygen consumption. All data used were from post-absorptive animals.

RESULTS

Body temperature during normal activity.—Normally active animals kept at room temperature (22.4°–25.4° C.) had body temperatures ranging from slightly less than 37° to as high as 41° C., with a mean of 38.8° C. (Fig. 1). To determine

temperature. The animals appeared normally active and unaffected by the change in environmental temperature.

Animals were maintained at T_A of 37.5°–40.5° C. for 24 hours to test their response to moderately high environmental temperatures. They showed a conspicuous elevation in T_B with a mean almost 2° C. higher than that of animals at room temperature. Animals maintained at T_A close to 35° C. also became hyperthermic and showed a mean T_B intermediate between that of animals held at room temperatures and those held at 39° C. There was no mortality in animals held at 35° C., but exposure to 39° C. for more than a few hours killed three out of the seven animals tested. At a high T_A the kangaroo mice did not salivate or pant; they merely sprawled out flat on the sand with legs extended and lower jaw and neck prone on the substrate. This prone posture alternated with brief bursts of intense activity characterized by repeated shifts in position and much digging and moving of sand.

gm.) is 1.8 cc O_2/gm/hr when the formula $M = 3.8W^{-0.27}$ is used (see Brody, 1945, and Morrison, Ryser, and Dawe, 1959). The observed basal metabolism of our kangaroo mice (mean, 1.3 ± 0.2 cc O_2/gm/hr) was about three-fourths of the predicted value. This relatively low figure is consistent with the observation on some other heteromyids (Dawson, 1955).

The only comparative data on the

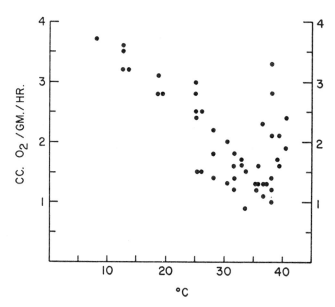

Fig. 2.—The relation of oxygen consumption to ambient temperature. Data obtained from ten animals. Each point represents the minimum level of oxygen consumption maintained by an animal for half an hour. Oxygen volumes are corrected to 0° C. and 760 mm. (Hg.) pressure.

Oxygen consumption.—The relation of oxygen consumption to T_A is summarized in Figure 2. There is no clearly defined zone of thermal neutrality, but oxygen consumption is minimal at about 35° C. The increase in oxygen consumption at temperatures above 35° C. is relatively more rapid than is the increase below this point of thermal neutrality. No differences in oxygen consumption were apparent between males and females.

The calculated metabolism of *Microdipodops pallidus* (mean weight, 15.2

energy metabolism of *Microdipodops* is that of Pearson (1948) on *M. megacephalus*. Pearson's data, obtained at temperatures near 24° C. from animals that were not post-absorptive, gave oxygen consumptions of 3.4–3.7 cc O_2/gm/hr. Pearson's measurements, as might be expected from the fact that he was not using post-absorptive animals, are higher than our determinations of 2.7 cc O_2/gm/hr at 25° C.

Hibernation and estivation.—No information on hibernation or estivation is available for *Microdipodops.* Hall (1946, p. 386) pointed out that kangaroo mice are often active above ground in temperatures many degrees below freezing, and Ingles (1954, p. 214) suggested that kangaroo mice probably do not hibernate.

Under laboratory conditions we found that kangaroo mice at any time of year

there are no conspicuous physiological differences between arousal from spontaneous dormancy and that from induced dormancy.

Animals dormant at room temperatures (estivating) started to arouse immediately upon being handled. The rate of increase in T_B varied but usually fell between 0.5° and 0.8° C. per minute. Usually within 20 minutes of the onset

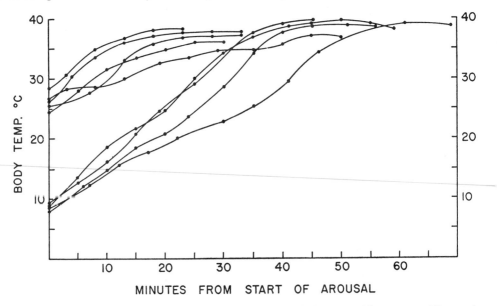

Fig. 3.—Increases in oral temperatures in nine kangaroo mice during arousal from torpor. All arousals took place in ambient temperatures between 23° and 26° C. Temperatures taken manually with thermocouples. The five upper animals were dormant at room temperature (22°–25° C.); the four lower animals were dormant at 5°–8° C.

will spontaneously become dormant at ambient temperatures ranging at least from 5° to 26° C. and can readily be induced to hibernate (or estivate) over this range of temperatures by reduction of food for 24 hours or less.

Body temperature and behavior during entry into torpor were not recorded, but the animals apparently entered torpor in the crouching posture normally used in sleeping. Dormant animals had body temperatures 1°–2° C. above ambient. Judging from the course of body temperature during arousal from torpor,

of arousal the animals attained their normal operating temperature, and within as little as 12–15 minutes from the start of arousal they appeared to behave normally, even though T_B approximated 30° C. Arousal from low temperatures was essentially the same as arousal from high temperatures (Fig. 3). However, animals arousing from low temperatures attained maximal body temperatures about 1° C. higher than did those arousing from room temperature.

Incidental to the measurement of T_B the relations of various types of behavior to body temperature were noted during nine arousals. Mice unsuccessfully attempted to right themselves when turned over at T_B between 16.1° and 18.2° C. and successfully righted themselves at T_B between 19.0° and 22.0° C. The first vocalizations were given at T_B between 24.7° and 28.6° C. Grain was available to the animals during arousal, and seven of the nine animals ate during arousal. The lowest T_B for eating was 25.5° C., and three animals ate at temperatures between 25° and 29.4° C. The mean T_B for onset of visible shivering for seven animals was 25.5° C. Two of the nine animals observed did not visibly shiver during arousal. Shivering usually stopped at a T_B of 34°–35° C.

DISCUSSION

The general features of thermoregulation in *Microdipodops pallidus* are similar to those of the related genus *Perognathus* in that both show well-developed patterns of hibernation and estivation, essentially normal behavior at T_B below 35° C., obligate hyperthermia at T_A above 35° C., and no apparent salivary response to elevated body temperature. *Microdipodops* differs from the related genus *Dipodomys* in that the latter does not readily become dormant at either high or low temperatures and does use salivation as an emergency thermoregulatory response (Schmidt-Nielsen and Schmidt-Nielsen, 1952).

In kangaroo mice, as in *Perognathus longimembris* (Bartholomew and Cade, 1957) and *Citellus mohavensis* (Bartholomew and Hudson, 1960), there appears to be no sharp physiological differentiation between hibernation and estivation. This underscores the point that the facultative hypothermia shown by mammals should not be thought of only as an adaptive response to low environmental temperatures; at least for small desert mammals the ability to become dormant and to decrease body temperature and metabolic activity may be more useful in the summer than in the winter, and it may be as important for water conservation as for energy conservation.

Kangaroo mice are unique among heteromyids in having conspicuous deposits of adipose tissue in the proximal third of the tail, which is considerably larger than either its base or its distal half. Hall (1946, p. 379) suggests that the fleshiness of the tail permits it to function in balancing. However, since these mice hibernate but do not show conspicuous seasonal deposits of subcutaneous fat over the body as a whole, it seems reasonable to suggest that the fat tail serves as a reserve of energy for use during periods of torpor. In the laboratory with food available in excess, many of the kangaroo mice showed a marked increase in tail diameter.

Our data (Fig. 1) show almost no indication of a discrete zone of thermal neutrality for the kangaroo mouse. Its critical temperature is unusually high for an animal living in an area characterized by cold winters. For months on end kangaroo mice can be active only at temperatures below thermal neutrality. Presumably, their energetic and thermal problems are reduced in cold weather by periodic episodes of torpor. It is of interest that we captured our animals on nights when environmental temperatures went below −10° C., and Hall (1946, p. 396) reports that these animals are often "active on nights when the temperature is so low as to freeze to a state of stiffness the bodies of mice caught in traps." Thus, although they can hibernate, they are also commonly active during subfreezing weather.

This species has remarkably shallow burrows, often no deeper than 4 inches (Hall, 1946, p. 396). Consequently, when high daytime temperatures occur, at least some members of the population may be exposed to temperatures near 35° C. It is possible, therefore, that the high point of thermal neutrality of this species allows a significant metabolic economy and a significant reduction in pulmocutaneous water loss during the severely hot desert summers.

Extrapolation of the plot of metabolism against ambient temperature below thermal neutrality does not intersect the abscissa within the usual range of body temperature (38°–39°C.) of kangaroo mice (Fig. 2). This means that, unlike some of the species considered by Scholander et al. (1950), and unlike the masked shrew, Sorex cinereus (Morrison, Ryser, and Dawe, 1959), the kangaroo mouse does not follow Newton's empirical law of cooling in a simple and direct manner. The failure to follow the pattern predicted by Newton's law of cooling may be related to the fact that kangaroo mice start to become hyperthermic as they approach their critical temperature (Fig. 1), and it suggests that the relation between skin and ambient temperature in this species differs from the usual pattern. It is of interest that Pearson's data (1960) for Reithrodontomys show a situation similar to that reported here for Microdipodops, that is, almost no zone of thermal neutrality, a high critical temperature, and a failure of the curve of metabolism against ambient temperature to intersect the abscissa at the usual body temperature. Although Pearson does not comment on this point, it appears that in Reithrodontomys as in Microdipodops the curve of metabolism against ambient temperature intersects the abscissa at a point above the lethal temperature for the species.

The apparent absence of a marked increase in salivation at high temperatures in Microdipodops correlates nicely with its strong tendency toward hyperthermia at high ambient temperatures. For animals living in a desert environment where water is usually in short supply, hyperthermia is a more advantageous response to heat than is evaporative cooling.

SUMMARY

Microdipodops pallidus occurs only on sparsely vegetated sand dunes in the desert parts of western Nevada and eastern California. In the absence of temperature stress body temperature, T_B, averages 38.8°C. There is no diminution of T_B with decreasing ambient temperature, T_A, at least to 8° C. However, hyperthermia is apparent at a T_A of 35° C. and at 39° C. T_B averages 40.5° C. Exposure for more than a few hours to 39° C. is often lethal. At high ambient temperatures kangaroo mice neither pant nor drool. They have no clearly defined zone of thermal neutrality; oxygen consumption is minimal at 35° C. and increases more rapidly at temperatures above this point than below it. Basal metabolism is 25 per cent less than that predicted on the basis of body size. Kangaroo mice are capable of both estivation and hibernation. In the laboratory they often become dormant at ambient temperatures ranging at least from 5° to 26° C. The rate of temperature increase during arousal at room temperature is 0.5°–0.8° C. per minute. Terminal body temperatures after arousal from low temperatures averaged about 1° C. higher than after arousal from room temperature. By the time the T_B of arousing animals reaches 30° C., their behavior appears normal. The thermoregulatory responses of kangaroo mice are compared with those of other desert heteromyids, and the ecological significance of their physiological capacities is discussed.

206

LITERATURE CITED

BARTHOLOMEW, G. A., and CADE, T. J. 1957. Temperature regulation, hibernation, and aestivation in the little pocket mouse, *Perognathus longimembris*. Jour. Mamm., 38:60–72.

BARTHOLOMEW, G. A., and HUDSON, J. W. 1960. Aestivation in the Mohave ground squirrel, *Citellus mohavensis*. Bull. Mus. Comp. Zool., 124:193–208.

BRODY, S. 1945. Bioenergetics and growth. New York: Reinhold Publishing Co.

DAWSON, W. R. 1955. The relation of oxygen consumption to temperature in desert rodents. Jour. Mamm., 36:543–53.

INGLES, L. G. 1954. Mammals of California and its coastal waters. Stanford, Calif.: Stanford University Press.

HALL, E. R. 1946. Mammals of Nevada. Berkeley: University California Press.

HALL, E. R., and LINSDALE, J. M. 1929. Notes on the life history of the kangaroo mouse (*Microdi-podops*). Jour. Mamm., 10:298–305.

LYMAN, C. P., and CHATFIELD, P. O. 1955. Physiology of hibernation in mammals. Physiol. Rev., 35:403–25.

MORRISON, P., RYSER, F. A., and DAWE, A. R. 1959. Studies on the physiology of the masked shrew *Sorex cinereus*. Physiol. Zoöl., 32:256–71.

PEARSON, O. P. 1948. Metabolism of small mammals, with remarks on the lower limit of mammalian size. Science, 108:44.

———. 1960. The oxygen consumption and bioenergetics of harvest mice. Physiol. Zoöl., 33:152–60.

SCHMIDT-NIELSEN, K., and SCHMIDT-NIELSEN, B. 1952. Water metabolism of desert mammals. Physiol. Rev., 32:135–66.

SCHOLANDER, P. F., HOCK, R., WALTERS, V., JOHNSON, F., and IRVING, L. 1950. Heat regulation in some arctic and tropical mammals and birds. Biol. Bull., 99:237–58.

Reprinted for private circulation from

PHYSIOLOGICAL ZOÖLOGY

Vol. XXXIV, No. 3, July 1961

Copyright 1961 by the University of Chicago

PRINTED IN U.S.A.

Counter-Current Vascular Heat Exchange in the Fins of Whales[1]

P. F. SCHOLANDER AND WILLIAM E. SCHEVILL. *From the Woods Hole Oceanographic Institution, Woods Hole, Massachusetts*

I T MAY BE a source of wonder that whales swimming about in the icy waters of the polar seas can maintain a normal mammalian body temperature. What prevents them from being chilled to death from heat loss through their large thin fins?[2] These are well enough vascularized to justify the question (fig. 1). One may conjecture that a whale may be so well insulated by its blubber that it needs such large surfaces to dissipate its heat. On the other hand, if heat conservation is at a premium, there must be some mechanism whereby the fins can be circulated without losing much heat to the water. One may point to two circulatory factors which would reduce the heat loss from the fin: *a*) slow rate of blood flow and, *b*) precooling of the arterial blood by veins before it enters the fin.

Bazett and his coworkers (1) found that in man the brachial artery could lose as much as 3°C/decimeter to the two *venae comitantes*. This simple counter-current exchange system is a mere rudiment compared to the multi-channelled arteriovenous blood vascular bundles which we find at the base of the extremities in a variety of aquatic and terrestrial mammals and birds. These long recognized structures have most recently been studied by Wislocki (2), Wislocki and Straus (3) and Fawcett (4).

The function of these bundles has long been a mystery. No matter what else they do, they must exchange heat between the arteries and veins, and it has been pointed out that they very likely play an important role in the preservation and regulation of the body heat of many mammals and birds (5).

In the present study we describe a conspicuous arteriovenous counter-current system in the fins and flukes of whales, which we interpret as organs for heat preservation.

Received for publication July 21, 1955.

[1] Contribution Number 807 from the Woods Hole Oceanographic Institution.

[2] In 'fin' we include the structures more specifically called flippers (pectoral fins), flukes (caudal fins) and dorsal fin.

MATERIAL

Two species of porpoises have been studied: *Lagenorhynchus acutus:* dorsal fin, tail-fluke, and flipper of an adult female collected 50 miles east of Cape Cod; *Tursiops truncatus:* dorsal fin and tail-fluke of a 4-month-old calf from Florida, supplied through the courtesy of the Marineland Research Laboratory.

Lagenorhynchus is a genus of fairly high latitudes, the southern limit of *L. acutus* being about latitude 41°N. on the New England coast and about 55°N. in the British Isles. It has been caught at least as far north as latitude 64°N. in west Greenland and Norwegian waters. *Tursiops* is found in lower latitudes, *T. truncatus* overlapping slightly with *L. acutus* and occurring south into the tropics.

DESCRIPTION

Figure 2 illustrates the vascular supply at the base of the dorsal, pectoral and caudal fins. It may be seen that all major arteries are located centrally within a trabeculate venous channel. This results in two concentric conduits, with the warm one inside. In addition to the circumarterial venous channels there are separate superficial veins, as seen in figure 2. The circumarterial venous channels are conspicuously thin walled compared to the simple veins, as may be seen in figure 3. When an artery was perfused with saline, the solution returned through both of these venous systems.

INTERPRETATION

Based on the anatomical findings and on the perfusion experiments, we interpret the artery within-vein arrangement as a heat-conserving counter-current system, as schematically presented in fig. 4. In such an arrangement the warm arterial blood is cooled by the venous blood which has been chilled in the fin. The result is a steep proximodistal temperature drop from the body into the appendage. The heat of the arterial blood does not reach the fin, but is short-circuited back into the body in the venous system. Body heat is therefore conserved at the expense of keeping the appendage cold. There is reason to believe that the analogous blood vascular bundle in the prox

27

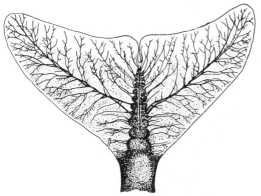

Fig. 1. Arterial supply to the flukes in the common porpoise (*Phocoena phocoena*), drawn from an x-ray picture by Braun (6).

mal part of the extremities of sloths serves a similar function, inasmuch as these animals can barely keep warm even in their warm environment (5). Cold extremities have been described in many arctic mammals and birds as important factors for conservation of body heat (7), but to what extent arteriovenous counter-current structures are present in these animals is not known.

The efficiency of heat exchange in a system like that diagrammed in figure 4 is related to the blood flow. The slower the flow, the more nearly identical will be the arterial and venous temperatures along the system, and the more efficient will be the heat conservation. At high rates of flow, warm blood will reach the periph-

ery and cool venous blood will penetrate into the body.[3]

It was shown by perfusion experiments on the detached fins that the arterial blood may return via the concentric veins, and/or through the separate superficial veins. As pointed out above, the concentric vein channels are very thin walled and weak compared to the thick-walled superficial veins (fig. 3). One may interpret these anatomical facts in the following way. If the animal needs maximal heat conservation, blood circulation through the fins should be slow, and the venous return should preferentially take place through the counter-current veins. But a slow rate of blood flow would need only weak venous walls, as actually found. If, on the other hand, the animal needed maximal cooling, as during exercise in relatively warm water, this would be most effectively accomplished by a high rate of blood flow through the fins, with venous return through the superficial veins and the least possible flow through the concentric veins. This would require the strong and thick walls of the superficial veins. One may even see the likelihood of a semiautomatic regulatory function in the concentric system, for when the artery is swelled by increased blood flow, the concentric veins will be more or less obliterated,

[3] The theory for a multichannel counter-current system has been elaborated in connection with the swimbladder in deep sea fishes (8).

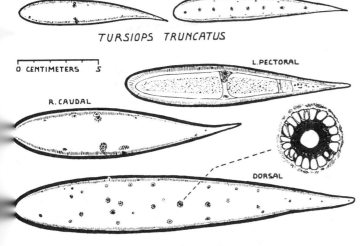

TURSIOPS TRUNCATUS

Fig. 2. Sections near base of fins and flukes of two species of porpoises. Each artery is surrounded by a multiple venous channel. Simple veins are near the skin (only the larger ones are indicated).

LAGENORHYNCHUS ACUTUS

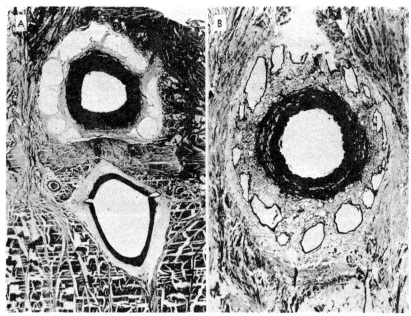

FIG. 3. Sections from *Tursiops truncatus*. (Courtesy of the Department of Anatomy, Harvard Medical School.) *A*. From tail-fluke. Upper: artery surrounded by thin-walled venous channels. Lower: superficial single thick-walled vein in the hypodermis. (× 9) *B*. From dorsal fin. Artery surrounded by thin-walled venous channels (× 12)

but will remain open when the diameter of the artery is reduced during decreased blood flow. Thus the anatomical findings fit logically into the simplest possible scheme of heat regulation in the fins.

There are a few observations available indicative of heat regulation in the fins of porpoises. Tomilin (9) made some observations on an east Siberian 'white-sided dolphin' on deck, and found that the fins could vary between 25° and 33.5°C, while the body varied only 0.5°. Schevill observed that the flukes in a Florida *Tursiops* out of water became about 10° warmer than the body surface itself. In both of these cases the animals were probably resisting overheating. On the other hand, Scholander (5) has recorded cold flippers in water-borne common porpoises (*Phocoena*).

The concentric counter-current system of an artery within a vein appears to be a peculiarly cetacean arrangement, and we have seen it only in the fins, flippers and flukes of these animals.[4] This is an impressive example of bioengineering, which, together with other

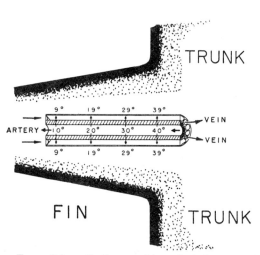

FIG. 4. Schematic diagram of hypothetical temperature gradients in a concentric counter-current system.

factors, adapts these homeotherms for a successful existence in a heat-hungry environment.

SUMMARY

The vascular supply to the fins and flukes of two species of porpoises, *Lagenorhynchus acutus* and *Tursiops truncatus*, is described. All major arteries entering the fins and flukes

---

[4] The present material is from odontocetes, but these structures have also been noted by Scholander in the tail flukes of a mysticete (fin whale).

are surrounded by a trabeculate venous channel. The arteries drain into these, but also into superficial simple veins. The artery within the venous channel is interpreted as a heat-conserving counter-current exchange system. The heat regulatory aspects of the two venous systems are discussed.

We wish to express our appreciation to Dr. F. G. Wood, Jr., and the Marineland Research Laboratory, Marineland, Fla., for providing the material of *Tursiops truncatus*, and to Dr. G. B. Wislocki of the Dept. of Anatomy, Harvard Medical School, Boston, Mass., for providing the histological sections and photographs.

REFERENCES

1. BAZETT, H. C., L. LOVE, M. NEWTON, L. EISENBERG, R. DAY AND R. FORSTER II. *J. Appl. Physiol.* 1: 3, 1948.
2. WISLOCKI, G. B. *J. Morphol.* 46: 317, 1928.
3. WISLOCKI, G. B. AND W. L. STRAUS, JR. *Bull. Mus. Comp. Zool. Harvard* 74: 1, 1932.
4. FAWCETT, D. W. *J. Morphol.* 71: 105, 1942.
5. SCHOLANDER, P. F. *Evolution* 9: 15, 1955.
6. BRAUN, M. *Zool. Anz.* 29: 145–149, 1905.
7. IRVING, L. AND J. KROG. *J. Appl. Physiol.* 7: 355, 1955.
8. SCHOLANDER, P. F. *Biol. Bull.* 107: 260, 1954.
9. TOMILIN, A. G. *Rybnoe Khozaistvo* 26: 50, 1950. (In Russian.)

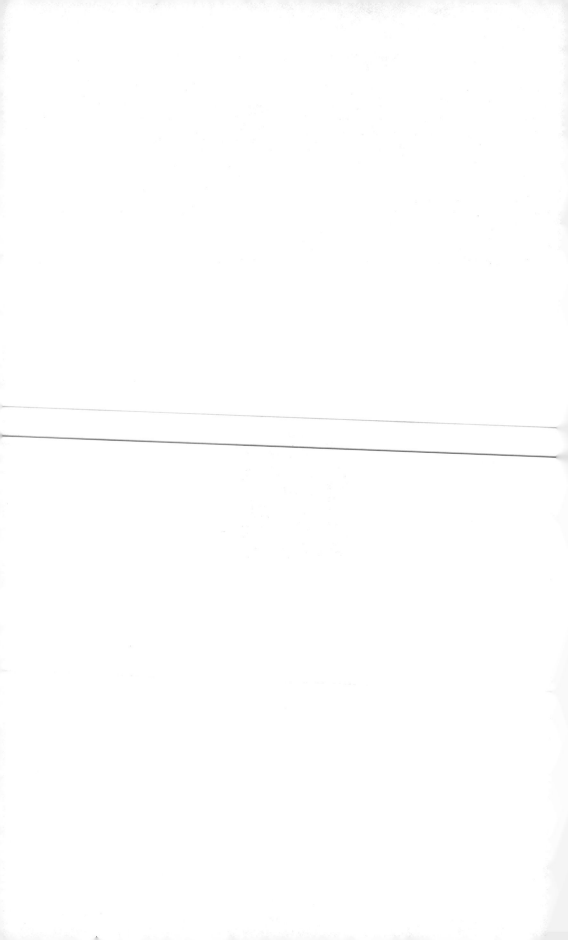

SECTION 3—REPRODUCTION AND DEVELOPMENT

Just as animal structures must be adaptive, so must reproductive and developmental patterns; the organism must be a functioning unit in its environment at all stages of its life cycle. In our selections, Millar discusses the evolution of litter size, and Sharman points out the adaptive value of some peculiarities of kangaroo reproduction. The interrelationships of reproductive and developmental patterns in the fisher are evident in the account by Wright and Coulter. Superfetation (or the fertilization of new ova during gestation, known in kangaroos, rabbits, and some rodents), delayed implantation (mainly in some mustelids, bears, and pinnipeds), and delayed fertilization (through sperm storage, known in some bats) all are interesting adaptive variations of the reproductive theme. Conaway reviews the adaptive significance of several reproductive patterns. A study of the reproductive adaptations of the red tree vole by Hamilton (1962), not reprinted here, related small litters, long gestation, delayed implantation during lactation, and slow development of young with the limited amount of energy available in the vole's food sources. The classic summary of mammalian reproduction by Asdell (as revised in 1964), Sadlier's monograph (1969) on ecology of reproduction and collections of contributions edited by Enders (1963) on delayed implantation and by Rowlands (1966) on comparative biology of mammalian reproduction also will repay study. The study of pheromones—chemicals produced by organisms that transmit olfactory signals—is a new and growing area combining reproductive and behavioral biology. Bronson's review is based mainly on laboratory experiments, but he discussed concepts of obvious importance to the mammalogist who would understand the physiological bases of phenomena such as the Bruce and Whitten effects, and their adaptive significance.

Two classic books in the field of development in which relative growth rates were considered at length are *On Growth and Form* by D'Arcy Wentworth Thompson (1942) and *Problems of Relative Growth* by Julian Huxley (1932). A detailed account (too long to include here) of one species in terms of relative growth and in comparison to several other species is the study by Lyne and Verhagen (1957) on *Trichosurus vulpecula,* an Australian brushtailed possum. Allen's paper (reprinted here) is of interest because it is one of the earliest to give serious consideration to variation with age and to possible relevance of such variation to systematic and other problems. Hall (1926) described at greater length than could be included here and in detail uncommon at that time the changes during growth of the skull in the California ground squirrel. Two among the many good studies of development of single species are by Layne (1960, 1966) on *Ochrotomys nuttalli* and *Peromyscus floridanus.* Clark's study of the Richardson ground squirrel is typical of many short descriptive papers in this area.

Although we have not included examples of methods of determining age other than the report of Wright and Coulter on the fisher, we must comment that age determination is important in many practical problems relating to wildlife management as well as in studies of population composition or of growth as such. Managers of deer herds, for example, can examine the teeth of hunter-killed animals using standards developed by Severinghaus (1949)

213

and later workers. Age determination by annuli was reviewed by Klevezal and Kleinenberg (1969)—see Adams and Watkins (1967) on its application to ground squirrels. Epiphyseal growth as observed in X-ray photographs and the use of lens weights are other means (see Wight and Conaway, 1962, on aging cottontails).

A short paper on maturational and seasonal molt in the golden mouse concludes our selection for this section. Studies of molt in furbearers are, of course, of special economic import, and knowledge of pelage differences related to age, sex, or season are of obvious use in most studies of mammalian populations.

The literature on mammalian reproduction and development is widely scattered. Journals such as GROWTH, DEVELOPMENTAL BIOLOGY, and BIOLOGY OF REPRODUCTION contain much of interest to mammalogists, but other journals, either more (e.g., ENDOCRINOLOGY) or less (e.g., THE JOURNAL OF EXPERIMENTAL ZOOLOGY) specialized, must also be consulted.

EVOLUTION OF LITTER-SIZE IN THE PIKA, *OCHOTONA PRINCEPS* (RICHARDSON)

John S. Millar[1]

Department of Zoology, University of Alberta, Edmonton, Alberta

Received May 30, 1972

Attempts to explain the evolutionary significance of litter-size have been relatively few, but varied in approach and scope. Lack (1948) considered that natural selection favors those animals producing the most offspring because they leave the most descendants. He suggested that an "upper limit is set by the number of young which the parents can successfully raise" but recognized that "there is an evolutionary alternative between producing more young, or fewer young which are better nourished and better protected." Several hypotheses have been based on Lack's basic premise that natural selection favors the production of maximum number of offspring. These generally view litter-size as only part of an overall reproductive strategy, but differ in the parameters considered important in determining litter-size. For example, "resources," length of breeding season, food supply, body size, altitude, latitude, mortality, population stability, and competition have all been suggested to influence litter-size directly or indirectly (Lord, 1960; Cody, 1966; Gibb, 1968; Smith and McGinnis, 1968; Spencer and Steinhoff, 1968).

During a study of the pika *Ochotona princeps* (Richardson) in southwestern Alberta, data on reproduction, mortality, and population density were obtained. Here these data are used to evaluate the significance of litter-size in pikas. Several aspects of the biology of these animals have been reported by Millar (1972*a*, *b*) and Millar and Zwickel (1972*a*, *b*).

METHODS

A total of 667 animals were collected from several study areas by shooting throughout the breeding seasons of 1968, 1969 and 1970. Ovaries of mature females were fixed in A.F.A. (alcohol-formalin, acetic acid), embedded in paraffin, and serially sectioned at 7–10μ. Corpora lutea and corpora albicantia were counted microscopically, and these counts were considered to be the litter-size at conception, or potential litter-size. Such counts may be biased by twinning of ova or polyovulation. Polyovulation and twinning of ova would result in females having more embryos than corpora lutea, but this situation was not observed. Twinning of ova would result in some embryos sharing a chorion with another, but again, none were found. Stage of gestation was determined from the size of embryos in collected females, and all embryonic losses occurred prior to mid-pregnancy (Millar, 1972*b*). The number of healthy embryos in late gestation was considered to be the litter-size at birth; the difference between the number of corpora lutea and healthy embryos in late stages of pregnancy provided an estimate of prenatal losses.

Discrete fat bodies were present in the interscapular, cardiac, and splenic regions of collected animals. These were removed and weighed, and mg fat per 100 gm body wt was used as an index of condition.

Age of collected animals was determined from histological sections of lower jaws. Adult mortality rates were based on age structure of the populations (Millar and Zwickel, 1972*a*).

Several marked populations were followed on one area during the three years

[1] Present Address: Department of Zoology, University of Western Ontario, London 72, Ontario.

TABLE 1. *Litter-size and number of litters per season of North American pikas. Litter-size based on counts of embryos.*

Region	Litter-size					Litters per Season	Source
	1	2	3	4	5		
California	1	6	8	,2	1	3–4	Severaid, 1955
		2	3			–	Grinnell et al., 1930
Nevada		1	5	2		–	Hall, 1946
Oregon				2		2‡	Bailey, 1936
			2			–	Roest, 1953
Colorado	1	2	1			2	Johnson, 1967
		1	2				Anderson, 1959
			1†				Dice, 1927
		3		1		2‡	Present study
Colorado and Utah		3	4	5		2	Hayward, 1952
Utah			1			–	Long, 1940
Washington					1	2‡	Dice, 1926
British Columbia			1†			–	Underhill, 1962
Alberta	8	38	33	1		2	Present study
Alaska*				1			Dixon, 1938
			2	1		2‡	Rausch, 1961, 1970

* *O. collaris.*
† born.
‡ based on scanty evidence.

of the study. Animals on 10 discrete rock slides were live-trapped and individually marked. Certain females were retrapped and observed as often as possible throughout each breeding season. Young pikas emerging from nests beneath the rocks were associated with particular females, counted, and marked. The number of young associating with a particular female was considered to be the litter-size at weaning.

Total populations were marked on 4 slides that were measured and mapped. Number of adults per unit area of rock slide provided estimates of population density.

RESULTS

Pikas in southwestern Alberta matured as yearlings (the spring followed their birth) and all females had the potential to produce two litters each breeding season. A comparison of breeding parameters among pikas throughout North America (Table 1) indicates that litter-size varies little and most populations (with the exception of California) have two litters per

year. Presumably, populations in areas with short summers do not have a sufficiently long period of favorable conditions to have animals maturing during the season of their birth (assuming growth rates similar to those in Alberta). A comparison of breeding parameters among several species of Asian pikas (Table 2) indicates that different species have quite different breeding patterns, and that most Asian species have higher fecundity than pikas in North America.

Mortality of Litters

Entire litters were frequently lost prior to independence (48% of 67 females known to be pregnant failed to produce weaned young), but females that were successful had a constant pattern of fecundity. For example, potential litter size did not vary in relation to season, year, age and type of habitat. It is the steady erosion of potential litter-size that is examined here. Mean litter-size was 2.64 at ovulation, 2.33 at birth, and 1.83 at weaning (Fig. 1). Thirteen per cent of all ova shed were lost

TABLE 2. *Summary of reproductive parameters of Ochotonidae.*

Species	Approximate Weight (grams)	Litters per Season	Mean embryo Counts	First Breeding	Source
1) *O. alpina*	130	2	2.2		Revin, 1968
2) *O. alpina*		2–3	3 *	yearling	Khmelevskaya, 1961
3) *O. alpina*		2	2–3**	yearling	Yergenson, 1939‡
4) *O. alpina*	115	1–2	2–6**		Kistchinsky, 1969
5) *O. daurica*		2	7.0	summer of birth	Nekypelov, 1954‡
6) *O. hyperborea*	110	1	4.8	yearling	Kapitonov, 1961
7) *O. macrotis*	180	2–3	5.0	yearling	Zimina, 1962
8) *O. macrotis*		2–4	6.0	summer of birth	Bernstein, 1964
9) *O. pallasi*		2–3	8.0	summer of birth	Shubin, 1956‡
10) *O. pallasi*			5.8	summer of birth	Chergenov, 1961‡
11) *O. pallasi*		3	6.0	summer of birth	Tarasov, 1950‡
12) *O. pusilla*	200	3–5	9.0	summer of birth	Shubin, 1965
13) *O. rufescens*	250		6.0		Puget, 1971
14) *O. rutila*	275	2–3	4.2	yearling	Bernstein, 1964
15) *O. princeps*	135	2	2.3	yearling	Present study
16) *O. princeps*	135	3–4	2.8	yearling	Severaid, 1955

* mode.

** range.

‡ not seen; cited by Bernstein, 1964.

before birth while losses between birth and weaning were estimated at 21% (Millar, 1972*b*). The extent of losses in litters of different size was evaluated by comparing the frequencies of litter-sizes at conception, birth and weaning. Prenatal mortality almost always involved only one embryo per litter, except when whole litters were lost. Assuming losses in all successful litters to involve only one offspring, an expected frequency was calculated by applying a a constant loss to each litter-size at the preceding stage. This expected frequency was then compared statistically (x^2) to the observed frequency. Prenatal, but not postnatal losses were compared directly in relation to initial litter size.

→

FIG. 1. Frequencies of litter-size of pikas in southwestern Alberta at ovulation (based on counts of corpora lutea and corpora albicantia), birth (based on counts of healthy embryos after day 13 of gestation), and weaning (based on counts of young emerging from the rocks).

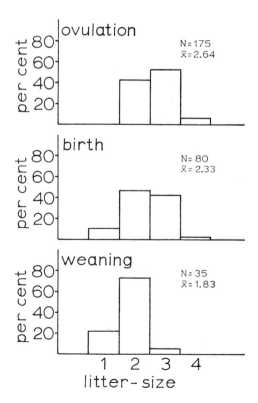

TABLE 3. *Mortality of embryos between conception and birth in relation to initial litter-size at conception based on counts of corpora lutea and corpora albicantia. Observed litter-size at birth based on counts of healthy embryos in late gestation. Expected frequencies at birth based on a uniform 13% loss in all litters during gestation and assumes that only one embryo is lost from any one litter.*

	Litter-size				
	1	2	3	4	Total
A. Observed Frequency at Conception	0	75	89	11	175
B. Percent Frequency at Conception	0	42.9	50.8	6.3	100
C. 13% loss [(13/100) × B]	0	5.6	6.6	0.8	13
D. Remaining (B – C)	0	37.3	44.2	5.5	87
E. Gain From Next Largest Litter (C)	5.6	6.6	0.8	0.0	13
F. Adjusted Frequency (D + E)	5.6	43.9	45.0	5.5	100
G. Expected Frequency at Birth [(F/100) × 80]	4.5	35.1	36.0	4.4	80
H. Observed Frequency at Birth	8	38	33	1	80
x^2	2.7222	0.2396	0.2500	2.6272	5.8390; $P > .100$; N.S.

A comparison of frequency of litter-sizes between conception and birth (Table 3) indicated that the frequency of litters at birth was not significantly different from the frequency expected if mortality was equal among all litter-sizes. However, a direct comparison of prenatal losses in relation to potential litter-size (Table 4) indicated that although there were no differences between females ovulating two and three ova, those producing four ova suffered significantly greater losses than those shedding three. Potential litters of four were uncommon, and heavy losses resulted

in almost no litters of four at birth. Similar differences in rates of mortality were apparent between birth and weaning. The frequency of litters at weaning was significantly different than expected if mortality was equal among all litter sizes at birth (Table 5). Litters of three were less common at weaning than expected, while litters of one and two were more common, indicating that greater mortality occurs in litters of three. Litters of three were common at birth (41% of 80 litters contained three young), but rare at weaning (6% of 35 weaned litters contained three young) indicating that most females could raise only two young. Conceiving more offspring than can be raised may be advantageous in case an offspring is lost for some other reason.

The fate of missing young is not known; only one dead nestling, estimated to be two weeks of age, was found.

TABLE 4. *Prenatal losses in successful pregnancies in relation to initial litter-sizes, based on differences between counts of corpora lutea and healthy embryos after day 13 of gestation (see Millar, 1972b). Tabulated as a percentage of females that have losses and a percentage of ova that are lost. Sample sizes in parentheses.*

Initial Litter-size	Prenatal Losses	
	% Females	% Ova
2	21.2 (33)	10.6 (66)
3	28.2 (39)	10.2 (117)
4	87.5* (8)	28.1** (32)

* Significantly higher than in females shedding 3 ova ($\chi^2 = 6.1871$, $P < .025$).
** Significantly higher than in females shedding 3 ova ($\chi^2 = 5.2825$; $P < .025$).

Index of Fat

The sizes of particular fat bodies are difficult to relate to the condition of animals because different fat bodies may be deposited or mobilized in response to changes in environmental (temperature, food supplies) conditions or physiological status (sex, age, breeding status) (Flux, 1971). The generalized fat index used here

TABLE 5. *Mortality of nestling pikas in relation to litter-size at birth. Observed litter-size at birth based on counts of healthy embryos in late gestation. Observed litter-size at weaning based on counts of young emerging from the rocks. Expected frequencies at weaning based on a uniform loss of 21% in all litters and assumes that only one nestling is lost from any one litter.*

| | Litter-size | | | | |
	1	2	3	4	Total
A. Observed Frequency at Birth	8	38	33	1	80
B. Percent Frequency at Birth	10.0	47.5	41.2	1.3	100
C. 21% Loss [(21/100) × B)]	2.1	10.0	8.6	0.3	21
D. Remaining (B − C)	7.9	37.5	32.6	1.0	79.0
E. Gain From Next Largest Litter (C)	10.0	8.6	0.3	0	18.9
F. Adjusted Frequency (D + E)	17.9	46.1	32.9	1.0	97.9
G. Adjusted Percent Frequency [(F/97.9) × 100]	18.3	47.1	33.6	1.0	100
H. Expected Frequency at Weaning [(G/100) × 35]	6.4	16.5	11.8	0.3	35
I. Observed Frequency at Weaning	8	25	2	0	35
χ^2	0.4000	4.3787	8.1389	0.3000	13.2176; $P < .005$

(mg cardiac, splenic, interscapular fat per 100 gm body weight) presumably reflects general fat levels. Fat animals are not necessarily healthy animals, but fat animals must be obtaining, or at least storing, more energy in relation to their requirements than thin animals. In the pika, females enter the breeding season with much higher fat indexes than males, but lose these reserves over the breeding season. A comparison of fat reserves in relation to pregnancy and lactation (Fig. 2) indicates that fat is deposited during pregnancy and drained during lactation. These drains occurred despite the presence of abundant food during the period of lactation, and did not vary in relation to littering period or type of habitat.

Population Density

This parameter is difficult to estimate, but pikas in southwestern Alberta exhibit several characteristics that indicate they are at or near saturation level.

For instance, overall population levels were relatively stable over the three years of the study. Although individual populations varied considerably (Table 6), yearly averages of adults per ha of rock were 7.5 in 1968, 5.8 in 1969 and 6.5 in 1970 (based on 0.8, 8.5 and 8.5 ha, respectively). Secondly, pikas appeared relatively immune to variations in environmental conditions; no "catastrophes" were recorded. Thirdly, although few populations were dense enough to exhibit responses to density, low populations produced more offspring per female to weaning than high populations, and any

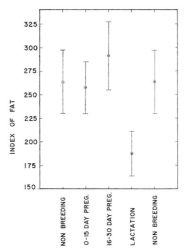

FIG. 2. Mean index of fat (mg interscapular, cardiac and splenic fat per 100 gm body wt) of mature female pikas in southwestern Alberta in relation to stages of pregnancy and lactation. Vertical lines denote two standard errors each side of the mean.

TABLE 6. *Population parameters for five discrete populations of pikas in southwestern Alberta. Density based on number of adult pikas per hectare of rock slide. Number of weaned young based on counts of young animals emerging from the rocks. Emigrants include only animals born in the populations and whose final location was known. Transients and immigrants include all young animals caught on the area that were known not to be born there.*

Popu-lations	Density	No. Ad. ♀ ♀	No. Weaned Young	No. Weaned Young per Female	No. Known Emigrants	No. Known Transients and Immigrants
1	4.0	5	9	1.80	3	1
2	4.1	6	14	2.33	–	1
3	5.4	6	14	2.33	3	7
4	10.9	12	13	1.08	–	–
5	22.0	6	2	0.33	–	1

populations that were still low after the breeding season could easily be replenished with immigrants (Table 6). These data indicate that the populations studied were probably at or near saturation level at all times.

DISCUSSION

Each female appears to be producing as many offspring as she can support. Females drain their energy reserves during lactation and large litters (three) appear to suffer higher mortality prior to independence than small litters (one and two). This limit of two offspring supported during lactation could be considered a local phenomena, but data from Colorado and Utah indicate that the same limit exists there. Counts of embryos from that area (Table 1) indicate that many females (16 of 26) gave birth to three or four young, while Krear (1965) noted that the number weaned was usually two.

The limit imposed on litter-size during lactation did not appear related to environmental conditions since weather was relatively mild and food was abundant during the breeding season. The limit was more likely physiological; possibly based on one or more of three factors: energy assimilation, drainage of energy reserves, and the rate of growth of the young. Small mammals such as bank voles (*Clethrionomys glareolus*), common voles (*Microtus arvalis*), mice (*Mus musculus*) and red squirrels (*Tamiasciurus hudsonicus*) increase their food intake during pregnancy and

lactation (Kaczmarski, 1966; Migula, 1968; Myrcha et al., 1969; Smith, 1968, respectively) and pikas likely do the same. Pikas also drain their reserves during lactation and apparently cannot support offspring through increased assimilation alone. Possibly, rather than draining their reserves below a critical level, or decreasing the size of each offspring, they sacrifice the size of the litter. The difference in energy expenditure between a female (mean weight 133 gm) supporting two and three offspring (weaning weight approximately 50 gm) would be considerable.

Few studies were found where survival of young or condition of females was documented for natural populations. Le Resche (1968) found greater mortality of twin moose calves (*Alces alces*) than single calves, and data collected by Markgren (1969) indicates a similar trend. Tree mice (*Phenacomys longicaudus*) support only a certain biomass of offspring, even when fed an abundance of natural foods (Hamilton, 1962) and young rabbits (*Oryctolagus cuniculus*) in small litters grow faster than those in large litters (Myers and Poole, 1963), indicating that milk is limited.

In general, the differential mortality in litters of different size during lactation appear to be related to the female's ability to support offspring. Prenatal mortality, however, occurred early in gestation when embryos were very small in relation to the size of the female, and mortality at that time was not likely related to any nutritional stress on the female. A possible explanation is that litter-size has been reduced over evolutionary time by limiting the capacity of the uterus, rather than by reducing the number of ova shed. This could operate by limiting the number of embryos implanted, as in the elephant shrew (*Elephantulus myurus*) (Horst and Gillman, 1941) or through limiting the number of implanted embryos carried to term, as in the alpaca (*Lama pecos*) (Fernandex-Baca et al., 1970). Although most evolved reductions in litter-size undoubtedly arise through changes in ovulation rates, some sort of restriction placed on litter-size by the uterus may be relatively common. Greater prenatal losses in large litters have been noted in white-tailed deer (*Odocoileus virginianus*) (Ransom, 1967) and European rabbits (*Oryctolagus cuniculus*) (Poole, 1960), and the trend is evident in other data presented for European rabbits (Brambell, 1942; Lloyd, 1963; McIlwaine, 1962), European hares (*Lepus europaeus*) (Flux, 1967) and snowshoe hares (*Lepus americanus*) (Newson, 1964).

Lack's (1948) hypothesis that females produce as many offspring as they can support appears to hold true for the pika. However, his suggestion that the most frequent litter-size at birth is the most productive is not supported by my data, and has been criticized on theoretical grounds by Mountford (1968). Litters of three were common at birth, but were generally reduced to two young at weaning. Such a system may appear inefficient, but the "wastage" of one offspring would be selected against only if the wastage contributed in some way to the death of the parent or surviving offspring. Perhaps, in pikas, there is an advantage in having an extra nestling available in case one is lost for some other reason before the critical period of lactation.

Spencer and Steinhoff (1968) suggested that animals with restricted breeding seasons have larger litters than those with extended breeding seasons to maximize the number of offspring produced during the

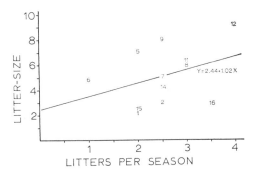

FIG. 3. Regression of mean litter-size against number of litters per breeding season of several species of pikas. Numbers refer to data in Table 2. A significant proportion of the variance of y is not explained by regression on x. ($F = 1.4289$; $P > .25$.)

females life span. From this, populations with short breeding seasons (hence fewer litters per year) should have the largest litters. A comparison of pika populations indicates no such trend (Fig. 3), while there is a trend (not significant) for large litters to be associated with long breeding seasons. Their assumption that maternal mortality varies directly with the size of the litter may be invalid (Tinkle, 1969). All behavioral or physiological attributes of a species can be related to the ability to survive or to reproduce. Since survival is a prerequisite to reproduction, it is likely that requirements for survival take priority over requirements for reproduction, and that females will sacrifice their litters before threatening their own survival. In the pika, this is seen as a loss of total litters at high population densities and under other adverse conditions (Millar, 1972*b*).

The trend in litter-size among species may be attributed, at least in part, to body size. Species with large litters are generally larger than those with small litters (Table 2).

Lord (1960) predicted a positive correlation between litter-size and latitude in non-hibernating prey species. This is not true of the pika; litter-size is relatively constant in North America (Table 1). My data do not, however, refute his suggestion

TABLE 7. *Reproduction and adult mortality in populations of pikas.*

Population	Adult Mortality*	Litter-size	Litters/ season	Sources
O. princeps (Colorado and Utah)	43.4	2.9	2	Present study; Johnson, 1967; Hayward, 1952
O. princeps Alberta	46.0	2.3	2	Present study
O. macrotis	58.2	6.0	2–4	Bernstein, 1964
O. rutila	64.8	4.2	2–3	Bernstein, 1964

* Based on per cent yearlings in mature populations (see Millar and Zwickel, 1972a).

that populations suffering low mortality produce relatively small litters because resources are limited. Mortality in Alberta was relatively low and litters were small (Table 7). The problem of determining the limited resource remains. Gibb (1968) considered food to be the important resource, but pikas are herbivores that feed on a wide variety of plants, and food appeared abundant in Alberta.

Cody (1966) suggested that successful offspring in saturated populations must be larger in order to be competitive. This may explain the partitioning of available maternal resources into only two offspring (rather than three or four smaller offspring). Perhaps the resource in question is space. Pikas are territorial (Kilham, 1958; Krear, 1965) and although detailed behavioral studies have not yet been done, competition for space likely occurs.

SUMMARY

Potential litter-sizes, mortality of embryos and nestlings, maternal fat reserves during pregnancy and lactation, and population parameters of pikas in southwestern Alberta are documented and used to evaluate several hypotheses on the significance of litter-size. Two, three, or sometimes four ova are shed at conception. Relatively heavy losses in litters of four occur during gestation. These losses are not likely caused by any nutritonal stress on females, and may be related to an evolutionary reduction in litter-size by limiting the capacity of the uterus. Relatively heavy losses in litters of three between birth and weaning, and low maternal fat reserves during lactation indicate that females produce as many offspring as they can support. Population appeared at or near saturation level, and the advantage in partitioning maternal resources into two, rather than three or four smaller offspring may be that the better nourished offspring are more successful at competing for space.

ACKNOWLEDGMENTS

This study was conducted under the supervision of Dr. F. C. Zwickel and was supported by funds from the National Research Council of Canada and the Department of Zoology and the R. B. Miller Biological Station, University of Alberta. R. A. MacArthur and H. Reynolds provided assistance during 1969 and 1970, respectively. Special thanks are due to F. C. Zwickel, J. O. Murie, S. C. Tapper, J. P. Ryder, A. E. Aubin and C. D. Ankney, for many valuable comments, suggestions, criticisms and heated arguments.

LITERATURE CITED

ANDERSON, S. 1959. Mammals of the Grand Mesa, Colorado. Univ. Kansas Publs. Mus. Nat. Hist. 9:405–414.

BAILEY, V. 1936. The mammals and life zones of Oregon. North Amer. Fauna No. 55:112–117.

BERNSTEIN, A. D. 1964. The reproduction by red pika (*Ochotona rutila* Sev.) in the Zailijsk Alutau (In Russian). Bull. Mosc. Soc. Nat., Biol. 69:40–48.

BRAMBELL, F. W. R. 1942. Intra-uterine mortality of the wild rabbit. *Oryctolagus cuniculus* (L.). Proc. Roy. Soc. B. 130:462–479.

CODY, M. L. 1966. A general theory of clutch size. Evolution 20:174–184.

DICE, L. R. 1926. Pacific coast rabbits and pikas. Occ. Papers Mus. Zool., Univ. Mich. 166:1–28.

———. 1927. The Colorado pika in captivity. J. Mammal. 8:228–231.

DIXON, J. S. 1938. Birds and Mammals of

Mount McKinley National Park, Alaska. Nat. Park Service, Fauna Ser. 3, 236 p.

FERNANDEZ-BACA, S., W. HANSEL, AND C. NOVOA. 1970. Embryonic mortality in the alpaca. Biol. Reprod. 3:243–251.

FLUX, J. E. C. 1967. Reproduction and body weights of the hare, *Lepus europaeus* Pallas, in New Zealand. N.Z.J. Sci. 10:357–401.

——. 1971. Validity of the kidney fat index for estimating the condition of hares: a discussion. N.Z.J. Sci. 14:238–244.

GIBB, J. A. 1968. The evolution of reproductive rates: are there no rules? Proc. N.Z. Ecol. Soc. 15:1–6.

GRINNEL, J., J. DIXON, AND J. M. LINSDALE. 1930. Vertebrate natural history of a section of northern California through the Lassen Peak region. Univ. Cal. Publ. Zool. 35:1–594.

HALL, E. R. 1946. Mammals of Nevada. Univ. Calif. Press. 710 p.

HAMILTON, W. J. 1962. Reproductive adaptations in the red tree mouse. J. Mammal. 43:486–504.

HAYWARD, C. L. 1952. Alpine biotic communities of the Uinta Mountains, Utah. Ecol. Monogr. 22:93–102.

HORST, C. J. V. D., AND J. GILLMAN. 1941. The number of eggs and surviving embryos in *Elephantulus*. Anat. Rec. 80:443–452.

JOHNSON, D. R. 1967. Diet and reproduction of Colorado pikas. J. Mammal. 48:311–315.

KACZMARSKI, F. 1966. Bioenergetics of pregnancy and lactation in the bank vole. Acta Theriol. 11:409–417.

KAPITONOV, V. I. 1961. Ecological observations on *Ochotona hyperborea* Pall. in the lower part of the Lena River. (In Russian; English summary). Zool. Zh. 40:922–933.

KHMELEVSKAYA, N. V. 1961. On the biology of *Ochotona alpina* Pallas. (In Russian; English summary). Zool. Zh. 40:1583–1585.

KILHAM, L. 1958. Territorial behavior in pikas. J. Mammal. 39:307.

KISTSCHINSKY, A. A. 1969. The pika (*Ochotona alpina hyperborea* Pall.) in the Kolyma highlands. (In Russian; English summary). Bull. Mosc. Soc. Nat. Biol. 74:134–143.

KREAR, H. R. 1965. An ecological and ethological study of the pika (*Ochotona princeps saxitilis* Bangs) in the Front range of Colorado. Ph.D. Thesis, Univ. of Colorado, Boulder. 329 p.

LACK, D. 1948. The significance of litter-size. J. Anim. Ecol. 17:45–50.

LE RESCHE, R. E. 1968. Spring-fall calf mortality in an Alaska moose population. J. Wildlife Manage. 32:953–956.

LLOYD, H. G. 1963. Intra-uterine mortality in the wild rabbit, *Oryctolagus cuniculus* (L.) in populations of low density. J. Anim. Ecol. 32:549–563.

LONG, W. S. 1940. Life histories of some Utah mammals. J. Mammal. 21:170–180.

LORD, R. D. 1960. Litter-size and latitude in North American mammals. Am. Midl. Nat. 64:488–499.

MARKGREN, G. 1969. Reproduction of moose in Sweden. Viltrevy 6:127–299.

McILWAINE, C. P. 1962. Reproduction and body weights of the wild rabbit *Oryctolagus cuniculus* (L.) in Hawke's Bay, New Zealand. N.Z.J. Sci. 5:324–341.

MIGULA, P. 1969. Bioenergetics of pregnancy and lactation in the European common vole. Acta Theriol. 14:167–179.

MILLAR, J. S. 1972a. Timing of breeding of pikas in southwestern Alberta. Can. J. Zool. 50:665–669.

——. Success of reproduction in pikas (*Ochotona princeps* Richardson) Fecudity of pikas in relation to the environment. (in preparation).

MILLAR, J. S., AND F. C. ZWICKEL. 1972a. Determination of age, age structure, and mortality of the pika, *Ochotona princeps* (Richardson). Can. J. Zool. 50:229–232.

——, AND ——. Characteristics and ecological significance of hay piles of pikas. Mammalia (*in press*).

MOUNTFORD, M. D. 1968. The significance of litter-size. J. Anim. Ecol. 37:363–367.

MYERS, K., AND W. E. POOLE. 1963. A study of the wild rabbit *Oryctolagus cuniculus* (L.) in confined populations. V. Population dynamics. C.S.I.R.O. Wildlife Research. 8:166–203.

MYRCHA, A., L. RYSZKOWSKI, AND W. WALKOWA. 1969. Bioenergetics of pregnancy and lactation in white mouse. Acta. Theriol. 14:161–166.

NEWSON, J. 1964. Reproduction and prenatal mortality of snowshoe hares on Manitoulin Island, Ontario. Can. J. Zool. 42:987–1005.

POOLE, W. E. 1960. Breeding of the wild rabbit, *Oryctolagus cuniculus* (L.) in relation to the environment. C.S.I.R.O. Wildlife Research 5:21–43.

PUGET, A. 1971. *Ochotona r. rufescens* (Gray 1842) in Afghanistan and its breeding in captivity. (In French; English summary). Mammalia 35:24–37.

RANSOM, A. B. 1967. Reproductive biology of white-tailed deer in Manitoba. J. Wildl. Manage. 31:114–123.

RAUSCH, R. L. 1961. Notes on the collared pika, *Ochotona collaris* (Nelson), in Alaska. Murrelet. 42:22–24.

——. 1970. Personal communication, College, Alaska.

REVIN, Y. N. 1968. A contribution to the biology of the northern pika (*Ochotona alpina* Pall.) on the Olekmo-Charskoe highlands (Yukatia). (In Russian; English summary). Zool. Zh. 47:1075–1082.

ROEST, R. I. 1953. Notes on pikas from the Oregon Cascades. J. Mammal. 34:132–133.

SEVERAID, J. H. 1955. The natural history of the pika (Mammalian genus *Ochotona*). Ph.D. Thesis, Univ. of Calif. 820 p.

SHUBIN, I. G. 1965. Reproduction of *Ochotona pusilla* Pall. (In Russian; English summary). Zool. Zh. 44:917–924.

SPENCER, A. W., AND H. W. STEINHOFF. 1968. An explanation of geographical variation in litter-size. J. Mammal. 49:281–286.

SMITH, C. C. 1968. The adaptive nature of social organization in the genus of three squirrels *Tamiasciurus*. Ecol. Monographs 38:31–63.

SMITH, M. H., AND J. T. McGINNIS. 1968. Relationships of latitude, altitude, and body size to litter size and mean annual production of offspring in *Peromyscus*. Res. Popul. Ecol. X:115–126.

TINKLE, D. W. 1969. The concept of reproductive effort and its relation to the evolution of life histories of lizards. Amer. Natur. 103: 501–516.

UNDERHILL, J. E. 1962. Notes on pika in captivity. Can. Field Nat. 76:177–178.

ZIMINA, R. P. 1962. The ecology of *Ochotona macrotis* Gunther dwelling in the area of the Tersky-Alutau mountain range. (In Russian; English summary). Bull. Mosc. Soc. Nat., Biol. 67:5–12.

Sonderdruck aus Z. f. Säugetierkunde Bd. 30 (1965), H. 1, S. 10—20

Alle Rechte, auch die der Übersetzung, des Nachdrucks und der photomechanischen Wiedergabe, vorbehalten.
VERLAG PAUL PAREY · HAMBURG 1 · SPITALERSTRASSE 12

The effects of suckling on normal and delayed cycles of reproduction in the Red Kangaroo

By G. B. Sharman

Eingang des Ms. 23. 12. 1963

Introduction

In non-lactating female marsupials the occurrence of fertilization, followed by immediate gestation of the embryo, does not delay the onset of the following oestrus. In those marsupials in which the gestation period is considerably shorter than the length of one oestrous cycle, such as *Didelphis virginiana* (Hartman, 1923) and *Trichosurus vulpecula* (Pilton and Sharman, 1962), oestrus recurs at the expected time if the young are removed at birth. In several species of Macropodidae, such as *Setonix brachyurus* (Sharman, 1955), *Potorous tridactylus* (Hughes, 1962) and the Red Kangaroo (Sharman and Calaby, 1964), the gestation period occupies almost the length of one oestrous cycle and oestrus is imminent at the time of parturition. Oestrus thus recurs just after the young reach the pouch (post-partum oestrus) presumably because pro-oestrus changes are initiated before the onset of the suckling stimulus. In all marsupials suckling of young in the pouch is accompanied by a lengthy period during which oestrus does not occur. This period is called the quiescent phase of lactation or, simply, the quiescent phase. It differs from seasonal anoestrus in that the ovaries and other reproductive organs respond to the removal of the suckling stimulus by resuming cyclic functions. Those marsupials in which post-partum oestrus occurs exhibit discontinuous embryonic development analogous to the delayed implantation which occurs in some eutherian mammals. If fertilization takes place at post-partum oestrus the resulting embryo assumes a dormant phase, at the blastocyst stage, and is retained as a dormant blastocyst during the quiescent phase. In these marsupials pregnancy (the interval between copulation at post-partum oestrus and parturition) is long and gestation of the embryo is interrupted by the dormant phase.

In the Red Kangaroo, *Megaleia rufa* (Desm.), the oestrous cycle averages 34 to 35 days and the gestation period is 33 days in length (Sharman and Calaby, 1964). Postpartum oestrus occurs, usually less than 2 days after the newborn young reaches the pouch, and a dormant blastocyst is found in the uterus of females, fertilized at postpartum oestrus, which are suckling young less than 200 days old in the pouch (Sharman, 1963). If the young is removed from the pouch suckling ceases and the dormant blastocyst resumes development: the young derived from it being born about 32 days after removal of the pouch young (RPY). This birth is followed by another postpartum oestrus or, if the female was not carrying a blastocyst, by a normal oestrus. Oestrus recurs at the same number of days after RPY irrespective of whether a delayed blastocyst was carried or not. The sequence of events from RPY to the next oestrus is called the delayed cycle of reproduction[1] to distinguish it from the normal reproductive cycle which follows oestrus. The delayed reproductive cycle may be divided into delayed gestation and delayed oestrus cycle according to whether a dor-

---

[1] The term "delayed cycle of reproduction" or "delayed (reproductive) cycle", was introduced by Tyndale-Biscoe (1963) to describe the resumption of ovarian activity, and the features associated with it, following removal of pouch young (RPY).

mant blastocyst does or does not complete development. If the young is retained in the pouch until it leaves in the normal course of events the delayed reproductive cycle occurs coincident with the latter stages of pouch life. The dormant phase of the blastocyst gives way to renewed development when the pouch young is a little over 200 days old and subsequent vacation of the pouch, at an average age of 235 days, is immediately followed by birth of another young (SHARMAN and CALABY, 1964). The young is suckled for another 130 days, that is until it is about a year old, after it leaves the pouch. During this period the normal reproductive cycle occurs if the pouch is not occupied. It is thus evident that, although the delayed reproductive cycle occurs after RPY and cessation af lactation, some factor other than the actual production of milk must be implicated for both delayed and normal cycles may also occur during lactation.

The aim of the experiments reported below was to determine the effect of the suckling stimulus on both normal and delayed reproductive cycles. Additional suckling stimulus was provided by fostering an extra young on to females already suckling a young-at-foot. The experimental approach was suggested by chance observations on a female Red Kangaroo which, while suckling her own young-at-foot, alternately fed the young of another female kept in the same enclosure. There are four teats in the pouch but the teat to which the young attaches after birth alone produces milk and its underlying mammary gland produces all the milk for the young from birth to weaning. The female's own young and the foster-young thus shared the products of a single mammary gland and used the same teat alternately. Some initial results, in so far as they were relevant to the theme of delayed implantation, were reported earlier in a review of that subject (SHARMAN, 1963).

Methods

The results presented consist of observations on a minimum of five reproductive cycles in the female Red Kangaroo in each of the following categories:

1. Normal cycle of reproduction, suckling one young-at-foot.
2. Normal cycle of reproduction, suckling two young-at-foot.
3. Delayed cycle of reproduction, suckling one young-at-foot.
4. Delayed cycle of reproduction, suckling two young-at-foot.

The results are compared with data on the normal and delayed cycles of reproduction in non-lactating females most of which have been published elsewhere (SHARMAN, 1963; SHARMAN and CALABY, 1964; SHARMAN and PILTON, 1964). In most cases the experimental females were pregnant or carrying dormant blastocysts so that cycles of normal or delayed gestation with subsequent post-partum oestrus were studied. The gestation periods and cycles were regarded as having been significantly lengthened when they occupied a time greater by the length of two, or more, standard deviations than similar cycles in control, non-lactating, females

Some difficulty was experienced in getting females to accept foster-young and only six females readily did so. The experiments were therefore done serially one female being used in two and two females in three experiments.

The animals were watched from a hide overlooking the enclosures and observed with binoculars. An initial watch was always done to find whether females accepted their potential foster-young. Thereafter prolonged watches were kept on some females to determine the amount of time spent suckling the young-at-foot.

Vaginal smears for the detection of oestrus and copulation were taken as reported previously (SHARMAN and CALABY, 1964).

Results

Effects of suckling on the normal cycle of reproduction

In thirteen non-lactating female Red Kangaroos forty-two intervals from oestrus to
the succeeding oestrus averaged 34.64 days with a standard deviation of 2.22 days
(34.64 ± 2.22 days). Twenty gestation periods in fourteen females lasted 33.00 ± 0.32
days (Fig. 1A). In five females, each observed for a single reproductive cycle while
suckling one young-at-foot, the intervals between two successive oestrous periods were
not different from those in non-lactating females (Fig. 1B). In another female (K32a)

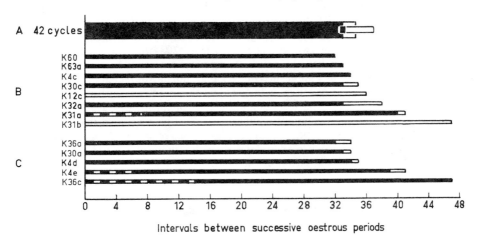

Fig. 1. Intervals between successive oestrous periods in nonlactating (control) female Red
Kangaroos (A), females suckling one young-at-foot (B) and females suckling two young-at-
foot (C). Black lines — continuous embryonic development, broken lines — approximate
periods of dormant phase in embryo induced and maintained by suckling young-at-foot, open
lines — no embryos present, bars inserted in A — standard deviations either side of mean.

the gestation period was not significantly different from that of control females but
oestrus did not occur until 5 days post-partum. This was the longest interval between
parturition and post-partum oestrus recorded but it is not regarded as significant. Two
cycles in female K31, one lasting 41 days and one 47 days, were abnormally long. The
41-day cycle is of special significance since the interval between copulation and birth
was 40 days. This differs so much from the gestation period in the control, non-lactating, females that it must be assumed that suckling of the single young-at-foot induced
a short quiescent phase in the uterus accompanied by a dormant phase of about 7 days
in the embryo. The 47-day cycle was over 12 days longer than the mean normal cycle
length and 7 days longer than the maximum cycle length. The female copulated at
oestrus but did not give birth so it is presumed that fertilization did not occur.

In three females already suckling one young-at-foot, which had another young-at-
foot fostered on to them at about the time of fertilization, the lengths of the reproductive cycles were not significantly different from those in control females. Two
females had significantly longer cycles than in control females. One of these (K36) was
used in three successive experiments while suckling the same two young-at-foot. In the
first of these (K36a) the extra suckling stimulus had no significant effect on the length
of the reproductive cycle. The second experiment concerned the delayed reproductive
cycle and is reported below. During the third experiment (K36c) the young were being
weaned but a highly significant result was obtained. The interval from copulation to

birth showed conclusively that a dormant phase had been induced and maintained in the embryo for about 14 days of the 47-day pregnancy. In the other female in which the cycle was prolonged (K4e) the embryo presumably had a dormant phase of about 6 days.

Effects of suckling on the delayed cycle of reproduction

In ten non-lactating females thirteen intervals from RPY to the succeeding oestrus were 34.46 ± 1.92 days. In seven of these females the delayed gestation period was 31.64 ± 0.65 days (Fig. 2A). There was no evidence that suckling one young-at-foot had any effect on the length of the delayed reproductive cycle (Fig. 2B). In one female (K12a) the interval from RPY to the following oestrus was 38 days but this falls short of the minimum interval accepted as significantly different.

All six females suckling two young-at-foot (Fig. 2C) were carrying a dormant blastocyst in the uterus when the pouch young were removed. In five of these the interval RPY to birth was significantly longer than in control females (Fig. 2C). The interval RPY to the next oestrus was longer than the mean for control non-lactating females in all six experimental females and in three of them (K4b, K30b, K36b) the difference from controls was highly significant. It must be concluded that the blastocysts of five of the above experimental females remained in the dormant phase for between 3 and 22 days longer after RPY than did those of control non-lactating females and females suckling one young-at-foot.

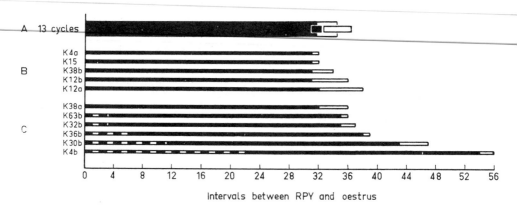

Fig. 2. Intervals between removal of pouch young (RPY) and the next oestrus in non-lactating (control) female Red Kangaroos (A), females suckling one young-at-foot (B) and females suckling two young-at-foot (C). Black lines — continuous embryonic development, broken lines — approximate periods of continued dormant phase of embryo maintained by suckling young-at-foot, bars inserted in A — standard deviations either side of mean.

The amount suckling in relation to occurrence of parturition and return to oestrus

Observations on the habits of the pouch young suggested that the stimulus causing withholding of the mother's reproductive cycles might be tactile and received via the teat. The young during the early stages of pouch life, when reproductive cycles were withheld, were suckled continuously and could not regain the teat if removed before the age of 6 weeks. Later young were able to take the teat back into their mouths but were seldom found free of the teat before the age of about 5 months. On the other

Table 1

Effects of suckling one and two young-at-foot on subsequent parturition and oestrus in Red Kangaroos

No. of female	Ages of young (days) Own young	Foster-young	No. of hours observed	Minutes of suckling Own young	Foster-young	Per day	Type of cycle	Occurrence of parturition	Occurrence of oestrus
K 60	255	—	13,5	44	—	78	normal	when expected	when expected
K 31 (a)	255	—	15,5	31	—	48	normal	late	late
K 31 (b)	269	—	96	203	—	51	normal	no young born	late
K 30 (c)	288	—	13,5	38	—	68	normal	when expected	when expected
All females suckling 1 young			138,5	316	—	53			
K 4 (e)	259	245	22,5	86	55	150	normal	late	late
K 63 (b)	309	309	15.5	27	31	90	delayed	late	when expected
K 36 (b)	322	315	78	192	83	85	delayed	late	late
All females suckling 2 young			116	305	169	98			

hand the pouch young present when the delayed reproductive cycle occurred apparently frequently released the teat as they were seen protruding their heads from the pouch to feed from the ground or leaving the pouch entirely (SHARMAN and CALABY, 1964).

Theoretically it was to be expected that if reproductive cycles resumed in response to a lowered suckling stimulus, as they did during the terminal stages of pouch feeding, then the cycles which occurred as soon as the young left the pouch should have been of normal length. Six of the eight cycles shown in Fig. 1B were the first which occurred after termination of pouch feeding. Four were of normal length but two cycles in one female (K 31 a, b) were lengthened by a significant amount. Observations on the habits of the young, just after they left the pouch permanently, showed that they frequently attempted to regain the pouch but were restrained from doing so by their mothers (SHARMAN and CALABY, 1964). In these cases they spent long periods with their heads in the pouch during which time they may have grasped the teat. It is also possible that the young, subjected permanently for the first time to the cooler environment outside the pouch, fed more frequently than they did during the terminal stages of pouch life. This would result in a greater suckling stimulus being exerted: at least during the initial stages of life outside the pouch.

A number of females suckling one or two young-at-foot were watched continuously for varying periods and the amounts of time spent suckling were recorded (Table 1). It was at once apparent that females feeding two young-at-foot spent nearly twice as much time suckling as did females with a single young-at-foot. The relationship between amount of suckling and interruption or resumption of the reproductive cycle is, however, not so obvious. Thus, in female K31, 48 and 51 minutes of suckling per day were associated with lengthening of the interval between successive oestrous periods and 48 minutes per day with inducing and maintaining a short dormant phase in the embryo. In two other females (K60, K30c) a greater amount of suckling apparently had no effect on the length of the cycle or on pregnancy. However, although the watches were done during the relevant cycles, they were not necessarily done at the critical period of the cycle when the suckling stimulus exerted its effect. This period could not be ascertained since no evidence of its occurrence was available until the females gave birth or returned to oestrus. The figures in Table 1 are thus to be regarded as no more than a guide to the amount of suckling which occurred at the critical period.

The most conclusive evidence about the effect of the suckling stimulus on the reproductive cycle came from the females from which pouch young were removed while they were suckling two young-at-foot (Fig. 2C). In one of these females (K38a) the suckling of two young-at-foot was without effect on the delayed reproductive cycle; in three (K32b, K36b, K63b) the delayed cycle began while two young were being suckled but in two others (K4b, K30b) the delayed cycle was only initiated when one of the suckling young-at-foot was removed. The interval from removal of the young-at-foot to completion of the delayed cycle was approximately the same (31—32 days) as from RPY to the completion of the cycle in the control females.

The two intervals between successive oestrous periods with intervening pregnancies which were observed in the same female (K36a, c) while suckling the same two young-at-foot call for some comment. Parturition and return to oestrus occurred when expected in the first cycle but were delayed significantly in a subsequent cycle when the young were much older and were being weaned (Fig. 1C). During this, latter, cycle one of the young frequently grasped the teat for periods of 10 minutes or more but when the female's pouch was examined it was found that no milk could be expressed from the teat and that the mammary gland was regressing. This was in contrast to the condition in other females suckling young-at-foot in which milk could usually be readily expressed. No watch was done to observe the amount of time the young spent sucking the dry teat as the significance of the observation was only realised after completion of the cycle. This cycle is, however, of particular significance because it appears likely that the suckling stimulus, in the absence of lactation, induced a quiescent phase in the uterus lasting some 14 days and a corresponding period of dormancy in the blastocyst.

Discussion

Delayed implantation in the Red Kangaroo is of the type usually referred to as lactation controlled delayed implantation. This description is adequate in so far as the delayed cycle of reproduction is initiated following removal of the pouch young and cessation of lactation. However, the delayed cycle also occurs during the seventh and eighth months of the 12-month lactation period. It therefore follows that, in these cases, the delayed cycle does not begin in response to the cessation of lactation or to the imminent cessation of lactation. The quiescent phase of lactation with asso-

ciated arrested development of the embryo is initiated during the early part of lactation while a small young is suckled continuously in the pouch but the normal reproductive cycle may, as has been shown above, occur during the latter part of lactation. It is thus much more likely that the amount of suckling stimulus which the
female receives at various phases of the lactation period is of paramount importance
in determining whether the normal reproductive cycle shall be interrupted or whether
the delayed cycle shall be initiated. The experiments reported above have shown that
in some females the normal cycle is interrupted and a quiescent phase of lactation,
with associated dormant phase of the embryo is induced by increasing the suckling
stimulus. It has also been shown that the stimulus of suckling of young, outside the
pouch, is capable of prolonging the quiescent phase of lactation and dormant phase
of the embryo.

Two other factors could be of importance in determining the time of onset of the
delayed cycle of reproduction: 1. Temporary or permanent vacation of the pouch.
2. Fall in milk yield. Temporary emergence from the pouch first occurs when the
young are less than 190 days old and permanent emergence at the average age of
235 days — that is a few days before the completion of the delayed cycle (SHARMAN
and CALABY, 1964) but the delayed cycle apparently begins when the young are a
little over 200 days old. Precise data on this point are difficult to obtain but assuming that the delayed cycle, once initiated, proceeds at the same rate in lactating
females as it does in females from which the pouch young are removed then it must
begin about 30 days before the young leaves the pouch. This is in agreement with the
massive amount of data obtained from Red Kangaroos taken in the field. The onset
of the delayed cycle can hardly occur in response to a fall in milk yield since it takes
place when the young is actively growing and when it is increasing rapidly in weight.
From the age of 200 days to the age of 220 days, during which period the delayed
cycle is resumed, the pouch young increase from about 2.5 to 3.5 kg in weight which
is not the expected result of a fall in milk yield. Furthermore removal of young from
the pouches of females which were suckling two young-at-foot must have been accompanied by a fall in milk yield yet under these circumstances the quiescent phase
of lactation with associated dormant blastocyst continued in five of six females
(Fig. 2C).

The importance of the suckling stimulus in marsupial reproduction was demonstrated by SHARMAN (1962) and SHARMAN and CALABY (1964) who transferred newborn young *Trichosurus vulpecula* and *Megaleia rufa* to the pouches or teats of non-
lactating, non-mated or virgin females of each of these species at the appropriate
number of days after oestrus. The suckling stimulus exerted by the young induced
the onset of lactation without the prior occurrence of pregnancy and oestrous cycles
were withheld while the foster-young were suckled in the pouch. SHARMAN and CA-
LABY (1964) were unable to demonstrate any behavioural differences between pregnant and non-mated female Red Kangaroos at the same number of days after oestrus
except that pregnant females repeatedly cleaned their pouches just before giving birth.
Other authors (HILL and O'DONOGHUE, 1913; HARTMAN, 1923; SHARMAN, 1955; PIL-
TON and SHARMAN, 1962) have drawn attention to the remarkable resemblances of
post-oestrous changes in pregnant females to those of non-mated females in various
species of marsupials. It is apparent, that whereas in polyoestrous eutherian mammals
hormones produced by the embryonic membranes modify the reproductive cycle and
prevent the recurrence of oestrus during pregnancy, no such mechanism has yet been
demonstrated in any marsupial. In those marsupials which do not have a seasonal
anoestrous period, such as the Red Kangaroo, the reproductive cycle is continuous
except when interrupted by the quiescent phase of lactation.

OWEN (1839—47) determined the gestation period (interval from mating to birth) of a lactating female Great Grey Kangaroo as 38–39 days. HEDIGER (1958) stated that K. H. WINKELSTRÄTER and E. CRISTEN in Zurich Zoo found gestation periods of 30 and 46 days in the same species and later, in the same paper, stated that a young was born on the forty-sixth day after mating in a lactating female Great Grey Kangaroo. However the dates quoted by HEDIGER show that the „gestation period" was actually 57 days. In non-lactating Great Grey Kangaroos Miss PHYLLIS PILTON (pers. comm.) found the gestation period was about 30 days and in the C.S.I.R.O. Division of Wildlife Research four gestation periods in three non-lactating females were 33 days 6 hours to 34 days 6 hours, 33 days 18 hours to 34 days 10 hours, 34 days to 34 days 17 hours and 34 days to 34 days 20 hours. It is apparent that, although the Great Grey Kangaroo does not have the same type of lactation controlled delayed implantation as occurs in the Red Kangaroo and other marsupials (SHARMAN, 1963), intervals between mating and birth in lactating females may be an unreliable guide to the gestation period. HEDIGER (1958) stated that exact gestation periods in kangaroos and other marsupials are difficult to determine because ovulation occurs several days after mating and spermatozoa can remain active in the oviduct for long periods. This may be true of the marsupial *Dasyurus viverrinus*, but HILL and O'DONOGHUE's (1913) work on this species has not been repeated and confirmed. Delayed ovulation and storage of spermatozoa do not occur in *Didelphis* (HARTMAN, 1923), *Setonix* (SHARMAN, 1955) or *Trichosurus* (PILTON and SHARMAN, 1962) and gestation periods in non-lactating females of these species can be determined with considerable accuracy. In the Red Kangaroo the intervals between mating and birth in some lactating females (Table 2) are not true gestation periods since they include

Table 2

Intervals from mating to birth and intervals from removal of pouch young (RPY) to birth in seven female Red Kangaroos subjected to different levels of suckling stimulus

No. of female	K 4	K 30	K 31	K 32	K 36	K 38	K 63
Intervals from mating to birth							
Non-suckling	33	—	—	33	—	—	33
Suckling 1 young	34	33	40	33	—	—	33
Suckling 2 young	34,39	33	—	—	32,47	—	—
Intervals from RPY to birth							
Non-suckling	32	32	—	—	—	32	—
Suckling 1 young	31	—	—	—	—	31	—
Suckling 2 young	54	43	—	35	38	32	35

a period of arrested development of the embryo. However, in thirteen non-lactating female Red Kangaroos one gestation period was 32 days, one was 34 days and eighteen were 33 days in length (SHARMAN and CALABY, 1964). The true gestation period, as in the species above, can therefore be determined with precision.

Perhaps failure to recognise the importance of the suckling stimulus accounts for the inaccuracy of some of the marsupial gestation periods given in International Zoo Year Book Vol. 1 (JARVIS and MORRIS, 1959). The list is incomplete and at least half of the figures given are wrong.

The occurrence of lactation controlled delayed implantation in marsupials was reported in 1954 (SHARMAN, 1954) and numerous papers have since appeared indicating that it is of widespread occurrence among kangaroo-like marsupials. Records of birth in captive female marsupials after long isolation from males, such as those reported by CARSON (1912) in the Red Kangaroo and, recently, by HEDIGER (1958) in Bennett's Wallaby, are readily explained in terms of the occurrence of delayed implantation.

I am indebted to Miss PAT BERGER, Mr. JOHN LIBKE and Mr. JAMES MERCHANT who helped with animal maintenance, handling and watching. The interest, assistance and advice on the manuscript given by my colleague Mr. J. H. CALABY is gratefully acknowledged.

Summary

In non-lactating female Red Kangaroos the oestrous cycle lasted about 35 days and the gestation period was about 33 days. Gestation did not interrupt the oestrous cycle. Postpartum oestrus, at which copulation and fertilization took place if the female was with a male, occurred just after parturition. Recurring reproductive cycles were replaced by the quiescent phase of lactation for up to about 200 days while the young were suckled in the pouch. If fertilization occurred at postpartum oestrus a dormant blastocyst was carried in the uterus during the quiescent phase of lactation. The delayed cycle of reproduction during which the hitherto dormant blastocyst, if present, completed development occurred following removal of young less than 200 days old from the pouch. If the young were retained in the pouch until they emerged in the normal course of events the delayed cycle of reproduction occurred coincident with the last month of pouch life and was completed a day or two after the young permanently left the pouch. Suckling of the young occupied one year: they were suckled for about 235 days in the pouch and for a further 130 days after leaving the pouch. The delayed cycle of reproduction could thus occur during, and long before the cessation of, lactation. Normal cycles of reproduction occurred during lactation if the pouch was not occupied.

The lengths of normal and delayed cycles of reproduction in females suckling one and two young-at-foot were compared with those in control, non-lactating, females. The results were as follows:

Normal cycle of reproduction

Females suckling one young-at-foot. Six cycles not significantly different from those of controls; two cycles significantly longer than in controls in one of which a dormant phase of about 7 days occurred in the embryo. Total: 8 cycles.

Females suckling two young-at-foot. Three cycles not significantly different from those of control females: two cycles significantly longer than in control females which included dormant periods of 6 and 14 days in the embryos. Total: 5 cycles.

Delayed cycle of reproduction

Females suckling one young-at-foot. No effect of suckling. Total: 5 cycles.

Females suckling two young-at-foot. One cycle not significantly different from those of control females. Five cycles longer than those of control females in which the dormant periods of the blastocysts were extended by 3, 3, 6, 11 and 22 days. In the two latter cycles resumption of development of the dormant blastocysts did not occur until removal of one of the suckling young-at-foot. Total: 6 cycles.

Observations showed that females with two young-at-foot suckled their young for about twice the length of time that females suckled a single young-at-foot. It was concluded that the suckling stimulus exerted by one or two young-at-foot could induce and maintain the quiescent phase of lactation and the associated dormant phase in the embryo. Available evidence suggested that the stimulus causing onset of the quiescent phase was tactile and received via the teat and that the delayed cycle of reproduction occurred, or the interrupted normal cycle was resumed, when the suckling stimulus was lessened.

It is suggested that some published gestation periods of marsupials owe their error to the failure of observers to appreciate the significance of concurrent suckling. Reported cases of female marsupials giving birth after long isolation from males can readily be explained as due to the occurrence of the delayed cycle of reproduction.

Zusammenfassung

Bei nichtsäugenden ♀♀ des Roten Riesenkänguruhs dauert der Oestrus-Cyclus rund 35 Tage, die Trächtigkeit rund 33 Tage. Trächtigkeit unterbricht den Cyclus nicht. Postpartum-Oestrus, bei dem Begattung und Befruchtung stattfanden, erfolgten unmittelbar nach der Geburt. Wiederkehr des Oestrus wurde durch eine Latenz während der Laktation bis zu 200 Tagen verhindert, während welcher das Junge im Beutel gesäugt wurde. Wenn beim Postpartum-Oestrus Befruchtung erfolgt war, enthält der Uterus während dieser Latenzperiode eine ruhende Blastocyste. Der verzögerte Cyclus der Fortpflanzung, während der die bisher ruhende Blastocyste (wenn sie vorhanden ist) ihre Entwicklung vollendet, tritt auf, wenn das Junge früher als 200 Tage nach der Geburt aus dem Beutel entfernt wird. Wenn die Jungen jedoch so lange im Beutel bleiben, bis sie ihn normalerweise verlassen hätten, fällt der verzögerte Cyclus der Fortpflanzung mit dem letzten Monat des Beutellebens zusammen und ist vollendet ein oder zwei Tage nachdem die Jungen den Beutel endgültig verlassen haben. Das Säugen dauert ein volles Jahr: die Jungen werden rund 235 Tage lang im Beutel und noch weitere 130 Tage bei Fuß gesäugt.

Der verzögerte Cyclus der Fortpflanzung kann also während und auch lange vor Beendigung der Laktation auftreten. Normaler Cyclus der Fortpflanzung tritt auf, wenn kein Junges im Beutel ist. Die Länge von normalen und verzögerten Cyclen der Fortpflanzung bei säugenden ♀♀ mit einem bzw. zwei Jungen bei Fuß wurde mit solchen bei nicht säugenden Kontroll-♀♀ verglichen. Die Ergebnisse waren:

Normaler Cyclus der Fortpflanzung

bei ♀♀, die 1 Junges bei Fuß säugten: 6 Cyclen waren nicht besonders verschieden von den Kontroll-♀♀. Zwei Cyclen waren bedeutend länger; bei einem davon machte der Embryo eine Ruhepause von etwa 7 Tagen durch. Im ganzen 8 Cyclen.

Bei ♀♀, die 2 Junge bei Fuß säugten: 3 Cyclen nicht besonders verschieden von den Kontroll-♀♀; 2 Cyclen bedeutend länger als bei den Kontroll-♀♀ mit Ruheperioden des Embryos von 6 und 14 Tagen. Im ganzen 5 Cyclen.

Verzögerter Cyclus der Fortpflanzung

bei ♀♀, die ein Junges bei Fuß säugten, ergab sich kein Einfluß des Säugens. Im ganzen 5 Cyclen

Bei ♀♀, die 2 Junge bei Fuß säugten, war 1 Cyclus nicht sehr verschieden von den Kontroll-♀♀. 5 Cyclen waren länger als bei den Kontroll-♀♀, bei denen die Ruhezeit der Blastocyste resp. 3, 3, 6, 11 und 22 Tage betrug. In letzteren beiden setzte die Weiterentwicklung nicht ein, bevor nicht eines der Jungen weggenommen wurde. Im ganzen 6 Cyclen.

Die Beobachtungen zeigten, daß ♀♀ mit 2 Jungen bei Fuß ihre Jungen doppelt so lange säugen, wie sie ein einziges gesäugt haben würden. Daraus wurde geschlossen, daß der Sauge-Stimulus, von einem oder zwei Jungen bei Fuß ausgelöst, sowohl die Ruhephase während der Laktation, als auch die damit gleichlaufende Ruhephase des Embryos einleitet und erhält. Die bisherige Erfahrung läßt annehmen, daß der Stimulus, der den Beginn der Ruhephase bewirkt, tactil ist und über die Zitze empfangen wird, und daß der verzögerte Cyclus der Fortpflanzung auftritt, oder der unterbrochene normale Cyclus wieder aufgenommen wird, wenn der Saugereiz sich vermindert.

Einige von anderer Seite veröffentlichte Daten über Trächtigkeitsdauern von Beuteltieren enthalten offenbar Fehler, da die betreffenden Autoren die Bedeutung gleichlaufenden Säugens nicht beachteten. Mitgeteilte Fälle, daß ♀ Beuteltiere auch nach langer Isolierung vom ♂ warfen, kann ohne weiteres durch das Auftreten des verzögerten Fortpflanzungs-Cyclus erklärt werden.

Literature

CARSON, R. D. (1912): Retarded development in a red kangaroo; Proc. zool. Soc. Lond. 1912, 234–235. — HARTMAN, C. G. (1923): The oestrous cycle in the opossum; Am. J. Anat. 32, 353–421. — HEDIGER, H. (1958): Verhalten der Beuteltiere (Marsupialia); Handbuch Zool. 8, 18 Lief 10(9), 1–28. — HILL, J. P., and O'DONOGHUE, C. H. (1913): The reproductive cycle in the marsupial *Dasyurus viverrinus.* Quart. J. micr. Sci. 59, 133–174. — HUGHES, R. L. (1962): Reproduction in the macropod marsupial *Potorous tridactylus* (Kerr); Aust. J. Zool. 10, 193–224. — MORRIS, D., and JARVIS, C. (Eds.) (1959): The International Zoo Year Book, Vol. 1; London. — OWEN, R. (1839–47): Marsupialia; In: The Cyclopaedia of Anatomy and Physiology, Vol. 3 (ed. R. B. Todd), London. — PILTON, P. E., and SHARMAN, G. B. (1962): Reproduction in the marsupial *Trichosurus vulpecula.* J. Endocrin. 25, 119–136. — SHARMAN, G. B. (1954): Reproduction in marsupials; Nature, Lond. 173, 302–303. — SHARMAN, G. B. (1955): Studies on marsupial reproduction. 3. Normal and delayed pregnancy in *Setonix brachyurus;* Aust. J. Zool. 3, 56–70. — SHARMAN, G. B. (1962): The initiation and maintenance of lactation in the marsupial *Trichosurus vulpecula;* J. Endocrin. 25, 375–385. — SHARMAN, G. B. (1963): Delayed implantation in marsupials; In: Delayed implantation (ed. A. C. Enders), Chicago. — SHARMAN, G. B., and CALABY, J. H. (1964): Reproductive behaviour in the red kangaroo in captivity; C. S. I. R. O. Wildl. Res. 9 (in press). — SHARMAN, G. B., and PILTON, P. E. (1964): The life history and reproduction of the red kangaroo (*Megaleia rufa*). Proc. zool. Soc. Lond. 142 (in press). — TYNDALE-BISCOE, C. H. (1963): The rule of the corpus luteum in the delayed implantation of marsupials; In: Delayed implantation (ed. A. C. Enders), Chicago.

Author's address: Dr. G. B. SHARMAN, C. S. I. R. O. Division of Wildlife Research, Canberra, A. C. T., Australia

BIOLOGY OF REPRODUCTION 4, 344–357 (1971)

Rodent Pheromones[1]

F. H. BRONSON

Department of Zoology, The University of Texas, Austin, Texas 78712

Received September 9, 1970

Social interactions, particularly among mammals, are often of a complex nature involving many specific cues and more than one sensory modality. An obvious challenge is the unraveling of such an array of stimuli and effects in order to elucidate specific stimulus–response systems; i.e., precisely defined cues operating through particular receptors to bring about known physiological events. One area of research that shows promise of being a rich source of such systems is that involving the chemical–olfactory modality. Olfactory signals are of widespread usage among mammals and assume major roles in a great many species. With respect to reproduction, it is now known that olfactory-mediated stimuli not only play a large role in the transfer of information that must necessarily precede insemination but that odors may also have relatively direct effects on anterior pituitary function itself. Male odors, for example, may alter the release of FSH, LH, ACTH, or prolactin in recipient females. There is the distinct possibility that precise identification of such compounds may well provide some of the best tools for elucidating the brain mechanisms and pathways normally acting as intermediaries between the environment of an animal and its reproductive behavior and physiology. The general purpose of this paper, then, is to review the subject of mammalian pheromones, particularly as they interact with reproductive processes. The bulk of experimental evidence on this subject concerns rodents with considerably

less experimentation outside of this order. Furthermore, the one species most heavily studied has been the house mouse. We can, therefore, most profitably discuss pheromonal functions within the framework of this one species supplementing the emerging generalities with information from other rodents when possible.

The term pheromone has been commonly used for externally voided substances that convey information between members of the same species. While known to exert their effects via olfaction, ingestion, and possibly absorption in insects (Barth, 1970), the prime and possibly the exclusive pathway used in mammals apparently involves odors and olfactory reception. Two general types of pheromones, modelled after entomological constructs, have been commonly recognized in mammals: (*a*) signalling pheromones that result in a more or less immediate change in motor activity on the part of the recipient animal and (*b*) priming pheromones that trigger neuroendocrine and endocrine activity. The response to a signalling pheromone (e.g., a sex attractant) may be positive or negative or there may be no response at all. A response, if it does occur, will be behavioral and is most apt to occur shortly after reception of the information. An example of a primer is a factor in male urine, the smell of which passively induces gonadotropin release in females and culminates in estrous behavior 48–72 hr later. Thus the two types of pheromones may both result in changes in behavior; however, in the case of the primer, the behavioral change is considerably delayed and requires hormonal mediation.

[1] This investigation was supported by Public Health Grant HD-04149 from the National Institute of Child Health and Human Development.

34

Informational content has been experimentally confirmed for several odor sources in mice: urine, preputial glands, coagulating glands, and plantar glands. Suspected sources should certainly include feces, generalized skin odors, nonspecialized sebaceous glands, sex accessory glands in addition to the coagulating glands, as well as Harderian, lacrimal, and submaxillary glands. Two problems seem paramount when discussing pheromonal function in mice regardless of the source of the odor: (a) no pheromone has been isolated and identified and (b) the term pheromone may even be misleading in many situations in which olfactory cues are known to be used. With respect to the latter problem, a prime function of odors among mice is certainly identification of species, sex, sexual state, and even individual. Several experimental studies utilizing conditioning techniques and good stimulus control have verified that mice can make such discriminations based solely on olfactory cues and, additionally, can even detect the difference between two members of the same highly inbred strain (Bowers and Alexander, 1967; Chanel and Vernet-Maury, 1963; Kalowski, 1967). The same general type of experiment has been successfully accomplished using rats but with less control over specific sensory modalities (Husted and McKenna, 1966). In all probability there is no such thing as a single pheromonal compound specifically tailored for individual recognition. Olfactory discrimination between individuals would more reasonably seem to involve the widest possible spectrum of odor sources and to include variations in both concentration and type of odor. Whether or not such individual recognition may be thought of as having a pheromonal basis (as opposed to simply the use of a variety of olfactory cues) depends upon the definition of the term pheromone (Gleason and Reynerse, 1969; Kirschenblatt, 1962; Whitten, 1966). It would seem most profitable to restrict the use of this term to situations where there seems a reasonable probability of isolating one or at least a restricted mixture of compounds that could, in turn, be synthesized and whose actions could then be reconfirmed experimentally. Secondly, it would seem necessary that any response to a pheromone should serve a reasonable biological function in a natural population. Following the first aspect of this argument, then, it is doubtful that the term pheromone can realistically have merit when referring to the melange of odors probably used in individual identification. Likewise, it is questionable if the term is useful when concerned with the probably large variety of animal odors that could influence the result of many psychological testing procedures (e.g., open-field emotionality and maze-running) unless the behavioral side of this can be more closely allied with a more naturalistic function (e.g., Douglas, 1966; Whittier and McReynolds, 1965).

Table 1 summarizes pheromones postulated for mice by various authors, i.e., the table actually presents a series of behavior patterns or physiological results which involve olfactory cues and in which both of the restrictions discussed previously seem possible of satisfaction. Since no pheromones have been isolated, the table can obviously only be used as a "best guess" for further research. Nevertheless the evidence for the probable existence of pheromones seems good for some of these effects. A postulated fear substance, for example, is based upon evidence that mice who are stressed by blowing air upon them, or by

TABLE 1

POSTULATED MOUSE PHEROMONES

Signalling pheromones:	Priming pheromones:
1. Fear substance	1. Estrus-inducer
2. ♂ sex attractant	2. Estrus-inhibitor
3. ♀ sex attractant	3. Adrenocortical activator
4. Aggression-inducer	
5. Aggression-inhibitor	

electroshock, excrete urine whose smell causes avoidance by other mice or which otherwise interferes with conditioning experiments (Carr, Martarano, and Krames, 1970; Muller-Velten, 1966; Sprott, 1969). Mice do learn to cease avoiding such odors after a period of time thus confirming the typical lack of stereotypy of mammalian responses to signalling pheromones. Importantly the phenomenon is species specific as tested within a three-species framework (Muller-Velten, 1966). Rats are also known to discriminate between the urine of shocked and unshocked rats (Morrison and Ludvigson, 1970; Valenta and Rigby, 1968). It seems reasonable, then, to suspect a relatively specific signalling pheromone in the urine of stressed animals which would function to communicate danger to other members of the same species in a natural population.

The presence of sex attractants is incompletely documented in mice. A female-originating pheromone acting as an attractant for males is listed in Table 1 largely on the basis of so many studies in other rodents (for rats, Carr and Caul, 1962; Carr et al., 1966; Carr, Krames, and Castanzo, 1970; Carr, Wylie, and Loeb, 1970; Le-Magnen, 1952; Stern, 1970; for voles, Godfrey, 1958; and for deermice, Moore, 1962). The opposite attraction, however, is better documented. Female mice become more active when in the presence of odor drawn from a cage of males (Ropartz, 1968a), and, additionally, Scott and Pfaff (1970) have shown that females prefer urine collected from intact males to that obtained from castrates. The male's preputial gland lipids have been implicated in this phenomenon (Bronson, 1966; Bronson and Caroom, 1971; Gaunt, 1968), with one report of such pheromonal activity in the free-fatty acid fraction (Gaunt, 1968). Considerable chemistry has been done on the normal male gland (Sansone and Hamilton, 1969; Spener et al., 1969), and it is probable that this will

be the first mouse pheromone to be isolated and identified. A word of caution is necessary, however, concerning the concepts of attraction and arousal as determined in the laboratory. What is actually known, with respect to the preputial gland, is that saline homogenates or lipid extracts of these glands are preferred by females in a two- or four-choice test situation and, additionally, that such preparations will more or less instantly awaken females from a sound sleep. It is not known, for example, whether the preputial lipids are attractive to both sexes nor is it known how other species react to them. Interestingly enough, there is a suggestion that the attraction for male urine may actually be stronger among diestrus females than for those in estrus (Scott and Pfaff, 1970). Additionally, the preputial gland has been implicated in male aggression (McKinney and Christian, 1970; Mugford and Nowell, 1970b). Thus, while there can be little doubt that the male's preputial contains one or more lipids acting in an informational context, the exact content of the message is still highly questionable.

Both "aggression-inducing" and "aggression-inhibiting" pheromones have been postulated in mice. The basis for assuming the presence of an inhibitor is the fact that males whose fur has been rubbed in female urine are not attacked as frequently as expected (Mugford and Nowell, 1970a). The general repellent action of the odor of one male upon another has been known for some time (Chanel and Vernet-Maury, 1963) and several workers have shown that the odor of strange male urine will elicit attack among normally compatible groups of males. Techniques for demonstrating this latter phenomenon have included the use of soiled bedding (Archer, 1968) and rubbing urine from a strange male onto the fur of one member of a usually or expected compatible pair of males (Mackintosh and Grant, 1966; Mugford and Nowell, 1970a). In addition, Ropartz has shown the lack both attack and response to attack in anosmic males (Ropartz, 1968b). Strange male urine is decidedly more effective in eliciting attack if obtained from an intact rather than from a castrate male (Mugford and Nowell, 1970a). Androgenized females elicit considerable aggression from normal males and it has been suggested (a) that these females were producing an androgen-dependent pheromone that resulted in attack and (b) that the preputial gland is the source of this pheromone (Mugford and Nowell, 1970b). Somewhat confusing, but indicative of the importance of the preputial gland, is the report that preputialectomized males fight more frequently than expected when housed with intact males (McKinney and Christian, 1970). There can be little doubt, then, that mice make ready use of urinary signals, and possibly a specific androgen-dependent pheromone from the preputial gland, in their assessment of a potentially aggressive encounter.

A correlated behavioral result of exposure to strange male odor is a general increase in motor activity. Importantly, no increase in activity is found if a group's own odor is passed back onto themselves in a control experiment (Ropartz, 1967a). Ropartz has implicated two odor sources in the activity change: a urinary factor traceable to the coagulating glands and a secretion by the plantar glands in the soles of the feet (Ropartz, 1967a). The former is apparently produced only among males found in groups while the latter is thought of as a type of individual identification. Of particular interest within this context is a possible primer effect; strange male odor causes an increase in adrenal weight and a decrease in adrenal ascorbic acid among isolated male mice (Archer, 1969; Ropartz, 1966). Ether extracts of coagulating glands cause an increase in adrenal weight (Ropartz, 1967a), but odors from cages of females also increase adrenal weight in males (Ropartz, 1967b). The full story on these interesting observa-

tions is, of course, yet to come, but the need to precisely measure changes in activity and aggression as well as to investigate the possible adrenal-activating primer at the same time in the same experimental program is obvious.

In summary of the status of mouse signalling pheromones, then, it can be said that experimental evidence is slowly accumulating to support the common sense concept that nocturnal animals rely heavily on signalling pheromones and olfactory reception for much of their social communication. Additionally, considering the manifold types of information possible or necessary in a mouse population, it can be expected that odors will be experimentally implicated in many other behavioral dimensions. Elucidation of both the chemistry and the correlated behavior of these phenomena should prove an interesting future area of research.

The major interest in pheromones for reproductive physiologists has been within the framework of the priming effects discovered since 1955. These odors have potent and obviously interesting effects: inducing or inhibiting estrus and ovulation, accelerating sexual maturity in young females, and blocking implantation (recent reviews include Bronson, 1968; Bruce, 1966, 1967, 1970; Gleason and Reynierse, 1969; Whitten, 1966; Whitten and Bronson, 1970). One of the best documented aspects of primer pheromone function in mammals is certainly in the control of the laboratory mouse estrous cycle. While typically thought of as possessing 4- to 5-day periodicity, cycles ranging up to 11–12 days may be easily produced and should be considered perfectly normal depending upon the olfactory environment in which the females are housed and the strain under consideration (Bronson, 1968). Two different phenomena interact to produce such variation: (a) the suppression of estrous cycling by a poorly understood odor that is produced among groups of females and (b) the acceleration or induction

F. H. BRONSON

of cycling by a factor in male urine. Cycles of relatively short duration, then, are obtained in some but not all strains by isolation or, in most strains, by exposure to male odor. Longer cycles result from grouping of females in the absence of any male odor. All-female suppression of estrus may take the form of a prolonged diestrus (Whitten, 1959) or spontaneous pseudopregnancy (van der Lee and Boot, 1956) and is only indirectly correlated with adrenocortical activity (Bronson and Chapman, 1968). The evidence that a priming pheromone is involved rests largely on reports that such suppression is alleviated by olfactory bulbectomy and is independent of vision, audition, and physical contact (Bianchifiori and Caschera, 1963; Mody, 1963; van der Lee and Boot, 1956; Whitten, 1959).

The presence of a pheromone has been better established as the stimulus associated with a male that can override the suppressive effects of all-female grouping and, hence, synchronize the attainment of estrus among a group of females (Whitten, 1956a). Male urine of the appropriate species induces estrous synchrony in both mice and deermice (Bronson and Marsden, 1964; Marsden and Bronson, 1964). The mouse pheromone is apparently of small enough molecular size to be transported at least eight feet on a 0.25-mph air current (Whitten et al., 1968). Additionally, it has been shown that just the presence of males in the same animal room is sufficiently stimulating to influence cycle duration in wild house mouse females (Chipman and Fox, 1966). Bladder urine has proven as effective in inducing estrus as externally voided urine, thus apparently ruling out the possibility that the preputial attractant and the estrus-inducing primer are the same compound (Bronson and Whitten, 1968). It should be noted here, however, that one laboratory has reported a relatively small but consistent degree of estrus induction by preputial homogenates (Albrecht, 1967; Gaunt, 1968). Several

possibilities exist for clarifying this apparent discrepancy; e.g., the preputial gland could concentrate to a small degree the particular pheromone under consideration or there may not be a highly specific pheromonal compound at all, only a mixture of odors all denoting a male and to which the female reacts. Important in this respect is the fact that the deermouse shows both the all-female grouping phenomenon and estrus induction by male urine, yet has no preputial glands (Bronson and Marsden, 1964). On firmer ground is the fact that the pheromone is either an androgen metabolite or the product of androgen-maintained tissue since castration removes it from urine and testosterone replacement returns it to castrates of either sex (Bronson and Whitten, 1968). Saline homogenates of mouse testes are ineffective in inducing estrus (Bronson, unpublished observations). An active fraction of urine has been obtained from a Sephadex column, but isolation and identification have not yet been accomplished (Whitten, personal communication).

A recent experiment determined threshold amounts of male urine necessary for estrus induction in deermice and two types of house mice, a wild stock and the highly inbred SJL/J strain. The results of this experiment may be used to demonstrate variation in the effectiveness of the urine-estrus response and to evaluate sensitivity to priming pheromones in general. The basic procedure involved exposing cages containing 8–10 adult females to one of various doses of male urine for 3 days and assessing the results by vaginal smears on the 3rd and 4th mornings. Females were tested only after a minimum of 2 weeks maintenance in clean cages in a male-free room to maximize the suppression of their estrous cycles. Fresh male urine was diluted appropriately to 1 ml (per cage to be tested) with saline mixed with antibiotic and an antioxidant (propyl gallate). Delivery of urine to females' cage was by way of syringe pumps with attache timers set to deliver 0.42·ml aliquots once every hour. Polyethylene tubing connected the 10-ml syringes to a Pasteur pipette taped to the wire lid of each cage. Urine, saline (control dose), and/or urine–saline dilutions thus ran down the inside of each cage onto the females' bedding once every hour. Deermouse females were tested with urine collected from mature males of the same species. Urine for testing both wild house mouse and SJL/J females was obtained from mature males of a stock resulting from crossing eight inbred lines of laboratory mice.

Doses of urine tested for the three types of females and results are shown in Fig. 1. Numbers of females represented by each point in Fig. 1 ranged from 40–60 for deermice, 38–80 for wild house mice, and 39–75 for SJL/J females. Lowest doses at which a significantly increased incidence of estrus occurred were 0.01, 0.1, and 1 ml/day/cage of females for deermice, wild house mice, and SJL/J females, respectively ($p < 0.01$ in each case when compared to control cages receiving only saline–antibiotic–antioxidant mixture). Estrus was thus induced in some female deermice by adding as little as 0.01 ml of male urine per day for a 2-day period to their bedding. Importantly, this response was set against a background of the presumably inhibiting odor typical of closely confined groups of females. At the other extreme was the lack of response among the inbred SJL/J females at doses lower than 1 ml/day; a quantity not much less than the output of a normal male in a 24-hour period. Whether or not such differences are traceable to quantitative differences in pheromone between the two types of urine or to the level of sensitivity of the female is, of course, unknown. It should be noted, however, that the comparison of the deermouse dose–response curve with those of the two types of mouse mice is somewhat suspect since deermouse females repeatedly chewed off the ends of the glass pipettes through which urine was being delivered in all cages tested

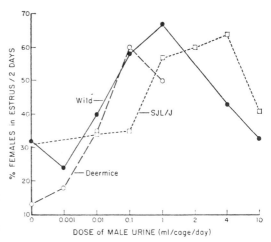

FIG. 1. Estrus-induction by dripping various doses of urine onto the bedding of cages containing deermice, SJL/J inbred mice, or wild house mice. (See text for explanation.)

at or above 0.01 ml/day. This brings up the interesting possibility of a sex attractant in deermouse male urine but does somewhat negate comparison of responsiveness; i.e., deermice were obviously being directly smeared with urine during their chewing activities. Finally it should be noted that species specificity has been established for the urinary primer pheromone within the limited framework of deermice, another inbred house mouse strain, and human male urine (Bronson and Whitten, 1968).

The picture emerging from the above studies, then, is that of a small molecule in male urine that is androgen dependent, species specific and, at least as tested in deermice in the laboratory, may act at remarkably low concentrations to initiate an estrous cycle via olfactory reception. The mechanism of action of this molecule is assumed to be by way of the hypothalamus and to cause the release of gonadotropin by the anterior pituitary. Adenohypophyseal acidophils decrease in female mice as a consequence of cohabitation with a male but at a time too late in the cycle to be a reflection of the initial effect of the pheromone (Avery, 1969). Bingel and Schwartz (1969) examined pituitary concentrations of LH during proestrus and estrus in mice and found no difference between cycles spontaneously initiated and those induced by the presence of males. The question of whether or not the initial action of the pheromone is to release both FSH and LH is still somewhat uncertain. Concentrations of FSH increased in both the plasma and pituitaries of ovariectomized females following 3 days of male exposure but changes in LH in this experiment were questionable (Bronson and Desjardins, 1969). A male may be substituted for an injection of either PMS or HCG in the normal injection regime used to induce ovulation in immature mice thus arguing for a total gonadotropin release by the male's pheromone (Zarrow et al. 1970). Hormonal controls on the action of the pheromone are likewise largely unknown. Bronson and Desjardins (1969) failed to find the expected change in pituitary and plasma FSH in male-exposed, ovariectomized females when females were given chronic injections of estradiol. Even though the dosage of estradiol used was relatively low, such a finding correlates well with Whitten's (Whitten, 1956a) earlier conclusions that the male is effective in inducing a cycle only when exposure takes place during metestrus or diestrus. A blockage of the release of FSH in response to the pheromone by the higher titers of estrogen at proestrus and estrus, therefore, seems at least possible. Finally, the same pheromone discussed above is undoubtedly the stimulus associated with a male that accelerates sexual maturation in young female mice since male-soiled bedding will also exert this effect (Vandenbergh, 1967, 1969).

The estrus-inducing compound discussed above, coupled with individual identification, provides adequate pheromonal basis to conceptualize a block to implantation resulting from exposure to a strange male; a phenomenon first described in laboratory mice by Bruce (1959). Removal of the stud from an inseminated female's home cage and replacement with another male results in a failure to implant and a return to estrus in both mice and deermice. This effect is absent in anosmic females and may be duplicated by exposing the inseminated female to pooled urine providing the urine is collected from a normally or experimentally androgenized mouse (Bruce, 1965; Bruce and Parrott, 1960; Dominic, 1965). Such a response to a male would seem to be similar to the estrus-induction phenomenon previously described, given: (a) accommodation on the part of the female to the pheromones of the original stud but not to those of the stranger, and (b) the hormonal responses to male odor possibly being somewhat different in inseminated females than they are during metestrus or diestrus. The former supposition would seem to be of prime importance since reexposure to the original stud male does not result in a blocked pregnancy. Additionally, the efficiency of blockage is enhanced if stud and strange males are of different strains (Parkes and Bruce, 1961). Thus the key concept would appear to be discrimination on the part of the female between the odors of the two males; a discrimination allowing her to cease responding to the stud after his original induction of her estrous cycle but to react to the new male by hormonal changes leading to a return to estrus at the expense of implantation. Such a process implies signalling pheromonal differences (individual identification of males) as well as sensory, central integrative, and steroidal influences on the part of the female, none of which are well understood. The obvious effect of strange male exposure is to prevent functionalization of the corpus luteum. Protection from the block may be obtained by injecting prolactin concurrently with male exposure in both mice and deermice (Bronson et al., 1969; Bruce and Parkes, 1960; Dominic, 1966) suggesting that the primary effect of the male is to enhance the normal tonic inhibtion of prolactin release. Chapman *et al.* (1970), however, found a decrease in pituitary LH (and, hence, presumably a release of LH) preceding any effect on pituitary prolactin content in inseminated females as a consequence of strange male exposure. To further confuse the picture, adrenalectomy protects against the block in one stock of laboratory mice (Synder and Taggert, 1967) and greatly increased levels of circulating corticosterone have been reported in inseminated deermice during the initial stages of strange male exposure (Bronson *et al.*, 1969). In the latter case, however, adrenalectomy offered no protection against an implantation failure and even pharmacological doses of ACTH did not decrease the probability of implantation. There is evidence on hand, then, that the odor of strange male urine acts at the level of the adenohypophysis to influence prolactin, LH, ACTH, and probably FSH also since this latter hormone is increased during normal estrus induction by urinary pheromone. One could wish at this point for assay of all four hormones in ovariectomized or immature females and in recently inseminated females following exposure to the cleanest urinary preparation available. Of particular interest is the possibility mentioned previously of a urinary primer acting on the adrenal cortex that may or may not have any effect on gonadotropin release.

Given the present status of our knowledge about mouse pheromones, it is interesting to ask several questions, the first of which concerns the number of pheromones actually denoted by past experimental work. Table 1 lists five suspected signalling and three possible priming pheromones. The list could be expanded by assuming Ropartz' verification of the effect of coagulating glands on general motor activity to be entirely separate from aggression induction. On the other hand, the list could be considerably shortened by applying Occam's razor and assuming the brain of a recipient animal

sufficiently flexible to use the same odor for different types of information and, hence, for different responses depending upon the situation. One could, for example, suspect that the aggression-inhibiting aspect of female urine is a sex attractant which overrides any potential aggression. Importantly, the estrus-suppressing primer could be a laboratory artifact reflecting high concentrations of a noxious compound in the bedding. It is known, for example, that ammonia is capable of inhibiting cycling in rats (Takewaki, 1949). Interesting in this regard is the fact that high doses of male urine are apparently ineffective in inducing estrus (see Fig. 1). Finally it could be stated that the concept of an adrenocortical activator needs further verification, particularly with respect to past social experiences of the test animals and in regard to biological function, before a specific primer is postulated.

Before grappling with the problem of the many diverse reports of olfactory-induced changes in aggression and motor activity, then, the list of potential pheromones could be considerably reduced. Likewise, it could be postulated that the compound(s) in the male's preputial gland acting to attract females are the same that elicit aggression when perceived by another male rather than a female. An argument could also be made that increased aggression, motor activity, and adrenal function are all related to simply the perception of strangeness. Male mice when housed together normally establish social hierarchies which are organized initially by fighting and often maintained thereafter by recognition of individual status in the hierarchy without overt aggression. Placing a mouse with a new individual odor among an already organized group or overriding the odor of some individuals by rubbing their fur with urine from strange males could certainly be expected to result in the resumption of fighting in an effort to reorganize the group. Increased adrenal function is known to accompany fighting among

352 F. H. BRONSON

males (Christian, 1971) and such an effect may be prolonged in a defeated mouse (but not in an undefeated mouse) just by the presence of a dominant male without physical contact (Bronson, 1967). It would thus seem that the melange of odors probably used in individual identification would be of prime importance as opposed to either a specific aggression-inducing compound or an adrenocortical-activating primer. Arguing against this unified concept are the androgen dependency of the aggression-induction signal, Ropartz' implication of the coagulating and plantar glands in activity changes, and the verification that strange male odor increases adrenal function in *isolated* males. Since not all odors associated with males are androgen dependent, one must almost postulate an androgen-dependent, core compound that carries the message "male" with increased activity and aggression being partially determined by the relative strangeness of the melange of associated odors. A similar concept has been applied by Whitten (1966) to conceptualize the pheromonal block to pregnancy by strange males. It could even be that one androgen-dependent compound occurs in both the preputial and the coagulating glands and, hence, leads to both glands being postulated as sources for effects on aggression and general motor activity. Finally, it should be noted that the fear and aggression-inducing reactions to male urine have been conceptually linked in mice by Carr, Martarano, and Krames (1970), who used both electroshock and physical defeat as stressors in the same study. These workers showed that the odor of a trained fighter is different from that of a trained loser and, secondly, that animals of different past social experiences may react differently to the two types of odors. There is thus the possibility that mice might secrete a urinary product that would vary in concentration with dominance status. Furthermore, since low ranking animals are commonly in a stressed condition, the urine of a

subordinate might be similar in some respects to the urine of a physically stressed animal. Some interesting speculations arise since both adrenal and testicular function are known to be altered by subordination and presumably, by electroshock treatment. Nevertheless, the key concept in aggression induction would still seem to be detection of first a male odor and, secondly, a strangeness and it is difficult to visualize this aspect of an animal as being denoted exclusively by a single compound varying in concentration. A signalling pheromone associated with social dominance has been suggested for rats (Krames et al. 1969).

Regardless of how future research resolves the above problems, it is probable that, as of this time, we are not justified in postulating as many specific pheromones as are listed in Table 1. A more realistic list would include three signalling pheromones and one primer: (a) a fear substance; (b) a female-originating signal which both attracts males and blocks attack when strangeness is detected; (c) a male-originating signal that attracts females and, when coupled with the many odors denoting strangeness, leads to aggression; and (d) a single primer that induces estrus, accelerates the attainment of sexual maturity and when coupled with strangeness, will induce estrus at the expense of implantation. A possible adrenocortical-activating primer is particularly interesting but needs further elaboration with respect to its function and interaction with other pheromones.

A second question concerns what might be expected in terms of pheromone chemistry. First of all the signalling–priming dichotomy is useful for discussing pheromone function but may be inadequate terms of chemistry; i.e., one pheromone could serve both functions. Secondly, the one reasonably well isolated mammalian pheromone, that of the tarsal gland black-tailed deer, is apparently a decided mixture of compounds (Brownlee et al.

244

1969; Muller-Schwarze, 1969). The major component causing licking and sniffing by other deer is *cis*-4 hydroxydodec-6-enoic acid lactone but several other lactones in the gland also result in this behavior and it is probable that the normal pheromone is a mixture of several lactones. Should mixtures prove common for mammalian pheromones, chemical isolation and identification with correlative biology will prove more difficult. On the other hand the pheromone in the boar's preputial gland, which acts on estrous females (Dutt *et al.*, 1959; Signoret, 1970; Signoret and deBuisson, 1961), might prove to be a simple androgenic steroid like andros-16-en-3-ol which structurally resembles civetone and muskone (Katkov and Gower, 1968; Sink, 1967). The latter possibility is intriguing for mouse pheromones because of the verification of androgen dependency for the sex attractant, aggression-inducer, and estrus-inducing primer. The probability of a high degree of species specificity in pheromones, however, argues against the widespread use of the more typical androgen metabolites as pheromones.

A question that is always paramount in biology is the generality of phenomena across species. Anecdotal information confirms the wide use of olfactory signals for a variety of communicative purposes in mammals (e.g., Wynne-Edwards, 1962). Reasonable verification of probable primer effects has been limited to reproductive phenomena and, additionally, to only mice, deermice, and two species of voles. Deermice, *Peromyscus aniculatus*, show all-female suppression as well as both the estrus-inducing and pregnancy-blocking actions of male urine (Bronson and Dezell, 1968; Bron and Marsen, 1964; Elefteriou *et al.*, 1962). The vole, *Microtus ochrogaster*, an induced ovulator, rarely cycles unless placed in the presence of males or male-soiled bedding, or when disturbed by cage changings (Christenson and Gier, 1971; Richmond and Conaway, 1969). Another vole, *M. agrestis*, shows the

strange male implantation block although olfactory mediation has not yet been verified (Clulow and Clarke, 1964). It seems not unreasonable to expect that the use of priming pheromones has much wider usage among mammals than that now documented. Male induction of estrus is known in some other species even through the exact sensory modality in use is unknown (e.g., the desert pocket mouse, *Perognathus penicillatus*, Wilken and Ostwald, 1968). Probably there is no way at present to generalize about the use of pheromones, priming or signalling, among mammals. Female mice of some strains cease cycling entirely following anosmia (Whitten, 1956b) while other species show only minor effects of the loss of the sense of smell (e.g., guinea pig, Donovan and Koprina, 1965). Surprisingly, the rat may well prove to be a typical, or average, mammal in this respect. Barnett (1963) has documented the importance of olfactory communication in wild rats. The rat's ability to display sexual behavior, however, is not completely dependent upon any one sensory modality (Beach, 1947). Experimental anosmia may result in some degree of depression of sexual behavior (Bermont and Taylor, 1960) but strong, direct effects of olfactory lobectomy appear only when combined with loss of another sensory modality; for example, puberty is delayed in female rats by a combination of anosmia and blinding but not by either procedure alone (Reiter and Ellison, 1970). Furthermore, a possible primer effect has been documented in the rat, the induction of estrus by a male (exact modality used is unknown), but this effect occurs only under conditions of starvation (Cooper and Haynes, 1967). The pregnancy-blocking response could not be demonstrated in rats (Davis and de Groat, 1969) but incidence of spontaneous pseudopregnancy is apparently influenced by social conditions (Alloiteau, 1961; Everett, 1963). In other words, the rat apparently does utilize olfactory stimuli to

354 F. H. BRONSON

support its reproduction and sexual behavior but is far from being dependent upon
them as long as other sensory modalities are
operative. In all probability, there will prove
to be many species like the rat just as there
are probably many that heavily utilize
pheromones, priming and signalling, and
some species relatively divorced from such
communication.

Finally, it is interesting to speculate that,
in addition to finding many more species
utilizing priming pheromones, we could also
expect to find their interaction with other
aspects of reproductive physiology. A largely
unexplored area of research is that dealing
with early olfactory imprinting (e.g., Mainardi, 1963). A third type of pheromone
classification has been evolved to cover this
phenomenon in which early exposure to
odors permanently alters the nervous system
in terms of adult response to odors. Other
possible areas of research include the response of the lactating female to odors.
Maternal behavior is completely absent in
anosmic mice (Gandelman et al., 1971) and
nonsocial odors have been shown to influence milk ejection in rats (Grosvenor and
Mena, 1967).

REFERENCES

ALBRECHT, E. D. (1967). Source of the pheromone
causing estrous synchronization in the laboratory
mouse. Thesis, University of Vermont.
ALLOITEAU, J. J. (1961). Le controle hypothalamique
de l'adenohypophyse. III. Regulation de le fonction
gonadotrope, femelle. Activite LTH. Biol. Med.
(Paris) 51, 250.
ARCHER, J. E. (1968). The effect of strange male odor
on aggressive behavior in male mice. J. Mammal.
49, 572–575.
ARCHER, J. E. (1969). Adrenocortical response to
olfactory stimuli in male mice. J. Mammal. 50, 836–
841.
AVERY, T. L. (1969). Pheromone induced changes in
the acidophil concentration of mouse pituitaries.
Science 164, 423–424.
BARNETT, S. A. (1963). "The Rat, A Study in Behavior." Aldine, Chicago.
BARTH, R. H. (1970). Pheromone–endocrine interactions in insects. Mem. Soc. Endocrinol. 18, 373–
404.

BEACH, F. A. (1947). A review of physiological and
psychological studies of sexual behavior in mammals. Physiol. Rev. 27, 240–307.
BERMONT, G., AND TAYLOR, L. (1960). Interactive
effects of experience and olfactory bulb lesions in
male rat copulation. Physiol. & Behav. 4, 13–17.
BIANCIFIORI, C., AND CASCHERA, F. (1963). The effect
of olfactory lobectomy and induced pseudopregnancy on the incidence of methyl-cholanthrene-
induced mammary and ovarian tumors in C3Hb
mice. Brit. J. Cancer 17, 116–118.
BINGEL, A. S., AND SCHWARTZ, N. B. (1969). Pituitary
LH content and reproductive tract changes during
the mouse oestrous cycle. J. Reprod. Fert. 19, 215–
222.
BOWERS, J. M., AND ALEXANDER, B. K. (1967). Mice
Individual recognition by olfactory cues. Science
158, 1208–1210.
BRONSON, F. H. (1966). A sex attractant function for
mouse preputial glands. Bull. Ecol. Soc. Amer. 47
182 (Abst.).
BRONSON, F. H. (1967). Effects of social stimulatio
on adrenal and reproductive physiology of rodent
In "Husbandry of Laboratory Animals" (M. I
Conalty, ed.), pp. 513–542, Academic Press, Lor
don.
BRONSON, F. H. (1968). Pheromonal influences o
mammalian reproduction. In "Reproduction an
Sexual Behavior" (M. Diamond, ed.), pp. 341–36
Indiana Univ. Press, Bloomington.
BRONSON, F. H., AND CAROOM, D. (1971). Preputi
gland of the male mouse: attractant function.
Reprod. Fert. 25, 279–282.
BRONSON, F. H., AND CHAPMAN, V. M. (1968). A
renal-oestrus relationships in grouped or isolate
female mice. Nature (London) 218, 483–484.
BRONSON, F. H., AND DESJARDINS, C. (1969). Relea
of gonadotrophin in ovariectomized mice after e
posure to males. J. Endocrinol. 44, 293–297.
BRONSON, F. H., AND DEZELL, H. E. (1968). Studies
the estrus-inducing (pheromonal) action of ma
deermouse urine. Gen. Comp. Endocrinol. 10, 33
343.
BRONSON, F. H., ELEFTHERIOU, B. E., AND DEZE
H. E. (1969). Strange male pregnacy block in de
mice: prolactin and adrenocortical hormones. B
Reprod. 1, 302–306.
BRONSON, F. H., AND MARSDEN, H. M. (1964). Ma
induced synchrony of estrus in deermice. G
Comp. Endocrinol. 4, 634–637.
BRONSON, F. H., AND WHITTEN, W. K. (1968). Oestr
accelerating pheromone of mice: assay, androge
dependency, and presence in bladder urine.
Reprod. Fert. 15, 131–134.
BROWNLEE, R. G., SILVERSTEIN, R. M., MULL
SCHWARZE, D., AND SINGER, A. G. (1969). Isolati

identification, and function of the chief component of the male tarsal scent in black-tailed deer. *Nature (London)* **221**, 284–285.

BRUCE, H. M. (1959). An exteroceptive block to pregnancy in the mouse. *Nature (London)* **184**, 105.

BRUCE, H. M. (1965). The effect of castration on the reproductive pheromones of male mice. *J. Reprod. Fert.* **2**, 138–142.

BRUCE, H. M. (1966). Smell as an exteroceptive factor. *J. Anim. Sci.* **25**, 83–89.

BRUCE, H. M. (1967). Effects of olfactory stimuli on reproduction in mammals. *Ciba Found. Study Group [Pap.]* **26**, pp. 29–42.

BRUCE, H. M. (1970). Pheromones. *Brit. Med. Bull.* **26**, 10–13.

BRUCE, H. M., AND PARKES, A. S. (1960). Hormonal factors in exteroceptive block to pregnancy in mice. *J. Endocrinol.* **20**, 29–30.

BRUCE, H. M., AND PARROTT, D. M. V. (1960). Role of olfactory sense in pregnancy block by strange males. *Science* **131**, 1526.

CARR, W. J., AND CAUL, W. F. (1962). The effect of castration in rats upon discrimination of sex odors. *Anim. Behav.* **10**, 20–27.

CARR, W. J., LOEB, L. S., AND WYLIE, N. R. (1966). Responses to feminine odors in normal and castrated male rats. *J. Comp. Physiol. Psychol.* **62**, 336–338.

CARR, W. J., KRAMES, L., AND CASTANZO, D. J. (1970). Previous sexual experience and olfactory preference for novel versus original sex partners in rats. *J. Comp. Physiol. Psych.* **71**, 216–222.

CARR, W. J., MARTORANO, R. D., AND KRAMES, L. (1970). Responses of mice to odors associated with stress. *J. Comp. Physiol. Psych.* **71**, 223–228.

CARR, W. J., WYLIE, N. R., AND LOEB, L. S. (1970). Response of adult and immature rats to sex odors. *J. Comp. Physiol. Psych.* **72**, 51–59.

CHANEL, J., AND VERNET-MAURY, E. (1963). Determination par un test olfactif des inter-attractions chez la souris. *J. Physiol. (Paris)* **55**, 121–122.

CHAPMAN, V. M., DESJARDINS, C., AND WHITTEN, W. K. (1970). Pregnancy block in mice: changes in pituitary LH and LTH and plasma progestin levels. *J. Reprod. Fert.* **21**, 333–337.

CHIPMAN, R. K., AND FOX, K. A. (1966). Oestrus synchronization and pregnancy blocking in wild house mice (*Mus musculus*). *J. Reprod. Fert.* **12**, 233–236.

CHRISTENSON, C. M., AND GIER, H. T. (1971). Induction of estrus in an induced ovulator. *Biol. Reprod.* (in press).

CHRISTIAN, J. J. (1971). Population density and reproduction efficiency. *Biol. Reprod. Suppl.* **3**.

CULOW, F. V., AND CLARKE, J. R. (1964). Pregnancy-block in *Microtus agrestis* an induced ovulator. *Nature (London)* **219**, 511.

COOPER, K. J., AND HAYNES, N. B. (1967). Modification of the oestrous cycle of the under-fed rat associated with the presence of the male. *J. Reprod. Fert.* **14**, 317.

DAVIS, D. L., AND DEGROOT, J. (1969). Failure to demonstrate olfactory inhibition of pregnancy ("Bruce effect") in the rat. *Anat. Rec.* **148**, 366 (Abst.).

DOMINIC, C. J. (1965). The origin of the pheromones causing pregnancy block in mice. *J. Reprod. Fert.* **10**, 469–472.

DOMINIC, C. J. (1966). Observations on the reproductive pheromones of mice. II. Neuroendocrine mechanisms involved in the olfactory block to pregnancy. *J. Reprod. Fert.* **11**, 415–421.

DONOVAN, B. T., AND KOPRINA, P. S. (1965). Effect of removal or stimulation of the olfactory bulbs on the estrous cycle of the guinea pig. *Endocrinology* **77**, 213–217.

DOUGLAS, R. J. (1966). Cues for spontaneous alternation. *J. Comp. Physiol. Psych.* **62**, 171–183.

DOUGLAS, R. J., ISSACSON, R. L., AND MOSS, R. T. (1969). Olfactory lesions, emotionality and activity. *Physiol. & Behav.* **4**, 379–387.

DUTT, R. J., SIMPSON, E. C., CHRISTIAN, J. C., AND BARNHART, C. E. (1959). Identification of preputial glands as the site of production of sexual odor in the boar. *J. Anim. Sci.* **18**, 1557–1558 (Abst.).

ELEFTHERIOUS, B. E., BRONSON, F. H., AND ZARROW, M. X. (1962). Interaction of environmental stimuli on implantation in the deermouse. *Science* **137**, 764.

EVERETT, J. W. (1963). Pseudopregnancy in the rat from brief treatment with progesterone: Effect of isolation. *Nature (London)* **198**, 694.

GANDELMAN, R., ZARROW, M. X., AND DENENBERG, V. H. (1971). Olfactory bulb removal eliminates maternal behavior in the mouse. *Science* **171**, 210–211.

GAUNT, S. L. (1968). Studies on the preputial gland as a source of a reproductive pheromone in the laboratory mouse (*Mus musculus*). Thesis, University of Vermont.

GLEASON, K. K., AND REYNIERSE, J. H. (1969). The behavioral significance of pheromones in vertebrates. *Psych. Bull.* **71**, 58–73.

GODFREY, J. (1958). The origin of sexual isolation between bank voles. *Proc. Roy. Phys. Soc. Edinburgh* **27**, 47–55.

GROSVENOR, C. E., AND MENA, F. (1967). Effect of auditory, olfactory, and optic stimuli upon milk ejection and suckling-induced release of prolactin in laboratory rats. *Endocrinology* **80**, 840–846.

HUSTED, J. R., AND MCKENNA, F. S. (1966). The use

356 F. H. BRONSON

of rats as discriminative stimuli. *J. Exp. Anal. Behav.* **9**, 677–679

KALKOWSKI, W. (1967). Olfactory bases of social orientation in the white mouse. *Folia Biol. (Krakow)* **16**, 69–87.

KATKOV, T., AND GOWER, P. B. (1968). The biosynthesis of 5α-androst-16-en-3-one from progesterone by boar testes homogenate. *Acta. Biochim. Biophys.* **164**, 134–136.

KIRSCHENBLATT, J. (1962). Terminology of some biologically active substances and validity of the term "pheromones". *Nature (London)* **195**, 916–917.

KRAMES, L., CARR, W. J., AND BERGMAN, B. (1969). A pheromone associated with social dominance among male rats. *Psychon. Sci.* **16**, 11–12.

LEMAGNEN, J. (1952). Les phenomenes olfacto-sexuels chez le rat blanc. *Arch. Sci. Physiol.* **6**, 295–332.

MACKINTOSH, J. H., AND GRANT, E. C. (1966). The effect of olfactory stimuli on the aggressive behavior of laboratory mice. *Z. Tierpsych.* **23**, 584–587.

MAINARDI, D. (1963). Eliminazione della barriera etologica all'isolamento reproduttino tra *Mus musculus domesticus e M. m. bactrianus* mediante axione sull'apprendimento infantile. *Rend. Inst. Lombardo Sci. Lett.* B **97**, 291–299.

MARSDEN, H. M., AND BRUNSON, F. H. (1964). Estrous synchrony in mice: Alteration by exposure to male urine. *Science* **144**, 3625.

MCKINNEY, T. D., AND CHRISTIAN, J. J. (1970). Effect of preputialectomy on fighting behavior in mice. *Proc. Soc. Exp. Biol. Med.* **134**, 291–293.

MODY, J. D. (1963). Structural changes in the ovaries of IF mice due to age and various other states: Demonstration of spontaneous pseudopregnancy in grouped virgins. *Anat. Rec.* **145**, 439–447.

MOORE, R. E. (1962). Olfactory discrimination as an isolating mechanism between *Peromyscus maniculatus fuginus* and *Peromyscus polionotus leucocephalus*. Doctoral dissertation, University of Texas, Austin.

MORRISON, R. R., AND LUDVIGSON, H. W. (1970). Discrimination by rats of conspecific odors of reward and non-reward. *Science* **167**, 904–905.

MUGFORD, R. A., AND NOWELL, N. W. (1970a). Pheromones and their effect on aggression in mice. *Nature (London)* **226**, 967–968.

MUGFORD, R. A., AND NOWELL, N. W. (1970b). The aggression of male mice against androgenized females. *Psychon. Sci.* **20**, 191–192.

MULLER-SCHWARZE, D. (1969). Complexity and relative specificity in a mammalian pheromone. *Nature (London)* **223**, 525–526.

MÜLLER-VELTEN, H. (1966). Über den Angstgeruch bei der Hausmaus (*Mus musculus* L.) *Z. Vergl. Physiol.* **52**, 401–429.

PARKES, A. S., AND BRUCE, H. M. (1961). Olfactory stimuli in mammalian reproduction. *Science* **134**, 1049–1054.

REITER, R. J., AND ELLISON, N. M. (1970). Delayed puberty in blinded anosmic rats: role of the pineal gland. *Biol. Reprod.* **2**, 216–222.

RICHMOND, M., AND CONAWAY, C. J. (1969). Induced ovulation and oestus in *Microtus ochrogaster*. *J. Reprod. Fert.* **6**, 357–376.

ROPARTZ, P. (1966). Contributions e l'etude du determination d'un effet de groupe chez les souris. *C. R. Acad. Sci. Ser. D* **262**, 2070–2072.

ROPARTZ, P. (1967a). Role des communication olfactives dans le comportment social des souri males. *Colloq. Int. Cent. Nat. Rech. Sci.* **173**.

ROPARTZ, R. (1967b). La seule odeur d'un groupe de souris femelles est capable d'induire un hyper trophie surrenalienne chez des males isoles. *C. R. Acad. Sci. Ser. D* **264**, 2811–2814.

ROPARTZ, P. (1968a). Mise en evidence d'une augmentation de l'activite locomotrices des groupes d souries femelles en response a l'odeur d'un group de males etranges. *C. R. Acad. Sci. Ser. D* **267**, 2341–2343.

ROPARTZ, P. (1968b). The relation between olfactor discrimination and aggressive behavior in mice *Anim. Behav.* **16**, 97–100.

SANSONE, G., AND HAMILTON, J. G. (1969). Glyceri ether, wax ester and triglyceride composition of th mouse preputial gland. *Lipids* **4**, 435–440.

SCOTT, J. W., AND PFAFF, D. W. (1970). Behavior and electrophysiological responses of female mic to male urine odors. *Physiol. & Behav.* **5**, 407–41

SIGNORET, J. P. (1970). Sexual behavior patterns i female domestic pigs (*Sus scrofa* L.) reared in iso lation from males. *Anim. Behav.* **18**, 165–168.

SIGNORET, J. P., AND DEBUISSON, F. DUMESNIL. (1961 Etude du comportement de la truie en oestrus. I Congr. Anim. Reprod. Artif. Insem., 4th.

SINK, J. D. (1967). Theoretical aspects of sex odor swine. *J. Theor. Biol.* **17**, 174–180.

SNYDER, R. L., AND TAGGART, N. E. (1967). Effects adrenalectomy on male-induced pregnancy block mice. *J. Reprod. Fert.* **14**, 451.

SPENER, F., MANGOLD, H. K., SANSONE, G., A HAMILTON, J. G. (1969). Long-chain alkyl aceta in the preputial gland of the mouse. *Act. Bioche Biophys.* **192**, 516–521.

SPROTT, R. L. (1969). "Fear communication" via o in inbred mice. *Psych. Res. Reports* **25**, 263–26

STERN, J. J. (1970). Responses of male rats to s odors. *Physiol. & Behav.* **5**, 519–524.

TAKEWAKI, K. (1949). Occurrence of pseudopre nancy in rats placed in vapor of ammonia. *Pr Jap. Acad.* **25**, 38–39.

VALENTA, J. E., AND RIGBY, M. K. (1968). Discri

nation of the odor of stressed rats. *Science* **161**, 599–601.

VANDENBERGH, J. G. (1967). Effect of the presence of a male on the sexual maturation of female mice. *Endocrinology* **81**, 345–349.

VANDENBERGH, J. G. (1969). Male odor accelerates female sexual maturation in mice. *Endocrinology* **84**, 658–660.

VAN DER LEE, S., AND BOOT, L. M. (1956). Spontaneous pseudopregnancy in mice. II. *Acta Physiol. Pharmacol. Neerl.* **5**, 213–214.

WHITTEN, W. K. (1956a). Modifications of the oestrous cycle of the mouse by external stimuli associated with the male. *J. Endocrinol.* **13**, 399–404.

WHITTEN, W. K. (1956b). The effect of the removal of olfactory bulbs on the gonads of mice. *J. Endocrinol.* **14**, 160–163.

WHITTEN, W. K. (1959). Occurrence of anoestus in mice caged in groups. *J. Endocrinol.* **18**, 102–107.

WHITTEN, W. K. (1966). Pheromones and mammalian reproduction. *In* "Advances in Reproductive Physiology" (A. McLaren, ed.), Vol. 1, pp. 155–177. Academic Press, New York.

WHITTEN, W. K., AND BRONSON, F. H. (1970). Role of pheromones in mammalian reproduction. *In* "Advances in Chemoreception" (J. W. Johnson, D. C. Moulton, and A. Turk, eds.). Appleton-Century-Crofts, New York.

WHITTEN, W. K., BRONSON, F. H., AND GREENSTEIN, J. A. (1968). Estrus-inducing pheromone of male mice: transport by movement of air. *Science* **161**, 584–585.

WHITTIER, J. L., AND McREYNOLDS, P. (1965). Persisting odors and a biasing factor in open-field research with mice. *Can. J. Psych.* **19**, 224–230.

WILKIN, K. K., AND OSTWALD, R. (1968). Partial contact as a stimulus to laboratory mating in the desert pocket mouse *Perognathus penicillatus*. *J. Mammal.* **49**, 570–572.

WYNNE-EDWARDS, V. C. (1962). "*Animal Dispersion in Relation to Social Behaviour,*" Oliver and Boyd, London.

ZARROW, M. X., ESTES, S. A., DENENBERG, V. H., AND CLARK, J. H. (1970). Pheromonal facilitation of ovulation in the immature mouse. *J. Reprod. Fert.* **23**, 357–360.

REPRODUCTION AND GROWTH IN MAINE FISHERS[1]

PHILIP L. WRIGHT, Montana Cooperative Wildlife Research Unit and Department of Zoology, University of Montana, Missoula

MALCOLM W. COULTER, Maine Cooperative Wildlife Research Unit, University of Maine, Orono

Abstract: New data concerning reproduction, aging techniques, and growth of fishers (*Martes pennanti*) were obtained from 204 specimens taken from October to April during 1950–64. All female fishers more than 1 year old were pregnant. The immature class consisted of juveniles in their first year. The period of delayed implantation lasted from early spring until mid- or late winter. Nine adult females taken in January, February, or March showed implanted embryos. Fishers in active pregnancy had corpora lutea 7 times the volume of those in the period of delay. Most litters are born in March, but some as early as late February and some in early April. Counts of corpora lutea of 54 animals taken during the period of delay and during active pregnancy averaged 3.35 per female. The number of embryos, either unimplanted or implanted, corresponded exactly with the number of corpora lutea in 18 of 21 animals. Two recently impregnated 1-year-old females, recognizable from cranial characters, had tubal morulae, confirming that females breed for the first time when 1 year old. Also confirmed are previous findings of Eadie and Hamilton that juvenile females can be distinguished from adults by open sutures in the skull throughout their first year. Juvenile males can be recognized in early fall by open sutures in the skull, absence of sagittal crest, immature appearance and lighter weight of bacula, unfused epiphyses in the long bones, and small body size. The sagittal crest begins to develop in December and often is well developed by March. The baculum grows slowly during the early winter, but by February there was some overlap with weights of adult bacula. Male fishers showed active spermatogenesis at 1 year. Open sutures were found in juvenile male skulls throughout the first year. Pelvic girdles of juveniles were distinguished by an open pubo-ischiac symphysis; adults of both sexes showed the two innominates fused into a single bone resulting from at least a partial obliteration of the symphysis. Mean body weights of animals weighed whole in the laboratory were as follows: adult males, 10 lb 12 oz; juvenile males, ? lb 7½ oz; adult females 5 lb 8 oz; juvenile females, 4 lb 11 oz.

After reaching an all-time low during the early part of the century, the fisher has made a remarkable recovery during the past 25 or 30 years in Maine (Coulter 1960) and in New York State (Hamilton and Cook 1955). The increase in abundance of this high quality furbearer in New York to the point that it could be legally trapped allowed Hamilton and Cook (1955) and late Eadie and Hamilton (1958) to discover significant facts from studying carcasses obtained from trappers.

[1] This study is a contribution from the Maine and the Montana Cooperative Wildlife Research Units, the University of Maine, the Maine Department of Inland Fisheries and Game, the University of Montana, the Montana Fish and Game Department, the U. S. Bureau of Sport Fisheries and Wildlife, and the Wildlife Management Institute cooperating. The study was supported by Grant GB-3780 from the National Science Foundation.

In Maine, the season was reopened in 1950, permitting collection of data and material from fishers trapped there. The purpose of the present paper is to present new information about reproduction, age determination, and growth of fishers, derived from study of Maine animals obtained between 1950 and 1964.

More than a dozen biologists and many wardens of the Maine Department of Inland Fisheries and Game collected material from trappers. Special thanks are due to Myron Smart, Biology Aide, who assisted in numerous ways throughout the entire study, and to Maynard Marsh, Chief Warden, who made arrangements for confiscated specimens to be processed at the Maine Unit. Numerous graduate assistants at the Maine Unit helped with processing carcasses and the preparation of skulls and bacula. We are indebted to Howard L. Mendall for editorial assistance and to Virginia Vincent and Alden Wright who made the statistical calculations. Margaret H. Wright did the microtechnique work. Elsie H. Froeschner made the drawings. Some of these findings were summarized in an unpublished Ph.D. dissertation presented by Coulter at the State University College of Forestry at Syracuse University.

FINDINGS OF PREVIOUS WORKERS

Hall (1942:147) published data from fur farmers in British Columbia showing that the gestation period in captive fishers ranges from 338 to 358 days and that copulation normally takes place about a week after the young are born. Enders and Pearson (1943) described the blastocyst of the fisher from sectioned uteri of trapper-caught animals and showed that the long gestation period is due to delayed implantation. It was assumed that the blastocysts remain inactive from spring until sometime during winter. De Vos (1952) studied fishers in Ontario and made preliminary attempts to establish an aging method based upon skulls of males and females and the bacula of males. Hamilton and Cook (1955) published information about the current status of fishers in New York State and described a technique for recovering the unimplanted blastocysts from fresh reproductive tracts by flushing them out with a syringe. Eadie and Hamilton (1958) provided additional data on the numbers of blastocysts in pregnant tracts and described cranial differences between adult and immature females.

MATERIALS AND METHODS

Coulter collected material in Maine from trapped fishers, starting in 1950 when the season was first reopened. The intensity of the collection varied over the years depending upon the legal regulations in effect. Data are available from 204 animals.

In addition to animals legally taken during the trapping season, Coulter obtained a number of animals both before and after the season, taken by trappers who were trapping other species, primarily bears and bobcats. Trappers who caught fishers accidentally were required to turn them over to the Department of Inland Fisheries and Game which in turn brought or sent them to the Maine Unit at Orono where they were autopsied by Coulter. Unskinned fishers as well as skinned carcasses were submitted to the laboratory. Whenever possible, weights were taken immediately before and after skinning to obtain an index for converting the weights of carcasses received from trappers to whole weights. During the trapping season carcasses were collected at trappers' homes. Usually the material was submitted in fresh condition; often it was frozen or thoroughly chilled when received at the laboratory. Because of the interest of the cooperators, most of the material was accompanied by collection dates, method of capture, locality, and other notes. These data together with measurements, weights, observations about the condition and completeness of the specimens, and a record of material saved for future study were entered on individual cards for each animal.

A special effort was made from the fall of 1955 to the spring of 1958 to obtain complete skeletons, and 59 such specimens were obtained. Coulter trapped a series of especially needed animals in late March and early April, 1957. Because of excellent cooperation by State Game Wardens and Regional Biologists, a good sample of specimens was available for study over a 6-month period from October to April.

This series of fishers is an unusually valuable one for discerning important aspects of the growth and the reproductive cycle of this mustelid. For example, nine females in active pregnancy were obtained, as well as several adult males in full spermatogenesis. Furthermore, the juvenile fishers were growing and maturing rapidly during the collecting period, and this fairly large collection has allowed us to reach significant conclusions concerning the onset of sexual maturity and the distinction between the age classes with more assurance than de Vos (1952) was able to do with more limited material.

The reproductive tracts of female fishers were removed in the laboratory and preserved in 10-percent formalin, in AFA, or in special cases, Bouin's fluid. The bacula of all the males were air-dried, as were the skulls of both sexes. Testes from a few representative males were fixed in formalin also. Coulter solicited the cooperation of Wright in 1955 and all of the material then available was shipped to him for further analysis and for histological work. Most of the skeletal material was cleaned by dermestid beetles in Montana.

This study was carried out without the aid of known-age specimens. Since the study was completed, three known-age animals have become available: an 18-month-old female in Maine which was in captivity for 1 year, and two females captured in central British Columbia, released in western Montana, and recaptured 6 years later. Study of these three animals in no way affects the findings presented in this paper. Evidence is presented to indicate that young-of-the-year animals can be distinguished from adults by studying either their skulls and skeletons or their reproductive tracts. Animals judged by these criteria to be less than 1 year old are, for convenience, referred to as juveniles even though in a few cases they may be almost 1 year old. Except for one criterion for distinguishing yearling females from older adult females, described by Eadie and Hamilton (1958:79–81) and confirmed here, no method of determining the relative ages of adults was discovered.

Wherever appropriate, standard deviations and standard errors have been calculated, but generally such figures are not presented here. When it is stated that a significant or highly significant difference exists, it is based upon the use of the t test.

FINDINGS

Female Reproductive Tracts

The reproductive tract of the female fisher is similar to that of other mustelids. The ovaries are completely encapsulated with only a small ostium through which a small portion of the fimbria extends. The ovary must be cut free from the bursa under a dissecting scope with a pair of fine scissors. The oviduct encircles the ovary as in other mustelids. The oviducts were not highly enlarged in any animals studied, since no estrous stages were seen. The uterus has common corpus uteri which allows embryo developing in one horn to migrate to the other horn. The uterine horns are 40–60 mm long in adult females in inactive pregnancy, and 2½–4 mm in diameter. Immature fishers show smaller uteri with horns about 30–40 mm long and 1½–2½ mm in diameter. No search was made for an os clitoridis.

The ovaries from each preserved tract were dissected from the fixed reproductive tract, blotted, and weighed. Each ovary from animals taken in fall or early winter was sliced macroscopically and the number of corpora lutea present determined by the use of a dissecting microscope. Of the 77 tracts handled in this way, 44 animals showed corpora and were thus judged to be adults. Thirty-three animals were without corpora and were judged to be immature. The average combined weights of the ovaries was 134.4 mg for adults and 76.5 mg for immatures. The average weight of the left ovaries (I—40.3 mg, A—70.0 mg) was greater than that of the right ovaries (I—36.2 mg, A—64.4 mg) in both immatures and adults, but no special significance is ascribed to this matter. The average number of corpora lutea from this series of 44 adult females was 1.68 in the right ovaries and 1.60 in the left; the average was 3.28 per adult female. Eadie and Hamilton (1958) reported that the mean number of corpora lutea in 23 adult New York fishers was 2.72. The difference in the average number of corpora lutea between the Maine and New York samples is highly significant. The distribution of the corpora lutea from all of the Maine, pregnant animals is shown in Table 1.

To determine the relationship between the number of corpora lutea in the ovaries and the number of blastocysts in the uteri, 11 tracts of adult females were studied in detail. After the ovaries were removed and sectioned by hand, uteri were selected that appeared to be the best preserved. These

Table 1. Distribution of corpora lutea in ovaries of pregnant Maine fishers.

No. of Corpora in Single Ovaries	No. of Cases	No. of Corpora in Both Ovaries of Individual Females	No. of Females
4	2	5	1
3	14	4	21
2	42	3	30
1	43	2	2
0	8		—
	—		54
Total 109*			

* One case in which only one ovary available.

entire uteri were dehydrated and cleared in wintergreen oil. Study of the entire cleared tract under a dissecting scope using transmitted light often revealed the location of blastocysts. Serial sections of each of these tracts were made to locate the blastocysts. As soon as all of the expected blastocysts were found, no further sectioning of that tract was done. In some cases the entire uterus was sectioned before all the blastocysts could be located, and in 2 of the 11 tracts, 1 potential blastocyst was not found. This represents a loss of only 6 percent, as there were 35 corpora in the ovaries of the 11 animals and 33 blastocysts were located. The technique of Hamilton and Cook (1955:30–31) of flushing the uteri for the blastocysts was not followed here since the tracts had been fixed in formalin.

The sectioned blastocysts were similar to those described by Enders and Pearson (1943:286). The extremely thick zona pellucida, 14.4 μ according to these authors, makes it possible to find the blastocysts in very poorly preserved material. None of the blastocysts studied was in better condition than those seen by Enders and Pearson, and the relative numbers of nuclei in the trophoblast and the inner cell mass for this species is still not known. In order to ob-

Table 2. Findings in nine reproductive tracts of female fishers in active pregnancy.

DATE KILLED	WEIGHT OF OVARIES (MG)		DISTRIBUTION OF CORPORA LUTEA		STATE OF UTERUS
	Right	Left	Right	Left	
January, 1965	—	—	–	–	3 embryos, 18 mm CR (Crown–Rump)
February 2, 1961	118	77	3	1	4 embryos, 2R, 2L, 17-mm swellings, embryo 8 mm CR
February 7, 1956	182	98	3	0	3 embryos, 1R, 2L, embryo 13 mm CR
February 21, 1964	110	73	2	1	3 embryos, 2R, 1L, embryo 18 mm CR
Late February, 1959	98	108	1	3	4 embryos, 2R, 2L, embryo 8 mm CR
March 3, 1959	179	138	3	1	4 fetuses, 2R, 2L, 2 males, 2 females, fetuses 53, 54, 55, 57 mm CR
March 11, 1965	—	—	2	1	3 fetuses, 2R, 1L, 3 males, fetuses 69, 71, 74 mm CR
March 13, 1956	92	92	2	1	3 early embryos, 2R, 1L, 7-mm swellings
March 20, 1957	121	137	1	2	3 fetuses, 2R, 1L, 3 females, fetuses 74, 80, 83 mm CR

tain such material, adult tracts would have to be preserved in a matter of minutes after the animal was killed.

Tracts of nine adult fishers in which there were implanted embryos were studied (Table 2). Studies of the marten and a weasel are of some value in estimating the times of parturition in these tracts. Jonkel and Weckwerth (1963:96–97) made a series of laparotomies on late-winter adult female marten (*Martes americana*) and determined that the interval between implantation and parturition was less than 28 days. In the long-tailed weasel (*Mustela frenata*), Wright (1948) showed that the postimplantation period lasted about 23 or 24 days. In estimating the parturition dates from the pregnant fisher tracts it is assumed that the period of active pregnancy is about 30 days. This seems reasonable on the basis of the larger size of the fisher in comparison with the marten and the weasel.

The female fisher with the largest fetuses, taken on March 20, would probably have borne young before April 1. The one with the earliest stages was taken on March 13, and it is estimated that her litter would not have been born until after April 1. The one with the 13-mm (crown-rump) embryos, taken on February 7, would have borne her

young before the end of February. Two recently captured females produced litters on March 2 and on March 20 at the Maine Unit. The evidence indicates that the majority of Maine fishers produce their litters during the month of March, but some do so as early as mid-February, and some as late as early April.

The ovaries of female fishers with implanted embryos were all serially sectioned. The ovaries are much larger than those in inactive pregnancy, the average combined weight being 231.9 mg as compared with 134.4 mg for the inactive group. The corpora lutea are markedly enlarged in active pregnancy as is generally known in mustelids with long periods of delayed implantation (Wright 1963:87). The corpora of three of these animals averaged 2,380, 2,917, and 3,057 μ in diameter, whereas corpora from two animals with unimplanted blastocysts averaged 1,387 and 1,219 μ. Although these corpora in animals with implanted embryos are more than seven times the volume of those with unimplanted embryos, the increased size of the ovaries is not due solely to the increase in corpus size.

In no case is the histological preservation of high quality, but the corpora lutea were readily seen and counted in all ovaries

There is a great deal of interstitial tissue in all of these ovaries, and in this they differ from weasel ovaries (Deanesly 1935:484) in which the interstitial tissue is most active in late summer but by implantation time shows considerable degeneration. There are also numerous small and medium-sized follicles in these fisher ovaries. In all cases the cells of the corpora lutea are highly vacuolated. Vacuolated cells in corpora are common in many mustelids during the period of inactive pregnancy. Eadie and Hamilton (1958:78) noted that their fisher corpora in ovaries in inactive pregnancy were highly vacuolated. Wright and Rausch (1955:348–350) describe vacuolated corpora in the wolverine (*Gulo gulo*) in inactive pregnancy, but during active pregnancy the vacuolation had disappeared. It appears then that vacuolated corpora lutea during active pregnancy is a condition not commonly seen in this group. We suppose that the corpora lutea of active pregnancy are secreting progesterone, whereas during the inactive period there may be no active secretion of progesterone. This is suggested by the urine analysis conducted in various stages of pregnancy by Ruffie et al. (1961) on the European badger (*Meles meles*) which has a similar reproductive cycle.

The number of embryos or fetuses in these eight animals averaged 3.38 and in each case the number of corpora lutea corresponded to the number of embryos; that is, there was seen here no loss of potential embryos that may have occurred during either the preimplantation period or the postimplantation period.

There was evidence of migration of embryos from one uterine horn to the other in five of the eight animals. Migration of embryos is well known in other mustelids. It apparently occurs largely during the process of spacing just before implantation. Only one example of migration was seen in all

11 tracts which were preserved during inactive pregnancy and which were sectioned to locate all of the blastocysts.

On a few occasions at the time of autopsy, Coulter observed darkened areas in the uteri which were apparently placental scars. After being fixed and cleared, most of these areas were no longer visible. Wright (1966:29) found that in the badger (*Taxidea taxus*) placental scars can readily be found in cleared tracts of parous females, provided the uteri were preserved at once after death. Placental scars are difficult to find, even in lactating badgers, in material that is not freshly preserved. It seems likely that the general level of preservation in these fisher tracts was not good enough to preserve placental scars.

Breeding Season

Earlier workers, Hall (1942:147), for example, indicate that the female fisher breeds soon after her litter is born; thus the gestation period may be as long as 51 weeks. Since no recently postparturient tracts were available for study, this particular point could not be confirmed from wild-caught animals. However, among specimens collected in late March and early April, 1957, two recently bred nulliparous females were obtained and the tracts preserved fresh. These two tracts are the best preserved in the entire series, and tuba embryos were found in each by serially sectioning the oviducts. Each animal had three corpora lutea, 2 R, and 1 L, and 3 morulae were found in one and 2 in the other. In the one taken on March 28, one morula had about 228 nuclei (Fig. 1A); the other embryos were of comparable development, but it was not possible to count the nuclei.

The animal taken on April 4 showed 2 morulae with 12 and 20 nuclei (Fig. 1B). No evidence was found of the expected

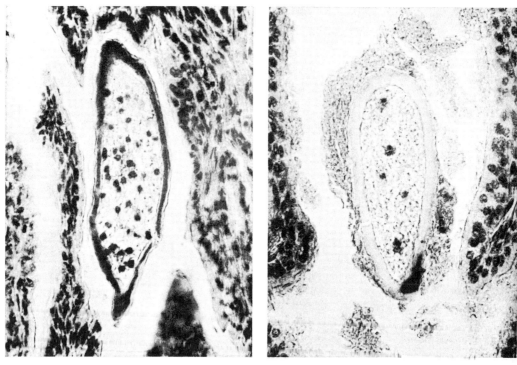

Fig. 1. Photomicrographs of tubal morulae from recently impregnated female fishers. (Left) Embryo of about 225 nuclei, from 1-year-old female taken on March 28. (Right) Embryo of about 12 nuclei from oviduct of 1-year-old female taken on April 4.

third embryo. The only mustelid possessing a long period of delay in implantation in which the rate of cleavage is known is the long-tailed weasel (Wright 1948). If the fisher has a comparable slow rate of cleavage, the March 28 animal was impregnated about March 18, and the April 4 specimen was impregnated about March 27. This is probably about the same time as recently parturient females would be impregnated. The young developing from these tubal embryos would normally have been born about 1 year later.

The ovaries of these nulliparous animals were largely masses of interstitial tissue, apparently of cortical origin. There were no graafian follicles of medium or large size. The small, almost fully formed corpora lutea with organized connective tissue centers also suggested that ovulation had oc-

curred some 8 or 10 days earlier. The luteal cells were not vacuolated. The medulla of these ovaries was discernible only as a small area adjacent to the mesovarium.

Both of these recently bred females, even though nulliparous, showed slight mammary development. In weasels, Wright (Unpublished data) has never seen mammary development associated with the summer breeding season. The nipples become conspicuous for the first time about the time of implantation.

Both of the fishers in question were judged to be 1 year old, on the basis of the development of both their skulls and skeletons. Another nulliparous female taken at the same time, March 27, was also judged to be 1 year old, but showed no sign of reaching estrus. This animal might have attained estrus within 2 or 3 weeks.

Table 3. Findings in tracts of male fishers taken in late winter and early spring.

DATE (1957)	WEIGHT OF COMBINED TESTES AND EPIDIDYMIDES (G)	PAIRED TESTIS WEIGHT (G)	PAIRED EPIDIDYMIS WEIGHT (G)	STATUS OF SPERM IN TESTES	STATUS OF SPERM IN EPIDIDYMIDES	BACULUM WEIGHT (MG)	ESTIMATED AGE OF ANIMAL	BODY WEIGHT
January 5	2.7	1.8	0.4	None	None	1262	Juv.	7 lb 3 oz
February 26	7.4	5.6	1.4	Active spermatogenesis	None	?	?	?
February or early March	6.3	4.8	1.1	None	None	1725	Juv.	10 lb 7 oz
March 1	6.3	4.8	1.0	Active spermatogenesis	Few	1252	Juv.	8 lb 5 oz
March 1	8.6	6.9	1.3	Abundant	Abundant	1550	Juv.	9 lb 12 oz
March 1–15	10.3	7.6	1.9	Abundant	Abundant	1562	Adult	—
March 17	11.3	8.6	1.9	Abundant	Abundant	1522	Adult	11 lb 5 oz
March 27	7.4	5.8	1.2	Abundant	Abundant	1921	Adult	8 lb 3 oz
March 27	13.0	9.8	2.2	Abundant	Abundant	2053	Adult	14 lb 6 oz
April 4	9.0	7.0	1.7	Abundant	Abundant	1800	Adult	9 lb 5 oz

Coulter has often noticed a definite change in travel pattern beginning in March and suspects that it is associated with breeding activities. Earlier, the animals are fairly solitary and travel in long routes in more or less direct fashion. But during March there are numerous cases of animals traveling together. The incidence of scent posts is much higher than in early or midwinter. At this season, reports are received of "dozens of fisher" in a given locality. Closer study shows that only two or three animals may be responsible for an unbelievable maze of tracks in a small area.

In the European badger, which may also have a gestation period of almost a full year, both Neal and Harrison (1958:115–116) and Canivenc and Bonnin-Laffargue (1963:121–122) present evidence for sterile matings occurring outside of the usual breeding season and ovulation in animals already in inactive pregnancy. Although no fishers were obtained during the period extending from early April until October, it is clear from the material at hand that ovulation occurs only during the breeding season, and there is no evidence of sterile matings.

Male Tracts

Since testes were generally inactive during the trapping season, they were not routinely saved from trapper-caught specimens. With a breeding season in March and April, it was obvious that late-winter animals would show transitional stages from the inactive early-winter condition to the active state in the breeding season. An effort was made, therefore, in the late winter of 1957 to preserve testes from available males. The results of the observations are included in Table 3.

The weights of the combined testes and epididymides were obtained after first stripping free the tunica vaginalis. Then the testes were further separated from the epididymides and both were weighed again. Thus, the total of the separated weights does not equal the combined weights because additional connective tissue and fat had been removed. Representative sections of testes from each animal and from the

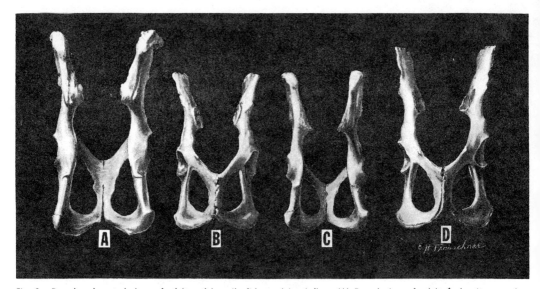

Fig. 2. Dorsal and ventral views of adult and juvenile fisher pelvic girdles. (A) Dorsal view of adult ♂ showing complete disappearance of the symphysis in a portion of the anterior half of pubo-ischiac junction. (B) Ventral view of adult ♀ showing almost complete disappearance of pubo-ischiac symphysis. There are conspicuous rugosities projecting from each side of the symphyseal line. (C) Dorsal view of juvenile ♀ showing complete separation of the innominates by symphyseal cartilage. (D) Ventral view of juvenile ♂ in which the two innominates are completely separated by a substantial symphyseal cartilage.

tail of the epididymis were prepared and stained.

The juvenile male taken on January 5 was aspermatic. By late February and early March three juveniles showed somewhat enlarged testes, but only one of these animals was in breeding condition. All of the adults taken from early March into early April were fully developed with abundant sperm in the tails of the epididymides. It would have been desirable to have tracts from additional males taken earlier in the winter. The results indicate, however, that adult males are fully active sexually during the breeding season; and the young males, now just 1 year old, are also apparently in breeding condition.

Skeletal Development

The series of 59 skeletons was studied with respect to the fusion of the epiphyses in each of the long bones and representative vertebrae. Sixteen specific sites were studied in addition to the status of fusion of the pubo-ischiac symphysis and certain sutures in the skulls.

Examination of the November and December skeletons showed striking differences between two groups, apparently juveniles and adults, in both sexes. All of the sutures studied were open in November and December males judged to be juveniles; and most of the sutures were only partly closed in comparable females thought to be juveniles. The obviously juvenile animals were smaller and showed many open sutures in the skulls. The bacula of the males in this group were small and weighed less than 1,000 mg, compared to an average of more than 2,000 mg for those with closed sutures. The ovaries of females regarded by skeletal criteria as juveniles were all without corpora lutea; the ovaries of all those classed as adults possessed corpora lutea.

The pubo-ischiac symphysis clearly remains open longer than most of the sutures

258

It was completely open in all animals that were regarded as less than 1 year old taken throughout the fall, winter, and early spring. It was at least partially obliterated, when viewed either dorsally or ventrally, in all animals regarded as more than 1 year of age (Fig. 2). The findings of striking differences in the fusion of this symphysis parallel those of Taber (1956), who described differences in this symphysis extending over several years in deer (*Odocoileus hemionus* and *O. virginianus*). The pubo-ischiac symphysis should be studied in other mammals in which aging criteria are needed.

Baculum

Weights of bacula are shown in Fig. 3, and drawings of representative types are shown in Fig. 4. The bacula of adults are more than 100 mm long, and they commonly weigh 2,000 mg or more. The fully mature baculum shows an elevated ridge near the proximal end that completely encircles the bone in a diagonal line when viewed from the side. The bacula of juveniles taken in the fall and early winter are much smaller. Although they show the typical splayed tip at the distal end, which is universally perforated by a small, round, or oval foramen, they do not show the enlarged proximal end typical of the adults. The series of bacula in Fig. 3 shows that those of the juveniles are growing rapidly during the winter months. By February some of them weigh as much as 1,600 mg (one 2,099 mg) and thus overlap the weight of those of adults. Two such bacula are shown in Fig. 4, F and G. Since the testes of juveniles in February were becoming active, it seems reasonable to assume that such animals were secreting androgen at high levels.

The fully adult baculum undoubtedly develops under the influence of androgen,

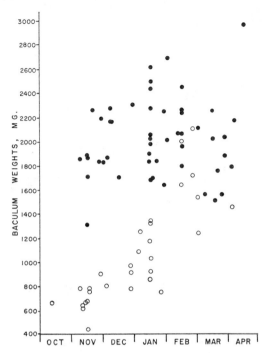

Fig. 3. Baculum weights. Adults are shown in solid dots, juveniles with open circles. The continued growth of the juvenile baculum during the winter months is clearly shown as is the overlap in weights of February, March, and April juveniles and adults.

as it probably does in all mustelids. This was demonstrated (Wright 1950) to be the case in the long-tailed weasel. Probably the fully adult type of baculum would develop by late spring in these year-old males, since Deanesly (1935:469) concluded that the adult baculum of the European stoat (*Mustela erminea*) develops to adult type within 1 month after the testes become active for the first time. Although Elder (1951:44) showed that bacula may continue to develop in succeeding years in sexually mature mink (*Mustela vison*), the lack of known-age fishers in this series does not make such conclusions possible here. The tentative conclusion reached by de Vos (1952), that bacula were not of value in distinguishing adult from juvenile fishers, resulted from failure to recognize changes

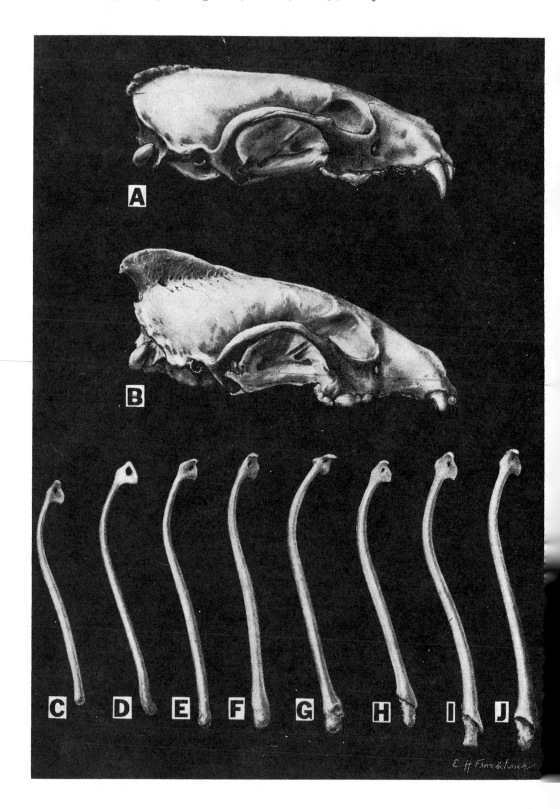

in the rapidly maturing skulls of juvenile male fishers during the late winter. This will be discussed further in a later section.

Skulls

The specimens were placed in four groups (adult males, juvenile males, adult females, and juvenile females) on the basis of reproductive condition and skeletal analysis, and 12 measurements were taken of each skull (see Wright 1953:78–79). Means, standard errors, and coefficients of variation were calculated for each group. It is clear from study of these statistics that the skulls of the juvenile animals in both sexes have not reached maximum growth. In many cases the differences between the means is statistically significant, but, because of overlap between the measurements in adults and juveniles, it is not possible to develop aging criteria based on measurement of a single skull parameter, with one exception to be discussed later.

The differences between the means of these measurements was generally much greater among males than among females. For example, the mean weight of adult male skulls was 70.6 g, whereas in juvenile males it was 53.9 g, a difference of some 20 percent. In female skulls, however, the adults average 32.1 g and the juveniles 31.1 g, a difference of only 3 percent.

The postorbital constriction becomes somewhat smaller with increased age in both sexes of fishers, as it does in other mustelids. Another striking difference between adult and juvenile skulls was seen in males where the zygomatic breadth averages 77.4 mm in adults and only 64.8 mm in juveniles. In spite of this 18 percent smaller measurement in juveniles, there is overlap. It is not possible to classify a male fisher as juvenile or adult solely on the basis of this measurement. The difference in zygomatic breadth would, in most cases, produce a broader appearing face on adult males than on juvenile males.

The sutures in the skulls of fishers, like those of all other mustelids, tend to disappear at a relatively young age (Marshall 1951:278, Greer 1957:322–323) as compared to the Ursidae, for example, where they persist for many years (Rausch 1961: 86, Marks and Erickson 1966:398). Juvenile male fishers taken in early fall (Fig. 5A) show almost all of the sutures unfused, but on specimens during March or April (Fig. 5C) almost all have completely disappeared. Eadie and Hamilton (1958:77) showed, in New York fishers from which they had reproductive tracts, that "All breeding females showed at least partial fusion of the temporal ridges . . . to form a sagittal crest, and [that] the maxillary-palatine sutures were completely fused. Non-breeding females showed the temporal ridges in various degrees of separation and had the maxillary-palatine sutures at least partly open. It is concluded that female fisher normally breed at the age of one year in the wild, and that these criteria will separate young-of-the-year from adults."

ig. 4. (A) Lateral view of skull of winter juvenile male, February, showing well developed sagittal crest and open zygomatic–maxillary suture. (B) Lateral view of skull of fully adult male showing typical tremendously developed sagittal crest and disappearance of zygomatic–temporal suture. The heavily worn teeth shown are not necessarily characteristic of adult shers. (C to J) Bacula of male fishers showing progressive changes with age, distal end to the top, the youngest to the eft and oldest to the right. C, D, and E are from juveniles, C taken October 12, D taken December 3, E taken January 5. and G are from late winter juveniles showing progressive changes toward the adult type with increased deposition of bone t the basal end. Both F and G were taken in February or early March. H, I, and J are selected adult bacula showing the haracteristic oblique ridge near the basal end and generally more massive appearance. H is from a smaller-than-average ale (body weight, 9 lb, 5 oz), I and J from larger-than-average males (I, carcass weight 10 lb; J, body weight, 14 lb oz).

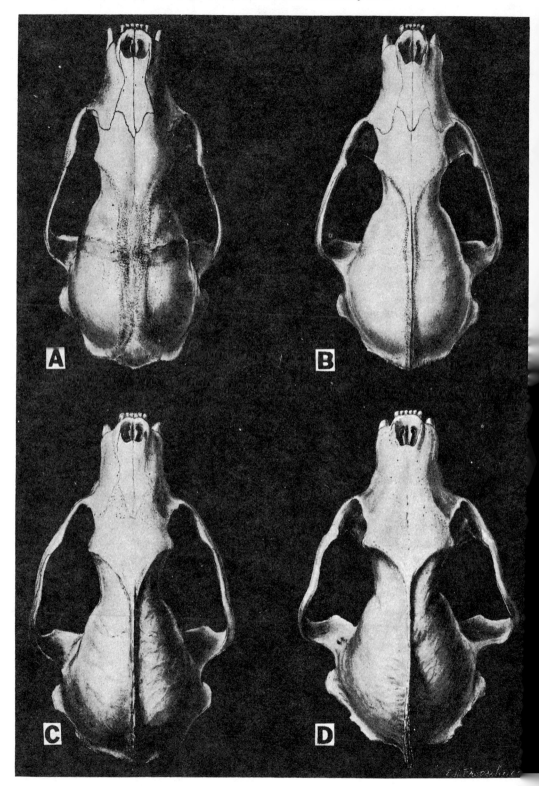

Our findings from study of 66 female fishers from Maine, from which comparable data were available, confirm in detail the findings of Eadie and Hamilton (1958). It is also clear that the maxillary-palatine suture is among the last, if not the last, to disappear.

These authors also describe a frequency distribution in the length of the sagittal crests in adult females, and reference to their Fig. 3 shows that there are two peaks of sagittal crest lengths which they tentatively regarded as representing a group of 1½-year-old females and another group of older females. When we plotted our data in comparable fashion, the line exactly paralleled theirs; and there is thus further evidence that such separation into young adults and older adults is possible. The distribution of the lengths of the sagittal crests of the adult Maine female fishers, plotted in the same fashion as did Eadie and Hamilton, is as follows: 0–10, 1; 11–20, 11; 21–30, 3; 31–40, 8; 41–50, 19; 51–60, 1.

The findings in the skulls and skeletons of the two recently bred nulliparous females, whose reproductive tracts were described in an earlier section, also provide significant evidence that the onset of breeding in female fishers occurs when they are 1 year old. In each case there was no sagittal crest, and the maxillary-palatine suture was partially open. Eadie and Hamilton (1958:79) found this suture closed in all New York fishers judged to be adults. In their collection, adult fishers, taken entirely in fall and winter, would have been at least 20 months old, whereas our two animals were almost exactly 1 year of age. One of these animals shows the pubo-ischiac symphysis still open; the other shows it partly closed. Further, the fact that during the fall and winter there is only one type of skull to be found in fishers that have not bred makes it virtually certain that wild Maine fishers are regularly impregnated at the age of 1 year and thus produce their first litters at the age of 2 years.

The sagittal crests of adult male fishers are extremely well developed as was mentioned by Coues (1877:65), and the degree of sexual dimorphism in skulls of fishers is greater than in any other American mustelid. All adult females develop sagittal crests, but even the most highly developed crests in females are almost vestigial compared with those of adult males. It is natural to suspect that with this tremendous development in mature males the crest might begin to develop earlier in juvenile males than in females. This is exactly the case, and sagittal crests were first seen in one of two juvenile males taken in December (Fig. 5B). By February, March, and April the crests of the juvenile class, now almost 1 year old, are well developed (Fig. 5C), as much so as they ever become in adult females.

In the female fishers it is clear that the sagittal crest develops first at the posterior end of the skull and grows progressively

Fig. 5. Dorsal view of male fisher skulls showing characteristic changes associated with development. (A) Juvenile male, October 12, showing narrow zygomatic breadth, all sutures in nasal region clearly open; the fronto-parietal sutures are partly fused. The poorly developed temporal lines are wide apart and thus there is no sagittal crest. (B) Juvenile male, December 3, showing disappearance of fronto-parietal suture, less conspicuous sutures in nasal region, and characteristic early development of sagittal crest running throughout the middle and posterior portions of the cranium. (C) Juvenile male in late winter, February, in which these naso-maxillary and maxillary-frontal sutures are barely visible, but the zygomatic-temporal sutures are still very distinct and the sagittal crest is better developed. (Same skull as shown in Fig. 4A). (D) Skull of adult male in which the entire dorsal skull is ankylosed into a single unit; no suture visible except for faint remains of posterior internasal suture. The characteristic highly developed keel-like sagittal crest of all adult males is clearly shown. (Same skull as shown in Fig. 4B).

forward over a period of months or probably years. In the male fisher the temporal lines move rapidly together during the winter months; and as soon as the crest is formed, it runs essentially the entire length of the dorsal region from the postorbital constriction to the inion, a distance of 50–60 mm. The sagittal crest continues to develop in adult males, and they have the crest developed to the extent of forming a "thin, laminar ridge" (Coues 1877:65). It is difficult to measure the extent of this ridge objectively; but since it extends posteriorly in fully adult males, one can use the method employed by Wright and Rausch (1955) on wolverines to subtract the condylobasal length from the greatest length of the skull. This is one accurate method of showing the posterior extension of this crest. This indirect measurement shows no overlap whatever between males classed as adults and those classed as juveniles. The mean for the former group is 11.9 mm and for the latter, 3.9 mm (see Fig. 4, A and B). Thus in male skulls if the difference between the greatest length of the skull and the condylobasal length is 6 or more mm (may be as much as 15 mm), the animal is an adult; if it is less than 6 mm, the animal is a juvenile.

Another reason for assuming that skulls of males with immature bacula, but with sagittal crests, are still in their first year of life is provided by data on the closure of sutures in the skull. The last sutures to close in males are the zygomatic-temporal, the naso-maxillary, the internasal, and the naso-frontal. In all of the skulls classed as adult, all of these sutures were closed, but in every male skull classed as juvenile, all four of these sutures were still open (Figs. 4 and 5).

On the basis of this evidence, it seems clear to us that males classed by de Vos (1952) as "adults" were in effect juveniles

as well as his "juvenile" class, and that only the animals he called "old adults" were adult males over 1 year of age.

It is concluded, therefore, that during the early winter, adult males can be separated from juvenile males by the occurrence of a well developed sagittal crest on adults; but by mid- or late winter only those males with all of the skull sutures closed are adults.

Body Weights

Both de Vos (1952) and Hamilton and Cook (1955) have provided body weights of wild-caught fishers, and both studies show that males often weigh twice as much as females. The latter indicate an average weight for males of 3,707 g (8 lb 3 oz) and 2,057 g (4 lb 9 oz) for females. De Vos's figures are roughly comparable. In both studies many of the body weights were estimated from carcass weights by applying a correction factor to skinned carcasses. (Most fisher specimens coming to biologists are likely to be carcasses skinned by trappers.) Hamilton and Cook (1955:21–22) state that the fresh carcasses average 80 percent of the unskinned weight. In the present study many fishers were confiscated and were available intact. Thus, it was possible to obtain a sample of weights taken directly from the entire unskinned carcasses allowing consideration of differences between adult and juvenile classes in both sexes.

Data obtained from those fishers which were weighed entire in the laboratory are shown in Table 4. The differences between the juveniles and adults in both sexes is highly significant although there is some overlap in each case. Furthermore, juvenile males are significantly heavier than the adult females. The available mean weight of adults are probably more satisfactory than those of the juveniles. Presumably the adults were no longer growing, but the juveniles were growing throughout the collection period from October to April. The sample is not large enough to allow a breakdown within the juvenile classes by month, but the smallest juveniles were taken in the fall.

The fact that weights of the juvenile males are 21 percent less than those of the adult males, while the weights of juvenile females are only 15 percent less than those of the adult females, further indicates that juvenile female fishers are more nearly full grown during the first winter of life than are the juvenile males.

In many cases, the fishers that were weighed whole were also weighed after skinning. This allowed determination of a correction factor. Thirty-nine animals were weighed both before and after skinning: 14 adult males, 5 juvenile males, 8 adult females, and 12 juvenile females. The carcasses averaged 81.9 percent of the whole weight; or, stated conversely, one could multiply the carcass weight by 1.22 to obtain an estimate of the entire adult weight. This latter conversion factor was applied to those animals that were weighed only after being skinned. Estimated entire body weights obtained in this fashion were comparable for both adult and juvenile males, but weights of females were significantly below the weights of those females weighed entire. For this reason, it was obvious that in the interval between skinning and weighing, many of the female carcasses had lost significant weight. It was therefore necessary to abandon any attempt to use the more numerous carcass weights for interpretation of possible growth rates in the juveniles or other weight changes that might exist between months.

Table 4. Body weights of Maine fishers weighed whole.

CLASS	NO. OF ANIMALS	MEAN BODY WEIGHT (OUNCES)	SE IN OZ	MAX.	MIN.
Adult ♂	23	172.1 (10 lb 12 oz)	±6.30	14- 6	7- 4
Juv. ♂	10	135.5 (8-7½)	±7.08	10- 8	6- 8
Adult ♀	13	88.2 (5-8)	±3.61	7-11	4- 8
Juv. ♀	17	75.0 (4-11)	±2.35	6- 8	3-13

DISCUSSION

This study indicates that in the fisher the adult class consists of all animals more than 1 year of age and that all animals of both sexes less than 1 year are sexually immature. Females older than 1 year normally are carrying unimplanted blastocysts throughout the year except during active pregnancy in late winter. The fisher, then, differs from all other American mustelids studied in this regard except the wolverine. The weasels, *Mustela erminea* and *M. frenata*, are similar in that the males reach sexual maturity in 1 year; but the females breed during their first summer and thus produce young at the age of 1 year (Wright 1963:83–84). In the marten, males also apparently reach sexual maturity in 1 year, but females may not breed until they are 2 years old, and thus two year-classes of immature females may be found in wild populations (Jonkel and Weckwerth 1963: 95–96). This has made further refinement of Marshall's (1951) original study of marten quite difficult.

In the female otter it appears that sexual maturity is delayed another year beyond that in the fisher and that there are two age-classes of immature otters (Hamilton and Eadie 1964:245). In the badger the same type of situation prevails as in the fisher except that some females breed precociously during their first summer and such females would produce litters at the age of 1 year, whereas most badgers produce their first litters at the age of 2 years (Wright 1966:42). Only in the wolverine (*Gulo gulo*) does it appear that a reproductive cycle like that of the fisher is found; but owing to a small sample of animals of the former species, the matter of age at sexual maturity is somewhat in doubt.

The recovery of the marten in Maine has been much slower than in the fisher (Coulter 1959) although both species originally occurred sympatrically in much the same habitat. The present study indicates that the potential rate of reproduction in the fisher is higher than in the marten. A large sample of winter-caught marten is not available from Maine, but such material was obtained from Montana. Wright (1963: 79) indicates that corpora lutea counts averaged 3.02 in a sample of 44 trapper-caught marten. The present study showed 3.28 for the fisher. Perhaps of greater significance, though, is the fact that some female martens (in Glacier National Park) (Jonkel and Weckwerth 1963) do not produce litters for the first time until they are 3 years old.

LITERATURE CITED

CANIVENC, R., AND M. BONNIN-LAFFARGUE. 1963. Inventory of problems raised by the delayed ova implantation in the European badger (*Meles meles* L.). Pp. 115–125. *In* A. C. Enders (Editor), Delayed implantation. University of Chicago Press, Chicago, Illinois. 309pp.

COUES, E. 1877. Fur-bearing animals: a monograph of North American mustelidae. Dept. Interior, Misc. Pub. 8, Washington, D. C. 348pp.

COULTER, M. W. 1959. Some recent records of martens in Maine. Maine Field Naturalist 15(2):50–53.

———. 1960. The status and distribution of fisher in Maine. J. Mammal. 41(1):1–9.

DEANESLY, RUTH. 1935. The reproductive processes of certain mammals. Part IX: Growth and reproduction in the stoat (*Mustela erminea*). Philos. Trans. Roy. Soc. London 225(528):459–492.

DE VOS, A. 1952. Ecology and management of fisher and marten in Ontario. Ontario Dept. Lands and Forests Tech. Bull. 90pp.

EADIE, W. R., AND W. J. HAMILTON, JR. 1958. Reproduction in the fisher in New York. New York Fish and Game J. 5(1):77–83.

ELDER, W. H. 1951. The baculum as an age criterion in mink. J. Mammal. 32(1):43–50.

ENDERS, R. K., AND O. P. PEARSON. 1943. The blastocyst of the fisher. Anat. Rec. 85(3): 285–287.

GREER, K. R. 1957. Some osteological characters of known-age ranch minks. J. Mammal. 38(3):319–330.

HALL, E. R. 1942. Gestation period in the fisher with recommendations for the animal's protection in California. California Fish and Game 28(3):143–147.

HAMILTON, W. J., JR., AND A. H. COOK. 1955. The biology and management of the fisher in New York. New York Fish and Game J. 2(1):13–35.

———, AND W. R. EADIE. 1964. Reproduction in the otter (*Lutra canadensis*). J. Mammal. 45(2):242–252.

JONKEL, C. J., AND R. P. WECKWERTH. 1963. Sexual maturity and implantation of blastocysts in the wild pine marten. J. Wildl. Mgmt. 27(1):93–98.

MARKS, S. A., AND A. W. ERICKSON. 1966. Age determination in the black bear. J. Wildl. Mgmt. 30(2):389–410.

MARSHALL, W. H. 1951. An age determination method for the pine marten. J. Wildl. Mgmt. 15(3):276–283.

NEAL, E. G., AND R. J. HARRISON. 1958. Reproduction in the European badger (*Meles meles* L.). Trans. Zool. Soc. London 29(2): 67–130.

RAUSCH, R. L. 1961. Notes on the black bear, *Ursus americanus* Pallas, in Alaska, with particular reference to dentition and growth. Z. Säugetier. 26(2):77–107.

RUFFIE, A., M. BONNIN-LAFFARGUE, AND R. CANIVENC. 1961. Les taux du pregnandiol urinaire au cours de la grossesse chez le Blaireau europeen. *Meles meles* L. Comptes rendus des séances de la Société de Biologie 155(4):759–761.

TABER, R. D. 1956. Characteristics of the pelvic girdle in relation to sex in black-tailed and white-tailed deer. California Fish and Game 42(1):15–21.

WRIGHT, P. L. 1948. Preimplantation stages in the long-tailed weasel (*Mustela frenata*). Anat. Rec. 100(4):593–607.

————. 1950. Development of the baculum of the long-tailed weasel. Proc. Soc. Expt. Biol. and Med. 75:820–822.

————. 1953. Intergradation between *Martes americana* and *Martes caurina* in western Montana. J. Mammal. 34(1):74–86.

————. 1963. Variations in reproductive cycles in North American mustelids. Pp. 77–97. *In* A. C. Enders (Editor), Delayed implantation.

University of Chicago Press, Chicago, Illinois. 309pp.

————. 1966. Observations on the reproductive cycle of the American badger (*Taxidea taxus*). Pp. 27–45. *In* I. W. Rowlands, Editor, Comparative biology of reproduction in mammals. Symposia Zool. Soc. London, No. 15. Academic Press, London. 527pp.

————, AND R. RAUSCH. 1955. Reproduction in the wolverine, *Gulo gulo*. J. Mammal. 36(3):346–355.

Received for publication August 22, 1966.

BIOLOGY OF REPRODUCTION 4, 239–247 (1971)

Ecological Adaptation and Mammalian Reproduction

C. H. CONAWAY

Caribbean Primate Research Center, University of Puerto Rico, San Juan, Puerto Rico 00905

Received September 9, 1970

In this paper some of the major variations in mammalian reproductive cycles are discussed from the viewpoint of their broad adaptive values. The variations involve qualitative differences in various aspects of the reproductive cycle and seem to be genetically fixed within a species. Research in the area of reproductive physiology has chiefly been confined to a few laboratory and domestic species which have been studied in great detail. As a result we have much information about these few forms, but perspectives may be distorted by this same wealth of information. It is often forgotten that these studies have been made on highly specialized forms which are usually the resultant of many years of domestication and artificial selection. Furthermore, the work has been done without regard to ecological or social context. For example, so much emphasis has been given to studies of the nonpregnant cycle, that we tend to lose insight into its significance in natural populations.

In natural populations the nonpregnant cycle is a rarity, and it is essentially a pathological luxury which cannot be tolerated. Even in relatively long-lived animals with low mortality rates a nonpregnant cycle is an exception. In a study done under the somewhat artificial conditions existing in the Cayo Santiago Island colony in Puerto Rico, only 6 of the 28 fertile mature female rhesus monkeys (*Macaca mulatta*) in a free ranging social group failed to conceive on the first estrus of the breeding season (Conaway and Koford, 1964). The remaining six conceived during their second estrous period.

For short-lived prey species the occurrence of a nonpregnant cycle is a disaster, which must be avoided if the individual is to contribute significant numbers of offspring to the population. Many small mammals in this category have only a few months of reproductive life. Any portion of this which is lost through a nonpregnant cycle can be critical. The only acceptable alternatives are either to safeguard against the occurrence of a nonpregnant cycle or to recover and recycle as quickly as possible. One of the basic variants in mammalian reproductive cycles seems to be in the methods which have evolved for handling the nonpregnant cycle or preventing its occurrence. Full understanding of these variations is impossible since we have detailed information about the nonpregnant cycle in fewer than 50 of the 1000 mammalian genera. This is not surprising since information must come from laboratory studies; however, it does indicate the importance of obtaining basic information on wide variety of forms.

CLASSIFICATION OF BASIC TYPES OF NONPREGNANT FEMALE REPRODUCTIVE CYCLES

Everett (1961) clearly defined pseudopregnancy and indicated sources of confusion in the usage of this term. As he proposed, the term pseudopregnancy will be used here to indicate the occurrence of any functional luteal phase in a nonpregnant cycle. The pseudopregnancy will be designated as "spontaneous" if the formation of a functional corpus luteum always follows ovulation. If activation of the corpus luteum does not obligatorily follow ovulation but r

23

quires a separate stimulation as in the laboratory mouse and rat, the phenomenon will be designated as "induced pseudopregnancy."

Within this framework mammalian female cycles may be broadly categorized as follows:

Type I

Both ovulation and pseudopregnancy are spontaneous. Sterile copulation does not alter the length of the progestational phase.

Subtype A. Medium length cycles (generally 2–5 weeks long).
The follicular phase is somewhat variable and may last from a few days to several weeks. This variation accounts for the major difference in lengths of cycles between species in this group. The luteal phase is relatively constant, lasting about 12–16 days. Cycles of this type appear to be the rule in ungulates, hystricomorph rodents, and higher primates.

Subtype B. Long cycles (over 5 weeks in length). The follicular phase is several weeks in length and the luteal phase is prolonged, lasting from 1 to 2 months. A period of anestrus follows the luteal phase. Cycles of this type occur in the dog and probably other canids.

Type II

Ovulation is induced; pseudopregnancy is spontaneous. When ovulation is induced by exogenous hormones, the length of the luteal phase does not differ from that following sterile copulation.

Subtype A. Medium length cycles (less than a month in length). Estrus is more or less behaviorally induced by proximity to the male or by social stimulation. The length of estrus may be extended in some species by continued behavioral stimulation. Synchronization of estrus in all members of a population through social facilitation occurs frequently. Cycles of this type seem characteristic of the Microtini, Lagomorpha, and some insectivores including the Soricidae.

A typical example of this type of cycle is shown by the cottontail rabbit (*Sylvilagus*

floridanus). In most areas the cottontail is a seasonal breeder with the breeding season starting during the spring and ceasing in late summer. The gestation period is 27 days and each female breeds during the postpartum estrus within 0.5 hr following parturition (Marsden and Conaway, 1963). During the breeding season a total of five to seven litters may be produced. Within any single population the time of onset of the breeding season may vary considerably between years; however, most of the females in that population usually conceive within a brief period each year and the population may remain in very close synchrony throughout the breeding season. Synchronization of estrus seems to be the result of behavioral displays which influence all members of the population.

The importance of behavioral induction of estrus is shown to an even greater degree in the similar cycle of the prairie meadow vole (*Microtus ochrogaster*). The onset and duration of estrus are largely dependent upon social stimulation. Proximity to the male is the most potent estrus-inducing stimulus, although various social and environmental stimuli are also somewhat effective. If stimulation is continuous, a state of constant estrus can be maintained for at least 30 days. Copulation at any time during estrus results in ovulation 10.5 hr later (Richmond and Conaway, 1969).

Subtype B. Long cycles (4–8 weeks in length). Estrus is more spontaneous and its duration is more fixed than in Subtype A. The luteal phase of the nonpregnant cycle persists from 4 to 6 weeks. This type of cycle is known to occur in the domestic cat and ferret.

Type III

Both ovulation and corpus luteum formation are spontaneous, but pseudopregnancy is induced via the release of luteotropin following copulation. In the short cycle (noncopulatory), no functional corpora lutea form and ovulation recurs after 4–7 days.

If sterile copulation occurs the resulting pseudopregnancy is similar to that of medium length cycles (Types I A and II A). This highly specialized cycle is known to occur in several groups within two rodent families, the Cricetidae and the Muridae.

If one examines the preceding types of female reproductive cycles with regard to ecological adaptation, several speculations seem warranted. Many small mammals, including many rodents, insectivores, and lagomorphs, may be characterized as staple small prey species. They have very short life-spans, often in the vicinity of 3–5 months. Mortality rates are extreme and annual production by an adult female which survives the reproductive season is very high (30–35 young per adult female per breeding season in cottontails). Sexual maturity occurs very early, usually at 1–2 months of age. High production rates are essential to the survival of the species and such animals cannot afford to be nonpregnant during the breeding season. Two different systems of safeguards appear to have been developed to minimize nonpregnant time. One of these is the Type II A cycle using induced ovulation and the second is the Type III cycle involving induced pseudopregnancy. In either case there is no protection against sterile copulation, since there is a pseudopregnancy of approximately 2 weeks. Sterile copulation, however, seems virtually unknown in natural populations.

Despite the fact that it has been so intensively studied and is so familiar, the Type III cycle seems to be of very restricted occurrence among mammalian species. It has been identified in a few members of each of two very large rodent families, the Cricetidae and the Muridae. These families have been separated for a considerable period of time and have shown some parallel evolution with the development of numerous ecologically equivalent species.

The family Muridae, commonly called the Old World rats and mice, consists of 101 genera distributed in seven or eight subfamilies. Within this assemblage, five species of *Rattus* and one of *Mus* clearly have been shown to have the Type III cycle. All of these are typical high production, small prey species. On the other hand, the large pseudomyid murid *Mesembriomys* of Australia has a Type I A cycle. The mean length of the nonpregnant cycle in this species is 26 days. Ovulation is spontaneous followed by spontaneous pseudopregnancy lasting about 14 days (Crichton, 1969).

Indirect evidence suggests that the Type I A cycle also occurs in other pseudomyid and perhaps hydromyid murids, although none of these forms can be considered as a high reproductive rate small prey species. All seem to have only a few litters per year and very few young per litter. They differentiated during the Miocene in the absence of placental carnivores (Simpson, 1961) and thus are not adapted to high levels of predation pressure. Within the genus *Rattus* there are several ecologically similar forest-dwelling species which have very low reproductive rates (Harrison, 1952) and apparently low mortality rates. It would be very useful to learn about the cycle of these forms which are closely related to the familiar *Rattus* species having the Type III cycle but ecologically very different and occupying a niche similar to that of *Mesembriomys*.

The Cricetidae is the second family of rodents in which the Type III cycle occurs. This family contains 99 genera divided among five subfamilies. The largest of these is the Cricetinae with 59 genera. In this subfamily at least six genera appear to have one or more species showing the Type III cycle. Also at least 2 of the 13 genera of the subfamily Gerbillinae appear to have this cycle. Several species of the genus *Microtus* in the subfamily Microtine, however, are known to have typical Type II A cycle. One might predict that still a third type of cycle might occur in this family since the Malagasy rat (subfamily Nesomyinae) are in many way ecologically similar to the Australian rats and therefore may have Type I A cycle.

Both the Type II A and Type III cycles seem to be characteristic of short life-span, high production prey species. Probably both types have developed idependently in several groups. The Type II A cycle is found in at least three orders, while the Type III cycle is known from some groups in only two highly specialized rodent families. The special adaptive significance of the Type III cycle is not clear. One possible suggestion is that those forms having the Type III cycle do not show the violent amplitude in population density cycles that characterize many Type II A forms. It is possible that since there is less estrous induction by behavioral stimulation in Type III spontaneous ovulators, the major reproductive outbursts and subsequent density-dependent die-offs are to some extent dampened in the Type III cycle. Many Type II A forms such as voles and hares are characterized by major cyclic fluctuations in population density.

If the short cycle (Type III) and medium length cycle with induced ovulation (Type I A) are associated with high turnover prey species, what can be said of the other cycle types? It seems that medium length cycles with spontaneous ovulation (Type I A) include in general the medium and larger herbivores, such as ungulates, hystricomorph rodents, and the omnivorous primates. These are long-lived prey species not subjected to the extreme mortality rates of smaller forms. Recovery from a nonpregnant cycle would be important but extreme rapidity would not be critical. Therefore these animals can afford a delay of several weeks before recycling.

The long cycle Types I B and II B seem restricted to large predators in which the emphasis may be on low production rather than high production. The question now arises: which is the primitive Eutherian pattern—the medium length or the long cycle? Is this the basic birth control mecha-

nism, or have other forms been forced to shorten a long cycle to increase production? One can only speculate about this at present, since information from so many groups is completely lacking.

One point which may apply is that in the primitive insectivore family Tenricidae, the length of pseudopregnancy seems prolonged and the cycle appears to be Type II B (C. H. Conaway and M. J. Hasler, unpublished). In the more advanced Soricids, however, the cycle is of the II A type (G. L. Dryden and C. H. Conaway, unpublished). These data would support the concept that long luteal life was the primitive pattern. On the other hand, the two major insectivore subgroups had already diverged by the middle Cretaceous (McKenna, 1969). Tenrecs and Carnivora were derived from palaeoryctoid Insectivora while the shrews and most other Cutherian mammalian orders trace to a leptictid-like insectivore stock. It may thus be that the medium-lived and the long-lived corpus luteum forms represent early and fundamental divergences not related to their present ecological adaptations.

INDUCED AND SPONTANEOUS OVULATION

As information is obtained about more forms it appears that induced ovulation is the more widespread phenomenon and that spontaneous ovulation occurs in a more restricted number of species. Since most of the common domestic and laboratory species are spontaneous ovulators, this point is often overlooked. It has also become increasingly apparent that, physiologically, both induced and spontaneous ovulation are the extremes of a single continuum. Some induced ovulators will ovulate under a variety of stimuli other than copulation. It is a common observation that the domestic rabbit will ovulate as a result of female–female mounting, as well as other stimuli. On the other hand, in spontaneous ovulators there is a copulatory LH surge similar to that found in induced ovulators (Taleisnik, Caligaris, and Astrada, 1966).

Induced ovulation appears to be of general occurrence in the primitive Eutherian order of Insectivora. Within the Rodenta, it occurs in some of the primitive Sciuromorpha while most of the more advanced Hystricomorpha seem to be spontaneous ovulators (Asdell, 1964). Also as previously noted, induced ovulation is of very widespread occurrence. Because of these reasons, it seems that induced ovulation may be regarded as the basic Eutherian pattern and spontaneous ovulation as a specialization. Other than Type III rodents, spontaneous ovulation occurs generally in ungulates (except Camelidae), primates, canid carnivores, and in hystricomorph rodents. It is sporadically distributed in other groups (Asdell, 1966).

Asdell (1966) concluded that because of the sporadic nature of the distribution of the two types of ovulation, no conclusion could be drawn regarding evolutionary trends of this character. For the ungulates, primates, and canids, at least, one can propose a common selective force favoring the development of spontaneous ovulation. All of these forms are characterized by having fairly complex social groups with often elaborate social structuring. Temporary pair bonding and consort relationships are common between an estrus female and a breeding male and the number of adult females usually exceeds the number of breeding males. Commonly breeding activity is restricted to a small segment of the adult male population which includes only animals of high social rank.

In a situation where the number of effective breeding males may be considerably less than the number of breeding females, it would be of great advantage to spread the estrous periods of females randomly over a period of time to insure conception in all females. Since induced ovulation seems to be accompanied by more or less synchroniza-

tion of estrus periods, a strong positive selective force favoring spontaneous ovulation and randomization of estrus periods would exist when the number of breeding males was limited. The female which could achieve estrus independently and avoid estrus synchronization would have a greater chance of conception.

This explanation, however, does not seem to fit all of the histricomorph rodents. While some of these species (viscacha and chinchillas) are colonial and may have a reduced number of breeding males, certainly others such as the porcupine are solitary. The occurrence of induced ovulation in the alpaca, llama, and probably other Camelidae (England et al., 1969; Fernandez-Baca, Madden, and Novoa, 1970) also would not be explained by this hypothesis.

Also some degree of estrus synchronization may occur in spontaneous ovulators. The Whitten effect may involve estrus synchronization (Whitten, 1956), but the importance of this in natural populations remains to be demonstrated. Kummer (1968) found that among hamadryas baboons in one male units having two adult females there was a high degree of estrous synchronization between the two females. Large units, however, showed less suggestion of synchronization. An extreme degree of estrus synchronization has been reported for Lemur catta where observed mating lasted only 1 week in a troop containing nine adult female lemurs (Jolly, 1966). In bands rhesus monkeys there is no evidence of estrous synchronization among the individual females (Conaway and Koford, 1964); however, there is some suggestion that the peak of mating activity may vary between troops (Koford, 1965).

DELAYED IMPLANTATION

Delayed implantation has clearly arisen independently a number of times. Not only does it occur in widely scattered species and groups within many orders but also in physiological controls vary. The selective forces favoring delayed implantation seem clear in most forms. The fundamental prerequisite is that the young be born at the optimum season of the year for survival and growth. In addition, within each species the length of the implanted gestation period is relatively rigidly fixed. Only heterothermic bats seem able to alter this significantly through delayed development (Bradshaw, 1962; Racey, 1969). Within these fundamental restrictions problems arise under several conditions.

In boreal areas solitary mammals may have difficulty in crowding mating, gestation, birth and rearing of the young to independency within the short growing season. A solution to this problem is in delayed implantation which allows almost complete flexibility of the time interval between mating and birth. Delayed implantation in northern mustelids, ursids, and roe deer would seem to fit this pattern.

The same forces apply when the time for mating, birth, and rearing of the young is behaviorally restricted to a brief period each year. Again all the reproductive events must be crowded into a short time interval. This is the problem faced by the colonial seals. The sexes are together for only a short period of time each year and must breed, give birth, and rear young during this interval.

Delayed implantation in marsupials seems to serve an entirely different function. Those forms showing the delayed implantation live in a severe and unpredictable environment. There is a prolonged nursing period during which the suckling young may frequently be lost as a result of severe environmental conditions. If this occurs, the loss of suckling stimulus causes an unimplanted blastocyst to break diapause and begin development. In this case implantation serves as a means to replace the lost suckling young as quickly as possible so that advantage can be taken of any improvement in environmental conditions.

At present I can offer no explanation of delayed implantation in New World Edentata or in the African bat *Eidolon*. Perhaps when more is learned about the natural history and ecology of these forms the significance of delayed implantation to them will become apparent.

SEASONAL BREEDING

It is becoming increasingly clear that at least some degree of seasonality in reproductive activity is almost universal in natural populations. This is true not only in the boreal and temperate zones but also in the tropics. Among natural populations one of the carefully documented studies which suggest continuous breeding is that done on the musk shrew (*Suncus murinus*) on Guam (Dryden, 1969).

The breeding season may not be regular, however, nor is it necessarily limited to one period each year. Desert species are often opportunistic breeders and breed irregularly following rainfall. In small rodents in the temperate zone there may be two distinct breeding seasons within the year. Several species of *Peromyscus* have a spring and fall breeding season with midsummer and midwinter cessations of reproduction (Hill, 1966). Generally, seasonality quickly and more or less completely disappears when species are brought into the laboratory. This is evident with small mammals.

Many studies have been directed toward a search for the environmental trigger which initiates reproductive activity. For many small mammals having short gestation periods it seems that a reverse approach may be more ecologically and physiologically sound. If these forms are regarded as potentially continuous breeders then the problem is to understand the environmental factor or combination of factors which depresses reproduction. Among such factors extreme deviations of temperature and aridity are certainly important. Extremely short photoperiod may have a depressing effect in some species, and lowered quality of nutrition is probably also effective. In a 3-year study in the Congo involving over 7000 specimens of several species, Dieterlen (1967) found both species and annual variations in the breeding seasons. He concluded that in general, the annual amplitude of reproductive periodicity was proportional to the degree of seasonal contacts. Small mammals seem to be able to adapt to reproduce under almost any set of environmental conditions, but deviation from these conditions inhibits reproduction.

If the viewpoint outlined above is adopted, then many variations in the reproductive season seen in natural populations can be explained. Instances of "unseasonal" early breeding in cottontail rabbits during a period of warm weather in midwinter (Hill, 1966) or extension of the breeding season and occasional continuous breeding throughout the winter in small rodents (Ashby, 1967; Krebs, 1964) could both be interpreted as indicating that those species were continuous breeders unless inhibited by adverse environmental conditions. Similarly the spring and fall breeding seasons of *Peromyscus* (Brown, 1964) would fit such a pattern. Voles have been found to continue to breed in irrigated fields after breeding had ceased in voles living in nearby nonirrigated areas (Bodenheimer, 1949). Again this seems to be best explained if one considers them as fundamentally continuous breeders but recognizes that reproduction may be inhibited by any of a variety of adverse factors.

In a regularly fluctuating environment, forms with longer gestation need to predict the optimum season for birth. Here it becomes imperative to use some regular and repeating event in the environment as a trigger for the breeding season. As has been suggested many times, variation in photoperiod is the most predictable changing factor in the environment and seems to be the clue used by a number of ungulates with six- to nine-month gestation periods.

POSTPARTUM ESTRUS

Postpartum estrus seems to have developed independently many times and its distribution is sporadic throughout many mammalian species. As Asdell (1964) has noted, it does not follow any phylogenetic pattern. Indeed it may not occur in all species of the same genus, as in *Peromyscus* (Asdell, 1964). Apparently the acquisition or loss of a postpartum estrus is one of the most easily made of the major reproductive adjustments. It can be a mechanism for increasing productivity and this seems its obvious function in short life-span, small mammals. It seems almost universal in those forms having Type II A or Type III cycles, while among Type I A forms it is common in the Hystricomorpha. Among other forms it is also known to occur in the mouse-sized pygmy squirrel (*Exilisciuru exilis*) of Malaysia (Conaway, 1968). This seems to emphasize the adaptive role of the postpartum estrus since it has not been reported from other squirrels or in fact from any other families of sciuromorph rodents. It also would be expected to occur in several species of *Perognathus* such as *P. parvus* and in other Heteromyinae which are typical small prey species, and in this respect similar to the pygmy squirrel.

A second group in which the postpartum heat has developed is the colonial seals. As discussed previously, in these forms the sexes are together for only a short period of time into which all reproductive events must be crowded. The postpartum estrus is integrated with delayed implantation. In marsupials also the postpartum heat is associated with delayed implantation and functions with it as an adaptation to an unpredictable environment as previously discussed.

DISCUSSION

This discussion has considered the major reproductive patterns which are relatively fixed within a species and change slowly over many generations. Another significant group of adaptive changes is the flexible quantitative changes that vary frequently within a species or even within an individual. These include variations in litter size, resorption rate, age at sexual maturity, number of litters, etc. These are the finer adjustments to specific environmental fluctuations and may be influenced by a wide variety of physical and behavioral variables. An excellent discussion of these has been given recently by Sadleir (1969).

Many of the speculations made in this paper may be erroneous. It does seem necessary to begin to try to understand the significance of the variations seen in reproductive cycles. Otherwise they become a meaningless and endless array. It also seems that after trying to establish some basic patterns we can then select appropriate species to test some of the hypotheses rather than haphazardly studying forms just because they are available and no one has worked with them. It would seem much more useful to study in detail one of the forest rats of the genus *Rattus* such as *R. sabanus* than to continue studying species of *Rattus* which are ecologically similar to the Norway rat. This might provide some evidence regarding the relative importance of adaptation versus phylogeny in establishing basic reproductive patterns.

REFERENCES

ASDELL, S. A. (1964). "Patterns of Mammalian Reproduction." Cornell Univ. Press, Ithaca, N. Y.

ASDELL, S. A. (1966). Evolutionary trends in physiology of reproduction. In "Comparative Biology of Reproduction in Mammals" (I. W. Rowlands, ed.), pp. 1–13. Academic Press, New York/London.

ASHBY, K. R. (1967). Studies on the ecology of field mice and voles (*Apodemus sylvaticus, Clethrionomys glareolus,* and *Microtus agrestis*) in Houghall Wood, Durham. *J. Zool.* **152,** 389–513.

BODENHEIMER, F. S. (1949). Problems of vole populations in the Middle East: Report on the population dynamics of the Levant vole (*Microtus guentheri*). *Res. Counc. Isr.* 77pp.

BRADSHAW, G. V. R. (1962). Reproductive cycle of the California leafnosed bat *Macrotus californicus. Science* **136,** 645–6.

BROWN, L. N. (1964). Reproduction of the brush mouse and white-footed mouse in the central United States. *Amer. Midl. Natur.* **72,** 226–240.

CONAWAY, C. H. (1968). Postpartum estrus in a sciurid. *J. Mammal.* **49,** 158–159.

CONAWAY, C. H., AND KOFORD, C. (1964). Estrous cycles and mating behavior in a free ranging band of rhesus monkeys. *J. Mammal.* **45,** 577–588.

CRICHTON, E. G. (1969). Reproduction in the Pseudomyine rodent *Mesembriomys gouldii. Aust. J. Zool.* **17,** 785–797.

DIETERLEN, F. (1967). Jahreszeiten und Fortpflanzungs Perioden bei den Muriden des Kivuss-Gebietes (Congo). *Z. Saug.* **32,** 1–44.

DRYDEN, G. L. (1969). Reproduction in *Suncus murinus. In* "Biology of Reproduction in Mammals." pp. 377–396. Blackwell, Oxford.

ENGLAND, B. G., FOOTE, W. C., MATTHEWS, D. H., CARDOZO, A. G., and RIERA, S. (1969). Ovulation and corpus luteum function in the llama (*Lama glama*). *J. Endocrinol.* **45,** 505–513.

EVERETT, J. W. (1961). The mammalian female reproductive cycle and its controlling mechanisms. *In* "Sex and Internal Secretions" (W. C. Young, ed.), 3rd Ed., Vol. II, pp. 497–555. Williams and Wilkins, Baltimore.

FERNANDEZ-BACA, S., MADDEN, D. H. L., AND NOVOA, C. (1970). Effect of different mating stimuli on induction of ovulation in the alpaca. *J. Reprod. Fert.* **22,** 261–267.

HARRISON, J. L. (1952). Breeding rhythms of Selangor rodents. *Bull. Raffles Mus.* **24,** 109–31.

HILL, E. P. (1966). Some effects of weather on cottontail reproduction in Alabama. *Proc. Annu. Conf. Southeastern Ass. Game Fish Commissioners,* **19,** 48–57.

JOLLY, A. (1966). "Lemur Behavior." Univ. of Chicago Press, Chicago.

KOFORD, C. B. (1965). Population dynamics of rhesus monkeys on Cayo Santiago. *In* "Primate Behavior" (I. DeVore, ed.), pp. 160–174. Holt, Rinehart, and Winston, New York.

KREBS, C. J. (1964). The lemming cycle at Baker Lake, Northwest Territories, during 1959–62. *Artic Inst. N. Amer. Tech. Paper No. 15,* 1–104.

KUMMER, H. (1968). "Social Organization of Hamadryas Baboons." Univ. of Chicago Press, Chicago.

MARSDEN, H., AND CONAWAY, C. H. (1963). Behavior and the reproductive cycle in the cottontail. *J. Wildl. Manage.* **27,** 161–170.

MCKENNA, M. F. (1969). The origin and early differentiation of Eutherian mammals. *In* "Comparative and Evolutionary Aspects of the Vertebrate Central Nervous System," pp. 217–240. New York Academy of Science, New York.

RACEY, P. A. (1969). Diagnosis of pregnancy and experimental extension of gestation in the pipistrelle bat, *Pipistrellus pipistrellus*. *J. Reprod. Fert.* **19**, 465–474.

RICHMOND, M., AND CONAWAY, C. H. (1969). Induced ovulation and oestrus in *Microtus ochrogaster*. *In* "Biology of Reproduction in Mammals," pp. 357–376. Blackwell, Oxford.

SADLEIR, R. M. F. S. (1969). "The Ecology of Reproduction in Wild and Domestic Mammals." Methuen, London.

SIMPSON, G. E. (1961). Historical zoogeography of Australian mammals. *Evolution*, **15**, 431–446.

TALEISNIK, S., CALIGARIS, L., AND ASTRADA, J. J. (1966). Effect of copulation on the release of pituitary gonadotropins in male and female rats. *Endocrinology* **79**, 49–56.

WHITTEN, W. K. (1956). Modification of the oestrus cycle of the mouse by external stimuli associated with the male. *J. Endocrinol.* **13**, 399–408.

Early Growth, Development, and Behavior of the Richardson Ground Squirrel (Spermophilus richardsoni elegans)

TIM W. CLARK[1]

Department of Zoology and Physiology, University of Wyoming, Laramie 82070

ABSTRACT: Data on growth and development were collected from five litters (30 individuals) of Richardson ground squirrels (*Spermophilus richardsoni elegans*) born in captivity. Hair first appeared on the head at 10 days of age and fully covered the body by day 25. Incisors erupted at 13-14 (lower) and 22-26 (upper) days of age. On day 14 trills were first noted. Eyes opened 21-24 days and ears opened 21-26 days after birth. Young were weaned at 28-35 days. Body weight increased at an instantaneous growth rate of 11.4% per day during the first weeks after birth and 2.0% at 10 weeks; 25% of body weight growth was completed by day 35, 50% by day 49, 75% by day 60 and 100% by day 100. Of the linear measurements foot length achieved adult size first (100% in 42 days), tail length next (100% in 56 days) and total length last (100% in 63 days).

INTRODUCTION

Twenty-eight species of ground squirrels (*Spermophilus* spp.) are currently recognized (Hall and Kelson, 1959). A review of the literature revealed early growth and development information on only 10 species. The existing data are not complete in all cases and much of the information was gathered in conjunction with studies on other phases of their biology.

The purpose of this paper is to describe the early growth, development, and behavior of the Richardson ground squirrel (*S. richardsoni elegans*) in southeastern Wyoming.

METHODS AND MATERIALS

Litters used in this study were obtained from live-captured pregnant females taken in April 1968 in the short grass prairie south of Laramie, Albany Co., Wyoming. (Squirrels were kept in individual, wire mesh cages (ca. 56 × 51 × 38 cm.) A wooden nest box (ca. 20 × 20 × 20 cm) with a 5 × 5 cm hole cut in one side near the top was placed in each cage. Surgical cotton was used for the nest material.

Unlimited amounts of a mixture of whole wheat, corn, alfalfa pellets, lettuce, cabbage, carrots, fresh green grass, small amounts of deer meet, and Purina dog chow were provided.

Data on growth and development were collected from five litters — a total of 30 individuals. The young were weighed and measured individually at various intervals. Individuals were carefully removed from the nest box and replaced after data collection. Since information is

---

[1] Present address: Department of Biology, Wisconsin State University, Medford Branch Campus, Medford 54451.

197

available on the weight of newly born Richardson ground squirrels (Denniston, 1957), and in an attempt to reduce the possibility of maternal cannibalism, it was decided not to measure and weigh all the young during the first two weeks. At 14 days after birth and at each 7-day interval thereafter, the young of all litters were weighed and measured. Young were toe-clipped for individual identification when two weeks old. The following measurements were taken (in mm) for all the pups, body length, tail length, and right hind foot and body weight in g. Total length was obtained by adding the tail length to the body length.

Percentage of adult size attained by *S. richardsoni* young was based on measurements of a series of adults of both sexes collected in the Laramie Basin of southeastern Wyoming throughout the seasonal activity cycle. Maximum (100%) of adult size was taken as the mean of the measurements recorded from 50 animals.

A system proposed by Brody (1945) to analyze measurements was used. Measurement values were plotted on a logarithmic scale, versus age on an arithmetic scale. On this plot, linear segments indicate periods when growth increments were a constant percentage of previous size. From these linear sections instantaneous growth rates were calculated from the formula:

$$IGR = (\ln M_2 - \ln M_1) / (t_2 - t_1)$$

where IGR is the instantaneous percentage growth rate for the unit of time in which t_2 and t_1 are expressed and $\ln M_2$ and $\ln M_1$ are the natural logarithms of the measurements made at t_1 and t_2.

RESULTS

PHYSICAL DEVELOPMENT AND BEHAVIOR

At two days after birth Richardson ground squirrels are reddish in color. Some blood vessels are visible beneath the relatively smooth and translucent skin. The internal organs are faintly visible in some pups. The pinnae are evidenced only by slight irregularities in the skin. Small vibrissae are present on the snout, the longest of the series extending posteriorly to about the anterior margin of the eye. The fused digits lack pigmentation. The tail is short and tapering. Audible squeaks are emitted. The pups can be sexed at this time, roll about quite actively with no special orientation, and cannot crawl or right themselves when placed on their backs.

By day 5 most pups have lost the stump of their umbilical cords. The toes are still fused, but the claws are fully separated and colored gray at the bases. The dorsal and lateral areas of the head and body are covered with gray pigment. The underparts are still pinkish. The eye region is very darkly pigmented. Pups are eliminating yellow feces at this time. They are barely able to pull themselves forward with their forelegs. Pups are able to right themselves with much apparent difficulty when placed on their backs.

TABLE 1.—Weight (g) and measurements (mm) of young *Spermophilus richardsoni elegans* from birth through 10 weeks of age

Age	N	Weight		Total length		Tail		Right hind foot	
		Mean ± SD	Range	Mean ± SD	Range	Mean ± SD	Range	Mean ± SD	Range
2 Days	6	5.9± 0.45	5.5-6.6	53.2± 2.99	49-57	10.2±0.75	9-11	7.5±0.54	7-8
6 Days	9	8.7± 0.62	7.8-9.9	61.3± 5.0	55-71	12.3±1.22	11-15	9.4±0.52	9-10
8 Days	6	11.7± 0.68	10.7-12.8	78.3± 3.14	74-82	13.1±1.72	11-16	11.7±0.51	11-12
2 Weeks	28	21.0± 6.19	10.7-28.1	102.9±11.27	80-122	19.7±3.12	15-25	17.9±2.68	13-23
3 Weeks	28	34.3± 6.23	23.3-44.5	138.7± 9.40	121-157	29.8±4.08	25-37	25.7±2.39	22-31
4 Weeks	28	48.9± 9.50	33.7-60.5	164.6±10.75	142-180	37.8±3.72	30-45	33.2±1.85	31-38
5 Weeks	28	80.5±20.1	58.5-109.0	191.6±15.92	175-230	48.3±5.82	35-60	36.4±2.11	35-41
6 Weeks	22	125.4±29.95	58.5-166.5	224.6±17.33	186-254	55.0±6.52	41-64	40.2±1.31	35-42
7 Weeks	22	166.1±28.31	122. -208.	250.0±11.44	233-268	64.4±5.14	58-74	40.7±1.18	39-43
8 Weeks	22	200.8±35.61	149. -256.	264.5±11.85	242-282	67.3±4.74	60-76	40.7±2.48	39-43
9 Weeks	22	240.8±30.67	199. -289.	271.4±11.85	250-295	67.5±4.77	62-76	41.3±1.91	39-44
10 Weeks	22	276.3±33.6	230. -329.	283.9±12.7	260-302	67.8±3.77	62-76	41.6±1.61	39-44

At six days the vibrissae are about 4 mm long. Short hair is present on the snout and cheeks. By day 8 all claws are completely pigmented.

At 11 days the lower incisors are visible just below the surface of the gums. They perforate the gum on day 12, and are well above the gum line on days 13 and 14.

At two weeks, the hair at the base of the nose has the characteristic rust-reddish adult color. The tail is covered with very short fine hairs, as is the body. The toes are fully separated and the pups can easily pull themselves forward with their forelegs. When placed on their backs they readily right themselves. The young utter a half-muted trill when disturbed.

At three weeks the upper incisors begin to appear through the gum line and by day 26 have erupted. The eyes of some pups are open on day 21 and by day 24 all young have their eyes open. Both eyes do not always open at the same time. The external auditory meatus is open by day 26. Young frequently leave the nest box shortly after the eyes open. The pups are not as vocal as before. They do not evert their anal glands or "bottle brush" (Bridgewater, 1966) when handled as is characteristic of adults.

At four weeks the young have acquired adult coloration. They

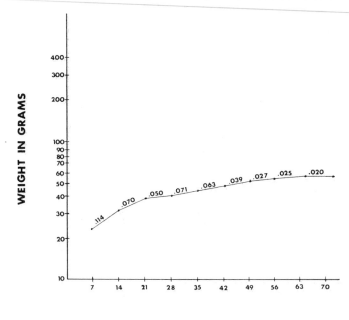

AGE IN DAYS

Fig. 1.—Growth in weight of young *Spermophilus richardsoni elegans* raised in captivity. Value above segments of curve are instantaneous growth rates. The value times 100 equals the percentage of increase per day during a particular age period.

run well and do not wobble while running. Pups urinate when handled and exhibit some resistance. A few pups can evert their anal gland and some attempt to bite. Solid brown feces are beginning to appear.

At five weeks the young are passing dark fecal pellets, apparently indicating a change from milk to a solid diet at the time of weaning (Neal, 1965). Tail-flicking was first noticed at this time.

INCREASE IN BODY WEIGHT

Weight (g) and measurements (mm) of young Richardson ground squirrels from birth through 10 weeks of age are given in Table 1. The mean weight at two days of age of Richardson ground squirrels was 5.9 g (SD = ± 1.5), with a range of 5.5 to 6.6 g. Weight of one-day-old Richardson ground squirrels as reported by Denniston (1957) averaged 5.96 g and for other *Spermophilus* from birth to one day old are: *S. spilosoma*, 3.8 g (Sumrell, 1949); *S. tridecemlineatus*, 2.6 g (Wade, 1927), *S. tereticaudus*, 3.7 g (range 2.7 to 4.7 g), and *S. harrisii*, 3.6 g (range 3.0 to 4.1 g) (Neal, 1965).

Figure 1 shows growth in body weight of young Richardson ground squirrels plotted on a logarithmic scale. The instantaneous growth rate (IGR) varied from week to week, being the most rapid immediately after birth and declining with age. During the first week after birth, weight increased about 11.4% per day, whereas at three weeks of age it increased 7.0% per day and at 10 weeks 2.0% per day. Neal (1965) found that the IGR of body weight in *S. tereticaudus* was 11.0% per day shortly after birth and dropped to 0.53% per day at 84 days of age and in *S. harrisii* was 9.3% per day shortly after birth and dropped to 0.16% per day at 77 days of age.

Figure 2 shows the rate at which adult weight and measurements were achieved in Richardson ground squirrels. Twenty-five per cent of weight growth is completed by the 35th day, 50% by day 49, 75% by day 60, 90% by day 75 and 100% by the 100th day. In *S. tereticaudus*, 25% of weight growth is completed by 34th day, 50% by day 50, 75% by day 68 and 90% by day 79, and in *S. harrisii*, 25% of weight growth is completed by day 38, 50% by day 56, 75% by day 75 and 90% by day 150 (Neal, 1965).

Weight and measurements taken from wild young at the time of their first emergence from their nest burrows, compared with weights and measurements of young Richardson ground squirrels raised in captivity, suggested that the wild young were about 5 to 6 weeks of age. As pointed out by Neal (1965), field weight is highly sensitive to food supply and environmental stress; growth therefore can be assumed to be somewhat more rapid in the laboratory than in the field.

INCREASE IN LINEAR MEASUREMENTS

Figure 3 presents the growth of the right hind foot, tail length and total length of young *Spermophilus richardsoni elegans* raised in captivity. The hind foot growth rate is generally the most rapid, with the tail growth rate next. During the first two weeks after birth the growth rate of the right hind foot was 7.76% per day; that of the total length was 5.49% per day and of the tail 5.48% per day. *Spermophilus tereticaudus* during the first two weeks of age gained 5.7% per day for tail growth, 4.9% per day for the hind foot growth and *S. harrisii* gained 4.9% per day for tail growth and 4.8% per day for the hind foot.

The hind foot attained 25% of maximum growth at 7 days, 50% at 16 days, 75% at 26 days, 90% at 35 days and 100% at 42 days. In total length the 25% point of maximum growth was obtained in 7 days, 50% in 20 days, 75% in 37 days, 90% at 46 days and 100% at 63 days. Twenty-five per cent of maximum tail growth was reached in 10 days, 50% at 26 days, 75% at 39 days, 90% at 45 days and 100% in 56 days.

In *S. tereticaudus,* the hind foot attained 50% of maximum growth in 22 days and 90% in 66 days, and in *S. harrisii,* the hind foot reached the 50% point in 23 days and the 90% in 65 days. Tail measurements in *S. tereticaudus* attain the 50% point in 27 days and 90% in 210 days, whereas *S. harrisii* attained the 50% in 30 days and 90% in 217 days (Neal, 1965).

DISCUSSION

Information on various aspects of growth and development of several species of *Spermophilus,* including *S. harrisii, S. townsendi, S. richardsoni, S. columbianus, S. undulatus, S. tridecemlineatus, S.*

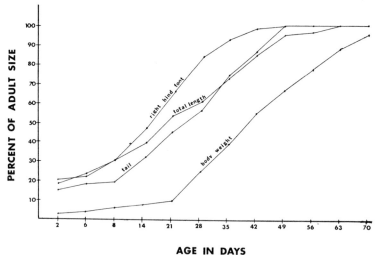

AGE IN DAYS

Fig. 2.—Increase in weight and measurements of young *Spermophilus richardsoni elegans* raised in the laboratory in terms of per cent adult weight and measurements at the indicated ages.

spilosoma, S. beecheyi, S. tereticaudus, and *S. lateralis* has been given by several authors (Neal, 1965; Svihla, 1939; Denniston, 1957; Shaw, 1925; Mayer and Roche, 1954; Bridgewater, 1966; Blair, 1942; Tomich, 1962; McKeever, 1964). These data, although not complete in all cases, permit at least a gross comparison of development and growth rates of a group of species of varying habits and habitats.

Juvenile ground squirrels emerge from their nest burrow with certain characteristics for survival already developed. At the time of emergence locomotor abilities, vocalizations, eye and ear opening, acquisition of teeth, ability to "bottle brush" the tail and evert the anal glands are usually accomplished. Ground squirrels that hibernate have a more rapid developmental rate during growth than non-hibernating ground squirrels of the same general size (Neal, 1965). Table 2 demonstrates this difference in growth rates and development between hibernators and nonhibernators. Acquisition of hair on head, eruption of lower and upper incisors, first trill, body fully covered

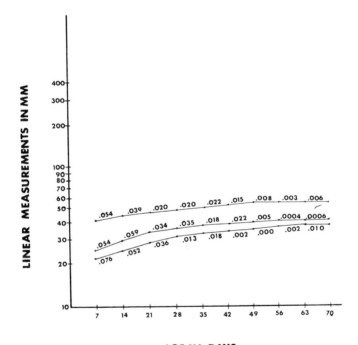

AGE IN DAYS

Fig. 3.—Growth in total length (top curve), tail (middle curve), right hind foot (bottom curve) and tail (bottom curve) of young *Spermophilus richardsoni elegans* raised in captivity. Values above segments of curve are instantaneous growth rates. The value times 100 equals the percentage of increase per day during a particular age period.

TABLE 2.—Comparison of the development and growth rates of young *Spermophilus* as reported in the literature with *S. richardsoni elegans*. Numbers are days since birth

DEVELOPMENTAL AND GROWTH FEATURES

Species	Hair on head	Lower incisors erupted	Upper incisors erupted	First trill	Fully covered with short hair	Eyes open	Ears open	Weaned	Hibernator or nonhibernator
S. harrisii harrisii (Neal, 1965)	14	21	28	28	21-28	29-34	21-28	49	NH
S. townsendi (Svihla, 1939)	8	16	….	16	….	19-22	….	"Soon after eyes open"	H
S. richardsoni elegans	10	13-14	22-26	14	25	21-24	21-26	28-35	H
S. columbianus (Shaw, 1925)	7	14-19	by 19	14	7	19-23	….	30	H
S. undulatus barrowensis (Mayer & Roche, 1954)	10	….	….	18	10	20	….	22	H
S. tridecemlineatus (Bridgewater, 1966)	4	14-17	16-19	12	8	21-23	….	29	H
S. spilosoma major (Blair, 1942)	….	….	….	34	….	27-28	….	48	NH
(Sumrell, 1949)	….	….	20-26	25-30	23-24	20-26	….	31	….
S. beecheyi (Tomich, 1962)	13	21	28	….	20	34-37	24-28	60-74	H?
S. tereticaudus neglectus (Neal, 1965)	4	7	21	….	21	25-27	21	35	H
S. lateralis (McKeever, 1964)	8	….	….	15	15	27-31	….	….	H

with short hair, eye and ear opening, and weaning appear to occur earlier in hibernating than in nonhibernating ground squirrels (Table 2).

Although data exist on only 10 of the 28 species of ground squirrels, the foregoing comparison indicates that hibernators develop at a relatively more rapid rate than nonhibernators. The relatively rapid growth and development of hibernating ground squirrels would appear to have adaptive significance. This not only probably reduces the length of time young must remain confined to the nest but also may allow for the earlier acquisition of the necessary motor and behavioral agility required to survive aboveground. An accelerated ontogeny would conceivably be a distinct advantage to hibernating ground squirrels that have to grow rapidly and store an adequate quantity of fat in a limited amount of time in order to survive the period of inactivity.

Acknowledgments.—I wish to acknowledge Drs. R. H. Denniston, C. A. McLaughlin, and J. A. Swatek for reviewing the manuscript.

REFERENCES

BLAIR, F. W. 1942. Rate of development of young spotted ground squirrels. *J. Mammal.*, **23**:342-343.

BRIDGEWATER, D. D. 1966. Laboratory breeding, early growth development and behavior of *Citellus tridecemlineatus* (Rodentia). *Southwestern Natur.*, **11**:325-337.

BRODY, S. 1945. Time relations of growth of individuals and populations, p. 487-574. *In* S. Brody, Bioenergetics and growth. Reinhold, New York. 1023 p.

DENNISTON, R. H. 1957. Notes on breeding and size of young in Richardson ground squirrel. *J. Mammal.*, **38**:414-416.

HALL, E. R. AND K. KELSON. 1959. The mammals of North America. Ronald Press, New York. 1083 p.

MAYER, W. V. AND E. T. ROCHE. 1954. Development patterns in the Barrow ground squirrel, *Spermophilus undulatus barrowensis*. *Growth*, **18**: 53-69.

MCKEEVER, S. 1964. The biology of the golden-mantled ground squirrel. *Ecol. Monogr.*, **34**:383-401.

NEAL, B. J. 1965. Growth and development of the round-tailed and Harris antelope ground squirrels. *Amer. Midl. Natur.*, **73**:479-489.

SHAW, W. T. 1925. Breeding and development of the Columbian ground squirrel. *J. Mammal.*, **6**:106-113.

SUMRELL, F. 1949. A life history study of the ground squirrel (*Citellus spilosoma major* (Merriam)). M.S. thesis, Univ. of New Mexico, Albuquerque.

SVIHLA, A. 1939. Breeding habits of Townsend's ground squirrel. *Murrelet*, **20**:6-10.

TOMICH, P. Q. 1962. The annual cycle of the California ground squirrel, *Citellus beecheyi. Univ. California Publ. Zool.*, **65**:213-282.

WADE, O. 1927. Breeding habits and early life of the thirteen-striped ground squirrel, *Citellus tridecemlineatus* (Mitchell). *J. Mammal.*, **8**:269-275.

SUBMITTED 21 AUGUST 1968   ACCEPTED 6 DECEMBER 1968

Article IX.—CRANIAL VARIATIONS IN NEOTOMA MICROPUS DUE TO GROWTH AND INDIVIDUAL DIFFERENTIATION.

By J. A. ALLEN.

PLATE IV.

In view of the stress naturally, and very properly, laid upon the importance of cranial characters in the discrimination of species in groups of closely-allied forms, it seems desirable to ascertain the character and amount of change in not only the general form of the skull but in the form of its separate bones due to growth, and also to determine the amount and kind of individual variation that may be expected to occur in skulls unquestionably of the same species. Having of late had occasion to examine a large amount of material relating to the genus *Neotoma*, the subject has been forcibly brought to my attention, and some of the results of a careful examination of a large series of skulls pertaining to several species of this genus are here presented. No attempt is made to treat the subject exhaustively, only a few special points being here presented.

As is well known to all experienced workers in mammalogy, the general contour of the brain-case, the relative size and form of individual bones, notably the interparietal, and the condition of the supraorbital and other ridges for muscular attachment, alter materially after the animal reaches sexual maturity; the deposition of osseus matter, the closing of sutures, the building out of crests and rugosities continuing throughout life, so that a skull of a very old animal may differ notably from that of an individual of the same species in middle life, and this latter from one just reaching sexual maturity.

The Museum has at present a large series of specimens of *Neotoma micropus* Baird, including ages ranging from nursling young to very old adults. They are mainly from three localities in the eastern coast district of Texas, namely, Brownsville, Corpus Christi, and Rockport. In order to avoid any complications that

233

might arise through geographic variation, only the specimens from
Rockport and Corpus Christi—localities less than twenty-five
miles apart, and similar in physical conditions—are here considered. There is not the slightest reason for questioning their conspecific relationship. The series selected to illustrate variations
due to age are, with one exception, from Rockport ; those figured
to show individual variation are all from Corpus Christi.

Variations due to Age.

General Contour.—The variation in the general form of the
skull resulting from growth is due mainly to the lengthening of
the several skull segments without a corresponding relative increase in the breadth of the skull. Hence in the young skull, in
comparison with an adult skull of the same species, the braincase is disproportionately large in comparison with the anteor-
bital and basal portions of the skull. This is well shown in
Plate IV, and in the subjoined table of measurements of three

Measurements and Ratios showing Cranial Variations due to Age

in *Neotoma micropus.*

	No. 5834, ♀ juv.	Ratio[1]	No. 4480, ♂ juv.	Ratio[1]	No. 4478, ♂ very old.	Ratio[1]
Occipito-nasal length	31	100	41	100	53	100
Length of nasals	10	32.3	14.5	35.4	22	41.5
Length of frontais	13	42	15	36.6	18	34
Length of parietals on median line	5	19.4	6	14.6	8	15
Greatest length of parietals	12	39	15	36.6	16	30.2
Length of interparietal	4.5	14.5	5.5	13.4	7	13.2
Length of brain-case	14	45.2	17	41.5	21	39.6
Greatest rostral breadth	5.5	17.7	6.3	15.4	6.5	12.3
Least interorbital breadth	6	19.4	6	14.6	6	11.3
Breadth of brain-case	16	51.6	19.5	45	20	38
Breadth of interparietal	11	35.5	10	24.4	7.5	14.2
Greatest zygomatic breadth	20 ?	64.6	23	56.1	30	56.6
Depth of skull at middle of palate	8	26	11	26.8	15	28.5
Depth of skull at front of basisphenoid	11	35.5	12	29.3	14	26.4
Length of tooth-row (crown surface)	8[2]	25.8	8	19.5	9	17
Length of incisive foramina	6	19.3	8.5	20.7	11.5	21.7
Width of incisive foramina	3	9.7	3	7.3	3.5	6.6
Length of palatal floor	5	16.1	7	17	7	13.2

[1] Ratio to occipito-nasal length.
[2] From No. 4482, ♀ juv., in which the last molar has just come into use.

specimens of *N. micropus* from Rockport, Texas.   No. 5834, ♀ juv., is a nursling so young that the last molar is still wholly enclosed in the jaw;[1] No. 4480, ♂ juv., though not quite full-grown, would pass as a 'young adult'; No. 4478, ♂ ad., is a very old male, with the teeth well worn down, and the fangs visible at the alveolar border.   Other specimens in the series furnish a complete series of gradations between the two extremes (Nos. 5834 and 4478).

In general contour (Figs. 1–11, Pl. IV), the young skull, in comparison with adults, is much more convex in dorsal outline,[2] very broad posteriorly, and very narrow anteriorly.   In comparing the relative length of the several skull segments the occipito-nasal length is taken as the basis, and the skulls will be referred to as *A* (=No. 5834), *B* (=No. 4480), and *C* (=No. 4478).

Rostral Segment.—In *A* the ratio of the rostral segment to the total length is 32.3 per cent.; in *B*, 35.4; in *C*, 41.5—giving a rapid *increase* in the ratio with age.

Frontal Segment.—In *A* the ratio of the frontal segment—*i. e.,* the distance between the naso-frontal and fronto-parietal sutures —to the total length is 42 per cent.; in *B*, 36.6; in *C*, 34—a considerable *decrease* in the ratio with age.

Parietal Segment.—In *A* the ratio of the parietal segment— *i. e.,* the distance from the latero-anterior angle of the parietal bone on either side to the occipito-parietal suture—to the total length is 39 per cent.; in *B*, 36.6; in *C*, 30.2—again a rapid decrease in the ratio.

Brain-case.—The length of the brain-case in *A* is 51.6 per cent. of the total length of the skull; in *B*, 45; in *C*, 38.

In each case the change in ratio is due to the disproportionate growth of the rostral portion of the skull.   Thus in *A* the nasals have a length of only 10 mm.; in *B* they have increased to 14.5 mm., and in *C* to 22 mm., while the total occipito-nasal length of

---

[1] The length of the tooth-row given in the table is taken from an older specimen (No. 4482, ♀ juv.), in which the last molar has reached the level of the others and is just beginning to show traces of wear.

[2] In Figs. 10 and 11 it should be noted that the greater flatness of the skull interorbitally, as compared with Fig. 6, is masked by the raised supraorbital borders in the older skulls when viewed in profile.

the skull has increased only from 31 mm. in *A* to 53 mm. in *C*. In other words, the nasal bones have increased in length 120 per cent., while the total length has increased only 77 per cent.

Transverse Breadth.—In respect to the breadth of the skull the variations with growth are much less than in its length. Thus the greatest diameter of the rostrum varies only from 5.5 mm. in *A* to 6.5 in *C*—an increase of about 20 per cent. in the breadth of the rostrum, against an increase of 120 per cent. in its length. The interorbital breadth remains nearly constant, being 6 mm. in all three of the skulls here compared. The width of the brain-case shows an increase of 25 per cent. against an increase in the total length of the skull of 77 per cent. The zygomatic breadth shows an increase of about 50 per cent., due almost wholly to the thickening and increased convexity of the zygomatic arches.

Vertical Depth.—In respect to the depth of the skull, the variations with age prove especially interesting, although only such as would be expected from the facts already given. For present purposes the depth of the skull is taken at two points, namely, (*a*) at the middle of the palatal region, and (*b*) at the posterior border of the basisphenoid (basisphenoid-basioccipital suture). The palatal depth increases markedly with age, correlatively with the growth of the rostrum ; the basisphenoidal depth changes but slightly after the molars have attained to functional development. Thus in *A* the basisphenoidal depth is 11 mm. ; in *B*, 12 mm. ; in *C*, 14 mm.—an increase of about 28 per cent. The palatal depth in *A* is 8 mm. ; in *B*, 11 mm. ; in *C*, 15 mm.—an increase of nearly 88 per cent.

Tooth-row.—The length of the upper tooth-row varies about 12 per cent., due almost wholly to the wearing down of the teeth, the length of the crown surface being much less, in slightly worn teeth, than the length taken at the alveolar border.

Interparietal.—The interparietal shows surprising modification with age, both as to size and form, but especially in respect to the latter. At early stages, as in *A*, this bone is more or less crescentic in shape, with the transverse diameter more than twice the antero-posterior diameter. Thus in *A* the two diameters are respectively 11 and 4.5 mm. ; in *B*, 10 and 5.5 mm. ; in *C*, 7.5 and 7 mm. In other words, the short, broad, convex sub-crescentic interparietal in *A* becomes transformed in *C* into a squarish, flat bone in which the two diameters are nearly equal, instead of the transverse being twice as great as the antero-posterior, as in *A*. This would be almost incredible were not the proof so abundantly furnished by the material in hand, where every stage of transition is shown. (Figs. 1–8, Pl. IV.) This change is coincident with the development of the raised supra-orbital borders and their prolongation backward as ridges to the parieto-occipital suture, and the flattening of the whole dorsal aspect of the post-rostral portion of the skull. In old age these ridges become confluent with the lateral edges of the interparietal which has now lost its postero-lateral moieties, partly apparently by absorption and partly by their being overgrown by the mediad posterior angle of the parietals. A sharp thin ridge for muscular attachment also extends back from the posterior base of the zygomatic arch. The interparietal at the same time develops a more or less prominent median angular projection at its posterior border, confluent with the median ridge of the supraoccipital. The contrast between these conditions, obtaining only in very old skulls, and their almost entire absence in skulls which have just reached sexual maturity, is strikingly great.

Supraoccipital.—The supraoccipital changes from a posteriorly convex, thin lamina of bone, in early life, to a thick, nearly vertical plate, with a strongly-developed median ridge produced into an angular spine at its superior border, and with a lateral ridge on either side about midway between the median line and its lateral borders ; these lateral ridges also each develop an angular rugosity or process about midway their length. The superior border is also produced into an incipient occipital crest.

Basioccipital.—The basioccipital becomes greatly altered by growth, as in fact is the case with the whole postpalatal region. In comparing stages *A* and *C* it is found that the distance across the occipital condyles increases only about 15 per cent., while the breadth of the anterior border increases 100 per cent., and the length about 50 per cent. (Figs. 12-14, Pl. IV.)

Basisphenoid.—The basisphenoid doubles in length, and its anterior third becomes differentiated into a narrow projecting neck.   The presphenoid at stage *A* is nearly hidden by the palatal floor. (Figs. 12-14, Pl. IV.)

Postpalatal Region as a whole.—This doubles its length with an increase in breadth of only about 50 per cent.   At stage *A* the postpalatal border terminates slightly behind the posterior edge of M.2 ; in stage 3 it holds very nearly the same position.   The distance between the postpalatal border and the front border of the auditory bullæ, compared with the total length of the skull, is as 1 to 9 in *A*, and as 1 to 5 in *C*.   In *A* the pterygoid hamuli reach the second fourth of the bullæ ; in *C* they terminate slightly in advance of the bullæ.   The bullæ themselves in *A* are more obliquely placed than in *C*, in relation to the axis of the skull, and are quite differently shaped.   Also the form of the foramen magnum has undergone much change.   These points are all well shown in Figs. 12-14 of the accompanying plate.

Incisive Foramina.—Consequent upon the growth of the rostral portion of the skull, the incisive foramina undergo marked change in form, and somewhat in position, as regards both their anterior and posterior borders.   In the stage designated as *A* they are short and broad, and extend relatively further both anteriorly and posteriorly than in stage *B* or *C*, their anterior border being nearer the base of the incisors, and their posterior border being carried back to or slightly behind the front border of the first molar.   Thus in *A* the length of the incisive foramina is 6 mm., with a maximum breadth of 3 mm., while in *C* the dimensions are respectively 11.5 and 3.5 mm.—a great increase in length with only slight increase in breadth.   At the same time the anterior border is considerably further from the base of the incisors, and the posterior border is slightly in advance, instead of slightly behind, the front border of the molars.

Spheno-palatine Vacuities.—In adults of *Neotoma micropus*, as in other species of the ' round-tailed ' section of the genus, there is a long, broad vacuity on each side of the presphenoid and anterior third of the basisphenoid, which Dr. Merriam has recently named[1] the '*spheno-palatine vacuities*,' and he has also called attention to the fact that they are not present in some forms of the 'bushy-tailed' section of the genus. It is therefore of interest in the present connection to note that these vacuities are absent at stage *A*, and are only partially developed at later stages (Figs. 12-14, Pl. IV). My attention was called to the matter by finding several nearly fully-grown skulls from Texas and northeastern Mexico with these vacuities either quite absent or represented by an exceedingly narrow slit, while I could find no differences in the skins or in other cranial characters that gave the slightest hint that the animals were not referable to *N. micropus*. Further examination of young skulls of undoubted *N. micropus* from Rockport and Corpus Christi, Texas, showed that the closed condition was in this species a feature of juvenility. It is thus of interest to find that a feature which proves to be merely a character of immaturity (and quite inconstant as well) in *N. micropus* is a permanent condition in *N. cinerea occidentalis*.[2]

In the development of these vacuities it appears that as the presphenoid increases in length it becomes reduced in width; at the same time, as the skull broadens, the edges of the ascending wings of the palatine bones become slightly incised. There is, however, much individual variation in this respect, as will be shown later.

Molars.—When the molars first cut the gum they have nearly the entire crown-surface capped with enamel. Very soon, even before the tooth has attained its full height, the enamel begins to disappear from the centers of the enamel loops, the capping remaining longer over the narrower loops than over the broader ones; it quickly disappears from all as soon as the crown-surface becomes subject to wear. In stage *A*, in which only M.1 and M.2 have appeared, and are less than one-third grown, the enamel walls of the loops nearly meet over the dentinal areas— quite meeting over the narrower portions, especially in the case of the middle transverse loop of each tooth. Some time before the age represented by *B* is reached, the crown-surface is worn to an

[1] Proc. Biol. Soc. Wash., VIII, p. 112, July, 1893.

[2] Unfortunately the outline figures here given (Figs. 12–15, Pl. IV,) fail to show clearly the points at issue.

even plane ; the tooth has reached its normal length, but the fluting of the sides still extends to the alveolar border.   As attrition goes on, with the advance of the animal in age, the crown-surface wears down, and the neck of the tooth appears above the alveolar border, till, especially in the upper molars, the fluted terminal and the smooth basal portions are of nearly equal extent ; but in old age (as in *C*) the smooth basal portion is the longer and the division of the root into fangs is clearly shown. With this wearing down the tooth increases somewhat in both width and length, but the pattern of the enamel folds undergoes but slight change until nearly the whole crown is worn away, except that the angles become gradually more rounded.

Résumé.—As already stated the change with age in the general form of the skull is due to the relatively disproportionate increase in length of the pre- over the post-orbital region, and the same disproportionate increase of the basal region as compared with the frontoparietal elements.   In the first case the rostrum becomes relatively greatly produced ; in the second the basioccipital and adjoining parts become so greatly enlarged as to change the entire aspect of the basal region of the skull.   Thus the occipital condyles, which in *A* terminate slightly in advance of the most convex portion of the supraoccipital, and are crowded up very close to the bullæ, form in *C* the most posterior part of the skull, with a considerable interval between them and the bullæ.   (Figs. 12–14, Pl. IV.)

INDIVIDUAL VARIATION.

In comparing a large series of skulls of the same species it quickly becomes apparent that no element of even the adult skull is constant, either as to form or relative size.   There is also much variation in the size of skulls of the same sex and approximately the same age.

Variation in Size.—Thus in *Neotoma micropus,* from the same locality, there are dwarfs and giants.   While the females average smaller than the males, size is by no means a safe criterion of sex.   Thus two old females, not appreciably different in age, from Corpus Christi, Texas, vary as follows : No. 2948, total length 51 mm., zygomatic breadth 26 mm. ; the corresponding dimensions in No. 2955 are 45 mm. and 24 mm. These are merely the extremes of a series of six specimens ; with a much larger series doubtless the difference would be considerably increased. A series of six old males, from the same locality and indistinguishable as to age, vary as follows : No. 2952, total length 50.5 mm., zygomatic breadth 27 mm. ; the corresponding dimensions in No. 2956 are 45 mm. and 25 mm.

Nasals and ascending branches of the Premaxillæ.—Ordinarily in *N. micropus* the nasals terminate in a gradually narrowed evenly rounded point, a little less than 2 mm. in front of the posterior termination of the ascending branches of the premaxillæ. The distance between the points of termination of the nasals and premaxillæ, however, frequently varies between 1.5 and 2.5 mm. ; more rarely from 1 to 3 mm. These extremes each occur in the ratio of about 10 per cent. of the whole, while probably 60 per cent. would not vary much from the normal average of about 2 mm. (See Figs. 1–8 and 16, 17, Pl. IV.)

The nasals, as already said, usually terminate in an evenly rounded point, but in several of the 50 skulls of *N. micropus* before me their posterior border forms a double point, *each* nasal terminating in a distinctly rounded point ; in one or two the posterior border is squarely truncate ; in others it is irregularly uneven. The ascending branches of the premaxillæ usually terminate in an obtusely V-shaped point, with a uniformly even outline, their breadth, however, being subject to variation ; in some specimens they terminate in a brush of irregular spiculæ. (Figs. 1–8 and 16, 17, Pl. IV.)

Frontals.—The posterior border of the frontals is subject to great irregularity, varying from a nearly transverse line (rounded slightly at the outer corners) to a gentle, rather even convexity, and thence to an acute angle, involving the whole posterior border. It is difficult to decide what outline is the most frequent, though the tendency seems to be greatest toward a well-pronounced rather even convexity. Figures 1–8 and 18, 19, Plate V, well show the variation in the position and direction of the fronto-parietal suture.

[*September, 1894.*]

Parietals.—The anterior outline of the parietals of course conforms to the posterior outline of the frontals, and must be equally variable. It hence follows that their length on the median line is also variable. Their posterior border is also subject to much variation in consequence of the great diversity in the form of the interparietal.

Interparietal.—In middle-aged specimens the interparietal tends strongly to a quadrate form, varying from quadrate to diamond shape, through a more or less marked median angular extension of both its anterior and posterior borders, and occasionally of its lateral borders as well. Often it forms a quadrate figure, in which each of its four sides is slightly convex ; again the corners are so much rounded, and the lateral breadth so much in excess of the antero-posterior, as to give a lozenge-shaped figure. In other cases it is distinctly shield-shaped ; in others it is hexagonal. In size the variation is fully 50 per cent. of what may be regarded as the average dimensions. These remarks have strict reference to fully adult specimens, and as nearly as can be judged these variations are not at all due to differences of age, which, as already shown, has so great an influence upon the size and form of this exceedingly variable element of the skull.[1] (Figs. 20–23, Pl. IV. Compare also the interparietal, as shown in Figs. 1–8.)

Ventral aspect.—The ventral aspect of the skull presents numerous points of variability, only a few of which will be here mentioned. The palate varies more or less in breadth, and especially in the development of the anterior palatal spine, which is sometimes slight, and sometimes so strongly produced anteriorly as to touch the vomer. The postpalatal border may be evenly concave, or present a slight median process. The presphenoid is very variable in size, being often an exceedingly slender rod of bone, and at other times very stout, the variation in thickness being nearly or quite 100 per cent. The anterior third of the basisphenoid shares in the same variability. As the

---

[1] As regards variation with age in the form of the interparietal, *Neotoma micropus* is only an example of what doubtless prevails throughout the genus, and even in many other genera as well. Yet in adult animals the form of this bone seems, as a rule, to be sufficiently constant to be of more or less taxonomic value. Thus in the *N. cinerea* group it may be said to be normally quadrate ; in the *N. fuscipes* group it is quite constantly shield-shaped. In *N. floridana*, however, and in the *N. mexicana* group, it seems to be nearly or quite as variable as in *N. micropus*, both as to size and shape.

ascending borders of the palatals are also variable in respect to the extent of their development, it follows that there is, even among adults, a wide range of variation in the size of the spheno-palatine vacuities.

Teeth.—Aside from differences due to age and attrition, the teeth vary in size to a considerable extent among individuals strictly comparable as to sex and age, some having a much heavier dental armature than others. But more particularly note-worthy in this connection is the variation in the color of the teeth, which seems strongly a matter of individuality. Although Dr. Merriam has recently placed *N. micropus* in his " *Neotoma leucodon* group,"[1] which has, among other alleged characters, " color of teeth white or nearly white," the teeth in *N. micropus* average blacker than in any other species of the genus known to me. Were this all it might be considered that *N. micropus* was erroneously referred to the ' *leucodon* group '; but unfortunately the range of individual variation in the color of the teeth in the large series at hand covers also the whole range of variation for the genus. Thus in some instances the molar teeth are intensely black from base to crown, while the crown-surface itself is strongly blackish, even the enamel loops, as well as the enclosed dentine being tinged with blackish ; in other cases the teeth are merely slightly tinged with brownish near the base and at the bottom of the sulci. These extremes are connected by a series of very gradual intergradations. In other words, among hun-dreds of skulls of *Neotoma*, those with the blackest teeth occur in *N. micropus*, as well as those in which the teeth are practically white.

In the suckling young the teeth are pure white ; before M.3 has come to wear, M.1 and M.2 have become more or less blackened ; in young adults, and in middle aged specimens, the teeth are often intensely black ; in old specimens, with the teeth much worn, the teeth average lighter than in the younger indi-viduals. There is, however, a wide range of variation in the color of the teeth in specimens of corresponding age, whether old or young. The black coloring consists to a large extent of a

---

[1] Proc. Biol. Soc. Wash., IX, p. 118, July 2, 1894.

superficial incrustation which tends to scale off in flakes in the prepared skull, and its absence apparently may be due sometimes to removal in the process of cleaning the skull for the cabinet. In other words, the blackness is to some extent an accidental or pathological condition, due probably more or less to the particular character of the food or to the health of the animal.

General Remarks.

The bearing of what has been stated above respecting variations in the form of the skull and of its principal elements due to age is of course obvious, the inference being that in animals which have reached sexual maturity variations due wholly to growth, in passing through adolescence to senility, may readily be mistaken, when working with very small series or with single specimens, for differences of subspecific or even specific importance. Not only do the individual bones vary in their outlines and proportions and in relative size, but the skull varies as a whole in its relative dimensions, including depth as well as length and breadth. There is beside this a wide range of purely individual variation, affecting every character that can be used in a diagnostic sense. Thus in a series of fifty skulls of *Neotoma micropus* it would be easy to select extremes, of even individual variation, that depart so widely from the average, in one or more characters, as to deceive even an expert, on considering these alone, into the belief that they must represent very distinct species ; yet in the present instance the proof that such is not the case is overwhelming. In *N. micropus* the coloration is remarkably constant, for a member of this genus, at all seasons and ages, so that the case is less complicated than it would be in many other species of the group, where the color of the pelage varies radically with season and age.

Personal criticism is not the purpose of the present paper, and it was not my intention at the outset to refer specifically to the work of any of my *confrères*. Since its preparation was begun, however, its *raison d'être* has perhaps been emphasized by the publication of two brochures of ' preliminary descriptions ' of species and subspecies of the genus *Neotoma*, numbering altogether 10 species and 8 subspecies, which added to the 22 species and subspecies previously standing practically unchallenged, makes, at the present writing, a total of 40 forms of the genus *Neotoma*. Of these no less than 26 have been described within the last nine months.[1] Without the material before me used by the original describers of these forms it would be presumptive to give an opinion respecting the merits of many of them. While the greater part may have some real basis, it is evident that others are almost unquestionably synonyms of previously-described forms, judging by 'topotypes' in this Museum, the brief diagnoses accompanying the names affording in these cases no characters that are in the least degree distinctive.

The genus *Neotoma* was chosen for treatment in this connection in preference to some other almost solely by chance, as the facts of variation above presented are not at all exceptional. In fact the common muskrat (*Fiber zibethicus*) would have shown a still more striking case of variability, as would also various species of many other genera. Yet describers of new species are constantly laying stress upon cranial differences that have not necessarily the slightest specific or even subspecific importance; and, so far as can be judged from their descriptions, they are entirely unconscious that such can be the case.

On the other hand, it is equally certain that such alleged characters may have the value assigned them; since it is now a well known fact that the extremes of purely individual variation in any character, external or internal, may exceed in amount the average differences that serve to satisfactorily distinguish not only well-marked subspecies, but even forms that are unquestionably specifically distinct. Hence it must often happen that the determination of the status of a species or subspecies originally described from one or two specimens, in groups especially susceptible to variation, must depend upon the subsequent examination of a large amount of material bearing upon this and its closely-related forms.

---

[1] For a list of the species and subspecies of *Neotoma* described prior to July 6, 1894, see Abstr. Proc. Linn. Soc. New York, No. 6, pp. 34, 35, July, 1894.

EXPLANATION OF PLATE IV.

Figures all Natural size.

Neotoma micropus *Baird.* Showing cranial variations due to age and individualism. (Unless otherwise stated, the specimens are from Rockport, Texas.)

Figs. 1–8. Dorsal aspect of skull, showing gradual change in form with age, and especially in the form and relative size of the interparietal. Fig. 1, No. 5834, ♀ juv. (suckling). Fig. 2, No. 2975, ♀ juv. (nearly sexually adult), Corpus Christi, Texas. Fig. 3, No. 5841, ♀ ad. Fig. 4, No. 4480, ♂ ad. Fig. 5, No. 2958, ♂ ad., Corpus Christi. Fig. 6, No. 4479, ♂ ad. Fig. 7, No. 4477, ♀ ad. Fig. 8, No. 4478, ♂ ad.

Figs. 9–11. Skull in profile, to show change of form with growth. Fig. 9, No. 5834, ♀ juv. (nursling). Fig. 10, No. 4480, ♂ ad. (rather young). Fig. 11, No. 4478, ♂ ad. (very old).

Figs. 12–15. Ventral aspect, showing variations in postpalatal region due to age. Fig. 12, No. 5834, ♀ juv. (nursling). Fig. 13, No. 5841, ♀ ad. (young adult). Fig. 14, No. 2958, Corpus Christi, ♂ ad. (very old). Fig. 15, No. 1456, *Neotoma cinerea occidentalis*, ♂ ad., Ducks, B. C. (for comparison with *N. micropus*).

Figs. 16, 17. To show extremes of individual variation in relative posterior extension of nasals and ascending branches of premaxillæ. Locality, Corpus Christi, Texas. Fig. 16, No. 2958, ♂ ad. Fig. 17, No. 2948, ♀ ad.

Figs. 18, 19. To show extremes of individual variation in posterior border of frontals. Locality, Corpus Christi, Texas. Fig. 18, No. 2949, ♂ ad. Fig. 19, No. 2951, ♂ ad.

Figs. 20–23. To show individual variation in the size and form of the interparietal. Specimens all from Corpus Christi, Texas. Fig. 20, No. 2949, ♂ ad. Fig. 21, No. 2948, ♀ ad. Fig. 22, No. 2952, ♂ ad. Fig. 23, No. 2945, ♀ ad.

NOTE.—If the Brownsville, Texas, series of specimens had also been included, the range of individual variation would have been considerably increased.

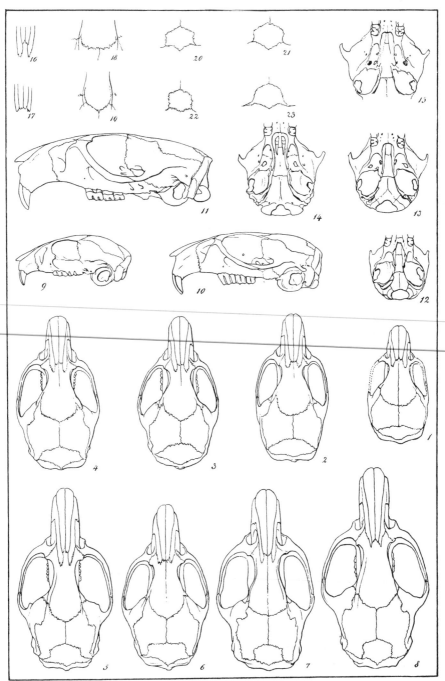

Neotoma micropus.

Figures nat. size.

MATURATIONAL AND SEASONAL MOLTS IN THE GOLDEN MOUSE, *OCHROTOMYS NUTTALLI*

Donald W. Linzey and Alicia V. Linzey

Abstract.—The adult pelage of the golden mouse (*Ochrotomys nuttalli*) is attained by a single maturational molt. Data on the post-juvenile molt were obtained from 96 young golden mice. This molt began on the ventral surface and spread dorsally, meeting in the dorsal midline. It then proceeded anteriorly and posteriorly. The average age at which male golden mice began molting was 36 days, whereas that of females was 38 days. The average duration of molt for the sexes was 29 days and 25 days, respectively. Golden mice undergo two seasonal molts—spring and fall. Data were obtained from 36 mice. The winter pelage was generally much darker than the summer pelage. Both spring and fall molts were more irregular than the post-juvenile molt, and the spring molt tended to be more irregular than the fall molt. Young golden mice born after 1 October and 8 April appeared to combine the post-juvenile and seasonal molt. Hair replacement was more irregular than during the normal post-juvenile molt.

During the course of a study on the ecology and life history of the golden mouse, *Ochrotomys nuttalli nuttalli*, in the Great Smoky Mountains National Park (Linzey, 1966), considerable data were obtained on pelage changes. The limited data presented by Layne (1960) have been the only published information concerning molt in this species.

Maturational Molt

The adult pelage of the golden mouse is attained after a single maturational molt. Data on the post-juvenile molt were obtained from 96 young golden mice. Eighty-four of these mice were raised in captivity. Data from the remaining 12 individuals were obtained from field observations.

The molt from the golden-brown juvenile pelage to the golden-orange adult pelage, although varying in details, followed a definite pattern (Fig. 1). The first indication of the beginning of the dorsal molt was the appearance of new golden fur along the line separating the golden-brown dorsal fur from the white fur of the ventral surface. The replacement of the juvenile pelage progressed dorsally on both sides and met on the dorsal midline forming a continuous band of new fur. The molt then proceeded anteriorly between the ears and onto the head, while posteriorly, it joined the molt proceeding dorsally near the thighs. By this time, new fur had appeared on the sides of the face and just anterior to the ears. The molt along the sides of the body had nearly been completed by this time. The last two areas in which the fur was replaced were the top of the head and the base of the tail. In some individuals, the new fur first appeared just in back of the front leg. It proceeded both posteriorly and dorsally and formed a band of new fur just behind the ears. The molt proceeding posteriorly then covered the remainder of the body.

This pattern of molt generally agrees with that described for *Peromyscus*

236

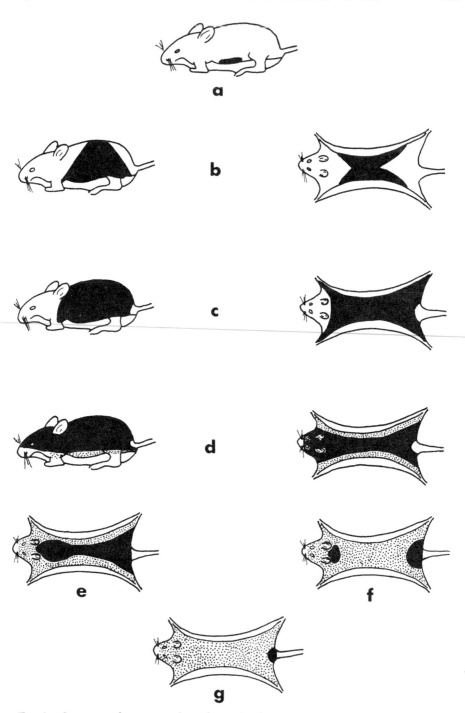

Fɪɢ. 1.—Sequence of post-juvenile molt on the dorsum in *Ochrotomys nuttalli*. Shaded portions represent areas of active hair replacement. Stippled areas represent adult pelage.

TABLE 1.—*Duration of post-juvenile molt and average age at beginning and ending of molt in 34 captive golden mice (Range of values in parentheses).*

	Males (15)	Females (19)
Duration	29 days (14–45)	25 days (12–49)
Beginning	36 days (33–42)	38 days (31–47)
Ending	64 days (51–87)	63 days (51–84)

truei (Hoffmeister, 1944), *Peromyscus gossypinus* (Pournelle, 1952) and *Peromyscus boylei* (Brown, 1963). It differs from that reported for *Peromyscus leucopus noveboracensis* (Gottschang, 1956).

Data on the beginning, ending, and duration of the post-juvenile molt on the dorsum in male and female golden mice are compared in Table 1. The average duration of molt for males was slightly longer than for females. The shortest time recorded was between 12 and 14 days, whereas the maximum time required was about 49 days. Approximately 3.5 weeks are required for most *Peromyscus leucopus noveboracensis* to attain their full adult coat according to Gottschang (1956). He recorded a minimum duration of 12 days for captive individuals and 10 days for one wild mouse to undergo the complete molt; the maximum number of days required was about 36.

In the field, animals undergoing various stages of maturational molt were recorded in June (1), July (1), August (2), and December (8). These mice were between 150 mm and 164 mm in total length (mean, 156 mm). In the captive population, male golden mice began molting when their total length was 149 mm, whereas females averaged 146 mm. At the completion of molt, their measurements averaged 163 mm and 160 mm, respectively. From these data, it appears that both wild and captive individuals molted at approximately the same body size, although it is not known whether they were the same age.

The youngest individuals in captivity to begin molting during the current study were 31 days of age. Layne (1960) recorded one young *Ochrotomys* molting at 31 days of age with the molt apparently being complete 10 days later. Molting was in progress in one four week old mouse, while in another of the same age, it had not yet begun (Layne, 1960). Collins (1918) reported that the transition from juvenile to post-juvenile pelage in *Peromyscus* usually began at 6 weeks and was completed about 8 weeks later. The earliest age at which *Peromyscus leucopus noveboracensis* began molting was 38 days (Gottschang, 1956). These were all males. The youngest female to begin molting was 40 days of age. Ninety-five per cent of his mice of both sexes started the pelage change between the ages of 40 and 50 days. Young *Peromyscus gossypinus* began molting when they were between 34 and 40 days of age (Pournelle, 1952).

Gottschang (1956) found that, in general, mice of the same sex in a single litter started molting simultaneously. However, in every case where a difference did occur, he found that the males started to molt first. In the current study, the males in 13 out of 21 litters containing mice of both sexes began molting before the females, whereas the females began molting first in three litters. The initiation of molt was simultaneous in the remaining five litters.

The progression of the ventral molt was studied in seven individuals (four males, three females). The white belly fur was dyed purple by the stain Nyanzol A (20 g per liter of water-hydrogen peroxide mixture in ratio of two to one) and replacement by new hairs was followed. The ventral molt began approximately 2–4 days before the dorsal molt. Hair replacement occurred first in the center of the belly and continued laterally, and then dorsally into the golden fur. Simultaneously, new hair appeared over the entire chest and abdomen. The last areas to acquire new pelage were the throat and the ventral bases of the hind limbs. The ventral molt was complete at about the time that the dorsal molt covered the entire back (Fig. 1).

SEASONAL MOLT

Mice of the genus *Peromyscus* are generally considered to undergo one annul adult molt in autumn (Collins, 1923). However, Osgood (1909) and Brown (1963) recorded two annual molts in *Peromyscus melanotis* and *Peromyscus boylei*, respectively.

Golden mice in the Great Smoky Mountains National Park apparently undergo two annual molts. These take place during the spring (April–June) and fall (October–December). The difference between summer and winter pelage was clearly distinguishable with the unaided eye. The winter pelage was much darker than the usual summer pelage, especially on the mid-dorsum. Osgood (1909) noted that winter specimens of *Peromyscus melanotis* possessed a paler colored pelage, whereas summer specimens were in a dark pelage. The fall molt of *P. boylei* was characterized by the replacement of a bright cinnamon-brown pelage by a more drab, brown winter pelage (Brown, 1963).

Nineteen of 21 adult golden mice in captivity underwent a fall molt between October 20 and December 24. A total of 10 adult golden mice were observed in the wild between December 12–17. Six of these were molting; four already had the winter pelage. The fall molt appeared to be more irregular than the post-juvenile molt. In several animals, it began near the hind leg, covered the rump and then progressed anteriorly to the head. Replacement of the hair was completed first over the posterior half of the body. This separated the two remaining areas of molt—the base of the tail and the head. The replacement of fur at the base of the tail was completed shortly thereafter. The final area of molt was on the head between the ears, and this sometimes required several weeks for completion. This is in contrast to the post-juvenile molt, where the last area of molt in all of the animals was at the base of the tail.

The spring molt must have occurred between 1 April and 15 June. All wild individuals observed between 26 March and 1 April 1964 still retained their winter pelage. By 15 June, all adult golden mice had either already completed their spring molt or were very near completion. Seventy-four per cent (23) of the adult individuals in the captive population molted during the spring. Of those molting, 83% (19) did so between 15 May and 30 June. As in the fall molt, the pattern was irregular. Hair replacement occurred in patches along the sides and across the shoulders, and a simultaneous molt of the entire dorsum took place in only five of 31 individuals (16%). In the cases where this molt was complete, it followed a more regular pattern, with hair replacement occurring last on the nape of the neck.

Gottschang (1956) noted no difference in the onset, progress or length of time required for the pelage change between spring-, summer-, or fall-born litters of *Peromyscus leucopus*. During the current study, however, golden mice born after 1 October and 8 April appeared to combine the post-juvenile molt and seasonal molt. The process of hair replacement was more irregular than during the normal post-juvenile molt. The molt began at a point just behind the front legs, as in the regular post-juvenile molt. It then proceeded dorsally and posteriorly at approximately equal rates. During the combined fall molt (post-juvenile plus fall molt), the replacement of hair at the base of the tail was completed prior to the completion of molt on the head in all cases. In this respect, this combined molt was more similar to the regular seasonal molt than to the regular post-juvenile molt. Upon completion of this molt, the mice had acquired the typical dark winter pelage. However, during the combined spring molt (post-juvenile plus spring molt), hair replacement was completed last at either the tail or head regions.

On the average, those animals born after 1 October began molt at a later age than did those animals born earlier in the breeding season. Males in this group began molting at an average age of 37 days, whereas spring and summer-born males began at 35 days of age. Females born after 1 October began molting at an average age of 43 days, while females born earlier in the season began molting at an average age of 37 days.

ACKNOWLEDGMENTS

We thank Dr. W. Robert Eadie of Cornell University for his advice and criticism of the manuscript. We gratefully acknowledge the financial assistance provided by The Society of the Sigma Xi and the cooperation of the National Park Service.

LITERATURE CITED

BROWN, L. N. 1963. Maturational and seasonal molts in *Peromyscus boylei*. Amer. Midland Nat., 70: 466–469.

COLLINS, H. H. 1918. Studies of normal molt and of artificially induced regeneration of pelage in *Peromyscus*. J. Exp. Zool., 27: 73–99.

———. 1923. Studies of the pelage phases and nature of color variations in mice of the genus *Peromyscus*. J. Exp. Zool., 38: 45–107.

GOTTSCHANG, J. L. 1956. Juvenile molt in *Peromyscus leucopus noveboracensis*. J. Mamm., 37: 516–520.

HOFFMEISTER, D. F. 1944. Phylogeny of the Nearctic cricetine rodents, with especial attention to variation in *Peromyscus truei*. Ph.D. thesis, Univ. California, 406 pp.

LAYNE, J. N. 1960. The growth and development of young golden mice, *Ochrotomys nuttalli*. Quart. J. Fla. Acad. Sci., 23: 36–58.

LINZEY, D. W. 1966. The life history, ecology and behavior of the golden mouse, *Ochrotomys n. nuttalli*, in the Great Smoky Mountains National Park. Ph.D. thesis, Cornell Univ., 170 pp.

OSGOOD, W. H. 1909. Revision of the mice of the American genus *Peromyscus*. N. Amer. Fauna, 28: 1–285.

POURNELLE, G. H. 1952. Reproduction and early post-natal development of the cotton mouse, *Peromyscus gossypinus gossypinus*. J. Mamm., 33: 1–20.

Division of Biological Sciences, Cornell University, Ithaca, New York. Accepted 16 January 1967.

SECTION 4—ECOLOGY AND BEHAVIOR

Animal behavior, or ethology, and ecology are varied and expanding fields of biology. The older term "natural history" is a concept that embraces both, and earlier "naturalists" were the pioneers of these fields. Ecology is of special importance to man owing to his increasing awareness of, and concern for, his own environment and such problems as the need to regulate human populations, to reduce pollution of air and water, and to conserve endangered species. The papers selected here can suggest to the reader some basic ecological principles that apply to man himself, as well as to other mammals. Our selections illustrate concepts such as territoriality and home range (applied to mammals in the classic paper by Burt), studies that provide sound theory and quantitative results based on large sample sizes (Caughley), and the application of experimental procedures (as in the manipulation of rats in city blocks reported by Davis and Christian or the tests run by McCarley in compartmented cages). The application of newer techniques such as Pearson's traffic counter for mouse runways, radio-tracking techniques used by Rongstad and Tester, and other types of modern technology, including radar (Williams *et al.*), and even earth-orbiting satellites (Craighead *et al.*, 1971, not included), all have contributed to advances in ecology and ethology. However the study by Estes and Goddard of the African wild dog will serve to remind the reader that careful observational methods such as were used so effectively by earlier field naturalists certainly have not been supplanted, but only expanded and supplemented.

A host of topics other than those we were able to include in our selection come to mind when the ecological literature is contemplated—topics such as food habits as learned from stomach contents or droppings, or the extensive literature on small mammal populations (grid live-trapping, employed by Congdon, is one of the most commonly used methods). Long-term cycles in populations and daily cycles in activity have had their share of attention also, as indicated in the papers by Tast and Kalela, and by Grodzinski.

Ecological problems may be approached at different levels of inclusiveness. The relationships of all species of plants and animals in an entire community may be studied; such a broad approach to entire ecosystems merges imperceptibly with problems concerning factors that limit distributions, hence to ranges of species and to faunal and zoogeographic problems (see Section 6). At a less inclusive level, the ecological relationships of a single species or population may be studied. These approaches are called autecology, or population ecology, as opposed to community or synecological studies. If we restrict ourselves further to the environmental relationships of individual animals, we find our studies, again by gradual stages, merge with those that are primarily physiological (see Section 2). Physiological techniques also enter directly into the study of ecosystems when energy flow is considered, as often is the case in recent studies (see, for example, Grodzinski *et al.*, 1970).

Just as ecological principles may be applied to the mammal, man, so may many ethological concepts, and part of the current interest in mammalian behavior (Ewer, 1968) stems from this possibility. For example, our selections concerning paternal (Barlow) and maternal (Rongstad and Tester) behavior (see also Rheingold, 1963) have obvious parallels in human society, and interpretations of observations are often controversial, as Barlow's reply attests. Other aspects of behavior illustrated by our selections include predation or foraging (Suthers, Estes and Goddard); orientation (Suthers, Layne), social dominance (Roberts and Wolfe), and reproductive behavior (Samaras). The last-mentioned paper is of particular interest in that the observations reported raise the issue of "altruistic" or "reciprocal" behavior, concepts of great current interest (see Wilson, 1975).

ECOLOGY and ECOLOGICAL MONOGRAPHS, published by the Ecological Society of America, and the JOURNAL OF ANIMAL ECOLOGY, by the British Ecological Society, are among the most important journals containing articles on mammalian ecology. Comparable journals in the field of ethology are ANIMAL BEHAVIOUR, ANIMAL BEHAVIOUR MONOGRAPHS, BEHAVIOUR (including Supplements), and ZEITSCHRIFT FÜR TIERPSYCHOLOGIE. In recent years a number of new ecological journals have been established, such as OIKOS and OECOLOGIA, in which papers on mammals occasionally appear.

Ecological and ethological texts have also burgeoned in recent years. Those we judge to be of particular interest to mammalogists include *Animal Behavior: An Evolutionary Approach* (Alcock, 1975), *Ecology: An Evolutionary Approach* (Emlen, 1973), *Animal Behavior* (Hinde, 1970), *Mechanisms of Animal Behaviour* (Marler and Hamilton, 1966), *Ecology*, by Krebs (1972), and Robert Ricklefs' *Ecology* (1973). Many other new books in these fields, as well as older classics, will reward the student, but they are far too numerous to mention.

MORTALITY PATTERNS IN MAMMALS

Graeme Caughley

*Forest Research Institute, New Zealand Forest Service, Rotorua, and Zoology Department,
Canterbury University, New Zealand*

(Accepted for publication December 8, 1965)

Abstract. Methods of obtaining life table data are outlined and the assumptions implicit in such treatment are defined. Most treatments assume a stationary age distribution, but published methods of testing the stationary nature of a single distribution are invalid. Samples from natural populations tend to be biased in the young age classes and therefore, because it is least affected by bias, the mortality rate curve (q_x) is the most efficient life table series for comparing the pattern of mortality with age in different populations.

A life table and fecundity table are presented for females of the ungulate *Hemitragus jemlahicus,* based on a population sample that was first tested for bias. They give estimates of mean generation length as 5.4 yr, annual mortality rate as 0.25, and mean life expectancy at birth as 3.5 yr.

The life table for *Hemitragus* is compared with those of *Ovis aries, O. dalli,* man, *Rattus norvegicus, Microtus agrestis,* and *M. orcadensis* to show that despite taxonomic and ecological differences the life tables have common characteristics. This suggests the hypotheses that most mammalian species have life tables of a common form, and that the pattern of age-specific mortality within species assumes an approximately constant form irrespective of the proximate causes of mortality.

Introduction

Most studies in population ecology include an attempt to determine mortality rates, and in many cases rates are given for each age class. This is no accident. Age-specific mortality rates are usually necessary for calculating reproductive values for each age class, the ages most susceptible to natural selection, the population's rate of increase, mean life expectancy at birth, mean generation length, and the percentage of the population that dies each year. The importance of these statistics in the fields of game management, basic and applied ecology, and population genetics requires no elaboration.

The pattern of changing mortality rates with age is best expressed in the form of a life table. These tables usually present the same information in a variety of ways:

1) Survivorship (l_x): this series gives the probability at birth of an individual surviving to any age, x (l_x as used here is identical with P_x of Leslie, Venables and Venables 1952). The ages

are most conveniently spaced at regular intervals such that the values refer to survivorship at ages 0, 1, 2 etc. yr, months, or some other convenient interval. The probability at birth of living to birth is obviously unity, but this initial value in the series need not necessarily be set at 1; it is often convenient to multiply it by 1,000 and to increase proportionately the other values in the series. If this is done, survivorship can be redefined as the number of animals in a cohort of 1,000 (or any other number to which the initial value is raised) that survived to each age x. In this way a kl_x series is produced, where k is the constant by which all l_x values in the series are multiplied.

2) Mortality (d_x): the fraction of a cohort that dies during the age interval x, x + 1 is designated d_x. It can be defined in terms of the individual as the probability at birth of dying during the interval x, x + 1. As a means of eliminating decimal points the values are sometimes multiplied by a constant such that the sum of the d_x values equals 1,000. The values can be calculated from the l_x series by

$$d_x = l_x - l_{x+1}$$

3) Mortality rate (q_x): the mortality rate q for the age interval x, x + 1 is termed q_x. It is calculated as the number of animals in the cohort that died during the interval x, x + 1, divided by the number of animals alive at age x. This value is usually expressed as $1,000q_x$, the number of animals out of 1,000 alive at age x which died before x + 1.

These are three ways of presenting age-specific mortality. Several other methods are available— e.g. survival rate (p_x), life expectancy (e_x) and probability of death (Q_x)—but these devices only present in a different way the information already contained in each of the three series previously defined. In this paper only the l_x, d_x and q_x series will be considered.

METHODS OF OBTAINING MORTALITY DATA

Life tables may be constructed from data collected in several ways. Direct methods:

1) Recording the ages at death of a large number of animals born at the same time. The frequencies of ages at death form a kd_x series.

2) Recording the number of animals in the original cohort still alive at various ages. The frequencies from a kl_x series.

Approximate methods:

3) Recording the ages at death of animals marked at birth but whose births were not coeval. The frequencies form a kd_x series.

4) Recording ages at death of a representative sample by ageing carcasses from a population that has assumed a stationary age distribution. Small fluctuations in density will not greatly affect the results if these fluctuations have an average wave length considerably shorter than the period over which the carcasses accumulated. The frequencies form a kd_x series.

5) Recording a sample of ages at death from a population with a stationary age distribution, where the specimens were killed by a catastrophic event (avalanche, flood, etc.) that removed and fixed an unbiased sample of ages in a living population. In some circumstances (outlined later) the age frequencies can be treated as a kl_x series.

6) The census of ages in a living population, or a sample of it, where the population has assumed a stationary age distribution. Whether the specimens are obtained alive by trapping or are killed by unselective shooting, the resultant frequencies are a sample of ages in a living population and form a kl_x series in certain circumstances.

Methods 1 to 3 are generally used in studies of small mammals while methods 4 to 6 are more commonly used for large mammals.

TESTS FOR STATIONARY AGE DISTRIBUTION

Five methods have been suggested for determining whether the age structure of a sample is consistent with its having been drawn from a stationary age distribution:

a) Comparison of the "mean mortality rate," calculated from the age distribution of the sample, with the proportion represented by the first age class (Kurtén 1953, p. 51).

b) Comparison of the annual female fecundity of a female sample with the sample number multiplied by the life expectancy at birth, the latter statistic being estimated from the age structure (Quick 1963, p. 210).

c) Calculation of instantaneous birth rates and death rates, respectively, from a sample of the population's age distribution and a sample of ages at death (Hughes 1965).

d) Comparison of the age distribution with a prejudged notion of what a stationary age distribution should be like (Breakey 1963).

e) Examination of the "l_x" and "d_x" series calculated from the sampled age distribution, for evidence of a common trend (Quick 1963, p. 204).

Methods a to c are tautological because they assume the sampled age distribution is either a kl or kd_x series; method d assumes the form of the life table, and e makes use of both assumptions. These ways of judging the stationary nature of a population are invalid. But I intend something more general than the simple statement that these five methods do not test what they are supposed to test. Given no information other than a single age distribution, it is theoretically impossible to prove that the distribution is from a stationary population unless one begins from the assumption that the population's survival curve is of a particular form. If such an assumption is made, the life table constructed from the age frequencies provides no more information than was contained in the original premise.

Mortality Samples and Age Structure Samples

Methods 4 to 6 for compiling life tables are valid only when the data are drawn from a stationary age distribution. This distribution results when a population does not change in size and where the age structure of the population is constant with time. The concept has developed from demographic research on man and is useful for species which, like man, have no seasonally restricted period of births.

Populations that have a restricted season of births present difficulties of treatment, some of which have been discussed by Leslie and Ranson (1940). Very few mammals breed at the same rate throughout the year, and the stationary age distribution must be redefined if it is to include seasonal breeders. For species with one restricted breeding season each year, a stationary population can be defined as one that does not vary either in numbers or age structure at successive points in time spaced at intervals of 1 yr. The stationary age distribution can then be defined for such populations as the distribution of ages at a given time of the year. Thus there will be an infinite number of different age distributions according to the time of census, other than in the exceptional case of a population having a constant rate of mortality throughout life.

The distribution of ages in a stationary population forms a kl_x series only when all births for the year occur at an instant of time and the sample is taken at that instant. This is obviously impossible, but the situation is approximated when births occur over a small fraction of the year. If a population has a restricted season of births, the age structure can be sampled over this period and at the same time the number of live births produced by a hypothetical cohort can be calculated from the number of females either pregnant or suckling young. In this way a set of data closely approximating a kl_x series can be obtained.

If an age distribution is sampled halfway between breeding seasons, it cannot be presented as a kl_x series with x represented as integral ages in years. With such a sample (making the usual assumptions of stability and lack of bias) neither l_x nor d_x can be established, but q_x values can be calculated for each age interval $x + \frac{1}{2}$, $x + 1\frac{1}{2}$. The age frequencies from a population with a continuous rate of breeding are exactly analogous; they do not form a kl_x series but can be treated as a series of the form

$$k\,(l_x + l_{x+1})\,/2$$

This series does not allow calculation of l_x values from birth unless the mortality rate between birth and the midpoint of the first age interval is known.

Because a sample consists of dead animals, its age frequencies do not necessarily form a mortality series. The kd_x series is obtained only when the sample represents the frequencies of ages at death in a stationary population. Many published samples treated as if they formed a kd_x series are not appropriate to this form of analysis. For instance, if the animals were obtained by shooting which was unselective with respect to age, the sample gives the age structure of the living population at that time; that the animals were killed to get these data is irrelevant. Hence unbiased shooting samples survivorship, not mortality, and an age structure so obtained can be treated as a kl_x series if all other necessary assumptions obtain. Similarly, groups of animals killed by avalanches, fires, or floods—catastrophic events that preserve a sample of the age frequencies of animals during life—do not provide information amenable to kd_x treatment.

A sample may include both l_x and d_x components. For instance, it could consist of a number of dead animals, some of which have been unselectively shot, whereas the deaths of others are attributable to "natural" mortality. Or it could be formed by a herd of animals killed by an avalanche in an area where carcasses of animals that died "naturally" were also present. In both these cases d_x and l_x data are confounded and these heterogeneous samples of ages at death can be treated neither as kd_x nor kl_x series.

Even if a sample of ages at death were not heterogeneous in this sense, it might still give misleading information. If, for instance, carcasses attributable to "natural" mortality were collected only on the winter range of a population, the age frequencies of this sample would provide ages at death which reflected the mortality pattern during only part of the year. But the d_x series gives the proportion of deaths over contiguous periods of the life span and must reflect all mortality during each of these periods.

It has been stressed that the frequencies of ages in life or of ages at death provide life-table information only when they are drawn from a population with a stationary age distribution. This age distribution should not be confused with the stable distribution. When a population increases at a constant rate and where survivorship and fecundity rates are constant, the age distribution eventually assumes a stable form (Lotka 1907 a, b; Sharpe and Lotka 1911). Slobodkin (1962, p. 49) gives a simple explanation as to why this is so. A stable age distribution does not form a kl_x series except when the rate of increase is zero, the season of births is restricted, and the sample is taken at this time. Hence the stationary age distribution is a special case of the stable age distribution.

THE RELATIVE USEFULNESS OF THE l_x, d_x AND q_x SERIES

Most published life tables for wild mammals have been constructed either from age frequencies obtained by shooting to give a kl_x series, or by determining the ages at death of animals found dead, thereby producing a kd_x series. Unfortunately, both these methods are almost invariably subject to bias in that the frequency of the first-year class is not representative. Dead immature animals, especially those dying soon after birth, tend to decay faster than the adults, so that they are underrepresented in the count of carcasses. The ratio of juveniles to adults in a shot sample is usually biased because the two age classes have different susceptibilities to hunting. With such a bias established or suspected, the life table is best presented in a form that minimizes this bias. An error in the frequency of the first age class results in distortions of each l_x and d_x value below it in the series, but q_x values are independent of frequencies in younger age classes. By definition, q is the ratio of those dying during an age interval to those alive at the beginning of the interval. At age y the value of q is given by

$$q_y = d_y/l_y$$

but

$$d_y = l_y - l_{y+1}$$

therefore

$$q_y = (l_y - l_{y+1})/l_y .$$

Thus the value of q_y is not directly dependent on absolute values of l_x but on the differences between successive values. If the l_x series is calculated from age frequencies in which the initial frequency

is inaccurate, each l_x value will be distorted. However, the difference between any two, divided by the first, will remain constant irrespective of the magnitude of error above them in the series. Thus a q_x value is independent of all but two survivorship age frequencies and can be calculated directly from these frequencies (f_x) by

$$q_x = (f_x - f_{x+1})/f_x$$

if the previously discussed conditions are met.

The calculation of q from frequencies of ages at death is slightly more complex:

by definition $q_y = d_y/l_y$

but $l_y = \sum\limits_{x=0}^{\infty} d_x - \sum\limits_{x=0}^{y-1} d_x$

therefore $q_y = d_y \div (\sum\limits_{x=0}^{\infty} d_x - \sum\limits_{x=0}^{y-1} d_x)$

$$= d_y / \sum\limits_{x=y}^{\infty} d_x$$

but the frequencies of ages at death (f'_x) are themselves a kd_x series and so $q_y = f'_y / \sum\limits_{x=y}^{\infty} f'_x$.

Thus the value of q at any age is independent of frequencies of the younger age classes. Although the calculated value of q for the first age class may be wrong, this error does not affect the q_x values for the older age classes.

The q_x series has other advantages over the l_x and d_x series for presenting the pattern of mortality with age. It shows rates of mortality directly, whereas this rate is illustrated in a graph of the l_x series (the series most often used when comparing species) only by the slope of the curve.

A LIFE TABLE FOR THE THAR, *Hemitragus jemlahicus*

The Himalayan thar is a hollow-horned ungulate introduced into New Zealand in 1904 (Donne 1924) and which now occupies 2,000 miles² of mountainous country in the South Island. Thar were liberated at Mount Cook and have since spread mostly north and south along the Southern Alps. They are still spreading at a rate of about 1.1 miles a year (Caughley 1963) and so the populations farthest from the point of liberation have been established only recently and have not yet had time to increase greatly in numbers. Closer to the site of liberation the density is higher (correlated with the greater length of time that animals have been established there), and around the point of liberation itself there is evidence that the population has decreased (Anderson and Henderson 1961).

312

The growth rings on its horns are laid down in each winter of life other than the first (Caughley 1965), thereby allowing the accurate ageing of specimens. An age structure was calculated from a sample of 623 females older than 1 yr shot in the Godley and Macaulay Valleys between November 1963 and February 1964. Preliminary work on behavior indicates that there is very little dispersal of females into or out of this region, both because the females have distinct home ranges and because there are few ice-free passes linking the valley heads.

As these data illustrate problems presented by most mammals, and because the life table has not been published previously, the methods of treatment will be outlined in some detail.

Is the population stationary?

Although it is impossible to determine the stationary nature of a population by examining the age structure of a single sample, even when rates of fecundity are known, in some circumstances a series of age structures will give the required information. This fact is here utilized to investigate the stability of this population.

The sample was taken about halfway between the point of liberation and the edge of the range. It is this region between increasing and decreasing populations where one would expect to find a stationary population. The animals came into the Godley Valley from the southwest and presumably colonized this side of the valley before crossing the 2 miles of river bed to the northeast side. This pattern of establishment is deduced from that in the Rakaia Valley, at the present edge of the breeding range, where thar bred for at least 5 yr on the south side of the valley before colonizing the north side. Having colonized the northeast side of the Godley Valley, the thar would then cross the Sibald Range to enter the Macaulay Valley, which is a further 6 miles northeast. The sample can therefore be divided into three subsamples corresponding to the different periods of time that the animals have been present in the three areas. A 10×3 contingency test for differences between the three age distributions of females 1 yr of age or older gave no indication that the three subpopulations differed in age structure ($\chi^2 = 22.34$; $P = 0.2$).

This information can be interpreted in two ways: either the three subpopulations are neither increasing nor decreasing and hence are likely to have stationary age distributions, or the subpopulations could be increasing at the same rate, in which case they could have identical stable age distributions. The second alternative carries a

TABLE I. Relative densities of thar in three zones

Zone	Number females autopsied	Mean density index[a]	Standard error
Godley Valley south	258	2.19	0.56
Godley Valley north	240	1.67	0.53
Macaulay Valley	115	2.66	0.69

$F_{2,36}$ for densities between valleys = 1.74, not significant
[a]Density indices were calculated as the number of females other than kids recorded as autopsied in a zone each day, divided by the number of shooters hunting in the zone on that day.

corollary that the subpopulations would have different densities because they have been increasing for differing periods of time. But an analysis of the three densities gives no indication that they differ (Table I). This result necessitates the rejection of the second alternative.

The above evidence suggesting that the sample was drawn from a stationary age distribution is supported to some extent by observation. When I first passed through the area in 1957, I saw about as many thar per day as in 1963-64. J. A. Anderson, a man who has taken an interest in the thar of this region, writes that the numbers of thar in 1956 were about the same as in 1964 (Anderson, pers. comm.). These are subjective evaluations and for that reason cannot by themselves be given much weight, but they support independent evidence that the population is stationary or nearly so.

Is the sample biased?

A sample of the age structure of a population can be biased in several ways. The most obvious source of bias is behavioral or range differences between males and females. For instance, should males tend to occupy terrain which is more difficult to hunt over than that used by females, they would be underrepresented in a sample obtained by hunting. During the summer thar range in three main kinds of groups: one consists of females, juveniles and kids, a second consists of young males and the third of mature males. The task of sampling these three groupings in the same proportions as they occur throughout the area is complicated by their preferences for terrain that differs in slope, altitude and exposure. Consequently the attempt to take an unbiased sample of both males and females was abandoned and the hunting was directed towards sampling only the nanny-kid herds in an attempt to take a representative sample of females. The following analysis is restricted to females.

Although bias attributable to differences in behavior between sexes can be eliminated by the simple contrivance of ignoring one sex, some age classes of females may be more susceptible than others to shooting. To test for such a difference, females other than kids were divided into two groups: those from herds in which some members were aware of the presence of the shooter before he fired, and those from herds which were undisturbed before shooting commenced. If any age group is particularly wary its members should occur more often in the "disturbed" category than is the case for other age groups. But a χ^2 test ($\chi^2 = 7.28$, df $= 9$, $P = 0.6$) revealed no significant difference between the age structures of the two categories.

The sample was next divided into those females shot at ranges less than 200 yards and those shot out of this range. If animals in a given age class are more easily stalked than the others, they will tend to be shot at closer ranges. Alternatively, animals which present small targets may be underrepresented in the sample of those shot at ranges over 200 yards. This is certainly true of kids, which are difficult to see, let alone to shoot, at ranges in excess of 200 yards. The kids have therefore not been included in the analysis because their underrepresentation in the sample is an acknowledged fact, but for older females there is no difference between the age structures of the two groups divided by range which is not explainable as sampling variation ($\chi^2 = 9.68$, df $= 9$, $P = 0.4$). This is not to imply that no bias exists—the yearling class for instance could well be underrepresented beyond 200 yards—but that

no bias could be detected from a sample of this size.

The taking of a completely representative sample from a natural population of mammals is probably a practical impossibility, and I make no claim that this sample of thar is free of bias, but as bias cannot be detected from the data, I assume it is slight.

Construction of the life table

The shooting yielded 623 females 1 yr old or older, aged by growth rings on the horns. As the sampling period spanned the season of births, a frequency for age 0 cannot be calculated directly from the number of kids shot because early in the period the majority had not been born. In any case, the percentage of kids in the sample is biased.

The numbers of females at each age are shown in Table II, column 2. Although the ages are given only to integral years each class contains animals between ages x yr — $\frac{1}{2}$ month and x yr $+ 2\frac{1}{2}$ months. Variance owing to the spread of the kidding season is not included in this range, but the season has a standard deviation of only 15 days (Caughley 1965).

Up to an age of 12 yr (beyond this age the values dropped below 5 and were not treated) the frequencies were smoothed according to the formula

$$\log y = 1.9673 + 0.0246x - 0.01036 \, x^2,$$

where y is the frequency and x the age. The linear and quadratic terms significantly reduced

TABLE II. Life table and fecundity table for the thar *Hemitragus jemlahicus* (females only)

1 Age in years x	2 Frequency in sample	3 Adjusted frequency	4 No. female live births per female at age x m_x	5 $1,000 \, l_x$	6 $1,000 \, d_x$	7 $1,000 \, q_x$
0	—	205[a]	0.000	1,000	533	533
1	94	95.83	0.005	467	6	13
2	97	94.43	0.135	461	28	61
3	107	88.69	0.440	433	46	106
4	68	79.41	0.420	387	56	145
5	70	67.81	0.465	331	62	187
6	47	55.20	0.425	269	60	223
7	37	42.85	0.460	209	54	258
8	35	31.71	0.485	155	46	297
9	24	22.37	0.500	109	36	330
10	16	15.04	0.500	73	26	356
11	11	9.64	}0.470	47	18	382
12	6	5.90		29		
13	3		}0.350			
14	4					
15	3					
16	0					
17	1					

[a]Calculated from adjusted frequencies of females other than kids (column 3) and m_x values (column 4).

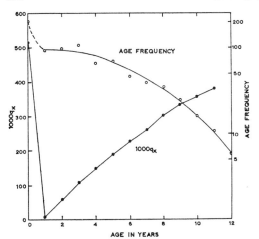

FIG. 1. Age frequencies, plotted on a logarithmic scale, of a sample of female thar, with a curve fitted to the values from ages 1 to 12 yr, and the mortality rate per 1,000 for each age interval of 1 yr ($1,000q_x$) plotted against the start of the interval.

variance around the regression, but reduction by the addition of a cubic term was not significant at the 0.05 level. There are biological reasons for suspecting that the cubic term would have given a significant reduction of variance had the sample been larger, but for the purposes of this study its inclusion in the equation would add very little. The improved fit brought about by the quadratic term indicates that the rate of mortality increases with age. Whether the rate of this rate also increases, is left open. The computed curve closely fitted the observed data (Fig. 1) and should greatly reduce the noise resulting from sampling variation, the differential effect on mortality of different seasons, and the minor heterogeneities which, although not detectable, are almost certain to be present. The equation is used to give adjusted frequencies in Table II, column 3.

The frequency of births can now be estimated from the observed mean number of female kids produced per female at each age. These are shown in column 4. They were calculated as the number of females at each age either carrying a foetus or lactating, divided by the number of females of that age which were shot. These values were then halved because the sex ratio of late foetuses and kids did not differ significantly from 1:1 (93 ♂ ♂ : 97 ♀ ♀). The method is open to a number of objections: it assumes that all kids were born alive, that all females neither pregnant nor lactating were barren for that season, and that twinning did not occur. The first assumption, if false, would give rise to a positive bias, and the second

and third to a negative bias. However, the ratio of females older than 2 yr that were either pregnant or lactating to those neither pregnant nor lactating did not differ significantly between the periods November to December and January to February ($\chi^2 = 0.79$, $P = 0.4$), suggesting that still births and mortality immediately after birth were not common enough to bias the calculation seriously. Errors are unlikely to be introduced by temporarily barren females suckling yearlings, because no female shot in November that was either barren (as judged by the state of the uterus) or pregnant was lactating. Errors resulting from the production of twins will be very small; we found no evidence of twinning in this area.

The products of each pair of values in columns 3 and 4 (Table II) were summed to give an estimate of the potential number of female kids produced by the females in the sample. This value of 205 is entered at the head of column 3. The adjusted age frequencies in column 3 were each multiplied by 4.878 to give the $1,000l_x$ survivorship values in column 5. The mortality series (column 6) and mortality-rate series (column 7) were calculated from these.

Conclusions

Figure 1 shows the mortality rate of females in this thar population up to an age of 12 yr. Had the sample been larger the graph could have been extended to an age of 17 yr or more, but this would have little practical value for the calculation of population statistics because less than 3% of females in the population were older than 12 yr.

The pattern of mortality with age can be divided into two parts—a juvenile phase characterized by a high rate of mortality, followed by a postjuvenile phase in which the rate of mortality is initially low but rises at an approximately constant rate with age.

Table II gives both the l_x and m_x series, and these two sets of values provide most of the information needed to describe the dynamics of the population. Assuming that these two series are accurate, the following statistics can be derived: generation length (i.e. mean lapse of time between a female's date of birth and the mean date of birth of her offspring), T:

$$T = \frac{\Sigma l_x m_x x}{\Sigma l_x m_x} = 5.4 \text{ yr};$$

mean rate of mortality for all age groups, $\bar{q}_x$:

$$\bar{q}_x = 1/\Sigma l_x = 0.25 \text{ per female per annum};$$

life expectancy at birth, e_0:

$$e_0 = \Sigma l_x - \tfrac{1}{2} = 3.5 \text{ yr}.$$

The last two statistics can also be expressed conveniently in terms of the mortality series by

$$\bar{q}_x = 1/\Sigma\ (x+1)\ d_x$$

and

$$e_0 = \frac{\Sigma\ (2x+1)\ d_x}{2}.$$

The relationship of the two is given by

$$\bar{q}_x = 2/(2e_0 + 1).$$

Life Tables for Other Mammals

The difficulty of comparing the mortality patterns of animals that differ greatly in life span can be readily appreciated. To solve this problem, Deevey (1947) proposed the percentage deviation from mean length of life as an appropriate scale, thereby allowing direct comparison of the life tables of, say, a mammal and an invertebrate. For such comparisons this scale is obviously useful, but for mammals where the greatest difference in mortality rates may be at the juvenile stage the scale often obscures similarities.

By way of illustration, Figure 2 shows $1,000q_x$ curves for two model populations which differ only in the mortality rate of the first age class. When the values are graphed on a scale of percentage deviation from mean length of life the close similarity of the two sets of data is no longer apparent. Thus the use of Deevey's scale for

comparing mortality patterns in mammals might result in a loss rather than a gain of information. In this paper, absolute age has been retained as a scale in comparing life tables of different species, although this scale has its own limitations.

Domestic sheep, Ovis aries.—Between 1954 and 1959, Hickey (1960) recorded the ages at death of 83,113 females on selected farms in the North Island of New Zealand. He constructed a q_x table from age 1½ yr by "dividing the number of deaths which have occurred in each year of age by the number 'exposed to risk' [of death] at the same age." An age interval of 1 yr was chosen and the age series 1½, 2½, 3½ etc. was used in preference to integral ages.

The q_x series conformed very closely to the regression: log $q_x = 0.156x + 0.24$, enabling him in a subsequent paper (Hickey 1963) to present the interpolated q_x values at integral ages. He also calculated q for the first year of life from a knowledge of the number of lambs dying before 1 yr of age out of 85,309 (sexes pooled) born alive.

These data probably provide the most accurate life table for any mammal. The $1,000q_x$ curve is graphed in Figure 3.

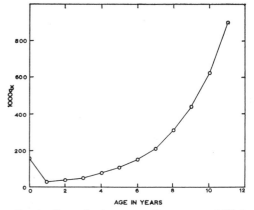

FIG. 3. Domestic sheep: mortality rate per 1,000 for each age interval of 1 yr ($1,000q_x$), plotted against the start of the interval. Data from Hickey (1963).

Dall sheep, Ovis dalli.—During his study on the wolves of Mount McKinley National Park, Murie (1944) aged carcasses of dall sheep he found dead, their ages at death being established from the growth rings on the horns. This sample can be divided into those that died before 1937 and those that died between 1937 and 1941. The former sample was used by Deevey (1947) to construct the life table presented in his classic paper on mortality in natural populations. Kurtén (1953) constructed a life table from the same

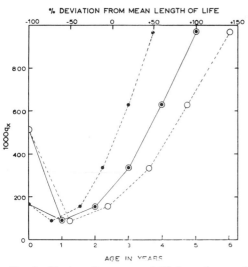

FIG. 2. The mortality rate per 1,000 for each year of life for two model populations that differ only in the degree of first-year mortality. These $1,000q_x$ values are each graphed on two time scales: absolute age in years (continuous lines) and percentage deviation from mean life expectancy (broken lines).

data, but corrected the underrepresentation of first-year animals resulting from the relatively greater perishability of their skulls by assuming that adult females produce 1 lamb per annum from about their second birthday. Taber and Dasmann (1957) constructed life tables for both males and females from the sample of animals dying between 1937 and 1941, and adjusted both the 0 to 1- and 1 to 2-year age frequencies on the assumption that a female produces her first lamb at about her third birthday and another lamb each year thereafter, that the sex ratio at birth is unity and that the loss of yearlings is not more than 10%.

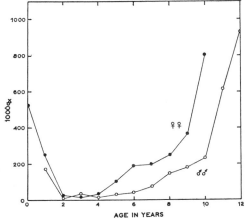

Fig. 4. Dall sheep: mortality rate per 1,000 for each age interval of 1 yr ($1,000q_x$), plotted against the start of the interval. Data from Murie (1944).

Figure 4 shows a version of this table constructed from the pre-1937 sample. The mortality of the first year class has been adjusted by assuming that the sex ratio at birth is unity, that 50% of females produce their first kids at their second birthday and that thereafter 90% produce kids each year. The figure of 50% fecundity at age 2 is borrowed from Woodgerd's (1964) study on the closely related *Ovis canadensis*, and the subsequent 90% fecundity is based on Murie's (1944) statement that twins are extremely rare. To allow for temporarily or permanently barren animals, 10% is subtracted from the potential fecundity.

This life table must be taken as an approximation. As Deevey (1947) has pointed out, the pre-1937 and 1937–41 samples differ significantly in age structure. The obvious conclusion is that the mortality rate by age was changing before and during the period of study. Consequently the age structure of the sample is likely to be only an approximation of the kd_x series. Furthermore,

the q_x values for age 1 yr are likely to have been biased by differential perishability of skulls, but no arbitrary adjustment has been made.

Man.—Most of the life tables available for man show that males have a higher rate of mortality than females. However, Macdonell's (1913) tables for ancient Rome, Hispania and Lusitania suggest that this might not always have been so and that in some circumstance the reverse can be true.

A $1,000q_x$ curve for Caucasian males and females in the United States between 1939 and 1941 is shown in Figure 5. The values are taken from Dublin, Lotka, and Spiegelman (1949).

Rat, Rattus norvegicus.—Wiesner and Sheard (1935) gave the ages at death of 1,456 females of the albino rat (Wistar strain) in a laboratory population. Their table begins at an age of 31 days, but Leslie et al. (1952) calculate from Wiesner and Sheard's data that the probability of dying between birth and 31 days was 0.316. Figure 6 gives a q_x curve constructed from these data.

Short-tailed vole, Microtus agrestis.—The ages at death of 85 males and 34 females were reported by Leslie and Ranson (1940) from a laboratory population of voles kept at the Bureau of Animal Population, Oxford. Frequencies for both sexes were pooled and the data were smoothed by the formula $f_x = f_0 e^{-bx^2}$ where f is the frequency of animals alive age x, and b is a constant. The computed curve closely fitted the data ($P = 0.5$ to 0.7). Figure 6 shows the $1,000q_x$ curve derived from the authors' sixth table.

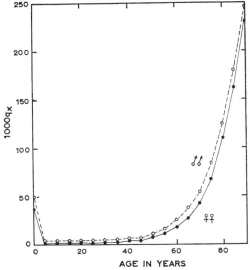

Fig. 5. Man in U.S.: mortality rate per 1,000 per year of age ($1,000q_x$). Data from Dublin et al. (1949).

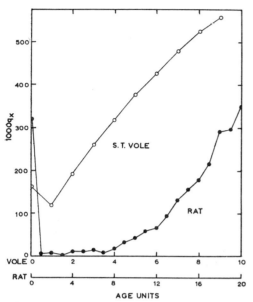

FIG. 6. Short-tailed voles and rats: mortality rate per 1,000 for each age interval $(1,000q_x)$, plotted against the start of the interval. Age interval is 56 days for voles and 50 days for rats. Rat data from Wiesner and Sheard (1935); vole data from Leslie and Ranson (1940).

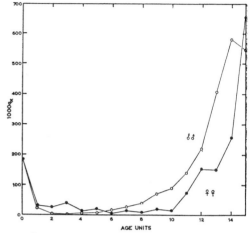

FIG. 7. Orkney vole: mortality rate per 1,000 for each age interval of 56 days $(1,000q_x)$, plotted against the start of the interval. Data from Leslie et al. (1955).

The pooling of mortality data from both sexes is strictly valid only when the two q_x series are not significantly different. Studies on differential mortality between sexes are few, but those available for man (Dublin et al. 1949, and other authors), dall sheep (Taber and Dasmann 1957, and this paper), the pocket gopher (Howard and Childs 1959) and Orkney vole (Leslie et al. 1955) suggest that although mortality rates certainly differ between sexes, the trends of these age-specific rates tend to be parallel. Consequently, this life table for voles, although based on presumably heterogeneous data, is probably quite adequate for revealing the gross pattern of mortality with age.

Orkney vole, Microtus orcadensis.—Leslie et al. (1955) gave a life table for both males and females in captivity from a base age of 9 weeks. In addition they gave the probability at birth of surviving to ages 3, 6, and 9 weeks, but did not differentiate sexes over this period. The q_x curve given here (Fig. 7) was constructed by calculating survivorship series for both males and females from these data, drawing trend lines through the points, and interpolating values at intervals of 8 weeks.

Proposed life tables not accepted

In the Discussion section of this paper the life tables discussed previously are examined in an

attempt to generalize their form. Only a small proportion of published life tables are dealt with, and any generalization from these could be interpreted as an artefact resulting from selection of evidence.

To provide the reader with the information necessary for reaching an independent conclusion, the published life tables not selected for comparison are listed below with the reason for their rejection. Only those including all juvenile age classes are cited. These tables are rejected only for present purposes because comparison of mortality patterns between species demands a fairly high level of accuracy for individual tables. The inclusion of a table in this section does not necessarily imply that it is completely inaccurate and of no practical value.

Tables based on inadequate data (i.e. less than 50 ages at death or 150 ages of living animals): *Odocoileus hemionus* (Taber and Dasmann 1957), *Ovis canadensis* (Woodgerd 1964);

Probable sampling bias: *Lepus americanus* (Green and Evans 1940), *Rupicapra rupicapra* (Kurtén 1953), fossil accumulation (Kurtén 1953, 1958; Van Valen 1964) *Balaenoptera physalus* (Laws 1962);

Age structure analyzed as a kd_x series: *Sylvilagus floridanus* (Lord 1961), *Odocoileus virginianus* and *Capreolus capreolus* (Quick 1963);

Death and emigration confounded: *Peromyscus maniculatus* (Howard 1949), *Capreolus capreolus* (Taber and Dasmann 1957, Quick 1963);

Sample taken between breeding seasons: *Odocoileus virginianus* (Quick 1963);

Form of life table, or significant portion of it, based largely on assumption: *Callorhinus ursinus* (Kenyon and Scheffer 1954), *Myotis mystacinus* (Sluiter, van Heerdt, and Bezem 1956), *Cervus elaphus* (Taber and Dasmann 1957), *Rhinolophus hipposideros, Myotis emarginatus,* and *Myotis daubentonii* (Bezem, Sluiter, and van Heerdt 1960), *Halichoerus grypus* (Hewer 1963, 1964);

Sample from a nonstationary population: *Sylvilagus floridanus* (Lord 1961);

Inadequate aging: *Gorgon taurinus* (Talbot and Talbot 1963);

Confounding of l_x and d_x data: *Rangifer arcticus* (Banfield 1955).

DISCUSSION

The most striking feature of the q_x curves of species accepted for comparison is their similarity. Each curve can be divided into two components: a juvenile phase where the rate of mortality is initially high but rapidly decreases, followed by a postjuvenile phase characterized by an initially low but steadily increasing rate of mortality. The seven species compared in this paper all produced q_x curves of this "U" or fish-hook shape, suggesting that most mammals share a relationship of this form between mortality rate and age. This conclusion, if false, can be invalidated by a few more life tables from other species. It can be tested most critically by reexamining some of the species for which life tables, although published, were not accepted in this paper. Those most suitable are species that can be adequately sampled, and accurately aged by growth rings on the horns or growth layers in the teeth (chamois, Rocky Mountain sheep, and several species of deer), or those small mammals that can be marked at birth and subsequently recaptured.

High juvenile mortality, characterizing the first phase of the q_x curve, has been reported also for several mammals for which complete life tables have not yet been calculated (e.g. for *Oryctolagus cuniculus* (Tyndale-Biscoe and Williams 1955, Stodart and Myers 1964), *Gorgon taurinus* (Talbot and Talbot 1963), *Cervus elaphus* (Riney 1956) and *Oreamnos americanus* (Brandborg 1955). Kurtén (1953, p. 88) generalized this phenomenon by stating that "the initial dip [in the survivorship curve] is a constitutional character in sexually reproducing forms at least . . .". This phase of mortality is highly variable in degree but not in form. Taber and Dasmann (1957) and Bourlière (1959) have emphasized the danger of

considering a life table of a population in given circumstances as a typical of all populations of that species. Different conditions of life tend to affect life tables, and the greatest differences between populations of a species are likely to be found at the juvenile stage. For example, the rate of juvenile mortality in red deer (Riney 1956) and in man differ greatly between populations of the same species.

The second phase—the increase in the rate of mortality throughout life—is common also to the seven species compared in this paper. However, although the increase itself is common to them, the pattern of this increase is not. Mortality rates have a logarithmic relationship to age in domestic sheep and to a less marked extent in the rat, the Orkney vole, and the dall sheep, whereas the relationship for the thar and the short-tailed vole appears to be approximately arithmetic. However, this difference may prove to be only an artefact resulting from the smoothing carried out on the data from these two species.

Despite these differences, the characteristics common to the various q_x curves dominate any comparison made between them. The similarities are all the more striking when measured against the ecological and taxonomic differences between species. Taxonomically, the seven species represent three separate orders (Primates, Rodentia, and Artiodactyla), and ecologically they comprise laboratory populations (rats and voles), natural populations (thar, dall sheep and man) and an artificial population (domestic sheep). The agents of mortality which acted on these populations must have been quite diverse. Murie (1944) reported that most of the dall sheep in the sample had been killed by wolves; most mortality in the thar population is considered to result from starvation and exposure in the winter; mortality of domestic sheep seems to be largely a result of disease, physiological degeneration, and possibly iodine deficiency in the lambs (Hickey 1963); whereas the deaths in the laboratory populations of voles and rats may be due to inadequate parental care and cannibalism of the juveniles, and perhaps disease and physiological degeneration in the adults. These differences suggest that the q_x curve of a population may assume the same form under the influence of various mortality agents, even though the absolute rate of mortality of a given age class is not the same in all circumstances. This hypothesis is worth testing because it implies that the susceptibility to mortality of an age class, relative to that of other age classes, is not strongly specific to any particular agent of mortality. A critical test would be to compare the life tables of two stationary populations of the same species, where only one population is subjected to predation.

Although no attempt is made here to explain the observed mortality pattern in terms of evolutionary processes, an investigation of this sort could be informative. A promising line of attack, for instance, would be an investigation of what appears to be a high inverse correlation between the mortality rate at a given age and the contribution of an animal of this age to the gene pool of the next generation. Fisher (1930) gives a formula for the latter statistic.

Bodenheimer (1958) divided expectation of life into "physiological longevity" ("that life duration which a healthy individual may expect to live under optimum environment conditions until dying of senescence") and "ecological longevity" (the duration of life under natural conditions). This study suggests that such a division is inexpedient because no clear distinction can be made between the effect on mortality rates of physiological degeneration and of ecological influences.

It is customary to classify life tables according to the three hypothetical patterns of mortality given by Pearl and Miner (1935). These patterns can be characterized as: 1) a constant rate of mortality throughout life, 2) low mortality throughout most of the life span, the rate rising abruptly at old age, and 3) initial high mortality followed by a low rate of mortality. Pearl (1940) emphasizes that the three patterns are conceptual models having no necessary empirical reality, but a few subsequent writers have treated them as laws which all populations must obey. None of these models fit the mortality patterns of the seven species discussed in this paper although Pearl's (1940) later modification of the system provides two additional models (high–low–high mortality rate and low–high–low mortality rate), the first of which is an adequate approximation to these data. For mammals at least, the simple threefold classification of mortality patterns is both confusing and misleading. The five-fold classification allows greater scope; but do we yet know enough about mortality patterns in mammals to justify the construction of any system of classification?

ACKNOWLEDGMENTS

This paper has greatly benefited from criticism of previous drafts by M. A. Bateman, CSIRO; P. H. Leslie, Bureau of Animal Population; M. Marsh, School of Biological Sciences, University of Sydney; J. Monro, Joint FAO/IAEA Div. of Atomic Energy; G. R. Williams, Lincoln College, and B. Stonehouse, Canterbury University, New Zealand; and B. B. Jones and W. G. Warren of this Institute. The equation for smoothing age frequencies of thar was kindly calculated by W. G. Warren. For assisting in the shooting and autopsy of specimens, I am grateful to Chris Challies, Gary Chisholm, Ian Hamilton, Ian Rogers and Bill Risk.

LITERATURE CITED

Anderson, J. A., and J. B. Henderson. 1961. Himalayan thar in New Zealand. New Zeal. Deerstalkers' Ass. Spec. Publ. 2.
Banfield, A. W. F. 1955. A provisional life table for the barren ground caribou. Can. J. Zool. 33: 143-147.
Bezem, J. J., J. W. Sluiter, and P. F. van Heerdt. 1960. Population statistics of five species of the bat genus Myotis and one of the genus Rhinolophus, hibernating in the caves of S. Limburg. Arch. Néerlandaises Zool. 13: 512-539.
Bodenheimer, F. S. 1958. Animal ecology today. Monogr. Biol. 6.
Bourlière, F. 1959. Lifespans of mammalian and bird populations in nature, p. 90-102. In G. E. W. Wolstenholme and M. O'Connor [ed.] The lifespan of animals. C.I.B.A. Colloquia on Ageing 5.
Brandborg, S. M. 1955. Life history and management of the mountain goat in Idaho. Idaho Dep. Fish and Game, Wildl. Bull. 2.
Breakey, D. R. 1963. The breeding season and age structure of feral house mouse populations near San Francisco Bay, California. J. Mammal. 44: 153-168.
Caughley, G. 1963. Dispersal rates of several ungulates introduced into New Zealand. Nature 200: 280-281.
——. 1965. Horn rings and tooth eruption as criteria of age in the Himalayan thar Hemitragus jemlahicus. New Zeal. J. Sci. 8: 333-351.
Deevey, E. S. Jr. 1947. Life tables for natural populations of animals. Quart. Rev. Biol. 22: 283-314.
Donne, T. E. 1924. The game animals of New Zealand. John Murray, London.
Dublin, L. I., A. J. Lotka, and M. Spiegelman. 1949. Length of life. Ronald Press, New York.
Fisher, R. A. 1930. The genetical theory of natural selection. Clarendon Press, Oxford.
Green, R. G., and C. A. Evans. 1940. Studies on a population cycle of snowshoe hares on the Lake Alexander area. II. Mortality according to age groups and seasons. J. Wildl. Mgmt. 4: 267-278.
Hewer, H. R. 1963. Provisional grey seal life table, p. 27-28. In Grey seals and fisheries. Report of the consultative committee on grey seals and fisheries. H. M. Stationary Office, London.
——. 1964. The determination of age, sexual maturity, longevity and a life table in the grey seal (Halichoerus grypus). Proc. Zool. Soc. Lond. 142: 593-623.
Howard, W. E. 1949. Dispersal, amount of inbreeding, and longevity in a local population of prairie deermice on the George Reserve, Michigan. Contrib. Lab. Vertebrate Biol. Univ. Michigan, 43.
Howard, W. E., and H. E. Childs Jr. 1959. The ecology of pocket gophers with emphasis on Thomomys bottae mewa. Hilgardia 29: 277-358.
Hickey, F. 1960. Death and reproductive rate of sheep in relation to flock culling and selection. New Zeal. J. Agric. Res. 3: 332-344.
——. 1963. Sheep mortality in New Zealand. New Zeal. Agriculturalist 15: 1-3.
Hughes, R. D. 1965. On the composition of a small sample of individuals from a population of the banded hare wallaby, Lagostrophus fasciatus (Peron & Lesueur). Austral. J. Zool. 13: 75-95.

Kenyon, K. W. and V. B. Scheffer. 1954. A population study of the Alaska fur-seal herd. United States Dep. of the Interior, Special Scientific Report—Wildlife No. 12: 1-77.

Kurtén, B. 1953. On the variation and population dynamics of fossil and recent mammal populations. Acta Zool. Fennica 76: 1-122.

———. 1958. Life and death of the Pleistocene cave bear: a study in paleoecology. Acta Zool. Fennica 95: 1-59.

Laws, R. M. 1962. Some effects of whaling on the southern stocks of baleen whales, p. 137-158. In E. D. Le Cren and M. W. Holdgate [ed.] The exploitation of natural animal populations. Brit. Ecol. Soc. Symp. 2. Blackwell, Oxford.

Leslie, P. H., and R. M. Ranson. 1940. The mortality, fertility and rate of natural increase of the vole (Microtus agrestis) as observed in the laboratory. J. Animal Ecol. 9: 27-52.

Leslie, P. H., U. M. Venables, and L. S. V. Venables. 1952. The fertility and population structure of the brown rat (Rattus norvegicus) in corn-ricks and some other habitats. Proc. Zool. Soc. Lond. 122: 187-238.

Leslie, P. H., T. S. Tener, M. Vizoso and H. Chitty. 1955. The longevity and fertility of the Orkney vole, Microtus orcadensis, as observed in the laboratory. Proc. Zool. Soc. Lond. 125: 115-125.

Lord, R. D. 1961. Mortality rates of cottontail rabbits. J. Wildl. Mgmt. 25: 33-40.

Lotka, A. J. 1907a. Relationship between birth rates and death rates. Science 26: 21-22.

———. 1907b. Studies on the mode of growth of material aggregates. Amer. J. Sci., 4th series, 24: 199-216.

Macdonell, W. R. 1913. On the expectation of life in ancient Rome, and in the provinces of Hispania and Lusitania, and Africa. Biometrika 9: 366-380.

Murie, A. 1944. The wolves of Mount McKinley. Fauna Nat. Parks U.S., Fauna Ser. 5.

Pearl, R. 1940. Introduction to medical biometry and statistics. 3rd ed. Philadelphia: Saunders.

Pearl, R., and J. R. Miner. 1935. Experimental studies on the duration of life. XIV. The comparative mortality of certain lower organisms. Quart. Rev. Biol. 10: 60-79.

Quick, H. F. 1963. Animal population analysis, p. 190-228. In Wildlife investigational techniques (2nd ed). Wildlife Soc., Ann Arbor.

Riney, T. 1956. Differences in proportion of fawns to hinds in red deer (Cervus elaphus) from several New Zealand environments. Nature 177: 488-489.

Sharpe, F. R., and A. J. Lotka. 1911. A problem in age-distribution. Phil. Mag. 21: 435-438.

Slobodkin, L. B. 1962. Growth and regulation of animal populations. Holt, Rinehart and Winston, New York.

Sluiter, J. W., P. F. van Heerdt, and J. J. Bezem. 1956. Population statistics of the bat Myotis mystacinus, based on the marking-recapture method. Arch. Néerlandaises Zool. 12: 63-88.

Stodart, E., and K. Myers. 1964. A comparison of behaviour, reproduction, and mortality of wild and domestic rabbits in confined populations. CSIRO Wildl. Res. 9: 144-59.

Taber, R. D., and R. F. Dasmann. 1957. The dynamics of three natural populations of the deer Odocoileus hemionus columbianus. Ecology 38: 233-246.

Talbot, L. M., and M. H. Talbot. 1963. The wildebeest in western Masailand, East Africa. Wildl. Monogr. 12.

Tyndale-Biscoe, C. H., and R. M. Williams. 1955. A study of natural mortality in a wild population of the rabbit Oryctolagus cuniculus (L.). New Zeal. J. Sci. Tech. B 36: 561-580.

Van Valen, L. 1964. Age in two fossil horse populations. Acta Zool. 45: 93-106.

Wiesner, B. P., and N. M. Sheard. 1935. The duration of life in an albino rat population. Proc. Roy. Soc. Edinb. 55: 1-22.

Woodgerd, W. 1964. Population dynamics of bighorn sheep on Wildhorse Island. J. Wildl. Mgmt. 28: 381-391.

Ann. Acad. Sci. fenn. A, IV Biologica: 186 (1971)

Comparisons between rodent cycles and plant production in Finnish Lapland

JOHAN TAST and OLAVI KALELA

Department of Zoology and Kilpisjärvi Biological Station, University of Helsinki
Helsinki, Finland

Tast, J. & Kalela, O. (1971). Comparisons between rodent cycles and plant production in Finnish Lapland. — Ann. Acad. Sci. fenn. A, IV Biologica 186, 1 –14.
The powerful synchronous fluctuations found in 1963 –1971 in the populations of four species of microtine rodents (*Microtus oeconomus*, *M. agrestis*, *Lemmus lemmus* and *Clethrionomys rufocanus*) living in a fell district in Finnish Lapland are compared with the annual variations of flowering frequency and associated features in the organology of certain plants.

During most of the summer these rodents feed on growing above-ground organs of plants belonging mainly to the field layer. They show a marked preference for generative plant organs, which probably contain compounds essential as rodent food. In late summer, autumn and winter buds and storage organs grown for winter form an increasing proportion of their diet.

Special attention is paid to the root vole (*Microtus oeconomus*). In the area studied, the cotton grass (*Eriophorum angustifolium*) is the most important food plant of the root vole throughout the year. In 1963 –1971 there were three distinct peaks in the flowering frequency of the cotton grass. These peaks were followed in winter by abundant occurrences of first-year rhizomes, very rich in food reserves, and in spring by an above-average amount of sprouting shoots. In the goldenrod (*Solidago virgaurea*), which is one of the favourite food items of root voles in spring and summer, the frequency of flowering shoots and of sprouting shoots in the following spring varied in fair synchrony with the comparable development of cotton grass.

Corresponding to the three periods of favourable food conditions, the populations of the root vole (and of the other rodent species) displayed three cyclic highs though one of them was small. The clearest correlation between food situations and population fluctuations was found in spring, which emphasizes the importance of wintering conditions.

1. Introduction

In most animal populations survival is influenced by a wide variety of factors. Rodent populations characterized by strong annual fluctuations in numbers are no exception to this rule. Thus the widely varying hypotheses presented to explain the fluctuations are not necessarily mutually exclusive.

Annual variations in rodent numbers are known to be particularly strong in arctic and adjacent areas. They often cover large unbroken areas such as Northern Lapland simultaneously and, within each area, the different species tend to fluctuate synchronously. Without ignoring the probability that widely varying factors always affect mortality and natality rates, it must be concluded that some basic climatic factor is at work directly or indirectly influencing the size of the populations.

Among the hypotheses of the latter type

those based on production biology can be traced back some thirty years. They imply that rodent cycles are correlated with variations in the quality or quantity of their vegetable foods. A review of these studies, including important papers by Braestrup (1940), Hustich (1942), Lauckhart (1957) and Svärdson (1957), has been published earlier (Kalela, 1962). The results of some feeding experiments concerning the summer food of *Lemmus lemmus*, *Microtus oeconomus* and *Clethrionomys rufocanus* in Finnish Lapland were referred to in the same connexion. Of the green plants that constitute the staple food of these species, growing organs proved to be distinctly preferred to fully grown ones. A conspicuous feature was the high position of generative organs in the order of preference, especially during the first half of the breeding season. After midsummer, increasing towards the winter, buds and storage organs for the following winter were included in the preferred food.

On all the above evidence, the general course of the rodent cycles was considered to result from the interaction of two complex factors: (1) random oscillations in meteorological conditions (summer temperatures), which caused variations in flowering frequency and, simultaneously, in the nutritional state of the food plants as a whole; (2) rhythms inherent in the food plants and in the rodent populations themselves. The effect of the latter group of factors is to smooth out the influence of climatic variation. However favourable the meteorological conditions have been in the meantime, most arctic perennial plants require an invigoration period of a few years to flower anew (Sørensen, 1941).

One of the starting points in formulating the above hypothesis was the annual variation in the body weight of rodents. For instance, studies carried out on *Clethrionomys rufocanus* at Kilpisjärvi in 1954—1960 (Kalela, 1957, 1962) revealed a distinct decrease in the body weight (most easily observed in males that had overwintered) for the years of cyclic low. Food shortage was considered the simplest explanation.

As many species of plants that belong mainly to the field layer are used as food by the rodents in question, the hypothesis implies a certain co-variation in their flowering, similar to what is known on the flowering and seeding of several species of trees in Southern Sweden (Svärdson, 1957; Bergstedt, 1965), England (Perrins, 1965), etc.

We do not know of any systematic records of the annual flowering frequency of field-layer plants. At the Kilpisjärvi Biological Station in Finnish Lapland (69° 03′ N, 20° 49′ E), year to year variations in the numbers of both flowering and sterile shoots of two plant species have been studied for several years. These two species are the narrow-leaved cotton grass (*Eriophorum angustifolium*) and the goldenrod (*Solidago virgaurea*).

Eriophorum angustifolium is one of the most abundant peatland plants in the study area. According to our feeding experiments and field observations, it is among the highest ranking food items of the root vole (*Microtus oeconomus*) throughout the year. Although our »cafeteria tests» have revealed plant species that root voles prefer even more, this species of cotton grass constitutes their principal diet in most of their regular habitats (Tast, 1964, 1966). During most of the summer they eat aboveground parts, at other times, they mainly consume succulent organs of *Eriophorum angustifolium* lying underground.

Solidago virgaurea grows in the forested habitats occupied by *Microtus oeconomus* during its population peaks. In such areas it is one of its favourite food items.

The present study deals mainly with the root vole. Other microtine rodents in the Kilpisjärvi area include the Norwegian lemming (*Lemmus lemmus*) and the field vole (*Microtus agrestis*) for which *Eriophorum angustifolium* constitutes an important food plant in summer, and the grey-sided vole (*Clethrionomys rufocanus*) which eats *Solidago virgaurea* during the same season, as also does *M. agrestis*. These four rodents are predominantly green-eaters, though in winter the root vole feeds mainly on underground organs of graminids and herbids (Tast, 1964, 1966). As pointed out by Koshkina (1957), *C. rufocanus* is also a fairly typical green-eater, in which respect it resembles the *Microtus* species more than its congeneric species. (For a survey of the plant food of these green-eaters, see Kalela, 1962.)

The fifth and rarest species of microtine

rodents living in our study area, the red vole (*Clethrionomys rutilus*), differs from the others being a seed-eater by preference. For reasons of ecological uniformity this species will be referred to only in passing. The cyclic peaks of the red vole are not strictly synchronous with those of the other species considered, which is another reason for studying this species in a paper of its own.

Other factors besides variations in food conditions will be referred to. Special attention will be paid to the probable effect of predators, particularly mustelids, on the cyclic declines in rodent numbers.

2. Fluctuations in the numbers of rodents

2. 1. *Material and methods*

The indices for the fluctuations of rodent numbers in 1963 – 1971 used in this study were obtained from the following snap trappings:

1 – 2. Line trapping (see Kalela, 1957, p. 13) performed in the subalpine birch region in June (Fig. 1) and from mid-August to September (Fig. 2). The traps were set in lines straight up the mountain slopes from the shore of Lake Kilpisjärvi to the timber line. The catches made in the different habitats covered by forest or scrub have been combined. The numbers of trap-nights generally varied from, say, 500 to 1700, but in the early summers of 1963 and 1966 they numbered only 180 and 340, respectively.

3. Snap trapping (see Tast, 1966, 1968) in Saananvankka, an alpine area near the timber line consisting of alpine meadows and grassy thickets, some of which are seasonally subject to flood (Fig. 3). The trapping was performed at the end of June and beginning of July. Traps were set in groups of five. The number of trap-nights varied from 150 to 600.

4. Snap trapping (see Tast, 1966, 1968), in a subalpine open bog area at the end of June and beginning of July (Fig. 4). Here two traps were set at each point. Trap-nights numbered about 100 per annum.

2.2. *Fluctuations*

Figs. 1 and 2 show the general course of the fluctuations in the total numbers of rodents in the subalpine region. Together they indicate both the annual variation in 1963— 1971 and the early summer to autumn trend for each year. A similar picture of the situation in the alpine region in early summer is given by Fig. 3, in which the years 1961 and 1962 are also included.

Although the data available on food conditions are the most representative in the case of the root vole (*Microtus oeconomus*), trapping data are the smallest for this species. Line trapping, as practised in the subalpine region (Figs. 1 and 2), is not suitable for catching this species, which is very patchy in its incidence owing to its specific demands on the habitat. Above the timber line the root vole lives close to the limit of its climatic tolerance, so there, too, the catches were small (Fig. 3), even though the vegetation in the area studied was suitable for the species.

Despite their smallness, the root vole catches shown in Figs. 1 – 3 indicate annual and seasonal trends similar to those of the other rodents. This is also apparent from the data in Fig. 4, which were collected in a small subalpine area of open bog representing a favourite habitat of this species. Thus in the survey of the annual and seasonal variation in the total number of rodents given below it seems justified to apply the results equally to *Microtus oeconomus*.

After a sharp dip in 1961, and most of 1962 (see Fig. 3), the rodent populations, including those of *M. oeconomus*, increased rapidly during the summer of 1963. They survived the next winter well and some increase occurred the following summer. Consequently a peak was reached late in 1964.

The populations collapsed during the snowy season of 1964/1965 and continued to decrease the following summer. The early part of 1966 was characterized by an extremely deep cyclic low. There was a slight recovery in the summers of 1966 and 1967 but this did not lead to a continuous rise in numbers. This can be seen from the situation in spring 1968: in all samples there was a drop in the numbers found that spring as compared with the previous spring. The difference was not statistically significant in the catch obtained by line-trapping (Fig. but was highly significant ($\chi^2 = 7.78$; P < 0.0 when more effective trapping methods we used (Figs. 3 and 4).

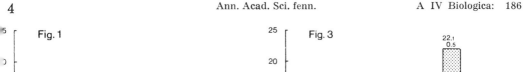

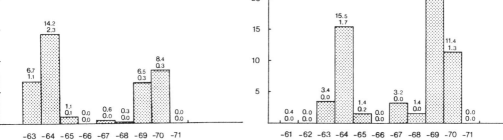

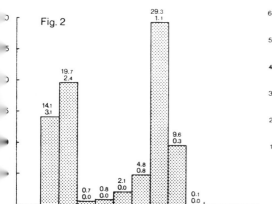

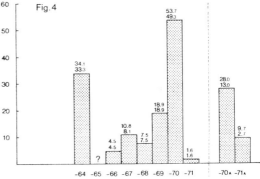

Fig. 1. Fluctuations in relative numbers of rodents in the subalpine birch region in 1963—1971, according to trapping in June. The upper figures indicate the total numbers of all green-eating rodents and the figures below them the *Microtus oeconomus* numbers per 100 trap-nights. Further explanation in text.

Fig. 2. Fluctuations in relative numbers of rodents in the subalpine birch region in 1963—1971, according to trapping from mid-August to mid-September. For further explanations, see Fig. 1 and text.

Fig. 3. Fluctuations in relative numbers of rodents in 1961—1971 in an alpine area slightly above the timber line, according to trappings performed in late June and early July. The upper figures indicate the total numbers of all green-eating rodents and the figures below them the *Microtus oeconomus* numbers per 100 trap-nights. Further explanation in text.

Fig. 4. Fluctuations in relative numbers of rodents in 1964—1971 in a subalpine open bog area, a favourite habitat of *Microtus oeconomus*. Most of the trapping was done in late June and early July, that marked 1970a and 1971a in August. No trapping in 1965 when, according to signs of rodent activity, all populations were very low. For further explanations, see Fig. 3 and text.

The next cyclic rise began in the summer of 1968, but high densities were reached only during the following summer, with a pronounced peak late in 1969. Taken as a whole (see below) the populations were fairly high in the spring of 1970 but decreased markedly in the course of the summer; *Microtus oeconomus* was a typical example. 1971 was characterized by a deep low in the populations of all the species.

The following differences between the species in their fluctuations can be noted:

In summer 1969 the *Microtus agrestis* population displayed an exceptionally pronounced cyclic peak followed by a crash before the advent of spring — i.e., slightly ahead of the decline found in most of the species.

Up to 1969 *Lemmus lemmus* occurred only sparsely in the study area. In that year it became abundant in the alpine region and migrated into the subalpine region late in the summer and in autumn. Its populations were high in spring 1970. The lemmings ceased to reproduce by July, but their young survived fairly well, so the population continued to increase during the summer of 1970. It collapsed, however, in late autumn, i.e., a few months later than the populations of most of the species.

3. Annual variations in plant production

3.1. General

The means of propagation and the growth of the two plant species studied can be outlined as follows:

Söyrinki (1938, 1939) found that *Eriophorum angustifolium* is able to produce viable seed in his study area — the fells of Petsamo (70° N, 31° E). The seedlings, however, serve mainly as a means of dispersal into new areas; in closed vegetation, particularly on wet peatland, this cotton grass is restricted almost exclusively to vegetative reproduction. In the habitats occupied by *Microtus oeconomus* at Kilpisjärvi we have never found any *E. angustifolium* seedlings. Of the 200 and more shoots of this plant dug up in 1971, for instance, all had been produced vegetatively.

The development of *E. angustifolium* from bud to flowering shoot takes some 3 to 5 years (Metsävainio, 1931; Tast, 1966). The shoots die out after flowering, but in the meantime new vegetative buds have grown and during the first summer these develop rhizomes up to 50 cm in length. The following spring the rhizomes reach the earth surface, where small shoots with a few leaves are seen. The leaves survive the winter and new ones develop every summer, their numbers thus increasing until the plant is able to flower.

The reproduction and growth of *Solidago virgaurea* differs:

According to Söyrinki (1938, 1939) seedlings of this species are common around Petsamo, though there is much local and annual variation in their frequency. Vegetative propagation, though it probably occurs, is of minor importance. Our observations at Kilpisjärvi corroborate these views.

The development of *Solidago virgaurea* from seedling to the flowering stage sometimes takes as long as ten years (Perttula, 1941; Söyrinki, 1954). After reaching this stage, the shoot is able to flower several times.

3.2. Methods

In 1963—1971 the numbers of fertile and sterile shoots of *Eriophorum angustifolium* were recorded in 75 plots measuring one square metre and located on open bog. Owing to human influence, the viability of this cotton grass has diminished in many of the study plots since the mid-1960's, and the numbers of both fertile and sterile shoots have dropped markedly.

A start was made with the recording of fertile and sterile shoots of *Solidago virgaurea* in 1965, two years later than for *Eriophorum*. The study plots, totalling 100 square metres, were situated in mesotrophic birch forests.

3.3. Fertile shoots

The total numbers of fertile shoots of *Eriophorum angustifolium* in the study plot were as follows:

1963	1964	1965	1966	1967	1968	1969	1970	197
779	470	181	666	56	53	188	71	3

As can be seen, annual differences are marked[1], though, for the reason stated above the figures for 1967 on are considerably lower than the earlier ones.

The total numbers of fertile *Solidago* shoots were as follows:

1965	1966	1967	1968	1969	1970	19'
60	192	132	91	211	39	1'

---

[1] In the tables presented in subsections 3.3 and 3.4, all year-to-year changes are statistically significant at a level of at least P < 0.01, with the following two exceptions: the change of fertile *Eriophorum* shoots from 1967 to 1968 and the change sterile *Solidago* shoots from 1965 to 1966, in which there are no significant differences.

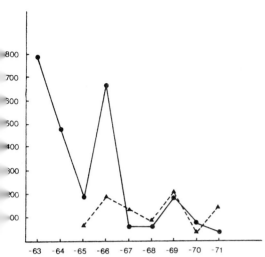

Fig. 5. Annual fluctuation in the numbers of fertile shoots of *Eriophorum angustifolium* (solid line) and *Solidago virgaurea* (broken line). The declining trend in *Eriophorum* is due to the deterioration of the habitat as a consequence of human activity. Further explanation in text.

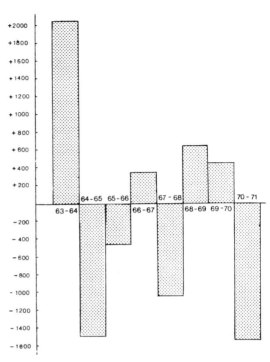

Fig. 6. Year-to-year changes in the numbers of sterile shoots of *Eriophorum angustifolium*, indicating the variation in the numbers of first-year shoots, which correspond to the numbers of first-year rhizomes the previous winter. The overall decrease is due to the deterioration of the habitat consequent to human activity. Further explanation in text.

The annual variations in the flowering frequency of the two plant species investigated tally fairly well with each other.[1] This indicates that there must be some common climatic factor affecting the flowering of these two species of different organologic types, which grow in different habitats.

The fluctuations in flowering frequency of both species are shown in Fig. 5.

3.4. Sterile shoots

The following table shows that in *Eriophorum angustifolium* sterile shoots, too, vary considerably in number from year to year:

1963	1964	1965	1966	1967	1968	1969	1970	1971
4111	6171	4686	4224	4575	3544	4183	4647	3087

---

[1] Between their flowering frequencies, as given in Fig. 5, there is a positive correlation (r = 0.43). To eliminate the effect of the decreasing trend in the numbers of *Eriophorum angustifolium* shoots since 1967, the method of dummy variables was used (for 1965 and 1966 a value of 0 was given, for other years 1). Then R = 0,87; statistical significance close to the 5 % level.

Fig. 6 shows the changes in the numbers of sterile shoots from one year to the next. The figures so obtained reflect the numbers of shoots aged about one year (which is the time required for a bud to develop to a shoot reaching above the ground). A plus sign indicates a more or less abundant development of new shoots, a minus sign the reverse.

There was a more or less abundant occurrence of the youngest shoots in 1964, 1967, 1969 and 1970, i.e. in the summers subsequent to or (1969) coinciding with peak occurrences of flowering shoots (cf. p. 5). In the summers 1965, 1966, 1968 and 1971 the youngest shoots were poorly represented. The marked dips in the numbers of sterile shoots are due to the fact that many shoots — particularly those less than two years old — die before reaching the flowering stage.

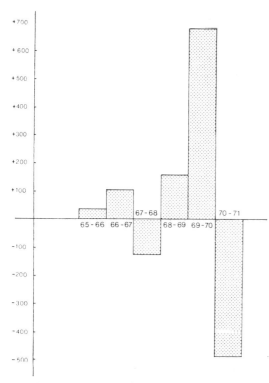

Fig. 7. Year-to-year changes in the numbers of sterile shoots of *Solidago virgaurea*, indicating the variation in the amount of first-year shoots. Further explanation in text.

As will be explained in Section 4, the data illustrated by Fig. 6 also denote the annual variations in the numbers of the youngest rhizomes, rich in nutrients. Indeed, from the standpoint of the present study, the most relevant conclusion can be drawn from these variations.

The numbers of sterile shoots of *Solidago* are summarized in the following table:

1965	1966	1967	1968	1969	1970	1971
520	555	660	535	694	1368	879

Fig. 7 shows year to year changes in the numbers of sterile shoots. Remarkably enough, new sterile shoots were abundant during the same years in the case of both *Eriophorum* and *Solidago*, in spite of the fact that those of *Eriophorum* are produced vegetatively, while those of *Solidago* are almost exclusively seedlings. The most

noticeable decreases were also found in the same years: 1968 and 1971.

To sum up, therefore, there is a co-variation in both flowering and the development of new shoots in the two plant species investigated, which differ considerably from each other organologically and with respect to habitat.

3.5. *Are the rodents themselves the cause of the fluctuations in the flowering of their food plants?*

Basing on his studies in the Taimyr Peninsula, Tikhomirov (1959) pointed out that the flowering of *Eriophorum angustifolium* varies greatly from year to year. He found that a year immediately following a rodent peak was characterized by poor flowering of cotton grasses. According to Tikhomirov this was due to the feeding activity of lemmings (*Lemmus obensis*) that had consumed most of the flower buds during their peak years. Similarly Pruitt (1966, 1968) assumes that, in his study areas in Alaska, the poor flowering of *Eriophorum angustifolium* in certain years was due to foraging by rodents — in this case *Microtus oeconomus*.

The above interpretation could well fit the observations we made in 1965 and 1970 but has no bearing on our data for 1967 because the microtine populations in 1966 were extremely low.

To throw more light on the subject, the following method was adopted: in the bog areas where the flowering of cotton grass was recorded, we surrounded six areas measuring 3 × 1 metres by rodent proof wire netting. Each such exclosure was matched as closely as possible with an adjacent control plot having the same dimensions and vegetative cover. The control plots were open to grazing by rodents mostly root voles and lemmings. The numbers of flowering plants were counted in three consecutive years. The results are as follows:

	1968	1969	1970
Exclosures	61	77	36
Control plots	64	92	33

$$\chi^2 = 0.93$$

There were no statistically significant differences within each year between the flowering frequencies of cotton grass in the exclosures and control plots. On the contrary

Fig. 8. An *Eriophorum angustifolium* stand examined in the present study. — Photo by A. Kaikusalo.

the flowering frequencies in both types of plots varied significantly and synchronously from one year to another.

These results indicate that, in our study area, rodent feeding is not a major cause of annual fluctuations in the flowering of cotton grass.

4. Variations in quality and quantity of the rodent food

Of the results obtained in the previous section, the following three in particular merit further consideration: (1) annual variation and fairly close synchronization of flowering frequency in the two plant species studied; (2) annual variation in the frequency of the youngest vegetative shoots — and youngest rhizomes — in *Eriophorum angustifolium;* (3) annual variation in the frequency of *Solidago virgaurea* seedlings.

4.1. General

With special reference to trees, Lauckhart 1957, p. 232) writes: »Botanists assume that

northern plants must manufacture and store food for several years to produce a single seed crop, and when the seed is produced, this crop utilizes all the stored nutrients from the twigs and stems. It seems logical that herbivores feeding on these plants would be influenced by these changes in stored nutrients.» The same applies to many dwarf shrubs and perennial herbs, *Solidago virgaurea* being an example of the latter.

In other perennial plants, such as *Eriophorum angustifolium*, the shoot dies after flowering and fruiting, and the growth of the plant continues from new buds. The usual explanation for the death of the shoot is nutritional: »death is seen as the result of metabolic patterns in which the flowers, fruits and seeds in some way compete so successfully with the rest of the plant for energy sources and other materials that death is the eventual result» (Hillman, 1964, p. 136; cf. Schumacher, 1967, p. 306 ff.).

From the point of view of the consumer, it is immaterial to which of these two types its food plants belong.

4.2. Summer food

In feeding root voles and lemmings with *Eriophorum angustifolium* and other species of the sedge family, we used leaves of shoots differing in age but belonging to the same shoot system (see Kalela, 1962). The leaves of the older (e.g. 4-year) shoots were preferred to those of the younger (e.g. 2-year) ones — a result that accords with the above conclusions based on the principle of an invigoration period (cf. p. 2)

As a further result of our experiments, all the three green-eaters (*Microtus oeconomus*, *Lemmus lemmus* and *Clethrionomys rufocanus*) studied for summer food showed a marked preference for generative plant organs (flower buds, flowers, developing and ripe fruits, moss sporogonia) to vegetative ones (Kalela, 1962). Even in good flowering years, however, the proportion of fertile shoots may be rather small (as was the case in the *Eriophorum* and *Solidago* stands studied by us), so it is worth pointing out that the generative organs of widely differing plants were eaten by the experimental animals. In fact, they readily accepted the generative organs (usually excluding the seeds) even of plants belonging to groups otherwise consumed little, if at all, by them (Kalela, 1962).

The above generative plant organs hardly play a major role as an energy source for the rodents in question; of decisive importance are compounds present in these organs which the rodents only need in small amounts. According to Löve & Löve (1945), generative plant organs are generally richer than vegetative organs in substances with estrogenic activity. The biochemical composition of pollens has been extensively studied for apicultural purposes (for a review, see Barbier, 1970). The pollens were found to be rich in essential amino acids, vitamins, trace elements and plant-growing substances. In Chauvin's experiments (1968), a diet containing small amounts (1—5 %) of pollen produced a marked increase in birth rates when fed to female mice.

Thus when examining, in the next section, the extent to which variations in flowering frequency of a certain food plant are reflected in the fluctuations of rodent numbers, two factors must be borne in mind: (1) the phenomenon of »flower-eating», (2) the increase in the nutritional state of the whole plant during the invigoration period that precedes flowering.

4.3. Food from autumn to spring: (a) seeds and seedlings

The above is mainly concerned with feeding conditions in summer. Passing on to the situation in other seasons, reference may be made first to some studies on certain rodent species known to be seed-eaters by preference.

The unpublished results of experiments carried out by Peiponen indicate that red voles (*Clethrionomys rutilus*) living in the subalpine region at Kilpisjärvi also have a marked preference for the flowers of herbs and dwarf shrubs. In late summer they change over to seeds, mostly of the same species. A related phenomenon can be seen in the case of red squirrels (*Sciurus vulgaris*) living in North European spruce forests: in a winter when the spruce bears little seed but plenty of flower buds, the squirrels feed almost exclusively on the latter, concentrating on male buds; the following winter they consume the spruce seed (Vartio, 1946; Lampio, 1948; Svärdson, 1957; Kalela, 1962). In Continental Europe good crops of beechmast and acorns generally give rise to outbreaks of bank vole (*Clethrionomys glareolus*) (Turček, 1960).

Obviously, assuming that the fructificative rhythm of plants is also reflected in the food conditions of green-eaters in seasons other than summer, the effect must be based on factors other than those mentioned above for seed-eaters.

We shall first consider the possible effect of variations in the abundance of seedlings

The stimulating effect of fresh greens on the reproduction of various herbivores has been pointed out by several authors. Recent studies on *Microtus montanus* (Pinter & Negus, 1965; Negus & Pinter, 1966; Pinter, 1968) have shown that young sprouting shoots of wheat contain hormone-like compounds which, even when given in minute quantities, significantly and rapidly stimulate various parameters in the reproductive physiology of this species. These authors tentatively postulate »plant estrogens» as the agents responsible for such an effect. Similarly, the great abundance of *Solidago* seedlings as found in our study area in certain springs (e.g. 1970, see p. 7) following profuse flowering is a factor to be considered in examining the spring food conditions of rodents.

In *Eriophorum angustifolium* stands, the amount of fresh greens (sprouting leaves of first-year and older shoots) has also been greater in springs with many new shoots (vegetative in this case) than in others.

4.4. Food from autumn to spring: (b) storage organs

Vegetative, not generative, propagation is the principal means of reproduction of most of the plants eaten by rodents in our study area.

The feeding habits of *Microtus oeconomus* provide an example of the way in which such plants are used as rodent food in autumn and winter (Tast, 1966; for other species, see Kalela, 1962). Already in August, root voles begin to consume underground organs of certain species of plants. As, among the latter, *Eriophorum angustifolium* is prominent in the area studied by us, attention will be concentrated on it.

In the vertical rhizome system of *Eriophorum angustifolium* an individual rhizome, when fully grown or almost fully grown at the onset of its first winter, is generally 20—40 cm long. There is a distinct difference between the shape of first-year rhizomes and that of older ones. The latter are soft, often almost hollow, and dark in colour. The light-coloured first-year rhizomes are compact, due to the abundance of reserve food.[1]

As each rhizome ends in a bud from which a shoot will grow, the amount of first-year rhizomes in winter is reflected by the number of first-year shoots found the following spring. Thus it can be concluded from Fig. 6 that there are sharp annual variations in the amount of the nutrient-rich first-year rhizomes of *Eriophorum angustifolium*.

There is little doubt that the winter food resources available to *Microtus oeconomus* depend to a great extent on this variation, just as the winter food resources available for the seed-eaters depend on the variations in the seed crop. In fact, the spring to spring fluctuations in the numbers of (root) voles clearly parallel the variations in the numbers of first-year rhizomes of the cotton grass (see Figs. 3 and 6).

Large quantities of nutrients are obviously required not only for an abundant formation of seeds (as in beech and goldenrod) but also for the rich formation of rhizomes of the type found in *Eriophorum angustifolium*. First-year rhizomes of the cotton grass have, in fact, been produced most abundantly in years of maximum flowering (1963, 1966 and 1969) or in the summer preceding such a year (1968) — i.e. in years which, according to the principle of an invigoration period, are characterized by a high nutritional state in the plant. A comparable situation was found in the Scotch pine (*Pinus silvestris*) studied by Hustich (1948, 1956, 1969) in Northern Lapland. The maximum years for growth in the length of annual shoots coincided with maximum intensity of female flowering. Those years were preceded by thermically favourable summers with maximum radial growth.

From the above, it can also be understood why, in *Eriophorum angustifolium* and *Solidago virgaurea* with their synchronously varying flowering frequency, there is a co-variation in the amount of first-year shoots in spring, even though they are produced by vegetative means in *Eriophorum*, and almost exclusively from seedlings in *Solidago*.

Subsections 4.3 and 4.4 give the idea that populations of green-eaters and seed-eaters

---

[1] On 15th August 120 *Eriophorum* shoots were dug up and dried at a temperature of 85°C. From this material the weights of first-year and older rhizomes were determined for 10 metres' rhizome length of each. The dry weight of the first-year rhizomes exceeded that of the older ones by 45 %.

do not necessarily differ in principle in their relation to the varying food supplies. The following example illustrates this view.

Both in *Microtus oeconomus*, a green-eater, and in *Clethrionomys rutilus*, a seed-eater, population growth is probably accelerated in a summer characterized by profuse flowering of the respective food plants. During the subsequent winter there are plenty of food resources which the increased populations can consume effectively and still survive well.

5. Fluctuations in rodent numbers in relation to food situations, with special reference to *Microtus oeconomus*

Eriophorum angustifolium can be considered the most important food plant of *Microtus oeconomus* in the study area throughout the year. *Solidago virgaurea* is also a good example of a plant species used as food by root voles in spring and summer.

Therefore, and with reference to the conclusions drawn in the previous section, there is some justification in simplifying the study of the population fluctuations of *Microtus oeconomus* in their relation to food situations as follows:

(1) The more or less synchronous variation in the flowering frequency of both these plant species will be used as a measure of the nutritional value of the summer food plants of the root vole. (2) The variation in the amount of the young *Eriophorum* rhizomes will be used as a measure of its winter food resources, bearing in mind, too, the possible significance of the consequent increase of sprouting shoots in spring. (3) The variation in the amount of *Solidago* seedlings will be used as an additional criterion of the food situation of root voles in spring.

Our material on plant production is obviously inadequate to throw light on the fluctuations of the other species of green-eating rodents in the area. But as far as it goes (see p. 2), the effects are similar for the populations of all the species studied.

Summer 1963 was characterized by a rich flowering. The following winter the food situation was good. In summer 1964 flowering was fairly abundant, though not as much as the previous summer. In 1963 and 1964 the numbers of root voles (and also the total numbers of rodents) grew to a high cyclic level. The subsequent rapid decline took place in winter 1964/1965, during which food was scarce, and continued in summer 1965, which was characterized by a very low flow-

ering frequency. The lowest point of the decline was reached in spring 1966.

The next good flowering season was summer 1966, which was followed by a winter and spring with fairly good food conditions. Owing to the extremely low level of the initial population in spring 1966, there could be no sharp increase in absolute numbers, though a certain increase could be observed up to spring 1967 (Figs. 1 and 2). In this case, the cyclic increase did not continue; the numbers of rodents dropped between spring 1967 and spring 1968. This is precisely what could be expected according to our hypothesis, because the flowering frequency in summer 1967 was low and the food situation the following winter and spring was poor.

Poor flowering continued in summer 1968 but the rich formation of buds for the next winter indicates that invigoration had occurred in *Eriophorum* shoots. The population increase which started that summer continued throughout the next one, leading to a cyclic peak late in 1969. Food conditions were good throughout the winter 1969/70 and continued so the following spring. The populations of most of the rodent species, including *Microtus oeconomus*, were high in spring 1970, but declined rapidly the subsequent summer when flowering was scanty. The food situation was very poor in winter 1970/71, and in spring and summer 1971 all the rodent populations were extremely low.

The above results can be summarized as follows:

During the nine years of the study period there were three peaks of flowering frequency with associated peaks in the amount of food in winter and spring. Corresponding to these maxima, there were two distinct highs in the numbers of *Microtus oeconomus* (and also in the other rodents). The rodent peak corrsponding to one of the periods of optimum food conditions was abortive owing to the short time (one year) available for the populations to respond to the improving conditions.

The clearest correlation between food conditions and population fluctuations in rodents occurred in spring.[1] The abortive peak, too, was visible only in the spring phase. In autumn it levelled out, resulting in a five-year cycle (1964—1969). Such situations possibly account for the fact that five-year cycles are fairly common in the population

fluctuations of rodents living in Northern Lapland (see Kalela, 1962).

The fact that the best correlation between food conditions and population fluctuations is found in spring (when few if any young of the year are included in the catch) stresses the importance of wintering conditions. According to the above analysis, however, the size of the spring populations results from a sequence of interconnected food conditions affecting both survival from autumn to spring and the reproduction rate in the preceding summer.

6. Synchronizing factors

Though the evidence still lies on a narrow basis, the results of this study accord with the view that annual variations in the quality and quantity of the vegetable food are the basic factor responsible for the cyclic nature of the fluctuations in the numbers of rodents in Lapland.

On the other hand, there is little doubt that a variety of other factors besides the basic one are at work. These factors, which have to be considered separately in each case, include: (1) over-exploitation of food resources, which accelerates the decline of the population; (2) stress and its consequences at peak densities; (3) climatic disasters which, directly or by suddenly deteriorating the quality of the food, cut down the number of rodents, regardless of the current population density.

Of special interest are the factors that account for the marked synchrony in fluctuations characteristic of the northern population of different rodent species living in the same area. Some of them have been discussed

in our previous review (Kalela, 1962). The most essential factor was considered to be the shortness of summer in Lapland, owing to which the assimilation periods of plant species tend to coincide, being subject to common fluctuations in climate.

Recent studies (see Maher, 1970, and the literature quoted by him) suggest that predation by mustelids plays a marked role in the declining phase of the fluctuations in arctic rodents. Unpublished observations made at Kilpisjärvi by Mr. Jussi Viitala, M.A., in connexion with his live-trapping of rodents, support this view. Mustelids (*Mustela erminea* and *M. rixosa*) increased rapidly throughout the cyclic rodent high in 1969 and the early summer of 1970. In his trapping area measuring 100×200 metres, no less than ten weasels were trapped from 17th July to 3rd August 1970 (the traps were too small for stoats). After June, when the rodent populations had begun to decline, the relative effect of predation by mustelids must have increased progressively, contributing effectively to the decline of all the populations, irrespective of species.

Acknowledgements. Our best thanks are due to Mr. Simo Lahtinen, M.A. (Institute for Economic Research, Bank of Finland) for his help on statistical problems. This study has been supported by grants from the National Research Council for Sciences.

---

[1] On comparing the fluctuations in the numbers of rodents at Saänanvankka, from where the longest series of annual catches are available (Fig. 3), with year-to-year changes in the numbers of first-year shoots (Figs. 6 and 7) illustrating food conditions in spring and previous winter, the following results are obtained: rodents and *Eriophorum* = 0.47, rodents and *Solidago* r = 0.52, and rodents and both plant species together R = 0.62; all these figures show positive correlation.

References

BARBIER, M. (1970). Chemistry and biochemistry of pollens. — In *Progress in phytochemistry.* Ed. by Reinhold, L. & Liwschitz, Y., vol. 2, 1—34. London.

BERGSTEDT, B. (1965). Distribution, reproduction, growth and dynamics of the rodent species Clethrionomys glareolus (Schreber), Apodemus flavicollis (Melchior) and Apodemus sylvaticus (Linné) in southern Sweden. — Oikos 16, 132—160.

BRAESTRUP, F. W. (1940). The periodic die-off in certain herbivorous mammals and birds. — Science, N.Y. 92, 354—355.

CHAUVIN, R. (1968). (Quoted according to Barbier, 1970.)

HILLMAN, W. (1964). The physiology of flowering. 164 pp. — New York — Chicago — San Francisco —Toronto — London.

HUSTICH, I. (1942). Något om perioder och växlingar i Lapplands växt- och djurvärld under senaste årtionden. — Memo. Soc. Fauna Flora fenn. 17, 200—209.

HUSTICH, I. (1948). The Scotch pine in northernmost Finland. — Acta bot. fenn. 42, 1—28.

HUSTICH, I. (1956). Notes on the growth of Scotch pine in Utsjoki in northernmost Finland. — Acta bot.fenn. 56, 1—13.

HUSTICH, I. (1969). Notes on the growth of pine in Northern Finland and Norway. — Ann. Univ. Turku A, II: 40, 159—170.

KALELA, O. (1957). Regulation of reproduction rate in subarctic populations of the vole Clethrionomys rufocanus (Sund.). — Ann. Acad. Sci. fenn. A, IV: 34, 1—60.

KALELA, O. (1961). Seasonal change of habitat in the Norwegian lemming, Lemmus lemmus (L.). — Ann. Acad. Sci. fenn. A, IV: 55, 1—72.

KALELA, O. (1962). On the fluctuations in the numbers of arctic and boreal small rodents as a problem of production biology. — Ann. Acad. Sci. fenn. A, IV: 66, 1—38.

KOSHKINA, T. V. (1957). Comparative ecology of the red-backed voles in the northern taiga. (Russ., with English summary.) — Bull. Mosc. Soc. Nat. Biol. Sect., 37, 3—65.

LAMPIO, T. (1948). Luontaiset edellytykset maamme oravataouden perustana. (Summary: Squirrel economy in Finland based on natural prerequisities.) — Suomen Riista 2, 97—147.

LAUCKHART, J. B. (1957). Animal cycles and food. — J. Wildl. Mgmt 21, 230—234.

LÖVE, A. & LÖVE, D. (1945). Experiments on the effects of animal sex hormones on dioecious plants. — Ark. Bot. 32 A, 1—60.

MAHER, W. (1970). The pomarine jaeger as a brown lemming predator in northern Alaska. — Wilson Bull. 82, 130—157.

METSÄVAINIO, K. (1931). Untersuchungen über das Wurzelsystem der Moorpflanzen. — Ann. bot. Soc. Vanamo 1, 1—417.

NEGUS, N. C. & PINTER, A. J. (1966). Reproductive responses of Microtus montanus to plants and plant extracts in the diet. — J. Mammal. 47, 596—601.

PERRINS, C. M. (1965). Population fluctuations and clutch-size in the great tit, Parus major. — J. Anim. Ecol. 34, 601—647.

PERTTULA, U. (1941). Untersuchungen über die generative und vegetative Vermehrung der Blütenpflanzen in der Wald-, Hainwiesen- und Hainfelsenvegetation. — Ann. Acad. Sci. fenn. A, 58: 1, 1—388.

PINTER, A. J. (1968). Effects of diet and light on growth, maturation, and adrenal size of Microtus montanus. — Am. J. Physiol. 215, 461—466.

PINTER, A. J. & NEGUS, N. C. (1965). Effects of nutrition and photoperiod on reproductive physiology of Microtus montanus. — Am. J. Physiol. 208, 633—638.

PRUITT, W. O., Jr. (1966). Ecology of terrestrial mammals. — In *Environment of Cape Thompson region, Alaska.* Ed. by Wilimovsky, N.J. & Wolfe, J.N. U.S.Atom. Energy Comm., Div. Tech. Inf. Ext. PNE-481, Chapter 20, 519—564.

PRUITT, W. O., Jr. (1968). Synchronous biomass fluctuations of some northern mammals. — Mammalia 32, 172—191.

SCHUMACHER, W. (1967). Physiologie. — In *Lehrbuch der Botanik,* pp. 204—379. Ed. by Denfer D., Mägdefrau, K., Schumacher, W. & Firbas, F. Stuttgart.

SVÄRDSON, G. (1957). The »invasion» type of bird migration. — Br. Birds 50, 315—343.

SØRENSEN, T., (1941). Temperature relations and phenology of the northeast Greenland flowering plants. — Meddr. Grønland 129, 1—311.

SÖYRINKI, N. (1938). Studien über die generative und vegetative Vermehrung der Samenpflanzen in der alpinen Vegetation Petsamo-Lapplands I. Allgemeiner Teil. — Ann. bot. Soc. Vanam 11: 1, 1—311.

SÖYRINKI, N. (1939). Studien über die generativ und vegetative Vermehrung der Samenpflanze in der alpinen Vegetation Petsamo-Lapplands

II. Spezieller Teil. — Ann. bot. Soc. Vanamo 14: 1, 1—406.

SÖYRINKI, N. (1954). Vermehrungsökologische Studien in der Pflanzenwelt der Bayerischen Alpen. I. Spezieller Teil. — Ann. bot. Soc. Vanamo 27: 1, 1—232.

TAST, J. (1964). Lapinmyyrä, Pohjois-Suomen tulvamaiden tuntematon jyrsijä. (Summary: The vole Microtus oeconomus — the unknown inhabitant of flooded lands in North Finland.) — Suomen Luonto 23, 59—67.

TAST, J. (1966). The root vole, Microtus oeconomus (Pallas), as an inhabitant of seasonally flooded land. — Ann. zool. fenn. 3, 127—171.

TAST, J. (1968). Influence of the root vole, Microtus oeconomus (Pallas), upon the habitat selection of the field vole, Microtus agrestis (L.), in northern Finland. — Ann. Acad. Sci. fenn. A, IV: 136, 1—23.

TIKHOMIROV, B. A. (1959). Interrelationships of the animal world and the vegetation cover of the tundra. (Russ. with English summary.) — Academy of Science, USSR, Komarov bot. Inst., 104 pp. Moscov-Leningrad.

TURČEK, F. J. (1960). Über Rötelmausschäden in den slowakischen Wäldern im Jahre 1959. — Z. angew. Zool. 47, 449—465.

VARTIO, E. (1946). Oravan talvisesta ravinnosta käpy- ja käpykatovuosina. (Summary: The winter food of the squirrel during cone and cone failure years.) — Suomen Riista 1, 49-74.

Presented for publication November 12th, 1971
Printed December 1971

TERRITORIALITY AND HOME RANGE CONCEPTS AS APPLIED TO MAMMALS

By William Henry Burt

TERRITORIALITY

The behavioristic trait manifested by a display of property ownership—a defense of certain positions or things—reaches its highest development in the human species. Man considers it his inherent right to own property either as an individual or as a member of a society or both. Further, he is ever ready to protect that property against aggressors, even to the extent at times of sacrificing his own life if necessary. That this behavioristic pattern is not peculiar to man, but is a fundamental characteristic of animals in general, has been shown for diverse animal groups. (For an excellent historical account and summary on territoriality, with fairly complete bibliography, the reader is referred to a paper by Mrs. Nice, 1941). It does not necessarily follow that this trait is found in all animals, nor that it is developed to the same degree in those that are known to possess it, but its wide distribution among the vertebrates (see Evans, L. T., 1938, for reptiles), and even in some of the invertebrates, lends support to the theory that it is a basic characteristic of animals and that the potentialities are there whether the particular animal in question displays the characteristic. Heape (1931, p. 74) went so far as to say:

"Thus, although the matter is often an intricate one, and the rights of territory somewhat involved, there can, I think, be no question that territorial rights are established rights amongst the majority of species of animals. There can be no doubt that the desire for acquisition of a definite territorial area, the determination to hold it by fighting if necessary, and the recognition of individual as well as tribal territorial rights by others, are dominant characteristics in all animals. In fact, it may be held that the recognition of territorial rights, one of the most significant attributes of civilization, was not evolved by man, but has ever been an inherent factor in the life history of all animals."

Undoubtedly significant is the fact that the more we study the detailed behavior of animals, the larger is the list of kinds known to display some sort of territoriality. There have been many definitions to describe the territory of different animals under varying circumstances. The best and simplest of these, in my mind, is by Noble (1939); "territory is any defended area." Noble's definition may be modified to fit any special case, yet it is all-inclusive and to the point. Territory should not be confused with "home range"—an entirely different concept that will be treated more fully later.

The territoriality concept is not a new one (see Nice, 1941). It has been only in the last twenty years, however, that it has been developed and brought to the front as an important biological phenomenon in the lower animals. Howard's book "Territory in Bird Life" (1920) stimulated a large group of workers, chiefly in the field of ornithology, and there has hardly been a bird life-history study since that has not touched on this phase of their behavior.

In the field of mammals, much less critical work has been done, but many of the older naturalists certainly were aware of this behavior pattern even though they did not speak of it in modern terms. Hearne (1795) apparently was thinking of property rights (territoriality) when he wrote about the beaver as

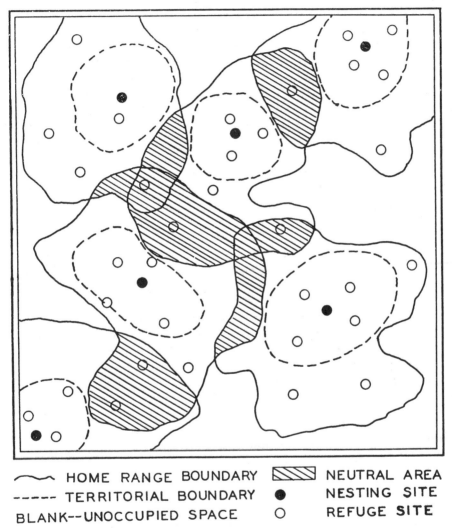

HOME RANGE BOUNDARY   NEUTRAL AREA

TERRITORIAL BOUNDARY   NESTING SITE

BLANK--UNOCCUPIED SPACE   REFUGE SITE

Fig. 1. Theoretical quadrat with six occupants of the same species and sex, showing territory and home range concepts as presented in text.

follows: "I have seen a large beaver house built in a small island, that had near a dozen houses under one roof; and, two or three of these only excepted, none of them had any communication with each other but by water. As there were beavers enough to inhabit each apartment, it is more than probable that each family knew its own, and always entered at their own door without having any further connection with their neighbors than a friendly intercourse" (in Morgan, 1868, pp. 308–309). Morgan (*op. cit.*, pp. 134–135), also writing of the beaver, made the following observation; "a beaver family consists of a male and female, and their offspring of the first and second years, or, more properly, under two years old. . . . When the first litter attains the age of two years, and in the third summer after their birth, they are sent out from the parent lodge." Morgan's observation was later confirmed by Bradt (1938). The works of Seton are replete with instances in the lives of different animals that indicate territorial behavior. In the introduction to his "Lives" Seton (1909) states "In the idea of a home region is the germ of territorial rights." Heape (1931) devotes an entire chapter to "territory." Although he uses the term more loosely than I propose to, (he includes home ranges of individuals and feeding ranges of tribes or colonies of animals), he carries through his work the idea of defense of an area either by an individual or a group of individuals. Not only this, but he draws heavily on the literature in various fields to support his thesis. Although the evidence set forth by Seton, Heape, and other early naturalists is of a general nature, mostly garnered from reports by others, it cannot be brushed aside in a casual manner. The old time naturalists were good observers, and, even though their techniques were not as refined as those of present day biologists, there is much truth in what they wrote.

A few fairly recent published observations on specific mammals serve to strengthen many of the general statements made by earlier workers. In speaking of the red squirrel (*Tamiasciurus*), Klugh (1927, p. 28) writes, "The sense of ownership seems to be well developed. Both of the squirrels which have made the maple in my garden their headquarters apparently regarded this tree as their private property, and drove away other squirrels which came into it. It is quite likely that in this case it was not the tree, but the stores that were arranged about it, which they were defending." Clarke (1939) made similar observations on the same species. In raising wild mice of the genus *Peromyscus* in the laboratory, Dice (1929, p. 124) found that "when mice are placed together for mating or to conserve cage space it sometimes happens that fighting takes place, especially at first, and sometimes a mouse is killed. . . . Nearly always the mouse at home in the cage will attack the presumed intruder." Further on he states, "However, when the young are first born, the male, or any other female in the same cage, is driven out of the nest by the mother, who fiercely protects her young." Similarly, Grange (1932, pp. 4–5) noted that snowshoe hares (*Lepus americanus*) in captivity "showed a definite partiality for certain spots and corners to which they became accustomed" and that "the female would not allow the male in her territory (cage) during late pregnancy and the males themselves were quarrelsome during the breeding season."

Errington (1939) has found what he terms "intraspecific strife" in wild muskrats (*Ondatra*). Much fighting takes place when marshes become overcrowded, especially in fall and winter during readjustment of populations. "But when invader meets resident in the tunnel system of one of [the] last lodges to be used in a dry marsh, conflict may be indeed savage." Gordon (1936) observed definite territories in the western red squirrels (*Tamiasciurus fremonti* and *T. douglasii*) during their food gathering activities. He also performed a neat experiment with marked golden mantled squirrels (*Citellus lateralis chysodeirus*) by placing an abundance of food at the home of a female. This food supply attracted others of the same species. To quote Gordon: "she did her best to drive away the others. Some of her sallies were only short, but others were long and tortuous. There were rather definite limits, usually not more than 100 feet from the pile, beyond which she would not extend her pursuit. In spite of the vigor and the number of her chases (one day she made nearly 60 in about 6 hours) she never succeeded in keeping the other animals away." This individual was overpowered by numbers, but, nevertheless, she was using all her strength to defend her own log pile. To my knowledge, this is the best observation to have been published on territorial behavior in mammals. I have observed a similar situation (Burt, 1940, p. 45) in the eastern chipmunk (*Tamias*). An old female was watched fairly closely during two summers. Having marked her, I was certain of her identity. "Although other chipmunks often invaded her territory, she invariably drove them away [if she happened to be present at the time]. Her protected area was about fifty yards in radius; beyond this fifty-yard limit around her nesting site she was not concerned. Her foraging range (*i.e.*, home range) was considerably greater than the protected area (territory) and occasionally extended 100 or more yards from her nest site." From live trapping experiments, plotting the positions of capture of individuals on a map of the area covered, I interpreted (*op. cit.*, p. 28) the results to mean that there was territorial behavior in the white-footed mouse (*Peromyscus leucopus*), a nocturnal form. When the ranges of the various individuals were plotted on a map, I found that "the area of each of the breeding females is separate—that although areas sometimes adjoin one another, they seldom overlap." Carpenter (1942) writes thus: "The organized groups of every type of monkey or ape which has been adequately observed in its native habitat, have been found to possess territories and to defend these ranges from all other groups of the same species." In reporting on his work on the meadow vole (*Microtus pennsylvanicus*), Blair (1940, pp. 154–155) made the statement "It seems evident that there is some factor that tends to make the females occupy ranges that are in part exclusive; .... Possibly there is an antagonism between the females, particularly during the breeding season, but the available evidence does not indicate to me that they have definite territories which they defend against all trespassers. It seems highly probable that most mammalian females attempt to drive away intruders from the close vicinity of their nests containing young, *but this does not constitute territoriality in the sense that the term has been used* by Howard (1920), Nice (1937), and others *in reference to the breeding territories of birds*." (Ital. mine.) To quote Howard (1920, pp. 192–193): "But the Guillemot is generally surrounded by other Guillemots, and the birds are often so densely packed along the ledges that there is scarcely standing room, so it seems, for all of them. Nevertheless the isolation of the individual is, in a sense, just as complete as that of the individual Bunting, for each one is just as vigilant in resisting intrusion upon its few square feet as the Bunting is in guarding its many square yards, so that the evidence seems to show that that part of the inherited nature which is the basis of the territory is much the same in both species." Blair, in a later paper (1942, p. 31), writing of *Peromyscus maniculatus gracilis*, states: "The calculated home ranges of all sex and age classes broadly overlapped one another. Thus there was no occupation of exclusive home ranges by breeding females. . . . That individual woodland deer-mice are highly tolerant of one another is indicated by the foregoing discussion of overlapping home ranges of all sex and age classes." Reporting on an extensive field study of the opossum, Lay (1942, p. 149) states that "The ranges of individual opossums overlapped so frequently that no discernible tendency towards establishment of individual territories could be detected. On the contrary, tracks rarely showed that two or more opossums traveled together." It seems quite evident that both Blair and Lay are considering the home range as synonymous with the territory when in fact they are two quite distinct concepts. Further, there is no concrete evidence in either of the above papers for or against territoriality in the species they studied. It is to be expected that the territory of each and every individual will be trespassed sooner or later regardless of how vigilant the occupant of that territory might be.

It is not intended here to give a complete list of works on territorial behavior. The bibliographies in the works cited above lead to a great mass of literature on the subject. The point I wish to emphasize is that nearly all who have critically studied the behavior of wild mammals have found this behavioristic trait inherent in the species with which they worked. Also, it should be stressed, there are two fundamental types of territoriality in mammals—one concerns breeding and rearing of young, the other food and shelter. These two may be further subdivided to fit special cases. Mrs. Nice (1941) gives six major types of territories for birds. Our knowledge of territoriality in mammals is yet too limited, it seems to me, to build an elaborate classification of types. Some day we may catch up with the ornithologists.

HOME RANGE

The home range concept is, in my opinion, entirely different from, although associated with, the territoriality concept. The two terms have been used so loosely, as synonyms in many instances, that I propose to dwell briefly on them here. My latest Webster's dictionary (published in 1938), although satisfactory in most respects, does not list "home range," so I find no help there. Seton (1909) used the term extensively in his "Lives" where he explains it as follows: "No wild animal roams at random over the country: each has a home region, even if it has not an actual home. The size of this home region corresponds somewhat with the size of the animal. Flesh-eaters as a class have a larger home region than herb-eaters." I believe Seton was thinking of the adult animal when he wrote the above. We know that young adolescent animals often do a bit of wandering in search of a home region. During this time they do not have a home, nor, as I consider it, a home range. It is only after they establish themselves, normally for the remainder of their lives, unless disturbed, that one can rightfully speak of the home range. Even then I would restrict the home range to that area traversed by the individual in its normal activities of food gathering, mating, and caring for young. Occasional sallies outside the area, perhaps exploratory in nature, should not be considered as in part of the home range. The home range need not cover the same area during the life of the individual. Often animals will move from one area to another, thereby abandoning the old home range and setting up a new one. Migratory animals have different home ranges in summer and winter—the migratory route is not considered part of the home range of the animal. The size of the home range may vary with sex, possibly age, and season. Population density also may influence the size of the home range and cause it to coincide more closely with the size of the territory. Home ranges of different individuals may, and do, overlap. This area of overlap is neutral range and does not constitute part of the more restricted territory of animals possessing this attribute. Home ranges are rarely, if ever, in convenient geometric designs. Many home ranges probably are somewhat ameboid in outline, and to connect the outlying points gives a false impression of the actual area covered. Not only that, it may indicate a larger range than really exists. A calculated home range based on trapping records, therefore, is no more than a convenient index to size. Overlapping of home ranges, based on these calculated areas, thus may at times be exaggerated. From trapping records alone, territory may be indicated, if concentrations of points of capture segregate out, but it cannot be demonstrated without question. If the occupant of an area is in a trap, it is not in a position to defend that area. It is only by direct observation that one can be absolutely certain of territoriality.

Home range then is the area, usually around a home site, over which the animal normally travels in search of food. Territory is the protected part of the home range, be it the entire home range or only the nest. Every kind of mammal may be said to have a home range, stationary or shifting. Only those that protect some part of the home range, by fighting or agressive gestures, from others of their kind, during some phase of their lives, may be said to have territories.

SIGNIFICANCE OF BEHAVIORISTIC STUDIES

I think it will be evident that more critical studies in the behavior of wild animals are needed. We are now spending thousands of dollars each year in an attempt to manage some of our wild creatures, especially game species. How can we manage any species until we know its fundamental behavior pattern? What good is there in releasing a thousand animals in an area large enough to support but fifty? Each animal must have so much living room in addition to other essentials of life. The amount of living room may vary somewhat, but for a given species it probably is within certain definable limits. This has all been said before by eminent students of wildlife, but many of us learn only by repetition. May this serve to drive the point home once more.

LITERATURE CITED

BLAIR, W. F. 1940. Home ranges and populations of the meadow vole in southern Michigan. Jour. Wildlife Management, vol. 4, pp. 149–161, 1 fig.

——— 1942. Size of home range and notes on the life history of the woodland deer-mouse and eastern chipmunk in northern Michigan. Jour. Mamm., vol. 23, pp. 27–36, 1 fig.

BRADT, G. W. 1938. A study of beaver colonies in Michigan. Jour. Mamm., vol. 19, pp. 139–162.

BURT, W. H. 1940. Territorial behavior and populations of some small mammals in southern Michigan. Miscl. Publ. Mus. Zool. Univ. Michigan, no. 45, pp. 1–58, 2 pls., 8 figs., 2 maps.

CARPENTER, C. R. 1942. Societies of monkeys and apes. Biological Symposia, Lancaster: The Jaques Cattell Press, vol. 8, pp. 177–204.

CLARKE, C. H. D. 1939. Some notes on hoarding and territorial behavior of the red squirrel *Sciurus hudsonicus* (Erxleben). Canadian Field Nat., vol. 53, no. 3, pp. 42–43.

DICE, L. R. 1929. A new laboratory cage for small mammals, with notes on methods of rearing Peromyscus. Jour. Mamm., vol. 10, pp. 116–124, 2 figs.

ERRINGTON, P. L. 1939. Reactions of muskrat populations to drought. Ecology, vol. 20, pp. 168–186.

EVANS, L. T. 1938. Cuban field studies on territoriality of the lizard, Anolis sagrei. Jour. Comp. Psych., vol. 25, pp. 97–125, 10 figs.

GORDON, K. 1936. Territorial behavior and social dominance among Sciuridae. Jour. Mamm., vol. 17, pp. 171–172.

GRANGE, W. B. 1932. Observations on the snowshoe hare, Lepus americanus phaeonotus Allen. Jour. Mamm., vol. 13, pp. 1–19, 2 pls.

HEAPE, W. 1931. Emigration, migration and nomadism. Cambridge: W. Heffer and Son Ltd., pp. xii + 369.

HEARNE, S. 1795. A journey from Prince of Wale's fort in Hudson's Bay, to the Northern Ocean. London: A. Strahan and T. Cadell, pp. xliv + 458, illustr.

HOWARD, H. E. 1920. Territory in bird life. London: John Murray, pp. xii + 308, illustr.

KLUGH, A. B. 1927. Ecology of the red squirrel. Jour. Mamm., vol. 8, pp. 1–32, 5 pls.

LAY, D. W. 1942. Ecology of the opossum in eastern Texas. Jour. Mamm., vol. 23, pp. 147–159, 3 figs.

MORGAN, L. H. 1868. The American beaver and his works. Philadelphia: J. B. Lippincott and Co., pp. xv + 330, illustr.

NICE, M. M. 1941. The role of territory in bird life. Amer. Midl. Nat., vol. 26, pp. 441–487.

NOBLE, G. K. 1939. The role of dominance in the life of birds. Auk, vol. 56, pp. 263–273.

SETON, E. T. 1909. Life-histories of northern animals. An account of the mammals of Manitoba. New York City: Charles Scribner's Sons, vol. 1, pp. xxx + 673, illustr., vol. 2, pp. xii + 677–1267, illustr.

——— 1929. Lives of game animals, Doubleday, Doran and Co., Inc., 4 vols., illustr.

Museum of Zoology, Ann Arbor, Michigan.

HIGH ALTITUDE FLIGHTS OF THE FREE-TAILED
BAT, *TADARIDA BRASILIENSIS*, OBSERVED WITH RADAR

TIMOTHY C. WILLIAMS, LEONARD C. IRELAND, AND JANET M. WILLIAMS

ABSTRACT.—Both search and height-finding radars were used to observe the airborne behavior of free-tailed bats, *Tadarida brasiliensis mexicana*, near several caves in the southwestern United States. Radar echoes from dense groups of bats covered areas as large as 400 square kilometers and rose to altitudes of more than 3000 meters. The presence of large numbers of bats within these areas was confirmed by visual observation from a helicopter. Bat flights appeared on radar at dusk and at dawn as a slowly expanding or contracting target, usually located near a known roost. The direction in which the echo expanded most rapidly was not due to drift of the bats by winds. This leading edge often moved at more than 40 kilometers per hour, indicating the capacity for rapid, well-directed, high altitude flight in these animals. Bats flying at such high altitudes must employ sensory systems other than echolocation for orientation and navigation.

Birds often have been reported flying high above the ground, and recent radar studies of bird migration have revealed that in some cases the majority of migratory birds often may be above 3000 meters (Richardson, 1972; Hilditch *et al.*, 1973). Both the respiratory physiology and the orientational abilities of birds appear especially adapted to high altitude flight (Schmidt-Nielsen, 1971; Griffin, 1969). In contrast, bats usually are considered to be restricted to flight at lower altitudes. One exception is *Tadarida brasiliensis mexicana*, the free-tailed or guano bat. These bats have been reported flying at high altitudes both at dawn and dusk as they enter or leave their caves in the southwestern United States (Campbell, 1925; Davis *et al.*, 1962; Constantine, 1967a). The exact altitude attained by these small animals is difficult to estimate under crepuscular light conditions, but Davis *et al.* (1962), using visual triangulation, estimated dawn flights to be as high as 3300 meters. In the present study, observations with radars, employing techniques initially devised for the study of bird migration, were made to gain further information on both the horizontal and vertical distribution of high altitude bat flights near San Antonio, Texas. A helicopter was used to verify the presence of large numbers of bats within radar targets.

807

Methods

Three types of radar were used in the present study—search, height-finding, and weather. Three search radars were used to gain information on the horizontal distribution of bats: a 10-centimeter wavelength, 400-kilowatt peak power, ASR-6 approach control radar located at San Antonio International Airport; a 10-centimeter, 100-kilowatt, MPN-14 approach control unit located at Randolph Air Force Base; and a 23-centimeter, 5-megawatt, FPS 91-A surveillance radar located at Lackland Air Force Base. The FPS-6 height-finding radar at Lackland Air Force Base (11-centimeter, 4-megawatt) gave information on the altitude of bat flights. Information concerning both height and horizontal distribution of the bats was obtained from the FPS-77 weather radar at Randolph Air Force Base (5-centimeter, 300-kilowatt). Both the Lackland Air Firce Base height-finder and the FPS-77 weather radar regularly determined the altitude of aircraft to within 200 meters. Although the accuracy for bat targets is not known, there is no reason to suspect an error of more than 200 meters. Further information on these radars may be obtained from the maintenance handbooks published regularly by the Federal Aeronautics Administration (surveillance radars) and the National Oceanic and Atmospheric Administration (weather radar), both in Washington, D.C.

The reader is referred to Eastwood (1967) for an excellent review of the use of radar in orientation research, and to Richardson (1972) for a discussion of the limitations of radars similar to those employed in the present study.

Data on horizontal distribution of the bats were recorded from the Plan Position Indicator (PPI) of the search radars with either a Polaroid 180 camera or with a time-lapse 8-millimeter camera, which recorded an entire night's data. Samples of such photographs are shown in Fig. 1. The radars were operated with linear polarization and minimal Sensitivity Time Control (STC) to maximize small targets. In most cases, Moving Target Indicator (MTI) was used to reduce ground return from the hills near the bat cave. A discussion of these "anti-clutter" circuits will be found in Richardson (1972). Data on the altitude of bats were recorded from the Range Height Indicator of the height-finding and weather radars with a Polaroid camera (Fig. 2), or by visual inspection of the display when the camera was not available. No "anti-clutter" circuits were used on the height-finding radars.

Observations were made from 20 to 29 October 1967, 6 June to 11 August 1968, and 19 July to 26 August 1971. Observations were made on a non-interference basis and, thus, other operations at the radar sites often prevented our obtaining continuous coverage of bat activity. Whenever possible, observations were started at least 1 hour before sunset or sunrise and continued for as long as bat activity was discernible on the radar displays.

All radars used in this study were subject to night-to-night changes in sensitivity due to changed atmospheric conditions or maintenance adjustments of the radars. At present we do not know the degree to which these sensitivity changes affect the size and shape

---

→

Fig. 1.—PPI photograph of bats emerging from several caves near San Antonio, Texas. Photographs are of the Lackland AFB FPS-91-A PPI with MTI. True compass bearings from the radar are indicated at the periphery of the display. Photographs were taken on 10 August 1968, A at 2045 hours, B at 2105 hours, and C at 2122 hours Central Daylight Savings Time. Ring at about one-fourth radius of lower figure is 32 kilometers range from the radar. White arrows in upper figure point to the location of known bat roosts: Frio (F), Val Verde (V), Nye (N) and Bracken (B), given in Davis et al. (1962), and a deserted railroad tunnel (R), described in Constantine (1967b). Note large targets in lower figures appear near these large roosts.

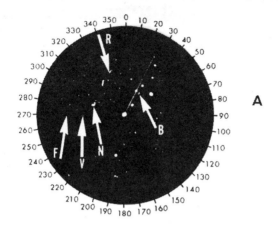

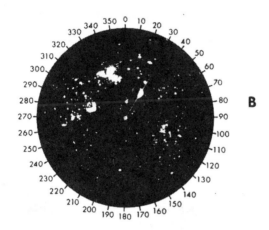

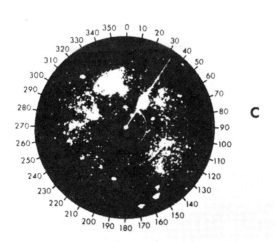

of the radar echoes seen on the PPI and RHI displays. For this reason, we have avoided comparing data on magnitude of bat targets from different radars or on different nights, and measurements of magnitude of radar echoes should be taken as minimum values. Direction, speed, and altitude data appeared to be less affected by changes in radar sensitivity.

Observations from a Helicopter

Visual observation from a helicopter was used to verify the presence of bats within the targets seen on radar. We used a small twin rotor fire rescue helicopter equipped with floodlights for this purpose. Bats could be reliably identified if they passed through a cylindrical area roughly 30 meters long and 15 meters in diameter, extending downward from the aircraft at about 45 degrees. Such observations were possible only if the airspeed of the helicopter was kept below 60 kilometers per hour. It was not possible to reliably count more than 30 animals per minute passing through the searchlight.

Weather Observations

Hourly surface weather observations were obtained from the 24th Weather Squadron at Randolph Air Force Base, located 17 kilometers south-southeast of Bracken Cave. No winds aloft data were available from the San Antonio area. Winds were, therefore, extrapolated from winds aloft data obtained from radiosondes at Victoria, Texas (160 kilometers southeast of Bracken Cave), and Midland, Texas (460 kilometers northwest of the cave). These stations made observations at 0000 and 1200 Greenwich Mean Time (0700 and 1900 hours local time). Personnel at the local weather bureau in San Antonio informed us that this technique should give winds aloft to within 20 degrees at San Antonio for windspeeds greater than 10 kilometers per hour.

RESULTS

Typical Appearance and Development of Bat Targets on Radar

Radar observations of Bracken Cave were made on 51 occasions, and approximately 3 hours of radar data were collected during each observation period. Thirty-five observation periods were conducted during the evening and 16 in the morning. Information from both search and height-finding radars was obtained on 14 occasions, exclusively from search radars on 26, and from height-finders alone on three. Bat targets were not observed on eight occasions due to heavy rainstorm in the area (see below).

Fig. 1 shows a series of photographs taken of the Lackland Air Force Base search radar PPI and illustrates the typical horizontal development of radar echoes that represent the emergence of bats from several large caves. The white arrows in Fig. 1A indicate the position of five large bat roosts, including Bracken Cave, reported by Davis et al. (1962), and by Constantine (1967b). This photograph was taken 20 minutes after sunset. At this time observers at Bracken Cave reported that a column of bats had emerged from the cave, attained an altitude of at least 150 meters, and was not less than 1.5 kilometers long. Note that small irregular targets can be seen near the roosts; the larger circular echoes, which were moving rapidly, represent aircraft.

Fig. 1B shows the appearance of the same radar display 20 minutes later. The bat echoes have now expanded and assumed a fan-shaped form. Of the 164 bat radar targets we recorded, including those near roosts other than

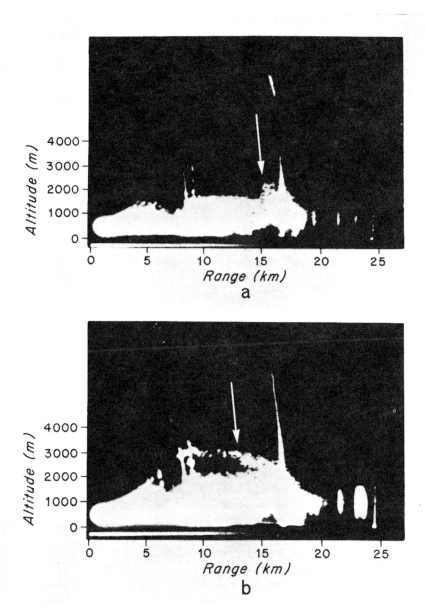

FIG. 2.—Altitudinal distribution of bats emerging from Bracken Cave. Photographs of the RHI display (see text) of the FPS-77 at Randolph Air Force Base, Texas. A, taken at 2110 hours Central Daylight Savings Time, shows the pattern of ground echoes in the area. The bats are the small mound rising toward the tip of the white arrow. The two vertical lines are range marks in the RHI display. B, taken at 2125 hours (CDST), illustrates the development of a layer of bats (white arrow) leaving the cave. Altitude in meters above sea level is given at left; subtract about 300 meters to obtain altitude above ground level. Range of target from the radar is given at bottom of each figure.

Bracken, 152, or 98 per cent, showed a rough fan shape at some time during their development. In some cases a bat echo expanded relatively rapidly in only one direction, in other cases the echoes expanded in two separate directions, possibly indicating two flight directions. At the time Fig. 1B was taken, observers at the Bracken Cave reported that the column of bats extended as far (more than 1.5 kilometers) and as high (about 300 meters) as they could see in the dim light. Also, in Fig. 1B numerous small targets may be seen developing to the southeast of the radar. These targets followed the same pattern of development as did larger targets, remaining essentially stationary and slowly expanding, probably indicating the location of a large number of relatively small bat roosts.

The bat targets continued to expand until they reached the size indicated in Fig. 1C, about an hour after sunset. They then gradually grew more diffuse until it appeared that a fine clutter covered almost the entire PPI screen. This diffuse activity often remained visible for 1 to 2 hours.

Both circumstantial and direct evidence indicated that these large radar echoes were due to bats. Such echoes are seen only during months when large numbers of *T. b. mexicana* are resident in the area, only at sunrise and sunset, and reliably only near large active bat roosts. The presence of bats in these large formations was confirmed by visual observation from a helicopter (see below).

Fig. 2 shows two photographs of the RHI display of the FPS-77 weather radar used in the height-finding mode with the radar beam directed toward Bracken Cave. These pictures illustrate the typical vertical dispersion of bat echoes. Bats emerging from the cave were first seen (Fig. 2A) as a small mound rising above the clutter due to ground return. In Fig. 2B, taken 15 minutes later, the column has risen to a height of about 3000 meters and has progressed about 9 kilometers toward the radar. The increased return near the ground in Fig. 2B probably represents low-flying bats. Data from height-finding radars repeatedly indicated that the bats were distributed as a layer approximately 350 to 500 meters thick and extending outward from the cave.

Greater detail than is shown in Fig. 1 was gained by decreasing the total area covered in a PPI display and observing the emergence from a single roost. Repeated visual inspection of the PPI and time-lapse movie films showed that the large targets, such as shown in Fig. 2, were actually made up of many smaller targets that we will term sub-targets. These sub-targets appeared near the location of the bat roost, then moved outward through the length of the large target and diminished and disappeared at its periphery. Within the large targets, the impression was of an ever changing mass of scintillating luminous dots similar in appearance to large scale passerine bird migrations viewed on radar (Drury and Nisbet, 1964), but confined to a discrete area. Due to the large numbers of these sub-targets and their scintillating nature, it was not possible to measure the exact speed of any single sub-target. But in all cases the majority of sub-targets moved toward those areas in which the large radar target was expanding most rapidly, although sub-targets could be seen leaving the main target at all points of the compass.

During morning activity, the sequence of events shown in Figs. 1 and 2 occurred in reverse order. One to 2 hours before sunrise, small scintillating targets would appear in the vicinity of bat roosts. These would increase in number until they formed radar targets similar to those in Fig. 1B. Over the next 1 to 3 hours the morning targets would slowly diminish in size, disappearing usually about 1 hour after sunrise. The small scintillating targets were similar to those described above but appeared near the edge of the large targets and moved toward the approximate position of the cave. The fanlike shape of the targets seen in the morning indicates that the bats tend to enter their roosts nonrandomly, with the major portion of the bat flight approaching the cave from a single direction.

The succession of events shown in Fig. 1 indicates the general pattern of development of bat echoes. Exceptions to this general pattern occurred on four occasions when two separate flights of bats were seen to emerge from a single cave, with intervals of from 20 to 60 minutes between flights. All observations during 1967 and 1968 were made in clear, or partially overcast weather when no rain was present in the area. Unusually heavy rainfall occurred in the study area during the summer of 1971. Radar observations made during rain revealed greatly reduced bat activity; heavy rainstorms were in the area on the eight occasions when bats failed to produce echoes near Bracken Cave. At this time circular polarization (CP) (see under Methods above) was often used to reduce the returns from rain clouds. Tests with CP at other times showed that this circuit reduced the size of a bat target on the PPI display by approximately 30 per cent. Because no targets at all were seen during the inclement weather mentioned above, it is unlikely that large numbers of bats emerged from Bracken Cave at that time.

Observations from a Helicopter

The presence of large numbers of bats within the areas indicated by the radar targets was verified by direct visual observation from a helicopter equipped with floodlights. On two evenings the aircraft was directed to fly across the large radar target near Bracken Cave. Within the radar target large numbers of bats would suddenly flash through the light beam more rapidly than the observer could count (more than 30 per minute) and then disappear, followed rapidly by another group. Beyond the perimeter of the radar targets, bats were far less numerous and were flying singly or in groups of less than 10. The greatest concentration of animals outside the radar targets was usually within 160 meters of the ground and these bats were often observed flying with the characteristic darting flight of feeding animals. These feeding flights were not seen in bats above 200 meters. Insects upon which the bats feed were common at less than 200 meters, passing like rain through the beam of the searchlights, but were rarely seen above 200 meters.

TABLE 1.—*Number of* Tadarida brasiliensis mexicana *seen from a helicopter at night in a 2-minute interval.*

Altitude above ground level (meters)	Counts made inside radar target	Counts made outside radar target
1500–1330	1	1
1300–1160	1	2
1160–1000	3	4
1000–830	8	1
830–660	13+	1
660–490	26+	1
490–320	8	2
320–160	*	2
160–0	*	15+

* 300 meters was the minimum altitude for night flight in this area.

Further data on the flight patterns of bats leaving Bracken Cave was gained by observing the bat flight from the helicopter before sunset. The aircraft was positioned about 1 kilometer west of the column of emerging bats at an altitude of approximately 330 meters. The thick column of bats that emerged from the cave first broke up into large groups of 10,000 to 1000 animals; these in turn divided into groups of 1000 to 100. These small groups may well have formed the small scintillating sub-targets seen to move across the large bat echoes on the search radar PPI displays. Presumably these small groups continued to disperse until, near the edge of the large radar echoes, they diminished below the critical density for detection by radar.

Thus, it appears that the large targets seen by radar were due to reflection of radar energy from groups of more than 100 animals flying in close proximity. The existence of such concentrations of small animals would explain the ability of radar to detect bats at ranges of up to 150 kilometers.

The vertical distribution of bats was investigated by moving the helicopter up or down in a spiral both within and outside the radar targets near Bracken Cave. Observations were restricted to less than 1500 meters above ground level. The most extensive probes were made between 2000 and 2200 hours (local time) on the night of 8 August 1968. Data from the two probes on this date, one inside the radar target, are listed in Table 1. The bats observed within the radar target were concentrated in a layer between 800 and 500 meters altitude. The slight increase in numbers of bats beyond the radar target at altitudes of 1000 to 1300 meters may indicate a continuation of the layer seen close to the cave but with the bats too widely dispersed to produce a radar echo. On the same night, height-finding radar indicated that the bat flight was less than 1000 meters above the ground. Incomplete separation of bat targets from ground return prevented accurate altitude measurement. A similar situation may be seen in Fig. 2B; a flight of bats below the high altitude flight blends into the ground return.

TABLE 2.—*Radar observations of bat targets* (Tadarida brasiliensis mexicana) *near Bracken Cave, Texas.*

Date	Radar	Direction of flight	Maximum extent (km)	Maximum speed (km per hour)	Maximum altitude/ Radar	Maximun rate of climb (meters per minute)
			Evening Activity			
1968						
7/13	A	NNE	15	36	2400/HL	93
7/14	A	NNW	10	33	2900/HL	70
7/18	A	S	10	15	2600/HL	60
7/24	W	NE	10	100	2000/W	10
7/27	W	NE	25	80	2900/W	8
8/1	A	NE	5	27	2700/HL	120
8/5	L	N	25	25	2100/HL	50
8/9	R	NW+NE	7	19	3100/W	100
8/11	W	E	12	45	2100/W	50
1971						
7/23	L	WNW	9	105	1550/W	9
			Morning Activity			
1968						
7/20	A	SW	8	—	1300/HL	—
8/2	A	NNE	10	—	3000/HL	—
8/3	A	N	5	—	3000/HL	—
8/4	A	N	12	—	2500/HL	—

Radar identification code: A = ASR-6 at San Antonio International Airport, L = FPS-91-A at Lackland Air Force Base, R = MPN 14 at Randolph Air Force Base, HL = FPS-6 at Lackland Air Force Base, W = FPS-77 at Randolph Air Force Base.

Variations in Bat Targets

Bat targets varied in size, direction of motion, speed of the leading edge, altitude, and rate of climb. Although bat emergence patterns from several caves were observed on the four nights we used the long range Lackland Air Force Base radar (see Fig. 1), the greater majority of our observations were made of the Bracken Cave, and, thus, the following analyses pertain only to that roost. Data obtained on all occasions when it was possible to obtain simultaneous search and height finding radar information are listed in Table 2. As discussed under Methods above, great care must be exercised in comparing data on size and density of radar targets due to large variations in radar sensitivity.

Direction.—Flight direction, as listed in Table 2, was taken as the direction of the principal axis of the radar target from the cave for evening observations, and toward the cave for morning observations. Flight directions from Bracken Cave recorded during all evening observation periods are shown in Fig. 3. Bat targets leaving this roost were photographed on 28 occasions; on four evenings

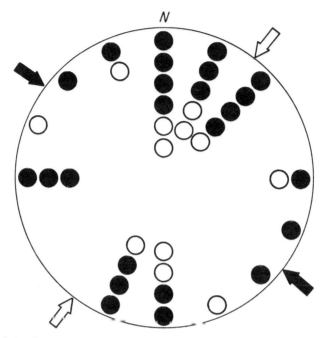

FIG. 3.—Flight directions taken by bats emerging from Bracken Cave, as shown on radar. Closed circles indicate that MTI was in use when flight direction was determined. Open circles indicate normal radar. Closed arrows show the area of expected maximum MTI suppression of targets for both the Lackland AFB FPS-91-A and the San Antonio Airport ASR-6 (because the two radars are located along a straight line joining them to Bracken Cave, areas of expected maximum suppression are the same for both units). Open arrows show the area of expected minimum suppression.

two separate flight directions were recorded, giving a total of 36 data points in Fig. 3. There did not appear to be any single preferred flight direction for exit from Bracken. A Chi-square test for goodness of fit (Batschelet, 1965) applied to this distribution of flight directions, however, indicated that the distribution was not uniform ($\chi^2_p = 15.33$, df = 5, $P < .01$). There appears to be a clustering of data points both to the northeast and to the southwest of the cave. This clustering was probably influenced by MTI circuitry, which tends to accentuate targets moving toward or away from the radar. The axes of expected maximum and minimum suppression are shown in Fig. 3. MTI circuitry was not, however, the sole determining factor of our observed flight directions. Most flight directions recorded on normal radar, when MTI was not in use, were included in the two clusters of data points to the northeast and to the southwest of the cave. A Chi-square test applied to the distribution of flight directions recorded when MTI was in use, with four groups, defined as a 90-degree sector of the unit circle bisected by either a line of maximum or minimum suppression, was not significant ($\chi^2_p = 4.33$, df = 3,

$P > .05$). The hypothesis that the distribution was uniform, therefore, could not be rejected on the basis of this sample.

The dispersal of bats from their caves in one or two principal directions might be due either to drift by winds aloft or to active flight in a preferred direction. The average difference between wind direction and flight direction for bats leaving the cave was 57 degrees, when all flight directions determined by search radar were considered and the wind direction was taken as that at 1500 meters. If we repeat the computations using only the nine cases of bats leaving the cave for which flight direction, altitude, and winds at that altitude are known (Table 2), the value is 68 degrees. Thus, it appears that the flight directions seen on radar are not simply due to wind drift of an ascending column of bats but reflect some degree of horizontal flight in a preferred direction.

The mean difference as calculated above for all bat flights returning to their roosts was 88 degrees, and for only those listed in Table 2 is 103 degrees. This indicates that at least in the final approach to their caves bats do not tend to fly either upwind or downwind.

Size of target.—As mentioned above, under Methods, changes in radar sensitivity produce probably small, but at present unknown, variations in the size of bat echoes on a PPI display and, thus, we cannot compare data from different radars or data from different nights in detail. The following limited analyses, however, appear warranted. Bat radar targets recorded from all roosts, including those shown in Fig. 1, varied from 5 to 25 kilometers in length and from 20 square kilometers to over 400 square kilometers in area. The mean length for evening activity was 17 kilometers with a standard deviation of 7 kilometers and for morning activity was 9 kilometers (standard deviation, 3 kilometers). These variations might reflect true differences in numbers or flow patterns of bats emerging from a cave (Herreid and Davis, 1966), or they might result from differential sensitivity of radars to bat flights at various altitudes. The correlation between length of echo and altitude for the observations in Table 2 was $+ .24$ ($P > .05$). The lack of a significant correlation between these two parameters indicates that altitude is not a major factor in the size of a target seen on the PPI display once a flight of bats rises into the radar beam. Thus, variations in the size of radar targets probably reflect variations in the numbers of bats aloft.

Speed.—An estimate of the horizontal flight speed of bats was gained by measuring the maximum rate of expansion over intervals of 5 to 10 minutes of the large targets on each night. The speed of the small scintillating targets mentioned above was in most cases greater than the expansion of the whole target. The speed of the leading edge of the bat targets emerging from Bracken Cave (measured over at least 5 kilometers) ranged from 7 to 105 kilometers per hour, with a mean of 40 and a standard deviation of 25 kilometers per hour (see also Table 2).

Some of the higher speeds recorded might be due to bats flying lower than the radar horizon for a period of time and then ascending, producing a sudden expansion of the radar echo. The majority of these observations are, however, lower than the maximum flight speed of 100 kilometers per hour for *T. b. mexicana* as estimated by Davis *et al.* (1962). Horizontal speed was negatively correlated with rate of climb. Rates of climb varied from 7 to 120 meters per minute, with a mean of 57 and a standard deviation of 40. The correlation between speed of the leading edge and rate of climb for the nine cases where both measures were available (Table 2) was –.85 ($P < .01$). This indicates, as would be expected, that bats displaying a high vertical velocity have a low horizontal velocity, and suggests that only our highest speeds represent bats in level flight.

Altitude.—For our calculations of the altitude of bat flights, we used the top of the major echo shown on the RHI display (Fig. 2). The majority of bats would be flying below this level, but a few small targets, presumably small groups of bats, were often found up to 500 meters above this level. During the course of four morning and 13 evening observation periods, the average maximum altitude reached by bat flights was 2300 meters (standard deviation, 600 meters), and the range was from 600 meters to 3100 meters above ground level. Approximate altitude above sea level may be obtained by adding 300 meters to these figures. Thus, high altitude flights appear to be common for *T. b. mexicana* both when entering and when leaving their roosts.

DISCUSSION

Radar has revealed that free-tail bats appear to leave and return to large roosts in massive high altitude formations, which may extend over an area greater than 400 square kilometers and rise to altitudes of more than 3000 meters. These formations are not constant from night to night, but change in both altitude and horizontal distribution. The speed of the leading edge of bat radar targets, the fact that the direction of motion of the leading edge does not appear to be due to prevailing winds, and the orientation of the majority of the small, relatively faster targets within the main target toward the leading edge, argues against an entirely random pattern of dispersion by *T. b. mexicana*, and for the conclusion that these bats make rapid, well-directed and perhaps goal-oriented flights.

The flight patterns of these bats are remarkable for both their large horizontal extent and their altitude. Both rapid sustained horizontal flight and flight at altitudes of 3 kilometers suggests that bats may be capable of the high, long-distance flights shown by many species of birds during migration (see Eastwood, 1967). This conclusion is supported by studies with *Myotis lucifugus*, which have revealed homing speeds of 32 kilometers per hour (Mueller, 1966), and by banding studies that have shown *T. b. mexicana* makes yearly migratory flights of 1300 kilometers (Constantine, 1967a). The reader is referred to Griffin (1970) for a review of bat homing and migration.

Physiological adaptations for high altitude flight in bats are at present unknown. Experiments by Thomas and Suthers (1973) have revealed that the average oxygen consumption of flying bats is similar to values obtained for birds by Tucker (1968), but bats lack air sacs and other adaptions believed important for high altitude flight of birds (Bretz and Schmidt-Nielsen, 1971; Schmidt-Nielsen, 1971).

The known orientation systems of bats appear inadequate for guidance at altitudes of 3000 meters. Olfaction and audition appear to operate only at short range (Twente, 1955). Griffin (1971) has shown that echolocation is restricted to a range of less than 300 meters. Wieder-orientierung, or spatial memory (Möhres and Oettingen-Spielberg, 1949; Neuweiler and Möhres, 1967) appears restricted to orientation within a fixed and probably learned path rather than for paths that change from night to night. Vision was shown to be used for homing of *Phyllostomus hastatus* by Williams *et al.* (1966) and Williams and Williams (1970), and this may be the sense used by *T. b. mexicana*. However, in the absence of large landmarks such as mountains or coastlines, visual piloting might prove difficult for animals with the limited visual acuity of bats (Suthers, 1966). The stars might also be used to obtain directional information as has been shown for birds by Sauer and Sauer (1960) and Emlen (1967). During our studies it was not possible to observe the airborne behavior of bats under totally overcast skies. Thus, the use of either the stars or sunset and sunrise for orientation remains a real but untested possibility.

Finally, we must acknowledge that there may be some as yet unknown sensory system used by bats for orientation. Observations of birds maintaining course without being able to see either the sky or the earth indicate that we have yet to understand the total sensory information available to flying animals (Bellrose, 1967; Griffin, 1969; Williams *et al.*, 1972).

The advantages of high altitude flights to bats are at present unclear. *Tadarida brasiliensis* did not appear to be feeding, and although Gressitt and Yashimoto (1963) reported finding insects at altitudes up to 5700 meters above the Pacific Ocean, our observations from a helicopter revealed few insects above 500 meters. Davis *et al.* (1962) suggested that *T. b. mexicana* may fly 100 kilometers or more in their nightly feeding flights. If bats were able to obtain tailwinds at high altitudes, the effort expended in reaching these altitudes might be offset by the reduced effort required for long distance flight. At present we lack sufficient physiological data for such calculations, and our data are insufficient to test whether the bats tended to choose the most favorable altitude for a given flight direction.

Acknowledgments

We are deeply indebted to the many men of the United States Air Force and the Federal Aviation Administration who assisted us in this research. Funds were provided by the United States Air Force Office of Scientific Research through a contract with the Office of Ecology, Smithsonian Institution, H. Buechner, Head, and through grant no. AFSOR-71-2123 to the Research Foundation of the State University of New York. Analysis of the data was also accomplished under National Science Foundation grants GZ-259 and GB-13246 to the Woods Hole Oceanographic Institution. We are indebted to Dr. D. R. Griffin and to Dr. C. H. Herreid for their helpful suggestions on the manuscript.

LITERATURE CITED

BATSCHELET, E. 1965. Statistical methods for the analysis of problems in animal orientation and certain biological rhythms. AIBS, Washington, D.C., 57 pp.

BELLROSE, F. 1967. Radar in orientation research. Proc. 14th Ornith. Cong., Blackwell, Oxford, pp. 281–309.

BRETZ, W. L., AND K. SCHMIDT-NIELSEN. 1971. Bird respiration: Flow patterns in the duck lung. J. Exp. Biol., 54:103–118.

CAMPBELL, C. A. R. 1925. Bats, mosquitoes and dollars. Stratford, Boston, 262 pp.

CONSTANTINE, D. G. 1967a. Activity patterns of the Mexican free-tailed bat. Univ. New Mexico Press, Publ. Biol., 7:1–79.

———. 1967b. Rabies transmission by air in bat caves. Publ. U.S. Public Health Serv., 1617:ix + 1–51.

DAVIS, R. B., C. F. HERREID II, AND H. L. SHORT. 1962. Mexican free-tailed bats in Texas. Ecol. Monogr., 32:311–346.

DRURY, W. H., AND I. C. T. NISBET. 1964. Radar studies of orientation of songbird migrants in southeastern New England. Bird Banding, 35:69–119.

EASTWOOD, E. 1967. Radar ornithology. Methuen and Co., London, 278 pp.

EMLEN, S. 1967. Migratory orientation in the Indigo bunting *Passerina cyanea*. Parts I and II. Auk, 84:309–342 and 463–489.

GRESSITT, J. L., AND C. M. YASHIMOTO. 1963. Dispersal of animals in the Pacific. Proc. 10th Pacific Sci. Cong., Bishop Mus. Press, Honolulu, Hawaii, pp. 283–292.

GRIFFIN, D. R. 1969. The physiology and geophysics of bird navigation. Quart. Rev. Biol., 44:255–275.

———. 1970. Migrations and homing of bats. Pp. 233–264, in Biology of Bats (W. A. Wimsatt ed.), Academic Press, New York, 1:xii + 1–406.

———. 1971. The importance of atmospheric attenuation for the echolocation of bats (Chiroptera). Anim. Behav., 19:55–61.

HERREID, C. H., AND R. B. DAVIS. 1966. Flight patterns of bats. J. Mamm., 47:78–86.

HILDITCH, C. D. M., T. C. WILLIAMS, AND I. C. T. NISBET. 1973. Autumnal bird migration over Antigua, W. I. Bird Banding, 44:171–179.

MÖHRES, F. P., AND T. OETTINGEN-SPIELBERG. 1949. Versuche über die Nahorientierung und das Heimfindevermögen der Fledermäuse. Verhandlung der Deutschen Zoologen in Mainz, pp. 248–252.

MUELLER, H. 1966. Homing and distance orientation in bats. Z. Tierpsychol., 23:403–421.

NEUWEILER, G., AND F. P. MÖHRES. 1967. Die Rolle des Ortsgedächtnisses bei der Orientierung der Grossblatt Fledermäuse, *Megaderma lyra*. Z. Vergl. Physiol. 57:147–171.

RICHARDSON, J. 1972. Temporal variations in the ability of individual radars in detecting birds. Field notes Associate Committee on Bird Hazard to Aircraft, Nat. Res. Council, Ottawa, Canada, 61:1–58.

SAUER, E. G. F., AND E. M. SAUER. 1960. Star navigation of nocturnal birds. Cold Springs Harbor Symp. Quant. Biol., 25:389–393.

SCHMIDT-NIELSEN, K. 1971. How birds breathe. Sci. Amer., 225:72–88.

SUTHERS, R. 1966. Optomotor responses by echolocating bats. Science, 152:1102–1104.

THOMAS, S. P., AND R. A. SUTHERS. 1973. The physiology and energetics of bat flight. J. Exp. Biol., in press.

TUCKER, V. A. 1968. Respiratory exchange and evaporative water loss in the flying Budgerigar. J. Exp. Biol., 48:67–87.

TWENTE, J. W. 1955. Aspects of a population study of cavern-dwelling bats. J. Mamm., 36:379–390.

WILLIAMS, T. C., AND J. M. WILLIAMS. 1970. Radio tracking of homing and feeding flights of a neotropical bat *Phyllostomus hastatus*. Anim. Behav., 18:302–309.

WILLIAMS, T. C., J. M. WILLIAMS, AND D. R. GRIFFIN. 1966. The homing ability of the neotropical bat *Phyllostomus hastatus*, with evidence for visual orientation. Anim. Behav., 14:468–473.

WILLIAMS, T. C., J. M. WILLIAMS, J. M. TEAL, AND J. W. KANISHER. 1972. Tracking radar studies of bird migration. Animal orientation and navigation: a symposium, NASA SP-262, Washington, D.C.

Department of Biology, State University of New York, Buffalo (present address of Ireland: Department of Psychology, Oakland University, Rochester, Michigan). Accepted 20 February 1973.

EFFECT OF HABITAT QUALITY ON DISTRIBUTIONS OF THREE SYMPATRIC SPECIES OF DESERT RODENTS

Heteromyid rodents *Dipodomys deserti* Stephens, *Dipodomys merriami* Mearns, and *Perognathus longimembris* (Coues) coexist in the Kelso Dune area, Mojave Desert, San Bernardino County, California. During summer 1970 these rodent species were trapped in adjacent dune and valley habitats during two periods of differing habitat quality. During the first or "before" time period (1 July to 15 August), habitat conditions were very poor due to a continuing local drought. Evidence supporting this judgement on habitat quality follows. I observed no annual plant growth or perennial seeding during visits to the area during winter and spring of 1969 to 1970; none of the rodents captured during the summer were reproductively active and none were juveniles. The "after" period occurred when the drought was temporarily broken on 15 August. An intense cloudburst drenched the area resulting in a period of rapid growth and flowering of first annual and then perennial plants. This paper compares distributions of the three rodent species in dune and valley habitats and changes in their distribution from "before" to "after" periods. This comparison provides insight into mechanisms of habitat selection and spatial separation of these species.

Field data on species dominance were obtained by baiting an area with wild bird seed to maximize rodent densities. Observations were then made on interspecific encounters from a platform using battery-powered red lights for illumination. In 11 interspecific encounters between kangaroo rats, *D. merriami* would retreat from the larger *D. deserti*. Retreat always began before any physical contact took place and the dominant individual showed little or no interest in pursuit. In four encounters with either of the kangaroo rat species, *P. longimembris* always retreated with no interest in pursuit being shown by the kangaroo rats. Individuals of each species were trapped and exposed to each other in the laboratory to gather supportive evidence on species dominance. Each encounter took place during early evening in a 1.0 by 1.8 meter (m) cage illuminated by a dim red light and center divided with plywood. A single individual of each of two species was placed in opposite ends of the divided cage and allowed five minutes to adjust before the divider was removed. All actions and postures were noted for ten minutes, or less if fighting seriously endangered one of the individuals. Five trials of each of the four combinations of species were tested. Both field and laboratory observations showed that *D. deserti* was by far the most aggressive species and the dominant rodent. *D. merriami*, the subordinate kangaroo rat, was dominant in all encounters with *P. longimembris*. Results of these observations were used to supplement distribution data obtained from the trap grid.

TABLE 1.—*Summary of recapture frequencies and total captures of three rodent species during summer 1970.*

Recapture Frequencies	Species		
	D. deserti	*D. merriami*	*P. longimembris*
1	7	3	10
2 to 5	5	3	8
6 to 10	1	3	1
11 to 15	2	0	0
Total number of individuals (N)	15	9	19
Total captures	61	37	39

A study plot was selected in an area where there was a sharp boundary between dune and valley habitats. A four hectare trapping grid consisting of 140 trap stations with 18 m between stations was placed half on dunes and half on valley floor. A single Sherman live trap was placed at each station and baited with wild bird seed and peanut butter. Dune and valley were trapped with a roving, alternating trap row procedure to help prevent rodents from becoming trap "shy" or trap "bums." Each trap was set late in the evening, checked and sprung before 0800 to prevent heat death of any captives. The area was trapped continuously from July through August 1970. Forty-three rodents were captured, marked by toe clipping, and released (Table 1).

Trapping data were analyzed using stepwise discriminant analysis (Dixon, 1970). For analysis, each trap station was given an X, Y co-ordinate. X axis trap rows ran parallel to the dune-valley boundary; Y axis trap rows were at right angles to the boundary. Y ordinates one through seven were located on valley floor, eight through fourteen on dunes. The X, Y co-ordinates defined the location of each rodent capture and were the basis upon which distributions were separated.

During this study *Dipodomys deserti* was almost entirely restricted to dune substrate during both time periods (Fig. 1). Previous work has indicated that this is optimum habitat for the desert kangaroo rat (Grinnell, 1922; Durrant, 1943; Butterworth, 1960; Eisenberg, 1963). Most workers consider *D. deserti* to be one of the most specialized kangaroo rats (Grinnell, 1922; Wood, 1935; Lidicker, 1960), and this specialization was reflected in the strong substrate preference which did not change with habitat quality. "Before" and "after" distributions of *D. deserti* were significantly different from the other two species distributions; however, separation increased in the "after" period (Table 2).

During the "before" period *Dipodomys merriami* was captured in both habitats, but its initial distribution was significantly different when compared to the "after" period (Fig. 1).

TABLE 2.—*Stepwise discriminant analysis F-matrix comparing three species distributions in two time periods.* * significant F-value at the 0.05 level.

	D. deserti	*D. merriami*
	BEFORE 15 AUGUST RAIN	
D. merriami	3.5006*	
P. longimembris	3.0916*	0.1001
	AFTER 15 AUGUST RAIN	
D. merriami	23.2952*	
P. longimembris	14.4996*	2.7976

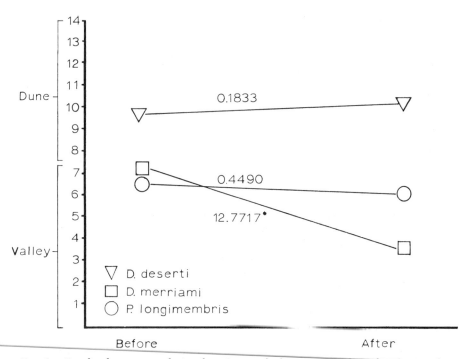

Fig. 1.—Results from an analysis of variance of changes in rodent distribution from "before" to "after" time periods. One through 14 on the Y axis represent trap rows. Symbols indicate mean Y axis distribution. * significant F-value at the 0.05 level.

The difference in between-period distribution of *D. merriami* was caused by its withdrawal from the dunes during the "after" period. Due to the extreme conditions during the "before" period, I suggest that withdrawal from the dunes during the period of plant growth and seed production was related to the increasing availability of food resources in areas not occupied by the dominant *D. deserti*. Lidicker (1960) found that although sandy soils are preferred, *D. merriami* will tolerate a wide range of soil types with the presence of open areas being the most important factor affecting its distribution. Open areas were present in both habitat areas of the trap grid and are, therefore, not considered to be important in determining the local distribution of Merriam's kangaroo rat.

During both periods *Perognathus longimembris* was found in both habitats almost equally (Fig. 1). Interaction with kangaroo rats does not appear to be a major factor in determining its distribution. Separation may be related to feeding behavior as was noted for *Dipodomys heermanni* and *Perognathus parvus* which took different seeds where they occur sympatrically (Smith, 1942). Brown and Lieberman (1973) have shown that *D. deserti*, *D. merriami*, and *P. longimembris* harvested seeds of different sizes; however, there was considerable overlap between the seeds harvested by *D. merriami* and *P. longimembris*. They also showed that the kangaroo rats forage primarily on the open ground away from vegetative cover, whereas *P. longimembris* concentrated its foraging under the cover of shrubs. For a detailed discussion of rodent species diversity on sand dunes and resource utilization by the species see Brown (1973) and Brown and Lieberman (1973).

The important point shown by this study is that investigators should be careful to note habitat conditions when making short-term samples of rodents for purposes of determining distribution or spatial relationships between species as these relationships may change with habitat quality.

I thank Dr. Richard MacMillen and my committee members for assistance with this research. This paper is part of a thesis submitted in partial fulfillment of the requirements for the degree of Master of Science at California State Polytechnic University.

LITERATURE CITED

BROWN, J. H. 1973. Species diversity of seed-eating desert rodents in sand dune habitats. Ecology, 54:775-787.

BROWN, J. H., AND G. A. LIEBERMAN. 1973. Resource utilization and coexistence of seed-eating desert rodents in sand dune habitats. Ecology, 54:788–797.

BUTTERWORTH, B. B. 1960. A comparative study of sexual behavior and reproduction in the kangaroo rats *Dipodomys deserti* (Stephens) and *Dipodomys merriami* (Mearns). Unpublished PhD dissertation, Univ. Southern California, 169 pp.

DIXON, W. J. 1970. BMD Biomedical Computer Programs. University of Calif. Press, Berkeley, 600 pp.

DURRANT, S. D. 1943. *Dipodomys deserti* in Utah. J. Mamm., 24:404.

EISENBERG, J. F. 1963. The behavior of heteromyid rodents. Univ. California Publ. Zool., 69:1–100.

GRINNELL, J. 1922. A geographical study of the kangaroo rats of California. Univ. California Publ. Zool., 24:1–124.

LIDICKER, W. Z. 1960. An analysis of intraspecific variation in the kangaroo rat *Dipodomys merriami*. Univ. California Publ. Zool., 67:125–218.

SMITH, C. F. 1942. The fall food of brushfield pocket mice. J. Mamm., 23:337–339.

WOOD, A. E. 1935. Evolution and relationships of the heteromyid rodents with new forms from the Tertiary of western North America. Ann. Carnegie Mus., 24:73–262.

JUSTIN CONGDON, *Box 6, Department of Zoology, Arizona State University, Tempe, 85281. Submitted 16 October 1973. Accepted 29 January 1974.*

CHANGES IN NORWAY RAT POPULATIONS INDUCED BY INTRODUCTION OF RATS

David E. Davis and John J. Christian

Division of Vertebrate Ecology, Johns Hopkins School of Hygiene and Public Health, Baltimore 5; Naval Medical Research Institute, Bethesda, Maryland

The introduction of aliens into an existing population of mammals may be followed by unexpected effects that relate to social structure and population composition. These effects were studied by introducing alien rats into stationary and increasing populations of rats in city blocks. This work is part of a continuing study of the mechanisms of change in vertebrate populations using Norway rats (*Rattus norvegicus*) in residential areas in Baltimore as experimental animals (Davis, 1953). These rats inhabit back yards, basements, and garages and feed on garbage. The human sanitary conditions in general are poor and remain unchanged for months at a time, so that the food supply of the rats has only slight seasonal variations. Other environmental conditions are similarly subject to little change for many months at a time. The constancy of these factors permits experiments on populations in a relatively stable environment. Finally, the population of rats in each block is essentially discrete and isolated, as rats rarely travel from one block to another (*see* Davis, 1953, for references).

METHODS AND PROCEDURES

The procedures followed to study the effects of introducing strange rats into a population consisted of taking some rats from a stationary or increasing population in one block and introducing them into a comparable population in another block and observing the resulting changes in the second population. The status (stationary or increasing) of the population was determined by estimates at bimonthly intervals for more than a year. Blocks that appeared to be either stationary or increasing were selected in October and monthly estimates made. From this group 4 stationary and 4 increasing populations were chosen. To get a base line for adrenal weights, six rats of one sex, weighing over 200 grams each, were removed from each block during the first experimental week (week 1). Alien rats were then introduced into each block during the third week and at the same time native rats were removed from the increasing populations. The details of these removals and introductions are contained in tables 2 and 3. Estimates were made during the sixth and eighth weeks, each followed by the removal of a small sample of rats for adrenal weights. The adrenal weights were expressed for each sample as the mean per cent of standard reference values (Christian and Davis, 1955). The rats to be introduced into a population were individually marked by toe-clipping prior to introduction, whereas the native rats were not marked. The details of the history of each population were complicated by the impossibility of introducing exactly the same number of rats into each block on exactly the same days, and the numerical population size also differed in each block.

A discussion of the likelihood of error is desirable when it is claimed that two populations differ in number, since the detection of changes is fundamental to the conclusions derived from these introductions. Some aspects of the census method were discussed by Brown, *et al.* (1955). However, the basic problem is that, even with trapping, the true number of rats in a block is not known. Nevertheless, a check on the validity of estimation can be made by comparing estimates before and after a trapping program. Suppose that an estimate of N_1 rats is first obtained, subsequently T rats are removed and a second estimate of N_2 rats is made. Obviously N_1 should equal $T + N_2$. A figure for percentage of error can be given as $\frac{N_1 - (T + N_2)}{N_1}$. For example, if the estimate for a block is 151 rats and then 81 are removed by trapping and an estimate of 63 is made, then $\frac{151 - (81 + 63)}{151} = 4.6$ per cent. Other procedures could be used such as $\frac{T - (N_1 - N_2)}{T}$ or $\frac{N_1 - T - N_2}{N_1 - T}$. The first procedure is preferred because it bases the calculations on N_1, which is the estimate that was used to determine the status of the block. A total of 50 populations was available to determine the extent of error. Each had been trapped during the past 6 years, and an estimate had been made before trapping and another within a month after cessation of trapping. Naturally, some changes can occur during the intervening month, but for practical reasons it is usually not possible to make an estimate promptly after the cessation of trapping. These blocks contained 1,707 rats by the estimates (N_1) and 1,502 were trapped. The number of rats per block

TABLE 1. — Distribution of Differences Among Estimates

Per cent Error[*]	Blocks		Total
	Positive	Negative	
0–9	13	6	19
10–19	6	11	17
20–29	4	4	8
30–39	2	3	5
40–49	1	0	1
Totals	26	24	50

[*] $\dfrac{N_1 - (T + N_2)}{N_1}$

varied from 15 to 182. The distribution of errors is given in Table 1. The percentage of error was independent of the number of rats in the population. From these differences the standard error of the difference can be calculated to be 10.7 per cent. This value can be used as an indication of the errors to be expected in estimates of population changes in blocks. For example, from Table 2 it is seen that the estimate (block 150128) before introduction was 116 and after was 89. The percentage difference is 23.3 which when divided by 10.7 gives a ratio of 2.2. This difference appears to be statistically significant.

RESULTS AND DISCUSSION

The histories of the populations are given by blocks in tables 2 and 3 and figures 1, 2, and 3. The terms "replacement" and "supplement" require clarification for this discussion. We mean by replacement that approximately the same number of rats was introduced as was removed. Supplement means that many more alien rats were introduced than were removed. A quantitative percentage might have been used to distinguish these two terms, but it would have been rather meaningless because (1) the size of the individual rats varies considerably, and (2) immediate mortality is probably high. Therefore, we really do not know the actual number of rats that produced the results. Another factor is that births and deaths are normally high in any population of rats. The average monthly death rate is about 20 per cent for stationary rat populations; therefore, their birth rate is also about 20 per cent. Comparable mortality and birth rates for increasing populations are prob-

TABLE 2. — RESULTS OF INTRODUCTION OF RATS INTO STATIONARY POPULATIONS

Block number	140338		140344		140118		150128	
Zero week is	Dec. 16		Dec. 16		Feb. 9		Feb. 9	
	Week	Rats	W	R	W	R	W	R
Population	−20	62	−20	30	−19	105	−18	122
Population	−13	42	−13	32	−11	98	−10	118
Population	− 5	40	− 5	38	− 6	87	− 6	120
Population	0	49	0	35	0	100	0	116
Rats removed	1	6M	1	6F	1	6F	1	6M
Rats introduced	3	10M	3	8F	2-6	23F	2-6	27M
Rats removed	—	—	—	—	—	—	—	—
Population	6	45	6	44	6	76	7	89
Rats removed	6	6M	6	6F	6	6F	7	6M
Population	8	56	8	23	11	63	10	86
Rats removed	8	22	8	12	11	27	10	36
	Week	Index[1]	W	I	W	I	W	I
	1	83.0	1	91.8	1	93.4	1	79.2
Adrenal size	6	84.1	6	85.3	6	90.4	7	71.4
	8	88.0	8	102.4	11	68.4	10	73.6

[1] Mean of the individual per cent of appropriate reference value for the sex and size of the rat.

ably about 15 per cent and 25 per cent per month respectively. It is unwise, under these circumstances, to attempt a precise measurement of numerical differences in the numbers of rats used.

The population estimates (Table 2) in the two replacement blocks 6 weeks after the introduction of rats showed (Fig. 1) that block 140338 was not significantly different from the previous estimate, while block 140344 had apparently increased (P is about .04). While the apparent difference in results in these 2 blocks might be due to sex (the females less disturbing) or to numbers (fewer introduced into 140344), no interpretation will be attempted for the reasons cited above. The two supplemented blocks declined significantly (P is about .04 for each) (Fig. 1).

We recognize that the replacement pair of populations was done in December 1953, and the supplemented pair in February 1954, and that differences might be due to some seasonal aspect. However, the only known seasonal change, an increase in breeding from December to February, would produce the opposite result.

Population growth ceased in all four increasing populations following the replacement of native with alien rats (figs. 2, 3,). In no block was the difference statistically significant. It apparently made no difference whether the sex of the introduced rats

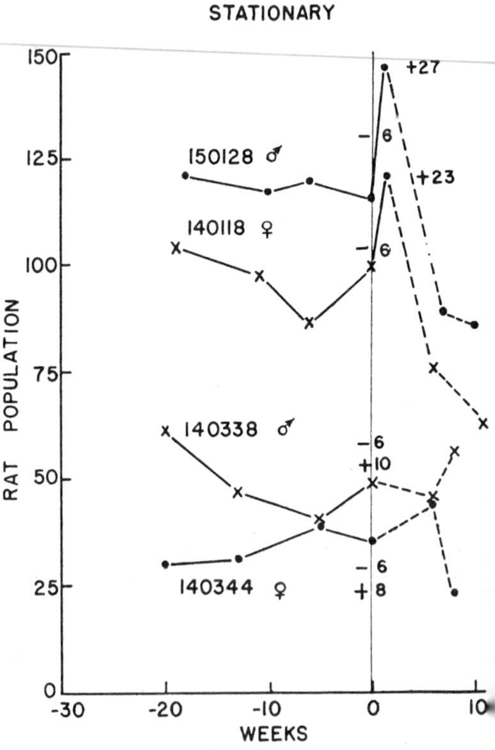

FIG. 1. The changes in four stationary populations for 20 weeks before introduction of rats (at 0 time and about 10 weeks after. The number added is indicated by a plus sign, the number removed by a minus sign.

Table 3. — Results of Introduction of Rats into Increasing Populations

Block number	140111		140134		140201		140222	
Zero week is	Dec. 16		Dec. 16		Feb. 9		Feb. 9	
	Week	Rats	W	R	W	R	W	R
Population	−20	110	−20	80	−25	86	−20	57
Population	−13	118	−13	88	−16	100	−14	62
Population	− 5	133	− 4	115	− 8	105	− 5	95
Population	0	150	0	140	0	135	0	90
Rats removed	1	6F	1	6F	1	6M	1	6M
Rats introduced	3	22F	3	28F	3	18M	3	20M
Rats removed	3	28F	3	22F	3	20M	3	18M
Population	6	167	6	130	6	130	6	85
Rats removed	7	6F	7	6F	7	6M	7	6M
Population	10	152	10	130	10	130	10	70
Rats removed	10	48	10	73	10	57	10	24
	Week	Index[1]	W	I	W	I	W	I
Adrenal size	1	92.1	1	91.5	1	104.4	1	93.8
	7	93.8	7	102.8	7	85.3	7	99.1
	10	101.5	10	90.4	10	91.1	10	84.6

[1] Mean of the individual per cent of appropriate reference value for the sex and size of the rat.

was male or female. The population from block 140222 (Fig. 2) may have become stationary just prior to the introduction of aliens, but the high rate of reproduction (4/6 mature females were pregnant) suggests that the population was increasing. The population in block 140111 increased numerically after the introduction, but the difference between the two estimates is within the error of estimate and does not indicate a change in population. It appears that introducing a number of alien rats may halt the growth of increasing populations.

The reader may have noticed that the total number of rats removed from the four increasing blocks was about 5 per cent greater than the number introduced, so that the rats in these populations were not replaced in the strict arithmetic sense of the word. However, considering the previously mentioned birth and mortality factors, it is not desirable to be more precise. All aspects considered, it is likely that the four increasing populations were somewhat reduced following replacement procedures. The four blocks (taken together) increased by 173 rats in the 20 weeks preceding replacement, so that they might have been expected to have had 600 rats 10 weeks after replacement instead of the observed 482, although the rate of increase would decline as the population increased.

On several occasions episodes have been noted that appear to be explainable on the

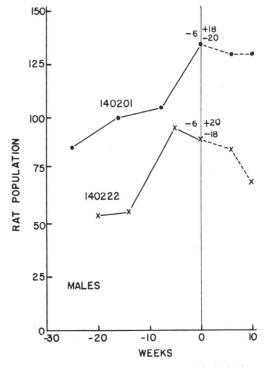

Fig. 2. The changes in two increasing blocks before and after the introduction of males (symbols as in Fig. 1.)

basis of introduction or actual immigration. In January 1946, about 60 rats were released in a block in one night as part of an experi-

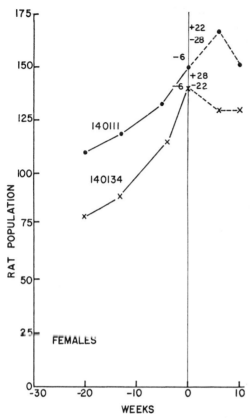

Fig. 3. The changes in two increasing blocks before and after the introduction of females (symbols as in Fig. 1.)

anisms in some way that causes the populations either to decline in numbers or stop growing.

The decrease is due in part to a decline in reproduction. Data are not available for the period immediately after introduction, as it is not feasible to follow the population changes and simultaneously to collect a number of rats for reproductive data. However, the large sample of rats collected from the blocks 8 to 11 weeks after introduction had a high reproductive rate (Table 4) and a low lactation rate. One would conclude from these data that the number of pregnancies was low immediately after the introduction. Only about 25 per cent of the females were lactating at 10 weeks, whereas normally about 40 per cent of the females of these rats are lactating (Davis, 1953). The high prevalence of pregnancy presumably resulted from their more or less simultaneous recovery from the effects of introduction. The decreases in rat populations obviously may have been due largely to mortality and movement, but data on this aspect are impossible to obtain under these conditions.

Table 4. — Reproductive Records of Local Rats Captured 8-11 Weeks after Artificial Immigration of Rats into Blocks

Population status	Number Mature Females	Per cent Pregnant	Mean Number Embryos	Per cent Lactating
Increasing	85	33.0	10.38	20.0
Stationary	66	31.8	9.63	28.8

Previous experiments have shown that the weight of the adrenal glands in rats responds to changes in population. An increase in adrenal weight parallels increases in population; the artificial reduction of a population also results in a decrease in adrenal weight (Christian, 1954; Christian and Davis 1955). The adrenal responses of the two sexes are parallel (ibid.). Experiments have indicated that changes in adrenal weight in response to changes in population result primarily from changes in cortical mass (Christian, 1955a, 1955b, 1956). To examine these problems, the adrenals of the rats from each block were removed and weighed. The observed adrenal weight for each rat was compared with a standard reference weight for the appropriate sex and size (length of head and body) of rat (Christian and Davis

ment on "homing" ability in rats. The block originally contained about 100 rats but within 3 weeks there were so few rats left in the block that the project was stopped. Calhoun (1948) noticed the same result when he introduced rats into blocks. These episodes, as well as miscellaneous observations, stimulated a test in 1947 of the idea that the introduction of rats into a population would result in its decline. Accordingly, rats were introduced over a period of 4 months into two populations that had just reached a level judged to be stationary (Davis, 1949). The introduction of 90 rats in one block and 101 in the other was accompanied by declines of about 25 per cent and of 40 per cent respectively. The populations increased after the introductions ended.

The present experiments suggest that the introduction of large numbers of rats into a population disrupts the population mech-

1955), and expressed as a per cent of the reference value. These percentages for the rats from each sample were averaged and the means are recorded at the bottom of tables 2 and 3. We have used the mean value of a given sample as the unit of measurement for comparing adrenal weight with population size (Christian, 1954; Christian and Davis, 1955).

The results indicate that, in the replacement stationary blocks (140338 and 140344), there was a small increase in adrenal weight after 8 weeks, while the populations apparently remained practically unchanged (tables 2 and 3, Fig. 1).

A mean decline in population size, paralleled by a decrease in adrenal weight in at least one of the two blocks, followed the addition of a large number of alien rats to stationary populations (blocks 140118 and 150128). The adrenal weights probably reflect largely the final results of population manipulation rather than the immediate effects, as the adrenal samples were obtained several weeks after the introductions or estimates. Therefore, the changes in adrenal weight probably reflect overall population changes rather than any immediate social strife resulting from the introductions. An experiment to collect samples a few days after the introductions is in progress and may show an increase in adrenal weight.

The adrenal glands of rats from the increasing blocks showed no consistent change, although population growth terminated (Table 3, figures 2 and 3). The replacement of rats in increasing populations had little effect on the adrenal weights of rats examined 10 weeks later.

The results reported here may be applicable to certain stocking programs. A routine part of many game-management programs has been the introduction of a number of animals into an area with the expressed hope of increasing the population either directly or eventually by reproduction. Indeed, such stocking has often been considered a panacea for all hunting problems. The present results, using rats as experimental animals, show that the disruption of a population following an introduction may actually produce a decline under certain conditions. Evidently the introduction of a number of animals may have disastrous results when a population is above the halfway point on a growth curve.

SUMMARY

Wild Norway rats (*Rattus norvegicus*) were introduced from one city block to another to simulate immigration. The population changes were determined by frequent estimates for about 20 weeks before introduction and 8 to 11 weeks thereafter. From two blocks with stationary rat populations, some rats were removed and then replaced by aliens. The populations remained stationary. In two blocks about four times as many rats were introduced as were removed. The populations declined about 25 per cent. In four blocks with increasing populations about one-fourth of each population was removed and replaced by alien rats from other blocks. The increase halted.

The reproductive rate 8 to 11 weeks after the introduction was normal for an increasing population, but the lactation rate was low, indicating that the decline in population growth was due in part to a decreased reproductive rate, and that the population was back to normal pregnancy rate in two months. The adrenal weights were also essentially normal for the population level two months after introduction.

REFERENCES

BROWN, R. Z., W. SALLOW, DAVID E. DAVIS, AND W. G. COCHRAN. 1955. The rat population of Baltimore 1952. Amer. J. Hyg., 61(1):89-102.

CALHOUN, J. B. 1948. Mortality and movement of brown rats (*Rattus norvegicus*) in artificially super-saturated populations. J. Wildl. Mgmt., 12(2):167-172.

CHRISTIAN, J. J. 1954. The relation of adrenal size to population numbers of house mice. Sc. D. dissertation, Johns Hopkins Univ., Baltimore.

———. 1955a. Effect of population size on the adrenal glands and reproduction organs of male mice in populations of fixed size. Amer. J. Physiol., 182(2):292-300.

———. 1955b. Effect of population size on the weights of the reproductive organs of white mice. Amer. J. Physiol., 181(3):477-480.

———. 1956. Adrenal and reproductive responses to population size in mice from freely growing populations. Ecology, 37(2):258-273.

——— AND D. E. DAVIS, 1955. Reduction of adrenal weight in rodents by reducing population size. Trans. N. Amer. Wildl. Conf., 20:177-189.

DAVIS, D. E. 1949. The role of intraspecific competition in game management. Trans. N. Amer. Wildl. Conf., 14:225-231.

———. 1953. The characteristics of rat populations. Quart. Rev. Biol., 28(4):373-401.

Received for publication October 24, 1955.

A TRAFFIC SURVEY OF *MICROTUS-REITHRODONTOMYS* RUNWAYS

By Oliver P. Pearson

Patient observation of the comings and goings of individual birds has long been one of the most rewarding activities of ornithologists. The development in recent years of inexpensive electronic flash photographic equipment has made it possible and practical for mammalogists to make similar studies on this aspect of the natural history of secretive small mammals. The report that follows is based on photographic recordings of the vertebrate traffic in mouse runways over a period of 19 months. Species, direction of travel, time, temperature and relative humidity were recorded for each passage. In addition, many animals in the area were live-trapped and marked to make it possible to recognize individuals using the runways.

THE APPARATUS

Two recorders were used. Each consisted of an instrument shelter and a camera shelter. Each instrument shelter was a glass-fronted, white box containing an electric clock with a sweep second hand, a ruler for measuring the size of photographed individuals, a dial thermometer and a Serdex membrane hygrometer. The ends of the box were louvered to provide circulation of air as in a standard weather station. This box was placed along one side of the runway, across from the camera shelter, so that the instruments were visible in each photograph (Plate I). The camera shelter was a glass-fronted, weather-proof box containing a 16-mm. motion picture camera synchronized to an electronic flash unit. In one of the recorders the camera was actuated by a counterweighted treadle placed in the mouse runway immediately in front of the instrument shelter (Plate I, bottom). An animal passing along the runway depressed the treadle, thereby closing an electrical circuit through a mercury-dip switch. This activated a solenoid that pulled a shutter-release pin so arranged that the camera made a single exposure. The electronic flash fired while the shutter was open. This synchronization was easily accomplished by having the film-advance claw close the flash contact. The camera would repeat exposures as rapidly as the treadle could be depressed, but at night about three seconds were required for the flash unit to recharge sufficiently to give adequate light for the next exposure.

169

The other recorder was actuated by a photoelectric cell instead of by a treadle. A beam of deep red light shone from the camera shelter across the runway and was reflected back from a small mirror in the instrument shelter to a photoelectric unit in the camera shelter. When an animal interrupted the light beam, the photoelectric unit activated a solenoid that caused the camera to make a single exposure, as in the other recorder.

To avoid the possibility of frightening the animals it would be desirable to use infra-red–sensitive film and infra-red light, but standard electronic flash tubes emit so little energy in the infra-red that this is not practical. Instead, I used 18 layers of red cellophane over the flash tube and reflector to give a deep red flash of light. Wild mice, like many laboratory rodents, are probably insensitive to deep red light. I found no evidence that the flash, which lasts for only 1/1000th of a second, frightened the mice. A muffled clunk made by the mechanism also seemed not to alarm the mice unduly.

When the camera diaphragm was set to give the proper exposure at night, daytime pictures were overexposed, since the shutter speed was considerably slower than 1/30th of a second. To reduce the daytime exposure, a red filter was put on the camera lens. The filter did not affect night exposures because red light from the flash passed the red filter with little loss. In addition, on one of the cameras the opening in the rotary shutter was reduced to give a shorter exposure.

Both recorders function on 110-volt alternating current. The treadle-actuated one could be adapted to operate from batteries. The units continue to record until the motion picture camera runs down or runs out of film. One winding serves for several hundred pictures. The film record can be studied directly by projecting the film strip without making prints.

The camera shelter and instrument shelter had overhanging eaves to prevent condensation of frost and dew on the windows. A small blackened light bulb was also kept burning in the camera shelter to raise the temperature enough to retard fogging on the glass. Animals were encouraged to stay in their usual runway by a picket fence made of twigs or slender wires. No bait was used.

A few individual animals could be recognized in the pictures by scars or molt patterns, but most had to be live-trapped and marked. Using eartags and fur-clipping I was able to mark distinctively (Plate I, bottom) all of the mice captured at any one station. The clipping remained visible for days or months depending upon the time of the next molt.

The apparatus produces photographic records such as those shown in the lower pictures in Plate I. These can be transposed into some form as Fig. 2.

THE STUDY AREA

The study centered around a grassy-weedy patch surrounding a brush pile in Orinda, Contra Costa County, California (Plate I). The runways wound through a 20 × 20-foot patch of tall weeds (*Artemisia vulgaris, Hemizonia* sp. and *Rumex crispus*) and under the brush pile. The weeds were surrounded by and somewhat intermixed with annual grasses. Oaks and other trees, as well as a house and planting, were 50 feet away.

Summer climate in this region is warm and sunny with official mean daily maximum temperatures rising above 80°F. in late summer. Official temperatures occasionally reach 100°, and temperatures in the small instrument shelters used in this study sometimes exceeded this. Nights in summer are usually clear and with the mean daily minimum temperature below 52° in each month. About 27 inches of rain fall in the winter and there is frost on most clear nights. The mean daily maximum temperature in January, the coldest month, is 54°, and the mean daily minimum 31°.

PROCEDURE

I placed the first recorder in operation on January 29, 1956, and the second on October 19, 1956. Except for occasional periods of malfunction and a few periods when I was away they continued to record until the end of the study on September 10, 1957. Approximately 778 recorder-days or 111 recorder-weeks of information were thus obtained. The monthly distribution of records was as follows: January, 54 days; February, 70; March, 90; April, 80; May, 84; June, 67; July, 52; August, 88; September, 48; October, 33; November, 52; and December, 60.

The recorders were placed at what appeared to be frequently used *Microtus* runways, usually situated on opposite sides of the weedy patch 20 to 30 feet apart. For one period of four months one of the recorders was placed at a similar weedy patch 70 yards away. Early in the study it was discovered that a neighbor's Siamese cat sometimes crouched on the camera shelter waiting for mice to pass along the exposed runway in front of the instrument shelter. Consequently, a 2½-foot fence of 2-inch-mesh wire netting was set up enclosing most of the weedy patch. This prevented further predation by cats at the center of the study area, although cats continued to hunt outside of the fence a few yards away from the recorders. The only other tampering with predation was the removal of two garter snakes on April 11, 1957.

RESULTS

Traffic in individual runways.—The recorders were operated at eighteen different stations. At seven of these apparently busy runways a traffic volume higher than a few passages per day never developed, and so the recorders were moved within two weeks. Perhaps the mice originally using these runways had abandoned them or had been killed shortly before a recorder was moved to their runway, or perhaps the disturbance of placing a recorder caused the mice to divert their activities to other runways. At the other eleven stations a satisfactory volume of traffic was maintained for three to more than twenty weeks. A station was abandoned and the recorder moved when the traffic had decreased to a few passages per day. Subsequently, I found that even this little activity does not indicate that the mice are going to abandon the runway, for on several occasions traffic in a runway dropped this low and then climbed again to high levels. At one recorder the total number of passages in consecutive weeks was 183, 84, 26, 75 and 203. The runway represented in Fig. 1 was one of those used most consistently, but even it shows marked daily and weekly fluctuations. It is probable that after a few weeks of disuse during the season when grass and weeds are growing rapidly, a runway would not be reopened, but during the rest of the year an abandoned runway remains more or less passable and probably more attractive to mice than the surrounding terrain.

Figure 1 summarizes the traffic in one of the busiest runways. On the first night there were an unusual number of records of harvest mice whose curiosity may have been aroused by the apparatus. Obviously they were not frightened away. After a short time traffic increased to a high level and remained high until the middle of November, when passages by *Microtus* decreased sharply. During the week before the decrease, seven marked individuals provided most of the *Microtus* traffic. One of these individuals, an infrequent passerby, disappeared at the time of the decrease, but the other six remained nearby for at least another week and continued to pass occasionally. Those *Microtus* that disappeared later were replaced by others so that even the infrequent passages in late November and early December were being provided by seven marked individuals. The decrease of *Microtus* traffic was caused, therefore, not by deaths but by a change in runway preference. Several of these same individuals were using another runway 20 feet away in mid-January, February and March.

Three to six marked *Reithrodontomys*, depending upon the date, were providing most of the harvest-mouse traffic in the runway represented in Fig. 1. The average number of passages per day of animals of all kinds was eighteen. In the ten other most successful runways, the average number of passages per day ranged from two to nineteen.

Figure 2 gives a detailed accounting of the traffic at a single recording station for six days. One can judge from this figure the kind of information (excluding

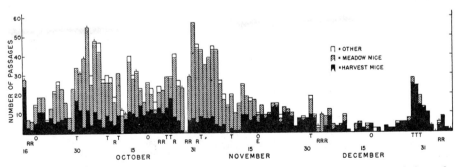

Fig. 1.—Traffic volume along one runway for 16 weeks. Meaning of symbols under the base line: *T*= live-trapping carried out for part of this day; *O*= full moon; *E*= total eclipse of the moon; *R*= rain. Columns surmounted by a vertical line represent days for which the recording was incomplete; the heights of the various segments of these columns should be considered minimum values.

temperatures and humidities) obtained with the recorders and can at the same time catch a revealing glimpse of an aspect of the biology of small mammals that has heretofore been revealed inadequately by trapping and other techniques. It may be seen that the mouse traffic was provided by one female and two male harvest mice and by three male, three female, and one or more unidentified meadow mice; together they gave between 15 and 24 passages each day. No individual passed more than eight times in one day. One harvest mouse (R2) seemed to spend the day to the left and to make a single excursion

PLATE I

Top: Camera shelter (foreground) and instrument shelter in position at a mouse runway on the study area. Bottom: The kind of records obtained with the recorder; *left*—a meadow mouse marked by clipping two strips of fur on the hips; *right*—a marked harvest mouse crossing the treadle.

to the right each night. Harvest mice first appeared in the evening between 6:37 and 7:22 and none passed after 6:26 in the morning. Five or six *Microtus* passed within a few hours (February 24), and there was nightly near-coincidence of *Reithrodontomys* and *Microtus*.

Traffic in all runways combined.—During the 111 recorder-weeks, the following passages of animals were photographed:

Meadow mouse, *Microtus californicus*	6,077
Harvest mouse, *Reithrodontomys megalotis*	1,753
Bird (see following account)	382
Brush rabbit, *Sylvilagus bachmani*	94
Shrew, *Sorex ornatus*	56
Peromyscus (see following account)	39
Fence lizard, *Sceloporus occidentalis*	33
Garter snake, *Thamnophis* sp.	17
Salamander (see following account)	11
Alligator lizard, *Gerrhonotus* sp.	10
House cat, *Felis domesticus*	6
Newt, *Taricha* sp.	5
Pocket gopher, *Thomomys bottae*	3
Gopher snake, *Pituophis catenifer*	3
Mole cricket, *Stenopelmatus* sp.	2
Ground squirrel, *Citellus beecheyi*	1
Weasel, *Mustela frenata*	1
King snake, *Lampropeltis getulus*	1
Racer, *Coluber constrictor*	1
TOTAL	8,495

On the basis of trapping results in this and in similar habitat nearby, large numbers of meadow mice and harvest mice were expected. The recording of at least 26 other species in the runways came as a pleasant surprise. Whereas all of these species would be expected to record their presence eventually, some of them are rarely seen or trapped near this location. After living five years on the study area, after doing considerable field work nearby, and after checking the recorders twice each day during the study, I have not yet seen a weasel or a ground squirrel within at least a mile of the study area. Weasels could easily escape detection, but large, diurnal ground squirrels must be very rare. The single individual recorded on August 31 may have been a young squirrel emigrating from some distant colony. Noteworthy absences were those of wood rats (*Neotoma fuscipes*), moles (*Scapanus latimanus*), and probably California mice (*Peromyscus californicus*), all of which were common within 100 feet of the recorders. An opossum (*Didelphis marsupialis*) was seen a few feet from one of the recorders but did not appear on the films. No house mice (*Mus musculus*) were detected in the photographs, although it is possible that some passages of *Mus* were listed as of *Reithrodontomys.* House mice were caught occasionally in houses nearby and in a field near a poultry house 200 yards away, but none was caught during frequent live-trapping near the recorders.

The total of 8,495 passages of animals gives an average of 11 passages per day in each runway. A patient, non-selective predator waiting for a single catch at runways such as these could expect, theoretically, a reward each 2.2 hours. The mean weight of animal per passage was about 31 grams, which would yield approximately 40 calories of food. This much each 2.2 hours would be more than enough to support an active mammal the size of a fox.

Meadow mouse.—The 6,077 *Microtus* passages were distributed throughout the day and night as shown in Fig. 3 (above). The hours of above-ground activity, however, were quite different in winter than in summer, so Fig. 3 is only a year-around average somewhat biased by the fact that more *Microtus* were recorded in the spring than in the other seasons. A more detailed analysis of the *Microtus* data will be given in a later report. By marking as many of the mice as possible, it was found that usually four or more individual *Microtus* were using each runway but rarely more than ten. On some occasions more than 60 *Microtus* passages were recorded at a single point in 24 hours.

Harvest mouse.—Harvest mice were almost entirely nocturnal (Fig. 3, center). They not only used the *Microtus* runways, but their passages were frequently intermixed with those of *Microtus* (Fig. 2). On fourteen occasions the two species passed within 60 seconds of each other, and on one occasion

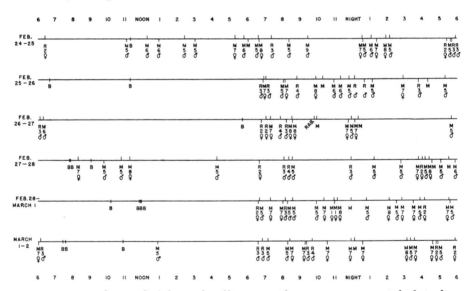

Fig. 2.—A sample record of the total traffic in a single runway over a period of six days. Marks above the base lines indicate passages from right to left, and marks below the base line passages from left to right. R represents *Reithrodontomys*; M, *Microtus*; B, bird (includes brown towhee, wren-tit, and song sparrow); and RAB, brush rabbit. Most of the mice are further identified by number and sex.

a 4-month-old male *Microtus* and a 5-month-old male *Reithrodontomys* appeared in the same photograph.

The history of one runway indicates that traffic by *Reithrodontomys* alone does not keep a *Microtus* runway open. One or more *Microtus* passed almost daily along this runway during February. At the end of the month the *Microtus* disappeared and two *Reithrodontomys* became active in the same runway. Despite an average of 3.3 passages per day by *Reithrodontomys* throughout March and up to mid-April, grass and weed seedlings grew up

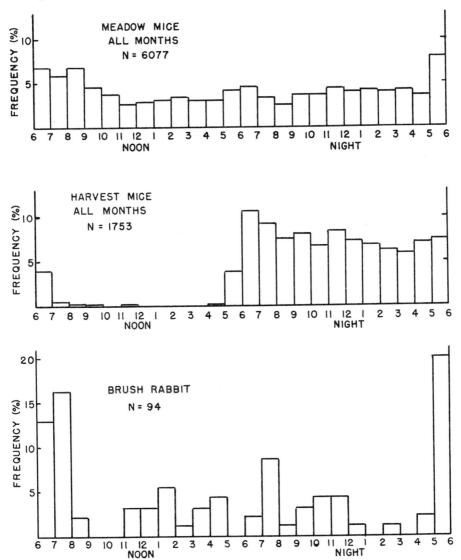

FIG. 3.—Distribution by hours of 6,077 passages of meadow mice (above); 1,753 passages of harvest mice (center); and 94 passages of brush rabbits (below).

in the runway and it began to look unused. By the end of April almost all traffic had ceased.

The *Reithrodontomys* data will be analyzed in a later report.

Birds.—Of the 382 bird records, at least 255 were of sparrows (at least 122 song sparrow; the remainder mostly fox sparrow, white-crowned sparrow and golden-crowned sparrow). Other birds recognized were wren-tit, wren, brown towhee and thrush. On several occasions birds, especially song sparrows, battled their reflections in the window of the instrument shelter. This caused long series of exposures. Each series was counted as a single passage. If the bird stopped for a minute or more and then returned to the battle, this was counted as another passage. All bird records were during daylight hours.

On three occasions a sparrow and an adult *Microtus* appeared in the same photograph. On one of these occurrences a song sparrow was battling its reflection when an adult, lactating *Microtus* came along the runway. The sparrow retreated about 12 inches toward the camera shelter and, as soon as the mouse had passed, returned to the runway.

Brush rabbit.—All except four of the records of brush rabbits were in June and July of 1957, a season when these animals, especially young ones, were abundant. Figure 3 (below) shows that they were most active in the early morning.

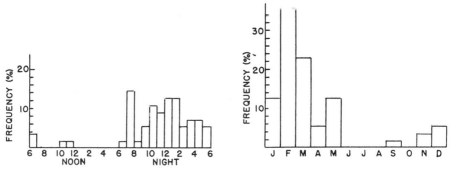

Fig. 4.—Distribution by hours of 56 passages of shrews (left) and distribution by months of 56 passages of shrews (right).

Shrew.—The dry, weedy habitat chosen was not favorable for shrews, and they were near the minimum weight necessary to depress the treadle of one of the recorders, so that some may have passed along the runway without making a record. The shrews were highly nocturnal (Fig. 4, left) and avoided the surface runways during the dry summer months (Fig. 4, right). Since captive specimens of *Sorex* are rarely inactive for more than one hour (Morrison, Amer. Midl. Nat., 57: 493, 1957), the scarcity of records in the daytime probably means only that the shrews were not moving above ground at this time. They may have been foraging along gopher, mole and *Microtus* tunnels during the daytime.

A shrew was marked on March 4, a few inches from one of the recorders. It was captured 15 feet away on May 30 and 5 feet farther away on June 23. It passed along the study runway five times in the 16-week interval between first and last capture: on March 13, 27, 31, and April 17, and possibly on April 10 (markings obscured). Another shrew was recorded on March 27. Unless baited traps attract shrews from a considerable distance, or the recorder repels them, a trapper setting traps in this runway for a few nights would have had small chance of recording the presence of this individual which apparently was nearby for at least 16 weeks.

Not a single shrew was recorded during the dry summer months of June, July and August. Nevertheless, on July 8 when I was checking the photo-electric recorder at 5:55 AM, a shrew emerged completely from a small hole in

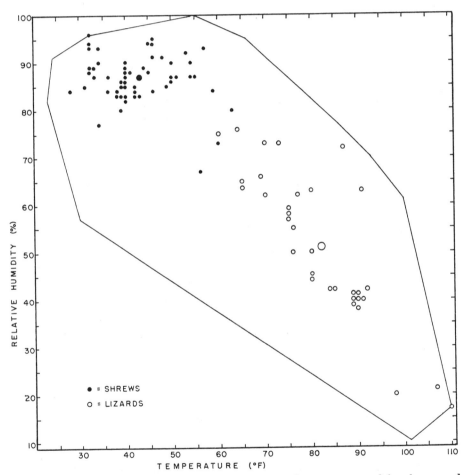

Fig. 5.—A comparison of the temperatures and humidities encountered by shrews and fence lizards in the runways. The larger circles show the position of the mean for each species. The large polygon encloses the range of temperatures and humidities available to the animals during the study.

the ground a few inches from the instrument shelter, twitched his nose rapidly for a few seconds, and retreated down the same hole. The air temperature was 54° and the relative humidity 78 per cent—normal for this season. Obviously shrews were present on the study area during some or all of the summer months but were not frequenting the surface runways.

Figure 5 shows the temperatures and relative humidities encountered above ground by the shrews on the study area compared with the total range of temperatures and humidities recorded throughout the study. By their nocturnal, winter-time activity shrews encountered the coldest, most humid conditions available in the region. In contrast, the similarly small, insectivorous fence lizards existing in the same habitat managed by their own behavioral patterns to encounter a totally different climate (Fig. 5). The mean of the temperatures recorded at the times of lizard passages was 39° warmer than that recorded for shrew passages, and relative humidity was 36 per cent lower.

This activity pattern of shrews differs from that reported by Clothier (Jour. Mamm., 36: 214–226, 1955) for *Sorex vagrans* in Montana. He found shrews there to be active "both day and night and throughout the year, even during extremely bad weather." It is important to understand, however, that he collected in damp areas near water, where the shrews may not have had to modify their activity to avoid desiccation. Extremely bad weather, for a shrew, is hot dry weather.

Peromyscus.—*Peromyscus truei* was abundant in brushy places and in houses nearby; *P. maniculatus* was scarce. Some of the *Peromyscus* records were clearly of *truei* and some may have been of *maniculatus,* but many could not be identified with certainty. No adult *P. californicus* was recognized although a few young ones may have passed and been listed as *truei.* All passages of *Peromyscus* were at night.

Salamander.—The record includes passages by both *Ensatina escholtzii* and *Aneides lugubris.* They were recorded in October, November, March and April. By being nocturnal and by avoiding the dry season, they encountered in these autumn and spring months about the same microclimate as shrews, but were recorded neither in the winter months nor at temperatures below 39°. A third species, *Batrachoceps attenuatus,* was common in the study area but is so small that it could not be expected to actuate either of the recorders. One *Batrachoceps* electrocuted itself underneath the treadle but has not been included in the records.

Comparison of traps and recorders.—The combination of live-trapping and photographing revealed a failure of small mammals to move between runways only a few feet apart. On several occasions meadow mice and harvest mice were live-trapped a few feet from one of the recorders, were released at the same place, and were recaptured a week or more later not more than a few feet away, yet during the intervening time they failed to pass the recorder. Conversely, some individual mice repeatedly recorded themselves on the films yet could never be induced to enter any of a large number of live traps placed in the same runway and in nearby runways. It is obvious that all mice present do not use all of the active runways close to their home, and it is also obvious that neither the recorders nor traps give a complete accounting of the mice present.

SUMMARY

A motion-picture camera synchronized to an electronic flash unit was used to record the passage of animals along meadow-mouse runways and to record the temperature, relative humidity and time at which they passed. More than 26 species used the runways during 111 weeks of recording. Meadow mice, harvest mice, sparrows, brush rabbits and shrews passed most frequently. The average traffic per day in each runway was 11 passages; on some days there were more than 60 passages. Rarely more than ten meadow mice or six harvest mice used a runway in any one period. Meadow mice and harvest mice used the same runways simultaneously. Traffic by harvest mice alone did not keep the runways open.

Meadow mice were active during the day and night; harvest mice were strongly nocturnal. Brush rabbits were active primarily early in the morning. Almost all shrews were recorded at night and in the winter months. Consequently, they encountered the coldest, most humid conditions available to them. In contrast, the similarly small, insectivorous fence lizards encountered a microclimate that was 39° warmer and 36 per cent less humid.

Neither traps nor recorders accounted for all the individuals living nearby.

Museum of Vertebrate Zoology, Berkeley, California. Received October 29, 1957.

PREY SELECTION AND HUNTING BEHAVIOR
OF THE AFRICAN WILD DOG[1]

RICHARD D. ESTES, Division of Biological Sciences, Cornell University, Ithaca, New York

JOHN GODDARD, Game Biologist, Ngorongoro Conservation Area, Tanzania

Abstract: African wild dog (*Lycaon pictus*) predation was observed in Ngorongoro Crater, Tanzania, between September, 1964, and July, 1965, when packs were in residence. The original pack of 21 dogs remained only 4 months, but 7 and then 6 members of the group reappeared in the Crater at irregular intervals. The ratio of males:females was disproportionately high, and the single bitch in the small pack had a litter of 9 in which there was only one female. The pack functions primarily as a hunting unit, cooperating closely in killing and mutual defense, subordinating individual to group activity, with strong discipline during the chase and unusually amicable relations between members. A regular leader selected and ran down the prey, but there was no other sign of a rank hierarchy. Fights are very rare. A Greeting ceremony based on infantile begging functions to promote pack harmony, and appeasement behavior substitutes for aggression when dogs are competing over meat. Wild dogs hunt primarily by sight and by daylight. The pack often approaches herds of prey within several hundred yards, but the particular quarry is selected only after the chase begins. They do not run in relays as commonly supposed. The leader can overtake the fleetest game usually within 2 miles. While the others lag behind, one or two dogs maintain intervals of 100 yards or more behind the leader, in positions to intercept the quarry if it circles or begins to dodge. As soon as small prey is caught, the pack pulls it apart; large game is worried from the rear until it falls from exhaustion and shock. Of 50 kills observed, Thomson's gazelles (*Gazella thomsonii*) made up 54 percent, newborn and juvenile wildebeest (*Connochaetes taurinus*) 36 percent, Grant's gazelles (*Gazella granti*) 8 percent, and kongoni (*Alcelaphus buselaphus cokei*) 2 percent. The dogs hunted regularly in early morning and late afternoon, with a success rate per chase of over 85 percent and a mean time of only 25 minutes between starting an activity cycle to capturing prey. Both large and small packs generally killed in each hunting cycle, so large packs make more efficient use of their prey resource. Reactions of prey species depend on the behavior of the wild dogs, and disturbance to game was far less than has been represented. Adult wildebeest and zebra (*Equus burchelli*) showed little fear of the dogs. Territorial male Thomson's gazelles, which made up 67 percent of the kills of this species, and females with concealed fawns, were most vulnerable. The spotted hyena (*Crocuta crocuta*) is a serious competitor capable of driving small packs from their kills. A minimum of 4–6 dogs is needed to function effectively as a pack. It is concluded that the wild dog is not the most wantonly destructive and disruptive African predator, that it is an interesting, valuable species now possibly endangered, and should be strictly protected, particularly where the small and medium-sized antelopes have increased at an alarming rate.

The habits of the African wild dog or Cape hunting dog (*Lycaon pictus*) have been described, sometimes luridly, in most books about African wildlife. Accounts by such famous hunters and naturalists as Selous (1881), Vaughan-Kirby (1899), and Percival (1924), repeated and embellished by other authors, have created the popular image of a wanton killer, more destructive and disruptive to game than any other African predator.

Because of its bad reputation, the wild dog was relentlessly destroyed in African parks and game reserves for many years. In Kruger National Park, for instance, it was shot on sight from early in the present century up until 1930 as part of an overall policy to keep predators down. In Rhodesia's Wankie National Park some 300 wild dogs were killed by gun and poison between 1930 and 1958.

Acceptance of modern concepts of wildlife management has finally brought an end to the indiscriminate destruction of wild dogs and other predators in most, if not all, African national parks. There is now

[1] Field work supported by the National Geographic Society; also by grants from the New York Explorers Club and the Tanzania Ministry of Agriculture, Forests and Wild Life.

52

a general awareness among game wardens of the predator's role in regulating populations, which perhaps began with Stevenson-Hamilton (1947), Warden of Kruger Park for almost 30 years, who related the alarming increase of impala (*Aepyceros melampus*) to the disappearance of the park's formerly large wild dog packs.

While the wild dog has benefited from more enlightened concepts of game management, its reputation, still based on popular writings and myth, remains unchanged. But recent scientific investigations indicate that a new and less-prejudiced evaluation of this species is long overdue. Kühme (1965) has studied the social behavior and family life at the den of a pack with young whelps on the Serengeti Plains, Tanzania. We have observed prey selection and hunting behavior in a free-ranging pack of adults and juveniles in nearby Ngorongoro Crater, a caldera with a floor area of 104 square miles that supports a resident population of around 25,000 common plains herbivores. The two studies together throw quite a different light on the habits, character, and predator–prey relationships of this highly interesting species.

We are indebted to Dr. B. Foster of the Royal College, Nairobi, and to G. C. Roberts of the Crater Lodge for reporting four kills and one kill respectively, that they witnessed in the Crater; also to Professors W. C. Dilger, O. H. Hewitt, and H. E. Evans of Cornell University for critical readings of the manuscript. Nomenclature follows Haltenorth (1963) for artiodactyls, and Mackworth-Praed and Grant (1957) for birds.

METHODS

Most observations were made from a vehicle; we each had a Land Rover and usually operated independently. The wild dogs, like many other African predators

in sanctuaries, paid very little attention to cars and could be watched undisturbed from within 30 yards or less. It was also feasible to keep pace with the pack during chases over the central Crater floor, either driving parallel to the leader at a distance of 100–200 yards or following behind and to one side so as not to get in the way of other pack members. To locate the pack initially, we often drove to an observation point on a hill and scanned the Crater with binoculars and a 20-power binocular telescope. When the pack was moving it could often be spotted at a distance of over 5 miles, and a number of chases and kills were clearly observed from a hilltop through the telescope.

RESULTS

Pack Composition

The pack that first entered Ngorongoro Crater in September, 1964, contained 21 animals, including 8 adult males, 4 adult females, and 9 juveniles. They remained more or less continually in residence through December, then disappeared and were presumed to have left the Crater. One juvenile female had died of unknown causes. During January, 1965, seven members of the same pack, 4 males, 1 female, and 2 juvenile males, reappeared; after a lapse of 5 months, apparently the same animals, minus one male, again took up residence, and have been observed off and on up to the present writing. In March, 1966, the female whelped but died 5 weeks later, leaving 8 male and 1 female pups. They were brought up by the 5 males, who fed them by regurgitation until they were old enough to run with the pack. However, the female and 4 male pups died, leaving an all-male pack of 9 in August, 1966.

While an all-male pack must be exceptional, there is other evidence to suggest that a high proportion of males may be common in this species. The pack Kühme studied consisted of 6 adult males and 2 adult females, which had 11 and 4 pups respectively, sex unreported. During 2–3 years of shooting in Kruger National Park, the ratio of males was 6 : 4, despite an attempt to select females (Stevenson-Hamilton 1947). We have no explanation to offer for the discrepancy, but if it is real and not normal, it might help explain the reported decline of wild dogs during recent years in many parts of Africa.

Social Organization

Leadership and Rank Hierarchy.—In the full pack of 21 and in the pack of 7, the same adult male was consistently the leader; he usually led the pack on the hunt, selected the prey, and ran it down. In the pack of 6, from which the above male was absent, the adult female was the leader. One of the males filled the position after her death.

Apart from the position of leader, we saw no indication of a rank order. Kühme concluded there was no hierarchy in the pack he observed, nor even a leader. The equality of pack members may partly explain the singularly amicable relations typical of the species. On the other hand, competition for food and females could easily lead to aggression; yet neither Kühme nor we ever saw a fight.

Food Solicitation and Appeasement Behavior.—Overt aggression and fighting are minimized through ritualized appeasement behavior derived from infantile food begging. Begging and appeasement appear in almost every contact between individuals, and particularly in situations where aggression would be most likely to occur—for instance, when animals are competing over a kill. However, we cannot comment on sexual competition, having seen none; we

observed sexual behavior on only two occasions, when one male mounted another repeatedly as the latter was feeding at a kill. Kühme also saw very little sexual behavior. When two animals were competing for the same piece of meat, each would try to burrow beneath the other, its forequarters and head flat to the ground and hindquarters raised, tail arched and sometimes wagging. The ears were flattened to the head and the lips drawn back in a "grin," while each gave excited twittering calls. As Kühme observed (p. 516), the dogs "tried to outdo each other in submissiveness."

In this way juveniles and even subadults manage to monopolize kills in competition with adults. The young thus enjoy a privileged position in the pack. Pups at the den successfully solicit any adult to regurgitate food by poking their noses into the corner of the adult's mouth, sometimes licking and even biting at the lips. Since all pack members contribute to feeding and protection of the young, the mother is not essential to their survival after the first few weeks.

Greeting Ceremony.—Whenever the pack became active after a rest period, and particularly if two parts of the pack were reunited after being separated, the members engaged in a Greeting ceremony (Fig 1), in which face-licking and poking the nose into the corner of the mouth played a prominent part. The ceremony thus appears to be ritualized food solicitation; the fact that Kühme actually saw regurgitation elicited by begging adults supports this interpretation. The Greeting ceremony in the wolf (*Canis lupus*), in which one takes another's face in its jaws, may have the same derivation.

As a prelude to greeting, dogs typically adopted the Stalking attitude (Fig. 2), with the head and neck held horizontally, shoulders and back hunched, and the tail usually

Fig. 1. Greeting ceremony.

Fig. 2. The Stalking attitude, here displayed by the pack leader while approaching a herd of gazelles.

hanging. Kühme (p. 512) interprets this posture as inhibited aggression; the same attitude is adopted when approaching potential prey and competitors of other species. The Stalking posture changed to greeting when dogs got close. In greeting-solicitation, as they licked each other's lips and poked the nose into the corner of the mouth, one or both crouched low, with head, rump, and tail raised stiffly (Fig. 1). Except for the raised head, this resembles the submissive posture displayed when two dogs are competing over food. The Greeting ceremony was also frequently performed while two dogs trotted or ran side by side.

Vocal Communication.—Although Percival (1924), Stevenson-Hamilton (1947), Maberly (1962), Kühme (1965), and others have given good descriptions of wild dog calls, the function of the calls has often been misinterpreted. This applies particularly to two of the three most frequently heard calls (Nos. 1 and 3):

1. *Contact call*—a repeated, bell-like "hoo." Often called the Hunting call, it has nothing to do with hunting as such, but is given only when members of a pack are separated. Though a soft and musical sound, it carries well for 2 or more miles. When members of the Ngorongoro pack were missing, an imitation of the Contact call would bring the rest to their feet, whereas there was at best only a mild reaction to imitations when the full pack was assembled.

2. *Alarm bark*—a deep, gruff bark, often combined with growling, given when startled or frightened. A good imitation near a resting pack elicited an immediate startled reaction.

3. *Twittering*—a high-pitched, birdlike twitter or chatter. The most characteristic and unusual vocalization, it expresses a high level of excitement. It is given in the prelude to the hunt, while making a kill, in mobbing hyenas or a pack member, and by dogs competing over food. Its primary function is evidently to stimulate and concert pack action. Kühme described this call (*Schnattern*) only in the context of the Greeting ceremony (p. 513).

Besides these vocalizations, whining may be heard during appeasement behavior and when pups are begging, and members of the pack sometimes yelp like hounds when close on the heels of their prey. Kühme (p. 500) further distinguishes an Enticing call (*Locken*) given by adults calling the young out of the den, and a Lamenting call (*Klage*) given by pups when deserted.

Olfactory and Visual Communication.—Wild dogs hunt primarily by sight and by daylight. We never saw them track prey by scent. Though they evidently have a good nose and may well use it for tracking in bush country, olfaction in this species seems to have a primarily intraspecific significance.

383

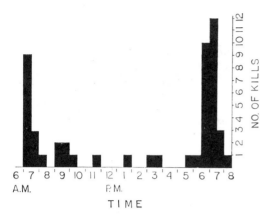

Fig. 3. Time distribution of 50 wild dog kills.

Wild dogs are renowned for their peculiarly strong, and to many humans disgusting, odor, which may emanate from anal glands but seems to come from the whole body. Sniffing under the tail, responsive urination and defecation are socially important activities. But the main role of the strong body odor may be to permit high-speed tracking of the pack by members that have lost visual contact. Lagging members seen running on the track taken by the rest of the pack sometimes appeared to be using their noses. Similarly, the white tail tip probably helps maintain visual contact in bush country, high grass, and under crepuscular conditions; in a species notable for every possible color variation, a white-tipped tail is the most constant and conspicuous mark.

Daily Activity Pattern

The Ngorongoro pack had two well-defined hunting periods each day (Fig. 3). That this periodicity is characteristic of the species may be inferred from Kühme's observations (p. 511), and from Stevenson-Hamilton's (1947) observations in Kruger National Park. In nine recorded instances, though, the Ngorongoro dogs killed between 8:30 AM and 3:30 PM, well outside the normal periods. Failure to kill during the regular hunting cycle is the likeliest explanation; it was more usual, however, for the pack then to wait until the following regular period. Wild dogs will also hunt on moonlight nights, as Stevenson-Hamilton noted. When the Crater dogs had not killed before dusk, the hunt was sometimes prolonged. The latest kill we recorded was at 7:32 PM, when it was fully dark.

Since they are capable of functioning as a pack and of hunting successfully after dark, the fact that wild dogs are so strongly diurnal may seem puzzling. But it may be explained by the fact that they hunt mainly by sight; it would be much more difficult to locate prey and single out a quarry at night. As to the regularity and brevity of their hunting cycles in early morning and late afternoon, this is partly a measure of their hunting efficiency, discussed below. Also, of course, these are the times in the day when diurnal animals, particularly herbivores, are most active and most approachable.

Apart from a certain amount of play and other social activities shortly before starting to hunt and immediately after feeding, pack members were usually active only while actually hunting. At other times they could often be found resting near or in the same place where they had settled after the morning or evening kill. When resting pack members customarily lay touching in close groups (Fig. 4). Generally speaking, the pack became active between 5:30 and 6:15 PM, and in the morning within 1 hour of dawn, remaining active for 1– hours. But where game is less plentiful than it is in Ngorongoro, and a pack must range more widely (Stevenson-Hamilton gives a range of at least 1,500 square miles for a Transvaal pack whose movements were reported over a period of years), a good deal of time between and during hunts must be spent in travel.

Fig. 4. Part of a resting pack, lying typically close together.

Hunting Behavior

Prelude to the Hunt.—Periods of activity were initiated by the actions of one or a few dogs apparently more restless than the others; rarely did the whole pack arise spontaneously at the start of an activity cycle. Typically, one dog would get up and run to a nearby group, nose the others and tumble among them until they responded. Within a few minutes the whole pack would usually become active. But if, as sometimes happened, the majority failed to respond to the urging of a few, then all would settle down to rest again. Sometimes, after a brief bout of general activity, the whole pack would lie down once more, even if it was past the usual time of hunting.

During the first 5 or 10 minutes after rousing, the pack members sniffed, urinated, defecated, greeted, and romped together. Play and chasing tended to become progressively wilder and reached a climax when the whole pack milled together in a circle and gave the twittering call in unison. As soon as this melee broke up, the pack usually set off on the hunt. Kühme (p. 522) interprets this performance (specifically the Greeting ceremony) as "a daily repeated final rehearsal for the behavior at the kill," wherein mutual dependence and friendliness are reinforced by symbolic begging, thus enabling the dogs to share the kill amicably. While this may be one function, the progressive buildup of excitement before hunting looked to us like nothing so much as a "pep rally," that served to bring the whole pack to hunting pitch. The behavior of domestic dogs urging one another to set off on a chase is somewhat similar.

The Mobbing Response.—During the milling preparatory to hunting, we sometimes saw what appeared to be incipient mobbing action toward a pack member, when up to half a dozen dogs would gang up on one, tumble and roll it but without actually biting it. Intensive play between two or three animals usually preceded and seemed to trigger a mobbing reaction in other members, who signaled their intentions by approaching in the Stalking posture. Percival (1924:48) reports seeing a pack mob and kill a wild dog he had wounded. The occurrence of "play" mobbing suggests that it could indeed become serious when an animal is maimed. On the other hand, sick and crippled pack members are often not molested: one very sick-looking old male in the large pack trailed behind the others for over a month before recovering, and though he kept usually a little apart, was tolerated at kills.

It is significant that basically the same mobbing behavior, at high intensity, is displayed when wild dogs kill large prey and when they harass spotted hyenas, their most serious competitor. It seems very likely, in fact, that mobbing is an innate response which governs pack action in hunting, killing, and mutual defense. It is perhaps the key to pack behavior in all animals that display it. That mobbing appears in play and can be released by a conspecific which is wounded or otherwise transformed from its normal self, supports the hypothesis that it is an innate response. It is also noteworthy that a wild dog removed from its pack apparently makes little effort to defend itself against attack. Selous (in Bryden 1936:24) reported that a wild dog caught by a pack of hounds shammed death and then escaped when he was about to skin it.

Hunting Technique.—Sometimes the pack would set off on the hunt at a run and chase the first suitable prey that was sighted. More often, there was an interval of 10–20 minutes during which the dogs trotted along, played together, and engaged in individual exploratory activity, stopping to sniff at a hole or a tuft of grass, then running to catch up with the rest. At this stage, when the hunt had started but before any common objective had been determined, individuals might forage for themselves. The observer would suddenly notice that a dog was carrying part of a gazelle fawn or a young hare (*Lepus capensis*), that must have been simply grabbed as it lay in concealment. Once during a moonlight hunt by a small pack that visited the Crater in 1963, individual dogs were seen to pick up at least two gazelle fawns and one springhare (*Pedetes surdaster*), a strictly nocturnal rodent, within ½ hour. Concealed small game such as this is apparently not hunted by the pack in concert.

Preparatory to the chase, there was frequently a preliminary stalking phase during which the pack approached herds of game at a deliberate walk, in the Stalking attitude (Fig. 2). The dogs appeared to be attempting to get as close as possible without alarming the game, and certainly the flight distances were much less than when the pack appeared running. The chase was launched the moment the game broke into flight. But game that began running at

more than 300 yards was generally not pursued. As far as we could tell, the prey animal was never singled out until after the pack, or at any rate the leader(s), had broken into a run.

In the pack of 21, juveniles and some adults usually lagged far behind, and often caught up 5–10 minutes after the kill was made. In the small pack, however, commonly all kept together and spread out on a front during the stalking phase. When all started running on a front, sometimes more than one dog picked out a quarry from the fleeing herd, whereupon the pack might split, some following one dog, the rest another. Kühme (p. 527) considered this the normal pattern and noted that often each animal acted for itself in selecting a quarry before all combined on a common goal. In this way the slowest prey tended to be selected. Selection by this method was exceptional for the Ngorongoro pack which had a definite leader; as a rule the lead dog made the choice and the rest of the pack fell in behind him. Nor did it appear that any effort was made to single out the slowest prey, although that would be difficult to observe clearly.

Again as a general rule, no attempt was made to carry out a concealed stalk, which would in any case be practically impossible by daylight on the short-grass steppe. But on one occasion the pack of six made use of a tall stand of grass to get near a group of Thomson's gazelles. On another hunt the pack apparently took advantage of a slight elevation in the expectation of surprising any game that might be out of sight on the far side. They moved deliberately up the slope, then broke into a run and swept at full speed over the crest on a broad front—but without finding any quarry that time.

When the leader had selected one of the fleeing herd, he immediately set out to run it down, usually backed up by one or two other adults who maintained intervals of 100 yards or more behind him, but might be left much further behind in a long chase. The rest of the pack lagged up to a mile in the rear. Discipline during the chase was so remarkable among all pack members that even gazelles which bounded right between them and the quarry were generally ignored. The average chase lasted 3–5 minutes and covered 1–2 miles. At top speed a wild dog can perhaps exceed 35 mph, and can sustain a pace of about 30 mph for several miles. Once when a chase had begun but no single quarry had yet been selected, a male in the pack of 21 broke away and proceeded to make a 5-mile circular sweep quite by itself, turning in bursts of speed when gazelles bounded off before him, but without ever singling one out. His average speed, as determined by pacing him in a vehicle, was approximately 20 mph.

In descriptions of wild dog hunting methods, much has been made of their intelligent cooperation in "cutting corners" in their prey, and particularly of their relay running, with fresh dogs taking the place of tired leaders. We concur that there is a basis for the first idea, but we saw no evidence whatever to support the contention that wild dogs run in relays. The truth is that wild dogs have no need to hunt in relays. The lead dog has ample endurance, if not the speed, to overtake probably any antelope, of which gazelles are among the fleetest. The fact that other members of the pack are able to cut corners on the prey is at least partly accounted for by the prey's tendency to circle instead of fleeing in a straight line. As explained later, some prey animals have a greater tendency than others of their species to do this. Of course, once overtaken, even a quarry that has been running straight is forced to start dodging if it is to avoid being caught

straightaway. Thus a dog running not too far behind the leader is well placed to cut corners when the quarry changes course, and it frequently happened that one of the followers made the capture. Most game, after a hard chase of a mile or two, was too exhausted by the time it began dodging to have any real chance of evading its pursuers.

Killing and Eating.—Wild dogs killed small game like Thomson's gazelles with amazing dispatch. Once overtaken, a gazelle was either thrown to the ground or simply bowled over, whereupon all nearby dogs fell on it instantly. Grabbing it from all sides and pulling against one another so strongly that the body was suspended between them, they then literally tore it apart (Fig. 5). It happened so quickly that it was never possible to come up to a kill before the prey had been dismembered. If it didn't go down at once, dogs began tearing out chunks while it was still struggling on its feet. We once saw a three-quarter term fetus torn from a Thomson's gazelle within seconds of the time it was overtaken and before it went down. As Kühme observed, there is no specific killing bite as in felids (Leyhausen 1965). When dealing with larger prey such as juvenile wildebeest and notably a female kongoni, the dogs slashed and tore at the hind legs, flanks, and belly—always from the rear and never from in front—until the animal fell from sheer exhaustion and shock. They then very often began eating it alive while it was still sitting up (Fig. 6). Self-defense on the part of a prey was never once observed; the kongoni, for example, did little more than stand with head high while the dogs cut it to ribbons, looking less the victim than the witness of its own execution.

In eating, the dogs began in the stomach cavity, after first opening up the belly, and proceeded from inside out. Entrance was

Fig. 5. The pack tearing apart a young gnu calf.

also effected through the anus by animals unable to win a place in the stomach cavity. While several dogs forced their heads inside and ripped out the internal organs, others quickly enlarged the opening in struggling for position. This resulted in skinning out the carcass, leaving the skin still attached to the head, which was seldom touched. Apart from these, the backbone and the leg bones, very little of a Thomson's gazelle would remain at the end of 10 minutes. In the pack of 21, if only part had managed to eat their fill, sometimes the rest went off to hunt again before the carcass was cleaned. They proceeded to chase and pull down another gazelle within as little as 5 minutes from the time of the previous kill, to be joined shortly by the other dogs. As each animal became satisfied it withdrew a little from the kill and joined others to rest, play, or gnaw at a bone it had taken along. Sometimes the pack stayed at the scene until the next hunting period; more often it withdrew to a nearby stream or waterhole and settled down there. Kühme never saw wild dogs drink. The Ngorongoro pack drank, though irregularly, before hunting and after eating.

Selection of Prey and Frequency of Kills

Table 1 summarizes prey selection b species, sex, and age in 50 recorded kill The 11 wildebeest calves were all taken i January during the peak calving seaso when the pack of seven dogs apparentl specialized on them; only kills of calv were seen by us or reported by Crater vis tors in this month. Thus the percentage calves in the total gives a biased picture prey selection during the rest of the yea With new calves excluded, the adjuste percentages, based on 39 kills, are as fc lows:

Thomson's gazelles	69 percent
Juvenile wildebeest	18 percent
Grant's gazelles	10 percent
Kongoni	one kill

Wright (1960:9) records a similar p ponderance of Thomson's gazelles in

Fig. 6. Dogs begin eating a yearling-class gnu while still alive, but evidently in a state of deep shock.

Table 1. Prey selection by species, sex, and age in 50 kills of the African wild dog.

PREY SPECIES	TOTAL No.	ADULT MALES	ADULT FEMALES	JUVENILE–SUBADULT	YOUNG*	PERCENT OF TOTAL KILLS
Thomson's gazelle	27	18	6	2	1	54
Wildebeest	18	0	0	7	11	36
Grant's gazelle	4	1	1	2	0	8
Kongoni	1	0	1	0	0	2

* Less than 6 months old.

kills on the Serengeti Plains (7 Thomson's gazelles, 1 wildebeest, 1 impala, and 1 reedbuck [*Redunca redunca*]), and notes that it is the staple diet of wild dogs in the Serengeti. Kühme also observed that wild dogs prey mainly on *Gazella thomsonii* and *G. granti*, and young wildebeest in the Serengeti.

In terms of actual preference, information from the Serengeti, where the Thomson's gazelle is by far the most numerous herbivore, is far less revealing than the figures from the Crater, where this species occurs in relatively small numbers. The status of the principal ungulates in Ngorongoro, based on an aerial count by Turner and Watson (1964), on two ground counts of the gazelles by the authors in collaboration with the Mweka College of Wildlife Management, and on our ground counts of the less numerous species, is as follows:

Wildebeest	14,000
Zebra	5,000
Thomson's gazelle	3,500
Grant's gazelle	1,500
Eland (*Taurotragus oryx*)	350
Waterbuck (*Kobus defassa*)	150
Kongoni	100
Reedbuck	100 (?)

The evidence suggests, then, that Thomson's gazelle is the preferred prey of the wild dog in East African steppe–savanna. In the miambo woodland (Brown 1965) that extends from mid-Tanzania into South Africa, where gazelles are not found, the main prey may be impala, followed by other medium- to small-sized antelopes and the young of large antelopes. In Kruger

Park, for example, of 88 identified wild dog kills, 85 percent were impala (Bourliere 1963). Stevenson-Hamilton listed other prey as reedbuck, bushbuck (*Tragelaphus scriptus*), duiker (*Sylvicapra grimmia* and *Cephalophus* spp.), and steinbok (*Raphicerus campestris*), also female waterbuck and kudu (*Tragelaphus strepsiceros*) when pressed by hunger. In Wankie Park, wardens' reports indicate a considerable toll of young kudu, eland, sable (*Hippotragus niger*), and tsessebe (*Damaliscus lunatus*). Instances where adult female and even adult male kudu were pulled down by wild dogs are also cited.

Bourliere states (1963:21) that "Carnivores actually only prey upon herbivores of about the same size and weight." While this generalization is open to dispute, it applies well enough to East African wild dogs preying on Thomson's gazelles. Where the main prey is impala, reedbuck, etc. that weigh in the 100–150 lb class, weight and size may be double or triple that of the wild dog. But the wild dogs of the East African steppe–savanna are smaller (also darker, with more black and less tan and white) than their counterparts in Central and South African woodland (Fig. 7). The average weight of the animals we have seen in East Africa would not exceed 40 lb; the members of a pack seen in Wankie Park, by comparison, looked to be a good 3 inches taller and 20 lb heavier. This consistent geographic size variation may be adapted to size of the principal prey species; specifi-

Fig. 7. Two specimens of the larger, lighter-colored wild dog of southern Africa, photographed in Wankie National Park

cally, wild dogs of the East African plains may be smaller as the result of specialization on Thomson's gazelle.

Kill Frequency.—Because of the difficulty of locating and relocating a free-ranging pack, our data for consecutive hunting periods are inadequate for defining the average kill frequency and average food intake per animal per day. Even when the pack was observed during the two daily hunting periods, it was rarely certain that it had not killed before, after, or between these periods. Nonetheless, because this type of information is badly needed, data covering consecutive hunting periods are presented in Table 2 as a rough average of kill frequency and meat available per animal per day.

The average frequency of two kills per day derived from the data for consecutive hunting periods agrees with our general observation that the pack usually killed during each period. To demonstrate this, on 28 hunts the pack performed as follows:

Chases	Kills	Failures	Did not chase
29	25	4	5

This indicates a success rate per chase of over 85 percent. As a further indication of efficiency in locating and running down

prey where game is plentiful, on eight occasions when the dogs were watched from the moment they left their resting place to the moment they killed, the mean time was only 25 minutes, with a range of 15–45 minutes. On five other occasions the pack failed to hunt seriously during the normal period; this was offset in the above figure by five periods during which the dogs chased and killed twice. The possibility that hunting activity and success might be reduced after having killed larger or more than one of the usual prey is not borne out by the six instances when the pack was observed during the next hunting period: four cases the pack killed again. There are some grounds for asserting, then, that wild dogs kill twice daily regardless of what their prey may be. Certainly they do not feed more than once from the same kill, at least not in Ngorongoro Crater, where the numerous scavengers dispose of all leftovers in very short order.

Meat Available per Animal per Day. The amount of meat available per wild dog per day works out at roughly 6 lb, assuming that 40 percent of the prey animal consists of inedible or unpalatable bone, skin, and stomach contents. Wright's (1960) calculation of 0.15 lb of food per day per pound

Table 2. Kill frequency and meat available per dog per day, based on observations of consecutive hunting cycles.

DATE	PREY	EST. WT. (LB)	NO. IN PACK	AVAIL- ABLE* MEAT/ DOG	MEAT/ DOG/DAY
1964					
Sept. 30	Juvenile wildebeest	125	21	3.6	3.6
Oct. 1	Thomson's gazelle (adult M)	60	21	1.7	
" " (adult F)		40	21	1.1	2.8
Nov. 11	Thomson's gazelle (adult F, including fetus)	50	21	1.4	
	2 Thomson's gazelles (adult F)	80	21	2.2	3.6
Nov. 12	Thomson's gazelle (adult M)	60	21	1.7	
	Kongoni (adult F)	250	21	7.4	9.1
Nov. 27	2 Thomson's gazelles (adult M)	120	21	3.4	
	Thomson's gazelle (subadult M)	50	21	1.4	4.8
Nov. 28	Grant's gazelle (subadult F)	90	21	2.6	
	Thomson's gazelle (adult M)	60	21	1.7	4.3
Dec. 5 (PM)	2 Thomson's gazelles (adult M)	120	12	6.0	
to					
Dec. 7 (AM)	Thomson's gazelle (juv. M)	40	21	1.1	3.5
1965					
Jan. 17 (PM)	4 wildebeest calves	180	7	15.5	7.8
to					
Jan. 19 (AM)					
July 16	2 Thomson's gazelles (adult M)	120	6	12.0	12.0

Kill frequency = 2 kills/day.
Meat available per dog per day: combined average = 6 lb; for pack of 21 = 4.5 lb; for pack of 7–6 = 9 lb.

* Available meat is based on 60 percent of carcass weight.

of dog also works out to 6 lb per day if the average weight of a dog is taken as 40 lb, but his figures are based on the total weight of the prey. In either case, two to three times as much food per day is available to wild dogs as is given to domesticated dogs of the same size. However, the number of dogs in the pack is an important factor. When there were 21 dogs, the amount of meat available per day was less than 5 lb per animal; in the pack of 7 and 6, each animal had approximately twice as much available meat. Since the small packs killed at the same rate, large packs are undoubtedly less wasteful.

Reactions of Prey Species

The reactions of game depended on the behavior of the wild dog pack. When the pack was at rest, all game would graze unconcernedly within 150 yards. When the dogs were walking or trotting, potential prey would stand until approached within 350–250 yards, or less if the pack was not headed directly toward them. When stalked, gazelles often stood watching until the pack came within 300–200 yards. But when the pack was running, gazelles, and wildebeest herds containing young, often acted alarmed at a distance of 500 yards, although again, individual animals not directly in the approach line might let the pack go by as close as 150 yards.

Gazelles.—The moment a running wild dog pack appeared on the plain, both gazelle species immediately reacted by performing the stiff-legged bounding display, with tail raised and white rump patch flashing, called Stotting or Pronking. Undoubtedly a warning signal, it spread wavelike in advance of the pack. Apparently in response to the Stotting, practically every

391

gazelle in sight fled the immediate vicinity.

Adaptive as the warning display may seem, it nonetheless appears to have its drawbacks; for even after being singled out by the pack, every gazelle began the run for its life by Stotting, and appeared to lose precious ground in the process. Many have argued that the Stotting gait is nearly or quite as fast as a gallop, at any rate deceptively slow. But time and again we have watched the lead dog closing the gap until the quarry settled to its full running gait, when it was capable of making slightly better speed than its pursuer for the first half mile or so. It is therefore hard to see any advantage to the individual in Stotting when chased, since individuals that made no display at all might be thought to have a better chance of surviving and reproducing. On theoretical grounds, then, it has to be assumed that the Stotting display offers an individual selective advantage which simply remains to be determined. Nor is this type of display confined to the gazelles: during the aforementioned kongoni chase, all six members of the herd began Stotting when the wild dog pack first headed in their direction, and the victim continued to Stot for some time after being singled out.

Table 1 shows that 67 percent of the Thomson's gazelles killed were adult males. This is evidently the result of territorial behavior. Because of attachment to territory, probably coupled with inhibition about trespassing on the grounds of neighboring rivals, territorial males tend to be the last to flee from danger. Moreover they show a greater tendency to circle back toward home, and these two traits together make them more vulnerable to wild dog predation than other members of the population. The same tendencies are displayed by females with young, concealed fawns, making them also more vulnerable.

Wildebeest.—Adult wildebeest, especially territorial bulls, show little fear of wild dogs, which is a good indication that they have little reason to fear them under normal circumstances (Fig. 8). While even territorial males will get out of the way of a running pack, they rarely leave their grounds, but merely trot to one side and turn to stare as the pack goes by. Bulls not infrequently act aggressively toward walking or trotting dogs, and may even make a short charge if the dogs give ground. In Rhodesia we have seen a pack of the larger variety of wild dogs chased by females and yearlings of the blue wildebeest (*C. t. taurinus*) which is also larger and perhaps generally more aggressive than the Western white-bearded gnu. But like zebras, all wildebeest will on occasion follow behind walking or trotting dogs, apparently motivated by curiosity, just as they will gather to stare at and follow lions (*Panthera leo*).

In hunting wildebeest, wild dogs are obviously highly selective. Having walked in the Stalking attitude to within several hundred yards or less and then run into the midst of a large concentration, the pack splits up and works through it, approaching one gnu after another only to turn away if it proves adult. Meanwhile the wildebeest mill and run in all directions, without ever making any effort to form a defensive ring —even when young calves are present. A defensive ring has been reported in some of the wild dog literature. Kühme (p. 528) observed something of the sort in large Serengeti concentrations, though they did not form any regular ring but simply crowded together in a milling mass. Individual females, on the other hand, defend their calves after being overtaken in flight. Against a pack, however, one wildebeest cannot put up any effective defense; while it confronts one or two, the rest go around and seize the calf.

Zebra.—The only other herbivore whose

ig. 8. Adults, and even a yearling gnu (4th from left) show little fear of running wild dogs, though they ran out of the
ay immediately after the picture was taken. The quarry is a young calf, visible as a light spot in the upper left.

eactions to wild dogs we observed in de-
ail, zebras are the least concerned about
hem, and do not hesitate to attack dogs
hat come too close. Wild dogs on their
art rarely stand up to them. Since the
nembers of a harem would probably co-
perate with the herd stallion to defend the
als, it would appear that wild-dog preda-
on on zebra is quite rare.

elations with Other Predators and
cavengers

Vultures.—Since wild dogs customarily
ill in early morning and late afternoon, the
arger vultures, the white-backed (*Pseu-
ogyps africanus*), Rüppells griffon (*Gyps
ippelli*), and lappet-faced (*Torgos trache-
otus*), whose activities are largely regu-
ted by the presence or absence of thermal
pdrafts, benefit rather little from their pre-
ation. Large vultures were more likely to
ppear at afternoon than morning kills. But
e two smallest species, the hooded and

Egyptian vultures (*Necrosyrtes monachus*
and *Neophron perenopterus*), were regularly to be found at wild-dog kills, a good
hour before other scavengers were even airborne. In addition to these vultures, other
regularly encountered scavengers included
the tawny eagle (*Aquila rapax*) and the
kite (*Milvus migrans*), while the uncommon white-headed vulture (*Trigonoceps
occipitalis*), the bateleur eagle (*Terathopius
ecaudatus*), and Cape rook (*Corvus capensis*) showed up infrequently.

On several occasions hooded vultures
were seen following a chase and landing
before the prey had even been pulled down,
shortly after full daylight. Aside from gleaning bits and pieces around the kill, vultures
had to wait until the dogs left before they
could feed on the carcass. But the kite successfully stole small pieces from the dogs
by swooping, grabbing, and mounting again
to eat on the wing. Although young animals sometimes stalked and ran at vultures

that approached close to the kill, the dogs were generally tolerant toward avian scavengers.

Jackals.—The Asiatic jackal (*Canis aureus*) was seen more frequently at kills than the black-backed jackal (*C. meso-melas*). Since the latter seemed to predominate at nocturnal kills by lions or hyenas, it may be that one is more nocturnal and one more diurnal in its habits. Also, the Asiatic jackal tended to behave more boldly and aggressively at kills. It would move closer to a feeding pack of dogs and take advantage of any opportunity to steal meat. When threatened by a dog, a little 15-lb jackal, coat fluffed, head down, and snarling, would stand its ground and snap ferociously if the dog continued to advance. Although it was pure bluff that quickly ended in flight if a dog attacked in earnest, it proved a surprisingly effective intimidation display in most encounters. But on the whole, wild dogs behaved almost as tolerantly toward jackals as toward vultures.

Spotted Hyenas.—Spotted hyenas, on the other hand, seriously compete with the dogs for their kills, attempting to play a commensal role against active resistance. In a place like the Crater, with an exceptionally large hyena population for such a small area, numbering some 420 adults (Kruuk 1966:1258), it is probably safe to say that wild dogs hunting singly or in twos and threes would very frequently lose their kills to hyenas, since this happened occasionally even to the pack of 21.

Hyenas actually stayed near the resting pack for hours at a time, evidently waiting for a hunt to begin. It was not unusual to see one or more of them slowly approach a group of dogs, then crawl to within a few yards and lie gazing at them intently, as though urging them to get started. Often several would wander between resting groups, sniffing the ground, consuming any

stools they found, and coming dangerously close, to stand staring with their short tails twitching—a sign of nervousness. Kühme (p. 534) reports an instance in which a hyena even touched a resting wild dog's face, meanwhile "whining friendly." Such boldness, particularly near the time when the pack was becoming active, often triggered the Mobbing response.

Hyenas, which weigh up to 150 lb, would be more than a match for wild dogs if they had the same pack (mobbing) instinct. Lacking it, they are nearly defenseless against a wild dog pack. With three to a dozen dogs worrying its hindquarters, the best a hyena can do is to squat down and snap ineffectively over its shoulder, while voicing loud roars and growls. On rare occasions a hard-pressed one would simply lie down and give up; a hyena we once saw crowded by a persistently curious group of juvenile wildebeest did the same thing. The spotted hyena seems on the whole to be notably timid by nature, as may be judged from the fact that mothers will often not even defend their offspring. Yet they are driven by hunger to take incredible and sometimes fatal risks.

Often, as under the above circumstances, they provoked attack by their own rashness. But in other cases the dogs seemed to go out of their way to harry hyenas encountered during the early stages of a hunt. Those unwary enough to let the pack get close could still usually get off entirely by cowering down and lying still. But those that stayed until the pack was close and then ran away were inviting pursuit and a good mauling. At the same time, hyenas following behind the pack were generally ignored. On one notable occasion, the pack of 21 took it in turns to mob the hyenas it happened upon in a denning area inhabited by more than 30 adults and cubs, many of which were foregathered as usual prior to the evening foraging. What was most surprising was that none, on this or any other occasion, attempted to take refuge underground. When hard-pressed, even half-grown pups bolted into nearby dense streamside vegetation, where the dogs did not follow. But presumably young pups were hidden in the dens, since one lactating female was reluctant to quit the immediate vicinity. She was repeatedly mobbed. Set upon by five or six dogs at a time, she would maintain a squatting defense as long as she could bear it, then break free to race for the nearest hole. Instead of going down it or backing into it, she threw herself into cup-shaped depressions next to the holes, which may or may not have been excavated by the hyenas themselves (territorial wildebeest also dig these depressions by pawing and horning the earth). In these she lay flat and tried to defend herself from the dogs, to whom only her back and head were exposed, while keeping up a steady volume of roars, growls, and staccato chuckles. Eventually she also took refuge in the bushes. Neither this hyena nor the next, which the dogs turned on its back and mauled for 2 minutes, bore any visible wounds. In fact, we have never known the dogs to kill or even seriously injure one. Either hyenas have exceedingly tough hides or else wild dogs are less in earnest about mobbing them than might appear.

Yet the degree to which hyenas are able to capitalize on wild-dog predation for their own benefit would justify a deep antagonism. They frequently drive away the last dogs on a kill unless the rest of the pack remains close by, and are quite capable of taking meat away from one or two dogs only a few yards removed from a kill where the rest are feeding. A more extraordinary example of this exploitation is the way hyenas take advantage of the wild dog's hunting technique: in the final moments of

Fig. 9. Hyenas appropriate a wild dog prey and begin eating it alive, while one of two dogs that caught it looks on, panting heavily from the chase. Hyena in foreground is half-grown. As shown in Fig. 6, the dogs reclaimed their prey when the rest of the pack arrived.

a chase, when only one or a few dogs are close to the quarry, hyenas have an opportunity to appropriate it before the rest of the pack arrives (Fig. 9). They attempted this with considerable regularity in the Crater, and we succeeded in recording one instance on film. In some cases it was a matter of chance that hyenas were near enough the scene of the capture to dash in at the decisive moment; in others up to three or four actually took part in the chase from the beginning. Though not as fast as the dogs, they were able to be in a position to intercept the quarry if it doubled back, or to grab it away from the dog(s) as soon as it was caught. When only two or three of them were on hand, the dogs hesitated to launch an immediate counterattack, particularly if more than one hyena was involved. But usually other pack members quickly appeared, joined together to mob the hyenas, and forced them to surrender the kill.

But sometimes the dogs were defeated by sheer numbers. Once when the leader of the pack of 21 had pulled down a juvenile wildebeest in a hyena denning area, some 40 hyenas closed in on the kill before the others could gather. Apparently intimidated by so many competitors, the dogs revenged themselves by mobbing stragglers, punishing them savagely. Twenty minutes later, while they were ranging for new prey, the hyenas pulled down an adult female wildebeest on their own, quite near the first kill. Their clamor drew the dogs back to the scene. But they did nothing this time but look on—there were now 60 hyenas!

CONCLUSIONS

Pack Function

Hunting is undoubtedly the primary function of the free-ranging pack. Wild-dog behavior is highly specialized and adapted for pack life by dint of the equal and exceptionally friendly relations between individuals, subordination of individual to group activity, discipline during the chase, and close cooperation in killing prey and mutual defense. It may, in fact, be seriously questioned whether a single wild dog could survive for long on its own. As demonstrated by the successful rearing of a litter after the mother died, feeding and protection of the young is another important pack function. The main selective advantages of the pack hunting unit may be summarized as follows:

1. Increased probability of success through cooperation, hence better opportunity to eat regularly at less cost in individual effort

2. More efficient utilization of food resources

3. Less disturbance of prey populations than would result if each animal hunted individually

4. Mutual protection against competitors (spotted hyenas) and possible predators (hyena, leopard [*Panthera pardus*], and lion)

5. The provision of food for infants at the den and the adults that remain with them when the pack is hunting, and for juveniles and sick or old adults unable to kill for themselves.

Effective Pack Size

We have presented evidence, though admittedly tentative, that large packs utilize prey resources more efficiently than small packs, with less waste. Competition from hyenas, where they are numerous, must exert a strong selective pressure in favor of large packs as well as for close cooperation at kills. While the observed tendency for small packs to keep closer together in the chase and at kills would tend to compensate somewhat for low numbers, there must be a minimum below which competition from hyenas, and reduced hunting and killing capability, would become a serious handicap. From our observations of both large and small packs, four to six would seem close to the minimum effective unit.

We believe that wherever wild dogs are reduced to such small packs, their ability to survive and reproduce may be endangered. This is not taking into account the possibility of a differential birth or mortality rate that results in a low ratio of females. If it represents a pathological condition, this alone could mean that the species is in serious trouble; a prompt investigation of reproduction and neonatal mortality is called for to find out to what extent an abnormally low percentage of females may be responsible for the apparent decline of the species in many parts of Africa.

Prey Relations

It seems clear that wild dogs are highly selective in the species they prey upon, specializing in East African steppe–savanna on Thomson's gazelles, and on wildebee calves during the gnu calving season. Considering their selectivity, their rate of killing, and the observed reactions of herbivores to them, it can only be concluded that wild dogs are by no means so wantonly destructive or disruptive to game as is commonly supposed. Kühme reached the same conclusion (p. 528). Indeed, until one comes to realize that plains game simply has no place to hide and no sanctuary where predators cannot follow, it is a recurrent surprise to note how short-lived and localized are disturbances due to predation.

In a prey population as small as that of Thomson's gazelles in Ngorongoro, if one assumes an average annual recruitment rate of roughly 10 percent, predation at the rate of only one a day obviously would reduce the population if maintained over a long period. There was, however, no evidence that the Thomson's gazelle population declined after wild dogs became resident in the Crater; our gazelle censuses in October, 1964, and May, 1965, showed no reduction that could not be accounted for by simple counting errors. Even an actual reduction would have no relevance to the overall situation, as Thomson's gazelles are the most numerous herbivores in their centers of distribution (the steppe–savanna from central Kenya to north-central Tanzania). On the Serengeti Plains, where the gazelle population is estimated at 800,000 and there are probably fewer than 500 wild dogs, predation by this species could have no appreciable effect. Indeed wild dogs are only one of nine predators on the Thomson's gazelle (Wright 1960), and not the most important one at that; jackals, which are numerous and specialize in catching new fawns, are probably the main predators.

Since wild dogs are nowhere numerous and everywhere apparently specialize on the most abundant small to medium-sized antelopes, it can be argued that more, not

fewer, of them are needed. The population explosion of impala in Kruger National Park and many other places where wild dog numbers have declined offers convincing evidence. The high percentage of territorial males in wild dog kills of Thomson's gazelles offers a more subtle example of how predation may benefit a prey species: in probably every gregarious, territorial antelope species, there is always a surplus of fit, adult and young-adult males which cannot reproduce for want of enough suitable territories, so that the removal of territorial males by predation is of perhaps major importance in opening up territories for younger and sexually more vigorous males.

In our judgment, the wild dog is an interesting, valuable predator whose continued survival may be endangered. We feel it should be strictly protected by law in all African states where it occurs, and that it should be actively encouraged, if this is possible, in every park and game reserve.

LITERATURE CITED

BOURLIERE, C. F. 1963. Specific feeding habits of African carnivores. African Wildl. 17(1): 21–27.

BROWN, L. 1965. Africa, a natural history. Random House, New York. 299pp.

BRYDEN, H. A. 1936. Wild life in South Africa. George G. Harrap Co. Ltd., London. 282pp.

HALTENORTH, T. 1963. Klassifikation der Säugetiere: Artiodactyla. Handbuch Zool. Bd. 8. 167pp.

KRUUK, H. 1966. Clan-system and feeding habits of spotted hyaenas (*Crocuta crocuta* Erxleben). Nature 209(5029):1257–1258.

KÜHME, W. 1965. Freilandstudien zür Soziologie des Hyänenhundes. Zeit. Tierpsych. 22(5): 495–541.

LEYHAUSEN, P. 1965. Über die Funktion der relativen Stimmungshierarchie. Zeit. Tierpsych. 22(4):395–412.

MABERLY, C. T. ASTLEY. 1962. Animals of East Africa. Howard Timmins, Cape Town. 221pp.

MACKWORTH-PRAED, C. W., AND C. H. B. GRANT. 1957. Birds of eastern and north eastern Africa. Series I, Vol. I. African handbook of birds. Longmans, Green and Co., London. 806[+ 40]pp.

PERCIVAL, A. B. 1924. A game ranger's note-book. Nisbet & Co., London. 374pp.

SELOUS, F. C. 1881. A hunter's wanderings in Africa. R. Bentley & Son, London. 455pp.

STEVENSON-HAMILTON, J. 1947. Wild life in South Africa. Cassell & Co., Ltd., London. 343pp.

TURNER, M., AND M. WATSON. 1964. A census of game in Ngorongoro Crater. E. African Wildl. J. 2:165–168.

VAUGHAN-KIRBY, F. 1899. The hunting dog. Pp. 602–606. *In* H. A. Bryden (Editor), Great and small game of Africa. R. Ward Ltd., London. 612pp.

WRIGHT, B. S. 1960. Predation on big game in East Africa. J. Wildl. Mgmt. 24(1):1–15.

Received for publication March 21, 1966.

HOMING BEHAVIOR OF CHIPMUNKS IN CENTRAL NEW YORK

Homing movements ranging from about 150 to 700 yards have been recorded for *Tamias* by Seton (LIFE HISTORIES OF NORTHERN ANIMALS, vol. 1: 341, 1909), Allen (Bull. N. Y. State Mus., 314: 87, 1938), Burt (Misc. Publ. Mus. Zool. Univ. Mich., 45: 45, 1940), and Hamilton (AMERICAN MAMMALS, p. 283, 1939). While engaged in other studies during the summer of 1952, I had the opportunity of making additional observations on the homing behavior of the eastern chipmunk, *Tamias striatus lysieri* (Richardson), on the campus of Cornell University at Ithaca, Tompkins County, New York.

Live-trapping was conducted from July 26 to August 3 in a tract of approximately 3 acres of hemlock and mixed hardwood forest bordering a small artificial lake. A maximum of 12 traps was employed. The chipmunks taken were sexed, aged (subadult or adult), marked by clipping patches of fur on various parts of the body, and transported in a cloth bag to one of six release points. The latter were situated in similar continuous habitat or in an area of campus buildings, lawns, shrubbery, and widely spaced trees adjacent to the woodland. An individual was considered as having homed when it was retaken within 115 feet of the original point of capture. Those chipmunks that returned and were recaptured were immediately released again in a different direction and usually at a greater distance. First releases averaged 675 feet (310–1,160) and second ones, 1,015 feet (500–1,570). Two animals that returned after second removals were liberated for a third time at distances of 1,130 and 2,180 feet. All distances given are calculated from the station where the animal was originally trapped. Since the mean home range size of chipmunks in this vicinity has been calculated as about .28 acre (Yerger, Jour. Mamm., 34: 448–458, 1953), it is assumed that in most, if not all, instances the removal distances involved were great enough to place the animal in unfamiliar territory beyond the boundaries of its normal range of movements.

A total of 18 individuals, consisting of five adult males, four adult females, two subadult males and seven subadult females, were marked and released a total of 29 times through July 30. Animals handled after this date are not included in the treatment of the data, since it is felt that there was insufficient opportunity for them to be retaken following their release. Seven of the chipmunks returned to the vicinity of original capture a total of ten times over distances varying from 430 to 1,200 and averaging 650 feet. In six of the ten returns the animals were retaken in the same trap in which they were initially caught. In two instances individuals were retrapped at stations 20 and 40 feet removed from the one where first taken, and in two others the individuals were recovered at a distance of 115 feet from the original site of capture.

Two adult males returned from 490 and 540 feet in two and three days, respectively, but were not recovered after second removals to 1,060 and 1,150 feet. Two other mature males trapped following their release at 310 and 750 feet had moved in a direction other than that of their original capture. An adult female was found in a trap a day after having been released at 775 feet. She returned a second time from 600 feet in two days. Another adult female was retrapped at the original trap station seven days following her release only 430 feet away. A single subadult male was recaptured in his original location the next day after his initial removal to 750 feet and two days after a second liberation at 1,200 feet. He was not retrapped subsequent to a third relocation 2,180 feet distant. Another young male was captured at a point 380 feet closer to its home area six days after being released 940 feet away. Three subadult females homed successfully. One returned from 580 feet the day after release, another from 650 feet in two days' time, and the third from 490 feet after an interval of five days. None were retaken following second liberations ranging from 940 to 1,570 feet. Two other subadult females were captured 100 and 120 feet closer to their original capture sites the day after having been released at 460 and 450 feet, respectively.

These limited data suggest that homing ability was restricted to rather short distances, only one individual being known to have returned from a point more than 775 feet away. The extent of these movements may be somewhat less than several reported by authors previously mentioned. However, because of the smaller home range size of chipmunks in this area as compared to other habitats in which the animals homed from more distant points, the actual distances moved over strange territory may be fairly comparable. The present results indicate no obvious differences in the proportion of adults and subadults homing nor in the average distance over which individuals in each of these age classes returned. The intervals of one to seven days between releases and recoveries, the relatively short distances involved, and the rather low proportion of returns (38.8 per cent) suggest that the animals may have returned to their home areas through random movements until familiar terrain was encountered. It should be mentioned, however, that the small number of traps employed might have been a factor in the low rate of recovery, since a chipmunk returning to its home region had a lower probability of being recaptured than would have been the case had more traps been present. This might also have tended to increase the apparent time taken to reach the home area following release. On the other hand, the use of a limited number of traps may have been advantageous in that there was less interference by traps with the normal activities and movements of the animals.—JAMES N. LAYNE, *Dept. of Biology, Univ. of Florida, Gainesville. Received December 1, 1956.*

COMPARATIVE ECHOLOCATION BY FISHING BATS

Roderick A. Suthers

ABSTRACT.—The acoustic orientation of two species of fish-catching bats was studied as they negotiated a row of strings or fine wires extending across their flight path. Orientation sounds of *Pizonyx vivesi* consisted of a steep descending FM sweep lasting about 3 msec. *Noctilio leporinus* used 8 to 10 msec pulses composed of an initial nearly constant frequency portion followed by a descending frequency modulation. The echolocation of small wires by *N. leporinus* differed from that of surface fish in that during wire avoidance no nearly constant frequency or entirely FM pulses were emitted, nor was the pulse duration markedly shortened as the barrier was approached. There was extensive temporal overlap at the animal's ear of returning echoes with the emitted cries when the bat was near the barrier—a strong contrast to the apparent careful minimization of such overlap during feeding maneuvers. *Noctilio* increased its average pulse duration about 2 msec when confronted with a barrier of 0.21 mm, as opposed to 0.51 mm, diameter wires. *Pizonyx* detected these wires well before pulse-echo overlap began, but at a shorter range than did *N. leporinus*, suggesting the latter species may have a longer effective range of echolocation.

At least two species of Neotropical bats have independently evolved an ability to capture marine or aquatic organisms. A comparison of the acoustic orientation of these animals is of particular interest in view of their convergent feeding habits yet strikingly different orientation sounds. *Noctilio leporinus* Linnaeus (Noctilionidae) catches fish by occasionally dipping its disproportionately large feet into the water as it flies low over the surface. Very small surface disturbances can be echolocated and play an important role in determining the locations of the dips (Suthers, 1965). Fish caught in this way are transferred to the mouth and eaten. *Pizonyx vivesi* Menegaux (Vespertilionidae) also possesses disproportionately large feet. Much less is known concerning the feeding behavior of this species, though it is reasonable to assume that it uses its feet in a manner similar to *N. leporinus* (but see Reeder and Norris, 1954). Extensive attempts to induce captive *P. vivesi* to catch pieces of shrimp from the surface of a large pool were unsuccessful. The following comparison is therefore based on the ability of these bats to detect small obstacles.

METHODS

The experimental animals consisted of two *P. vivesi*, selected as the best flyers of several collected in the Gulf of California, and one *N. leporinus* captured in Trinidad. The research was conducted at the William Beebe Memorial Tropical Research Station of the New York Zoological Society in Trinidad.

The bats were flown in a 4 × 15 m outdoor cage described elsewhere (Suthers, 1965). The test obstacles consisted of a row of strings or wires 2.5 m long which were hung at 55 cm intervals across the middle of the cage. Four sets of obstacles were used: 2 mm diameter strings, 0.51 mm, 0.21 mm, and 0.10 mm diameter wires, respectively. The bats were forced to pass through this barrier in order to fly the length of the cage. Each

79

flight was scored as a *hit* or a *miss* according to whether any part of the animal touched the test obstacles. Movement of the larger wires was easily visible following even gentle contact, but lateral illumination of the barrier was necessary in order to score flights through the row of 0.10 mm wires. The wires were occasionally shifted laterally about 20 cm across the width of the cage in order to reduce the possibility that the bats might learn their location.

An attempt was made to test each animal on two or more sets of obstacles per night, though this was not always possible. Cases in which the bat was making unusually frequent landings or was particularly reluctant to fly are omitted. Also excluded are flights on which the barrier was approached very near the upper ends of the wires, along either wall of the cage, or at an angle to the row of obstacles which was decidedly smaller than 90°. Experiments with *P. vivesi* no. 1 were terminated by its sudden death, which occurred before the 0.10 mm diameter wire was available. Tests with this size wire were therefore performed with a second healthy *Pizonyx*.

A series of flights through the 0.51 mm and 0.21 mm diameter wires was photographed with a 16 mm sound motion picture camera, while simultaneous two-channel tape recordings of the orientation sounds were obtained from microphones placed on opposite sides of the barrier. The position of the flying bat relative to the barrier was calculated by comparing the arrival time of each orientation sound at either microphone and also by matching the image of the bat in each frame of the film with rectified orientation sounds on the optical sound track. Details of these methods and the instrumentation are described elsewhere (Suthers, 1965). The overall frequency response of the recording system was approximately uniform between 15 and 100 kc/sec.

A total of 45 flights by *P. vivesi* and *N. leporinus* was tape recorded and photographed. Sixteen of these were discarded for reasons listed above, or because the bat did not fly on a straight path between the microphones, or because of a poor signal-to-noise ratio on one of the channels. The remaining 29 flights were analyzed and pulse intervals (the silent period from the end of one pulse to the beginning of the next) were plotted against the distance of the bat from the barrier (see Fig. 2). The animal's position was determined to within an accuracy of about ± 10–15 cm at a distance of two meters from the wires and ± 5–10 cm in the immediate region of the wires.

RESULTS AND CONCLUSIONS

Obstacle avoidance scores are given in Table 1. The greater success of *P. vivesi* in avoiding 0.10 mm wires may reflect its shorter maximum wingspan of 40 cm, compared to 50 cm for *N. leporinus*. Audio monitoring of the rectified orientation sounds emitted by these species during their flights indicated

TABLE 1.—*Percent of flights through barrier on which bat missed obstacles spaced at 55 cm intervals. Total number of flights in parentheses. Maximum wingspan of* P. vivesi *is about 40 cm; that of* N. leporinus *is about 50 cm.*

Bat	OBSTACLE DIAMETER (MM)			
	2	0.51	0.21	0.10
Pizonyx vivesi (1)	94%	83%	51%	
	(163)	(416)	(232)	
Pizonyx vivesi (2)		71%		37%
		(151)		(74)
Noctilio leporinus	91%	76%	60%	20%
	(207)	(203)	(217)	(55)

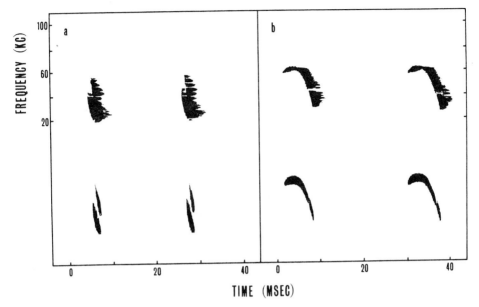

Fig. 1.—Sound spectrographs of orientation sounds emitted by *Pizonyx vivesi* (a) and *Noctilio leporinus* (b) when approaching wire obstacles. A pair of consecutive pulses, reproduced at two different filter settings of the sound spectrograph, is shown for each species. The narrow band filter setting (top) best indicates the frequency spectrum of the cries, whereas pulse duration and temporal relationships are more accurately shown using a wide band filter (bottom).

that approaches to the three larger diameter obstacles were accompanied by increases in the pulse repetition rate, whereas no such increase was noted during approaches to the 0.10 mm wires. This suggests that these latter wires were too small to be detected at an appreciable distance and that tests using them may indicate chance scores.

Tape recordings of flights between 0.51 and 0.21 mm diameter wires showed that these two species used distinctly different kinds of orientation sounds in detecting the obstacles (Fig. 1). When approaching the barrier at a distance of about 2 m, *P. vivesi* emitted ultrasonic pulses with a duration of about 3 msec at a mean repetition rate of 10 to 20 per sec. Each of these was frequency modulated (FM), sweeping downward from about 45 kc/sec to 20 kc/sec and accompanied by a second harmonic. The slightly lower starting frequency (36 kc/sec) reported by Griffin (1958) may be due to the lower sensitivity to high frequencies of microphones available at that time. At a similar distance from the barrier *N. leporinus* produced pulses with a duration of about 8 to 10 msec at comparable repetition rates. These sounds, however, were composed of an initial portion at a nearly constant frequency of about 60 kc/sec followed by an FM sweep down to 30 kc/sec. Neither species made any pronounced change in the frequency structure of its pulses as it approached and negotiated

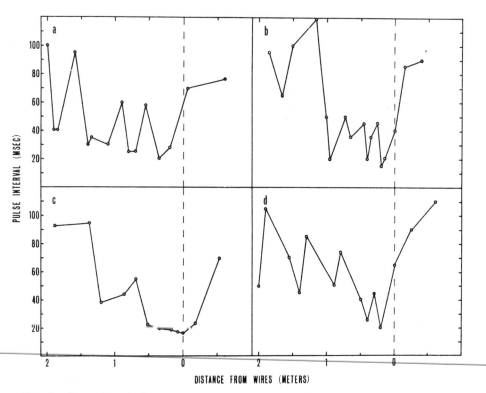

FIG. 2.—Examples of changes in orientation pulse intervals during flights by fishing bats through a barrier of fine wires spaced at 55 cm intervals across their flight path. Each dot represents one orientation sound: (a) *Noctilio leporinus* flying between 0.51 mm diameter wires; (b) *Pizonyx vivesi* flying between 0.51 mm diameter wires; (c) *N. leporinus* flying between 0.21 mm diameter wires; (d) *P. vivesi* flying between 0.21 mm diameter wires. On flights a, b, and d, the bat did not touch the wires. On the flight shown in c the wires were hit by the bat. Vertical dashed line indicates position of the wires.

the barrier. The pulse repetition rate was increased, however, to about 30 or 35 per sec. The use of a single pulse type by *N. leporinus* contrasts with its echolocation during normal cruising and feeding when constant frequency and entirely FM pulses are also employed (Suthers, 1965).

Fig. 2 gives examples of alterations in pulse intervals during one flight by each species through a barrier of 0.51 mm and of 0.21 mm diameter wires. The possible significance of the tendency to alternate long and short pulse intervals during the approach to the barrier is not known. It was not possible to reliably distinguish *hits* from *misses* on the basis of these graphs.

The minimum average distance of detection was estimated by calculating the point at which the bat began to shorten the pulse intervals. *Pizonyx vivesi* and *N. leporinus* must have detected the 0.51 mm wires at an average distance from the barrier of at least 110 and 150 cm, respectively, and the 0.21 mm

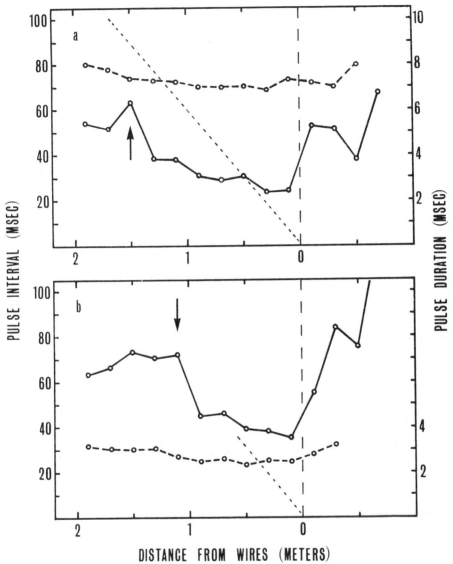

FIG. 3.—Mean pulse intervals (solid line) and mean pulse duration (broken line) of *Noctilio leporinus* (a) and *Pizonyx vivesi* (b) during approaches to the 0.51 mm diameter wires spaced across the flight path at 55 cm intervals. The bat is flying from left to right. Vertical dashed line indicates the position of the wires. Arrows indicate estimated minimum mean distance of detection as judged by progressive shortening of the pulse intervals. Dotted diagonal line shows the distance at which pulse-echo overlap will first occur for any given pulse duration. Echoes from the wires of pulses whose mean duration lies above this line will overlap with the emitted pulse by an average amount equal to their vertical distance above the line. Each point represents the mean interval or duration of pulses emitted in the adjacent ± 10 cm. Intervals for *N. leporinus* are averages of five flights; *P. vivcsi* intervals, of seven flights. All pulse durations are averages of three flights.

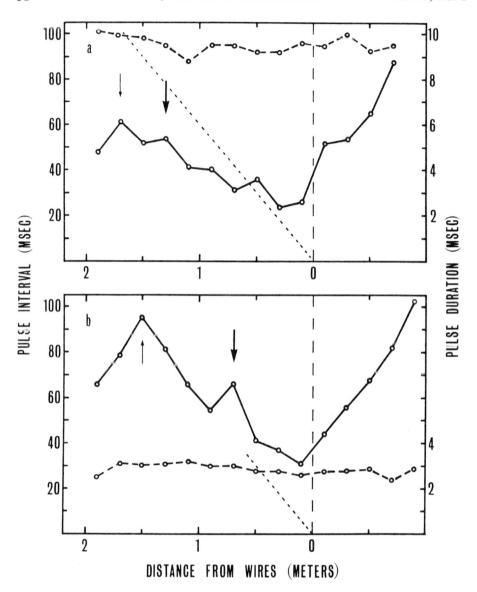

Fig. 4.—Mean pulse intervals (solid line) and mean pulse duration (broken line) of *Noctilio leporinus* (a) and *Pizonyx vivesi* (b) during approaches to the 0.21 mm diameter wires spaced across the flight path at 55 cm intervals. For explanation see legend of Fig. 3. Possible alternate interpretations of the point at which a progressive decrease in pulse intervals first appears are indicated by small arrows. The more conservative estimates denoted by the large arrows have been used in the text. Pulse durations and intervals of *N. leporinus* are averages of 10 flights; those of *P. vivesi* are averages of seven flights.

wires at an average of at least 70 and 130 cm, respectively (Figs. 3 and 4). Pulse durations did not markedly shorten as the barrier was approached. Thus at close ranges the echoes returning from the wires must have overlapped extensively with the emitted pulse. In the case of *N. leporinus* this overlap may have begun on the average when the bat was still 130 and 170 cm from the 0.51 and 0.21 mm wires, respectively (Figs. 3 and 4).

The data do not exclude the possibility that the start of pulse-echo overlap and the start of a progressive reduction in the pulse intervals by *N. leporinus* may occur simultaneously or be closely synchronized. The pulses of *P. vivesi* must have overlapped with their echoes during the last 40 and 50 cm of the approach to the 0.51 and 0.21 mm wires, respectively (Figs. 3 and 4). It seems clear that *P. vivesi* began to decrease its pulse intervals well before the first pulse-echo overlap occurred.

Noctilio leporinus regularly emitted longer pulses when approaching the 0.21 mm wires than when approaching the 0.51 mm wires. The significance of this difference is not known, although it is possible that the earlier initiation of pulse-echo overlap, or the increased duration of overlap at a given distance, when longer pulses are used, in some way facilitated detection of the finer wires. If this is true, however, why is overlap minimized with such apparent care during the detection of small cubes of fish muscle tissue projecting above the water surface (see below)? Since the difference in pulse duration as a function of wire diameter was already present when the bat was two meters from the barrier, either *leporinus* must have determined something about the wire diameter at a distance of more than two meters, or it must have remembered what kind of wires it had to detect and adopted a suitable pulse duration prior to their detection.

Details of the echolocation of *P. vivesi* during feeding are not known. Pulse-echo overlap during wire avoidance by *N. leporinus*, however, contrasts strongly with its apparent careful avoidance during catches of stationary 1 cm² cubes of fish muscle tissue projecting above the surface of the water. Pulse lengths under these conditions were progressively shortened as if to avoid pulse-echo overlap until the bat was 30 cm or less from the food (Suthers, 1965). Thus in the case of *N. leporinus*, at least, information concerning the position and nature of small wire obstacles is probably received in the presence of overlap, whereas most of this information regarding potential food must be obtained without such overlap. It has yet to be determined whether or not pulse-echo overlap is actually utilized by the bat.

It has been suggested (Pye, 1960; Kay, 1961) that possible nonlinearities in the ear may allow bats to utilize beat notes arising from pulse-echo overlap as a means of determining distance. Three species of chilonycterine bats have subsequently been found to maintain an overlap during the pursuit and catching of *Drosophila* (Novick, 1963, 1965; Novick and Vaisnys, 1964). *Myotis lucifugus* (Cahlander *et al.*, 1964) and *N. leporinus*, on the other hand, appear to minimize overlap when catching tossed mealworms or fish, respectively.

Should pulse-echo overlap be utilized by N. *leporinus* in determining its distance from the wires, then some basically different method, such as the temporal delay of the returning echo (Hartridge, 1945), must be employed in determining the range of potential food.

Since the constant frequency portion of the Doppler-shifted echo would at first overlap with the FM portion of the call, and later with part of both the constant frequency and FM portions of the call, any resulting beat note would have a complexly varying frequency structure from which it would be difficult for the bat to determine its distance from the barrier.

One would like to know if there is a significant difference in the range of echolocation for these fishing bats. *Noctilio leporinus* emits very loud pulses with a peak-to-peak sound pressure of up to 60 dynes/cm^2 at a distance of 50 cm from the mouth (Griffin and Novick, 1955). The intensity of sounds emitted by P. *vivesi* has not been measured, although the shorter range at which they can be detected on an ultrasonic receiver suggests they are less intense than those of *Noctilio*. M. *lucifugus*, a vespertilionid closely related to *Pizonyx*, can detect 0.46 mm diameter wires at 120 cm and 0.18 mm wires at 90 cm (Grinnell and Griffin, 1958), thus comparing favorably with fishing bats in this respect. Peak-to-peak sound pressures of this species have been measured at 12 dynes/cm^2 at 50 cm (Griffin, 1950). Sound intensity of the emitted pulse, however, is but one of a number of physical and physiological factors which must play important roles in determining the range of such a system for acoustic orientation.

ACKNOWLEDGMENTS

I wish to thank Prof. Donald R. Griffin, Drs. H. Markl, N. Suga, and D. Dunning for helpful criticism and assistance. Drs. R. E. Carpenter, G. W. Cox, and A. Starrett gave valuable assistance in obtaining live P. *vivesi*. Appreciation is also expressed to the San Diego Society of Natural History for use of the Vermillion Sea Station at Bahia de los Angeles, Baja California, and to the New York Zoological Society for the use of the William Beebe Memorial Tropical Research Station in Trinidad. The cooperation of the Mexican, Trinidadian, and United States governments in the transit of bats is gratefully acknowledged. This work was supported by grants from N.I.H., The Society of the Sigma Xi Research Fund, and the Milton Fund of Harvard University.

LITERATURE CITED

GRIFFIN, D. R. 1950. Measurements of the ultrasonic cries of bats. J. Acoust. Soc. Amer., 22: 247–255.
———. 1958. Listening in the dark. Yale Univ. Press, New Haven, 413 pp.
GRIFFIN, D. R., AND A. NOVICK. 1955. Acoustic orientation of neotropical bats. J. Exp. Zool., 130: 251–300.
GRINNELL, A. D., AND D. R. GRIFFIN. 1958. The sensitivity of echolocation in bats. Biol. Bull., 114: 10–22.
HARTRIDGE, H. 1945. Acoustic control in the flight of bats. Nature, 156: 490–494.
KAY, L. 1961. Perception of distance in animal echolocation. Nature, 190: 361–362.
NOVICK, A. 1963. Pulse duration in the echolocation of insects by the bat, *Pteronotus*. Ergebnisse Biol., 26: 21–26.

————. 1965. Echolocation of flying insects by the bat, *Chilonycteris psilotis*. Biol. Bull., 128: 297–314.

NOVICK, A., AND J. R. VAISNYS. 1964. Echolocation of flying insects by the bat, *Chilonycteris parnelli*. Biol. Bull., 127: 478–488.

PYE, J. D. 1960. A theory of echolocation by bats. J. Laryng. Otol., 74: 718–729.

REEDER, W. G., AND K. S. NORRIS. 1954. Distribution, habits, and type locality of the fish-eating bat, *Pizonyx vivesi*. J. Mamm., 35: 81–87.

SUTHERS, R. A. 1965. Acoustic orientation by fishing bats. J. Exp. Zool., 158: 319–348.

The Biological Laboratories, Harvard University, Cambridge, Massachusetts (present address: Department of Anatomy and Physiology, Indiana University, Bloomington, Indiana 47401). Accepted 11 November 1966.

SOCIAL INFLUENCES ON SUSCEPTIBILITY TO
PREDATION IN COTTON RATS

A recent study (Summerlin and Wolfe, 1971) demonstrated that socially dominant cotton rats (*Sigmodon hispidus*) are more exploratory and less neophobic (Chitty and Shorten, 1946; Chitty and Kemp on, 1949; Barnett, 1958) than subordinate ones. This suggested the possibility that dominant animals might be more exposed to being detected by predators under certain conditions. If animals of known rank are selected by predators, this should provide information on whether predation selects for or against aggressiveness in this species. Christian (1970) synthesizes the ideas of previous authors and points out several biological advantages that dominant animals have over lower ranking ones, such as occupying the more favorable habitats and being more successful breeders. If subordinate animals are more susceptible to predation, another selective mechanism favoring social dominance would be demonstrated. If, however, dominant individuals are taken more frequently, a possible mechanism might be demonstrated which selects against unilateral increase in aggressiveness in natural populations.

Cotton rats were trapped live in Oktibbeha, Lowndes, Leflore, and Bolivar counties, Mississippi. Animals were housed singly in the laboratory and exposed to the natural light cycle by way of windows in a room maintained at $21 \pm 3°C$. Individuals were marked for visual recognition by hair clipping.

Dominance relationships in groups of three males were determined in a glass front cage 1.2 by 0.6 by 0.6 meters (m). Three nest boxes were provided. Food (Purina laboratory chow) and water were provided *ad libitum*. Observations were made in a darkened room with a 20-w red light bulb illuminating the cage. Dominance rank in each group was determined by observing the outcome of agonistic encounters for a minimum of three hours. In decreasing order of intensity, fights, chases, threats, and avoidance were recorded as categories of agonistic behavior. Frequency of encounters and reversals of dominance generally decreased with time. Observations were continued on each group until a stable ranking could be determined.

Predation studies took place in a 3.7 by 7.3 m room illuminated by windows on three sides. The cement floor was covered with straw. Three nest boxes were provided on one end of the room. Observations were made from outside the building. Groups of three rats were placed in the room 12 hours (hr) before releasing the predator. Food and water were provided during the pretest and testing periods.

TABLE 1.—*Order of capture of cotton rats as related to social rank, by two types of predators*

Rank	Experiment 1 Group								Experiment 2 Group							
	A	B	C	D	E	F	G	H	A	B	C	D	E	F	G	H
							Cat									
I	1	1	1	1	2	1	2	1	1	1	1	1	2	1	1	2
II	2	2	2	3	1	2	1	3	3	2	2	2	1	2	3	3
III	3	3	3	2	3	3	3	2	2	3	3	3	3	3	2	1
							Hawk									
I	–	–	–	–	–	–	–	–	–	–	–	–	–	–	1	–
II	2	2	1	2	2	2	1	2	1	2	2	2	2	1	–	2
III	1	1	2	1	1	1	2	1	2	1	1	1	1	2	2	1

In the first predation test a female feral house cat was used as the mammalian predator. It was kept in a cage adjacent to the observation room and released via a sliding door manipulated from the outside. The cat could not observe the rats before being let into the observation room. It always stalked and killed all rats in the group before eating any of its prey.

A female red-tailed hawk (*Buteo jamaicensis*) was the avian predator. It remained loose in the observation room and stayed on a perch 1.2 m off the floor at the end of the room opposite the rat nest boxes. Rats were placed in the room at nightfall. Predation of the first rat occurred soon after dawn; a second animal was usually taken around noon.

The same procedure was repeated six months later in a larger room (9.2 by 9.2 m), using a male cat and a different female hawk. The first series is designated experiment 1, the second experiment 2 (Table 1).

Activity of individual cotton rats was measured for 20 minutes in a 0.9 by 0.9 m open field box. Interruptions of infrared beams passing diagonally across the box were monitored by two photo-electric cells and recorded on an Esterline-Angus multiple event recorder.

The influence of a predator on activity was tested by exposing the rats for 4 minutes to a crane-hawk (*Geranospiza nigra*) on a perch above the open box and measuring activity immediately after its removal.

As a control, activity was measured with a stuffed paper bag displayed in the same manner as the hawk. This was done prior to the activity test with the predator. Four groups of rats (12 individuals) were also tested, after ranking, in the empty activity box. There was an interval of 3 to 4 hr between all activity tests.

The mammalian predator selected dominant animals more frequently than subordinates. Conversely, the avian predator selected the subordinate animals more frequently (Table 1). An analysis of variance of ranked data variables indicated (1) no difference between prey

TABLE 2.—*Effect of the presence of a predator on activity of cotton rats. Activity of each animal is expressed as number of infrared beam interruptions during a 20 min test.*

Group	Dominant animals			Subordinate animals		
	Control	Predator	% Reduction	Control	Predator	% Reduction
A	236	152	36	81	8	90
B	182	115	37	35	13	63
C	183	115	37	113	2	98
D	103	64	38	54	8	85
E	245	176	28	36	0	100
F	105	77	27	22	0	100
G	125	75	40	30	4	87

selection of cats in experiments 1 and 2 ($F = 0.467$; $P < 0.01$), (2) no difference between prey selection of hawks in experiments 1 and 2 ($F = 0.438$; $P < 0.01$), and (3) significant difference in the rank of the prey selected by the two predators ($F = 38.07$; $P < 0.01$).

Our activity data confirmed the work of Summerlin and Wolfe (1971) which indicated a higher level of spontaneous activity for dominant animals. Activity in both dominant and subordinate animals was reduced after exposure to the avian predator (Table 2). A comparison of activity in the presence of the predator to control activity (Table 2) indicates that the movements of subordinate animals were inhibited more than were movements of dominant animals ($P < 0.05$; Wald-Wolfowitz runs test, Siegel, 1956).

Activity of dominant cotton rats was previously found to be inhibited less than that of subordinate rats by the presence of a strange object placed in the activity box (Summerlin and Wolfe, 1971). The object used as a control in this study did not elicit a decrease in activity of either the dominant or subordinate rats at the 0.05 probability level. The difference is probably due to difference in method. We placed the object on the rim of the exploratory box 1.0 m above the floor, whereas in the previous study it was placed on the floor.

Differences in hunting and capture techniques seem to be the most plausible explanation for the differential prey selection of the two predators. The cat typically entered the enclosure, assumed a crouched position, and chased the first rat that moved. As the predation results and activity scores indicate, this was almost invariably a high-ranking rat.

The hawk was much slower to strike, usually remaining immobile for a relatively long period after the rats became active in the morning before making its first kill. Similar observations were made by Sparrowe (1972) who reported that wild-trapped sparrow hawks (*Falco sparverius*) took some time in looking over and apparently evaluating each prey-capture opportunity before attacking in an experimental situation.

Several mechanisms have been suggested (Christian, 1970; Hutchinson and MacArthur, 1959; Ripley, 1959, 1961; Wilson, 1971) which could partially account for the lack of continually increasing aggressive tendency in natural populations due to the many apparent selective advantages of being dominant. Our results indicate the possibility of an additional factor. Differences in predator selection might prevent a unidirectional increase in aggressive tendency. This may especially hold true for genera such as *Sigmodon* in which dominant animals are more exploratory and thus more exposed to potential predators.

While increased predation at the dominant end of the social spectrum might contribute to imposing upper limits on the aggressiveness of a population, selection of subordinate animals by the hawk seems to fit the more classical predation hypothesis of Errington (1946, 1967) and others, as these animals would generally be weaker and less fit. In our experiments animals from the middle rank were taken first only 22 per cent of the time, whereas the dominant was taken first 40 per cent and the lowest-ranking individual 38 per cent overall. Predation involving predators using different hunting techniques and selecting from opposite ends of the dominance hierarchy, along with other factors, such as the survival value of exploratory behavior (Barnett, 1963; Metzgar, 1967), might produce a stabilizing selection which maintains an optimum genotype for aggressive tendency in certain prey species. It should be pointed out that the relative fitness of aggressive genotypes may fluctuate with population characteristics. Chitty (1967, 1970) suggests that more aggressive animals are less highly selected to survive local hazards (presumably including predators) and are at a disadvantage in the absence of crowding, but that these individuals are more fit at times of high density.

Also, the data illustrate that results of studies using different predators may be significantly different. Generalization from studies using only one predator species should be formulated with caution, as animals with different hunting techniques may select from different social strata in prey populations. This could have significant influence on studies designed to determine the role of predation in population regulation or other ecological or evolutionary phenomena.

We thank Michael P. Farrell, department of Agricultural and Experimental Statistics, MSU, for aid with the statistical analysis. This work was supported by a research grant from the National Science Foundation (NSF GB 8603).

LITERATURE CITED

BARNETT, S. A. 1958. Exploratory behaviour. British J. Psychol., 49:195–201.

———. 1963. The rat. A study in behaviour. Aldine Publishing Co., Chicago, 288 pp.

CHITTY, D. 1967. The natural selection of self-regulatory behavior in animal populations. Proc. Ecol. Soc. Aust., 2:51–78.

———. 1970. Variation and population density. Symp. Zool. Soc. London, 26:327–333.

CHITTY, D., AND KEMPSON, D. A. 1949. Prebaiting small mammals and a new design of live trap. Ecology, 30:536–542.

CHITTY, D., AND SHORTEN, D. 1946. Techniques for the study of the Norway rat (*Rattus norvegicus*). J. Mamm., 27:63–78.

CHRISTIAN, J. J. 1970. Social subordination, population density, and mammalian evolution. Science, 168:84–90.

ERRINGTON, P. L. 1946. Predation and vertebrate populations. Quart. Rev. Biol., 21:144–177, 211–245.

———. 1967. Of predation and life. Iowa State University Press, Ames, Iowa, xii + 277 pp.

HUTCHINSON, G. E., AND MACARTHUR, R. H. 1959. On the theoretical significance of aggressive neglect in interspecific competition. Amer. Nat., 93:133–134.

METZGAR, L. H. 1967. An experimental comparison of screech owl predation on resident and transient white-footed mice (*Peromyscus leucopus*). J. Mamm., 48:387–391.

RIPLEY, S. D. 1959. Competition between sunbird and honeyeater species in the Mollucan Islands. Amer. Nat., 93:127–132.

———. 1961. Aggressive neglect as a factor in interspecific competition in birds. Auk, 79:366–371.

SIEGEL, S. 1956. Nonparametric statistics. McGraw-Hill, New York, 312 pp.

SPARROWE, R. D. 1972. Prey-catching behavior in the sparrow hawk. J. Wildlife Mgmt., 36:297–308.

SUMMERLIN, C. T., AND WOLFE, J. L. 1971. Social influences on exploratory behavior in the cotton rat, *Sigmodon hispidus*. Commun. Behav. Biol., 6:105–109.

WILSON, E. O. 1971. Competitive and aggressive behavior. Pp. 183–217, *in* Man and beast (J. F. Eisenberg ed.), Smithsonian Institution Press, Washington, D.C., 401 pp.

MICHAEL W. ROBERTS AND JAMES L. WOLFE, *Department of Zoology, Mississippi State University, Mississippi State, Mississippi 39762. Submitted 23 July 1973. Accepted 1 May 1974.*

ETHOLOGICAL ISOLATION IN THE CENOSPECIES *PEROMYSCUS LEUCOPUS*

HOWARD MCCARLEY

Department of Biology, Austin College, Sherman, Texas

Accepted December 30, 1964

Peromyscus leucopus and *P. gossypinus*, constituting the cenospecies *Peromyscus leucopus*, have diverged genetically so that they have different morphological and adaptive norms. Genetic isolation, however, is apparently not complete because interspecific hybridization may occur (Dice, 1937; McCarley, 1954a). The present paper is a report of an ethological mechanism that helps maintain the genetic distinctness of the two species.

Previous studies by Dice (1940), Calhoun (1941), and McCarley (1954b, 1963) showed that *leucopus* and *gossypinus* were generally ecologically separated in areas of sympatric distribution: *leucopus* in upland woods and *gossypinus* in lowland woods. Overlapping frequently occurs, however, during the winter and spring reproductive seasons (McCarley, 1963). Consequently, ecological separation alone would not be adequate to account for the few recorded examples of natural interspecific hybrids in this cenospecies (Howell, 1921; McCarley, 1954a).

Work done by McCarley (1953) and Bradshaw (1957) using the procedures of Blair and Howard (1944) suggested that continued species separation of *leucopus* and *gossypinus* may, in part, depend on ethological, or species discrimination mechanisms. Experiments were begun in 1959 using techniques modified from the procedures of Blair and Howard (1944). These tests utilized three individuals, one male and two females or one female and two males. If males were to be tested, a male was placed in one of the two middle compartments of a four-compartmented cage and was free to move between these two compartments. A *leucopus* female was confined to one end compartment and a *gossypinus* female to the other end compartment. A reciprocal

arrangement of mice was used when females were tested. Each combination of three mice was observed daily, usually early in the morning, for not less than 5 nor more than 11 days. Observations were discontinued randomly. If the mouse being tested was observed nesting next to the mouse of its own species, it was recorded as a positive observation, otherwise as a negative observation. Only mice in breeding condition were used. Sympatric mice were from Leon County, Texas. Allopatric *leucopus* were from Bryan and Tillman counties, Oklahoma; allopatric *gossypinus* were from Nacogdoches County, Texas.

The results of these association experiments are summarized in table 1. Sympatric *leucopus* females and *gossypinus* males and females demonstrated a significant positive association with members of their own species of the opposite sex. Sympatric *leucopus* males associated with females of their own species more frequently than with *gossypinus* females but the deviation from the expected was insufficient to produce a significant χ^2 value. Table 1 also presents the results of tests utilizing allopatric stocks of *leucopus* and *gossypinus*. Allopatric mice, in this instance, did not associate with members of their own species significantly more often than with members of the other species. (In the case of allopatric *leucopus* females, five of the six tested showed a preference for individuals of the opposite species.) This suggests that existing isolating mechanisms are being reinforced (Koopman, 1950) in sympatric areas.

McCarley (1963) pointed out that in areas where *leucopus* and *gossypinus* are sympatric, the general restriction of *leucopus* to upland habitats (as opposed to the situation in allopatric

TABLE 1. *Results of discrimination tests using three mice in a four-compartmented cage*

	No. of tests	No. of individuals tested	Positive observations	Negative observations	χ^2 values
Sympatric *leucopus* males	36	19	113	78	3.010
Sympatric *leucopus* females	34	16	330	65	88.000
Sympatric *gossypinus* males	54	18	233	126	15.605
Sympatric *gossypinus* females	20	12	126	46	18.604
Allopatric *leucopus* males	14	7	69	36	4.830
Allopatric *leucopus* females	9	6	25	69	10.297
Allopatric *gossypinus* males	20	10	81	52	2.925
Allopatric *gossypinus* females	12	5	54	64	0.423

331

414

areas where *leucopus* occupies both uplands and lowlands) was the result of the presence of *gossypinus* in lowlands. The presence of an ethological mechanism in the form of species discrimination would support this hypothesis.

This study was supported by Grants No. G-8919 and G-19387 from the National Science Foundation. In addition to Austin College, facilities at the University of Oklahoma Biological Station and Southeastern Oklahoma State College were provided while I was in residence at these institutions.

Literature Cited

BLAIR, W. F., AND W. E. HOWARD. 1944. Experimental evidence of sexual isolation between three forms of mice of the cenospecies *Peromyscus maniculatus*. Contrib. Lab. Vert. Biol., Univ. Michigan, No. 26: 1–19.

BRADSHAW, W. N. 1957. Reproductive isolation in the *Peromyscus leucopus* group of mice. M.A. Thesis, Univ. of Texas, Austin.

CALHOUN, J. B. 1941. Distribution and food habits of mammals in the vicinity of the Reelfoot Lake Biological Station. Proc. Tenn. Acad. Sci., **6**: 207–225.

DICE, LEE R. 1937. Fertility relations in the *Peromyscus leucopus* group of mice. Contrib. Lab. Vert. Gen., Univ. Michigan, No. 4: 1–3.

——. 1940. Relations between the wood mouse and the cotton-mouse in eastern Virginia. J. Mammal., **21**: 14–23.

HOWELL, A. H. 1921. A biological survey of Alabama. U. S. Dept. of Agric. Bur. Biol. Surv., North Amer. Fauna, No. 45.

KOOPMAN, K. F. 1950. Natural selection for reproductive isolation between *Drosophila pseudoobscura* and *Drosophila persimilis*. EVOLUTION, **4**: 135–148.

MCCARLEY, HOWARD. 1953. Biological relationships of the *Peromyscus leucopus* species group of mice. Ph.D. Thesis, Univ. of Texas, Austin.

——. 1954a. Natural hybridization in the *Peromyscus leucopus* species group of mice. EVOLUTION, **8**: 314–323.

——. 1954b. The ecological distribution of the *Peromyscus leucopus* species group in eastern Texas. Ecology, **35**: 375–379.

——. 1963. The distributional relationships of sympatric populations of *Peromyscus leucopus* and *P. gossypinus*. Ecology, **44**: 784–788.

INFLUENCE OF FOOD UPON THE DIURNAL ACTIVITY OF SMALL RODENTS

ВЛИЯНИЕ КОРМА НА ЕЖЕДНЕВНУЮ АКТИВНОСТЬ МЕЛКИХ ГРЫЗУНОВ

WLADYSLAW GRODZIŃSKI

Department of Animal Genetics and Organic Evolution, Jagiellonian University, Kraków, Poland

The majority of mammals have an endogenous rhythm of diurnal activity which can be strongly influenced however by exogenous factors. The great variability of daily activity of these animals in the field can be explained by the action of such external factors as weather, food and population interdependencies (Aschoff, 1957; Harker, 1958).

The quality of food and possibility of access to it for herbivorous rodents change considerably in the year cycle. Many workers stated the relation between rodent activity and food conditions (Hatfield 1940, Naumov 1948, Saint-Girons 1959) but few of them turned their attention to the kind of food. That is why these investigations try to determine the influence of food caloricity upon the diurnal activity of small rodents.

Material and Methods

For experiments, 3 species of rodents were chosen (*Apodemus agrarius* Pall., *Clethrionomys glareolus* Schr. and *Microtus agrestis* L.); they are rather omnivorous and active both in darkness and in light — though normally with a certain preponderance of night activity (Miller 1955, Naumov 1948). *Apodemus agrarius* Pall. and *Clethrionomys glareolus* Schr. were caught in the field, but *Microtus agrestis* L. had been bred in a laboratory for several years. 28 individuals were used for the experiments but with regard to considerable individual variability, mean figures based on 3—5 specimens (males) are listed.

Efforts were made to obtain information on dependence of diurnal activity upon food in all its particulars and for this aim several methods were used. In the first line food preference of the above mentioned rodents was investigated, then from food willingly consumed three diets for each species were combined: "A" — a high caloric diet, "C" — a low caloric one and the medium diet — "B". "A" diets were composed of foods of higher concentrated value (oleaginous seed, hazelnuts, grain, insects). "C" diets consisted of bulky food (vegetables, fruits and new-mown grass). The medium "B" diets were composed of both concentrated and bulky foods. The animals were kept for a period of 10 days on each of the diets in various sequences (A—B—C, B—A—C, C—A—B).

134

Diurnal activity was recorded on Palmer's and Zimmerman's slow rotary kymo-
graphs and on Jaquet's polygraph by means of two-directional electro-magnetic
recorders. They were connected with electro-magnetic contacts installed in the cages
in the passage between nest and run. Activity phases could be read from the records
in minutes, as "exits" from, and "entrances" into, the nest were registered in a different manner. This installation could register the activity of animals in 18 cages
simultaneously.

Metabolism of the investigated rodents was defined in terms of oxygen consumption
(Pearson 1947), measured in Kalabukhov's automatic apparatus (spirometers)
(Skvartzov 1957).

Analyses of the structure of the alimentary canal of the three species have bean
carried out recently, in order to determine the proportion of intestine length to the
resorbing areas (number and shape of villi).

Results

Experimental rodents respond in an extremely rapid and brisk manner to any
change in the kind of food. Mean diurnal food consumption, calculated for 1 animal of
average weight, was 7—10 times higher on low caloric diet than on high caloric ones
(Table I). When comparing the diurnal consumption shape of these various foods in
terms of their caloric values, the differences are considerably smaller. The effect could
be seen immediately on the body weight of the animals, which after 10 days of diet
"A" increased in weight by 5—20%, and as a rule dropped in weight on diet "C".

Tab. 1. Diurnal food consumption on three different diets.

Species of rodents	Diets	Food in g per day	Capacity in g			Kcal
			Proteins	Fats	Carbo-hydrates	
Apodemus agrarius Pall. (19 g.)	A	2·5	0·35	1·50	0·28	14·90
	B	10·1	0·43	0·09	3·21	15·32
	C	16·4	0·09	0·05	2·12	8·47
Clethrionomys glareolus Schr. (22 g.)	A	2·1	0·27	0·85	0·48	9·98
	B	8·8	0·27	0·17	2·36	11·85
	C	20·5	0·14	0·06	2·27	9·36
Microtus agrestis L. (26 g.)	A	6·0	0·69	0·89	1·87	18·28
	B	26·5	0·77	0·29	2·86	17·36
	C	41·9	0·84	0·21	2·10	13·41

135

Caloricity of the food had a marked and rapid influence on the total amount of diurnal activity and on its pattern. The sum of this activity on the normal "B" diet amounted to about 148 minutes on the average for *Apodemus agrarius* Pall., 185 minutes for *Clethrionomys glareolus* Schr., and about 282 minutes for *Microtus agrestis* L. These numbers are close to the results obtained by other workers, except that the sum of diurnal activity for *Microtus agrestis* L. is visibly lower. This might be the results of breeding in laboratory conditions for several generations. In the highly caloric diet "A" the sums of activity were slightly lower — 135', 156' and 208', respectively, while on the low caloric diet "C" they were much higher: 198', 265' and 388', resp.

The pattern of daily activity underwent notable changes. This phenomenon, however, was slightly different in each of the three investigated species.

Apodemus agrarius Pall. (Fig. 1a). fed the normal diet "B" had a typical twofold pattern of nocturnal activity, an important preponderance of activity (71,6%) at night. When fed the highly caloric diet "A", nearly the whole 24-hours activity occurred at night and the activity pattern changed into a single one with an definite maximum in the middle of the night. However, on the low caloric diet "C" *Apodemus agrarius* Pall. was mainly active at night (72%), but produced a manifold activity pattern, with several peaks and three maxima late in the evening, after midnight and at dawn.

Clethrionomys glareolus Schr. (Fig. 1b) fed the concentrated diet "A" showed a definite preponderance of nocturnal activity (76·3%), the activity starting in the evening and lasting nearly all night long. When on diets "B" and "C" it became active both in the light and in darkness so that the preponderance of nocturnal activity diminished considerably (to 59 and 62·3%). For animals fed the medium diet "B", a twofold, mixed pattern of activity, with one broad maximum in the afternoon and a second higher maximum at night was maintained, typical for *Clethrionomys glareolus* Schr. On the bulky diet "C", however, the pattern became polycyclic, both during the day and at night.

Microtus agrestis L. (Fig. 1c) fed the unphysiologically concentrated diet "A" produced a twofold nightly pattern of activity, characterized by an interval of several hours between the higher evening maximum and the lower one at daybreak. When on the medium diet "B", the share of activity in the daytime increased slightly (up to 34·7%) and showed several additional phases. On the whole, the twofold night pattern was maintained with the sole difference that the two peaks drew nearer to each other and became larger. On the low caloric diet "C" the activity increased both in light and darkness, producing a typical polycyclic pattern.

Patterns and sums of activity, so unlike at different diets, resulted from the changes in the number of activity phases during 24-hours and also in the length of duration of the phases themselves. In animals fed a low caloric food the amount of activity

136

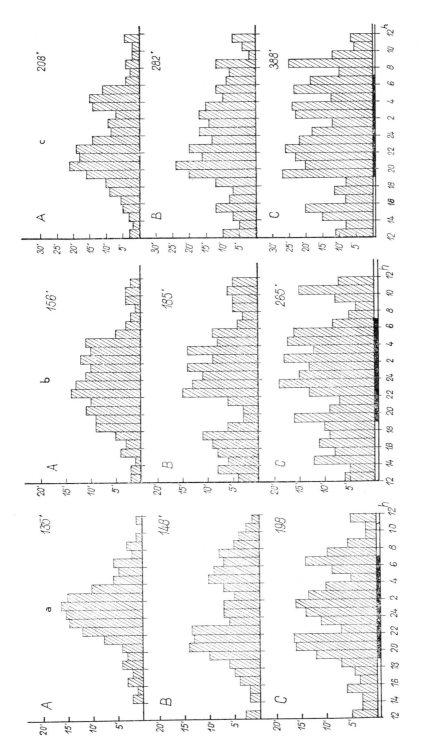

Fig. 1. Diurnal activity of (a) *Apodemus agrarius* Pall., (b) *Clethrionomys glareolus* Schr. and (c) *Microtus agrestis* L. on three different diets. A, B and C = diets of different caloric values.

137

419

phases augmented as a rule and their duration increased; the reverse was observed in animals on a highly concentrated diet. *Clethrionomys glareolus* Schr., for example, had 6—8 activity phases when on diet "A", and 10—13 mostly larger phases when on diet "C". After a change of diet, visible differences in activity appeared after 1—2 days, but the new pattern was consolidated only in 4—5 days. Therefore activity diagrams presented on figures 1a, b and c are based on mean values for the last 5 days of feeding on the diet.

Both the quality and quantity of the food eaten had a marked influence upon the metabolism of the animals. On the low caloric diets "C" the rate of their metabolism increased as a rule when compared to that on the mean diets "B", while on the high caloric diets "A" there was a certain descent in the rate of metabolism; e. g., *Microtus agrestis* L. after being fed the diet "B" showed at 20°C an oxygen consumption 0·097 ccm/g/min, on the diet "C" as much as 0·128 ccm/g/min, and on the diet "A" about 0·090 ccm/g/min on the average.

Discussion

Finally, two questions of a more general nature must be considered: (1) what is the factor limiting the plasticity of activity towards the action of different foods, and (2) can the results of laboratory experiment, presented here, be of any significance in the field conditions?

The structure of the alimentary canal determines the range of nutritional possibilities of a rodent. In different species of small rodents, considerable differences appear in the length and proportion of intestines and also in the structure of the stomach (Naumov 1948). The majority of mice feeding mainly on foods with high nutritive value, have a long small intestine, while voles, living mostly on bulky food, possess a well developed caecum and large intestine. According to our suppositions, this difference is largely the difference in the size of resorbing areas, that is in the amount and dimensions of the villi.

Rodents chosen for our experiments have a not very specialized alimentary canal. *Apodemus agrarius* Pall. has the shortest caecum in relation to the length of the small intestine (5 : 43 cm), the caecum of *Microtus agrestis* L. is the longest (12 : 30 cm), while in *Clethrionomys glareolus* Schr. both caecum and intestine are very long (11 : 53 cm). The activity pattern of *Clethrionomys glareolus* Schr. showed the greatest variability in response to different diets, the activity pattern of *Apodemus agrarius* Pall. being the most compact ant that of *Microtus agrestis* L. more polycyclic. Thus it appears that the alimentary canal limits both food preference and plasticity of diurnal activity of a species.

The application of results presented above for explaining changes in the activity of rodents in natural communities, though very tempting, might be difficult and even risky. In nature, food interdependencies of rodents are much more complicated;

138

not only quality of food, but the possibilities of access to it change within the year cycle, as well as the caloric demand of animals, their habit of storing reserves in nests, etc. Diurnal activity of the animals is subject to important oscillations in the year cycle (Kleitman 1949). This was also found in numerous rodents (Ostermann 1956). The mechanism of these oscillations is based on endogenic neuro-hormonal changes (Aschoff 1957) occuring in animals in the year cycle. These seasonal changes are a stabilized reaction to the year cycle in a temperate climate. Food as a factor with a direct influence on metabolism and diurnal activity may constitute here an important bridge between the physiological clock and the external conditions of environment.

LITERATURE

Aschoff, J., 1957: Aktivitätsmuster der Tagesperiodik. *Die Naturwiss.*, *44 : 361 — 367.*

Harker, E. J., 1958: Diurnal rhythms in the animal kingdom. *Biol. Rev.*, *33 : 1 — 52.*

Hatfield, D. M., 1940: Activity and food consumption in *Microtus* and *Peromyscus. J. Mammal., 21 : 29 — 36.*

Kleitman, N., 1949: Biological rhythms and cycles. *Physiol. Rev.*, *29 : 1 — 30.*

Miller, R. S., 1955: Activity rhythms in the wood mouse, *Apodemus sylvaticus* and the bank vole, *Clethrionomys glareolus. Proc. Zool. Soc. Lond., 125 : 505 — 519.*

Naumov, N. P., 1948: Oczerki sravnitelnoi ekologii myszevidnych gryzunov. *Izd. AN. SSSR. Moskva — Leningrad.*

Ostermann, K., 1956: Zur Aktivität heimischer Muriden und Gliriden. *Zool. Jb. (Physiol.), 66 : 355 — 388.*

Pearson, O. P., 1947: The rate of metabolism of some small mammals. *Ecology 28 : 127 — 145.*

Saint Girons, M. Ch., 1959: Les caractéristiques du rythme nycthéméral d'activité chez quelques petits mammifères. *Mammalia 33 : 245 — 276.*

Skvartzov, G. N., 1957: Usoveršcenstvovannaja metodika opredelenija intenzivnosti potreblenija kisloroda u gryzunov i drugich mielkich životnych. *Gryzuny i Borba s nimi, 5 : 424 — 432.*

GALAPAGOS SEA LIONS ARE PATERNAL

GEORGE W. BARLOW

Department of Zoology and Museum of Vertebrate Zoology
University of California, Berkeley, California 94720

Received July 21, 1973

The publication of a note (Barlow, 1972) describing the behavior of some territorial bulls of the Galapagos Islands sea lion (*Zalophus californianus*) toward large intruding sharks has drawn an unexpectedly wide response. This is probably because it is the first published report of a paternal role in a pinniped. One response was a brief article by Edward H. Miller (1974), questioning that conclusion. Miller raises five major points, to which I will reply in turn:

(1) *Distinctiveness of territories.* Miller seems to suggest that I described distinct territorial boundaries. To the contrary, I noted that the territorial boundaries of the bulls were indistinct; I relied, rather, on the spacing of the bulls. Miller extended this point to imply that, since the boundaries must be imprecise, it should be difficult for the sea lions to localize their territories. The inference is not altogether justified. While the territories lacked boundaries recognizable to the observer, and while the sand beach was deficient in distinctive features, the territories lay in a small enclosing cove whose shore had a low profile with outstanding landmarks behind it, facilitating general localization. The underlying reef must have provided some clues, too. Furthermore, the bulls were close to shore (25–45 m), although Miller takes this to mean they were far out.

The thrust of Miller's argument here, however, is that since the boundaries appeared to be indistinct, some overlap was expected. The implication is that it is not then noteworthy that the bulls ignored one another's territories when they converged on the shark. This misses the essential point that territoriality was recognized largely from the spacing of the bulls in a consistent topographical pattern. Furthermore, the mobbing-like response to the shark did not occur at the confluence of a number of territories. Rather, the territories were linearly arrayed, and the bulls farther away had to swim across the territories of the neighbors nearer to the shark.

Finally, when the bulls approached the shark, the distances between the bulls was much less than they otherwise tolerated.

(2) *Philopatry.*—Miller misconstrues my point about philopatry. I never said the bulls return to hold the same territories (though they might), for I have no such information. Rather, I deduced that philopatry should prove to be the case in the sense that most members of the rookery return to the same small cove.

For the evolution of paternal behavior, as described in my earlier paper, it would not be necessary that a given territorial bull return and protect the pups the following year (see below). It should be adequate if the bull, or related bulls, and also the cow simply came back to the same rookery. This follows since several bulls mob the shark, not just one, and since the females and the pups probably move about through the territories (Peterson and Bartholomew, 1967) but are nonetheless philopatric.

(3) *Bulls are indiscriminately territorial.*—Miller argues that it is generally true of pinnipeds that a territorial bull responds aggressively to any and all objects that roughly resemble in size, shape, and behavior, any intruding rival bull. That pinnipeds respond aggressively to heterospecific intruders is interesting but, taken alone, relatively uninformative. One must ask whether the intruders are competitors for breeding sites or food, or whether they are potential predators.

If the intruders are predators, as Miller suggests they could be at one point, then the larger question looms as to why a bull should respond aggressively to a predator? Aggression toward a predator involves considerable risk and energetic cost, for which there must be a counterbalancing gain. The bull maximizes its own chance of survival by avoiding or fleeing from the predator.

When it comes to the question of whether otariid bulls can distinguish between different types of intruders, Miller seems to argue both pro and con. He takes me to task for assuming that bulls can distinguish between rivals and sharks. Then he cites literature reporting their excellent visual acuity (see also Schusterman, 1973). He concludes that even if they can discriminate, they respond in the same way to all intruders of a similar size, shape, and manner. I found, however, that when I was SCUBA diving within the territory of the Galapagos bulls that they inspected but never bothered me; this was so even when pups and females were nearby. But when I was on land and near the females and young, the bulls were very aggressive. One interpretation might be that estrous females occur only on land. However, Peterson and Bartholomew (1967) reported that the California sea lion copulates in the water.

While I cannot explain the difference in responses to humans in the water as opposed to on land, two things are clear. First, bulls distinguish among intruders. Second, the responses differ with the setting of the encounter.

Pursuing the argument further, if sea lions are unable to distinguish between dangerous sharks and large sea lions, then why did all the small sea lions dash frantically out of the water when the shark turned into the rookery (Barlow, 1972)? I find it difficult to believe that a mammal as intelligent as a sea lion lacks the ability to recognize a large shark as the dangerous predator it is.

The crux of the differences in interpretation, however, lies in a part of the observations not made clear in my original article due to editorial deletion. It concerned the response of the territorial bulls to the return of other sea lions from the ocean. This desideratum led Miller to write that "Implicit assumptions made by Barlow (op. cit.) are: that bulls have the ability to distinguish sharks from potentially challenging males. . . ." I observed a number of cases in which one to three or four large sea lions of both sexes returned from the sea. Territorial bulls oriented briefly toward them as they swam in past them, and then on to the beach, but otherwise showed little response. Thus the territorial bulls also behaved in a dramatically different fashion toward different nonhuman intruders: sea lions in passage were scarcely noticed, but large sharks were turned away at the periphery and mobbed if they tried to penetrate the rookery. This was therefore an explicit observation, not an implicit assumption.

(4) *Behavior shared by many otariids.*—Miller notes that the elements of behavior I reported have been seen in several pinnipeds. Since these behaviors evolved in other contexts, it is said they cannot serve another function. Examples of widespread tendencies among pinnipeds that Miller noted are herding of other individuals, philopatry, repulsion of extraspecific as well as intraspecific territorial intruders, several territorial males joining to repulse intruders, and desisting

from the pursuit of an intruder when it has departed from the territory. Such observations are actually supportive of those I made. They show that otariids have already available to them the necessary behavioral repertoire to respond to sharks as reported. This is the familiar preadaptation argument.

Miller also seems to have overlooked the point that there are novel considerations inherent in the situation in the Galapagos Islands that have led to the evolution of the shark response. First, the sea lions there find themselves unable to keep their body temperature down during the frequently hot days on exposed beaches (Peterson and Bartholomew, 1967), such as where I made my observations. Consequently, they spend much of their time in the cool water just off the beach, as do even the young and the females. Second, the presence of sharks at these tropical latitudes presents a hazard: Eibl-Eibesfeldt and Hass (1959: 738) published a photograph of a Galapagos sea lion pup that had just had its back laid open by a shark and had crawled out onto the beach where it bled to death.

(5) *What is meant by a paternal role?*—By paternal role I mean only that the behavior of the male increases the likelihood of the survival of the pups. Miller confuses the issue by substituting the term paternal "care." I explicitly avoided that word because care implies a directly observable service-rendering relationship between one animal and another. Some examples might clarify the issue. Male marmoset monkeys transport and groom their own infants and hence render a tangible service to their offspring (Mitchell, 1969). Male baboons, in contrast, show little care of infants, but they do protect them from predators (Mitchell, 1969). We might argue as to whether the male baboon shows paternal *care*, or to what degree. Even more extreme is the male patas monkey; he provides no direct care of his offspring, but he is of considerable importance in warning the group when a predator approaches (Hall, 1965). The male patas has a paternal role because his behavior improves the survival of his young, even though he apparently provides no direct care.

Miller reasons that since in many pinnipeds the bulls sometimes kill pups, or otherwise ignore them, they cannot possibly be paternal. The larger issue has been missed! The behavior going on when the pups are killed must be more important than the loss of a few pups. Such deaths occur during aggressive interactions between bulls. This is a crucial moment in a bull's behavior. If a bull does not keep out rival males then it risks begetting no pups whatsoever instead of the approximately 80 *male* offspring over a lifetime, or the 16 per season that Bartholomew (1970) has estimated. These high stakes have generated intensive intrasexual competition, producing huge bulls that are morphologically and behaviorally adapted to overt combat. One expression is an extraordinarily high level of aggressiveness. The cost is a few pups trampled in rushes, and fewer yet killed through redirected biting. (If the pups in these species were not sired by the bulls that kill them, then the genetic consequences of killing them will be proportional to the degree of relationship.)

Miller seems to suggest that since bulls are clearly not paternal on the beach, and may even be inimical to the pups there, they could not be paternal in the water. This, and the preceding, implies an understanding by the animal of its own behavior. Lorenz has effectively dispelled such a view of animal behavior. A germane example is the pair of warblers who fed their nestling in the nest, but failed to respond to the same nestling out of, but next to, the nest (Lorenz, 1935). The parallel in the Galapagos sea lion is that bulls have evolved a specific response to intruding sharks, irrespective of the bulls' response to pups on land or elsewhere.

While it is not central to the argument, it is worth noting that in the Galapagos when the pups and bulls are in the water they are spatially separated most of the time. The bulls are more toward the periphery and the pups are mostly on the edge of the beach. Even when together in the water there is little danger of treading on the pups. So while territorial combat in more terrestrially disposed pinnipeds may result in the loss of some pups, such casualties are minimized in the Galapagos (I observed one intense territorial fight between two bulls and it was done entirely in the water). But I must not overstate the case: where and when conditions permit, such as the shade of cliffs or trees, or overcast skies, then young, females, and bulls may assemble on the beach.

There is still room for argument about the use of the term "paternal" in the strict sense. After all, the bull may not be the father of the pup in question but its uncle, cousin, etc. But since the bulls act collectively, to a degree, and since the pups cluster, this point becomes moot. Semantics aside, the important point is that the bull's behavior improves his inclusive fitness (Hamilton, 1971) by promoting the survival of related pups. Contrary to Bartholomew's suggestion, therefore, the role of the bull after insemination cannot be ignored.

Conclusions

The essential argument remains: if the bulls protect the pups, an algorithm for the evolution of such behavior must include a component for the degree of relationship between the bull that turns away the shark and the pups that benefit from this behavior. Since the pups sharing his genes move about through the territories, a bull is likely to be protecting his genetic investment by repelling sharks that intrude into any part of the rookery. And because births issue from inseminations of the previous year, this hypothesis requires a large element of philopatry. One could even argue that reciprocal altruism (Trivers, 1971) might have been involved in the origin of the response.

It remains a challenge to the student of pinniped biology to incorporate this interesting facet of the behavior of the Galapagos sea lion into the elegant model proposed by Bartholomew (1970). This will require closer study of pinnipeds in general, not just of the sea lions in the Galapagos Islands. Particular regard should be given to the degree of philopatry and hence kinship within the various rookeries and the risk of injury from sharks, as well as to the effectiveness and cost of the behavior of the territorial bulls. It would also be worthwhile to examine the strength of the bulls' responses to intruding sharks in relation to the presence, and age, of pups.

I am grateful to Edward H. Miller for giving me the opportunity to reply to his article. Thanks are due to the members of the convivial Berkeley Animal Behavior luncheon group for their comments on my reply.

Literature Cited

Barlow, G. W. 1972. A paternal role for bulls of the Galapagos Islands sea lion. Evolution 26:307–308.

Bartholomew, G. A. 1970. A model for the evolution of pinniped polygyny. Evolution 24: 546–559.

Eibl-Eibesfeldt, I., and H. Hass. 1959. Erfahrungen mit Haien. Z. Tierpsychol. 16:733–746.

Hall, K. R. L. 1965. Behaviour and ecology of the wild patas monkey, Erythrocebus patas, in Uganda. J. Zool. 148:15–87.

Hamilton, W. D. 1971. Selection of selfish and altruistic behavior in some extreme models, p. 59–91. In J. F. Eisenberg and W. S. Dillon (eds.), Man and beast: Comparative social behavior. Smithsonian Institution Press, Washington, D. C.

Lorenz, K. 1935. Der Kumpan in der Umwelt des Vogels. J. Ornithol. 83:137–213, 289–413.

Miller, E. H. 1974. A paternal role in Galapagos sea lions. Evolution 28:473–476.

Mitchell, G. D. 1969. Paternalistic behavior in primates. Psychol. Bull. 71:399–417.

Peterson, R. S., and G. A. Bartholomew. 1967. The natural history and behavior of the California sea lion. Spec. Pub. No. 1, Amer. Soc. Mammal., 79 p.

Schusterman, R. J. 1973. A note comparing the visual acuity of dolphins with that of sea lions. Cetology (15):1–2.

Trivers, R. L. 1971. The evolution of reciprocal altruism. Quart. Rev. Biol. 46:35–57.

REPRODUCTIVE BEHAVIOR OF THE GRAY WHALE
ESCHRICHTIUS ROBUSTUS, IN BAJA CALIFORNIA

WILLIAM F. SAMARAS[1]

ABSTRACT: Patterns of movement and repetitive behavior of gray whales in a specific area of Scammon's Lagoon suggest precopulative ritual. Individuals move into a specific area of the lagoon in the early morning and conspicuously engage in spyhopping, which is associated with the gray whale's courtship behavior. Whales expose their heads above the surface and remain motionless in this position for at least ten seconds before submerging. A lull in activity in the late morning and early afternoon is followed by intensified pairing and copulation.

ay whales, *Eschrichtius robustus* (Lilljeborg, 61), spend the summer months feeding pri- rily on amphipods and thereby acquire a ck blanket of blubber in the North Bering and ukchi Seas in preparation for their four- nth long, 7000 mile journey south to their ter quarters among the lagoons and shallow, tected bays of central and southern Baja ifornia (Rice and Wolman, 1971). It is in se lagoons that most of the calving takes ce. The gestation period for *E. robustus* is roximately 13 months with birth occurring m the middle of December into early February ice and Wolman, 1971). The round trip of ht months, beginning in October and ending May, gives these whales the distance record migrating mammals (Fig.1).

rom 29 January to 2 February 1973, I erved gray whales in the Laguna Ojo de bre (also known as Scammon's Lagoon), ich is one of the primary breeding and ing areas on the west coast of Baja California, ico, (Latitude 27°58′ N, Longitude 114°16′ . Entrance to Scammon's Lagoon is from the th through a narrow, shallow channel averag- 4 fathoms in depth (Fig. 2a). The main nnel is flanked by shallow bars whose presence idenced by onshore swells breaking over them. n after entering the lagoon at 0630 on 29 ary spouting whales were sighted in every

direction. Within a period of 15 minutes it was estimated that there were in excess of 50 animals in an area of approximately one square mile. Because of the large assemblage of whales just inside the entrance bar, I designated it the staging area (Fig. 2b).

STAGING AREA BEHAVIOR

Most of the whales in the staging area appeared to be milling around without any general direction of movement, and most of those observed through binoculars were large, ranging in size from 30 to 40 feet in length. Prior observations by Scammon (1874) indicated that the prepon- derance of whales in this part of the lagoon are, in the main, unattached males, although it is not known how he determined this. It appears reasonable that this assemblage of whales also includes sexually mature or oestrative females. Most of the incoming, pregnant females enter the lagoon and continue into the calving area often referred to as the nursery (Fig. 2c).

[1] Carson High School, 22328 S. Main St., Carson, California 90745, and Museum Associate in Mam- malogy, Natural History Museum of Los Angeles County, 900 Exposition Blvd., Los Angeles, Cali- fornia 90007.

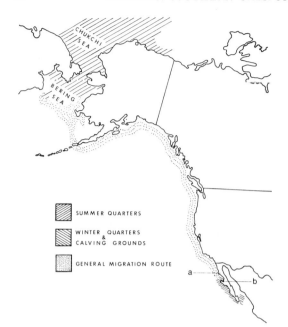

Figure 1. Seasonal distribution and migration route of the gray whale, *Eschrichtius robustus*: a, Baja California, Mexico; b, Scammon's Lagoon.

SUMMER QUARTERS

WINTER QUARTERS & CALVING GROUNDS

GENERAL MIGRATION ROUTE

Respiratory–diving behavior.—Within the staging area, the whales were swimming slowly, exhibiting their typical respiratory–diving behavior, which is initiated by the crown of the head and paired blow-holes breaking the surface of the sea. Exhalation is signified by a visible, heart-shaped, vaporous spout and a very audible blow. The whale then begins to round-out, exposing its dorsum to at least the primary dorsal ridge (sometimes called the dorsal bump) then submerges a few feet beneath the surface for one to two minutes and again resurfaces to exhale and inhale. This procedure is repeated, on the average, three to four times. After the last blow in the respiratory series or blast, as it was called by whalers (Slijper, 1962), the round-out is much more pronounced and deliberate. The entire dorsum and caudal peduncle arches above the surface exposing its entire array of dorsal ridges. The sequence ends with its flukes breaking above the water in an arching motion and re-entering at about a 45 degree angle. This action initiates the deep-dive phase.

Normally, the whale will remain submerged for an average of from five to eight minutes, but in the lagoon the mean is from three to five minutes. Upon resurfacing after the deep-dive,

the animal repeats the entire respiratory–dee dive sequence. Gray whales will seldom devia from this respiratory behavioral pattern unle frightened or agitated in some manner.

SOCIAL INTERACTION AREA

From the staging area the whales gradually swa south into the lagoon preferring the shallow waters along the west bank. Very few anima were observed moving down the deeper mi channel or along the east bank. Depths along t west bank vary from a few feet to less th five fathoms as determined by fathomet readings. Captain Scammon's observations duri the late nineteenth century confirm the fact th the gray whale has a penchant for very shallc water and generally will forsake the deeper wate of the lagoon.

The lagoon is not much more than a mile wi for a distance of some 10 miles inland from entrance. It then broadens off Brushy Isla (Isla Broza) (Fig. 2f). Within this broac portion of the lagoon, off the northern tip Brushy Island, much of the behavior which v interpreted as social orientation and interacti took place (Fig. 2e). Gilmore (1960) has si gested that gray whales are segregated by a sex, and reproductive condition in various ar of Scammon's Lagoon. His "outer" and "in mediate" areas essentially correspond to staging and social interaction areas, respectiv

Singly and in pairs and daily from dawn u about 1100, a line of whales moved into section of the lagoon. Sauer (1963) recor similar behavior for gray whales within time frame in Boxer Bay at St. Lawrence Isla Alaska. The whales moved south, gener preferred the western side of the lagoon, cros gradually towards the east bank, turned, mo north for a mile or two, then crossed again the west side. This general counterclockv pattern of movement was also observed by S in Boxer Bay and Fay (1963) at Kangee, a the southern coast of St. Lawrence Island. overall pattern resembled a convection cur (Fig. 2d). Within the area described, the wh deviated from their respiratory-diving cycle; the sic pattern being altered almost completely. T stayed near the surface, respired at irregular i vals, then executed deep-diving procedure caudal peduncles arched and flukes expo

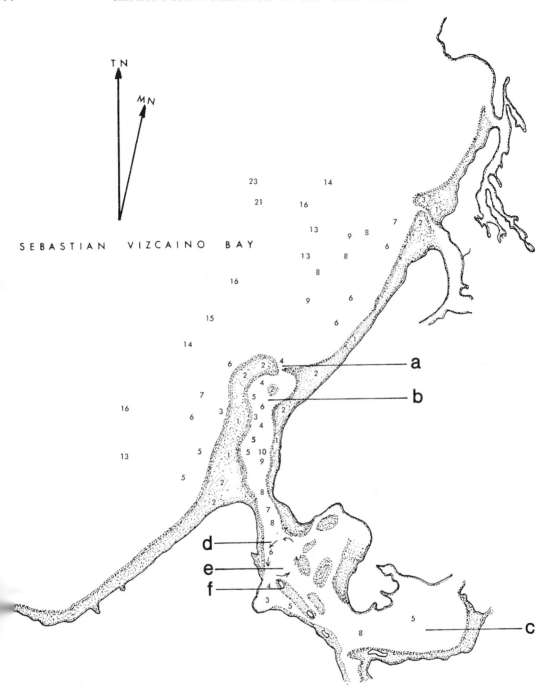

Figure 2. Scammon's Lagoon, Baja California, Mexico: a, entrance; b, staging area; c, the nursery; d, arrows indicate preferred direction of gray whale movement in the social inter-action area; e, social interaction area; f, Brushy Island. Depth is indicated in fathoms.

ever, rather than staying submerged for four ve minutes, they surfaced again in one or minutes. Interspersed with these deviations a constant display of massive, triangular heads poised erect above the sea's surface. Heads were visible in all directions around the lagoon. At any one time during the hours mentioned no fewer than half-a-dozen gray whale heads

SPYHOPPING   POSTURES

MODE 1

TRIANGULAR PROFILE

MODE 2

ARCUATE PROFILE

Figure 3. Spyhopping postures assumed by the gray whale, *Eschrichtius robustus* in Scammon's Lagoon. A parturating female was observed exposing her head as in Mode 2.

could be seen bobbing up in this section of Scammon's Lagoon. This perplexing behavioral trait is called spyhopping (Fig. 3).

Observations indicated that there was a general chronological order coupled with specific behavioral activities. The early morning hours to approximately 1100 included the basic pattern of whales circling within the social interaction area, a great deal of spyhopping and occasional breeching. Whales remained surfaced for prolonged periods of time, blowing slowly and rolling over exposing lateral surfaces and flippers. Females were observed nursing newborn calves.

During these hours the whales were easily approached by small boats propelled by oars. My experiences approaching dolphins of the genera *Globicephala*, *Delphinus*, and *Orcinus* in small boats showed that the high-frequency sound of an outboard motor and the cavitation shockwave created by the propeller irritated or frightened them. Evasive maneuvers consisted of

erratic course changes after deep-diving an swimming submerged at an increased rate c speed along with prolonged deep-dive period: From all indications, they appeared to ignore boat powered by oars or under sail. I have foun that grays can often be approached rather close' by larger boats powered by diesel engines whic are throttled back. The diesel produces a mu lower frequency sound. Although the above true for adult whales, yearlings or adolescen are difficult to approach in anything but a sm; boat under oars.

Copulatory behavior.—There was a noticeab change in the attitude and general behavior of t' whales beginning in the late morning and lasti until early afternoon, ca. 1100–1300 hrs. Duri this interval there was an evident hiatus surface and above-surface activity. The wha became skittish and unapproachable. T anticipatory-phase behavior was manifested by t crown of the head and blow-holes barely breaki water, followed by rapid exhalation and inhalati along with an increase in swimming speed a circling. Gradually, some of the animals beg to form trios. This episode was a prelude actual mating activity during the afternoon a early evening from about 1300 hours to dusk

The mating triad consists of two males, which at least one is sexually mature, and oestrative female. The role of the assisting m or helper bull is not fully understood but seems to act as a stabilizing agent keeping copulating pair together in a venter-to-ver attitude (Figs. 4 and 5). Just prior to intromiss as well as during copulation there was a g deal of flaying about of caudal peduncles flukes, rapid rolling and fluke-stands (flu extended vertically above the lagoon's surfa especially on the part of the assisting m Mating is a prolonged affair among gray wh: lasting well over an hour in many instan During the four days spent in Scammon's Lag observing gray whale behavior, five confir mating sequences were observed. The ph graphs of the copulating whales (Fig. 4) taken in the social interaction area off Br Island by photographer Bill Philbin. At appr mately 1600 hrs. on 30 January 1973, the 13 skiff used was unavoidably drawn into a ma interlude by the turbulence generated by copulating pair and the circling of the he bull. The mating took place at the surface full view, and lasted well over an hour. Du

Figure 4. (Top), two copulating gray whales observed in Scammon's Lagoon. Note the everted, cone-shaped vulva of the female. The penis of the male was between 5.5–6 feet in length. (Bottom), the semi-flaccid penis is seen here just as intromission was terminated as the pair rolled with the venter toward the surface and the female submerged. Anterior is to the right in each photograph.

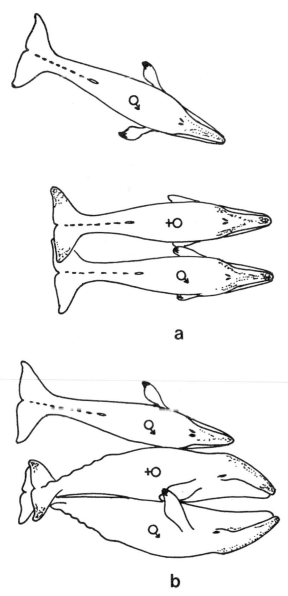

Figure 5. A mating triad of gray whales, *Eschrichtius robustus*: a, dorsal view of the lateral or side-by-side, precopulative swimming posture with the helper bull (top) generally swimming off to one side and somewhat behind the mating pair (bottom); b, dorsal view of the venter-to-venter copulative posture of the mating pair while in a horizontal position with the helper bull in a stabilizing status.

copulation venter-to-venter was the initial posture with flukes aligned and caudal peduncles rigid and extended, forward motion had ceased and the whales were horizontal. On one occasion during our involvement with this mating triad, one of the males (it was impossible to determine

whether it was the mating or helper bull) execute a fluke-stand not more than a meter off ou starboard side. This consisted of 15 feet of th caudal peduncle rigidly extended perpendicula to the lagoon's surface with its eight foot wid flukes spread straight out. This position wa maintained for at least 10 seconds before th whale submerged beneath the skiff. Fluke-stanc like this were frequently sighted in the socia interaction area during mating. They have als been observed during mating interludes along th Southern California coast.

Only once during mating sightings in the lagoc was any deviation from the mating-triad patter observed. Leatherwood and Philbin informe me that they witnessed a full-scale copulatic involving only two adult gray whales. A ca with a distinctive brown patch on its right si was seen among the mating animals and occasion was forced from between the copulati pair. In all probability, the calf was orphan or abandoned seeking a mother for sustenanc This assumption was reinforced some time la when it was again sighted some distance aw from the mating site and alone. Later, the inf became stranded in a shallow inlet on the ebbi tide. Three stranded calves were sighted in t outer portions of Scammon's Lagoon by the c of January.

OBSERVATIONS OF MATING
OUTSIDE OF THE LAGOON

On 26 February 1973, I observed a northbou mating trio one mile southeast of Whites Po Palos Verdes Peninsula. The whales remai surfaced for most of the 45 minutes that were able to stay with the animals before t became skittish and terminated the epis No other gray whales were within the vici to confuse observations. All three of the anin were large with estimated lengths of 35–40 f The mating pair swam side-by-side for appr mately 100 yards before abruptly rolling over c their sides, the male on his right, the female her left. Intromission was accomplished below the surface as they merged venter-to-ver The helper bull slowly swam around and u the copulating pair. Gradually, the mating wh rolled gently away from each other until they again side-by-side, but with ventral surfaces and exposed above the sea's surface (Fig. 4). penis was still inserted in the female's ge

rifice but retracting. At this time, the helper
bull swam obliquely in towards the now upside-
down pair, turned and came up broadside against
the copulating male, nudging him gently over
with a firm and deliberate action. The duration
of this episode was approximately 10 minutes.
I observed this procedure and general sequence
three times before the whales deep-dove and
ended copulation. In all three instances the same
male copulated with the female. Both mating
animals were about the same length. Only once
during the mating episode was I able to identify
which member of the mating pair the helper bull
actually assisted.

DISCUSSION

Two distinct modes of vertical, head-above-surface
posturing became manifest after many hours of
observing spyhopping behavior. The first and
most obvious visual-observing posture was the
whale's head smoothly rising above the lagoon's
surface and presenting a nearly vertical, *triangular
profile* (Fig. 3), remaining motionless and vigilant
for a number of seconds; and then submerging
again, scarcely disturbing the sea's surface. In
the second mode, the animal displayed an *arcuate
profile* with the head held obliquely to the plane
of the lagoon's surface (Fig. 3). The characteristically convex cranium and pronounced pro-
trusion of the distal tip of the rostrum over the
terminus of the mandibula is accentuated by the
concave-shape of the throat region. The posture
of the submerged portion of the body was unob-
servable.

Because of the consistency in the pattern of
spyhopping, it is probable that the gray whales are
not only visually orienting themselves within
their physical surroundings (Scammon, 1874)
but also observing each other. Spyhopping
behavior may well serve as a means of sex identification. Physically, the female is somewhat
larger than the male (Scammon, 1874; Rice and
Wolman, 1971). Walker (1971) maintains that
the female has a distinctly narrower cranial
profile than the male. Otherwise, there does not
appear to be any other external method of
discerning between the sexes except for external
genital differences. Subtle differences between
sexes may be expressed in the mode of posturing
during spyhopping. The male may characteristically display his sex by assuming one of the

aforementioned spyhopping postures with its
distinctive profile and the female, the other.
Spyhopping is the single most obvious behavioral
feature observed in the social interaction area
during the morning hours and preceding mating.
Spyhopping and mating have been observed all
along the California coast during both legs of the
migration. Spyhopping has been interpreted as
orientation behavior involved in migratory navigation (Scammon, 1874); as a method for establishing location and direction from learned reference points, such as prominent headlands (Walker,
1971 and Daugherty, 1972); as a method of
feeding whereby gravity is used to aid food to
pass down into the digestive tract (Walker,
1971); and, possibly, as a vigilance stance. I
believe that spyhopping plays an important role
in the social behavior and sexual interaction of
the gray whale.

ACKNOWLEDGMENTS

I am grateful to Donald R. Patten and Shelton P.
Applegate, both of the Natural History Museum of
Los Angeles County, and to John C. Ljubenkov and
Robert Osborn of the Cabrillo Marine Museum, San
Pedro, California for their constructive comments
and editing of the manuscript. I am especially indebted to the American Cetacean Society and the
Los Angeles City School District for their support of
this research. I also wish to thank William Philbin
for allowing me to use his exceptional photographs,
and to the many individuals who relayed their observations and information, especially Steve Leatherwood of the Naval Undersea Research and Development Center, San Diego.

LITERATURE CITED

Daugherty, A. E. 1972. Marine mammals of California. California Dept. Fish Game, 87 pp.

Fay, F. H. 1963. Unusual behavior of gray whales
in summer. Psychologische Forschung, 27:175–
176.

Gilmore, R. M. 1960. A census of the California
gray whale. U.S. Fish Wildl. Serv., Spec. Sci.
Rep., Fish, 342:1–30.

Rice, D. W., and A. A. Wolman. 1971. The life
history and ecology of the gray whale (*Eschrichtius robustus*). Amer. Soc. Mamm., Spec. Publ.
No. 3, viii + 142 pp.

Sauer, E. G. F. 1963. Courtship and copulation of the gray whale in the Bering Sea at St. Lawrence Island, Alaska. Psychologische Forschung, 27:157–174.

Scammon, C. M. 1874. The marine mammals of the northwestern coast of North America. John H. Carmany and Co., San Francisco, 319 pp.

Slijper, E. J. 1962. Whales. Basic Books Inc., New York, N.Y., 475 pp.

Walker, T. J. 1971. The California gray whale comes back. Natl. Geog. Mag., 193:394–41

Accepted for publication February 14, 1974.

BEHAVIOR AND MATERNAL RELATIONS OF YOUNG SNOWSHOE HARES[1]

ORRIN J. RONGSTAD, University of Minnesota, Minneapolis[2]

JOHN R. TESTER, University of Minnesota, Minneapolis

Abstract: The behavior of four young snowshoe hares (*Lepus americanus*) and the activities of a female during the time she was caring for these young were determined by using radiotelemetry. The female was with her young only once each day for only 5 to 10 minutes. The time of returning was remarkably constant, and appeared to be related to a certain light intensity during the evening twilight period. The movements of a second female during a period when she was known to have had two litters was also determined. It appeared that this family also assembled only once a day but later in the twilight period. The young spent the day in separate hiding places and got together only about 5 to 10 minutes before the female returned to nurse them; they then dispersed and remained alone until the following evening. Weekly home ranges of the young varied from 1.6 acres to 4.9 acres during the second to sixth week of life; they then increased rapidly to about the size of their mother's home range (approximately 20 acres).

Little is known of the behavior and maternal relations of young snowshoe hares living in the wild. Severaid (1942) reported on studies of penned hares, but information on wild hares has been difficult to obtain because of the secretive nature of the young. Recent development of small radio transmitters has made it possible to investigate movements and behavior of young hares without observing them. This paper reports on the activities of a female snowshoe hare and her litter of four young, and on movements of a second female during a period when she was known to have had two litters.

We thank V. B. Kuechle for construction and maintenance of the electronic aspects of the project and for help in the field, D. B. Siniff for assistance with computer processing of data, L. B. Keith and J. J. Hickey for a critical reading of the manuscript, and the many personnel of the Radio-tracking Project who assisted in the field and in the laboratory.

[1] Supported by the U. S. Atomic Energy Commission, COO–1332–58, and by PHS Training Grant No. 5 TO1 GM01779–01 from the National Institute of General Medical Sciences. Computer time was provided by the University of Minnesota Numerical Analysis Center.

[2] Present address: Department of Wildlife Ecology, University of Wisconsin, Madison.

METHODS

The study was conducted during the summer of 1966 on the University of Minnesota's Cedar Creek Natural History Area (93° 12′ W. longitude, 45° 24′ N. latitude), which has an automatic radio-tracking system (Cochran et al. 1965). Hares were tagged with collar-type radio transmitters similar to those described by Mech et al. (1965). Each transmitter broadcasted on a different frequency. The smallest transmitters, which weighed about 7.5 grams and had an expected battery life of 10 days, were used on young between 5 and 21 days of age. Although we did not place transmitters on the young until they were 5 days old, we believed that 2-day-old animals could carry them satisfactorily. On hares between 3 weeks and 3 months of age, we used transmitters weighing about 22 grams and having an expected life of 70 to 80 days. The transmitter for adults and older young weighed 35–42 grams and had an expected life of 150 to 180 days. To insure against loss of data because of premature battery failure, we recaptured the animals and changed batteries at 7, 50, and 100 days, respectively, for the three transmitter sizes. Maximum range was about 0.5 mile for the small transmitters and just over 1 mile for the largest.

The automatic tracking system had two permanent towers located 0.5 mile apart, and animal locations were determined by triangulation. Angular directions were read to the nearest whole degree. The effect of an animal's location, in relation to the position of the towers, on the accuracy of locations was discussed by Heezen and Tester (1967). In our study, the female with four young was in a position where the error polygon due to recording locations to the nearest degree was about 20 × 50 feet, and the other hare was in an area where the error polygon was about 40 × 40 feet.

Therefore, a location that we considered a point was actually within an area of the described size.

Although it was possible to obtain locations every 45 seconds, we found that a fix every 15 minutes usually provided satisfactory information on hare movements. During the day, when little movement occurred, we recorded fixes at 1-hour intervals unless the film record of the signal indicated extensive movements. To investigate relationships between individuals, every available fix was utilized during the time the animals were near each other. All times are given as Central Standard Time.

We had initially hoped that by plotting on a map the daily movements of a female, we would see a pattern develop that would tell us when and where she had young. We then intended to locate and radio-tag the young so that their movements could be recorded. One female (No. 220) was pregnant when captured, and we could predict through embryo palpation the approximate date of birth. Despite this information, we were unable to locate her young.

After this initial failure, we captured another pregnant female (No. 225) and put her in a small temporary pen (6 × 19 feet) in an unfamiliar area, but in a place we knew was used by other hares. She had a litter of four young on the night of July 12. On July 18 we placed transmitters on the female and her four young. Since all the transmitters were operating properly on the next afternoon, we quietly raised the sides of the pen and allowed the animals to leave without any disturbance.

In recapturing an animal, its exact location was first determined by using a portable receiver. The animal was then captured by chasing it into a drive net (Keith et al. 1968:802–803); in some cases, young were caught with a dip net. Because the drive net mesh was designed to capture adults, three thicknesses of net were used to capture young.

Home ranges were measured by dividing a map of the study area into 0.1-acre squares (66 × 66 feet) and determining the position of squares containing fixes and the number of fixes in each square. Home-range boundaries were arbitrarily delimited by searching alternately along the horizontal and vertical axes from squares containing fixes. Squares with fixes that were separated along either axis by not more than two vacant squares were considered within the home range; others were excluded. "Home-range size was then determined by summing the squares within the boundary" (Rongstad and Tester 1969:367). This method, developed by Siniff (unpublished), has an advantage over simply connecting external locations and measuring the enclosed area, in that it gives both intensity of use and area and is readily adapted to computer analysis. It may, however, give a slightly exaggerated home range because a single location in a square at times adds the entire 0.1 acre. The method could also eliminate a square in which an animal spent considerable time if that square were separated from other squares by more than two vacant ones.

RESULTS AND DISCUSSION

Activities of Female 225

Although female 225 had been trapped, moved to a strange area, and held in captivity during parturition, she did not abandon her young. We lost radio contact with one young when it was 12 days old. A weasel (*Mustela* sp.) killed another when it was 25 days old. The remaining two lived until March 1967, when one was killed by a red fox (*Vulpes fulva*) and the other moved out of range of the receiving towers and could not be relocated.

Fixes obtained by telemetry were used to determine when No. 225 was with her young and how far away she went when she left them. The young did not remain together but did stay in the general vicinity of the pen. The most striking discovery was that the female was with her young only once each day for only 5 to 10 minutes, and that this time was remarkably constant from day to day. The female would often spend the rest of the day as far as 900 feet from the young. It is probable that female 225 left her young as soon as they finished nursing. Three young hares observed by Severaid (1942:35) finished nursing on their own accord in 11 minutes. Zarrow et al. (1965:1836) found that suckling time in Dutch-belted rabbits ranged from 2.7 to 4.5 minutes, with a mean of 3.4 minutes.

We have found no references as to the number of times that young snowshoe hares nurse in each 24-hour period. Several varieties of domestic rabbits (Zarrow et al. 1965, Venge 1963) and wild rabbits (*Oryctolagus cuniculus*) in Australia (K. Myers, personal communication) nurse their young only once a day. It may be that this is a characteristic of all hares and rabbits.

To determine the distance female 225 was from her young throughout the day, we programmed the computer to calculate the distance between her and the pen site (Fig. 1). The pen site was selected because this was the approximate place where she met her litter each night.

Movements of No. 225 for the 9 days shown in Fig. 1 were essentially the same as for days not illustrated. The time that she returned to the young each day became earlier as the summer progressed (Fig. 2), and approximately followed the end of nautical twilight (time when the sun is 12° below the horizon). Nightly variations in meeting times may have been due to different light intensities caused by cloud cover. An example of this occurred on August 6

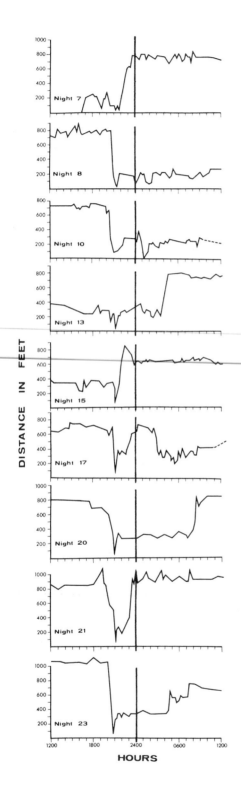

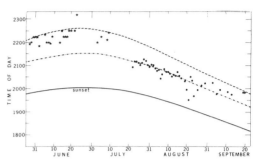

Fig. 2. The time that snowshoe hare female 220 (stars) and female 225 (dots) returned to their young. The top line is the time that the sun is 18° below the horizon (end of astronomical twilight); the middle line is the time the sun is 12° below the horizon (end of nautical twilight).

when the family got together at 8:27 PM, 20 minutes earlier than on any previous night. A thunderstorm that evening caused darkness to come rapidly and early. The meeting time became more varied as the young grew older.

Female 225 continued meeting with her young for an unusually long time, perhaps because the litter we were studying was her last of the year. Severaid (1942:35) reported that hares normally wean their young at between 25 and 28 days post partum, but he reported two instances where the last litters of the summer were nursed for almost 56 days. We recaptured No. 225 when her young were 71 days old; at this time milk could easily be squeezed from her teats. Also, the hair around the teats was matted and curled, suggesting that young were still nursing (Keith et al. 1968). The film record confirmed that this female and her two remaining young were still meeting nightly. It could not be determined from the film record exactly when the young were weaned. Milk could still

←

Fig. 1. Distance between female 225 and her young for various 24-hour periods during the time the young were nursing. Young were born the night of July 12, designated Night 1.

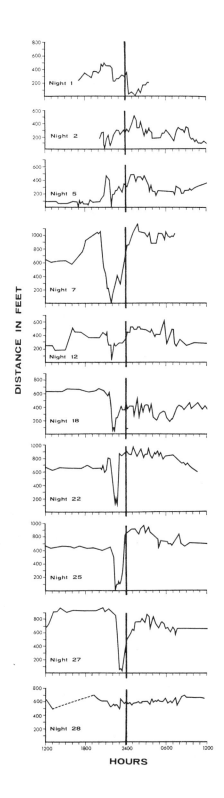

DISTANCE IN FEET

HOURS

be squeezed from the teats on October 11, but hair around the teats was no longer matted and curled; thus, nursing had probably stopped by this time.

Activities of Female 220

After noting how little time female 225 spent with her young, we reexamined the movement record of female 220 minute by minute and found that there was one place to which she also had returned each day. The plot of distance between this location and that of female 220 (Fig. 3) appeared similar to the plot for No. 225 (Fig. 1). We had originally failed to detect the place where the young of No. 220 were nursed, because we had sampled her location only every 15 minutes. This sampling interval often missed the time the family was together.

Female 220 first went to this location at 1:20 to 1:35 AM on May 27 and probably had her young at this time. Severaid (1942: 34) observed a female giving birth to four young in 20 minutes, making the 15 minutes No. 220 spent in the area a realistic time period for parturition. This female may have returned to the young more than once each night during the first, second, and possibly the fifth nights; otherwise she returned only once each night. The return was between 9:50 and 10:30 PM until night 27 (June 22), when she came to the general area at 11:00 PM and stayed 30 minutes. We had caught her with a drive net at 12:30 PM that same day to check for pregnancy; this handling may have affected her

←

Fig. 3. Distance between female 220 and her young for various 24-hour periods during the time the young were nursing. Young were probably born on May 27, designated *Night* 1. Night 28 was apparently the night of weaning. Dotted line indicates a period when a signal was received from only one tracking tower, but the direction of this bearing was such that the animal could not have gone to the young.

437

Table 1. Home range (in acres) of the young born on July 12 to snowshoe hare female 225. The numbers in parentheses indicate the number of locations on which the home range was based.

TIME PERIOD	MALE 227	FEMALE 228	MALE 229	FEMALE 230
July 19–26	1.8 (80)	1.6 (112)	2.4 (151)	2.1 (177)
July 27–Aug. 2	3.1[a](122)	3.4 (251)	2.7 (126)	3.4 (141)
Aug. 3–9		2.7 (318)	4.9 (310)	2.2[b](198)
Aug. 10–16		2.7 (316)	4.7 (354)	
Aug. 17–31		4.1 (144)	4.9 (234)	
Sept. 1–15		11.4 (382)	10.0 (397)	
Sept. 16–30		17.2 (667)	14.9 (572)	
Oct. 1–10		20.4 (769)	13.6 (822)	
Jan. 4–24		14.1 (735)	23.0 (841)	

[a] Transmitter quit on August 1.
[b] Killed on morning of August 8.

movements that night. The next night she did not approach within 300 feet of the nursing area, nor did she ever venture closer than this for the next 30 days. We do not know if the failure to return on the 28th night was part of a normal weaning process or was caused by our handling.

Female 220 had her next litter on either the night of July 2 or 3, and, although her transmitter operated only intermittently, she apparently returned to these young daily for only 16 days. We caught her on the 16th day, so we do not know if the trauma of capture caused her to abandon her young or if the young had died or been killed by a predator. This female was pregnant for a third time, but was accidentally killed before giving birth.

The timing of No. 220's meetings with her young was slightly less regular than that of female 225 and was a little later at night (Fig. 2). Her meeting times were closely related to the end of astronomical twilight (time when the sun is 18° below the horizon).

The range of movements of female 220 during the later stages of her two pregnancies became somewhat restricted. Her home range was 32.0 acres for the 20 days from May 1 to the morning of May 19, and only 8.1 acres for the 8 days prior to parturi-

tion on May 27. Her home range again expanded to 29.0 acres for the next 26 days and again decreased to about 8 acres prior to the birth of her next litter. In comparison, the home range of female 225 was 21.5 acres for the first month and 18.2 acres for the second month, after release from the pen.

No evidence is available from other female snowshoes to indicate that movements became restricted during later stages of pregnancy. Three hares that appeared to have had litters during April 1967, for example, did not show this tendency.

Activities of Young

During the first morning of life the four young of female 225 were huddled together under a small shelter in their pen. After we measured them they scattered, and two squeezed through the 1-inch-mesh poultry wire pen and had to be caught and returned. To prevent further escapes, a small 0.25-inch mesh pen was built within the larger pen. The female readily hopped into this pen to feed the young. Severaid (1942: 31) reported that young hares are generally found scattered about the pen by the third or fourth day, and if disturbed, they will scatter by the first day. From Severaid's observations and ours on penned hares, w

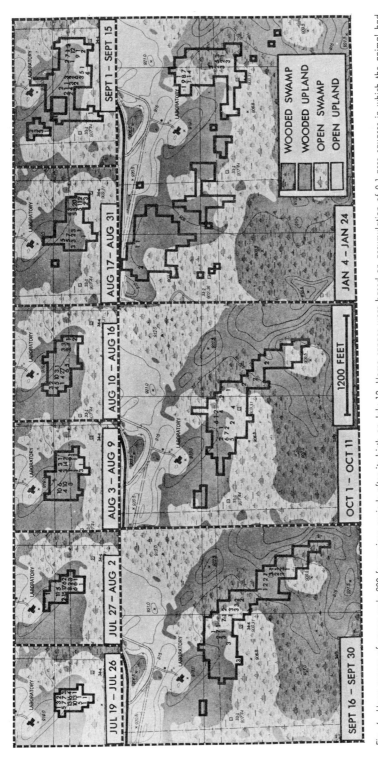

Fig. 4. Home ranges of young male 229 for various periods after its birth on July 12. Home ranges were based on accumulation of 0.1-acre squares in which the animal had been located. Numbers within the home range represent the percentage of time spent in that particular square. Squares in which the animals spent less than one percent of time were not labeled.

conclude that young hares remain together for only 1 to 4 days and thereafter scatter in the area around the place of birth.

On the day after release from the pen (day 8), all young were located with a portable receiver. Each was in a separate hiding place, and was separated from the others by as much as 60 feet. All were caught without a chase. On July 22, 3 days after release, the young were spaced about the same as on July 20 and were again caught without a chase. However, two ran after being released. On July 25 the young were caught, and batteries on the transmitters were changed. At this time, young were separated by up to 100 feet, and only one was caught without a chase. One had to be chased for at least 500 feet before it could be captured. On August 2, the transmitter on one animal stopped; the other three were captured (one without a chase) for a battery change. The daytime resting places of the three were still farther apart than on July 25. A weasel killed one of these young on the night of August 7–8; the other two lived through the winter.

Even though these young hares were carrying radio transmitters and could be captured without a chase, they were difficult to find. They remained motionless and were usually in small forms, concealed by vegetation.

Based on locations obtained with the tracking system, home ranges of the young of female 225 were between 1.6 and 2.4 acres during their second week of life (Table 1). Home ranges increased slightly during the third, fourth, and fifth weeks and showed major increases in the sixth and seventh weeks. After 8 weeks, home-range sizes were about the same as for adults. We do not know how the long nursing period affected home ranges. However, one young hare, estimated to be 25 days old when trapped, had a home range of 5.8

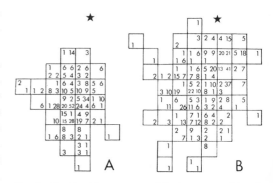

Fig. 5. Combined home ranges of the four young of female 225 for their 2nd (A) and 3rd week of life (B). Numbers in squares indicate the number of times a particular animal was located in that square. Upper left number is for male 227, upper right for female 228, lower left for male 229, and lower right for female 230. The stars indicate the same reference point for both home ranges.

acres for 10 days after capture. Another, estimated to be 60 days old when trapped, had a home range of 8.7 acres for 10 days after release, and 15.8 acres for the next 7 days. These home-range estimates are similar to those for the pen-released young of female 225.

For young male 229, changes in home-range size, shape, and intensity of use are shown in Fig. 4. The home ranges for its littermate (female 228), which lived through the winter, were approximately the same size (Table 1), and both animals lived in the same general area until November 14; at that time, hare 228 moved to a brush lowland about 1,500 feet east of the area it had previously used. The January home range, therefore, did not overlap its previous home ranges.

The entire area used by the four young during their 2nd and 3rd weeks of life was 2.8 acres and 4.8 acres, respectively (Fig 5A, B). Even at this early age, the young seemed to have spaced themselves. When we looked at only the five most intensively used 0.1-acre squares for each young, we found that the 0.5 acre used by No. 229 and the 0.5 acre used by No. 230 were not used intensively by any other animal and that No. 227 and No. 228 had only two squares in common.

Our data indicate that the young assembled each night about 5 to 10 minutes before the female returned, even on August 26 when, as described above, the female was 20 minutes earlier than on any previous night. The same mechanism was probably triggering the timing of the female and her young. The gathering place did not appear to be rigidly fixed, apparently varying by 30 to 50 feet between nights, but this was difficult to determine precisely because of limitations in the tracking system.

LITERATURE CITED

COCHRAN, W. W., D. W. WARNER, J. R. TESTER, AND V. B. KUECHLE. 1965. Automatic radio-tracking system for monitoring animal movements. BioScience 15(2):98–100.

HEEZEN, K. L., AND J. R. TESTER. 1967. Evaluation of radio-tracking by triangulation with special reference to deer movements. J. Wildl. Mgmt. 31(1):124–141.

KEITH, L. B., E. C. MESLOW, AND O. J. RONGSTAD. 1968. Techniques for snowshoe hare population studies. J. Wildl. Mgmt. 32(4):801–812.

MECH, L. D., V. B. KUECHLE, D. W. WARNER, AND J. R. TESTER. 1965. A collar for attaching radio transmitters to rabbits, hares, and raccoons. J. Wildl. Mgmt. 29(4):898–902.

RONGSTAD, O. J., AND J. R. TESTER. 1969. Movements and habitat use of white-tailed deer in Minnesota. J. Wildl. Mgmt. 33(2):366–379.

SEVERAID, J. H. 1942. The snowshoe hare; its life history and artificial propagation. Maine Dept. Inland Fisheries and Game. 95pp.

VENGE, O. 1963. The influence of nursing behaviour and milk production on early growth in rabbits. Animal Behaviour 11(4):500–506.

ZARROW, M. X., V. H. DENENBERG, AND C. O. ANDERSON. 1965. Rabbit: frequency of suckling in the pup. Science 150(3705):1835–1836.

Received for publication September 25, 1970.

SECTION 5—PALEONTOLOGY AND EVOLUTION

If there be one unifying principle that pervades all of biology, it is that of evolution. Not only is this evident in consideration of the papers here reproduced (which range from one that deals in part with intrapopulational variation up to those concerned with higher taxonomic categories), but it also is evident in the contents of virtually all other papers chosen for inclusion in this anthology. The few selections in this section, then, provide but a glance at some aspects of mammalian evolution.

The short paper by Reed nicely illustrates some problems that increases in knowledge have generated in our own part of the evolutionary tree. Linked inseparably with the evolutionary process is the fossil record, which is unusually good for some groups of mammals and provides much of the raw data for phylogenetic considerations. For papers relating to paleontology, we have chosen one (Wilson) that alludes to the importance of sound geographic and stratigraphic data and that ties in with the historic record, one (Radinsky) that deals with evolution and early radiation of perissodactyls, and two on rodents, one a paper by Zakrzewski revising the muskrat tribe, and the other a modern treatment of the major groups of the extremely complex order Rodentia by Wood (see also Wood, 1959).

For a recent analysis of one major group of rodents that includes good discussions of relationships, see Rowlands and Weir (1974). An excellent short overview of rodent evolution was written by Wilson (1972).

Two areas of recent interest that should be mentioned are the use of observable wear facets in attempts to understand function of teeth and cusps (for example, Crompton and Sita-Lumsden, 1970), and the elucidation of plate tectonics and moving continental land masses as significant factors in evolution and zoogeography (for example, see McKenna, 1973).

The study by Guthrie compares evolutionary change in molar teeth and relates this to variability in different measurements, using both fossil and Recent species of *Microtus*, and thus stresses the on-going evolutionary process. The elephants are represented by a good fossil record, and show major changes in dentition that led to better grinders, just as in the microtines studied by Guthrie. These changes and their functions in elephants were analyzed by Maglio (1973). Variability within populations of mammals has been summarized by Yablokov (1974).

The paper by Jansky is interesting because it provides an excellent example of evolutionary trends in features other than those directly related to "hard anatomy."

Nadler *et al.* studied the evolution of one species of ground squirrel in a relatively small area by examining chromosomes. The study by Smith *et al.* demonstrated striking differences between survival of wild- and laboratory-reared mice in a wild environment, even though it failed to reveal any effect of color on natural selection, which was the main hypothesis in initiating the study. Some earlier and later studies have revealed selective effects of color in small mammals (for example, Kaufman, 1974).

Species have not evolved in isolation but in ecosystems. The paper by Heithaus *et al.* examines the coevolution of certain tropical bats and plants.

The literature of mammalian evolutionary and paleontological studies is widely scattered. Aside from the journals and bibliographic sources mentioned in the Introduction, the interested student should consult EVOLUTION, the JOURNAL OF PALEONTOLOGY, and the recently initiated PALEOBIOLOGY. He should also be aware of the *Bibliography of Fossil Vertebrates, 1969-1972,* compiled by Gregory *et al.,* 1973 (as well as earlier volumes in the same series) and the News Bulletin of the Society of Vertebrate Paleontology.

Romer's textbook, *Vertebrate Paleontology* (1966) and Simpson's (1945) *The Principles of Classification and a Classification of Mammals* are especially recommended as sources of considerable information on the fossil history and evolution of mammals, and we would be remiss not to mention also Zittel's (1891-93) classic *Handbuch der Palaeontologie* (volume 4, Mammalia). Two substantial longer papers on systematics and evolution of special groups are Dawson's (1958) review of Tertiary leporids, and Black's (1963) report on the Tertiary sciurids of North America. Extensive paleofaunal studies of note are many: those by Hibbard (1950) on the Rexroad Formation from Kansas and by Wilson (1960) on Miocene mammals from northeastern Colorado serve as excellent examples.

South African Journal of Science
Suid-Afrikaanse Tydskrif vir Wetenskap

The Association, as a body, is not responsible for the statements and opinions advanced in its publications. Die Vereniging is nie, as 'n liggaam, verantwoordelik vir die verklarings en opinies wat in sy tydskrifte voorkom nie.

| Vol./Deel 63 | JANUARY 1967 JANUARIE | No. 1 |

THE GENERIC ALLOCATION OF THE HOMINID SPECIES HABILIS AS A PROBLEM IN SYSTEMATICS

CHARLES A. REED

THE recent controversial discussion, in *Current Anthropology* (Oct. 1965) and elsewhere, concerning the correct generic placement of the Lower Pleistocene hominid species *habilis* (Leakey, Tobias, and Napier, 1964), depends for its solution upon which one of two kinds of philosophy of systematics is followed. None of the participants in the discussion have emphasized this particular aspect of the issues, but an understanding of these concepts is basic to both argument and solution.

If one is impressed with the phylogenetic approach to the study of fossils, stressing the implications of those evolutionary innovations found in them which place a particular group at the beginning of a new evolutionary line, leading in time to new adaptive possibilities, then the classification will be vertical ('classification by clade'). Utilizing this approach to zoological systematics the investigator will emphasize the importance of the new evolutionary direction (the new adaptive plateau being approached), by placing his fossils in the taxon with the advanced forms derived from them. Leakey, Tobias, and Napier did exactly this when they placed the population *habilis*, from Bed I of Olduvai Gorge, Tanzania, in the genus *Homo* (Fig. 1).

The alternate approach to systematics is "classification by grade," wherein the investigator emphasizes in his taxonomic system,

as he emphasizes in his own thinking about the material, the greater or lesser degree of morphological likenesses between two populations which have essentially reached, at the generic or specific levels, a considerable similarity. Obviously, the individuals of *habilis* are anatomically more similar to individuals of *Australopithecus africanus* that they are to ourselves as *Homo sapiens*, or even to individuals of the mid-Pleistocene taxon *H. erectus*. Robinson (1965a, b) and separately Howell (1965), seeing clearly this essential anatomical similarity between *africanus* and *habilis*, wish to emphasize what to them is a clear closeness of biological relationship by placing the two populations together in the same genus, *Australopithecus* in this instance.*

The issues involved have roots deep in the history of post-Darwinian systematics, particularly as practised by palaeontologists. Simpson (1961) has summarized the problems with a suggestion for a solution which attempts (although in my opinion not

---

* The mentioning of two genera, but only two, as comprising the known Quaternary hominids is done on the basis of the general usage of the authors involved in the controversy presently being considered, and with the view that *Paranthropus* is probably best considered as a sub-genus of *Australopithecus*. We must not forget, however, that Mayr (1950) advocated that all Quaternary hominids be included in *Homo*, a practice followed only intermittently thereafter but espoused in at least two recent textbooks (Brace and Montagu, 1965; Buettner-Janusch, 1966). There is also another possible point of view, the one that *habilis* be included within *Homo erectus*, probably as a subspecies, although Tobias (1965b) has indicated that on the basis of present evidence this is a conclusion with which he could not agree.

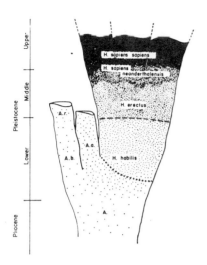

Fig. 1: Phylogeny and classification of the Family Hominidae, as presently understood (after Tobias, 1965a). The dotted line represents the boundary in time and between the taxa Homo **and** Australopithecus **as conceived on the basis of classification by clade; the dashed line represents the same concepts on the basis of classification by grade.**

Our problems with the systematics emerge irrevocably from the pattern of a continuous flow of genes, generation by generation, and from the occasional divisions of a population's gene pool into separate evolutionary streams.

The vertical type of classification based on clades is possible only if a population has proved its survival value by becoming the ancestral type of a new lineage, and if we have found a good record of these happenings. Thus if the population *habilis* had become extinct during the period of the formation of Bed I at Olduvai Gorge, its evolutionary potential would be unrecognizable and its remains would most certainly be classified with *Australopithecus* by whatever subsequent intelligent being was doing the paleontology. The *Homo*-ness of *habilis* lies in those characters which we can recognize as being important in initiating the lineage *Homo* only because we have a record of that lineage. Until, however, we had as complete a record of that lineage as we finally now have, systematics by clade was not possible.

A bit of an analogy, involving non-hominid lineages with which we are not personally involved, may help to clarify the principles. Thus the phylogenies of two super-families, those of the horses (Equoidea) and of the tapirs (Tapiroidea), diverged early in the Eocene. The first-known individual fossils of each of these two super-families are extremely similar, but each—to the eye of the expert—indicates its affinities to its known descendants by what might appear to be, but is not, a trifle of dental pattern (Radinsky, 1963). Where the fossil record is as complete as with these perissodactyls, the solution of the systematic problems has typically been to include in different clades (families or super-families) different populations which on the basis of similarity of anatomical form would be grouped at the grade level as closely related genera or as species in the same genus. If, at this Eocene level of evolution, one of these ancestral groups. such as *Hyracotherium* (ancestral to all later "horses" *sensu lato*),

successfully) to combine the two approaches. An earlier paper by myself (Reed 1960), as based on publications listed in its bibliography, states these particular issues in a shorter article and also points out the logical consequences of accepting either system, that "by clades" or the contrasting one, "by grades."

Neither system is necessarily correct, nor either wrong; they simply are based on two different, and in my opinion mutually exclusive, approaches to the systematic organization of biological populations in a time-continuum. For this reason, systematics remains an art and is not a science, depending upon the opinion of trained investigators for decisions which eventually are or are not followed by larger numbers of people who are interested in the fossils and the phylogeny, but have neither the time nor training to study the materials in detail.

had become extinct, no palaeontologist would be capable of recognizing its potential "horse-ness" and *Hyracotherium* would today be classified as a primitive tapir. Conversely, if *Homogalax*, the earliest of the tapiroid line, had become extinct without issue, undoubtedly it would today be classified as an Eocene equid.

In general, as the gaps in the fossil record of any lineage have been filled, the tendency has been, often without any realization of the philosophy of the systematics involved, to shift from a horizontal (grade) type of classification to the vertical (clade) type, and the recent flurry of published opinions as to the formal position of the species *habilis* illustrated a repetition of this historical pattern. Tobias (1965c) has stated that there is general agreement as to the meaning of the morphological data and the validity of the evolutionary position of the fossils included in the population *habilis* from Bed I at Olduvai Gorge; if precedent has any value as a guide, we may safely assume that *habilis* will remain in *Homo*.

In general, the Primates have been classified on the principle of grades, typical of groups with an incomplete fossil record and thus lacking well-defined lineages. As more fossils are found and the phyletic pattern becomes clearer, various parts of the sub-order (grade) Prosimii will become continuous with at least two lineages (platyrrhine and catarrhine) of the suborder (grade) Anthropoidea, and slowly the present pattern of the systematics will change.

Exactly this sort of change, to the surprise of some, is what is occurring in the Hominidae, due to the filling of the gaps priorly existing between the groups called Australopithecinae and Homininae. We should realize also that, as now defined, the names applied to extinct populations of *Homo* remain as grade concepts, as has already been stated clearly by Tobias and von Koenigswald (1964). Thus, if and when human fossils are found to fill the near-void now existing between the latest *erectus* and the earliest acknowledged neandertals, the whole present taxonomic scheme will necessarily be changed from the horizontal to the vertical. Perhaps that agonizing re-appraisal will be easier then—as indeed I hope it will be now at the *habilis* level—if we realize that it is inevitable.

REFERENCES CITED

BRACE, C. L. and M. F. ASHLEY MONTAGU (1965): *Man's evolution: An introduction to physical anthropology.* New York, The Macmillan Company.

BUETTNER-JANUSCH, JOHN (1966): *Origins of Man: Physical Anthropology.* John Wiley and Sons, Inc. New York.

HOWELL, F. CLARK (1965): *Early man.* New York: Life Nature Library, Time Incorporated.

LEAKEY, L. S. B., TOBIAS, P. V. and NAPIER, J. R. (1964): A new species of genus *Homo* from Olduvai Gorge. *Nature 202:7-9.* (Reprinted 1965 in *Current Anthropology, 6:424-27*).

MAYR, ERNST (1950): Taxonomic categories in fossil hominids. *Cold Spring Harbor Symposia in Quantitative Biology 15:109-18.*

RADINSKY, LEONARD (1963): Origin and evolution of North American Tapiroidea. Peabody Museum of Natural History, *Yale University, Bulletin 17,* 1-106.

REED, CHARLES A. (1960): Polyphyletic or monophyletic ancestry of mammals, or: What is a class? *Evolution 14,* 314-22.

ROBINSON, J. T. (1965a): *Homo 'habilis'* and the australopithecines. *Nature 205,* 121-24.

——— (1965b): Comment on "New discoveries in Tanganyika: Their bearing on hominid evolution," by Phillip V. Tobias. *Current Anthropology 6,* 403-6.

SIMPSON, GEORGE GAYLORD (1961): *Principles of animal taxonomy.* New York: Columbia University Press.

TOBIAS, PHILLIP V. (1965a): Early man in East Africa. *Science, 1949,* 22-33.

——— (1965b): *Homo habilis. Science 149,* 918.

——— (1965c): New discoveries in Tanganyika: Their bearing on hominid evolution. *Current Anthropology 6,* 391-99.

TOBIAS, P. V. and VON KOENIGSWALD, G. H. R. (1964): A comparison between the Olduvai hominines and those of Java and some implications for hominid phylogeny. *Nature 204,* 515-18. (Reprinted 1965 in *Current Anthropology 6,* 427-31).

DEPARTMENTS OF ANTHROPOLOGY AND
  BIOLOGICAL SCIENCES,
UNIVERSITY OF ILLINOIS AT CHICAGO CIRCLE,
CHICAGO,
ILLINOIS,
U.S.A.

447

FOSSIL ONDATRINI FROM WESTERN NORTH AMERICA

Richard J. Zakrzewski

ABSTRACT.—The cranium of *Pliopotamys minor* (Wilson) from the Hagerman local fauna is described. The cranial elements show a mosaic of shapes found in other distantly related arvicoline genera. The size of the cranium and the ratio of certain cranial measurements most closely approximate the extant genera *Neofiber* and *Arvicola*, which suggests a similar structural grade for the three genera. A partial cranium of a relatively advanced ondatrine recovered from deposits equivalent to those which contain the late Pliocene Benson local fauna is reported from Arizona. The stage of evolution of this animal is similar to species previously described from the early Pleistocene. These relationships suggest an early radiation of Ondatrini in the American southwest. Two partial lower jaws of a primitive ondatrine are reported from the San Joaquin formation of California. The nature of the specimens precludes generic assignment.

Isolated teeth and mandibles of arvicolines are common in deposits of late Cenozoic age. Remains of complete crania, however, are rare. Only one cranium of an extinct genus of arvicoline, *Cosomys*, has been described from North America (Wilson, 1932); therefore, the find of an almost complete cranium of *Pliopotamys minor* (Wilson, 1933) from the Hagerman local fauna, late Pliocene of Idaho, is significant. Additionally the partial cranium of an ondatrine from Arizona, found in deposits equivalent to those that contain the Benson local fauna (late Pliocene), and two partial lower jaws of a primitive ondatrine from the San Joaquin formation of California are discussed. Specimens of Ondatrini from these latter two sites have not been reported.

MATERIAL

The cranium of *Pliopotamys minor* (Idaho State University Museum, no. 17123) was found by John A. White and myself in May 1969, at the United States Geological Survey Cenozoic locality 20765 (for complete locality data see Hibbard, 1959). The specimen is slightly compressed dorsoventrally. The left zygomatic arch, auditory bullae, and all the teeth except the upper right first molar are missing.

I recently (1969) reviewed the systematic position of *Pliopotamys* and considered it to be ancestral to *Ondatra*. To find out more about the relationships of *Pliopotamys* I have compared the cranium of ISUM 17123 with crania from Recent species of *Ondatra*, *Neofiber*, *Microtus*, *Neotoma*, and *Arvicola*.

The specimen from Arizona (Frick Collection, American Museum of Natural History, no. 24649) is a partial cranium with a portion of the frontal, palatine, maxillaries, and all the molars except the upper left third molar present. The fossil was found 3.5 miles south of Benson, Arizona. I compared it with the cranium from the Hagerman local fauna and with isolated teeth of *Pliopotamys meadensis* Hibbard, 1938, and *Ondatra idahoensis* Wilson, 1933.

The specimens from California (University of California, Museum of Paleontology, nos. 32952 and 57958) are two partial mandibles with the first two molars present in both; however, the anterior loop on the lower first molar in each of the specimens is broken off. The specimens were found in the San Joaquin formation (UCMP locality V3520). I

284

compared these mandibles with mandibles of *Pliopotamys minor* and *Cosomys primus* from the Hagerman local fauna, and with the holotype of *C. primus* from the Coso Mountain local fauna of California.

The tribe Ondatrini, as defined by Kretzoi (1969), is used here for convenience in discussion.

CRANIUM FROM HAGERMAN

The size of the cranium in *Pliopotamys minor* is approximately equal to that of the woodrat, *Neotoma lepida* Thomas, 1893. The cranium of *Pliopotamys*, however, is typically arvicoline in structure. Arvicoline affinities are evident especially in the shape of the zygomatic plate, which expands ventrally and laterally rather than anteriad as in *Neotoma*, and in the configuration of the postorbital processes that extend at right angles to each frontal bone.

Viewed dorsally (Fig. 1B), the nasals of *Pliopotamys minor* are short, expand anteriorly, and flare slightly lateroventrally; they most closely resemble those of *Ondatra*.

The frontals of *Pliopotamys minor* end anteriorly in a W-shaped suture pattern; a pattern that resembles those in *Ondatra* and *Neofiber*. Posteriorly the shape of the frontal most closely resembles that of *Arvicola* in that the suture is concave anteriorly. The frontals of *Pliopotamys* possess low interorbital crests, like those in the other arvicoline genera I examined. In *P. minor* the crests abut just posterior to the least interorbital constriction. Anterior to this constriction the crests separate and continue forward until they end at the juncture of the premaxillaries and nasals. Just anterior to the interorbital area a small fossa is present between the crests making them appear well developed. Even though the development of the crests can vary with the ontogenetic age of the individual, I found a similar configuration and development of the fossa only in *Ondatra* and the ondatrine from Arizona. Posteriorly the crests separate and follow the lateral sutures of the parietals as temporal ridges to the end of the cranium. I found this configuration in *Arvicola* as well, but the interorbital crests in *P. minor* begin to separate more anteriorly.

The parietals of *Pliopotamys minor* resemble those of *Neotoma cinerea*. The interparietal is shaped like that in *Microtus ochrogaster*.

In lateral view (Fig. 1A), even though some dorsoventral compression has taken place, the profile of the cranium in *Pliopotamys* was lower than in extant arvicolines, being more like that in *Neotoma*.

The portion of the maxillary in which the alveoli are situated is separated from the rest of the cranium by the optic and sphenopalatine foramina. This portion of the maxillary in *Pliopotamys* can be considered shallow or narrow when compared to extant arvicolines. A similar condition exists in *Neotoma*, this reflects the fact that in both *Pliopotamys* and *Neotoma* the molars are mesodont and have well-developed roots.

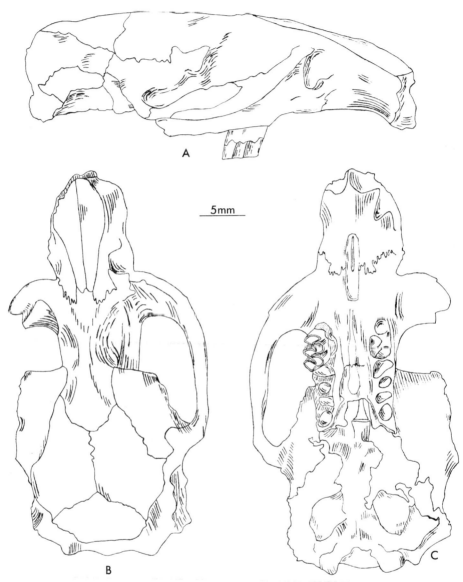

Fɪɢ. 1.—Cranium of *Pliopotamys minor* (Wilson); A, lateral view; B, dorsal view; C, ventral view.

In ventral view (Fig. 1C) the incisive foramina of *Pliopotamys minor* are as long as they are in *Neotoma lepida*. However, the length of incisive foramina varies with the ontogenetic age of the individual as has been pointed out by Quay (1954) for extant arvicolines and by Hibbard and Zakrzewski (1967) and Zakrzewski (1969) for extinct arvicolines.

TABLE 1.—*Measurements (in millimeters) of the cranium (ISUM 17123) of* Pliopotamys minor

Parameter	Measurements	Parameter	Measurements
Length of alveolar molar row	10.0	Diastema length	11.7
Greatest length of skull	38.5	Length of palatal bridge	9.1
Basal length	36.6	Breadth of braincase	14.6
Basilar length	34.0	Zygomatic breadth	21.4*+
Palatal length	21.8	Breadth of rostrum	7.7*
Palitar length	19.5	Least interorbital breadth	6.2*

* = estimate
\+ = width from midline to edge of right zygomatic arch multiplied by two

The lateral palatal grooves in *Pliopotamys* are moderately deep and the maxillary walls which form the outermost portion are vertical. These configurations most closely resemble those found in *Ondatra*. These grooves extend from the posterior end of the palate forward into the incisive foramina.

The median ridge is narrow, elevated throughout its entire length, and widens posteriorly. These configurations were observed in maxillaries of other *Pliopotamys* I examined earlier (Zakrzewski, 1969). The median ridge of *Neofiber* resembles that of *P. minor* in that it is elevated throughout its entire length. However, in *Pliopotamys* the ridge is solid throughout its length, whereas in *Neofiber* there is a slight groove running down the central portion of the median ridge to a position just posterior to the first molars, where the central part of the ridge becomes elevated into a narrower ridge.

Because the tympanic bullae are missing from the cranium the pterotic bones are exposed. They are globular in shape. Although part of the pterotic bone toward the midline of the skull is damaged, it does not appear as flattened and fanned out as it is in *Ondatra*. The overall shape of the pterotic in *Pliopotamys* most closely resembles that of *Neofiber*. Due to postdepositional damage the other portions of the cranium cannot be accurately described.

Measurements of 12 parameters of the skull were taken from the cranium of *Pliopotamys* using the definitions of Hershkovitz (1962), and are listed in Table 1. These same measurements were taken from small samples of extant *Ondatra zibethica, Neofiber alleni, Arvicola terrestris, Microtus ochrogaster,* and *Neotoma cinerea*. The relationship of these measurements is shown (Fig. 2) by means of a ratio diagram (Simpson *et al.*, 1960).

Apparently *Pliopotamys* possessed characteristics that exist today in distantly related arvicoline groups, as well as some that exist in the unrelated *Neotoma*. However, *Pliopotamys* has more characters in common with the extant amphibious arvicolines, especially *Ondatra*. In the measured parameters *Pliopotamys* tends to parallel the medium-sized amphibious arvicolines, *Neofiber* and *Arvicola* (Fig. 2A). This relationship might reflect a similar structural or functional grade for the three taxa. Willam A. Akersten of the Los Angeles County Museum (personal communication) stated that the post-

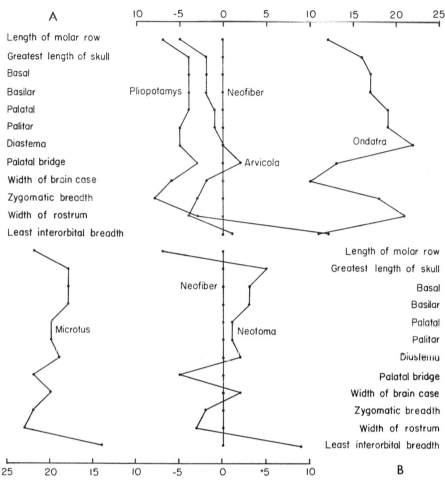

Fig. 2.—Ratio diagrams modified from Simpson *et al.* (1960) comparing various cranial measurements of six rodent genera. Logs of the means of dimensions in *Neofiber alleni* are assumed to be zero, while the differences between the log of the mean in *N. alleni* and the species being compared are plotted to the positive (+) or negative (−) sides of the zero line. A, comparison of *N. alleni* to *Arvicola terrestris*, *Ondatra zibethicus*, and *Pliopotamys minor*. B, comparison of *N. alleni* to *Microtus ochrogaster* and *Neotoma cinerea*.

cranial skeleton of *P. minor* exhibits aquatic adaptations similar to *Neofiber*, and that *Ondatra* is much better adapted to an aquatic existence than either of the other taxa. The curve (Fig. 2A) for *Ondatra* only loosely corresponds to any of the other taxa. The curve (Fig. 2B) for *Neotoma cinerea* (a terrestrial cricetine) parallels that of *Microtus ochrogaster* (a terrestrial arvicoline). The relationship suggests, again, that the curves are reflecting a similar structural or functional grade of the taxa that are compared.

The similarities between *Pliopotamys* and *Arvicola* are most likely due to parallelism, as Kowalski (1970) indicated that *Arvicola* has evolved from

Mimomys, the most common of the European fossil genera. The similarities between *Pliopotamys* and *Neofiber* could also represent parallelism. Recently Hibbard (personal communication) has redescribed an ondatrine from Kansan deposits in Texas which he considers to be ancestral to *Neofiber*. The relationship of this new ondatrine to *Pliopotamys* has not been established. Perhaps the two lines diverged from a common ancestor prior to the attainment of the *Pliopotamys* grade, or the *Neofiber* lineage might have evolved from *Pliopotamys* fairly early in the history of the group and remained relatively conservative in its aquatic adaptations.

SPECIMEN FROM ARIZONA

The ondatrine from Arizona is at an intermediate stage of evolution between *Pliopotamys meadensis* from the Dixon local fauna (Nebraskan) of Kansas and *Ondatra idahoensis* from the Grand View local fauna (Aftonian) of Idaho. This assumption is based on the relative development of the dentine tract on the lingual side of the first alternating triangle on the upper first molar. There is also a dentine tract on the posterior surface of the posterior loop of the upper third molar on the specimen from Arizona, which is not found in *P. meadensis*.

Roots of the teeth in the ondatrine from Arizona are not as developed as those of *Pliopotamys minor*, based on the configuration of that portion of the maxillary that contains the alveoli. This area is intermediate in development between that found in *P. minor* and *Neofiber*. The median area of the palate in the specimens from Arizona does not have a ridge, and thus resembles the median area of the palate in *Ondatra*. However, possibly the palate was damaged. The specimen from Arizona probably represents a new species of arvicoline, but the absence of specimens in advanced stages of ontogenetic development makes even generic assignment difficult at this time.

Hibbard's (1959) discussion of the importance of dentine tracts in the taxonomy of ondatrines apparently has led to the misconception among some workers (for example, Shotwell, 1970) that *Pliopotamys* is separated from *Ondatra* chiefly by the former's lack of well-developed dentine tracts. Although there is a high correlation in the taxa thus far described between poorly-developed or no dentine tracts in *Pliopotamys* and well-developed dentine tracts in *Ondatra*, this is not the chief basis for separation of the two forms. The chief character by which the two genera are separated is the presence or absence of cement in the re-entrant angles. Cement is present at sometime in the ontogeny of an individual in *Ondatra* and absent in *Pliopotamys*. In addition to the lack of cement and shorter dentine tracts, *Pliopotamys* is generally smaller and has better developed roots. However, if *Pliopotamys* is ancestral to *Ondatra* (as Hibbard and I believe) then a continuum, or temporal cline, in these parameters should exist that will document the evolutionary change of one genus into the other.

453

Clines within these genera have been demonstrated (Semken, 1966; Zakrzewski, 1969), as has overlap in size between the two (Zakrzewski, 1969: fig. 11). Therefore, the presence or absence of cement in these intermediate forms would appear to be the best criterion for taxonomic assignment at the generic level. However, a problem exists even with this parameter; as first pointed out by Hibbard (1959), in primitive species of *Ondatra* cement is not formed in the re-entrants until well into the adult or old adult stages. I believe the development of the dentine tract in the ondatrine from Arizona to be almost intermediate between those of the most advanced species of *Pliopotamys* and the most primitive species of *Ondatra* so far known, but the individual is only in the young adult stage of ontogenetic development; therefore, I do not know to which of the above two ondatrine genera the specimen should be assigned. A similar problem exists with the specimen that Hibbard (1959) assigned to *Ondatra* from the Borchers local fauna (Aftonian) of Kansas. This specimen (an isolated lower first molar) is immature and lacks cement in the re-entrant angles. However, in the development of dentine tracts and size it compares favorably with individuals of *O. idahoensis* from the Grand View local fauna of Idaho and is correctly assigned.

Because of the stage of evolution of the specimen from Arizona and the fact that it was not found at the type locality of the Benson local fauna (late Pliocene), I thought it might be a member of a younger local fauna that inhabited the area subsequent to the Benson local fauna. However, Dr. Richard H. Tedford of the American Museum of Natural History (written communication) stated that, based on field work by Ted Galusha, the site from which the specimen was obtained is at the same stratigraphic level as the type locality of the Benson local fauna, and that both sites are well below the lowest Curtis Ranch local fauna sites of middle Pleistocene age. These facts suggest that an early radiation of the Ondatrini occurred in the American southwest. Hibbard (1959) and Bjork (1970) suggested that *Pliopotamys* might have been a western form until mountain glaciation resulted in their movement onto the plains in the Pleistocene.

The Benson local fauna is considered to be approximately temporally equivalent to the Rexroad local fauna of southwestern Kansas, whereas the Hagerman local fauna is considered to be no older than the former two and probably slightly younger, but still late Pliocene in age (Zakrzewski, 1969). If these temporal associations are correct, then the relationship between *Pliopotamys minor* from the Hagerman local fauna and the ondatrine from the Benson local fauna is another example of a more primitive member of a lineage being present in the former fauna and a more advanced member in the latter fauna. I (1969) reported the presence of a pocket gopher with rooted teeth, *Pliogeomys parvus*, from the Hagerman local fauna, whose characteristics, with the exception of roots and size, are similar to those of *Geomys minor*, the type of which was described by Gazin (1942) from the Benson local fauna.

Pliopotamys is not known from the Rexroad local fauna, but *Ogmodontomys*, which is present, might have been its ecological equivalent. The latter genus becomes extinct on the plains shortly after the appearance of *Pliopotamys*.

SPECIMENS FROM CALIFORNIA

When I examined these specimens a few years ago, the presence of a well-developed fourth triangle on UCMP 57958 suggested that the tooth might be assignable to the genus *Pliopotamys*. The specimens from California approach *Pliopotamys* in length of re-entrant angles and width of tooth; however, the occlusal length is much shorter and the enamel is thinner. In UCMP 57958 the fourth alternating triangle is triangular rather than knob-like, as in most of the four-triangled *Cosomys* from UM-Ida.la-65 of the Hagerman local fauna (Zakrzewski, 1969: fig. 7E). Nor is there any modification of the alternating triangle into a prism fold or enamel ridge as is generally found in *C. primus*. The third external re-entrant angle does not appear to be very deep, but, as the specimen is broken just anterior to this point, not much certainty can be attached to this observation. UCMP 32952 resembles *Cosomys* in that it possesses an enamel ridge on the fourth triangle. The ondatrines from California are much more robust than the four-triangled *Cosomys* from UM-Ida.la-65. They are about the length of the three-triangled *Cosomys* from the other localities, but the width of UCMP 57958 falls outside the observed range of 252 measured specimens. The type of *C. primus* compares more favorably with the three-triangled *Cosomys* from the Hagerman local fauna.

The two specimens from California appear to be more closely related to each other (on the basis of thinness of enamel and the development of re-entrant angles, which are long and narrow) than either is to *Pliopotamys* or *Cosomys*. UCMP 57958 is much wider than UCMP 32952, but this may be due to ontogenetic differences between the individuals. Possibly these specimens represent another species of *Cosomys* that was more advanced than *C. primus*, or a small species of *Pliopotamys*, or another genus of arvicoline. The imperfect condition of the specimens precludes a definitive judgement at this time.

ACKNOWLEDGMENTS

For the loan of and permission to publish on specimens in their care, I am indebted to John A. White (Museum, Idaho State University), Richard H. Tedford (Division of Vertebrate Paleontology, American Museum of Natural History), and Donald E. Savage (Museum of Paleontology, University of California). I thank Jerry R. Choate (Museum of the High Plains, Fort Hays Kansas State College) and Claude W. Hibbard (Museum of Paleontology, University of Michigan) for the loan of specimens in their care and for critically reading the manuscript. David P. Whistler (Division of Vertebrate Paleontology, Los

Angeles County Museum of Natural History) made available the type of *Cosomys primus* for study.

The drawings were penciled by Lisa Hansen (Idaho State University). This study began at Idaho State University while I was on a Postdoctoral Fellowship sponsored jointly by Idaho State University and the Los Angeles County Museum of Natural History (NSF-GB5116).

Literature Cited

Bjork, P. R. 1970. The carnivora of the Hagerman local fauna (late Pliocene) of southwestern Idaho. Trans. Amer. Phil. Soc., n.s., 60:1–54.

Gazin, C. L. 1942. The late Cenozoic vertebrate faunas from the San Pedro Valley, Ariz. Proc. U.S. Nat. Mus., 92:475–518.

Hershkovitz, P. 1962. Evolution of neotropical cricetine rodents (Muridae), with special reference to the phyllotine group. Fieldiania: Zool., 46:1–524.

Hibbard, C. W. 1959. Late Cenozoic microtine rodents from Wyoming and Idaho. Papers Michigan Acad. Sci., Arts and Letters, 44:3–40.

Hibbard, C. W., and R. J. Zakrzewski. 1967. Phyletic trends in the late Cenozoic microtine *Ophiomys* gen. nov., from Idaho. Contrib. Mus. Paleo., Univ. Michigan, 21:255–271.

Kowalski, K. 1970. Variation and speciation in fossil voles. Symp. Zool. Soc. London, 26:149–161.

Kretzoi, M. 1969. Skizze einer Arvicoliden—Phylogenie—Stand 1969. Vert. Hung., Mus. Nat. Hist. Hung., 11:155–193.

Quay, W. B. 1954. The anatomy of the diastemal palate in microtine rodents. Misc. Publ. Mus. Zool., Univ. Michigan, 86:1–41.

Semken, H. A., Jr. 1966. Stratigraphy and paleontology of the McPherson *Equus* Beds (Sandahl local fauna), McPherson County, Kansas. Contrib. Mus. Paleo., Univ. Michigan, 20:121–178.

Shotwell, J. A. 1970. Pliocene mammals of southeast Oregon and adjacent Idaho. Bull. Mus. Nat. Hist., Univ. Oregon, 17:1–103.

Simpson, G. G., A. Roe, and R. C. Lewontin. 1960. Quantitative Zoology. Harcourt, Brace, and World Inc., New York, rev. ed., 440 pp.

Wilson, R. W. 1932. *Cosomys*, a new genus of vole from the Pliocene of California. J. Mamm., 13:150–154.

Zakrzewski, R. J. 1969. The rodents from the Hagerman local fauna, upper Pliocene of Idaho. Contrib. Mus. Paleo., Univ. Michigan, 23:1–36.

Sternberg Memorial Museum and Department of Earth Sciences, Fort Hays Kansas State College, Hays, 67601. Submitted 6 April 1973. Accepted 29 August 1973.

TYPE LOCALITIES OF COPE'S CRETACEOUS MAMMALS

Robert W. Wilson
Museum of Geology
South Dakota School of Mines and Technology, Rapid City

ABSTRACT

It is generally stated in paleontological literature that J. L. Wortman found the types of two species of Late Cretaceous mammals in unknown parts of South Dakota. These species, subsequently described and named by E. D. Cope, are **Meniscoessus conquistus** (probably the first Cretaceous mammal to be found and described), and **Thalaeodon padanicus**. They are the only Cretaceous mammals of published record from the state.

Review of some neglected sources of information leads to the conclusion that: (1) the type of **Meniscoessus conquistus** came from Dakota Territory, but not necessarily from South Dakota, and (2) E. D. Cope, rather than Wortman, found the type of **Thlaeodon padanicus,** and this specimen came from Hell Creek beds along the Grand River approximately four miles southeast of Black Horse.

E. D. Cope named and described two genera of Cretaceous mammals: these were the multituberculate *Meniscoessus* in 1882, and the marsupial *Thlaeodon* in 1892, with type species *M. conquistus* and *T. padanicus* respectively. Cope credited J. L. Wortman with the discovery of *Meniscoessus conquistus,* but said nothing about the type locality. In his description of *Thlaeodon padanicus,* he said nothing about either the discoverer or the place of discovery, except to state that the upper and lower jaws were found about one hundred feet apart, but probably pertained to a single individual. At a considerably later time, G. G. Simpson (1929) and others have stated that the type specimens of both *M. conquistus* and *T. padanicus* were found by Wortman in the "Laramie" [Lance] of South Dakota, but that no other locality data were available.

The Museum of Geology of the South Dakota School of Mines and Technology has been exploring the Hell Creek (Late Cretaceous) of South Dakota for mammals.[1] In an attempt to gain clues as to where Wortman might have found his specimens, I searched such literature as was available to me with care. As a result, I have reached tentative conclusions at variance with those of Simpson.

In respect to *Meniscoessus conquistus* not much can be said with assurance. A note by Wortman (1885, p. 296) states that Hill (Russell?) and Wortman found the type in the summer of 1883 (*sic,* but

[1] Work supported by National Science Foundation grant G23646

surely 1882) in Dakota. Because the division of the Territory into the present states of North and South Dakota did not take place until 1889, the question arises as to how it is known that the locality was in what is now South Dakota if nothing is known about the details of the locality. The only slight clue I can uncover is that a year after Wortman's finding of *Meniscoessus,* Cope, himself, was exploring the Cretaceous of the Dakota Territory. In a letter to his wife dated August 28, 1883 (Osborn, p. 306), and written at what is seemingly now Medora, North Dakota, he says in describing local outcrops: "This is the formation from which Wortman got the *Meniscoessus.*" This sentence can be taken literally as simply that the specimen came from Cope's Laramie Formation, or with more license that he meant these are the outcrops from which the specimen came. In the same letter, he wrote that he planned to go 30 miles south where the "badlands are said to be exceptionally bad." If he were following Wortman's footsteps at this point, he would have been approximately 45 miles north of the state line. After proceeding this far south along the Little Missouri, Cope went southeastward to White Buttes before turning back to Medora. White Buttes was his closest approach to South Dakota on this trip of several days, and he was then still about 30 miles from South Dakota. It may be that in the general area bounded by Medora, Marmath, and Bowman, North Dakota, Wortman found the type of *Meniscoessus,* but even if he did not, it is highly uncertain that the discovery was made in the South Dakota of today. As a matter of fact, most of the outcrops south of the state line for some miles may be somewhat too high in the geologic section for *Meniscoessus.*

In respect to the type locality of *Thlaeodon padanicus,* there are several bits of evidence suggesting (1) that Cope rather than Wortman found the specimen, and (2) that it was in fact found in South Dakota along the south bank of the Grand River southeast of Black Horse. These lines of evidence are itemized below.

1. Nowhere in the account published in 1892 in the American Naturalist does Cope credit Wortman with discovery of *Thlaeodon padanicus.*

2. The Indian name for the Grand River is Padani, and hence the specific name *T. padanicus* is a broad hint as to locality.

3. In the year of its discovery, Cope prospected along the Grand River. Wortman was also in South Dakota, but was occupied by collecting in the Big Badlands to the south, and such Cretaceous collections as he made seemed to have been in the Lance Creek area of Wyoming. In any case, even before the summer of 1892, he had left the employ of Cope, and was working for the American Museum of Natural History.

4. In a letter to his wife dated July 17, 1892 (Osborn, p. 431),
   Cope says, "We made noon camp on the bank of Grand R. and
   then climbed the bluffs on the S. side leaving the Rock Creek
   and this subagency to the N. We followed this high land,
   driving through the Grass, sometimes with, sometimes without trail. We had great distance views, fine air, and plenty
   of flowers. During the afternoon we crossed Five (sic, for
   Fire) Steel Creek, which comes in from the South. As evening
   approached thunderclouds arose in the W. and I began to
   think of camp. Oscar however drove on, and the Sioux boy
   kept ahead. As it grew late we turned down a low hill to the
   left and climbed a low bench at the foot of an opposite hill.
   I saw a low bare bank and lying around white objects. I told
   Oscar to let me get out, as I thought I saw bones. Sure enough
   the ground was covered with fragments of Dinosaurs, small
   and large, soon we found water and stopped for camp."
   ingly thought; see 1931, p. 415).

5. In the letter above-mentioned (Osborn, p. 443), Cope states
   his results as, "In the 3 days I collected I got 21 species of
   vertebrates, of which 3 are fishes, and all the rest reptiles
   except one mammal. This is a fine thing, the most valuable
   I procured, and new as to species at least; and it throws important light on systematic questions." This mammalian
   specimen is not otherwise accounted for in collections if it is
   not the type specimen of *T. padanicus* (as H. F. Osborn seem-

Reference to a geological map (Firesteel Creek Quadrangle,
South Dakota State Geological Survey) shows that the closest exposures from whence these bones could come after the Firesteel
crossing is in the vicinity of section 25, T. 20N, R. 22E, or sections 29
and 30, T. 20N, R. 23E. A good skeleton of *Anatosaurus* in the Museum of Geology collections is from the southwest corner of the
SW¼ of section 25, T. 20N, R. 21E. The type of *Thlaeodon padanicus*
surely came from somewhere in the area of these localities.

LITERATURE CITED

Cope, E. D., 1882, Mammalia in the Laramie Formation. Amer. Nat., v. 16,
    pp. 830-831.

——————————, 1892, On a New Genus of Mammalia from the Laramie
    Formation. Amer. Nat., v. 26, pp. 758-762, pl. xxii.

Osborn, H. F., 1931, Cope: Master Naturalist. Princeton Univ. Press, xvi
    plus 740 pp., 30 figs. Princeton.

Simpson, G. G., 1929, American Mesozoic Mammals. Mem. Peabody Mus.
    Yale Univ., v. 3, pt. 1, xv plus 235 pp., 62 text-figs., 32 pls.

Wortman, J. L., 1885, Cope's Tertiary Vertebrata. Amer. Jour. Sci. (3), v. 30,
    pp. 295-299.

THE ADAPTIVE RADIATION OF THE PHENACODONTID CONDYLARTHS AND THE ORIGIN OF THE PERISSODACTYLA[1]

Leonard B. Radinsky

Department of Biology, Brooklyn College, Brooklyn, New York

Accepted March 28, 1966

The mammalian order Condylarthra includes a heterogeneous assemblage of small- to medium-sized archaic omnivores and herbivores. Most families in the order flourished in the Paleocene and became extinct early in the Eocene. A few lineages, however, developed crucial adaptations which led to their emergence as new orders of mammals, one of which was the Perissodactyla. The origin of the Perissodactyla is better documented than that of any other order of mammals and provides an excellent opportunity to study the emergence of a major taxon.

Dental evidence indicates that perissodactyls were derived from the condylarth family Phenacodontidae. To view in proper perspective the evolutionary changes which led to the origin of the Perissodactyla, it will be necessary to survey the adaptive radiation of the Phenacodontidae.

The oldest true phenacodontid condylarth, *Tetraclaenodon*, first appears in faunas of middle Paleocene age, and by the beginning of the late Paleocene appears to have radiated into three main groups, represented respectively by *Phenacodus*, *Ectocion*, and an as yet unknown proto-perissodactyl. Forms transitional between *Tetraclaenodon* and *Phenacodus* (primitive species of *Phenacodus*), and between *Tetraclaenodon* and *Ectocion* (the genus *Gidleyina*) are known, but no intermediates between *Tetraclaenodon* and the most primitive known perissodactyl, the early Eocene genus *Hyracotherium*, have yet been found. However, *Tetraclaenodon* is the most advanced form which is

still unspecialized enough to have given rise to *Hyracotherium*. (The occurrence of incipient mesostyles in a small number of *Tetraclaenodon* specimens does not preclude this possibility; the alternative hypothesis, that proto-perissodactyls and *Tetraclaenodon* were independently derived from a still more primitive common ancestor, requires an additional complicating factor—an independent acquisition of molar hypocones by perissodactyls and phenacodontids.) Thus, in the absence of evidence to the contrary, *Tetraclaenodon* may be considered directly ancestral to perissodactyls. The major morphological changes involved in the evolution of the *Tetraclaenodon* stock into *Phenacodus*, *Ectocion*, and *Hyracotherium*, fall into two functional categories, one concerned with mastication and the other with locomotion.

MASTICATION
Dentition

The main changes involved in the evolution of the phenacodontid dentition occur in the molar teeth. The molars of *Tetraclaenodon* (see Fig. 1) are low-crowned, with low, obtuse cusps. The first and second upper molars are advanced over the primitive tritubercular molar pattern by the addition of a fourth main cusp, the hypocone. There are two relatively large intermediate cuspules, the protoconule and metaconule, and broad anterior and posterior cingula. The third upper molar is smaller than the second and lacks a hypocone. In the lower molars the paraconid has been reduced, leaving two main anterior cusps, the protoconid and metaconid, and a prominent anterior ridge, the paralophid. There are

[1] This work was supported in part by National Science Foundation Grant GB-2386.

408

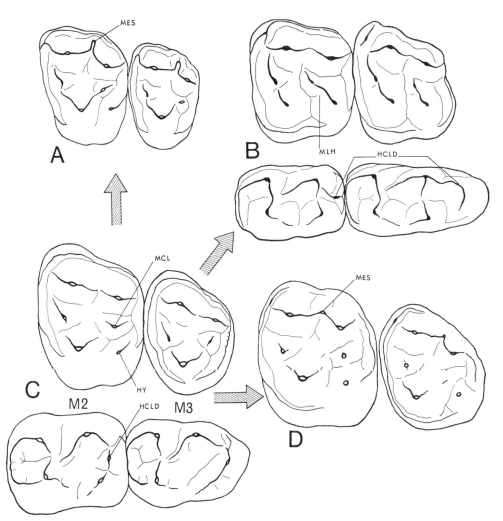

FIG. 1. Second and third molars of A. *Ectocion*, B. *Hyracotherium*, C. *Tetraclaenodon*, and D. *Phenacodus*. Lower molars of *Ectocion* and *Phenaodus* have the same basic cusp pattern as is seen in *Tetraclaenodon* and are therefore omitted. All about × 3. Abbreviations: HY, hypocone; HCLD, hypoconulid; MCL, metaconule; MES, mesostyle; MLH, metaloph.

three posterior cusps, a large hypoconid and slightly smaller entoconid and hypoconulid. The third lower molar is narrower posteriorly than the second. The wear facets on the molars of *Tetraclaenodon* suggest that both crushing and shearing occurred in mastication, with the emphasis apparently on crushing.

The teeth of *Phenacodus* are very similar to those of *Tetraclaenodon*, having low, obtuse cusps and ridges. The main

differences are the development of a small mesostyle on the upper molars and the enlargement of the posterior cingulum into a hypocone on the third upper molar. The upper molars are relatively long (anteroposteriorly) and narrow. As in *Tetraclaenodon*, the broad low cusps are more adapted for crushing than shearing. The addition of a hypocone on the third upper molar increases the surface area available for chewing. The mesostyle is not large

enough to add significantly to the ectoloph area.

In molars of *Ectocion* the cusps are
relatively higher and more acute and the
ridges connecting cusps are more prominent than in *Tetraclaenodon* or *Phenacodus*. The ectoloph is higher relative to the
lingual cusps and is folded into a prominent mesostyle. The upper molars are
relatively short and wide. The third upper molar does not develop a hypocone.
On the lower molars the paraconid is lost
and the paralophid no longer extends to
the metaconid (as it does in *Phenacodus*).
The high, narrow cusps and ridges provide
steep occlusal surfaces, indicating relatively more shear and less crushing than
occurred in *Tetraclaenodon* or *Phenacodus*. The prominent mesostyle increases
the length of ectoloph available for vertical shear against the labial sides of
ridges on the lower molars.

The molars of *Hyracotherium*, like
those of *Ectocion*, have relatively higher
and more acute cusps and ridges than do
those of *Tetraclaenodon* or *Phenacodus*.
However, *Hyracotherium* is even more
advanced in this respect than is *Ectocion*,
for the crests connecting cusps are better
developed. An important modification in
cusp pattern has been brought about by
the loss of the protcone-metaconule connection, an anterior shift of the metaconule and the development of a hypo-
cone-metaconule crest. These changes result in a cusp pattern with two oblique
tranverse crests (an anterior protoloph
and posterior metaloph) separated by a
lingually open valley. Correlated with the
changes in upper molar pattern, in the
lower molars the hypoconulid has been
posteriorly displaced, leaving the posterior
sides of the hypoconid and equally large
entoconid clear for shear against the anterior side of the metaloph above. Another
new feature in the dentition of *Hyracotherium* is the enlargement of the third
molars. In *Hyracotherium* the upper third
molar has a hypocone and is as large as
the second molar. The third lower molar
is larger than the second, owing to the

great enlargement of the hypoconulid. (In
the first and second lower molars the
hypoconulid is reduced.) However, even
excluding the enlarged hypoconulid, the
third lower molar is still as large as the
second. Finally, the lower molars of
Hyracotherium are narrower relative to
the uppers than is the case in the phenacodontids.

The changes in cusp pattern and tooth
proportions in evolution from *Tetraclaenodon* to *Hyracotherium* indicate an increase in the amount of shearing (especially along transverse crests) and a corresponding decrease in the amount of
crushing in mastication. A shift toward
increased shearing also occurred in *Ectocion*, but in that genus the emphasis was
on vertical ectoloph shear. The enlargement of the third molars in *Hyracotherium*
provided greater occlusal surface and could
have been brought about simply by a
slight posterior shift of the molarization
field. The greatly enlarged hypoconulid of
the third lower molar served the function
in occlusion of a paralophid and presumably developed in correlation with the
molarization (and enlargement) of the
posterior half of the upper third molar.
The relatively narrower lower molars of
Hyracotherium required a greater degree
of transverse jaw movement for complete
occlusion with the uppers than was necessary in *Tetraclaenodon*.

Jaw Musculature

The structure of the lower jaw, known
for *Phenacodus*, *Ectocion*, and *Hyracotherium* (see Fig. 2), provides information on
the relative proportions of the main components of the jaw musculature. In mandibles of *Hyracotherium* the coronoid
process is relatively smaller and the angle
relatively larger than in *Phenacodus* or
Ectocion. In addition, the posterior border of the angle is thicker and more heavily scarred (from insertions of the external masseter and internal pterygoid
muscles) in *Hyracotherium*. These differences suggest that the masseter and internal pterygoid muscles were relatively

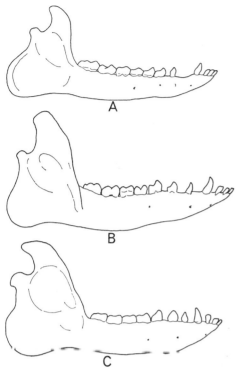

FIG. 2. Lower jaws of A. *Hyracotherium* (after Simpson, 1952), × ½; B. *Ectocion* (Yale Peabody Mus. no. 21211), × ⅔; C. *Phenacodus* (Princeton Univ. no. 14864), × ⅓. All in lateral view.

larger, and the temporalis, which inserts on the coronoid process, relatively smaller in *Hyracotherium* than in the phenacodontids.

In living ungulate herbivores the masseter-pterygoid complex is larger than the temporalis, while in carnivores the opposite is true (Becht, 1953, p. 522; Schumacher, 1961, pp. 143, 180). In carnivores, jaw movement is almost entirely confined to adduction, for which the temporalis is well suited, but in ungulates and many other herbivores transverse movement is important in mastication, and for transverse movement the deep part of the masseter and the internal pterygoid are more efficient than the temporalis (Smith and Savage, 1959, p. 297). Thus the relatively larger masseter and internal pterygoid musculature indicated by the jaw structure of *Hyracothe-*

rium suggests increased specialization for lateral jaw movement in mastication. This specialization of the jaw musculature correlates with the narrower lower molars and predominance of transverse shear indicated by the molar cusp patterns of *Hyracotherium*.

LOCOMOTION

Much of the postcranial skeleton is known for *Hyracotherium*, *Phenacodus* and, to a lesser degree, *Tetraclaenodon*, but that of *Ectocion* is largely unknown. Therefore the following discussion of locomotory adaptions will deal mainly with the first three genera.

Vertebral Column

Slijper (1946, p. 103) pointed out that with decreasing mobility of the vertebral column in ungulates the longissimus dorsi shifts its insertion posteriorly from lumbar to sacral vertebrae and consequently the neural spines of the lumbar vertebrae become less cranially, and even caudally, inclined. In *Phenacodus copei* (Amer. Mus. Nat. Hist. no. 4378) the lumbar neural spines are inclined cranially about 15 degrees from vertical. Kitts (1956, p. 21) states that the neural spine of the last lumbar vertebra of *Hyracotherium* is less cranialy inclined than that of *Phenacodus*. No specimen of *Hyracotherium* available to me preserves lumbar neural spines, but in *Heptodon posticus*, an early Eocene tapiroid similar in morphology to *Hyracotherium*, the neural spine of the last lumbar vertebrae (Mus. Comp. Zool. no. 17670) is inclined cranially about five degrees from vertical. This difference from the condition in *Phenacodus* suggests that the vertebral column in early perissodactyls was somewhat less flexible than that of *Phenacodus*.

Kitts (1956, p. 20) states that the zygapophyses of the lumbar vertebrae of *Hyracotherium* are embracing, but his illustration (*loc. cit.*, fig. 3) shows what appears to be a relatively flat prezygapophysis, similar to the condition in *Phenacodus*.

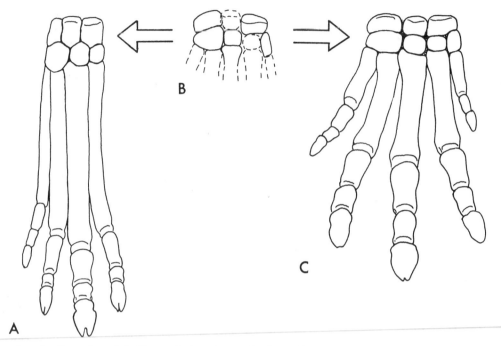

Fig. 3. Front feet of A. *Hyracotherium* (composite from Kitts, 1957, and Osborn, 1929, fig. 700), × ⅔; B. *Tetraclaenodon* (composite from AMNH nos. 2468 and 2547a), × 1; C. *Phenacodus* (AMNH no. 2961), × ⅓.

Forelimb

In *Tetraclaenodon* the humerus has a prominent deltoid crest, with the deltoid tubercle located on the distal half of the shaft, and a large medial epicondyle, with an entepicondylar foramen. The proximal end of the radius is about twice as wide as it is deep (anteroposteriorly) and articulates with the ulna along a wide flat facet, indicating loss of the ability to supinate. The carpus (see Fig. 3) is relatively low and wide, and has been called "alternating"; that is, in dorsal view the scaphoid rests partly on the magnum and the lunar partly on the unciform. The amount of overlap, however, is slight. Facets on the distal row of carpals indicate that there were five digits; except for the proximal head of the third metacarpal, the metacarpus is unknown.

The humerus, radius, and ulna of *Phenacodus* are similar to those of *Tetraclaenodon*, except that the deltoid crest of the

humerus is slightly weaker and the deltoid tubercle is higher on the shaft. The carpus of *Phenacodus* has been described as being of the serial type, i.e., with the scaphoid resting solely on the trapezoid and trapezium, and the lunar only on the magnum. This arrangement occurs in the large species of *Phenacodus*, *P. primaevus*, but in the small species *P. copei* (AMNH no. 16125), the lunar overlaps the unciform to about the same degree (which is very little) as in *Tetraclaenodon*.

The less prominent deltoid crest and higher deltoid tubercle suggest that the forelimb of *Phenacodus* was relatively less powerful but perhaps capable of more rapid movement than that of *Tetraclaenodon*. The small medial displacement of the lunar and scaphoid, resulting in loss of the lunar-unciform and scaphoid-magnum articulations in large species of *Phenacodus*, suggests a slight increase in importance of the ulna in weight support.

The forelimb of *Hyracotherium* differs from that of *Tetraclaenodon* in the following features: humerus with shorter and less prominent deltoid crest and more proximally located deltoid tubercle, greatly reduced medial epicondyle (with consequent loss of the entepicondylar foramen), and sharper intercondyloid ridge (= capitulum); radiohumeral index of about 1.0 compared to 0.8 in *Tetraclaenodon* and *Phenacodus*, ulna with narrower, less massive, more symmetrical olecranon; carpus relatively higher and narrower, with more extensive articulations between elements; cuneiform smaller and scaphoid displaced laterally to extensively overlap unciform and magnum, respectively; unciform, magnum, and scaphoid with larger posterior tuberosities; first digit lost and trapezium reduced to a tiny nubbin; remaining metacarpals relatively longer and thinner (the ratio of the length of the third metacarpal to the humerus is 1 : 2 compared to about 1 : 3 in *Phenacodus* and probably also *Tetraclaenodon*); fifth metacarpal relatively smaller.

All of these differences indicate increased specialization for running in *Hyracotherium*. The elongation of distal limb segments (radius and metacarpals) and reduction of lateral digits increases the length of stride and makes the limb a more effective lever. The reduction of the medial epicondyle probably correlates with the decreased importance of the pronator teres (which originates on that epicondyle), for the manus is fixed in a permanently pronated position, and may also correlate with the decrease in importance of the ulna as a weight-bearing element of the forearm. The latter change is indicated by the reduction in size of the cuneiform and lateral displacement of the lunar and scaphoid, which increases the relative size of the area of manus under the radius. The alternating arrangement of the carpals and more compact carpus make the wrist less flexible but better for resisting stresses. The larger posterior tuberosities on several of the carpals indicate more powerful flexor musculature. The

sharper intercondyloid ridge on the humerus restricts lateral movement at the elbow joint. The weaker deltoid crest, higher deltoid tubercle, and narrower and less asymmetrical olecranon are features associated with increased cursoriality. Thus, in a complex of features, the forelimb of *Hyracotherium* is more specialized for running than is that of *Tetraclaenodon* or *Phenacodus*.

Hind Limb

In *Tetraclaenodon* the greater trochanter of the femur is only slightly higher than the head, the lesser trochanter is very weak, and the third trochanter is large and located about two-fifths of the way down the shaft. The cnemial crest of the tibia is relatively large and extends about halfway down the shaft, the grooves for the astragalus are broad and very shallow, and the medial malleolus and distal end of the fibula (lateral malleolus) are large and massive. The astragalus has a relatively flat, low, and wide trochlea with a foramen, a relatively long neck, and a dorsoventrally flattened, convex head. The posterior astragalocalcaneal articulation is only slightly rounded. The calcaneum has a large peroneal tubercle and the ectocuneiform a large plantar process. The pes is pentadactyl, with the lateral toes slightly reduced (see Fig. 4).

The hind limb of *Phenacodus* is similar to that of *Tetraclaenodon*, differing in the following features: femur with larger lesser trochanter; tibia with weaker cnemial crest, smaller medial malleolus, and slightly deeper grooves for astragalus; fibula relatively slimmer, with smaller distal end; astragalus with a slightly relatively higher, narrower, and more deeply grooved trochlea, a slightly more curved posterior astragalocalcaneal facet, no astragalar foramen, and a deeper (dorsoplantarly) head; first and fifth metatarsals slightly more reduced.

The enlarged lesser trochanter of the femur suggests a stronger iliopsoas, an adductor of the femur. The reduction of the cnemial crest suggests reduced power

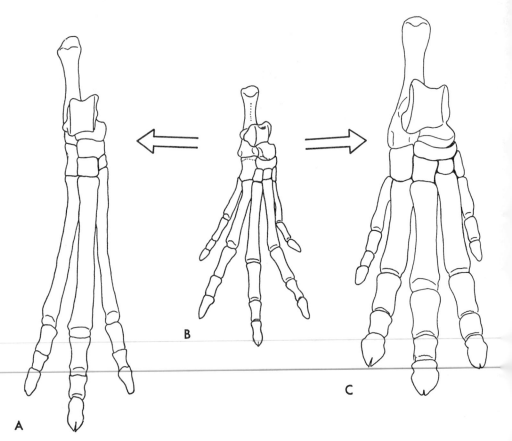

Fig. 4. Hind feet of A. *Hyracotherium* (from Kitts, 1956), × ½; B. *Tetraclaenodon* (from Matthew, 1897), × ½; C. *Phenacodus* (AMNH no. 293), × ⅓.

but increased speed in the hind limb. The more deeply grooved astragalar trochlea helps restrict lateral movement at the upper ankle joint and reduces the necessity for large lateral and medial malleoli. The loss of the astragalar foramen allows a slightly greater arc of rotation of the astragalus on the tibia. The more curved posterior astragalocalcaneal facet and deeper astragalar head may be related to a more digitigrade posture, which is suggested by the reduction of the lateral toes. In all of these features the hind limb of *Phenacodus* is slightly more specialized for running than is that of *Tetraclaenodon*.

The hind limb of *Ectocion* is known only from an astragalus and part of a calcaneum (AMNH no. 16127). The astrag-

alus (see Fig. 5) differs from that of *Tetraclaenodon* in having a slightly higher and narrower tibial trochlea with a slightly deeper groove and no astragalar foramen, a more anteriorly directed posterior calcaneal facet, a wider neck with a high anteroposteriorly oriented ridge at the dorsolateral corner, and a slightly flatter and deeper navicular facet. The high dorsolateral ridge probably marks the attachment of a strong lateral astragalocalcaneal ligament, which suggests restriction of rotation between astragalus and calcaneum. This interpretation is supported by the less oblique posterior calcaneal facet and the flatter head (the latter indicates less movement between astragalus and navicular). These features suggest a slight

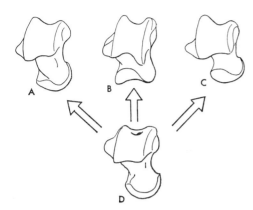

FIG. 5. Astragali of A. *Ectocion* (AMNH no. 16127), B. *Hyracotherium*, C. *Phenacodus*, D. *Tetraclaenodon*. Not to scale.

loss of freedom for lateral movement in the tarsus of *Ectocion* compared with the condition in *Tetraclaenodon*. The anterior end of the calcaneum is as wide in *Ectocion* as in *Tetraclaenodon*, suggesting that the pes of *Ectocion* was pentadactyl.

The hind limb of *Hyracotherium* differs from that of *Tetraclaenodon* in the same features mentioned for *Phenacodus*, but to a greater degree and with additional modifications. The latter include: femur with higher greater trochanter and more proximally located third trochanter; cnemial crest of tibia does not extend as far distally; first and fifth digits lost and remaining metatarsals relatively longer (length of third metatarsal/femur = 0.50 in *Hyracotherium* compared to 0.35 in the phenacodontids); tarsus relatively narrower and more compact, and astragalus, calcaneum, and navicular modified to eliminate the possibility of lateral movement of the foot.

The higher greater trochanter (which provides better leverage for the gluteal muscles, important abductors of the femur), more proximally located third trochanter, shorter cnemial crest, and longer metatarsals, plus the modifications noted in *Phenacodus*, are cursorial specializations of *Hyracotherium* which occur also in other running mammals. The loss of the first and fifth toes and the great

elongation of the remaining metatarsals are not unusual cursorial adaptations in later forms but are extremely progressive features for an early Eocene mammal. They require a compact, relatively rigid tarsus and it is in modifications of the tarsus to provide a stable ankle joint that *Hyracotherium* was unique.

The interpretation of tarsal mechanics in extinct animals is necessarily limited by lack of knowledge of the tarsal ligaments, for the ligaments may be as important as the bone articulations in restricting movement. Thus the degree of tarsal movement inferred from the bones alone represents the maximum amount possible and in life the actual amount of movement may have been considerably less.

The configurations of the tarsal articulations in *Tetraclaenodon* suggest that lateral movements of the foot (eversion and inversion) were possible, resulting from a combination of rotation at the lower ankle joint (between astragalus and calcaneum) and transverse tarsal joint (between astragalus and navicular). The posterior astragalocalcaneal articulation is only gently curved and the astragalonavicular articulation resembles a shallow ball-and-socket joint. In *Hyracotherium* the posterior astragalocalcaneal articulation is bent into a right angle and is more vertically oriented, restricting rotation at the lower ankle joint, and the astragalonavicular articulation is saddle-shaped (with the distal end of the astragalus concave mediolaterally), allowing a small amount of dorsoplantar rotation but no lateral movement. The saddle-shaped astragalonavicular articulation is unique to the Perissodactyla and a diagnostic feature of the order.

The redistribution of weight necessitated by the loss of the lateral toes and relative enlargement of the middle digit in *Hyracotherium* is reflected in the narrower, more compact tarsus, in which the cuboid and calcaneum are narrower (the peroneal tubercle of the calcaneum is lost), the neck of the astragalus shorter, wider, and deeper, and the head more closely appressed to the calcaneum, and the entocuneiform reoriented so that the vestigial first metatarsal is located behind the ectocuneiform and third metatarsal where it serves as attachment for deep flexor muscles and as a brace for the tarsus (Radinsky, 1963). The plantar process of the ectocuneiform is lost, its function apparently having been usurped by the reoriented vestige of the first metatarsal. Thus virtually the whole tarsus of *Hyracotherium* was remodeled to provide the stability required by the loss of lateral toes and great elongation of the metatarsus. Versatility was sacrificed for increased efficiency in running.

DISCUSSION

Absolute dating of the early Tertiary (Evernden *et al.*, 1964) indicates that evolution from *Tetraclaenodon* to *Hyracotherium* took place in less than five million years. Considering the magnitude of the morphological changes involved, the speed of that transition indicates a considerably higher rate of evolution in late Paleocene proto-perissodactyls than occurred during most of the subsequent 55 million years of perissodactyl evolution. This fact, coupled with the evidence of a major adaptive radiation of perissodactyls at the beginning of the Eocene, suggests that the origin of the Perissodactyla coincided with a shift to a new adaptive level.

The two major areas of specialization of the earliest perissodactyls, as far as the paleontological evidence indicates, are in mastication and locomotion, and there is evidence of experimentation among the condylarths in both of these fields. The dentition of *Phenacodus* is essentially a conservative continuation of the basic *Tetraclaenodon* pattern, while that of *Ectocion* is specialized for vertical shear. The molars of *Ectocion* are more specialized for vertical shear than are those of *Hyracotherium*, but are less specialized for transverse shear. In the closely related meniscotheriid condylarths, *Meniscotherium* has teeth which are more specialized for vertical shear than those of *Hy-*

racotherium and at least as specialized, although in a somewhat different way, for transverse shear. The specialization for transverse shear is also reflected in the mandible of *Meniscotherium*, which has a relatively large angular process and small coronoid process. This experimentation in dentition among condylarths suggests that a variety of ecological niches for medium-sized browsers was open at the beginning of the late Paleocene.

Phenacodus, Ectocion, and *Meniscotherium* appear to have been only slightly more specialized for running than was *Tetraclaenodon*, although the astragalus of *Ectocion* suggests that lateral movement at the ankle joint may have been restricted by ligaments. In *Hyracotherium*, however, a radical and unique remodeling of the ankle joint prevented lateral movement and made possible a precocious elongation of the metatarsals and reduction of the lateral digits. Other specializations for running are evident in the forelimb of *Hyracotherium*.

During the early Eocene perissodactyls underwent an extensive radiation while phenacodontid and meniscotheriid condylarths became extinct. Since the meniscotheriid dentition was at least as specialized for shear as was that of *Hyracotherium* it would seem that the masticatory specialization was less important for the success of the Perissodactyla than the adaptations for running. The early perissodactyls were considerably more specialized for running than were the contemporary predators, while the condylarths were not. It is surely no coincidence that the other major order of medium-sized to large herbivores, the Artiodactyla, appeared at the same time as the Perissodactyla, with their main adaptive feature a cursorial modification of the ankle joint (see Schaeffer, 1947). Thus it would seem that predator pressure, resulting in a major cursorial specialization, was the critical selective force involved in the origin of the Perissodactyla. Unfortunately there is no direct evidence of the ecological factors involved, for the faunas in which the condylarth-perissodactyl transition took place have not yet been discovered. The absence of perissodactyls in known late Paleocene faunas and their sudden appearance in abundance at the beginning of the Eocene suggests migration from an unknown area. Thus early perissodactyls may have originated isolated from, and perhaps under different selective pressures than, other descendant lineages of the middle Paleocene *Tetraclaenodon* stock.

SUMMARY

The middle Paleocene phenacodontid condylarth genus *Tetraclaenodon* gave rise to three late Paleocene groups, represented by *Phenacodus*, *Ectocion*, and an as yet unknown proto-perissodactyl. The main morphological changes indicated by the fossil evidence of this evolutionary radiation are specializations for mastication and locomotion. Molars of *Phenacodus* are very similar to those of *Tetraclaenodon*, with low broad cusps apparently mainly adapted for crushing. Teeth of *Ectocion* have prominent W-shaped ectolophs, an adaptation for vertical shear, while molars of *Hyracotherium*, the most primitive known perissodactyl, are specialized for both vertical and transverse shear. *Phenacodus* and *Ectocion* show little specialization for running over the primitive ambulatory condition of *Tetraclaenodon*, but the limbs of *Hyracotherium* display major cursorial modifications, including a unique remodeling of the ankle

which prevented lateral movement at that joint and made possible a precocious elongation and narrowing of the metatarsus.

LITERATURE CITED

BECHT, G. 1953. Comparative biologic-anatomical researches on mastication in some mammals. Proc. Kon. Nederl. Akad. Wetensch., (C) **56**: 508–527.

EVERNDEN, J. F., D. E. SAVAGE, G. H. CURTIS, AND G. T. JAMES. 1964. Potassium-argon dates and the Cenozoic mammalian chronology of North America. Amer. Jour. Sci., **262**: 145–198.

KITTS, D. 1956. American *Hyracotherium* (Perissodactyla, Equidae). Bull. Amer. Mus. Nat. Hist., **110** (1): 5–60.

MATTHEW, W. D. 1897. A revision of the Puerco fauna. Amer. Mus. Nat. Hist. Bull., **9** (22): 259–323.

OSBORN, H. F. 1929. Titanotheres of ancient Wyoming, Dakota, and Nebraska. U.S. Geol. Surv. Monograph 55 (2 vols.): 1–953.

RADINSKY, L. B. 1963. The perissodactyl hallux. Amer. Mus. Novit., **2145**: 1–8.

SCHAEFFER, B. 1947. Notes on the origin and function of the artiodactyl tarsus. Amer. Mus. Novit., **1356**: 1–24.

SCHUMACHER, G. H. 1961. Funktionalle Morphologie der Kaumuskulatur. G. Fischer, Jena, 262 pp.

SIMPSON, G. G. 1952. Notes on British hyracotheres. J. Linn. Soc. Zool., **42**: 195–206.

SLIJPER, E. J. 1946. Comparative biologic-anatomical investigations on the vertebral column and spinal musculature of mammals. Verh. Konink. Nederl. Akad. Wetensch. afd. Natuurk., (2) **42**: 1–128.

SMITH, J. M., AND R. J. G. SAVAGE. 1959. The mechanics of mammalian jaws. School Sci. Rev., **141**: 289–301.

GRADES AND CLADES AMONG RODENTS

ALBERT E. WOOD

Biology Department, Amherst College, Amherst, Massachusetts

Accepted September 30, 1964

As has been pointed out many times, the rodents are the most abundant and successful mammalian order. Their evolution has been channeled into a single major direction by the development, as an initial modification, of ever-growing, gnawing incisors, with associated changes in skull and jaw muscles. Subsequent evolution has involved a great deal of parallelism within the order, making it very difficult to disentangle the convergent and parallel changes from those that are truly indicative of phyletic relationship. The similarity in complexity of the evolutionary pathways among rodents to those among actinopterygians, and particularly teleosts, has also been pointed out.

Work by various authors has indicated that the evolution of the actinopterygians consists of the sequential attainment of a series of morphological stages, or grades (as in Huxley, 1958), each of which has been derived from the preceding one several independent times by a series of parallel trends. The classification of actinopts at the supraordinal level involves a series of taxa that are currently agreed to represent such polyphyletic grades rather than monophyletic units or clades (Schaeffer, 1956, p. 202).

The rodents were, classically, divided into three suborders on the basis of the structure of the jaw musculature and associated osteological differences—the Sciuromorpha, Myomorpha, and Hystricomorpha (Simpson, 1945). All recently proposed classifications of the order (Lavocat, 1956; Schaub, 1958, p. 691–694; Simpson, 1959; and Wood, 1955a and 1959), adopt the multiplicity of major groups postulated by Miller and Gidley (1918) or Winge (1924), and agree that the three classic suborders are not monophyletic clades, but rather, taken as a whole, represent a grade

that is an advance over the primitive rodent grade. The classic suborders represent alternative expressions of an advanced rodent grade, and may well have been achieved approximately simultaneously. The various clades within the order are still not clearly recognizable, and much work remains to be done before rodent cladal classification is stabilized to everyone's satisfaction, though considerable progress is being made.

There is no direct evidence as to the type of jaw muscles in the still unknown ancestral rodents that lived during the Paleocene. However, Edgeworth (1935, pp. 73 75), in discussing the primitive mammalian jaw musculature, indicates that a major part of it consists of an embryological single muscle mass, divisible into the *Temporalis, Zygomaticomandibularis,* and *Masseter.* The *Zygomaticomandibularis* is usually divided into anterior and posterior portions by the masseteric nerve. The masseter may be single or be divisible into two or more layers, with no clear indications as to which is the primitive condition.

Among students of rodent anatomy there have been many varying interpretations of the jaw musculature. Usually, the *Zygomaticomandibularis* has been considered to be part of the masseter (*Masseter medialis* of Tullberg, 1899, pp. 61–62; *Masseter profundus* of Howell, 1932, pp. 410–411), but sometimes it is treated as a separate muscle (Lubosch, 1938, p. 1068; Müller, 1933, pp. 14–24). The two parts of the masseter of Edgeworth are the *Masseter lateralis superficialis* and *Masseter lateralis profundus* of Tullberg, or the *Masseter superficialis* and *Masseter major* of Howell. Lubosch (1938, fig. 930) and Müller (pp. 19–20) also consider the anterointernal portion of what is usually called the masseter to be a distinct muscle, the *Maxillomandibularis*.

In the following discussion, the masseter is considered to consist of three parts— the *Masseter superficialis*, arising from the anterior end of the zygoma or the side of the snout and inserting on the ventral border of the angle (= *Masseter lateralis superficialis*); the *Masseter lateralis*, arising from most of the length of the lateral surface of the zygoma and inserting on the ventral part of the angular process (= *Masseter lateralis profundus*; *Masseter major*); and the *Masseter medialis*, arising from the medial side of the zygoma, whence it has sometimes spread to the medial wall of the orbit or forward through the infraorbital foramen, and inserting on the dorsal portion of the masseteric fossa of the jaw (= *Masseter profundus*; *Zygomaticomandibularis*; *Maxillomandibularis*). These are illustrated in Figs. 1–4.

The separation of evolutionary grades among the rodents can best be done on the basis of: (1) the incisor pattern and structure; (2) the structure of the jaw muscles and the associated areas of the skull and jaws; and (3) the general pattern and height of crown of the cheek teeth. These can be used as general clues to evolutionary grades throughout the order. The discussion below will largely be limited to these sets of criteria. On the other hand, the separation and identification of the clades must involve the use of all available data, and must not select one set of structures as the most critical one, with other criteria neglected.

GRADE ONE—PROTROGOMORPH RADIATION

The initial recorded rodent radiation, known from the Eocene but presumably having gotten well started in the later Paleocene, involved animals that had already acquired the basic gnawing adaptations.

The incisors were ever-growing, with the enamel limited to an anterior band, giving the perpetual chisel-edge that characterizes the Rodentia. The upper incisor was re-

curved, the worn surface being nearly vertical, and the lower incisor acted against it by moving upward and forward. The enamel cap had extended around the edges of the incisor, on both medial and lateral faces, to brace it better against the stresses of gnawing. The incisor enamel is of constant distribution on the incisor cross section, once the animal reached its adult size. Histologically, the incisor enamel in the Eocene members of the group is of the type called pauciserial by Korvenkontio (1934, p. 97, and fig. 1), in which the enamel is made up of irregular bands, ranging from a single row of enamel prisms, to as many as three or four rows of prisms. A change to the uniserial type of enamel (*op. cit.*, p. 130) has taken place in members of this radiation by the Oligocene.

As in all known rodents, there were no pre- or postglenoid processes, the glenoid fossa being elongate and slightly inclined from rear to front, so that the jaw could be moved backward bringing the cheek teeth into occlusion, or forward bringing the incisors together and separating the cheek teeth, vertically.

The dental formula had been reduced to the most primitive that is still found in living rodents. namely I_1^1, C_0^0, P_1^2, M_3^3. The cheek teeth were low-crowned and cuspidate in the earliest family (Paramyidae) or higher crowned and crested in derived families (Ischyromyidae, Sciuravidae), but were always based on a pattern of no more than four transverse crests. Occasional Eocene forms plus most later ones had hypsodont or even ever-growing cheek teeth (Cylindrodontidae, Aplodontoidea). Locomotion was largely scampering (or arboreal scampering), though some derivatives of this group had developed burrowing locomotion (Cylindrodontidae, Mylagualidae), and some may have been saltatorial (Protoptychidae).

The angle of the lower jaw was essentially in the same vertical plane as the rest of the jaw, as is usual among mammals. Specifically, it is usually in the plane of the incisive alveolus (sciurogna-

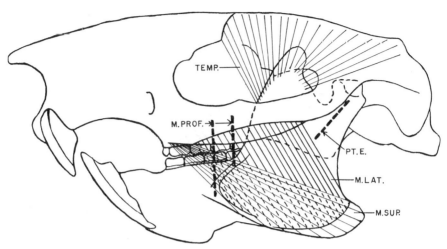

Fig. 1. Skull of the Eocene protrogomorph *Ischyrotomus*, with the jaw musculature restored, × 1. Abbreviations: M. LAT.—*Masseter lateralis*, dashed portions lying beneath *Masseter superficialis*; M. PROF.—dashed lines indicating the course of the *Masseter profundus*; M. SUP.—*Masseter superficialis*; PT. E.—dashed line indicating course of *Pterygoideus externus*; TEMP.—*Temporalis*.

thous), though occasionally (*Reithropara-mys*—Wood, 1962, fig. 41F) it has shifted to a position just laterad of the alveolus (incipiently hystricognathous).

The chief components of the jaw musculature were the temporal, the pterygoid, and the masseter. All showed a certain amount of differentiation (Fig. 1). In a form such as *Ischyrotomus*, the temporal was a large, fan-shaped muscle, arising in a semicircle from the frontal and parietal, and inserting on the coronoid process. Although the anterior fibers had a forward component and the posterior ones a backward component, its primary function was to raise the jaw, which pivoted about the condyle. The internal pterygoid, arising on the inner side of the pterygoid fossa and inserting on the inner surface of the angle (Wood, 1962, fig. 69B), pulled the jaw toward the midline as well as closing it. The external pterygoid (Fig. 1 PTE) arose on the external pterygoid process and inserted on the medial surface of the condyle. It helped to pull the jaw mesiad, but very largely served to slide the condyle forward and ventrad, along the glenoid cavity, to disengage the cheek teeth and bring the incisor tips into contact. The jaw was moved back again by the com-

bined action of the temporal and the digastric.

In *Ischyrotomus* the areas of origin and insertion of the *Masseter superficialis, M. lateralis*, and *M. medialis* are readily separable (Fig. 1). The *Masseter medialis* arose from the medial surface of the zygoma and inserted on the dorsal surface of the masseteric fossa of the lower jaw. It pulled the jaw nearly straight upward. There was the beginning of a differentiation of this muscle into two portions, the anterior inserting on the masseteric tuberosity by a separate tendon. It seems probable that these parts were separated by the masseteric nerve. The *Masseter lateralis* arose from a fossa extending most of the length of the zygoma, and occupying the ventral third of the arch. It inserted over much of the lateral surface of the angle, and pulled the lower jaw laterally, upward, and slightly forward. The most superficial of the three divisions of the masseter was the *Masseter superficialis*, which arose from the masseteric fossa on the base of the maxillary portion of the zygoma, immediately laterad of the upper premolars, and inserted along the ventral margin of the jaw all the way to the angle.

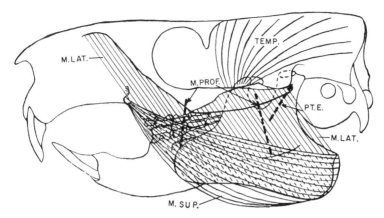

Fig. 2. Skull of the sciuromorphous sciurid *Marmota*, × 1. Abbreviations as for Fig. 1.

It was the major element in pulling the lower jaw forward, and hence in gnawing.

The functional activity of the jaws was composed of three parts (Becht, 1953, p. 515). A vertical or transverse movement, with the condyle toward the posterior end of the glenoid cavity, was used in the chewing activities of the check teeth. This would have involved the use of the main part of the temporal, the two inner parts of the masseter, and the internal pterygoid, and is the usual mammalian chewing activity. If the condyle were moved forward to the anterior end of the sloping glenoid cavity, the cheek teeth would be disengaged, and the same combination of muscles plus the *Masseter superficialis* would provide the motion of the lower incisor against the upper, resulting in gnawing. The third component, the shift from the first position to the second, would be brought about by the anterior portion of the temporal, the external pterygoid, and the *Messeter superficialis*; the reverse by the posterior portion of the temporal and the digastric.

The members of this grade include nearly all of the pre-Oligocene rodents of North America and Asia and some of those of Europe (none being known from the rest of the world). Several lines survive into the Oligocene or early Miocene, and the Aplodontoidea occur from the Oligocene to the present, mostly in North America, although some aplodontids are present in Palaearctica. This grade seems to include forms so related that they may be considered to be a clade, the Suborder Protrogomorpha.

GRADE TWO—SECOND RADIATION

Gnawing in the method outlined above was effective and presumably more efficient than that of the multituberculates or any of the other gnawing groups that were competing with the rodents in the Eocene. But the gradual filling of the available niches resulted in greater intra-ordinal competition and increased selective value for more efficient use of the incisors, which was brought about by a series of changes involving the muscles of mastication, the skull structure, the incisors, and the cheek teeth.

The modifications of the masseter muscle and the concomitant skull changes were the most prominent alterations leading to Grade Two. These changes involved either the *Masseter lateralis* or the *Masseter medialis* or both, the *Masseter superficialis* remaining essentially unchanged.

The *Masseter lateralis* may shift forward and upward, behind and median to the origin of the *Masseter superficialis*, onto the front of the zygomatic arch (Fig. 2). The shift was beginning in the ischyomyids *Titanotheriomys* (Wood, 1937, pp. 194–195, pl. 27, fig. 1, 1a, 1b) and *Ischy-*

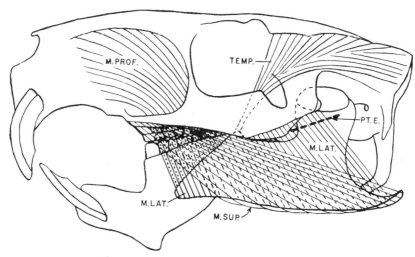

FIG. 3. Skull of the hystricomorphous caviomorph *Myocastor*, × 1. Abbreviations as for Fig. 1. Ventral part of *M. profundus* dotted.

romys troxelli (*op. cit.*, p. 191; Burt and Wood, 1960, p. 958), where the muscle was below, instead of lateral to, the infraorbital foramen. This process continued, with the muscle origin moving forward and upward along the anterior face of the zygoma, passing lateral and dorsal to the infraorbital foramen, eventually reaching almost to the top of the snout and forward onto the premaxillary. This pattern characterizes the sciuromorphous rodents—the Sciuridae, Castoroidea, and Geomyoidea. This shift of origin has changed the direction of pull of the anterior part of the *Masseter lateralis* by 30 to 60°, so that it essentially parallels the *Masseter superficialis*, greatly strengthening the forward component of masseteric action (Fig. 2).

In other rodents, the anterior part of the *Masseter medialis* has spread from the inner surface of the zygoma (or, perhaps, from the medial margin of the orbit) forward through the enlarged infraorbital foramen onto the snout (Fig. 3). In extreme cases, its origin extends as far forward as the premaxilla, almost reaching the posterior end of the external nares (*Hydrochoerus, Pedetes, Thryonomys*). This gives an almost horizontal resultant to the contraction of this muscle, and strongly aug-

ments the horizontal action of the *Masseter superficialis*. This pattern characterizes the hystricomorphous rodents—the Caviomorpha; the Dipodoidea, Theridomyoidea, and Thryonomyoidea; and the Anomaluridae, Ctenodactylidae, Hystricidae, and Pedetidae.

The Bathyergidae have developed perhaps the most massive masseters of any of the rodents, although there seems to have been very little shifting of the muscles (Tullberg, 1899, p. 78). The *Masseter medialis* has a broad expanse on the median side of the orbit (perhaps associated with the reduction of the eyes) and is confluent with the anterior end of the *Temporalis* (Tullberg, *op. cit.*, p. 75, and pl. 2, figs. 8–10, 17–18). In most members of the family, no part of the *Masseter medialis* passes through the small infraorbital foramen, but in *Cryptomys* (= *Georychus coecutiens*, Tullberg, 1899, p. 79) a small portion just edges through the foramen (*op. cit.*, pl. 2, fig. 17). Landry (1957, pp. 66–67) has argued that the small size of the infraorbital foramen and the limited forward extent of the *Masseter medialis* are secondary modifications of a hystricomorphous pattern, and that in spite of their differences, this family is relatively

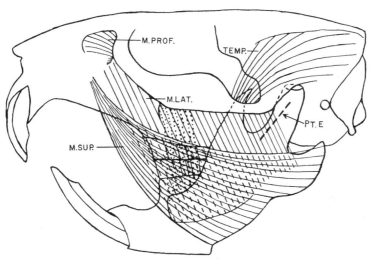

Fig. 4. Skull of the myomorphous cricetid *Ondatra*, × 1.5. Abbreviations as for Fig. 1. Ventral part of *M. profundus* dotted.

closely related to the Hystricidae. Most authors would not accept this conclusion. Since the earliest known bathyergids, from the Miocene of Kenya, were essentially identical in masseteric structure to living forms (Lavocat, 1962, p. 292), it is impossible to be certain of the direction of evolutionary change in this group. However, the *Masseter lateralis* seems to be in the process of spreading forward and upward onto the anterior side of the snout. This, together with the enlarged expanse of the *Masseter medialis* on the mesial side of the orbit, seem to be jaw muscle migrations sufficient to place these forms in Grade Two.

The expansion of the *Masseter medialis* onto the medial as well as lateral side of the orbit in bathyergids (Tullberg, 1899, pl. 2) and in *Castor* (*op. cit.*, pl. 22, fig. 9), putting it in an ideal position to expand through the infraorbital foramen if that opening were large enough, was probably a structural antecedent of the hystricomorphous pattern. Whether or not it indicates any close relationship between these forms and any histricomorphous rodents is arguable.

Finally, in the myomorphous rodents, both the *Masseter lateralis* and the *Mas-*

seter medialis have migrated, combining the features of the sciuromorphous and hystricomorphous groups (Fig. 4). This pattern characterizes the Muroidea, Spalacoidea, and Gliroidea. Such a type of masseter gives the greatest anteroposterior component of any of the types of rodent jaw musculature, with the possible exception of the paca (*Cuniculus*). It is perhaps not a coincidence that this pattern is found in the Muroidea, the most successful and cosmopolitan of all rodents.

At the same time that these changes in the masseter were occurring, the temporal muscle withdrew in most forms from the anterior area where it originated in *Ischyrotomus*, and is restricted in its origin to areas behind the tip of the coronoid process. In such forms it serves to raise the lower jaw and close the mouth or joins with the digastric and part of the *Masseter medialis* to move the jaw backward. However, the temporal keeps its anterior area of origin in the Bathyergidae and in some of the Rhizomyidae. Whether the conditions in these two families are primitive or secondary is unknown. The reduction of the temporal muscle continued in many rodents, especially those with enlarged auditory bullae (Howell, 1932, p. 411), so that in some it eventually became reduced to an exceedingly minute slip (Tullberg, 1899, pl. 9, figs. 8–9, *Ctenodactylus*; pl. 10, figs. 8–9, *Pedetes*; pl. 12, *Dipus* and *Alactaga*; and pl. 23, figs. 18–20, *Dipodomys*).

All of the sciuromorphous and myomorphous rodents and a number of the hystricomorphous ones (Theridomyoidea, Anomaluridae, Ctenodactylidae, and Pedetidae) have an angular process of the sciurognathous type, with the angle in the plane of the incisive alveolus. This is undoubtedly the primitive condition. In the other hystricomorphous rodents, the angle has shifted until it arises quite markedly laterad of the incisor. This would make the *Masseter lateralis* and *M. superficialis* more nearly vertical. This hystricognathous arrangement is fully developed in the earliest known (early Oligocene) members of the South American subordinal clade Caviomorpha (Wood and Patterson, 1959, p. 289) and of the African clade Thryonomyoidea (Wood, MS. 1), as well as in the Hystricidae, apparently of south Asiatic origin (Lavocat, 1962, pp. 292–293), and in the Bathyergidae.

Associated with these changes in the jaw muscles, but not necessarily occurring at precisely the same time, nor necessarily functionally correlated, there have been changes in the incisors, involving both their angulation and their histology. The lower incisors have usually become arcs of larger circles, so that they are more nearly horizontal, with the tips moving anteroposteriorly against the upper incisors. The upper incisors have tended to become either larger or smaller arcs, so that the tips tend to point either forward (true usually of burrowing forms), or slightly backward as is true of most living rodents. The former of these adjustments increases the ability to use the incisors as digging implements, with a corresponding increase in the rate of growth of the incisors, which reaches almost 0.5 cm per week in the lower incisors of geomyids (Manaro, 1959). The second change brings the enamel blades of

the upper and lower incisors more nearly into direct opposition than was true in Grade One.

Changes also took place in the histology of the incisor enamel. The pauciserial type has been modified, in members of Grade Two, in two different directions. In the uniserial type (Korvenkontio, 1934, p. 227), the lamellae are regular, and made up of one row of prisms each, with the prisms oriented in opposite directions in successive lamellae. This pattern is found in the Sciuridae, Castoridae, Geomyoidea, Gliridae, Muroidea, Spalacidae, Dipodoidea, and Anomaluridae among members of Grade Two, and in *Aplodontia*, *Meniscomys*, and *Ischyromys* among the members of Grade One (Korvenkotio, 1934, table on pp. 116–123).

The situation among the Theridomyoidea is most instructive. In the middle Eocene to Oligocene Pseudosciuridae, which are fully hystricomorphous in the infraorbital structure, the incisors are still pauciserial. The same is true of the more primitive members of the Theridomyidae, such as *Theridomys*. In more advanced theridomyids, there is a complete transition to the uniserial type of enamel. In *Issiodoromys* [= *Nesokerodon*] *minor*, Korvenkontio describes the enamel as "pauci-uniserial" (*op. cit.*, p. 116). He further describes that of *Protechimys gracilis* as pauciserial, and that of *Archaeomys laurillardi* as uniserial. These two forms are currently recognized as being two species of *Archaeomys* (Schaub, 1958, figs. 48–49). So in the Theridomyoidea, the transition from Grade One to Grade Two has occurred later in the incisor enamel than it did in the jaw musculature, the two apparently being completely independent.

A different type of enamel modification occurs in what Korvenkontio (*op. cit.*, p. 130) calls the multiserial type. Here each lamella is formed of four to seven identically oriented rows of prisms, the lamellae lying at an angle of about 45° to the surface of the enamel. Successive lamellae have the prisms oriented in opposite directions (Korvenkontio, 1934, pl. 8, figs. 3, 5, 7). This occurs in the Caviomorpha, and the Bathyergidae, Ctenodactylidae, Hystricidae, and Pedetidae.

Finally, there are likely to be differences in cheek-tooth formula or pattern associated with the change to Grade Two from Grade One. Primitively, the rodent cheek-tooth formula was P_1^2 and M_3^3, although some members of Grade One have lost P^3. This tooth has been preserved today only in *Aplodontia* and among the Sciuridae. In many rodents (most Caviomorpha, Anomaluridae, Castoroidea, Ctenodactylidae, Geomyoidea, Gliroidea, Hystricidae, and probably Pedetidae), P_4^4 have been retained. In such caviomorphs as the Echimyidae (Friant, 1936) and Capromyidae (Wood and Patterson, 1959, p. 324) and in the living African Thryonomyoidea (Wood, 1962, p. 316–317), the permanent premolars have been suppressed and the deciduous premolars are retained throughout life. This may also be true for the Pedetidae (Wood, MS. 2). According to Schaub (1958, p. 678), the reverse of this process occurs, with the elimination of the deciduous tooth in many hystricomorphous forms. Finally, the Muroidea and Spalacoidea have lost all the premolars and the Dipodoidea have almost reached this stage.

In summary, in Grade Two, there is a tendency to reduce the length of the tooth row, probably an adaptation permitting greater contrast between the gnawing and chewing activities, and therefore greater specialization in each. Usually, the loss of these teeth occurred at times when there are still gaps in the paleontological history of the groups. However, the loss of P^3 occurs within the known history of the Eomyidae (Wood, 1955b) and Gliridae (Schaub, 1958, figs. 201, 203), and the presence of P^4 is variable in living members of the Dipodidae (Schaub, 1958, p. 792).

Although the loss of cheek teeth brought about greater specialization of gnawing and chewing activities, it may have interfered with the functional activities of chewing,

since in almost all members of Grade Two there has been a tendency secondarily to elongate the cheek teeth by developing an additional transverse crest (mesoloph or mesolophid) in the middle of the teeth, making them five-crested in contrast to the four-crested pattern found in Grade One. This five-crested stage seems certainly to have developed independently in many lines, and therefore is no better than any other single criterion in determining the phylogenetic relationships (clades) among the rodents.

The changes in the jaw musculature look as though they are indicative of genetic relationships (i.e., clades), and were so used by most authors as far back as Brandt (1855) or even earlier, until fairly recently, giving three suborders of rodents, the Sciuromorpha, Hystricomorpha, and Myomorpha (see Simpson, 1945).

However, the use of other criteria for rodent classification complicated this apparently simple pattern. Tullberg (1899), for example, showed that rodents could be divided into two groups on the basis of the way in which the angle of the lower jaw originated—the Sciuragnathi, in which the angle arises in the plane of the alveolus of the lower incisor, and the Hystricognathi, in which it arises lateral to this plane. The hystricognathous forms include only those that are more or less hystricomorphous, whereas the sciurognathous ones may be sciuromorphous, myomorphous or hystricomorphous.

With an increase in the detailed studies of rodent paleontology since 1920, the chance that any of the three Brandtian suborders represents a clade has become progressively smaller, and students of fossil rodents have universally abandoned them at present.

The Sciuromorpha may be considered to be typical. The sciuromorphous condition was achieved by the squirrels (Sciuridae) in a transition, which is as yet not completely documented but that seems very probable, from a mid-Eocene paramyid such as *Uriscus* (Wood, 1962, p. 247;

Black, 1963, p. 229). A similar trend, not carried so far, is seen in the Oligocene ischyromyids, *Titanotheriomys* (Wood, 1937, pp. 194–195) and some species of *Ischyromys* (Burt and Wood, 1960, p. 958). These forms could not be in the ancestry of the squirrels, as their cheek-tooth pattern is much more advanced than is that of the squirrels.

The sciuromorphous Geomyoidea (including the extinct Eomyidae as well as the Geomyidae and Heteromyidae) seem to have many fundamental similarities especially in the basicranium (Wilson, 1949, pp. 42–48; Galbreath, 1961, pp. 226–230), to the myomorphous Muroidea (Muridae, Cricetidae), and have probably come from a common source. Whether this source was a sciuromorphous form, among some of whose descendants the *Masseter medialis* shifted forward, or whether it was a pro-trogomorphous form, and one group of descendants shifted the *Masseter lateralis* alone and the other shifted both branches of the muscle simultaneously, is completely unknown. It seems rather probable, however, that the Geomyoidea and the Muroidea are descended from some member of Grade One that would be included among the Sciuravidae. The jaw mechanism of the beavers (Castoridae) and their Oligocene to Miocene relatives, the Eutypomyidae, is almost identical to that of the squirrels, except for the expansion of the *Masseter medialis* onto the median side of the orbit. At present there is no evidence as to the pre-beaver ancestry of this group. The tooth structure of the Castoroidea is completely different from that of any of the other sciuromorphous rodents, which has led Schaub to include them, with the Theridomyoidea and Hystricoidea, in his Infraorder Palaeotrogomorpha (1958, p. 694). This association seems unnatural. It is possible that there is a special relationship of the beavers with either the ischyromyids or the sciurids, although the presence of five-crested teeth in both upper and lower jaws of the beavers makes this seem very unlikely.

The evidence that masseteric structure represents a grade is equally clear among the hystricomorphous rodents. These include the Old World porcupines (Hystricidae); the African Oligocene to Recent Thryonomyoidea (Cane Rats, Rock Rats, and Phiomyidae); the isolated African families Anomaluridae, Bathyergidae, Ctenodactylidae, and Pedetidae; the European Eocene to Oligocene Theridomyoidea; the South American Caviomorpha; and, as already indicated, the Dipodoidea. The lines of descent of most of these groups are either not clear or are unknown. The South American forms are a natural unit, the Suborder Caviomorpha of Wood and Patterson (1959, p. 289) or the Infraorder Nototrogomorpha of Schaub (1958, p. 720). It seems certain that these rodents have evolved in isolation in South America since the late Eocene or early Oligocene, when at least some members of the group were fully hystricomorphous and all were hystricognathous, and that they have had no connections with any other hystricomorphous forms during that period. On the basis of the available evidence, the most reasonable explanation for them is that they represent derivatives of a North American Grade One stock, that managed to reach South America by island hopping during the late Eocene, either via Middle America (Simpson, 1950, p. 375; Wood, 1962, p. 248; Wood and Patterson, 1959, p. 401–406), or via the West Indies (Landry, 1957, p. 91, who believed that these were hystricomorphs from the Old World; Wood, 1949, p. 47). The African Thryonomyoidea are clearly derived from the Oligocene to Miocene Phiomyidae (Lavocat, 1962, p. 289), whose Oligocene members (Wood, MS. 1) show no signs of relationship with any other group of hystricomorphous rodents, and can only (at present) be considered as an independent line derived from unknown protrogomorphs. The Hystricidae (all that seems to be left of the old Hystricomorpha) seem to have had a south Asiatic origin and differentiation, whence they spread, in the late Miocene or early Pliocene, to Europe and Africa. The Bathyergoidea are, unfortunately, very poorly known as fossils, though they occur in the African Miocene (Lavocat, 1962, p. 290). Certain Mongolian Oligocene fossils that have sometimes been referred to this family (Matthew and Granger, 1923, p. 2–5; Landry, 1957, pp. 72–73) have generally been agreed probably to be late members of the Grade One Cylindrodontidae.

The other hystricomorphous groups are all sciurognathous. The Dipodoidea (Dipodidae, Zapodidae) are extremely close to the cricetids in tooth pattern—so close, in fact, that many Miocene and Pliocene zapodids were originally referred to the Cricetidae (Schaub, 1930, pp. 616–617, 627–629; Wood, 1935b, *Schaubeumys*; Hall, 1930, *Macrognathomys*). The skeletal and myological differences between the Muroidea and Dipodoidea also seem to be relatively minor, and the Dipodoidea almost certainly belong to the same clade as do the Muroidea and Geomyoidea, which may be called the suborder Myomorpha.

The Theridomyoidea are an Eocene–Oligocene group, not known outside of Europe. The earliest members of the superfamily are close to the Paramyidae in cheek-tooth structure (Wood, 1962, p. 170) and in enamel histology (Korvenkontio, 1934, pp. 96–97), but are already fully hystricomorphous. It was long customary to consider them ancestral to the Caviomorpha, with the descendants, among other things, becoming hystricognathous. This interpretation is easily read into Schaub's classification, although he specifically states that current knowledge is not adequate to demonstrate such a relationship (1958, p. 693). But the closest resemblances to the theridomyoid tooth pattern are *not* found in the earliest caviomorphs as should be the case if they were genetically related (Wood and Patterson, 1959, pp. 400–401). Current work makes it equally improbable that there is a theridomyoid–thryonomyoid relationship

(Wood, ms. 1). The earliest known Anomaluridae are from the Miocene of Africa. There is no good evidence indicating relationship between them and any other group of rodents. It is conceivable that they are related to the Theridomyoidea, but there is no real evidence for such a relationship. The Ctenodactylidae, now exclusively African, have been shown by Bohlin (1946, pp. 75–146) to be abundant in the Oligocene of central Asia, and are known from Africa only since the late Miocene (Lavocat, 1962, p. 289). Work in progress (Dawson, 1964) rather strongly suggests an independent derivation of this family within central Asia from members of Grade One, though the jaw muscle transitions have not been worked out.

Finally, the Pedetidae are in many ways the most isolated of all rodents. They have lived in Africa since the Miocene (Stromer, 1926, pp. 128–134; MacInnes, 1957), and have a tooth pattern which is only very slightly reminiscent of that of any other rodents. They probably (with no evidence) represent an independent derivation from members of Grade One (Wood, ms. 2).

Schaub (various sources, especially 1958) completely abandoned the use of the zygomasseteric structure or that of the angle, in the subordinal classification of rodents, and relied only on the cheek-tooth pattern. He argued extensively (1958, p. 684, 691–694) that either the five-crested pattern ("plan *Theridomys*") originated only once, in the Theridomyoidea, and that all other five-crested forms are descended from them, or that his suborder Pentalophodonta, including these forms, is a natural group (clade) in that it contains those forms, and only those forms, that have achieved a five-crested pattern as a result of parallelism. As he stated (*op. cit.*, p. 693), our current knowledge of the detailed phylogeny of the rodents is still inadequate to permit us to make positive statements of the exact ancestry of most of the families of what are here included in Grade Two. Schaub further stated: "Il me parait aussi évident que l'idée de ce plan fondamental qui nous permet de révéler sinon tous, mais presque tous les parallélismes, peut servir comme base utilisable de la classification, tandis qu'on ne peut pas placer la même confiance dans celles qui s'appuie sur les structures zygo-massétériques et la configuration de l'angle mandibulaire" (*op. cit.*, p. 693).

The current conclusion of most students of fossil rodents is that there is no simple key to separating clades from grades within this complex order, and that no one set of criteria (tooth patterns, zygomasseteric structure, type of angle, fusion of ear ossicles, incisor histology, etc.) may be relied upon. Parallelisms and convergences are so abundant that only an analysis of all possible criteria can give reliable evidences of cladal unity (Lavocat, 1962, p. 288).

From the analysis of the features that are used to separate members of Grade Two from those of Grade One (jaw musculature; angle of the jaw; incisor position; incisor histology; cheek-tooth formula and pattern), it seems quite clear that these features evolved independently of each other. Hystricomorphous forms can be either hystricognathous or sciurognathous; any clade of Grade Two can include forms with high-crowned, as well as low-crowned, cheek teeth; and the changes in incisor histology seem to have taken place independently of all the others. This situation is not surprising and should not cause insurmountable difficulties in classification. It merely emphasizes that the grades must not be interpreted as clades, and that a key, based on grade characters, may be useful but is still only a key.

GRADE THREE—HYPSODONTY AND PATTERN MODIFICATION

The third grade in rodent evolution is not as clear-cut as are the first two. It is represented by those members of Grades One or Two that have developed extremely hypsodont or ever-growing cheek teeth.

These have developed independently many times, in almost all clades of rodents, as adaptations to grazing or burrowing modes of living. Among protrogomorphs, the burrowing cylindrodonts, the perhaps steppe-living protoptychids, the aplodontids and the mylagaulids all become very hypsodont. There is a definite trend toward hypsodonty in burrowing squirrels (*Cynomys*) and in some of the Old World ground squirrels. The burrowing geomyids and the desert-living saltatorial heteromyids have ever-growing cheek teeth. Extremely high crowns also characterize most of the Caviomorpha except for the New World porcupines (Erethizontidae); the Thryonomyoidea, the Bathyergidae, Ctenodactylidae, and Pedetidae in Africa; the Spalacidae and Rhizomyidae; the Castoridae; and the Microtinae among the Cricetidae.

Perhaps the suppression of the premolars and retention of the deciduous teeth, discussed above, are also features of this grade. On theoretical grounds, it would seem that a good explanation might be that the wear of the cheek teeth was so rapid that selection for increase of height of dP_4^4 was very strong, resulting in teeth that would last, proportionately, as long as in low-crowned ancestral forms. A long-growing tooth of this sort would be capable of increasing its horizontal dimensions, thus eliminating the primary adaptive reason for the replacement of deciduous teeth by permanent ones—the fact that the baby jaws were not big enough for adult-sized teeth. However, in the only case where the details of the suppression of P_4^4 by retained dP_4^4 are known (Phiomyidae, Wood, ms. 1), this change is taking place in animals some of which are still low-crowned while others are, at most, mesodont.

Two types of ever-growing teeth have developed among rodents. Usually, there has been growth of the pattern-bearing portion of the crown, so that the pattern is preserved with wear—at least in considerable part. This has resulted in cheek teeth that lose the details of cusp arrangement early in life, but in which a characteristic pattern is quickly achieved, and retained for the rest of the animal's lifetime. Such patterns are found in most caviomorphs, the Thryonomyoidea, the Theridomyidae, Bathyergidae, Ctenodactylidae, Pedetidae, Rhizomyidae, Castoridae, Spalacidae, and Microtinae.

In some rodents, however, there is little or no growth of the pattern-bearing portion of the crown, but rather a strong unilateral hypsodonty of the basal part of the crown. This arrangement usually results in the reduction of the enamel to one or a few transverse plates on each tooth, alternating with dentine (or occasionally cement) prisms. Such pattern developments are most characteristically developed in the Geomyidae (Merriam, 1895; Wood, 1936) and Heteromyidae (Wood, 1935a). Similar developments are present in Mongolian Oligocene cylindrodonts (Schaub, 1958, fig. 156), in several cases among caviomorphs (Wood and Patterson, 1959, p. 333 *et seq.*; figs. 9A, 14C, 16B, 23A), and in advanced theridomyids (Schaub, 1958, figs. 45, 49, 51, and 55).

The tendency to elongate the cheek teeth, discussed above under Grade Two, has been continued in a considerable number of forms by developments at the front end of the anterior cheek teeth (the anterocone and anteroconid), or by additions at the rear of the last tooth. The former is especially characteristic of the Microtinae, the latter of the Hydrochoeridae.

There can be no possible doubt that these high-crowned or ever-growing cheek teeth have been acquired independently in the various clades that are involved.

GRADE ZERO—THE BASIC LEVEL

The evidence suggests that the Paleocene rodent differentiation was based on a distinctly more primitive level of gnawing ability than that seen in later forms. This radiation, while essentially hypothetical, can be fairly well characterized, and is here called Grade Zero.

Among middle Eocene and later rodents,

the incisors universally have an enamel cap that covers the entire front face, and that curves around onto the buccal and lingual sides of the tooth for a short distance, serving to lock the enamel firmly onto the dentine. Among some of the earliest rodents of the Family Paramyidae, however, the locked-on pattern of enamel had not been achieved and the enamel merely forms a strip extending across most (but not all) of the width of the front edge of the tooth. As a result, there would have been danger of chipping or breaking off pieces of the enamel strip. This pattern shows up well in the late Paleocene *Paramys atavus* (Wood, 1962, fig. 21 B, C), and is also suggested in many individual specimens of several early Eocene paramyids, which seem to represent the last remnants of this Paleocene radiation. The early development of the *Leptotomus* incisor pattern, with the enamel extending over a very large part of the tooth, may also be derived from such a basic condition.

While there is no evidence one way or the other, it would seem entirely possible that the rodents of Grade Zero had a complete enamel cap on the unworn incisors, as did the multituberculates, and had merely achieved extreme unilateral hypsodonty. At some unknown time during the Paleocene, the rodents achieved a level where the incisors, including the enamel strip, became ever growing. Since the few known late Paleocene rodent incisors are all fragments, it cannot be determined when this condition was reached, though these incisor fragments seem to belong to ever-growing teeth similar (in this respect) to those of Grade One. This suggests that this type of tooth began to be acquired not later than middle Paleocene.

In the early rodents or their immediate precursors there was a reduction from the primitive placental formula of $I_3^3 C_1^1 P_4^4 M_3^3$ to that characteristic of the early Eocene Paramyidae, $I_1^1 C_0^0 P_2^2 M_3^3$. This almost certainly had taken place well before the end of the Paleocene, and presumably had begun before the enlargement of the incisors was completed.

The difference between the jaw musculature of Grade One (Fig. 1) and that of primitive mammals was presumably not very great, if Edgeworth's figures (1935, fig. 692a, b, p. 459) of the musculature of *Dasyurus* are any criterion. Here the *Masseter superficialis* has the same anteroposterior alignment as in *Ischyrotomus*, and the *Masseter lateralis* and *Masseter medialis* (*Masseter profundus* and *Zygomaticomandibularis* of Edgeworth) have an almost vertical alignment. Thus, the *Dasyurus* pattern of jaw musculature seems preadaptive for the beginnings of gnawing rodents, and therefore probably is essentially what was found in Grade Zero.

The skull structure of Paleocene rodents is completely unknown. But it seems probable that the development of free anteroposterior movement of the condyle of the lower jaw occurred *pari passu* with the development of extremely hypsodont to ever-growing incisors and the reduction of the dental formula discussed above, and that the structure of the condyle and glenoid fossa, of the incisors, of the cheek teeth and of the jaw muscles evolved as a unit complex.

Discussion

The analysis of rodent morphological evolution, given above, involves the interpretation of the classical suborders, the Sciuromorpha, Myomorpha, and Hystricomorpha, as representing alternative expressions of a major and a secondary adaptive level in the order, here called Grade Two and Grade Three. In all cases, they seem clearly *not* to be clades. The Protrogomorpha, as defined by Wood (1959, p. 170) are a closer approach to being both a clade and a grade. This suborder does not quite coincide with a grade because some members, while not having achieved any of the specializations of Grade Two, have reached a level of dental complexity that is here considered indicative of Grade Three. Whether the Protrogomorpha, as

here delimited, can be considered to represent a clade, is perhaps arguable. Certainly the Ischyromyoidea are a clade. Certainly the Aplodontoidea are derived from them, but most authors consider that the same is true of all the other rodents as well. However, the Protrogomorpha, as here defined, are related forms that have structural features in common, permitting the group to be satisfactorily defined.

Black has recently (1963, pp. 126–128) argued that the Sciuridae, because of their primitive dentition, should be returned to the Protrogomorpha, where Wood once included them (1955a). The suborder could then be defined as members of Grade One plus certain groups that had not gone very far in evolving into Grades Two or Three. It seems better, however, for the present to use the break between Grade One and Grade Two as a fundamental division in rodent classification, and hence to eliminate the Sciuridae from the Protrogomorpha. A major reason why Black considers that the squirrels can no longer be separated from the members of Grade One is that *Miosciurus* and *Protosciurus*, from the early Miocene, have zygomasseteric structures that have not fully achieved the sciuromorphous pattern. However, his description (1963, pp. 136, 140) and figures (*op. cit.*, pls. 3, 6) show that the masseter had already begun its migration in these forms, so that, technically, they belong to what is here called Grade Two. Naturally, there had to have been a transition from Grade One to Grade Two, and the transitional forms would be hard to place with exactitude, but it seems best to consider all the known Sciuridae as members of Grade Two.

The rest of the cladal classification of rodents must still remain largely as indicated by Simpson (1959) and Wood (1959, p. 172). The main changes that are required at the present time involve certain African rodents. The Phiomyidae are clearly not Protrogomorpha, but are hystricomorphous forms ancestral to the Thryonomyoidea, to which superfamily they should be referred. There seems to be even less justification than formerly (Wood, 1955a) for placing the Hystricidae close to any other known families.

All the available evidence suggests that the level of Grade Two has been achieved many times independently. Instead of the three suborders that were formerly recognized, it seems better to recognize at least eleven clades that have independently passed from Grade One to Grade Two. Which of these should be considered suborders and which merely families or superfamilies is, for the moment, largely a matter of convenience (Simpson, 1959; Wood, 1959).

A cladal classification of rodents, based on current knowledge, is as follows:

Order Rodentia
  Suborder Protrogomorpha
    Superfamily Ischyromyoidea
      Paramyidae, Sciuravidae, Cylindrodontidae, Protoptychidae, and Ischyromyidae
    Superfamily Aplodontoidea
      Mylagaulidae and Aplodontidae
  Suborder Caviomorpha
    Superfamily Octodontoidea
      Octodontidae, Echimyidae, Ctenomyidae, Abrocomidae, and Capromyidae
    Superfamily Chinchilloidea
      Chinchillidae, Dasyproctidae (incl. Cephalomyidae), Cuniculidae, Heptaxodontidae, and Dinomyidae
    Superfamily Cavioidea
      Eocardiidae, Caviidae, and Hydrochoeridae
    Superfamily Erethizontoidea
      Erethizontidae
  Suborder Myomorpha
    Superfamily Muroidea
      Cricetidae (incl. Melissiodontidae Schaub) and Muridae (incl. Gerbillidae Stehlin and Schaub)
    Superfamily Geomyoidea
      Geomyidae, Heteromyidae, and Eomyidae
    Superfamily Dipodoidea
      Dipodidae and Zapodidae
    Superfamily Spalacoidea
      Spalacidae and Rhizomyidae
    Superfamily Gliroidea
      Gliridae and Seleveniidae
  Clades not in suborders:
    Family Sciuridae (incl. Eupetauridae Schaub and Iomyidae Schaub)
    Superfamily Castoroidea

      Castoridae and Eutypomyidae
    Superfamily Theridomyoidea
      Pseudosciuridae and Theridomyidae
    Family Ctenodactylidae (incl. Tataromyidae Bohlin)
    Family Anomaluridae
    Family Pedetidae
    Family Hystricidae
    Superfamily Thryonomyoidea
      Phiomyidae (incl. Diamantomyidae Schaub), Thryonomyidae, and Petromuridae
    Family Bathyergidae

The Family Pellegriniidae of Schaub is based on a single species of completely unknown affinities, which should not be considered a family until more is known about it.

Summary

Rodent evolution can be envisioned as involving three relatively clear-cut evolutionary levels, here called Grades One, Two, and Three. The first involves well-developed gnawing animals, with a primitive mammalian jaw musculature. Grade Two includes those animals that have modified the jaw musculature in one of several ways that formerly were used as the basis for rodent subordinal classification. There were also changes in the dentition, especially in the development of cheek teeth with five transverse crests, rather than ones with no more than four crests as in Grade One. Changes occurred in numerous other parts of the skeleton and dentition, although these were probably not correlated with each other. Grade Three includes those rodents with very high-crowned or even ever-growing cheek teeth, in which there is sometimes the same type of limitation of the enamel that occurred during the Paleocene on the incisors. Grade Three also includes forms in which there has been a marked secondary increase in the length of the cheek teeth. A hypothetical Grade Zero is imagined for the rodents of the second half of the Paleocene.

Only Grade One comes close to approximating a clade. The Protrogomorpha, as here defined, include the members of Grade One and some forms that have reached Grade Three without going through Grade Two. The cladal classification of the rodents still requires the recognition of numerous independent lines, showing no evidence of interrelationship later than in members of Grade One. Only two or possibly three clades can be recognized that require units larger than the superfamily— the Protrogomorpha, the Caviomorpha, and perhaps the Myomorpha. The other rodents fall into nine familial or superfamilial clades.

ACKNOWLEDGMENTS

A discussion of the similarities and differences between rodent and teleost evolution prompted Schaeffer to make the oral suggestion that an analysis of evolutionary grades and clades among rodents would be very useful to students of the evolution of other groups. This has led to the preparation of the present review. This study was assisted by grants from the National Science Foundation and from the Marsh Fund of the National Academy of Sciences.

LITERATURE CITED

BECHT, G. 1953. Comparative biologic-anatomical researches on mastication in some mammals. Konink. Nederl. Akad. Wetensch., Proc., C, **56**: 508–527.

BLACK, C. C. 1963. A review of the North American Tertiary Sciuridae. Bull. Mus. Comp. Zool., **130**: 109–248.

BOHLIN, B. 1946. The fossil mammals from the Tertiary deposit of Taben-buluk, Western Kansu. Part II. Simplicidentata, Carnivora, Artiodactyla, Perissodactyla, and Primates. Rept. Scientific Exped. North-western Prov. China under leadership of Dr. Sven Hedin, 6, Vertebrate Paleontology, no. **4**: 1–259.

BRANDT, J. F. 1855. Beiträge zur nähern Kenntniss der Säugethiere Russlands. Mém. Acad. Imp. Sci. St.-Pétersbourg, ser. 6, **9**: 1–375.

BURT, A. M., AND A. E. WOOD. 1960. Variants among Middle Oligocene rodents and lagomorphs. J. Paleont., **34**: 957–960.

DAWSON, M. 1964. Late Eocene rodents (Mammalia) from Inner Mongolia. Amer. Mus. Novitates, **2191**: 1–15.

EDGEWORTH, F. H. 1935. The cranial muscles of vertebrates. Cambridge Univ. Press. ix + 493 pp.

FRIANT, M. 1936. Interprétation des dents jugales chez les lonchérinés. Vidensk. Med. Dansk. nat. For. Kjøbenhavn, **99**: 263–266.

GALBREATH, E. C. 1961. The skull of Heliscomys, an Oligocene heteromyid rodent. Trans. Kansas Acad. Sci., **64**: 225–230.

HALL, E. R. 1930. Rodents and lagomorphs from the later Tertiary of Fish Lake Valley, Nevada. Univ. California Publ. Geol., **19**: 295–312.

HOWELL, A. B. 1932. The saltatorial rodent Dipodomys: the functional and comparative anatomy of its muscular and osseous systems. Proc. Amer. Acad. Arts and Sci., **67**: 377–536.

HUXLEY, J. S. 1958. Evolutionary processes and taxonomy with special reference to grades. Uppsala Univ. Arssks., **1958**: 21–38.

KORVENKONTIO, V. A. 1934. Mikroskopische Untersuchungen an Nagerincisiven unter Hinweis auf die Schmelzstruktur des Backenzähne. Historische-phyletische Studie. Ann. Zool., Soc. Zool.-Bot. Fennicae Vanamo, **2**: 1–274.

LANDRY, S. O., JR. 1957. The interrelationships of the New and Old World hystricomorph rodents. Univ. California Publ. Zool., **56**: 1–118.

LAVOCAT, R. 1956. Réflexions sur la classification des rongeurs. Mammalia, **20**: 49–56.

———. 1962. Réflexions sur l'origine et la structure du groupe des rongeurs. Problèmes Actuels de Paléontologie, Colloques Internationaux, Centre National de la Recherche Scientifique, no. **104**: 287–299.

LUBOSCH, W. 1938. Muskeln des Kopfes: Viscerale Muskulatur (Fortsetzung). D. Säugetiere, in Handbuch der Vergleichenden Anatomie der Wirbeltiere. Bolk, Göppert, Kallius, and Lubosch (eds.), Berlin and Vienna, Urban and Schwarzenberg, **5**: 1065–1106.

MACINNESS, D. G. 1957. A new Miocene rodent from East Africa. British Mus. (Nat. Hist.), Fossil mammals of Africa, **12**: 1–35.

MANARO, A. J. 1959. Extrusive incisor growth in the rodent genera Geomys, Peromyscus, and Sigmodon. Quart. J. Florida Acad. Sci., **22**: 25–31.

MATTHEW, W. D., AND W. GRANGER. 1923. New Bathyergidae from the Oligocene of Mongolia. Amer. Mus. Novitates, **101**: 1–5.

MERRIAM, C. H. 1895. Monographic revision of the Pocket Gophers, family Geomyidae (exclusive of the species of Thomomys). North Amer. Fauna, **8**: 1–258.

MILLER, G. S., JR., AND J. W. GIDLEY. 1918. Synopsis of the supergeneric groups of rodents. J. Washington Acad. Sci., **8**: 431–448.

MÜLLER, A, 1933. Die Kaumuskulatur des Hydrochoerus capybara und ihre Bedeutung für die Formgestaltung des Schädels. Morph. Jahrb., **72**: 1–59.

SCHAEFFER, B. 1956. Evolution in the Subholostean fishes. Evolution, **10**: 201–212.

SCHAUB, S. 1930. Fossile Sicistinae. Eclog. geol. Helvetiae, **23**: 615–637.

——. 1958. Simplicidentata, *in* Traité de Paléontologie, Jean Piveteau (ed.), Paris, Masson et Cie., **6**(2): 659–818.

SIMPSON, G. G. 1945. The principles of classification and a classification of mammals. Bull. Amer. Mus. Nat. Hist., **85**:xvii + 1–350.

——. 1950. History of the fauna of Latin America. Amer. Scientist, **38**: 361–389.

——. 1959. The nature and origin of supraspecific taxa. Cold Spring Harbor Symposia on Quant. Biol., **24**: 255–271.

STROMER, E. 1926. Reste Land- und Süsswasser-bewohnender Wirbeltiere aus den Diamantenfeldern Deutschsüdwestafrikas. Pp. 107–153 *in* E. Kaiser, Die Diamantenwüste Südwestafrikas, vol. 2, Berlin, Dietrich Reimer.

TULLBERG, T. 1899. Ueber das System der Nagethiere: eine phylogenetische Studie. Nova Acta Reg. Soc. Scient. Upsala, ser. 3, **18**: v + 1–514.

WILSON, R. W. 1949. On some White River fossil rodents. Publ. Carnegie Inst. Washington, **584**: 27–50.

WINGE, H. 1924. Pattedyr-Slaegter. Vol. 2, Rodentia, Carnivora, Primates. Copenhagen, H. Hagerups Forlag.

WOOD, A. E. 1935a. Evolution and relationships of the heteromyid rodents with new forms from the Tertiary of western North America. Ann. Carnegie Mus., **24**: 73–262.

——. 1935b. Two new genera of cricetid rodents from the Miocene of western United States. Amer. Mus. Novitates, **789**: 1–3.

——. 1936. Geomyid rodents from the middle Tertiary. Amer. Mus. Novitates, **866**: 1–31.

——. 1937. The mammalian fauna of the White River Oligocene. Part II. Rodentia. Trans. Amer. Phil. Soc., n.s., **28**: 155–269.

——. 1949. A new Oligocene rodent genus from Patagonia. Amer. Mus. Novitates, **1435**: 1–54.

——. 1955a. A revised classification of the rodents. J. Mammal., **36**: 165–187.

——. 1955b. Rodents from the Lower Oligocene Yoder formation of Wyoming. J. Paleont., **29**: 519–524.

——. 1959. Are there rodent suborders? Syst. Zool., **7**: 169–173.

——. 1962. The early Tertiary rodents of the Family Paramyidae. Trans. Amer. Phil. Soc., n.s., **52**: 1–261.

——. MS. (1). The rodents of the Fayûm.

——. MS. (2). Unworn teeth of the African rodent, *Pedetes*.

WOOD, A. E., AND B. PATTERSON. 1959. The rodents of the Deseadan Oligocene of Patagonia and the beginnings of South American rodent evolution. Bull. Mus. Comp. Zool., **120**: 281–420.

CHROMOSOMAL DIVERGENCE DURING EVOLUTION
OF GROUND SQUIRREL POPULATIONS
(RODENTIA: *SPERMOPHILUS*)

Charles F. Nadler, Robert S. Hoffmann, and Kenneth R. Greer

Abstract

Nadler, C. F., R. S. Hoffmann, and K. R. Greer (Dept. Medicine, Northwestern Univ. Med. School, Chicago, 60611; Mus. Nat. History, Univ. Kansas, Lawrence, 66044; and Montana Dept. Fish and Game, Bozeman, 59715) 1971. Chromosomal divergence during evolution of ground squirrel populations. Syst. Zool., 20:298–305.—Two subspecies of Spermophilus richardsonii, occurring in the Madison River valley of southwestern Montana differ in chromosome number, S. r. aureus being 2n = 34, and S. r. richardsonii, 2n = 36. The zone of contact between them is 20–25 miles long; only one of 21 colonies investigated contained chromosomal hybrids (2n = 35). Cytological and morphological differences suggest that aureus and richardsonii passed through a period of divergence while isolated from one another, and that the present contact is secondary. [Chromosomes; evolution; Ground Squirrels.]

INTRODUCTION

Geographic variation of chromosomes occurs in contiguous, potentially interbreeding populations of several rodent species complexes including *Spermophilus richardsonu* (Nadler, 1964a, b; 1968a) *S. townsendii* (Nadler, 1968b) *Sigmodon hispidus* (Zimmerman and Lee, 1968), *Thomomys talpoides* (Thaeler, 1968) and *Spalax ehrenbergi* (Wahrman, et al., 1969). Prior to discovery of karyotypic differences, these various populations were considered conspecific because major cranial or pelage differences were lacking. The recognition of chromosomal divergence in these species complexes makes them valuable models for investigating the evolution of isolating mechanisms that maintain the observed cytological differentiation.

Among the four presently-recognized subspecies of *Spermophilus richardsonii* Sabine (Hall and Kelson, 1959), *S. r. richardsonii* Sabine and *S. r. aureus* Davis have diploid numbers of 36 and 34 respectively; their ranges meet in southwestern Montana. *Spermophilus r. elegans* Kennicott (2n = 34) from Wyoming and Colorado (Nadler, 1964a) and *S. r. nevadensis* Howell (2n = 34) from Nevada (Nadler, 1964b) are geographically isolated, but their somatic chromosomes are indistinguishable from S.

r. aureus. At one time Davis (1939) separated *elegans, nevadensis* and *aureus* from *S. richardsonii* as a different species, *S. elegans.* However, recently most authors have followed Howell (1938) in considering the four populations conspecific.

Objectives of the present study were to 1) locate the zone of contact between 2n = 34 and 2n = 36 populations, 2) identify possible intergradation and its characteristics, and 3) evaluate the systematic status of the *S. richardsonii* complex.

During 1968 and 1969 ground squirrel were trapped from 21 colonies at 13 localities in Gallatin, Madison, and Broadwater counties, southwestern Montana (Fig. 1). The chromosomes of 114 specimens were analyzed from cell suspensions of femoral bone marrow.

RESULTS

All specimens from localities designated by an open circle in Fig. 1 have 2n = 3 and a karyotype of 30 biarmed (metacentric and submetacentric) and 4 acrocentric autosomes, a submetacentric X, and a subtelocentric Y chromosome equal in size to the smaller two autosomal pairs (Fig. 2a). In north-central Montana *S. r. richardson* are cytologically indistinguishable from these animals except for an acrocentric

29

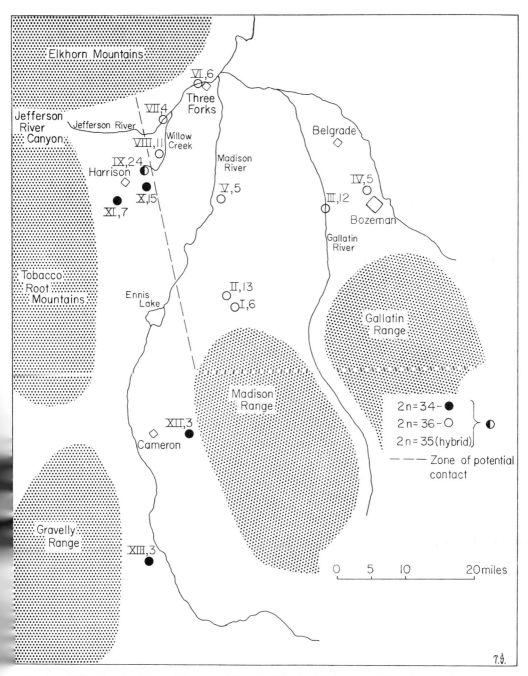

FIG. 1.—Collecting localities of *Spermophilus* in Gallatin, Madison and Broadwater Counties, Montana. Open circles (○) indicate colonies inhabited with animals with 2n = 36, and solid circles (●) denote 2n = 34 colonies. The half-solid circle, represents a colony (IX) with 2n = 34, 2n = 36, and 2n = 35 (hybrid) animals. Roman numerals designate colony and arabic numbers indicate sample size from the locality.

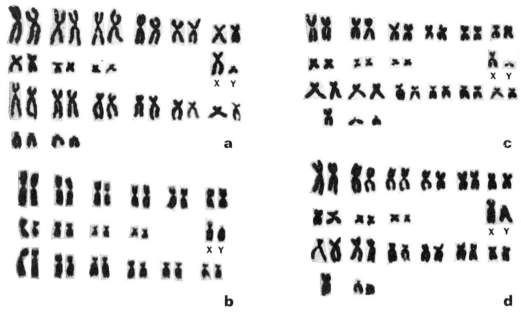

FIG. 2.—a) Karyotype of male *Spermophilus* (2n = 36); b) Karyotype of male *Spermophilus* (2n = 34); c) Hybrid male karyotype (2n = 35) with a small Y chromosome derived from 2n = 36 parent; d) Hybrid male karyotype (2n = 35) with large Y chromosome derived from 2n = 34 parent.

rather than subtelocentric Y chromosome (Nadler, 1964a).

A 2n = 34 characterizes all animals from localities indicated by solid circles in Fig. 1. They are chromosomally similar to *S. r. aureus* from Idaho (Nadler, 1968a), *S. r. elegans* (Nadler, 1964a), and *S. r. nevadensis* (Nadler, 1964b). Karyotypes are composed of 32 biarmed autosomes (Fig. 2b); and compared with the 2n = 36 karyotype, a smaller submetacentric X chromosome with more medially placed centromere, and a larger acrocentric Y chromosome.

Colony IX, located 4.0 miles from colony VIII (2n = 36; N = 11) and 2.0 miles from X (2n = 34; N = 15) (Fig. 1) displayed evidence of hybridization; one animal had 2n = 36, 19 animals had 2n = 34, and four animals had hybrid karyotypes of 2n = 35 (one adult and one juvenile male, and two adult females, one of which was lactating). Hybrid karyotypes reflected a haploid contribution from 2n = 36 and 2n = 34 parents and thus exhibited 15 sets of biarmed autosomes, 2 acrocentric autosomes from the

2n = 36 parent, and unmatched metacentric from the 2n = 34 parent, and a set of sex chromosomes. One male (Fig. 2c) had the small Y chromosome typical of 2n = 36 whereas the other had the large acrocentric Y found in 2n = 34 ground squirrels (Fig. 2d). Female hybrids possessed two X chromosomes of unequal size, resembling the X of 2n = 36 and of 2n = 34 specimens.

The observation of an unmatched metacentric in hybrids and a reexamination of the number of metacentric autosomes now leads us to postulate that chromosomal evolution between 2n = 36 and 2n = 34 resulted from a single Robertsonian centric fusion or fission involving equal-sized acrocentrics, without an added pericentric inversion as was previously thought (Nadler, 1964a). Differences in size and centromere location in the largest autosome pair, the X and the Y chromosome also accompanied karyotypic divergence.

DISCUSSION

The zone of potential contact between 2n = 36 and 2n = 34 populations appears

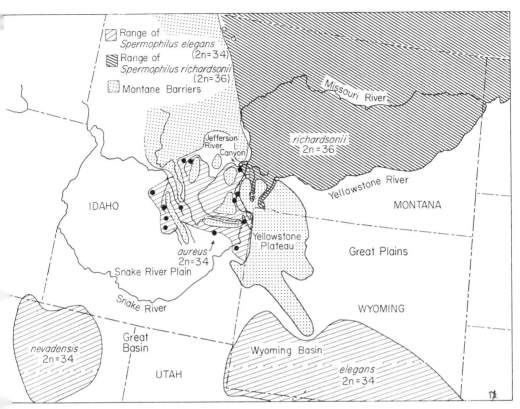

Fig. 3.—Map of Montana-Wyoming-Idaho border area, showing partial ranges of taxa discussed, and certain physiographic features of the region. Open spots indicate localities for *Spermophilus r. richardnii*; solid spots indicate localities for *S. r. aureus*.

be only 20–25 miles long. It is restricted the lower Madison River valley (Figs. 1,) by barriers of unsuitable montane habit; the Yellowstone Plateau complex to e south and east, of which the Madison nd Gallatin Ranges comprise the northest corner; and the Tobacco Root and Elkrn Mountains to the north, cut through the precipitous Jefferson River Canyon Fig. 3). Ground squirrels of the *S. richdsonii* complex inhabit valley bottoms nd foothills in southwestern Montana and jacent Idaho (Durrant and Hansen, 54), and not mountain meadows, as *S. r. egans* does in Colorado (Hansen, 1962; chleitner, 1969). Mountain meadows in e Montana-Wyoming-Idaho border area e instead inhabited by *S. armatus*, a ound squirrel similar to *S. r. aureus* and

S. r. elegans both morphologically (Davis, 1939), and cytologically (Nadler, 1966). The montane barriers to *S. richardsonii* may thus contain both habitat and biotic (competitive) components (Durrant and Hansen, 1954; Hoffmann, et al., 1969).

Only one of 21 colonies studied contained identifiable (2n = 35) hybrids, and within this colony the hybrid frequency, in 1969, was only 4/24. The narrowness of the zone of potential contact and infrequency of hybrids suggests that the situation is not now one of chromosomal polymorphism, wherein more than one karyotype is found within a single, potentially panmictic population, such as has been described for certain rodents (Matthey, 1963a, b). There is evidence that *aureus* and *richardsonii* ground squirrels have occupied the general area for at least a century (Merriam, 1891) and the contact zone for over 50 years; specimens of both taxa collected in 1917-18 are in the collection of the Department of Zoology and Entomology, Montana State University, Bozeman. They may, of course, have existed here much longer. Despite the long period of contact, and the considerable dispersal powers of ground squirrels (Hansen, 1962), hybridization is rare. This suggests either that behavioral or other barriers are inhibiting hybridization, or that some sort of selection is acting against the hybrids relative to the parental types. Hybrid fertility has not yet been tested, but one adult female hybrid collected was lactating, evidence that a fertility barrier to gene exchange may not be complete.

The isolated ranges of the three phenetically similar 2n = 34 taxa (*nevadensis, elegans, aureus*) (Fig. 3) suggests that the present distributional disjunctions have resulted from fragmentation of a formerly continuous range across the Snake River Plain of southern Idaho. The present-day absence of this ground squirrel on the Snake River Plain may be due to post-glacial changes in the physical environment and/or displacement by *S. townsendii*. Such a distribution pattern, centering on the northern Great Basin and adjacent desert plateaus and basins, contrasts with the present-day continuous distribution of *S. r. richardsonii* (2n = 36) on the northern Great Plains. The distributional patterns of a number of species of mammals suggests that the Great Basin and Great Plains were two important centers of differentiation during the Pleistocene (Hoffmann and Jones, 1970).

The question of which ground squirrel taxon is closer to the ancestral population cannot be answered unequivocally. The disjunct, relict distribution of the 2n = 34 ground squirrels argues for their being the older group (Hoffmann and Jones, 1970). Moreover, the oldest known fossil of the *richardsonii* complex is assigned to *S. r. elegans* (2n = 34); it occurs in sediments deposited during the Kansan glacial period (Hibbard, 1937). Much of the area presently occupied by *S. r. richardsonii* was either covered by glacial ice, or supported a boreal coniferous forest during the last glacial period, the Wisconsin (Lemke, et al., 1965; Wells, 1970). These factors must have restricted suitable ground squirrel habitat within the present range of *S. r. richardsonii* to a relatively small area from central Montana perhaps to western North Dakota. To the north was the ice sheet, to the east, coniferous forest, and to the west, the forested mountains of the Continental Divide.

There are, remarkably, no *S. r. richardsonii* at present known from south of the Yellowstone River in Montana or south and west of the Missouri River in North and South Dakota. There is thus a large area of apparently suitable habitat in the unglaciated area of southeastern Montana, southwestern North Dakota, northeastern Wyoming, and western South Dakota where the species does not occur; neither are there closely-related species which might serve as ecological equivalents. The Yellowstone-Missouri system thus seems to limit the distribution of *S. r. richardsonii* along the southern edge of its range. However, the upper Yellowstone has only recently become a major river due to late-Wisconsin capture of the drainage of the Yellowstone Plateau, much of which had formerly drained westward into the Snake River. Prior to that event, the main tributaries of the Yellowstone were the Bighorn and Powder rivers, and above the Bighorn the Yellowstone was probably quite small (Fenneman, 1931; Lemke et al., 1965).

All of these data suggest that *S. r. richardsonii* has occupied its present range on the northern Great Plains as the glacial ice and coniferous forests retreated in postglacial time, and that it spread from an area of origin north and west of the Yellowstone River *after* the Yellowstone River achieved its present size.

Two models could account for such an origin and spread. The first involves splitting, one or more times, of an ancestral population into separate units by forested barriers that developed in a glacial period (Wisconsin or earlier). During glacial periods, a barrier of continuous boreo-montane forest (Findley and Anderson, 1956) evidently separated the Great Basin and the Snake River Plain from the northern Great Plains (Fig. 3). If the ancestral form of the *richardsonii* complex were separated into two or more populations isolated on either side of this barrier, these might differentiate, undergo chromosomal changes and eventually develop a partially effective isolating mechanism. The population on the northern Great Plains was subjected to extreme conditions during glacial maxima, during which its range was sharply contricted. Such a population, subjected to "catastrophic selection," might be induced through forced inbreeding into chromosomal reorganization, as Lewis (1966) describes. Such a "founder" population (Mayr, 1963) would then have the opportunity to spread rapidly into the large area of the northern Great Plains made available by post-glacial retreat of the ice sheets. At the same time, during inter- or post-glacial periods, when non-forested connections were re-established between the Snake River Plain and the northern Great Plains, ground squirrels could then spread over low passes and come into contact again.

The second alternative is that the "founder" population arose in early post-Wisconsin time from ground squirrels crossing the Continental Divide from the southwest to occupy the area north of the Yellowstone River. Chromosomal reorganization in such a small isolated, newly established colony would not then be lost through hybridization with the more numerous ground squirrels having the ancestral chromosomal pattern. This "founder" colony could then proceed to occupy the empty niche available as the northern Great Plains became ice-free. The first model is supported by the present greatly restricted introgression between 2n = 34 and 2n = 36 populations of ground squirrels, and X and chromosome, as well as autosomal re-

patterning, which suggests a secondary contact between the two forms differentiated as a result of prolonged isolation; the present contact between 2n = 34 and 2n = 36 ground squirrels may well have been established after the Wisconsin glacial period. On the other hand, the absence of this ground squirrel on the Great Plains south of the Yellowstone-Missouri system supports the idea of post-Wisconsin differentiation proposed in the second model.

CONCLUSIONS

The biogeographic evidence thus suggests that a 2n = 34 ground squirrel from the Great Basin was ancestral, and that the Great Plains 2n = 36 form evolved from it by a dissociative process in which a metacentric chromosome divided to form two acrocentrics. The fact that true telocentrics were not formed argues against simple fission through the centromere. We are aware that there is a considerable body of evidence that centric fusion is the more common phenomenon in mammals; indeed, definitive cytological evidence for fission is lacking. Hence the possibility that *Spermophilus* evolution has proceeded from higher to lower chromosome numbers (Nadler, 1966) must be considered.

At present, the integrity of the two forms appears to be maintained naturally, by either selection against hybrids and/or an unknown isolating mechanism. In the zone of contact, hybrids are rare compared to parental types, and it is convenient to regard the two forms as semispecies (*sensu* Short, 1969). If further study verifies that gene flow is as sharply restricted between the two populations as now appears to be the case, and if this restricted gene flow is due to natural selection acting against hybridization between the two chromosomal forms, and not merely to temporal or geographic factors, then the distinctness of the 2n = 34 and 2n = 36 forms should be recognized by nomenclatural changes. Taxonomically, the 2n = 34 populations would then be referable to *Spermophilus elegans*; 2n = 36 populations retain the name S.

richardsonii. The bacula of the two forms are also distinctive; Burt (1960) stated "the structural differences . . . [are] greater than normally found between species." Additionally, fleas of the two taxa are different; *S. richardsonii* harbors *Oropsylla rupestris,* a flea of northern affinities, whereas *S. elegans* has *O. idahoensis,* a species common to a number of western *Spermophilus* (Jellison, 1945).

ACKNOWLEDGMENTS

We thank Jeff Greer, Karl R. Hoffmann, and Nancy Nadler for assistance in obtaining specimens; Kathleen Harris and Charles F. Nadler, Jr. provided technical assistance. R. Jackson and P. V. Wells, both of the University of Kansas, suggest valuable improvements in the manuscript, although they do not agree with all of our interpretation. The study was supported by NSF Grants GB 3251 and GB 5676X.

REFERENCES

BURT, W. H. 1960. Bacula of North American mammals. Misc. Publ., Mus. Zool., Univ. Mich., No. 113. 75 pp.

DAVIS, W. B. 1939. The recent mammals of Idaho. Caxton Printers, Caldwell, Idaho. 400 pp.

DURRANT, S. D., AND R. M. HANSEN. 1954. Distribution patterns and phylogeny of some western ground squirrels. Syst. Zool., 3:82–85.

FENNEMAN, N. M. 1931. Physiography of western United States. McGraw-Hill, New York and London. 534 pp.

FINDLEY, J. S., AND S. ANDERSON. 1956. Zoogeography of the montane mammals of Colorado. Jour. Mamm., 37:80–82.

HALL, E. R., AND K. R. KELSON. 1959. The mammals of North America. Ronald Press, New York, 1:xxx + 1–546, + 1–79.

HANSEN, R. M. 1962. Dispersal of Richardson ground squirrel in Colorado. Amer. Midl. Nat. 68:58–66.

HIBBARD, C. W. 1937. Notes on some vertebrates from the Pleistocene of Kansas. Trans. Kans. Acad. Sci., 40:233–237.

HOFFMANN, R. S., AND J. K. JONES, JR. 1970. Influence of late-glacial and post-glacial events on the distribution of Recent mammals on the northern Great Plains. Pp. 355–394 *in* Pleistocene and Recent environments of the Central Plains. Dort, W., Jr., and J. K. Jones, Jr., eds. Univ. Kansas Press, Lawrence. 433 pp.

HOFFMANN, R. S., P. L. WRIGHT, AND F. E. NEWBY. 1969. The distribution of some mammals in Montana. I. Mammals other than bats. J. Mammal., 50:579–604.

HOWELL, A. H. 1938. Revision of the North American ground squirrels, with a classification of the North American Sciuridae. N. Amer. Fauna, 56. 256 pp.

JELLISON, W. L. 1945. Siphonaptera: the genus *Oropsylla* in North America. J. Parasit., 31: 83–97.

LECHLEITNER, R. R. 1969. Wild mammals of Colorado. Pruett Publ. Co., Boulder, Colo. XIII. 254 pp.

LEMKE, R. W., W. M. LAIRD, M. J. TIPTON, AND R. M. LINDVALL. 1965. Quaternary geology of northern Great Plains. pp. 15–27 *in* The quaternary of the United States, H. E. Wright, Jr., and D. G. Frey, eds. Princeton Univ. Press, Princeton, N. J. 922 pp.

LEWIS, H. 1966. Speciation in flowering plants. Science, 152:167–172.

MATTHEY, R. 1963a. Polymorphisme chromosomique intraspécifique et intraindividual chez *Acomys minous* Bate (Mammalia-Rodentia-Muridae) Chromosoma, 14:468–497.

MATTHEY, R. 1963b. Polymorphisme chromosomique intraspécifique chez un Mammifèr *Leggada minutoides* Smith (Rodentia-Muridae) Rev. Suisse Zool., 70:173–190.

MAYR, E. 1963. Animal species and evolution Harvard Univ. Press, Cambridge. xiv + 797 pp

MERRIAM, C. H. 1891. Results of a biological reconnoissance of south-central Idaho. N. Amer Fauna, 5. 127 pp.

NADLER, C. F. 1964a. Chromosomes and evolution of the ground squirrel, *Spermophilus richardsonii.* Chromosoma, 15:289–299.

NADLER, C. F. 1964b. Chromosomes of the Nevada ground squirrel *Spermophilus richardsonii nevadensis* (Howell). Proc. Soc. Exper Biol. Med. 117:486–488.

NADLER, C. F. 1966. Chromosomes and systematics of American ground squirrels of the subgenus *Spermophilus.* J. Mamm., 47:579–59

NADLER, C. F. 1968a. Chromosomes of the ground squirrel, *Spermophilus richardsonii aureus* (Davis). J. Mamm., 49:312–314.

NADLER, C. F. 1968b. The chromosomes of *Spermophilus townsendi* (Rodentia: Sciuridae and report of a new subspecies. Cytogenetic 7:144–157.

SHORT, L. L. 1969. Taxonomic aspects of avian hybridization. Auk, 86:84–105.

THAELER, C. S. 1968. Karyotypes of sixteen populations of the *Thomomys talpoides* complex of pocket gophers (Rodentia-Geomyidae Chromosoma, 25:172–183.

WAHRMAN, J., R. GOITEIN, AND E. NEVO. 1969. Mole rat *Spalax*: evolutionary significance of chromosome variation. Science, 164:82–83.

WELLS, P. V. 1970. Postglacial vegetational history of the Great Plains. Science, 167:1574–1582.

ZIMMERMAN, E. G., AND M. R. LEE. 1968. Variation in chromosomes of the cotton rat, *Sigmodon hispidus*. Chromosoma, 24:254–250.

Manuscript received June, 1970

VARIABILITY IN CHARACTERS UNDERGOING RAPID EVOLUTION, AN ANALYSIS OF *MICROTUS* MOLARS

R. D. GUTHRIE

University of Alaska, College, Alaska

Accepted October 31, 1964

Information amassed by animal breeders has aided considerably the understanding of the genetic changes that accompany phenotypic population changes through time. In spite of genetic inferences from these artificial selection experiments, there are few studies of genetic and phenotypic changes in characters evolving under natural conditions. Because of the scarcity of statistically adequate series of fossils and the incompleteness of knowledge of phylogenetic patterns, the contributions of paleontology to the understanding of evolutionary dynamics have been far below its potential. However, as phylogenies become better known and series are emphasized rather than types, it is increasingly possible to study the detailed behavior of evolving characters. Findings of these studies, in turn, permit a more critical evaluation of our theoretical models.

One of the critical areas of evolutionary research is the behavior of the intrapopulational variation of a character when it is undergoing change. An understanding of the changes in genetic variation as the population moves from one mean to another is central to any investigation involving evolutionary mechanics. Lerner (1955) listed as one of the significant landmarks of population genetics the discovery of the great genetic reserves in natural populations, yet this high potential genetic variation is usually associated with relatively low phenotypic variation. According to our present concepts, sustained intensive directional selection would decrease and eventually exhaust this residual store of genetic variance. In reality the situation is never brought to this extreme since evolution, even at its most rapid pace, is slow compared to changes produced by artificial selection. However, the problem of the elimination of genetic variance does have meaning at its intermediate stages. It is the assumption of many evolutionary thinkers that as the population responds to the pressures of directional selection the genetic and phenotypic variation immediately decreases, discouraging further evolutionary changes proportionally. The findings of this study lead me to take issue with this assumption.

Empirical documentation supporting a reduction of phenotypic variation in evolving populations has been discussed by Simpson (1953) and Bader (1955), although, in their material, the decreases in phenotypic variation were slight. Since evolutionary change in both cases was taking place at only a moderate pace, an examination of a more rapidly evolving group would theoretically provide greater clarification as the interrelationships would be accentuated by the more intense selection pressures exerted over a shorter period of time. This study is an examination of such a rapidly evolving group. The variation of a suite of evolving characters has been compared to the variation of their more stable homologues.

One of the best examples of rapid evolution documented in the mammalian record has been chosen for this investigation. The setting for this rapid radiation is the late Pliocene and Pleistocene, a time of major ecological upsets, rapid introduction of new habitats, periodic invasions of new territory, and novel associations of faunas. The microtine rodents changed so rapidly during this time that they are used as one of the better markers for correlation of the Pleistocene stages (Hibbard, 1959). Microtines are well represented in the fossil record, and as a result of their generally high population densities, where present, fossils are usually abundant. The microtines have undergone a major adaptive shift from the seed–fruit diet of the typical cricetine to a bark–grass diet. This change has been accompanied by a characteristic increase in the complexity of the dentition, which is the most durable portion of a mammal and also the part most frequently preserved. The microtines have developed in this short period of time a tooth complexity comparable to that which the Equidae achieved throughout the entire Tertiary. Bader (1955) suggested about two million years as the average duration of a species of oreodont. This length of time would be too conservative for genera of microtines.

Preliminary studies indicated that the teeth and the areas of the particular teeth which are undergoing phylogenetic change (more variable interspecifically and intergenerically) are also those which are more variable intraspecifically and intrapopulationally. Two abundant species of *Microtus* that represent two minor grades of tooth complexity were selected, the extinct *M. paroperarius* from the Kansan glaciation and the recent species *M. pennsylvanicus*, first known from the Illinoian.

It should be emphasized that, unlike studies of fossil material which compared the variation between rapidly and slowly evolving lines for a variety of characters, this study was a comparison of characters within populations. The variation of tooth characters that are undergoing rapid evolution was compared with the variation of their serial homologues which are maintaining a fundamentally stable morphology. The hypothesis examined was that highly variable characters are not *ipso facto* vestigial. Quite the contrary, some of these characters have recently been, or are yet being, subjected to directional positive selection. Stated in another way, characters undergoing directional selection do not exhibit the expected phenotypic trend toward homogeneity; rather, they retain the same magnitude of variation or even increase that magnitude. A correlate of this statement is that those characters which are more variable between groups at a lower taxonomic level are also more variable within these groups.

As it is difficult to speak of selection intensity in wild populations, a phylogenetic unidirectional change in a mean will be equated in the ensuing discussions with selection response. This implied association does not necessarily follow since migration, inbreeding, and distortion of the gene pool due to random fluctuations alone may also cause a movement of the population mean. In the case of the microtine tooth variations, these exceptions to the assumption are probably not involved. The tooth evolution follows a syndrome of related adaptive changes of which increased tooth complexity is but one facet. According to our present knowledge, only selection can be held responsible for directional change of this type and magnitude.

EVOLUTION OF MICROTINE MOLARS

Most of the radiations involving grazing mammals began in the Miocene with the formation of the temperate and boreal grasslands. For some unknown reason the microtine radiation, involving a dietary shift from the fruiting part of the plant to the vegetative part, lagged until the late Pliocene. As in many other radiations involving the exploitation of a coarser diet, the low-crowned tuberculate teeth changed into complex high-crowned prismatic teeth to compensate for the increased rate of attrition.

The microtine molar crown consists of a wide enamel loop at one end with alternating left and right triangles following. These triangle-like extensions are termed *salient angles* and the troughs between are the *re-entrant angles* (Fig. 1). The crown pattern of the upper molars is oriented posteriorly (the loop on the anterior part of the tooth) while the crown pattern of the lower molars is just the reverse. Except for this reversal the tooth pattern of the uppers and lowers is fundamentally the same so that M^2 has approximately the same shape as M_2 except that the loop of

UPPERS   LOWERS

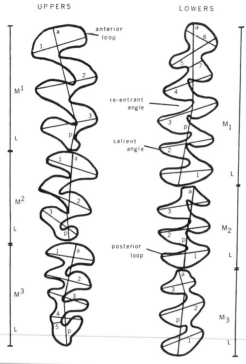

FIG. 1. A pictorial representation of the 42 measurements taken on the upper and lower teeth in two species of *Microtus*. Width measurements are numbered serially from the loop. Anterior and posterior lengths of each tooth are designated by (a) and (p) respectively, and the entire length of each tooth by (L).

the former is anterior and that of the latter posterior. In the upper molars the enamel border of the salient angles is convex on the anterior edge and concave on the posterior; in the lower teeth the pattern is reversed. Moving the teeth anterior–posteriorly produces a self-sharpening system of opposed shearing blades.

Microtine molars have become more complex by the addition of salient angles and in the more advanced forms the teeth are quite elaborate. Phylogenetically the uppers add on to the posterior margins of the teeth and the lowers to the anterior. As a consequence, the posterior margin of M^3 and the anterior margin of M_1 are the most variable between taxa. There have been numerous changes in all of the molar crowns although M^1, M_2, and M_3 are more constant than any of the other teeth. M_3 does vary in form intergenerically; perhaps this is a result of the position of the incisor root as it arcs past M_3. In some genera the incisor passes between M_3 and the two anterior molars and in other genera it does not. The addition of triangles is accomplished in M^3 and M_1, as illustrated in Fig. 2, by an increased penetration of the re-entrant angles in the trefoil or the primordium at the variable end of the tooth. In the other molars the addition of triangles is accomplished by a lateral pinching off, phylogenetically speaking, of a bud from the last triangle (see M^2 in Fig. 2). M^3 and M_1 maintain a labile primordium at the changing end, whereas this analogous area in the other molars abuts against the stable loop of the following tooth and cannot maintain such a variable structure, but has to resort to the use of the last salient angle if new angles are to be added.

The addition of salient angles has taken place throughout the late Pliocene and Pleistocene, but it would be naive to consider the whole subfamily as being constantly driven unidirectionally by a bombardment of selection pressures toward a new adaptive peak. Some groups within the subfamily have become stabilized intermediates between the two adaptive extremes. There is almost a whole generic continuum, even in the living forms, from the simple crushing bunodont dentition to a complex continuously growing hypsodont type. Within the various lines of descent there have been irregular increases in the rate of acquiring tooth complexity. Also there has been a varied differential between lines in the attainment of complex hypsodont molars. Microtine evolution is comparable to the evolution of horse cheek teeth through the Tertiary, where the more progressive grazers were often flanked by browsing groups with dentition of an ancestral pattern.

It is not intended to be implied that the teeth are the only or even the major characters undergoing change. Emphasis has been put on dentition in this treatment as

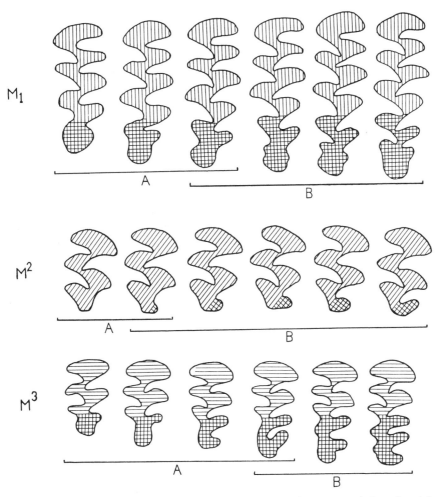

FIG. 2. A semischematic illustration of the extent of tooth crown variations found in the two species of *Microtus*: (A) *M. paroperarius*, (B) *M. pennsylvanicus*. The relatively stable areas are marked with parallel lines, and the variable areas are cross-hatched (see Fig. 1 for orientation).

it is one of the few characters which is consistently preserved in the fossil record. Although character choice in the fossil microtines is limited by default, it would have been difficult to have found a more suitable index of adaptive change.

METHODS, MATERIALS, AND MEASUREMENTS

Samples of multiple series were used in this study to investigate the horizontal (intraspecific) and vertical (phylogenetic)

species uniformity of the differential tooth variations. The main comparison is of individual variation within each series and not between series. The material is treated as four samples. The first sample represents the extinct *M. paroperarius*, which occurs only as a fossil. Samples two, three, and four are of the Recent meadow vole, *M. pennsylvanicus*. Sample two is one series with the sexes combined and the last two samples are another series with the sexes treated separately. These two species probably represent one evolutionary line; at least *M. pennsylvanicus* had to pass through the morphological stage represented by *M. paroperarius*.

The series of *M. paroperarius* was obtained from the collections of the University of Kansas Museum of Natural History. This species was first described by Hibbard (1944) and was considered in more detail, including a qualitative analysis of the intra-populational variation, by Paulson (1961). The sample was collected by Hibbard from several localities in Meade County, Kansas. These localities all belong to the Cudahy Fauna, which lies just below the Pearlette ash, a petrographically distinct volcanic ash. The Pearlette ash is a widespread Pleistocene marker of the non-glaciated areas in central and western North America and serves to delineate a contemporaneous fauna over a considerable territory. Hibbard (1944) considers the Cudahy Fauna to be late Kansan in age.

It was necessary to use teeth from several localities in order that a statistically adequate sample could be acquired. The series of *M. paroperarius* was taken as a not-too-serious deviation from an approximated population sample since the localities were all within one county and stratigraphically contemporaneous.

M. paroperarius is represented by single teeth, although a few remained attached to mandible fragments. The majority of the teeth came from K. U. localities 10 and 17, but a small number were from Locality No. 20. The individual tooth morphology was so characteristic that the individual molars could be easily identified as to upper or lower first, second, or third molars and separated as to left or right. The sexes were not distinguishable. The measurements of the left and right teeth were combined to increase the sample size. There was a positive correlation between the frequency of the teeth in the collection and their size. M_3 was the smallest and most fragile tooth and M_1 was the largest. There were fewer M_3's than any other tooth in the sample (31) and the M_1's were the

most numerous (58). This numerical disparity could have been due either to the fact that a more robust structure would better survive preservation or that, as fossils, a larger individual fragment would be more likely to be detected than a smaller one.

The second sample, of the Recent *M. pennsylvanicus*, was obtained from the Carnegie Museum collections through the Chicago Museum of Natural History. This sample was originally collected from the Pymatuning Swamp, Crawford County, Pennsylvania, an area 15 miles long by three miles wide. Goin (1943) included a qualitative review of the M^3 variations of this sample and discussed the locality in more detail. Fifty individuals were used, 25 males and 25 females. The sexes in sample two were combined as in the first sample (*M. paroperarius*). The teeth in the second sample, unlike those of *M. paroperarius*, were all in place in the jaws.

Samples three and four are, respectively, males and females of *M. pennsylvanicus*. There were 40 males and 42 females. This series was borrowed from the University of Michigan Museum of Zoology and was originally collected near the city of Lyndhurst, Ohio. The sexes were treated separately to eliminate the variable of sexual dimorphism and to see what changes this dimorphism brought about in the patterns of tooth variations.

In this study I treated the teeth as prismatic structures with no ontogenetic variation. This assumption is true for all practical purposes once the individual has passed the early juvenile age. Juveniles can be culled from *Microtus* samples by the criteria of overall small skull size, lack of suture closure, and lack of parallel-sided molars. The molars continue to grow throughout the life of the adult individual, maintaining an almost constant crown pattern.

I treated the tooth crown as if it were a two-dimensional surface. This procedure is also not precisely correct. The upper tooth-row surface wears to a slight convex profile and the lower conforms to this with a concave profile of the same magnitude. The mean of the greatest distance that the arc deviates from a straight line, intersecting the terminal ends of the arc, is 0.25 mm or 0.041 of the distance of the straight line. From the lateral view the teeth are also curved; the M_1^1's have their concave sides anterior and M_3^3's posterior. The M_2^2's have only a slight curvature. In most of the teeth there is a dorsoventral twist, so *Microtus* molars may be considered in form as segments of a broad helix.

The teeth of this genus are quite small, the whole tooth-row being only about 6 mm long in *M. pennsylvanicus*. To cope with the problem of measuring teeth of this size in detail, photographs of the tooth crown of the individual teeth in *M. paroperarius*, and of the whole tooth-row in *M. pennsylvanicus*, were taken through a dissecting microscope. The crown was first oriented at right angles to the ocular, then the camera was mounted and brought into focus. All pictures were taken through the same ocular at the same magnification. These were then enlarged and developed under the same conditions, including film, paper, and enlarger magnification. A note on the technique (Guthrie, in preparation) includes approximations of the errors in the technique at the various steps.

The measurements were then taken from the pictures with a dial micrometer reading to the nearest 0.1 mm. With the picture enlargement of 31.8×, this resulted in measurements to the nearest 3.3 microns. The measurements were quite repeatable. The exterior edge of the enamel was used in all measurements. Pictures of both left and right sides were taken of *M. pennsylvanicus*. The side with the picture of highest contrast was used, and, if there was any question, measurements were taken on both sides. Rarely was there a break or crack on both sides so that no measurement could be taken.

Measurements were made as illustrated in Fig. 1. The measurements on the whole were well defined. The only possible ex-

ceptions were the anterior part of M_1 and the posterior part of M^3. However, this is a function of their variability in form. Several measurements were used on the anterior part of M_1 and posterior part of M^3, but no one expresses adequately the variation in shape.

The width measurements for each tooth are numbered serially from the loop. Consequently, the uppers are numbered from anterior to posterior and the lowers from posterior to anterior. The total length is designated by L and the anterior and posterior lengths by a and p, respectively. Forty-two measurements were taken on each individual, 20 measurements on the uppers and 22 on the lowers.

DISCUSSION OF MOLAR VARIATIONS

The variation in M_1, M^2, and M^3 is represented in Fig. 2. The teeth viewed from left to right depict the nature and extent of the shape variations present in these samples. In reality, this variation does not fall into discrete classes as portrayed in Fig. 2; rather, each tooth in the figure represents a point along the variation continuum. The most variable portions are cross-hatched to facilitate the comparisons. Notice that in M_1 the rounded primordium on the lower part, actually the anterior part of the tooth, is utilized to construct new salient angles by the penetration of re-entrant angles into its lateral margins.

In the M^2 a new salient angle is formed by the budding off of the extreme posterior part of the crown, and varies in these samples all the way from absence to almost the size of the other salient angles. *M. paroperarius* has only a slight suggestion of this bud in some individuals, with most not having it at all. In *M. pennsylvanicus* this rudimentary stage is present only at a low frequency, most of the individuals having a well-developed salient angle.

The cross-hatched area in the posterior portion of the M^3 behaves differently than the cross-hatched area in the M_1. M^3 increases its number of salient angles phylogenetically by dropping a bud posteriorly

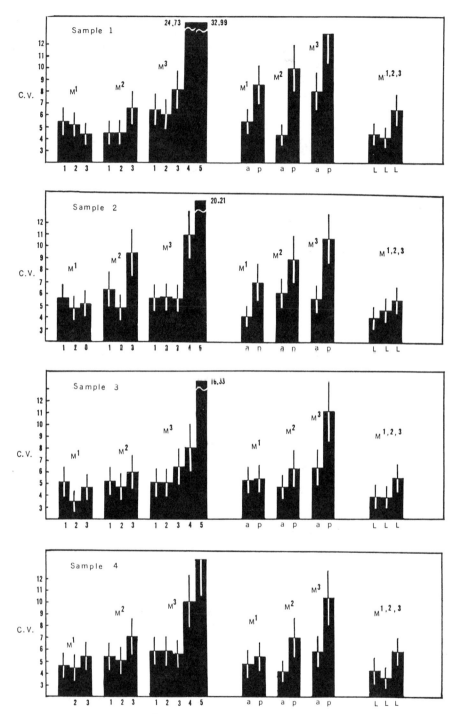

FIG. 3.   Coefficients of variation (C.V.) of the upper molars of *M. paroperarius* (sample 1) and *M. pennsylvanicus* (samples 2–4); samples are identified in text. The tongue inserts are equal to two standard errors in each direction. The measurements at the base of each histogram correspond to those in Fig. 1.

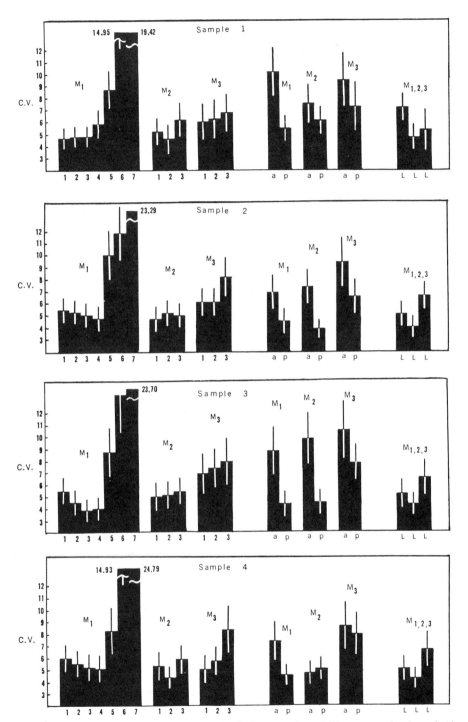

FIG. 4. Coefficients of variation (C.V.) of the lower molars of *M. paroperarius* (sample 1) and *M. pennsylvanicus* (samples 2–4); samples are identified in text. The tongue inserts are equal to two standard errors in each direction. The measurements at the base of each histogram correspond to those in Fig. 1.

and enlarging it lingually. However, on the labial side, the penetration of the re-entrant angles and the outgrowth of the salient angles act in a manner much the same as in the M_1. There is very little difference in principle in the mode of addition of salient angles in any of these teeth, only slight variations in detail.

These cross-hatched areas are the ones that vary most between species. For example, the M^3 tooth pattern at the extreme right in Fig. 2 is present in only one individual in the samples of *M. pennsylvanicus*, but is the most common tooth pattern in *M. chrotorrhinus*. Komarek (1932) reports a specimen of *M. chrotorrhinus* which has one less angle in the M^3 than usual. This specimen would correspond to the most common *M. pennsylvanicus* pattern. In addition to *M. pennsylvanicus*, several other species of *Microtus* have hints of the posteriolingual bud on the M^2, and in *M. californicus* it is of creditable magnitude (Hooper and Hart, 1962). A further discussion of the intrageneric variations in *Microtus* is given by Hooper and Hart in the preceding reference.

There is some overlap in shape between the fossil *M. paroperarius* and the recent *M. pennsylvanicus*. Referring to Fig. 2, in M_1 the third pattern from the left, in M^2 the second, and in M^3 the fourth are common to both species. However, it must be kept in mind that the discrete patterns illustrated here are only chosen points along a continuum.

The 42 different measurements are represented by histograms in Figs. 3 and 4. The most striking pattern is the high variation in the width measurements in the anterior part of M_1 and the posterior part of M^3. Although this varies slightly in magnitude between samples, the general pattern is much the same. The width measurements of M^1 have relatively low coefficients of variation, all under six. The M^2 width measurements also have relatively low coefficients of variation. The width measurement number three of M^2, which includes the incipient angle, has a larger coefficient

of variation than any of the other width measurements of either M^1 or M^2. This is the incipient angle which is predominantly present in *M. pennsylvanicus* and expressed in some individuals of *M. paroperarius* as a rudimentary bump.

In every case in the upper molars the anterior length is less variable than the posterior length, Fig. 3, (a) and (p) respectively. In the case of M^1 in samples three and four, which represent males and females from one series, the difference between (a) and (p) is not outstanding. The difference between the coefficients of variation of the anterior and posterior length is greatest in M^3, which has no overlap at two standard errors in either direction. The entire length measurements (L) of M^1 and M^2 appear to have about the same magnitude of variation. The length measurement of M^3 has a larger variation in all cases than either M^1 or M^2. It will be remembered that the upper molars add to the tooth complexity from the posterior margins. From the findings here it may also be stated that these phylogenetically variable posterior areas of the uppers have the greater intrapopulational variability.

The uniformity of the four samples would seem to increase with the order in which they are listed, as there are progressively fewer collecting restrictions imposed. The fossil *M. paroperarius* sample was taken from several localities and with some temporal variation involved. The second sample, of *M. pennsylvanicus*, was taken over a wider territory than samples three and four, which were collected near a small city. Since there is a high interpopulational variation in *M. pennsylvanicus*, even within the same subspecies (Snyder, 1954), the difference in uniformity of the collecting restrictions might be thought to affect the relative amount of within-sample variation. With but one or two exceptions, the measurements did not show this expected variational gradient between samples. There also proved to be no pattern differences of appreciable magnitude between the two sexes of *M. pennsylvanicus*.

In the uppers the measurements of *M. paroperarius* tend to be more variable than the samples of *M. pennsylvanicus*, especially the posterior part of M^3 where the coefficient of variation is about double, at least in the width measurements. In the width measurements of the phylogenetically more stable teeth M^1 and M^2 there is no notable difference in magnitude between *M. paroperarius* and *M. pennsylvanicus*.

The M^2 widths have a relatively low to moderate variation, with a coefficient of variation of about six or less, and no outstanding pattern within the tooth. M_3 width measurements tend to be more variable than those of the M_2 with the anterior width measurements having the greater variation. The coefficients of variation are very large in the anterior part of M_1 (note width measurements five, six, and seven). Another peculiarity of M_1 in *M. pennsylvanicus* is that the width measurements in the midsection of this tooth are less variable than either the anterior or posterior ones. Some of the other teeth show this to a minor degree (note M^1 and M^2). In the lowers the anterior length measurements (a) are more variable than the posterior length (p) in every case except the M_2 of sample four. Unlike the uppers, the lowers add on to the anterior margins of the teeth, and we may conclude from the coefficients of variation in Fig. 4 that these anterior areas of the lower molars also have the greatest variation.

In both the posterior lengths (p) and the whole lengths (L) there is a trend toward greater variation in an anterior to posterior direction in both the uppers and lowers. This is not so well marked in the anterior length (a) measurements.

Of the measurements of the entire tooth length, the length (L) of M_3 is the most variable in *M. pennsylvanicus* while the length (L) of M^1 is the most variable in *M. paroperarius*. This is a case where the patterns produced by the length variations (L) are somewhat misleading. In *M. pennsylvanicus* M^3 is the upper tooth with the most variation, which both the width and the length measurements suggest. M_1, on the other hand, is the most variable tooth in form among the lowers. This is evident in the width measurements but does not show up in the length (L) measurements of *M. pennsylvanicus*. Although M_1 is the most variable lower tooth it has developed a long stable posterior area which dampens the variations occurring at the anterior part of the tooth, thereby producing a deceptively low coefficient of variation for the entire tooth length. This effect is not present to the same degree in the M_1 of *M. paroperarius* (see Fig. 2). At this early phylogenetic stage the tooth has a relatively smaller stable posterior section.

M_3 has a relatively higher variability than the other phylogenetically more stable teeth M^1, M^2, and M_2. It is the one tooth that crosses over the incisor root and has a limited role in adding to the crown complexity of the tooth-row, and may even be in a state of reduction in this particular genus. In some other genera of microtines, *Dicrostonyx* for example, the incisor root does not cross over in this fashion and the M_3 has developed a more complex crown pattern. Also, it is not reduced in size laterally as it is in *Microtus*. These facts suggest that the peculiar relationship of M_3 to the incisor places some limitations on its potential for increased complexity.

In many of the cricetines both the upper and lower third molars have undergone considerable reduction; this is not the case in *Microtus*. Some individuals of *M. pennsylvanicus* have a longer M^3 than M^1.

In summary then, a quantification in these two species of the molar variability reveals an overall pattern of higher variation in the posterior parts of the upper molars and the anterior parts of the lowers. The greatest amount of variation is present in the anterior end of M_1 and the posterior end of M^3. A direct positive correspondence exists between those areas of the teeth which are changing phylogenetically and those which exhibit a greater magnitude of variation.

Supporting Evidence

The significance of a positive association between the rapidly evolving tooth characters and a relatively high variability in *Microtus* is dependent upon its general applicability. This may be either a special case or an expression of a more general phenomenon. The following is a presentation of evidence supporting its more general nature.

In the microtines this association is not limited to the *M. paroperarius–pennsylvanicus* line, but rather it is a common feature of the whole group. *Dicrostonyx* has the most complex crown pattern of the subfamily. *D. torquatus*, the species represented in the second phase of the last glaciation (Zeuner, 1958), has a variable expression of new salient angles on the posterior margin of M^1 and M^2 and the anterior margin of M_2 and M_3. These salient angles are highly variable in their occurrence, grading to complete absence in some individuals. The characteristic species of the last glaciation, phase one (early Wisconsin), was *D. henseli*, which did not possess the salient angle or bud as did *D. torquatus*. This bud seems to be a nascent character developing through the last glacial age. *D. groenlandicus*, a recent species, has this character present in all individuals. *D. hudsonius*, a species with a distribution presently limited to the Hudson Peninsula, is a living relict representative of the *D. henseli* tooth pattern of the early part of the last glaciation. *D. torquatus* exists as the modern Old World collared lemming. Thus there is a chronological and geographical representation of the stages of development of this salient angle. The fossil *D. henseli* and the recent *D. hudsonius* do not have the salient angle. *D. torquatus*, both modern and fossil, has a varied expression of the salient angle from absent to fully present (Hinton, 1926). In populations of *D. groenlandicus* all individuals have it. Some taxonomists give these forms only subspecific status; however, the principle dealt with here remains valid.

Kurtén (1959) suggested that the average rate of mammalian evolution during the Pleistocene was relatively higher than during the Tertiary. His analysis of the variability in several rapidly evolving groups, widely separated taxonomically, revealed an increase in the coefficient of variation in more lines than a decrease. Although his study did not deal in detail with the specific characters which are changing (he used an average of several measurements), it did serve to illustrate that rapidly evolving populations do not all tend toward morphological uniformity. On the contrary, it suggested the opposite. Wright, in the discussion at the end of Kurtén's paper, proposed that recombination is responsible for this amplification of potential variability.

Skinner and Kaisen (1947) noted that while there are few diagnostic patterns in the evolution of *Bison* cheek teeth, there is a general trend toward the molarization of P_4. The metastylid and median labial root of the P_4 increase in frequency through time. In early fossil *Bison* these characters are virtually absent and in modern ones almost universally present. The increases in the complexity of P_4 seem to have occurred over a relatively short period of time during the late Pleistocene. Since these evolving areas range from absent to fully developed in some populations during this period of incipiency, the variability is greater than that of the analogous areas of neighboring teeth.

Simpson (1937) discussed a sample of 33 Eocene notoungulates, *Henricosbornia lophodonta*, which he considered to be from one population, since their variation is normally distributed and they are from the same horizon and locality. These were originally described by Ameghino as belonging to 17 species, seven genera, and three families, principally on the basis of the variation present in the upper third molar. The variations present within this primitive form are characteristic of later species, genera, and families with which Ameghino was familiar. Here is an example of a considerable amount of variation in one population, the elements of which are later characteristic of higher taxa. It would be consistent with the evidence to assume that the tooth is undergoing evolutionary change in a manner which contributes to the types characteristic of later higher taxa.

Hooper's (1957) study of the dentition of *Peromyscus* gives supporting evidence to the main thesis proposed here of rapid evolution being accompanied by high phenotypic variation. A series of *P. maniculatus* from Distrito Federal, Mexico, for example, has highly variable molars. The mesoloph and mesostyle patterns found in this one series resemble the common patterns of the other 17 species of *Peromyscus* studied. In other words, the mesostyle and mesoloph patterns observed in 17 species of *Peromyscus* are also seen in this single series. *P. maniculatus* is first known from the Wisconsin age and has expanded its distribution over a considerable part of North America. It is considered to be one of the "younger" species of *Peromyscus* (King, 1961), and therefore has recently undergone evolutionary change at the species level.

The occurrence of the crochet in horse teeth is another example of an incipient character that is highly variable in the same population (Simpson, 1953; Stirton, 1940). The acquisition of this plication is one of the first features in a general trend toward increased tooth complexity. The crochet, an anastomosing ridge between metaloph and protoloph, shows up in the *Miohippus–Parahippus* line. It is also present in some species of *Archeohippus* and sometimes in the milk teeth of *Hypohippus* (Stirton, 1940). The incipient crochet juts out as a peninsula or pier from the metaconular part of the metaloph toward the protoloph. The degree of its development is extremely variable, from absence to a small spur extending halfway across, to a complete connection between the two lophs. The crochet varies both in frequency and extent between populations and within them, occurring in its various stages of representation in individuals of the same species at one locality.

Butler (1952), speaking of the molarization of premolars in Eocene horses, stated that the metaconule evolving in the premolars is most variable at the intermediate stages of molarization.

Wood's (1962) discussion of the tooth cusp variations in the early paramyid rodents showed that the hypocone is added to the tooth by two basically different means. In some forms it is derived from an enlargement of the posterointernal cingulum; in others it originates as a division of the protocone. Wood attributed these two distinctly different means of achieving fundamentally the same end product to a general selection toward the development of a posterointernal cusp irrespective of the nature of its origin. The addition of the fourth cusp, hypocone, is a common phenomenon in many lines during this part of the Tertiary, and seems to be correlated with the exploitation of more demanding food substances. Wood stated, "There is no question but that all of these variants may occur within a single genus and sometimes within a single species." Here again, when a directional selection pressure is being applied, more phenotypic variation is exhibited in the incipient than in the nonincipient cusps.

The lower third premolar is used to characterize various genera of fossil rabbits. Hibbard (1963) observed much variation within a primitive rabbit genus, *Nekrolagus*, and found at a low frequency a pattern of the P_3 that is characteristic of modern genera. The common tooth pattern of *Nekrolagus* is also found at a very low frequency in some modern genera. This comparative study documents a chronological frequency change in which the early fossil populations have the incipient characters represented at a low frequency and the modern populations at a high frequency. Here is another case in which there is a high variation associated with incipient characters, and the axis of this variation is parallel to phylogenetic change.

Another opportunity to try the hypothesis is on the results of artificial selection experiments. If the hypothesis does approximate the real condition, the character that is artificially selected for or against should behave in a manner similar to the evolving characters that have just been discussed. That is, characters undergoing artificial selection could be expected not to experience a decrease in their phenotypic variation, but to maintain or even increase the variation.

MacArthur (1949) selected for large and small size in mice using the weight at 60 days as a measure of size. In the unselected control the coefficient of variation was 11.1. However, in the strain selected for large size it was 12.8, and in the small line 14.3.

Falconer (1955) also selected for large and small size in mice using the sixth week weight as a measure of size. He stated, "The phenotypic variability, also, does not reflect the expected decline of genetic variance, and in addition reveals a striking and unexpected change in the small line." He further reported that the large line showed a slight increase in variation over the whole course of the experiment, although it remained relatively low compared to the variation of the small line. The coefficient of variation in the small line increased to about double the original value between the seventh and ninth generations and remained at this high level. The realized heritability remained substantially constant up to the point at which response ceased. This phenomenon, he suggested, was due to the release of genetic variation through recombination.

In their selection experiments for wing length in *Drosophila*, Reeve and Robertson (1953) found that the coefficients of variation at the twentieth to seventy-ninth generations were all below two in the unselected strain and all two or above in the selected strain. The strain selected for long wings showed an increase of about 50 per cent in total variance. They attributed this entirely to an increase in additive genetic variance, which rose about two and one-half times, while the absolute amount of other genetic variance remained about the same. This led them to suppose that selection for long wing length would be far more effective in the selected than in the unselected stock.

Clayton and Robertson (1957), selecting for low and high bristle number in *Drosophila*, concluded that "Selection had by no means led to uniformity, but in some cases even magnified the total variation."

Robertson (1955) selected for thorax length in three stocks of *Drosophila* with about the same initial amount of variation. The coefficient of variation in the small lines increased immediately in the first generations and was higher than the control in all three lines, although there were between-strain differences in the pattern of increase in variation. In the large lines the variation fluctuated around that of the control stock. Thus, in the large strains the changes in response to selection occurred without appreciable change in the coefficient of variation, while the variation of the small line increased.

Although the changes in variation accompanying selection response in these experiments do not behave in a completely uniform manner, they do maintain and usually increase the initial magnitude of variation. Thus, evidence supporting the association between directional selection and a constant or increased variation is found both in rapidly evolving groups and in artificial selection experiments in which the degree of variational change has been recorded.

The Theory and Model

The most frequently employed explanation for an inordinate amount of variation is vestigiality. In such a case the characters under consideration are not becoming more complex phylogenetically but are decreasing in pattern complexity. Morphological characters which are in the process of reduction or elimination exhibit more variation than do their more functional homologues. This high correlation between vestigial and highly variable characters no doubt influenced Hinton (1926) to believe the microtines to be, in tooth form, degenerate descendants of the multituberculates and consequently undergoing reduction in tooth complexity. However, there is a time gap in the fossil record of some 35 million years between the multituberculates and microtines. The concept of the vestigial nature of microtine teeth has been perpetuated by some mammalogists (Goin, 1943; Hall and Kelson, 1959). But the position that microtines did not arise from a cricetine stock and have not undergone a general increase in tooth complexity is untenable. Not only does the fossil record support an increase in microtine tooth complexity, but there is an almost complete continuum of recent intermediate forms between the Microtinae and Cricetinae. Vestigiality can be discounted as an explanation of the variation differential in the other examples as well, as these characters are also increasing in complexity.

Lately, much attention has been given to the loss of buffering capacity against environmental stress as the genome tends toward homozygosity (Lerner, 1954). Since directional selection reduces the amount of individual heterozygosity, the loss of buffering would result in a greater magnitude of individual deviation from the mean, increasing the phenotypic variation of that population. This process may be the cardinal factor involved in an explanation of the phenomenon of an increase in variation accompanying directional selection. However, there is some discouraging evidence against an explanation of this nature. (1) The increase of phenotypic variation becomes evident early in artificial selection (Robertson, 1955) before an appreciable amount of genetic variance could have been lost by selection. (2) A character in which selection has considerably altered the mean can often be returned with little difficulty to the original mean by reversed selection. This reversal could not take place if the population had reached a relatively homozygous level for that particular character. (3) A correlate of the latter is that often a substantial heritable component is still present after the mean has been considerably altered by selection (Lerner, 1958). (4) Bader (1962) showed that, in tooth form, inbred mice exhibit slightly less phenotypic variation than wild populations; and the outcrossed heterozygote is less variable than either. (5) If the tooth variation discussed here in *Microtus* is non-genetic, it is difficult to explain the phylogenetic increase in tooth complexity, since the most important cause of evolutionary change is selection acting upon heritable variation. From some preliminary crosses of microtines (Steven, 1953; Zimmermann, 1952), it does seem that these variations are heritable. In at least one species of *Microtus* (*M. arvalis*) there is also a geographic cline in the frequency of tooth complexity. The variations were classed into two discrete types (simple or complex); the frequency of "complex" ranges from five per cent to 95 per cent in the cline from northern to southern Europe (Zimmermann, 1935).

The accumulated evidence from breeding experiments suggests, contrary to the "wild type" or normality concept, that there is considerable heterozygosity underlying the relatively coherent facade of the phenotype. The variation expressed in the phenotype is only a fraction of the total possible variation present (Mather, 1956). There is a diversity of opinion as to the mechanisms involved in the maintenance of this large amount of potential variability. The position that the balanced additive factors maintain the stored variability has much evidence in its favor in terms of its general applicability to evolution at the intrapopulational level. Stated in more detail, this position asserts that there exist balanced systems of linked heterozygous polygenes structurally associated and maintained by selection and perhaps also by decreased recombination.

Delayed responses to selection are best accounted for on the basis of genetic linkage. A rather common phenomenon in experimental breeding is for a selected strain to reach a plateau of response only to have it resume progress after a period of relaxation of the selection pressures. The most plausible explanation of this phenomenon is linkage disassociation; the various elements are unable to segregate out immediately because of linkage restrictions (Mather, 1949). The ineffectiveness of experiments to reduce the variation by selection for intermediates (Lerner's type II selection, Lerner, 1958; Falconer, 1957), and the ineffectiveness of selection for the extremes to alter the variation (type III selection, Falconer and Robertson, 1956) both suggest that the additive genetic material resides in balanced linkage groups.

Structural change, which often inhibits crossing over, may establish an isolation of segments of the chromosome where crossing over is likely to occur only with configurations of that same type; however, the general importance of this mechanism is still not clear. As well as promoting these devices that inhibit recombination, selection can operate directly to maintain these blocks intact (Lerner, 1958) and this is probably the most important mechanism. Carson (1959) reports that most natural inversions are heterotic when removed from nature to the laboratory culture, and that strains derived from a single pair of wild flies retain with extreme tenacity most of their initial inversion variability.

The advantages of a system of balanced linkage groups are multiple. The population can maintain a high degree of heterozygosity in many individuals without the rigorous selection required if these elements were segregating at random. The close linkage association also serves as a buffer against random fluctuations away from the optimum. And perhaps most important, it holds genetic material in reserve, thereby maintaining an evolutionary plasticity.

There is evidence that integrated chromosome segments are important in the association or correlation of continuously distributed characters, and that they behave in a manner similar to single independent genes acting pleiotropically. To resolve or disassociate the correlation of two characters by selection would produce strong evidence for linkage. Such disassociation has been accomplished (Mather and Harrison, 1949; Mather, 1956). Correlation due to pleiotropy is, of course, more resistent to evolutionary change than the more labile system of linkage groups. Linkage groups can originate or be disposed of by the selection for various recombination and structural patterns. It would be a slow process for the population to await a new mutation at one locus which acted upon the desired characters in exactly the right magnitude.

Selection can maintain a frequency of balanced genetic material within each chromosomal block or "internally" at levels that insure a considerable proportion of "relationally" balanced, or heterozygous, individuals in the population. As long as this block remains intact it will carry reserves of variability which may be released and made available for segregation by crossing over. With selection against the crossovers, this residual genetic variability can be maintained (Lerner, 1958). In order to maintain the internally balanced linked groups a selection intensity would be required equivalent to the frequency of crossovers which deviate from the balanced configuration (Falconer, 1960).

The increase in variation of evolving characters may be further enhanced when an interbreeding population experiences the stress of two selective optima. This condition would occur in most evolutionary changes when the group is partially exploiting two adaptive zones. Thus, a character in transition may be expected to experience some reduction in stabilizing selection along its axis of change.

The high variation usually associated with vestigiality can also be accounted for in the context of this theory. A vestigial character is in essence an evolving character, as reduction plays a great part in evolutionary change. According to the explanation given for the greater amount of variation in evolving characters, the stored variability is maintained in a linked system by stabilizing selection. When this balance is altered by directional selection, the variability is released. Due to its decreasing functional role, the variation of a vestigial character would also be compounded by a decrease in stabilizing selection.

Carson (1955), in his discussion of the genetic composition of marginal populations, surmised that, since the marginal populations contain fewer inversions than central populations, the more stringent selection on the periphery is against the heterotic groups which predominate in the central population. These findings are in agreement with the idea expounded here, that directional selection away from the mean is selection for the breakdown of present linkage configurations. Carson further reported that when strong artificial selection was applied to both marginal and central population lines, the marginal lines showed the greater initial response. This difference would exist if the genetic material has been made available for segregation in the marginal populations by the breakdown of the linkage groups.

Reeve and Robertson (1953), selecting for wing length in *Drosophila*, found that the selected strain showed an increase in additive genetic variance of 250 per cent, all other genetic variance remaining about the same. They further suggest that selection for long wing length would be more effective in the selected than in the unselected stock. Robertson (1955) states: "Selection generally leads to an increase in variance which appears to be largely due to the increased effects of genetic segregation and this constitutes an aid to selection progress."

This release of additive genetic variation provides a mechanism whereby directional selection, in continuously distributed polygenic systems, increases its own resolving power. Selection against the mean and its present balance situation is selecting against the present linkage configurations, which results in a breakdown of these integrated

units. The genetic components are then released and made available for novel segregants hitherto unavailable. The consequence of this is an increase in phenotypic variation, which is heritable in an additive fashion. As the amount of variation is a determining factor of the effectiveness of selection, in conjunction with selection intensity and heritability, further selection gains are facilitated.

To set up a simple visual model of this theory let us suppose, as is expressed in Fig. 5, that there is a series of loci with alleles acting in an additive fashion either to the left or right of the mean. Loci a, b, c, and d control the size of character X and e, f, g, and h control Y. The contribution of each allele is specified. Further, suppose these are balanced "internally" and "relationally," with an equal frequency of each linkage group. The mean will be considered as zero with the deviations from it in both positive and negative directions. A stabilizing selection for the mean would cull out deviants, the crossovers, from this configuration. The genetic material present is potentially able to produce an individual representative of any point in the figure, but this particular linkage configuration limits the phenotypes to a coherent cluster around the mean. The broken circle represents a variation of two standard deviations from the mean, if each locus were acting individually with an equal frequency of each allele. The linked configuration, however, would produce a population with a lower variation, expressed here at two standard deviations by the solid inner circle.

If a new adaptive optimum $(\overline{X}_2)$ were created with a consequent directional selection of moderate magnitude exerted on the distribution, the linkage groups would be selected against by a selection for the crossovers in the direction of the new adaptive optimum, resulting in a partial breakdown of the coherent phenotype.

A structural association of the loci controlling the two characters (Fig. 6) would result in their correlation. The points all fall along the diagonal axis between +2

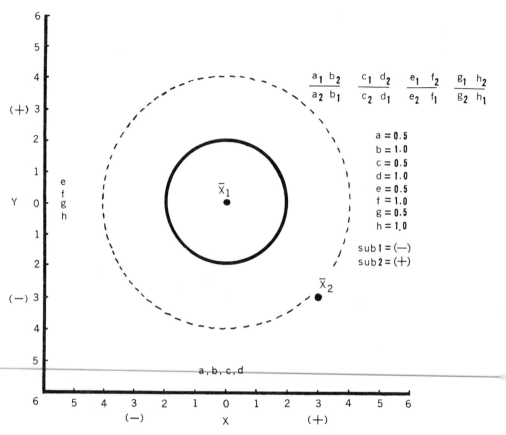

Fig. 5. An elementary model of the non-correlated case of two characters X and Y, where low phenotypic variation is maintained by selection for the linkage configuration represented in the upper right. With equal frequencies of each linkage group the variation of the population, at two standard deviations, would be circumscribed by the solid circle. The dashed line represents the same loci with no linkage. Selection for $\overline{X}_2$ would increase the variation as the linkage configuration would be selected against.

and −2 units, as shown by the ellipse. If one were to think in terms of the major axis of variation as size, this provides a relatively constant individual shape throughout a population in which the individuals are varying in size. As in Fig. 5 it will be noted that an imposed directional selection will produce an increase in variation. Even directional selection parallel with the main axis of size will increase the variation. The greatest increase in variation, however, would be produced by a selection pressure at right angles to the principal axis of variation, toward $\overline{X}_2$, which would be selection for shape changes. The long-term effect of

this type of selection would be twofold: (1) an increase in phenotypic variation, and (2) a decrease in the correlation of characters X and Y. Unlike the situation in Fig. 5, if a selection pressure is exerted on only one character (perpendicular to the scale of the other), the second character is also initially affected. However, it is an inherent mechanism of the model that the linkage which provides the correlation of the characters will be selected against when only one character is subjected to directional selection. This system then would further contribute to evolutionary plasticity.

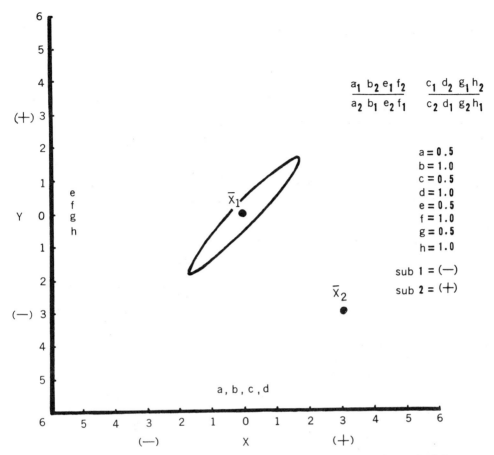

FIG. 6. Same as Fig. 5 except that characters X and Y are now correlated due to the linkage association. This new linkage configuration maintains a coherence differentially on the size and shape axes. Selection toward $\overline{X}_2$ would considerably increase the variation along the shape axis.

I do not wish to imply that the theory expressed here accounts for all the various behavior exhibited by residual genetic variation. Rather, I have investigated one aspect, the association of directional selection and the maintenance or increase of the initial phenotypic variation, and have hopefully offered a plausible explanation, which will be further explored soon by breeding experiments with *Microtus*.

SUMMARY

This is a study of the intrapopulational variability present in the dentition of two species of *Microtus*, and the more general questions arising from it. The central thesis is that quantitative characters undergoing rapid evolution do not show the decline in phenotypic variation predicted by our present evolutionary concepts. On the contrary, the variation is maintained and usually increased.

Of the two species used, the fossil species is thought to be ancestral to the modern meadow vole; thus the study materials comprise an evolutionary line with two grades of tooth complexity represented. In the molar crowns of both species, the areas which are changing phylogenetically are those which vary most within the population. Evidence from other sources in which characters are undergoing directional selection, both evolutionary and artificial, suggests that a greater variation in characters undergoing directional selection is a general condition.

A theory to account for association between rapidly evolving characters and a relatively higher amount of phenotypic variation is that the coherence of the population around the mean is due to balanced heterozygous linkage groups and that with the application of directional selection this organization is partially broken down. The genetic variation is then released and made available for recombination. The relatively high variability associated with vestigial characters is also fitted into the context of the theory. The theory suggests that directional selection on continuously distributed characters increases its own effectiveness.

ACKNOWLEDGMENTS

I wish to express my appreciation to Dr. E. C. Olson, University of Chicago, the chairman of my graduate committee, for his encouragement and assistance with the presentation. Thanks are also due to Dr. Vernon Harms and Dr. Brina Kessel, University of Alaska, for their helpful criticisms of the manuscript. My deepest gratitude goes to Dr. R. S. Bader, University of Illinois, for the many stimulating discussions which were to form the nucleus of my interests in evolutionary mechanisms. I wish to thank also those in charge of collections at the University of Kansas Museum of Natural History, University of Michigan Museum of Zoology, Carnegie Museum, and Chicago Museum of Natural History for the use of the specimens.

LITERATURE CITED

BADER, R. S. 1955. Variability and evolutionary rate in oreodonts. Evolution, 9: 119–140.

——. MS. Phenotypic and genotypic variation in odontometric traits of the house mouse. (Manuscript submitted for publication.)

BUTLER, P. M. 1952. Molarization of premolars in the *Perissodactyla*. Proc. Zool. Soc. London, 121: 819–843.

CARSON, H. L. 1955. The genetic characteristics of marginal populations of *Drosophila*. Cold Spring Harbor Symp. Quant. Biol., 20: 276–285.

——. 1959. Genetic conditions which promote or retard the formation of species. Cold Spring Harbor Symp. Quant. Biol., 24: 87–105.

CLAYTON, G. A., AND A. ROBERTSON. 1957. An experimental check on quantitative genetical theory II. Long term effects on selection. J. Genetics, 55: 152–180.

FALCONER, D. S. 1955. Patterns of response in selection experiments. Cold Spring Harbor Symp. Quant. Biol., 20: 178–196.

——. 1957. Selection for phenotypic intermediates in *Drosophila*. J. Genetics, 55: 551–561.

——. 1960. Introduction to quantitative genetics. Ronald Press, New York.

FALCONER, D. S., AND A. ROBERTSON. 1956. Selection for environmental variability of body size in mice. Z. indukt. Abstamm.-u. Vererblehre, 87: 385–391.

GOIN, O. B. 1943. A study of individual variation in *Microtus pennsylvanicus pennsylvanicus*. J. Mamm., 24: 212–223.

GUTHRIE, R. D. MS. A technique for detailed biometrical analysis of two-dimensional crowns. (Manuscript submitted for publication.)

HALL, E. R., AND K. R. KELSON. 1959. The mammals of North America. Ronald Press, New York.

HIBBARD, C. W. 1944. Stratigraphy and vertebrate paleontology of Pleistocene deposits of Southwestern Kansas. Geol. Soc. Amer. Bull., 55: 707–754.

——. 1959. Late Cenozoic microtine rodents from Wyoming and Idaho. Papers Michigan Acad. Sci., Arts, and Letters, 44: 3–40.

——. 1963. The origin of the P3 pattern of *Sylvilagus, Caprolagus*, and *Lepus*. J. Mamm., 44: 1–15.

HINTON, M. A. C. 1926. Monograph of the voles and lemmings (*Microtinae*) living and extinct. Richard Clay, Suffolk.

HOOPER, E. T. 1957. Dental patterns in mice of the genus *Peromyscus*. Univ. Michigan Mus. Zool., Misc. Publ., No. 99.

HOOPER, E. T., AND B. S. HART. 1962. A synopsis of recent North American microtine rodents. Univ. Michigan Mus. Zool., Misc. Publ. No. 120.

KING, J. H. 1961. Development and behaviorial evolution in *Peromyscus*. Pp. 122–147 in W. F. Blair (ed.), Vertebrate Speciation. Univ. of Texas Press, Austin.

KOMAREK, R. V. 1932. Distribution of *Microtus chrotorrhinus*, with description of a new subspecies. J. Mamm., 13: 155–158.

KURTÉN, B. 1959. Rates of evolution in fossil mammals. Cold Spring Harbor Symp. Quant. Biol., 24: 205–215.

LERNER, I. M. 1954. Genetic homeostasis. John Wiley, New York.

——. 1955. Concluding survey. Cold Spring Harbor Symp. Quant. Biol., 20: 334–340.

——. 1958. The genetic basis of selection. John Wiley, New York.

MacArthur, J. W. 1949. Selection for small and large body size in the house mouse. Genetics, **34**: 194–209.

Mather, K. 1949. Biometrical genetics. Methuen, London.

——. 1956. Polygenetic mutation and variation in populations. Proc. Royal Soc. London, **145**: 293–297.

Mather, K., and B. J. Harrison. 1949. The manifold effect of selection. Heredity, **3**: 1–52, 131–162.

Paulson, R. G. 1961. The mammals of the Cudahy fauna. Papers Michigan Acad. Sci., Arts, and Letters, **46**: 127–153.

Reeve, E. C. R., and F. W. Robertson. 1953. Studies in quantitative inheritance II. Analysis of a strain of *Drosophila melanogaster* selected for long wings. J. Genetics, **51**: 276–316.

Robertson, F. W. 1955. Selection response and the properties of genetic variation. Cold Spring Harbor Symp. Quant. Biol., **20**: 166–177.

Simpson, G. G. 1937. Supra-specific variation in nature and in classification. Amer. Nat., **71**: 236–276.

——. 1953. The major features of evolution. Columbia Univ. Press, New York.

Skinner, M. F., and O. C. Kaisen. 1947. The fossil *Bison* of Alaska and preliminary revision of the genus. Bull. Amer. Mus. Nat. Hist., **89**: 127–256.

Snyder, D. P. 1954. Skull variation in the meadow vole (*Microtus pennsylvanicus*) in Pennsylvania. Ann. Carnegie Mus., **33**: 201–234.

Steven, D. M. 1953. Recent evolution in the genus *Cleithrionomys*. Symp. Soc. Exp. Biol., **7**: 310–319.

Stirton, R. A. 1940. Phylogeny of the North American *Equidae*. Univ. California Publ., Bull. Dept. Geol. Sci., **25**: 165–198.

Zeuner, F. E. 1958. Dating the past. Methuen, London.

Zimmermann, K. 1935. Zur Rassenanalyse der mitteleuropäischen Feldmause. Arch. Naturgesch., N. F. 5.

——. 1952. Die simplex-Zahnform der Feldmaus, *Microtus arvalis*. Pallas. Verh. Deut. Zool. Ges., Freiburg.

Sonderdruck aus Z. f. Säugetierkunde Bd. 32 (1967) H. 3, S. 167—172
VERLAG PAUL PAREY · HAMBURG 1 · SPITALERSTRASSE 12
© 1967 Verlag Paul Parey, Hamburg und Berlin

Evolutionary adaptations of temperature regulation in mammals[1]

By L. Jansky

Eingang des Ms. 25. 10. 1966

Generally speaking, adaptations may take place either during individual life of animals (acclimations and acclimatizations), or they may be specific to certain species (evolutionary adaptations) (Hart 1963b). They may be realized by different mechanisms with different degree of efficiency, however the aim of all adaptations is essentially the same — to reduce the dependence of animals on environmental conditions and thus to increase their ecological emancipation. The study of physiological mechanisms of adaptations is therefore of great ecological importance since it helps us to elucidate physiological processes influencing limits of distribution of different species and having a profound effect on the quality or density of animal populations. The comparison of individual and evolutionary adaptations permits us to trace the evolutionary progressive physiological processes and to contribute to the problems of phylogeny.

In lowered temperatures mammels tend to lose heat. Theoretically, they can prevent hypothermia either by increasing heat production in the body or by reducing heat loss from the body to the environment. Heat production is realized by shivering; heat conservation may be manifested by reducing the body surface, by improving its insulation qualities and by decreasing the body—air temperature gradient according to formula:

$$H = K \frac{T_B - T_A}{I} \quad (1) \quad (\text{Hart, 1963b})$$

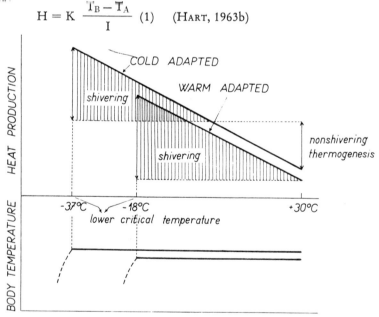

Fig. 1. Scheme of heat production of rats adapted to warm (30° C) and cold (5° C) environments. According to Hart & Jansky, 1963

[1] Presented at the 40th meeting of the German Mammalogical Society in Amsterdam.

L. Jansky

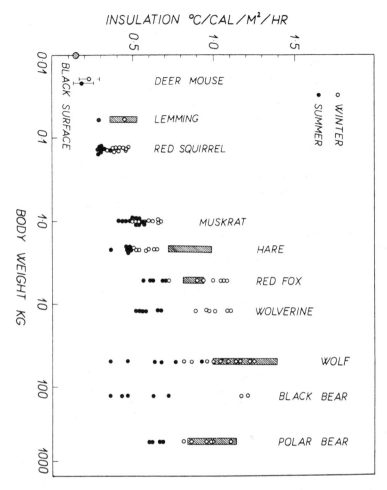

Fig. 2. Seasonal changes in fur insulation in various mammals (HART, 1956)

(H = heat production, K = a constant representing the body surface area, T_B = body temperature, T_A = air temperature, I = insulation qualities of the body surface.)

Similarly, the adaptations of temperature regulation to cold can be realized either by increasing the capacity of heat production or by mechanisms leading to reduction of heat loss from the body. The adaptation to cold appears as a shift of the lowest temperature limit animals can survive (lower critical temperature).

In our earlier work we have shown that the individual adaptations are manifested predominantly by an increased capacity of heat production owing to the development of a new thermogenetic mechanism — called nonshivering thermogenesis (HART, JANSKY 1963). Physiological background of this phenomenon consists in an acquired sensitivity of muscular tissue to thermogenetic action of noradrenaline liberated from sympathetic nervous endings (HSIEH, CARLSON 1957). Nonshivering thermogenesis potentiates heat production from shivering and in rats shifts the lower critical temperature for about 20° C (from −18° C down to − 37° C; Fig. 1).

Mechanisms controlling heat loss by changes in body surface area or by changes in body-air temperature gradient are not common in individual adaptations. On the other hand it is well known, that certain species can improve body insulation in winter season. However, this phenomenon becomes functionally justified only in animals of greater size (size of fox and larger; Fig. 2. Hart 1956).

The individual adjustments with the aid of nonshivering thermogenesis are encountered both in acclimations under laboratory conditions and in seasonal acclimatizations induced in the same species under natural conditions. They are undoubtelly very efficient and biologically important. On the other hand, from the ecological point of view, they have also their negative side. The increased heat production results in higher demands for energy restitution in the body, which is attained in cold adapted animals by an increased food consumption. As a result, individuals adjusted this way become more dependent on the quantity and availability of food and they are forced to use more effort to provide it. The reduced dependence of animals on temperature factors is thus substituted by increased dependence on food factors.

Contrary to individual adaptations, in evolutionary adaptations mechanisms leading to the reduction of the heat loss are greatly emphasized. Their importance consists in the fact that they save energy for the organism and have lower demands to its restitution in the body. This fact is obviously evolutionary very important — in the processes of phylogeny there occurs natural selection of those individuals that are less impeded by the lack of food, often occuring in nature.

Evolutionary adaptations are realized in the first place by an increased insulation of the body cover (fur, Fig. 3). This adjustment, typical for arctic animals, can reduce the heat loss so efficiently, that even considerably reduced ambient temperatures (down to − 50° C) do not result in an increased heat production in larger animals. (Fig. 4; Scholander et al. 1950a, b). The same role plays a thick layer of subcutaneous fat which appears in some mammals, such as seal and swine. The insulation qualities of this fat layer can be increased by an active restriction of the blood flow to this area. This results in superficial hypothermia, which also efficiently prevents the heat loss (Irving 1956). Animals endowed with superficial hypothermia have normal thermogenetic abilities. However, compared to the species from tropical regions with little insulation and to arctic species with great surface insulation they show a reduced sensitivity of afferent sensory input to temperature stimuli (Fig. 5).

A tendency to reduce heat loss by reduction of the body surface area may be considered as another type of evolutionary adaptations. This phenomenon occurs in animals living permanently in cold climate, which are generally larger and have shorter body appendages than animals from tropical zone (Bergmann's and Allen's rules). Both the validity and the physiological significance of these rules have been recently questionend by several workers, however.

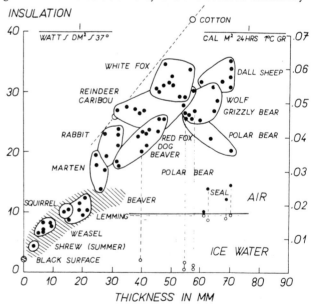

Fig. 3. Insulation in relation to winter fur thickness in arctic and tropical mammals (Scholander et all., 1950 b)

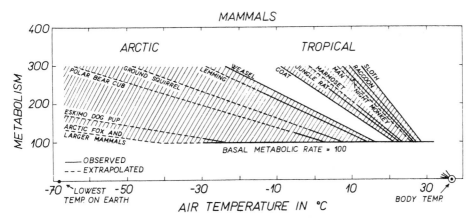

Fig. 4. The effect of environmental temperature on metabolism of arctic and tropical mammals
(SCHOLANDER et all., 1950 a)

The reduction of heat loss by changing the body-air temperature gradient can be
realized either by active choice of higher environmental temperature or by considerable lowering of body temperature.

It is generally recognized that the active choice of the environmental temperature
occurs by seasonal migrations and by changes in patterns of daily activity. It was
found that different species of
voles and shrews transfer the
peak of daily activity to warmer part of the day in a cold
weather (JANSKY & HANÁK
1959).

The mechanisms leading to
reduction of body-air temperature gradient by lowering of
body temperature are especially
developed in hibernators. According to the latest view hibernation is not considered as a
lack of temperature regulation
rather as a special adaptation of
thermogenetic processes. There
are two reasons for that: first,
hibernators have the same capacity of heat production as other
hemeotherms of similar size
(see JANSKY, 1965) and second,
the entering, the arousal and
the deep hibernation are under
remarkably precise physiological
control (see LYMAN, 1963).

This indicates a leading role
of central nervous system in
controlling hibernation, which
is adapted to hypothermal con-

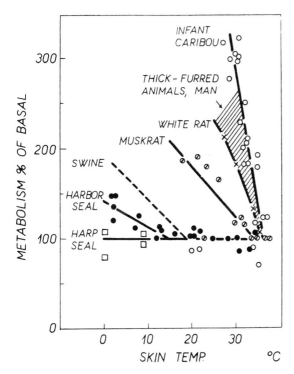

Fig. 5. Heat production as a function of skin temperature under fur of the back for a series of mammals
(HART, 1963 a)

ditions and it is functional at all levels of body temperature. This adaption has certainly its metabolic background, however only little is known about this phenomenon so far.

The control of entering into hibernation is realized by the active inhibition of shivering heat production by signals from subcortical centres of the brain. Simultaneously with the decrease in shivering an active inhibition of the activity of the sympathetic nervous system also takes place, which is manifested by the reduction of heart rate and by vasodilatation. These changes facilitate the lowering of body temperature of animals which is realized successively in the form of "undulating" cooling so the organism can slowly prepare to hypothermia (Fig. 6). Nervous control of hibernation persists in deep hypothermia as evident from the sensitivity to thermal and other stimuli. The arousal from hibernation is equally an active process, very efficiently controlled, so that organism can produce a great amount of heat in minimum of time. The coordination of thermogenetic processes depends also on the activity of nervous centres. Characteristic of awakening is the preponderence of sympathetic

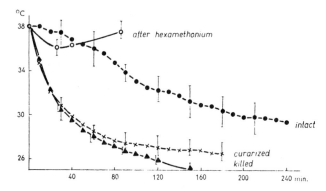

Fig. 6. Changes in body temperature of the bat *Myotis myotis* during entering hibernation (JANSKY, HÁJEK, 1961)

nervous system, leading to vasoconstriction and to an increase in heart rate. The main source of heat in awakening is again constituted by shivering. However, nonshivering heat production was also found during arousal and also the rapidly beating heart, working against a high pressure, may contribute a certain amount of heat.

Summary

On the basis of all mentioned data we conclude that the adaptations of temperature regulation to cold may be realized either by an increased ability to produce heat or by reducing the heat loss. While the individual adaptations are manifested chiefly metabolically as evident from an increased capacity of heat production, the inherited adaptations are realized mainly by mechanisms leading to the heat loss reduction (e. g. increased insulation by fur or by superficial hypothermia, reduction of body surface area, active choice of environmental temperature and lowering the body temperature). The control of the mentioned adjustments consists in the changes in function of the central and sympathetic nervous systems inducing changes in intensity of the energy metabolism (individual adaptations), changes in the plasticity of vasomotor mechanisms and in heat production of hibernators during entering into and awakening from hibernation (evolutionary adaptations). Morphologically based adjustments (improvement of insulation by fur) appearing in both evolutionary and individual adaptations forms the connecting link between both types of adaptations.

Zusammenfassung

Aus allen erwähnten Daten folgern wir, daß die Adaptationen der Temperaturregulierung bei Kälte entweder durch die erhöhte Wärmeproduktion oder durch die Verringerung des Wärmeverlustes erreicht werden. Während die individuellen Adaptationen hauptsächlich metabolischer Art sind, was durch die erhöhte Kapazität der Wärmeproduktion in Erscheinung tritt, findet man erbliche Adaptationen zumeist in Form von Mechanismen, die eine Verringerung des Wärmeverlustes bewirken (z. B. erhöhte Isolierung durch das Fell oder durch oberflächliche

Hypothermie, Verringerung der Körperoberfläche, aktive Wahl der Umgebungstemperatur und Absinken der Körpertemperatur). Die Steuerung der erwähnten Anpassungen beruht auf Veränderungen in der Funktion des zentralen und des sympathischen Nervensystems, welche Veränderungen in der Intensität des Energiestoffwechsels (individuelle Adaptationen) hervorrufen, weiterhin Veränderungen in der Plastizität der vasomotorischen Mechanismen und in der Wärmeproduktion von Winterschläfern beim Einritt in den Winterschlaf und beim Erwachen (evolutive Adaptationen). Morphologische Adaptationen (Verbesserung der Isolierung durch das Fell), die sowohl als evolutive und auch als individuelle Adaptationen vorkommen, stellen die Verbindung zwischen beiden Typen der Adaptation her.

Literature

HART, J. R. (1956): Seasonal changes in insulation of the fur. Can. J. Zool. **34**: 53—57.
— (1963a): Surface cooling versus metabolic response to cold. Fed. Proc. **22**: 940—943.
— (1963b): Physiological responses to cold in nonhibernating homeotherms. Temperature — Its Measurements and Control in Science and Industry **3**: 373—406.
HART, J. S., and JANSKY, L. (1963): Thermogenesis due to exercise and cold in warm and cold acclimated rats. Can. J. Biochem. Physiol. **41**: 629—634.
HSIEH, A. C. L., and CARLSON, L. D. (1957): Role of adrenaline and noradrenaline in chemical regulation of heat production. Amer. J. Physiol. **190**: 243—246.
IRVING, L. (1956): Physiological insulation of swine as bare-skinned mammals. J. Appl. Physiol. **9**: 414—420.
JANSKY, L. (1965): Adaptability of heat production mechanisms in homeotherms. Acta Univ. Carol.-Biol. 1—91.
JANSKY, L., and HÁJEK, I. (1961): Thermogenesis of the bat *Myotis myotis* Borkh. Physiol. Bohemoslov. **10**: 283—289.
JANSKY, L., and HANÁK, V. (1959): Studien über Kleinsäugerpopulationen in Südböhmen. II. Aktivität der Spitzmäuse unter natürlichen Bedingungen. Säugetierkundliche Mitteilungen **8**: 55—63.
LYMAN, C. P. (1963): Homeostasis in Hibernation. Temperature — Its Measurement and Control in Science and Industry **3**: 453—457.
SCHOLANDER, P. F., HOCK, R., WALTERS, V., JOHNSON, F., and IRVING, L. (1950a): Heat regulation in some arctic and tropical mammals and birds. Biol. Bull. **99**: 237—271.
SCHOLANDER, P. F., WALTERS, V., HOCK, R., and IRVING, L. (1950b): Body insulation of some arctic and tropical mammals and birds. Biol. Bull. **99**: 225—236.

Author's address: L. JANSKY, Ph. D., Department of Comparative Physiology, Charles University, Prague 2, Viničná 7, ČSSR

Ecology (1974) 55: pp. 412–419

BAT ACTIVITY AND POLLINATION OF *BAUHINIA PAULETIA*: PLANT-POLLINATOR COEVOLUTION[1]

E. Raymond Heithaus[2]

Department of Biological Sciences, Stanford University, Stanford, California 94305

Paul A. Opler

Organization for Tropical Studies, Apartado 16, Ciudad Universitaria,
Costa Rica, Central America

Herbert G. Baker

Department of Botany, University of California, Berkeley, California 94720

Abstract. The relationship between the pollination biology of a tropical plant, *Bauhinia pauletia*, and the foraging strategies of the nectarivorous bats visiting it was studied. At least two bat species are pollen vectors, *Phyllostomus discolor* and *Glossophaga soricina*. *Artibeus jamaicensis* and *Sturnira lilium* were also captured near *Bauhinia* flowers. Larger bats (*P. discolor*) drain flowers of nectar and forage in groups, while smaller bats (*G. soricina*) make brief visits and forage independently. These foraging strategies should optimize energetic gain for the bats and promote outcrossing for the plant. *Bauhinia pauletia* is self-compatible, but is found where conditions favor outcrossing. Andromonoecism (the presence of hermaphrodite and male flowers) in this species appears to be an adaptation to pollination by large pollinators that also promotes outcrossing.

Key words: Andromonoecism; behavior; Bauhinia; Chiroptera; coevolution; energetics; pollination.

INTRODUCTION

Bats are important pollinators in tropical communities, with at least 500 species of neotropical plants wholly or partly dependent on bats for pollination (Vogel 1969). Most of the literature on this subject has been devoted to describing instances of bat pollination and summarizing floral characteristics conducive to chiropterophily (Faegri and van der Pijl 1966). Some effort has gone into analysis of pollen feeding of glossophagine bats (Alvarez and Quintero 1969, Howell 1972).

The importance of bats as pollinators goes beyond the large number of species involved. Janzen (1970) has argued that the evolutionary plasticity afforded by outcrossing is extremely important in tropical communities, and Bawa (*in press*) found a high proportion of genetically self-incompatable tree species in one dry, lowland, tropical community. Others claim that wet tropical forest trees ought to be capable of inbreeding because of the large distances between trees (Baker 1955, Federov 1966), but such widely dispersed plants could be outcrossed if bats act as long-distance pollen vectors (Baker 1973).

The effectiveness of bats in outcrossing depends both on plant flowering strategies and the foraging patterns of bats. This paper studies one species of chiropterophilous plant and the foraging patterns of the bats visiting it, in the Tropical Dry Forest life zone (Holdridge 1967) in Costa Rica.

Bauhinia pauletia Pers. (Caesalpiniaceae) is found from southern Mexico to Trinidad. In Costa Rica it is uncommon in the Pacific lowlands, occurring in distinct small patches in savanna or pasture habitats (Standley 1937), while elsewhere it occurs in sparse to moderate stands (Wunderline, *pers. comm.*)

METHODS

Floral biology

This study was conducted 1.5 km west of Cañas, Guanacaste Province, Costa Rica. A 100 by 15 m patch of *Bauhinia pauletia* containing more than 20 individuals was observed casually in December 1970 and intensively from October 11, 1971, to January 1, 1972.

Two types of flowers may be found on individual shrubs, hermaphroditic and functionally staminate with an aborted gynoecium (i.e., plants are andromonoecious). The flowers are borne on a woody inflorescence that retains floral scars after unfertilized flowers fall off.

Flowers were observed in situ to determine the time of opening and anthesis and the pattern of nectar production. The relative frequency of staminate and hermaphroditic flowers was determined for 11 nights, distributed from October 16 to December 12 by counting flowers along a transect. The reproductive success of 81 marked inflorescences was followed by comparing the number of floral scars to the

[1] Manuscript received February 10, 1973; accepted July 27, 1973.

[2] Current address: Department of Biological Sciences, Northwestern University, Evanston, Illinois 60201.

number of fruits produced and multiplying by the percent of the flowers that set pods when hand-pollinated. These inflorescences were observed at 10- to 12-day intervals, October 15–December 12 to determine the rate of flower production and the reproductive success through time.

Breeding system studies of *Bauhinia pauletia* were conducted on November 20 and December 2. Buds were opened by hand at 1700 hr, anthers of hermaphroditic flowers were removed, and paper bags were placed over the flowers. Between 1900 and 2030 hr, the treated flowers were self-pollinated, cross-pollinated, or rebagged without further treatment. Twenty to 25 flowers distributed over several individuals were used for each treatment. Treated inflorescences were checked after 10 to 20 days to determine the extent of fruit set.

Plant voucher specimens were collected and deposited at the Missouri Botanical Garden, St. Louis, Missouri; Field Museum of Natural History, Chicago, Illinois; and Universidad de Costa Rica, San Jose, Costa Rica.

Pollinator activity

The behavior of flower visitors was studied by direct observation and photography. Bats were captured with nylon mist nets set near the shrubs on October 11 and 16, and November 16. Bats were tested for the presence of pollen, identified, banded with numbered aluminum bands, and released. Pollen loads were sampled by rubbing captured bats with a glycerin jelly preparation that was later mounted on glass slides as described by Beattie (1971). Pollen was identified by comparison with known types of pollen collected in Guanacaste.

Visual observations were made on November 4 and 5 between 1830 and 2130 hr to ascertain whether visitation rates varied with time, position of the flower, or type of bat. On November 4, three observers watched different sections of the patch containing several shrubs and 30–60 open flowers. Visitation times to the nearest 10 min and the size of the bat visitor ("large" vs. "small") were recorded. Observations were made by moonlight when possible or with the aid of dim flashlights.

Six observers repeated visual observations on November 5, from 1810 to 2125 hr. Each person recorded visits in a 17 m section of the patch containing roughly 40 to 70 flowers. The following information was recorded for each visit: size of the bat, relative flower position (top of canopy vs. low in canopy), and time of the visit to the nearest 5 min.

Since bat activity might be influenced by lunar cycles, stigmas were examined for pollen deposition on two mornings, once after a full moon and once after a new moon (December 3 and 17). Twenty-six and 11 stigmas respectively were examined, and the number of pollen grains touching the receptive portion of each stigma was counted.

RESULTS

Bauhinia pauletia flowers opened nearly synchronously between 1815 and 1900 hr and were receptive for only one night. Individual flowers usually opened in less than 5 sec and some pollen was immediately available. Each inflorescence produced a pair of flowers every 2nd or 3rd day, with a mean interval between successive antheses of 2.7 days ($N = 489$, $SD = .576$).

Only five of the 10 stamens in a flower produced pollen. The remaining five stamens were atrophied and lacked productive anthers. Anther dehiscence began in the bud at 1800 hr, but pollen was not fully exposed until approximately 30 min after opening. Anthers were loosely attached to the stamens and often were detached during the night by pollinator activity or the wind. Little pollen was available by 0500 hr on the morning following opening.

Nectar flow began just before flower opening and continued at an approximate rate of 0.5 ml/hr for the first several hours. No measurements were made between 2330 and 0500 hr. Nectar accumulated in the calyx, and small amounts moved up the basal portion of the stamen filaments by capillary action.

All flowers began to droop and wilt the morning following anthesis, and those which had not been pollinated dropped after 1 to 3 days. Pollinated flowers remained and incipient pod formation could be noted a few days following pollination.

Flowers were either hermaphroditic or functionally staminate. Each night 25% to 60% of the flowers contained degenerate pistils that dropped from the flower at anthesis. On any particular night, a single plant produced predominantly one type; however, a pair of flowers on one inflorescence sometimes included both types. Individuals were not always consistent from night to night in the production of one or the other flower type. The rate of flower production and the percentage of hermaphroditic flowers on 10 nights covering a 2-mo period is shown in Fig. 1. The proportion of hermaphroditic flowers increased from October 16 to November 4, then decreased through the rest of the flowering period. The small increase seen on December 12 was not significant (proportionality test). The initial, highest, and lowest values, however, are significantly different from each other (proportionality test, $P > .95$). The mean percentage of hermaphroditic flowers for all nights was 42.0% ($N = 869$).

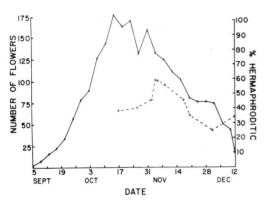

FIG. 1. The rate of flower production of 81 inflorescences and the proportion of hermaphroditic flowers produced in a patch of *Bauhinia pauletia* near Cañas, Costa Rica. Solid line = No. of flowers. Dashed line = % of hermaphroditic flowers.

The rate of flower production was greatest during October, then gradually declined (Fig. 1).

The results of the breeding system experiments are summarized in Table 1. All treatments resulted in some fruit set, with cross-pollinated flowers having the same success as selfed flowers and with control flowers showing the least success. The difference between fruit set in the self-pollinated group and the cross-pollinated group was not significant ($\chi^2 = 0.824$, df = 1). The fruit set of control flowers was significantly lower than that of selfed flowers ($\chi^2 = 5.017$). These data are based on observations of pod development 10 days after treatment. Wind and animal damage to treated flowers precluded following development of pods to maturity. Pod development was initiated by 58.8% of the artificially pollinated flowers.

Mature pods were produced by 12.0% of the 2373

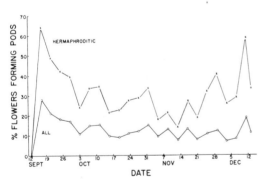

FIG. 2. The proportion of flowers producing pods through time. Solid line = % of all flowers forming pods. Dashed line = % of estimated No. of hermaphroditic flowers forming pods.

TABLE 1. Breeding system experiments

Treatment	N	Pods set	% Success
Cross	21	14	66.7
Self	30	16	53.4
Control	19	4	21.1

flowers that were observed but not treated. These flowers represent the output of 81 inflorescences over a 3½-mo period; however, single inflorescences produced flowers for several weeks only. Since only 43.3% of the flowers were estimated to be hermaphroditic and thus capable of producing fruits, pollination success can only be expressed by fruit set success for potentially reproductive flowers. Since 27.8% of the hermaphroditic flowers produced pods, the proportion of untreated flowers producing pods was roughly half the proportion (58.8%) of artificially pollinated flowers that showed incipient pod development. All observed ripe pods contained mature seeds, although these were often destroyed by insects (Bruchidae).

The temporal pattern of pod formation for all flowers on the monitored inflorescences is shown by Fig. 2. The estimated success of hermaphroditic flowers, based on the estimated proportion of hermaphroditic flowers and assuming a constant fertility rate of 58.8% is also shown. Fruiting success appears to be high initially, to decrease, then to increase somewhat at the end of the flowering period. Estimates of pollination success for the period prior to October 11 are based on extrapolation of known flowering rates. The pitfalls of extrapolation are acknowledged, but we feel it is justified by the consistent rates of flower production by individual inflorescences that we observed.

Many animals were attracted to *Bauhinia pauletia* flowers soon after anthesis, including small numbers of moths (Noctuidae, Pyralidae, and Sphingidae) and many leaf-nosed bats (Phyllostomatidae). Observations and photographs indicated that noctuids and pyralids withdrew nectar from the flowers while clinging to the calyx (Fig. 4A) and seldom contacted the stamens or pistil. Sphingids were potential pollinators but few were present. They contacted the floral reproductive parts, but we saw no more than one or two visits per night of observation. However, bats were commonly seen visiting flowers; some bees were observed collecting pollen early in the morning.

Sixty-seven bats were captured in mist nets near *B. pauletia*, 66 of them in 24 net-hours during October. Three subfamilies of Phyllostomatidae were represented: Phyllostomatinae: *Phyllostomus discolor* Wagner, $N = 40$; Glossophaginae: *Glossophaga soricina* Pallas, $N = 13$; and Stenoderminae: *Artibeus*

TABLE 2. Pollen load type

	Bauhinia	Crescentia	Bauhinia Crescentia	Unknown 23 Bauhinia	Zero
Bat					
Phyllostomus discolor	10	1	7	0	0
Glossophaga soricina	0	3	7	1	1
Artibeus jamaicensis	2	0	0	0	4
Sturnira lilium	1	0	0	0	0
Uroderma bilobatum	0	0	0	0	1

jamaicensis Leach, $N = 8$; *Sturnira lilium* E. Geoffroy, $N = 5$; and *Uroderma bilobatum* Peters, $N = 1$. We banded 41 bats on October 11, and 26 on October 16. Only one of these, a male *P. discolor*, was recaptured near *B. pauletia* (five nights after banding).

The distribution of pollen among the bat species is summarized in Table 2. On 32 of 48 bats tested (84.2%) large amounts of pollen ("pollen loads") adhered to the wings, head, and body. *Phyllostomus discolor* and *G. soricina* carried pollen most frequently (91% of 32 pollen loads were carried by these species). Pollen grains from both *B. pauletia* and *Crescentia* spp. (Bignoniaceae) were abundant, while one unidentified type was also present. Both *Crescentia alata* HBK and *Crescentia cujete* L. were in flower nearby, but their pollen was indistinguishable. Mixed *Bauhinia* and *Crescentia* pollen loads were common, especially on *G. soricina. Phyllostomous discolor* carried *Crescentia* pollen only in mixed species pollen loads, and even then *Crescentia* was uncommon relative to *Bauhinia* pollen. In contrast, *G. soricina* often carried *Crescentia* and *Bauhinia* pollen in equal amounts, so this species appeared to visit *Crescentia* flowers more often than *P. discolor*.

The capture and pollen load data indicate that *P. discolor* and *G. soricina* were the principal bat visitors to *B. pauletia*. This is supported by photographs of bats approaching or feeding at flowers. It was possible to identify bats on the basis of size, shape of the head, form of the flight membranes, and color in 30 photographs. *Phyllostomus discolor* (Fig. 5B and 5C) was identified in 13 pictures and *G. soricina* (Fig. 5D) in 17. No other species could be identified.

On November 4 and 5 during 27 man-hours of observation 1147 visits to flowers by bats were recorded. It was possible to distinguish bats of two sizes, large and small. The four common bat species netted near the flowers were also clearly large or small. Their forearm lengths are an index of differences in total sizes: *P. discolor* = 59.2–63.2 mm; *A. jamaicensis* = 54.9–60.6 mm; *G. soricina* = 33.1–39.1 mm; and *S. lilium* = 40.5–43.6 mm (Goodwin and Greenhall 1961).

Most visits were made by large bats on both nights (Table 3). The ratio of 4:1 (large:small) in observed visits is close to the ratio obtained from netting results for *P. discolor* and *G. soricina* (3.1:1.0).

We noted a number of other differences between large and small bats in addition to frequency of visits. These included variations in the floral visitation behavior, movement patterns within the plant vicinity, and height of flowers visited. Large bats (924 observations, 9 photographs) invariably grasped a section of branch beneath an open flower with their feet, then bent their heads into the calyx. The wings were usually spread and not used for support. The bat's weight bent the thin branches so that nectar was drained from the inverted flowers. Visits lasted 1–3 seconds (Fig. 4C and 4D).

Small bats, on the other hand, generally hovered in front of a flower and appeared to lap the nectar (Fig. 4D). Such visits lasted less than 1 sec. Small bats sometimes clung to flowers with their thumbs and feet, but flowers were not inverted by their weight. The duration of this clinging visit was also very short, less than 1 sec. The hovering visit was much more common, in a ratio of 12:1 based on 13 photographs.

Flowers visited once by small bats were very likely to be visited several more times. For example, one flower was visited 14 times in 30 min. Since the bat that was visiting this flower approached from the same direction and angle each time, the impression created was that the same bat was making repeated visits.

Observations revealed that large bats are most prone to visit flowers classed as "high" in the canopy. Of 924 visits, 87.6% were to high flowers and the

TABLE 3. Ratio of large and small bat visits

	Nov. 4		Nov. 5	
	%	N	%	N
Large	80.1	219	80.5	705
Small	19.9	52	19.5	171

TABLE 4. Patterns of visitation for large bats

	Mean	Range	SD	N
No. of visits per peak*	14.63	2–61	13.57	49
Duration of peaks (min)	10.9	5–25	5.83	49
Interval between peaks (min)	21.6	10–55	10.78	43
No. of peaks per section	8.2	7–9	.7528	6

* Peak = any concentration of visits of which at least three occur within one 5-min interval.

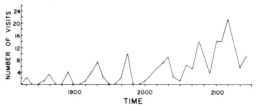

FIG. 3. The distribution of visits by large bats to 40–70 B. pauletia flowers, November 5.

remainder to "low" flowers. Small bats visited 59.7% high flowers and 40.3% low flowers in 223 visits. Both types of bat visited higher flowers more often (χ^2, $P > .99$), but large bats visited the high flowers almost to the exclusion of low flowers. It appeared that the large bats visited inflorescences that would not drop into foliage during their visits.

Social behavior during foraging also differed for large and small bats. Large bats appeared to arrive in groups to feed. Groups of two to six bats were often seen flying in a linear formation over the flowers. Vocalizations were commonly heard as these groups flew by. While it was impossible to count all bats feeding in an area at any given time, we estimated that groups consisted of from two to 12 individuals. In contrast, small bats appeared to forage alone.

The pattern of visitation to flowers in restricted sectors of the Bauhinia patch supports the hypothesis that large bats feed in groups. The pulsed nature of visits was striking (Fig. 3). After a period of active flower visitation in one section, very few or no visits were likely for up to 55 min (average interval between peak visitation periods = 21.6 min, N = 43). The average period of active visitation was 10.9 min (N = 49), consisting of an average of 14.4 visits. An average of 8.2 periods (range = 7–9) of active visitation were observed in each of the six sections observed for 3 hr on November 5. These patterns are summarized in Table 4. The distribution of

flower visits by large bats through the evening is exemplified by Fig. 3 (from Observer 2, November 5). Note that activity reached a peak late in the evening, but that the visits were pulsed throughout the observation period.

Observations on November 4 gave similar results, but the data were not lumped with those of November 5 because a different time interval was used. The pulsed visitation pattern was generally the same, and there were five to seven visitation periods observed.

No such patterns were discerned for small bats. Repeated visits in particular sectors were usually to the same flower and seemed to be by the same bat. No groups of small bats were ever seen flying over the flowers.

Finally, analyses of pollen loads suggested another difference in visitation patterns. Glossophaga soricina (a small bat) carried Bauhinia pauletia pollen only in mixed pollen loads with Crescentia pollen. Small bats carried either pure or mixed Crescentia pollen. Phyllostomus discolor (a large bat) had pure B. pauletia pollen loads on 55% (10) of the bats tested, and on only 41% (7) were very small amounts (< 5% of all pollen) of Crescentia pollen found.

The differences in the flower-visiting behavior between large and small bats are summarized in Table 5.

The bat activity level around B. pauletia varied considerably on different nights. This is reflected by the low netting success of November 16 com-

TABLE 5. Patterns of visitation to Bauhinia pauletia by large and small bats

Large	Small
(Probably Phyllostomus discolor)	(Probably Glossophaga soricina)
Visits high flowers	Visits high and low flowers
Grasps branch beneath flower, pulling it down	Hovers
Drains nectar well	Laps small amounts of nectar
Visits in groups in pulses	Visits singly, flies independently
Visits same flower infrequently	Returns to same flower at short intervals
Carries 55% pure Bauhinia pollen loads; 45% mixed pollen loads (all mixed loads > 90% Bauhinia)	Carries no pure Bauhinia pollen loads; 100% Crescentia or mixed loads

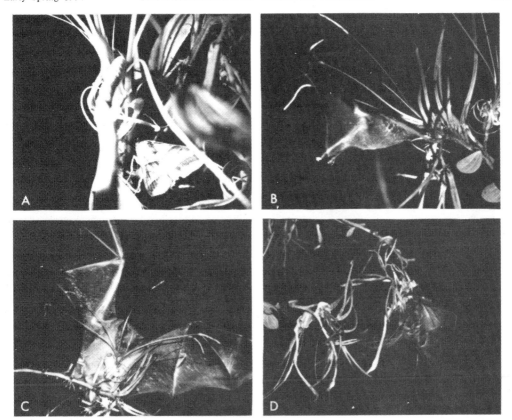

Fig. 4. *Bauhinia pauletia* flower visitors. (A) Noctuid moth approaching base of flower. (B) *Glossophaga soricina* hovering while feeding. (C) *Phyllostomus discolor* reaching for branch below a flower. (D) *Phyllostomus discolor* hanging while feeding.

pared to October 11 and 16. Differences in moonlight may have influenced activity (Crespo et al. 1972). For example, bat activity was nearly absent during 20 min of observation on December 2 (full moon), but was estimated to be light to moderate on December 16 (new moon). Apparent differences in bat activity during these nights were reflected in the amounts of pollen deposited on stigmas. The mean number of grains adhering to the stigmatic surfaces of pistils collected on December 3 was 23.2 ($N = 26$, SD $= 42.22$), while on December 17 the avearge number of grains per stigma was 294 ($N = 11$, SD $= 152.4$). This difference was highly significant (χ^2, $P > .99$).

DISCUSSION AND CONCLUSIONS

This study reveals that coevolution between plants and pollinators can be more complex than the exchange of food for pollinator services. Bats not only effect pollination, they also promote outcrossing in *B. pauletia* as discussed later. Furthermore, two dis-

tinct floral resource utilization strategies are seen in bats, and these strategies may be related to their evolution of social behavior.

Breeding systems which promote outcrossing allow for greater evolutionary plasticity than inbreeding systems (Stebbins 1970). It has been suggested that genetic variability is important to plants subjected to changing selective pressures such as herbivores or seed predators that are also evolving (Janzen 1970). Plant breeding systems in the tropical dry forest region are typified by the prevalence of self-incompatibility systems among trees (Bawa *in press*), a large number of dioccious species (Bawa and Opler *pers. comm.*), and the frequent occurrence of other strategies that promote outcrossing (e.g., heterostyly, protandry). This supports the hypothesis that maintenance of plasticity is generally important there. *Bauhinia pauletia* is not a likely exception to this generality. As extensive insect damage to seeds was observed, *B. pauletia* appears to be subjected to the type of changing selection discussed by Janzen

(1970). Also, the sympatric bat-pollinated *B. ungulata* ensures outcrossing through genetic self-incompatibility (Bawa *pers. comm.*).

Although *B. pauletia* is genetically self-compatible, self-pollination (autogamy) of flowers is improbable. The apparent correlation between pollen deposition on stigmata and bat activity (on two nights in December) suggests that pollen movement depends greatly on animal transport. The strategy of andromonoecism seems also to be associated with dependence on animal transport of pollen. Self-pollination of flowers during a bat visit is improbable, since bats tend to push the stamens and pistil apart (Fig. 4B–4D) so anthers are unlikely to contact the stigma. Although small bats tended to visit particular flowers repeatedly, mixed pollen loads on *G. soricina* suggest that these small bats fly from flower to flower in a "trapline." Probably these bats visit a series of flowers repeatedly, and their visits decrease the probability of autogamy. The movement of groups of large bats from one area to another clearly promotes outcrossing. Pollen transfer among flowers of the same plant (geitonogamy) is not precluded by bat behavior, but it is clear that bats cause much if not most crossing.

The simultaneous production of male and hermaphrodite flowers on the same individual (andromonoecism) is rare, and its significance has not been investigated adequately. Carr et al. (1971) report andromonoecism in *Eucalyptus* spp. of the series *Corymbosae* but make no attempt to relate the phenomenon to pollination systems. It clearly reduces inbreeding, since functionally male flowers cannot be self-pollinated. We feel andromonoecism has additional adaptive value because it increases the ratio between pollen and ovules (or pollen and stigmatic surface). Large quantities of pollen may be necessary to ensure deposition of sufficient pollen to fertilize up to 30 ovules per stigma, especially since pollen is scattered over the entire ventral surface of bats. The stigmatic surface area is .008 cm^2, while the ventral surface area of just the wings of an average glossophagine bat is 144 cm^2 (Findley et al. 1972). This means the ratio of wing area (of an average small bat) to stigma surface area is 18,000:1. The need for large amounts of pollen in this system are obvious. Andromonoecism, therefore, may be associated with specialization for pollination by pollen vectors that are very large relative to stigma size.

The problem of utilization of large pollen vectors has been solved by other bat-pollinated plants in three basic ways. For example, *Pseudobombax barrigon* (Bignoniaceae) also maintains a high ratio of pollen to stigma area, but by producing hundreds of stamens per flower. Pollen may also be placed on a bat selectively as in *Bauhinia ungulata*, which deposits pollen primarily on a bat's abdomen. Finally, larger stigmata, such as found in *Ochroma pyramidale* (Con. ex Lam.) Urban (Bombaceae), can greatly reduce the relative difference between pollinator and stigma-surface areas.

This explanation of andromonoecism can be extended to other pollination systems. *Aesculus californica* (Spach) Nutt. (Hippocastanaceae) is pollinated by butterflies, but it has a minute stigmatic surface (Benseler 1968). The stigmas of eucalypts are also very small, especially relative to the large number of ovules per flower.

Bauhinia pauletia nectar is used by at least two species of bats that have evolved different resource utilization strategies. Circumstantial evidence suggests that the great majority of the large bats were *Phyllostomus discolor*, while small bats were *Glossophaga soricina*. *Artibeus jamaicensis* and *Sturnira lilium*, characterized by reduced uropatagia easily distinguished in most photographs, were not seen. In addition, fewer than one-third of the captured *A. jamaicensis* carried pollen. Since we found that *A. jamaicensis* and *S. lilium* visited other flowers in the study region, we cannot assume that they never visited *B. pauletia*; however, they are certainly less important to *B. pauletia* pollination than *P. discolor* and *G. soricina*.

Resource utilization patterns appear to be related to aspects of the social behavior of these bats. Visitation by *P. discolor* is clearly coordinated, as indicated by the pulsed pattern illustrated in Fig. 4. This pattern and the formation of bats flying over the plants suggests a social organization that includes feeding groups. If so, more than one group was involved on November 4 and 5, as observers at opposite ends of the *B. pauletia* patch simultaneously watched peaks of visitation on 14 occasions. Therefore, it is impossible to speculate on the cohesiveness of these assemblages. Group foraging by *Phyllostomus hastatus* (Goodwin and Greenhall 1961) and *Myotis adversus* (Dwyer 1970) has been observed. As *Artibeus jamaicensis* and *Carollia perspicillata* may also forage in groups, this strategy may be more widespread than previously thought. In contrast, *G. soricina* forages singly around *Bauhinia pauletia*.

Energetic considerations are of extreme importance in plant-pollinator relationships (Heinrich and Raven 1972, Wolf et al. 1972) and may account for these bat foraging differences. Since the large bat *P. discolor* always drains nectar from *B. pauletia* flowers the reward in nectar in successive visits is correlated with the time between visits. Short times between repeated visits to the same flower will net the visitor little nectar. By visiting flowers in groups *P. discolor* may decrease the probability of quick successive visits to the same flower. Furthermore, bats might learn to allow the optimal time between visits to the same group of flowers for sufficient replenishment of nectar and to return before competing groups discover the new resource.

In contrast, individuals of the smaller bats spend a very short time at each flower, probably lapping small amounts of nectar at each visit. With the slow depletion of nectar, repeated visits to the same flower are not energetically disadvantageous. In fact, there may be some advantage to learning the position of particular flowers and then repeatedly utilizing these sources, because search time would decrease relative to feeding time. Independent foraging appears to be favored when nectar is not drained with each visit.

These bat-foraging strategies may parallel finch-foraging strategies in deserts. Cody (1971) found that the formation of foraging flocks of mixed finch species was related to food abundance, the birds forming flocks as food became less abundant. He hypothesized that renewable resources temporarily depleted are most efficiently utilized by flocks that can learn to vary the time between returns to the same feeding points. *Phyllostomus discolor* foraging may fit this pattern. Independent foraging of finches was observed when resources were abundant. This may be analogous to the independent foraging of *G. soricina*, which does not immediately deplete the floral nectar supply and thereby has an abundant resource.

In summary, *Bauhinia pauletia* is a genetically self-compatible plant living in conditions where maintenance of genetic variability is an asset. Bats pollinate its flowers and promote outcrossing. Andromonoecism increases the probability of pollen transfer from these large pollinators to relatively small stigmata and also decreases the probability of autogamy. Two kinds of bat have evolved different nectar-resource-utilization strategies, but both behaviors tend to result in cross-pollination of *B. pauletia* flowers.

ACKNOWLEDGMENTS

This work was partially financed by OTS Pilot Studies Grant N-70-58, NSF Grants GB-7805 and GB-25592 (Herbert G. Baker and Gordon W. Frankie, Principal Investigators) and a series of NSF grants to Peter H. Raven. We especially thank Patricia A. Heithaus for assistance during this study. Sandra Opler assisted in field observations. John T. Doyen, Theodore H. Fleming, Richard Holm, Peter H. Raven, and an anonymous reviewer read early drafts and offered many helpful suggestions.

LITERATURE CITED

Alvarez, T., and L. González Quintero. 1969. Análisis polinico del contenido gástrico de murciélagos Glosso-

phaginae de Mexico. An. Esc. Nac. Cienc. Biol. Mex. **18**: 137–165.

Baker, H. G. 1955. Self-compatibility and establishment after long distance dispersal. Evolution **9**: 347–348.

———. 1973. Evolutionary relationships between flowering plants and animals in American and African tropical forests, 350 p. *In* B. J. Meggers, E. Ayensu, and W. D. Duckworth [eds.] Tropical forest ecosystems in African and South American: a comparative review. Smithson. Inst. Press, Wash., D.C.

Bawa, K. 1974. Breeding systems of tree species of a lowland tropical community: evolutionary and ecological considerations. Evolution. (in press).

Beattie, A. J. 1971. A technique for the study of insect-borne pollen. Pan-Pac. Entomol. **47**: 82.

Benseler, R. W. 1968. Studies in the reproductive biology of *Aesculus californicus* (Spach.) Nutt. Ph.D. Thesis (Botany), Univ. California, Berkeley.

Carr, S. G. M., D. J. Carr, and F. L. Ross. 1971. Male flowers in Eucalypts. Aust. J. Bot. **19**: 73–83.

Cody, M. L. 1971. Finch flocks in the Mojave Desert. Theor. Pop. Biol. **2**: 142–158.

Crespo, R. F., S. B. Linhart, R. J. Burns, and G. C. Mitchell. 1972. Foraging behavior of the common vampire bat related to moonlight. J. Mammal. **53**: 366–368.

Dwyer, P. D. 1970. Foraging behavior of the Australian large-footed myotis (Chiroptera). Mammalia **34**: 76–80.

Faegri, K., and L. van der Pijl. 1966. The principles of pollination ecology. Pergamon Press, New York. 291 p.

Federov, A. A. 1966. The structure of the tropical rainforest and speciation in the tropics. J. Ecol. **54**: 1–11.

Findley, J. S., E. H. Studier, and D. E. Wilson. 1972. Morphological properties of bat wings. J. Mammal. **53**: 429–444.

Goodwin, G. G., and A. M. Greenhall. 1961. Review of the bats of Trinidad and Tabago. Bull. Am. Mus. Nat. Hist. **122**: 191–301.

Heinrich, B., and P. H. Raven. 1972. Energetics and pollination ecology. Sciences **176**: 597–602.

Holdridge, L. R. 1967. Life zone ecology. Trop. Sci. Cent., San Jose, Costa Rica. 206 p.

Howell, D. J. 1972. Physiological adaptations in the syndrome of chiropterophily with emphasis on the bat *Leptonycteris* Lydekker. Diss. Abstr. Int. B. Sci.

Janzen, D. H. 1970. Herbivores and the number of tree species in tropical forests. Am. Nat. **104**: 501–527.

Standley, P. C. 1937. Flora of Costa Rica. Field Mus. Nat. Hist. Publ. Bot. Ser. **18**(1–4). 1616 p.

Stebbins, G. L. 1970. Adaptive radiation of reproductive characteristics in Angiosperms. I. Pollination mechanisms. Ann. Rev. Ecol. Syst. **1**: 307–326.

Vogel, Von S. 1969. Chiropterophilie in der neotropischen Flora. Neue Mitteilungen III. Flora, Abt. B **158**: 289–323.

Wolf, L. L., F. R. Hainsworth, and F. G. Stiles. 1972. Energetics of foraging: rate and efficiency of nectar extraction by hummingbirds. Science **176**: 1351–1352.

A C T A   T H E R I O L O G I C A

VOL. XIV, 1: 1—9.   BIAŁOWIEŻA   5.IV.1969

Michael H. S M I T H, Ronald W. B L E S S I N G, James L. C A R M O N
and John B. G E N T R Y

Coat Color and Survival of Displaced Wild and Laboratory Reared Old-field Mice

[With 1 Table and 4 Figures]

Wild and laboratory reared mice from central Florida and South Carolina were released into enclosures on the significantly darker South Carolina soils. The lighter southern mice disappeared at the same rate as the darker northern form, as did the males and females from both localities. Prior experience with field conditions was associated with a large selective advantage; the laboratory reared mice disappeared much faster than the wild mice. The correlation between soil and pelage color implies a selective advantage for mice to match the soil, but this advantage must be relatively small because the light form did not disappear at a higher rate than did the darker mice. The relationship between reflectivity and wavelength was linear for both pelage and soil samples. This probably means that the evolution of the dorsal pelage color in this species takes place by modifications in the slope or the intercept of the pelage line.

I. INTRODUCTION

The old-field mouse, *Peromyscus polionotus* W a g n e r, shows considerable morphological variation throughout its range (S m i t h, 1966). The adaptive significance of many of these variations has been implied by correlational techniques, but no attempt has been made to experimentally determine under field conditions the selection coefficients for any of these traits. Despite a lack of data, it can still be argued that the observed polymorphisms are not random but rather represent evolutionary responses to differences in the local environments. Following this reasoning, we would expect mice currently living in one locality to be better adapted for survival in this particular area than mice that are taken another area and released at the first site. Of course, it is necessary to assume that the areas are separated to such an extent

1

that genetic exchange between the populations is negligible, the environments at the two sites are significantly different, and finally that the populations have persisted in these areas for a sufficient length of time for selection to produce differences between them. Under these conditions, two populations would normally be expected to differ in more than one way. For our purposes in this paper, we will concentrate only on differences in coat color of the old-field mouse in relation to the color of their soil background.

II. MATERIALS AND METHODS

Old-field mice were captured in the field or reared in the laboratory and then released into two outdoor enclosures. The enclosures, each containing approximately two acres, were adjacent to one another in field 3—412 on the Savannah River Plant in Aiken County, South Carolina, USA. The vegetation in area 3—412 is typical on an old-field (O d u m, 1960; G o l l e y & G e n t r y, 1965), and *P. polionotus* occurs abundantly in this area (C a l d w e l l, 1964; D a v e n p o r t, 1964). C a l d w e l l & G e n t r y (1965) described an enclosure similar to the ones used in this study.

The wild mice were collected from other fields on the SRP and also from Citra in Marion County, Florida. Laboratory reared mice were taken from two different colonies; one was located at the University of Georgia and the other at the Savannah River Ecology Laboratory (see C a r m o n, G o l l e y & W i l l i a m s, 1963, and S m i t h, 1966 for details concerning the maintenance of the colonies). The University of Georgia colony, which was in its seventh generation in the laboratory, was derived from a wild stock collected in the 3—412 area. The other colony came from stock collected in the Ocala National Forest just east of Citra, Florida and was in its fifth to seventh generation in the laboratory. The laboratory animals were from three to six months of age.

Each experiment consisted of releasing 10 adult males from each area into one of the enclosures and 10 adult females from each area into the other enclosure. Pregnant females were excluded from all experiments. The wild mice were held in the laboratory for about 10 days before their release. Experiments consisted entirely of wild mice or of laboratory reared mice; the two types were never released in the enclosures at the same time. Two experiments were conducted with laboratory reared mice and two with the wild mice; a total of 160 mice, 80 wild and 80 reared in the laboratory, were released. The order of the experiments was randomly selected; it was laboratory, wild, wild, laboratory. Sex, locality, and wild versus laboratory origin were the three factors being tested. The various types of mice were placed into the enclosures in a random order with the restriction that each factor must be tested twice in each enclosure.

Each mouse was toe clipped and then released. Populations were censused by live trapping for one day each week for a month. At the end of this time the survivors were removed and another experiment started. The first experiment was started on January 12, 1967 and the last one on May 2, 1967. There were 134 traps per enclosure, and they were distributed in a grid with 30 feet between the traps.

Reflectance was recorded for the mid-dorsal pelage and for the surface and subsurface soil where the mice were captured. Reflectance was measured between 400 to 700 mµ (violet to red) using the Bausch and Lomb Spectronic 505 Recording Spectrophotometer with its visual reflectance attachment in a way similar to the method used by Sealander, Johnston & Hamilton (1964). Flat skins were used in determining the reflectance of the pelage. The mice used in this part of the work were collected at the same time and areas at those released in the enclosures. Soil samples were taken from areas alongside the burrows in which the mice were captured. The reflectance at eight equally spaced points from 400 to 700 mµ was used for the statistical analyses. Reflectivity is given as a percentage of the amount of light reflected from a pressed white magnesium carbonate standard.

III. RESULTS

There was a linear relationship between wavelength (X) and amount of light (Y) reflected from each soil sample or from the pelage of each mouse. Each sample was analyzed separately, so a series of values for

Table 1.

Results of linear correlation and regression analyses of the relationship between reflectivity and wavelength between 400 to 700 mµ for individual soil and pelage samples from Florida and South Carolina.

Statistical Parameters	FLORIDA SOIL			SOUTH CAROLINA SOIL		
	Pelage	Surface	Subsurface	Pelage	Surface	Subsurface
Sample Size *	12	10	10	19	15	15
Range of r **	.907—.995	.992—.998	.990—.998	.968—.990	.956—.994	.951—.991
Range of a ***	4.49—11.94	16.27—20.99	12.66—18.68	2.68—8.53	9.98—14.99	8.03—14.88
$\bar{a} \pm$ S.E.	6.75±.75	17.17±.44	14.12±1.64	4.36±.37	11.41±.32	11.07±.46
Range of $b \cdot 10^{-4}$	1.30—2.68	3.71—7.55	8.43—10.72	1.89—5.32	3.08—6.87	3.35—8.72
$\bar{b} \pm$ S.E. $\cdot 10^{-4}$	1.67±.01	4.61±.03	9.33±.02	2.88±.02	4.63±.02	5.48±.04

* The number of observations for each regression equals 8,
** All r values were significant at the 0.01 level,
*** a is given in per cent reflectivity at 400 mµ.

the correlation coefficient (r), the intercept (a) and the slope (b) were generated (Table 1). A linear model accounted for the majority of the variability in all cases (82.3 to 99.6%).

The pooled data for reflectivity of the pelages and surface soils are given in Figs. 1 and 2, respectively. Surface soils from Florida and South Carolina are obviously different in reflectivity at the various

M. H. Smith *et al.*

wavelengths but the same is not true for the pelages. However, it would
be an error to conclude from Fig. 1 that the relationship between reflectivity of the pelage and wavelength is necessarily the same in the
two areas, since the pooled data consists of a series of independent
straight lines. In South Carolina b is larger for a given a than in Florida. In other words, the ratio of reflected red to violet light is larger
for a given amount of reflected violet light in South Carolina than in

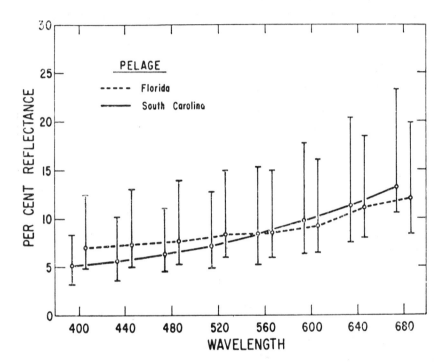

Fig. 1. The range and mean reflectance values for the mid-dorsal pelage of
Peromyscus polionotus from Florida and South Carolina as a function of wavelength from the violet (400 mμ) to the red (700 mμ) end of the spectrum. Reflectance is given as a percentage of the amount of light coming off a pressed white
magnesium carbonate standard.

Florida. The distribution of the b and a values for the pelages, surface
soils, and subsurface soils from South Carolina and Florida is given in
Fig. 3. The relationships between b and a differs in each case for the
two areas.

The disappearance rates of the four groups are given in Fig. 4. There
were no significant differences between the disappearance rates of Flo-
rida and South Carolina mice ($\chi^2 = 0.18$), or of males and females ($\chi^2 = 0.18$), or in the two enclosures ($\chi^2 = 2.2$), but wild mice survived better
than laboratory reared mice ($\chi^2 = 12.84$).

IV. DISCUSSION

The linear mathematical model accounted for more than 82 per cent of the variability between reflectivity and wavelength for every soil and pelage sample. The similarity of the relationship for the two types of samples was most likely the result of selection for the mice to match their background. D i c e (1947) showed that mice that contrast with their background were captured by owls more often than those that blended in with their background. However, the exact way in which

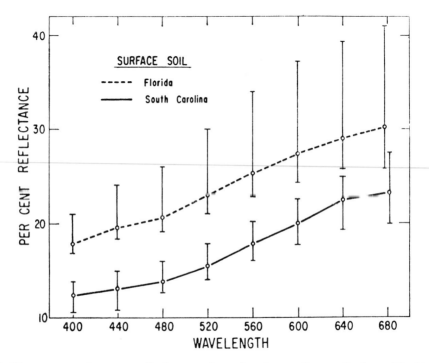

Fig. 2. The range and mean reflectance values for dry surface soils from Florida and South Carolina as a function of wavelength from the violet (400 mμ) to the red (700 mμ) end of the spectrum. Reflectance is given as a percentage of the amount of light coming off a pressed white magnesium carbonate standard.

populations respond to selection of this kind was not known. It now appears that there are only two parameters that are subject to modification by selection. These are the slope (b) and the intercept (a) of the reflectance line. This statement probably has general applicability for many, if not all, species of small mammals; we recently found a significant linear correlation between reflectivity and wavelength for six species picked at random from the University of Georgia's museum (manuscript in preparation).

The gross similarity of the mathematical relationships goes beyond mere linearity. Slopes and intercepts were positive for all samples. The largest intercepts for the pelage lines were associated with the largest

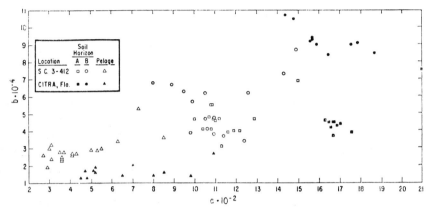

Fig. 3. The slope (*b*) plotted against the intercept (*a*) of the linear relationship between reflectance and wavelength for pelages, surface soils (A), and subsurface soils (B) from South Carolina and Florida.

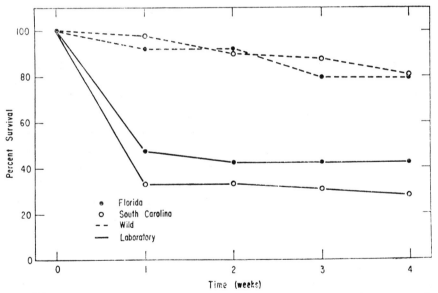

Fig. 4. Disappearance rates of wild and laboratory reared old-field mice from Florida and South Carolina released in South Carolina.

intercepts for the soil lines. Despite these similarities there were differences in the reflectivity of the pelages and soils at the two localities. The differences include the slopes of the pelage and subsurface soil lines and the intercepts of the surface soil lines. Both the mice and the backgrounds upon which they occur are different in the two areas, and thus there is reason to expect the light Florida mice to disappear faster then the dark South Carolina mice when released on the darker northern soils. The Florida and South Carolina mice did not disappear at different rates. Since the reason(s) for this contradiction is not readily apparent from the data, it is necessary to examine some of the assumptions underlying our reasoning concerning differential survival.

First, we must assume that there was sufficient selection pressure to cause an observable difference in the survival rates of the various groups during the study period of one month. The relatively high disappearance rate of the laboratory reared mice was probably the result of a high selection pressure. Most of the mice disappear during the first week after introduction, and the survival curves are almost horizontal after this time even for intervals longer than one month (S c h n e l l, 1964; G o l l e y & G e n t r y, in press). Second, predators must use the visual cues associated with the dorsal pelage to locate the mice. There is no way to test this assumption with the available data. However, considering the relatively high survival rate of the wild mice, it seems unlikely that many of the mice escaped from the enclosures. Thus, their disappearance is probably synonymous with their death by predation or disease. The latter is not probably since none of the mice appeared to be sick when captured in the traps.

Potential predators include several species of large mammals, raptorial birds and snakes. Large snakes do go in and out of the enclosures over the sheet metal fence, and many of these species are known to eat *Peromyscus* on the Savannah River Plant (D u e v e r, 1967). Since the relative importance of each of the potential predators and of the various senses used in their hunting behavior is not known, it could be concluded that the predators of the mice in the enclosures were not using the visual cues associated with the dorsal pelage to locate their prey. This conclusion is inconsistent with the overall correlation between soil and pelage color found in this species by ourselves and H a y n e (1950). It seems more likely that there is only a slight selection pressure for modifying the slope and intercept of the reflectivity-wavelength relationship of the dorsal pelage in relation to similar values for the soil. Under these conditions it would take a relatively long period of time to produce the differences in the dorsal pelage that are used to distinguish between the various subspecies (S c h w a r t z, 1954).

As the selection coefficient (s) approaches 0, the sample size needed to adequately estimate s approaches infinity. Our sample size was probably too small to detect a slight difference in the survival rates of the mice from the two localities. Prior experience with field conditions was associated with a large selective advantage. Thus learning is apparently very important to the survival of mice under natural conditions. S m i t h (1967) presented evidence for a different disappearance among the sexes based on sex ratios among young and adult mice, but the difference would have been slight and accordingly undetectable considering our sample size. These data are consistent with the conclusions made above.

Acknowledgements: The senior author was supported during the preparation of the manuscript by an Atomic Energy Commission Contract AT(40—1)—2975. Laboratory reared mice from South Carolina were also provided by funds from this grant. Two of the junior authors were supported by an AEC Contract AT(38—1)—310 while conducting the field work in South Carolina. Both contracts are with the University of Georgia. Travel funds to collect mice in Florida were provided by an NSF Grant GB5140. We would like to thank Drs. R. J. B e y e r s and J. W. G i b b o n s for critically reading the manuscript.

REFERENCES

1. C a l d w e l l, L. D., 1964: An investigation of competition in natural populations of mice. J. Mammal., *45*: 12—30.

2. C a l d w e l l, L. D. & G e n t r y, J. B., 1965: Interactions of *Peromyscus* and *Mus* in a one-acre field enclosure. Ecology, *46*: 189—192.

3. C a r m o n, J. L., G o l l e y, F. B. & W i l l i a m s, R. G., 1963: An analysis of the growth and variability in *Peromyscus polionotus*. Growth, *27*: 247—254.

4. D a v e n p o r t, L. B., Jr., 1964: Structure of two *Peromyscus polionotus* populations in old-field ecosystems at the AEC Savannah River Plant. J. Mammal., *45*: 95—113.

5. D u e v e r, A. J., 1967: Trophic dynamics of reptiles, in terms of the community food web and energy intake. M. S. thesis, Univ. Georgia, 89 p.

6. D i c e, L. R., 1947: Effectiveness of selection by owls of deer mice which contrast in color with their background. Contrib. Lab. Vert. Biol. Univ. Mich., *34*: 1—20.

7. G o l l e y, F. B. & G e n t r y, J. B., 1965: A comparison of variety and standing crop of vegetation on a one year and a twelve year abandoned field. Oikos, 15, 2: 185—199.

8. G o l l e y, F. B.: Response of rodents to acute gamma radiation under field conditions. Proc. Sec. Nat. Symp. Radioecology, Ann Arbor, Mich. [in press].

9. H a y n e, D., 1950: Reliability of laboratory-bred stocks as samples of wild populations, as shown in a study of variation of *Peromyscus polionotus* in parts of Florida and Alabama. Contrib. Lab. Vert. Biol. Univ. Mich., *46*: 1—56.

10. O d u m, E. P., 1960: Organic production and turnover in old field succession. Ecology, *41*: 34—49.

11. S c h n e l l, J. H., 1964: An experimental study of carrying capacity based on the disappearance rates of cotton rats (*Sigmodon hispidus komareki*) introduced into enclosed areas of natural habitat. Ph. D. diss., Univ. Georgia, 46 p.

12. S c h w a r t z, A., 1954: Old-field mice, *Peromyscus polionotus,* of South Carolina. J. Mammal., *35*: 561—569.

13. S e a l a n d e r, R. K., J o h n s t o n, R. F., & H a m i l t o n, T. H., 1964: Colorimetric methods in ornithology. Condor, *66*: 491—495.

14. S m i t h, M. H., 1966: The evolutionary significance of certain behavioral, physiological, and morphological adaptations of the old-field mouse, *Peromyscus polionotus*. Ph. D. diss., Univ. Florida, 186 p.

15. S m i t h, M. H., 1967: Sex ratios in laboratory and field populations of the old-field mouse, *Peromyscus polionotus*. Researches Population Ecology, 9: 108—112.

Received, June 8, 1968.

Mailing address:

Savannah River Ecology Laboratory SROO, Box A Aiken, South Carolina, USA 29801

Department of Zoology and Institute of Ecology University of Georgia

Computer Center University of Georgia Athens, Georgia, USA 30601

Michael H. SMITH, Ronald W. BLESSING, James L. CARMON i John B. GENTRY

UBARWIENIE FUTERKA A PRZEŻYWANIE DZIKICH I LABORATORYJNYCH
PEROMYSCUS POLIONOTUS

Streszczenie

Dzikie i hodowane w laboratorium *Peromyscus polionotus* (W a g n e r, 1843) były wpuszczone do zagród w Południowej Karolinie. Chociaż gleby w Karolinie są ciemniejsze niż na Florydzie (Fig. 2), to jednak kolor futerka badanych populacji myszy nie różnił się (Fig. 1). Zależność pomiędzy zdolnością odbijania światła a długością fali była liniowa zarówno dla skórek jak i dla próbek gleby (Tabela 1). Myszy z Florydy wydawały się na oko jaśniejsze i miały mniejszy kąt nachylenia prostej regresji przy danej stałej wielomianu w porównaniu do myszy z Południowej Karoliny (Fig. 3). Naprzekór tym różnicom, jaśniejsze osobniki ginęły w takim samym tempie jak ciemniejsze z południa. Dotyczy to zarówno samców jak i samic z obu miejscowości (Fig. 4). Wcześniejsze doświadczenia w warunkach terenowych były związane z dużym zróżnicowaniem w selekcji; hodowane w laboratorium myszy ginęły znacznie szybciej w porównaniu z dzikimi (Fig. 4).

Korelacja między barwą gleby a barwą futerka zwierzęcia sugeruje istnienie selektywnej dominacji myszy przystosowanych do gleby. Dominacja musi być względnie mała, ponieważ jasne formy nie giną w większym stopniu niż ciemne. Ewolucja w ubarwieniu grzbietowej strony futerka u tego gatunku zachodzi prawdopodobnie poprzez modyfikację kąta nachylenia linii obrazującej zależność pomiędzy zdolnością odbicia światła od powierzchni futerka a długością fali lub też przez zmianę stałej wielomianu równania regresji.

SECTION 6—ZOOGEOGRAPHY AND FAUNAL STUDIES

Studies of faunas, both of local areas and of broad regions, have contributed substantially to the literature in mammalogy. From the earliest contributions to the present, papers and books dealing with faunistics have included much information on systematics, ecology, distribution, ethology, and reproduction, among other topics; the sobriquet "natural historian" implied an interest in all these fields and more.

Darlington's *Zoogeography* (1957) and Udvardy's *Dynamic Zoogeography* (1969) are good sources of general information on that subject; Hesse *et al.* (1937) is a substantial and still useful earlier reference. Insular biogeography was aptly dealt with by Carlquist (1965) and in a more quantitative and theoretical way by MacArthur and Wilson (1967). Other general treatises that should be called to the attention of the beginning student include Matthew's (1939) *Climate and Evolution* and Dice's (1952) *Natural Communities.*

Among the major faunal catalogues are Allen (1939) for Africa, Ellerman and Morrison-Scott (1951) for the Palearctic, Miller and Kellogg (1955) and Hall and Kelson (1959) for North America, Cabrera (1958, 1961) for South America, and Troughton (1965) and Ride (1970) for Australia. At the regional level, recent treatments of the faunas of the whole of Canada (Banfeld, 1974) and the eastern half (Peterson, 1966), the East African countries of Kenya, Tanzania, and Uganda (Kingdon, 1971, and subsequent volumes), West Africa (Rosevear, 1965, 1969), Rhodesia, Zambia, Malawi and Botswana (Smithers, 1966, 1971), the Soviet Union (Ognev, 1962, and subsequent volumes), China and Mongolia (Allen, 1938, 1940), Arabia (Harrison, 1964, and subsequent volumes), are excellent examples, as are many of the state lists published for North America (for example, Hall, 1946; Jackson, 1961; Baker and Greer, 1962; Jones, 1964; Anderson, 1972; Armstrong, 1972; Lowery, 1974; Bowles, 1975; Findley *et al.*, 1975). Hall's *Mammals of Nevada* stands out in completeness from most points of view, whereas Armstrong's study of mammals in Colorado is especially good with respect to zoogeographic analyses and Lowery's book on Louisiana is particularly appealing to the specialist and nonspecialist alike. In terms of smaller geographic areas, Harper (1927) on the Okefinokee Swamp, Johnson *et al.* (1948) on the Providence Mountains of California, Anderson (1961) on the Mesa Verde of Colorado, Foster's (1965) study of the Queen Charlotte Islands, and Turner's (1974) paper on the Black Hills illustrate that substantial information can be gleaned from the study of a geographically restricted fauna. These publications as well as several reproduced here clearly indicate that the serious student of faunistics must be broadly trained in the discipline of mammalogy.

Because of interest in faunal studies over the years, it was inevitable that certain "laws"—directed at overall explanations for natural phenomena associated with distribution and variation—would emerge. These have been of two basic sorts, various "ecological rules" such as those proposed by Allen, Bergmann, and Gloger, and the biogeographic systems proposed on a worldwide scale by Wallace and others and applied more specifically to North America by Merriam (Life-zones), Shelford (Biomes), and Dice (Biotic

Provinces). Space does not permit the reproduction of the lengthy papers dealing with these subjects.

Four selections in this section (Kurtén, Guilday, Koopman and Martin, and Findley and Anderson) deal with various zoogeographic problems related to Pleistocene, subfossil, and Recent faunas, whereas another (Brown) examines the nonequilibrium insular nature of mammals isolated on mountain tops. The short paper by Hansen and Bear contrasts pocket gophers from different environments in Colorado. D. E. Wilson's contribution compares bat faunas from six different zoogeographic regions using trophic groupings. Those of Hagmeier and J. W. Wilson III, are attempts from different points of view to analyze distributional patterns of North American mammals, both based on initial compilations from one of the faunal catalogues (Hall and Kelson) cited above.

| Eiszeitalter und Gegenwart | Band 14 | Seite 96-103 | Öhringen/Württ., 1. September 1963 |

Notes on some Pleistocene mammal migrations from the Palaearctic to the Nearctic

Von Björn Kurtén, Helsingfors

Mit 2 Abbildungen im Text

S u m m a r y. The following dates for mammalian migrations from the Palaearctic to the Nearctic are suggested: Smilodontine sabre-tooths, Elster (Kansan); black bears, Elster (Kansan); brown and grizzly bears, Würm (Wisconsin); wolverine, Saale (Illinoian). Correlation between the mammalian faunas in the Nearctic and the Palaearctic should be based on compilation of many additional migration items.

Z u s a m m e n f a s s u n g. Die Einwanderungen einiger paläarktischer Säugetiere ins Nearktikum werden folgendermaßen datiert: Smilodontine Säbelzahnkatzen Elster (Kansan). Schwarzbären Elster (Kansan). Braunbären, bzw. Grizzlybären Würm (Wisconsin). Vielfraß Saale (Illinoian). Für eine befriedigende Korrelation zwischen den eiszeitlichen Säugetierfaunen des nearktischen und paläarktischen Raumes müßten noch eine Reihe von Einwanderungsbeispielen analysiert werden.

The correlation between the Pleistocene mammalian faunas of North America and Europe is at present a topic of much informal discussion and controversy, but on which relatively little has been published. The knowledge of Pleistocene faunal evolution in both areas is still incomplete, more so in the Nearctic, but a fairly coherent succession has recently been worked out by Hibbard and his associates (Hibbard, 1958; Taylor & Hibbard, 1959) and related to the standard glacial-interglacial sequence. Perhaps it may still be said that knowledge is too incomplete to permit a stage-by-stage correlation on faunal evidence. Nevertheless the topic is of such interest that contributions to it should be welcomed. It appears to the present writer that the most hopeful method to attack the problem is the detailed study of case histories of intercontinental migration by means of taxonomic and evolutionary analysis. Hence I have chosen to offer some notes on a number of Carnivora, summarized from a comparative study in progress. They are intended to supply some preliminary correlation items of the desired kind. It may perhaps be hoped that specialists on this and other groups would analyse other instances of intercontinental migration in the Pleistocene, of which many would be available.

It may be assumed that migration between the Old and the New World occurred exclusively across the Bering Strait, as far as the mammals are concerned, and that it took place only when there was a land bridge; that is to say, during a glacial phase. Immigrants are therefore assumed to have migrated in the preceding cold phase, if their first occurrence is in an interglacial fauna. First occurrence in a cold fauna is taken to signify an immigration during the same cold phase. In either case, however, it is a prerequisite that a probable ancestor should be known to have existed, at the given time, in the area of assumed origin.

Errors are bound to arise sometimes because our knowledge is incomplete and later discoveries may reveal that the immigrant was actually present at an earlier date. A more detailed analysis of the evolving populations may help us to avoid errors of this kind. If it can be shown that the assumed ancestor and the immigrant form an essential morphological continuum, this may be taken as additional indication that the correct time of migration has been found. A morphological gap between the two will suggest that part of the evolutionary sequence is missing and may have taken place in either area. I have endeavoured to pay full attention to this factor in the examples to follow.

A. Sabre-tooths of the genera *Megantereon* and *Smilodon*

The close relationship between the European *Megantereon* and the American *Smilodon* was recognized by Schaub (1925). Unfortunately, however, Matthew (1929) made the error of synonymizing the European *Homotherium* with *Smilodon*. With the discovery of excellent material of the sabre-tooth *Dinobastis* in the late Pleistocene of Texas (Meade, 1961) it has become clear that this form, not *Smilodon*, is the American ally of the European *Homotherium*, whereas *Megantereon* is allied to *Smilodon*. They form two quite distinct groups of sabre-toothed cats, which may be termed the tribes *Homotheriini* and *Smilodontini*.

Schaub (1925) pointed to the detailed similarity in the postcranial skeletons of *Megantereon* and *Smilodon*. The neck, for instance, is much elongated, and the distal segments of the heavy limbs are shortened. In the Homotheriini the neck is less elongated and the limbs are not shortened distally, indeed the forearm is extremely long. The dentitions and skulls (fig. 1) are also quite distinctly constructed in the two groups. The *Smilodontini* have dirk-like, very long upper canines, which were evidently used for stabbing exclusively; they are relatively and absolutely larger in *Smilodon*, but this is only a specialized character and does not obscure the essential similarity. The lower canines were reduced and form flanking elements in the transverse incisor row. The cheek teeth are not much modified from the normal feline type, except for a progressive reduction of the anterior elements and of the protocone (internal cusp) in the upper carnassial. The skull profile tends to be triangular, with a high occiput, only slightly overhanging the occipital condyles. The post-orbital processes are well set off with a marked constriction behind, separating them from the small but globular braincase. The development of the mastoid processes is correlated with that of the head depressors and is more pronounced in the advanced *Smilodon* with its enormous sabres.

In contrast, the Homotheriini have relatively short, flattened sabres with crenulated, sharp cutting edges all along the front and back; wear facets show that they were used in biting, as well as stabbing and slashing. The incisors form a semicircle, unlike the

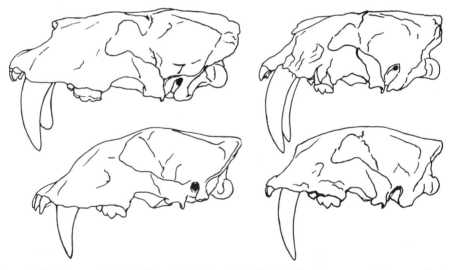

Fig. 1. Skulls of Homotheriini (left) and Smilodontini (right) in side view. Upper left, *Homotherium crenatidens*, Perrier, Villafranchian. Lower left, *Dinobastis serus*, Friesenhahn (Texas), Wisconsin. Upper right, *Smilodon neogaeus*, Arroyo Pergamino (Argentina), Pampean. Lower right, *Megantereon megantereon*, Senèze, Villafranchian. Not to scale.

7 Eiszeit und Gegenwart

smilodont transverse row. The cheek teeth are highly modified, extremely thin slicing blades. The skull is long and low with an arched profile, the frontal region is broad without distinctly set off processes, and the braincase is separated from the occipital plane by a marked constriction.

The two tribes probably arose independently from orthodox feline cats by divergent evolution along quite different adaptive paths. The first stage in the development of sabre-like upper canines is seen in the present-day *Felis nebulosa.* The idea of an iterated evolution of sabre-toothed cats gains in credibility from the fact that independent evolution of sabre-toothed carnivores has been demonstrated in the Marsupialia and Creodonta.

The history of both groups is mainly or entirely contained in the Pleistocene (including the Villafranchian), and most or all of the Eurasian and American Pleistocene sabre-tooths may be referred to one group or the other.

The two main types of homotheres, *Homotherium* and *Dinobastis,* occur in both hemispheres. The former are only early Pleistocene (Villafranchian) in Europe, but may occur later in America *(Ischyrosmilus?).* The latter are middle and late Pleistocene *(Dinobastis serus,* North America; *Dinobastis latidens,* Europe; *Dinobastis ultimus,* China).

The smilodont cats show a more definite evolutionary trend and indicate an intercontinental migration at a rather narrowly defined point in time.

All the Eurasian forms are referred to the genus *Megantereon.* The earliest forms, apart from some possible Indian ancestors in the Pliocene, occur in the earliest Villafranchian in Europe (Villafranca d'Asti, Etouaires). They are relatively small, of perhaps puma size. There is evidence of a gradual size increase, and the late Villafranchian forms (Senèze, Val d'Arno, Nihowan) are somewhat larger. All of the European Villafranchian

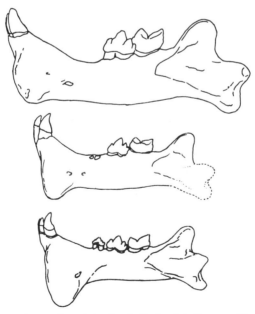

Fig. 2. Evolution of lower jaw and teeth in the Smilodontini. Bottom, *Megantereon megantereon,* Senèze, Villafranchian, with P$_3$ and large jaw flange. Centre, *Smilodon gracilis,* Port Kennedy Cave, Yarmouth, retaining P$_3$ (alveolus), flange somewhat reduced. Top, *Smilodon neogaeus,* Lapa Escrivania, Brazil, late Pampean, size greatly increased, P$_3$ lost, flange reduced. All to the same scale.

populations may, however, be referred to a single evolving species, *M. megantereon*. At the close of the Villafranchian, the line became extinct in Europe, but in Asia (and Africa) it survived into the middle Pleistocene. The last representative of this line in Asia, known at present, is *Megantereon inexpectatus* from Locality 1 (the Peking Man site) of Choukoutien. The date of this form is late Elster or early Hoxnian. *M. inexpectatus* is larger than *M. megantereon*, showing that the phyletic growth continued, and in fact it is about the same size as the earliest American Smilodons.

Our next glimpse of this evolutionary line comes from the New World with the earliest known members of the genus *Smilodon*. They come from Port Kennedy Cave out of deposits that appear to be Yarmouth in age (HIBBARD, 1958). As fig. 2 shows, these forms resemble advanced *Megantereon* more than advanced *Smilodon* in many characters. They still retain a well developed P$_3$ (alveolus in the figured specimen) and have a distinct inner cusp (protocone) in the upper carnassial. On the other hand, the reduction of the dependent flange on the lower jaw has already got under way in the early *Smilodon*. The trend in this character, within *Smilodon*, shows well enough that the genus evolved from a large-flanged ancestor like *Megantereon*. Apparently the sabre still bit inside the lower lip in *Megantereon*, and had to have a sheath supported by the jawbone. In *Smilodon* the sabre bit outside of the lower lip, and the sheath could be reduced. If this character is made the key character of the two genera, the Port Kennedy Cave form should go into *Smilodon* in spite of numerous resemblances to *Megantereon*. The development of the jaw flange in the Choukoutien form is unfortunately unknown (TEILHARD, 1945).

In *Smilodon* of Illinoian age, from the Conard Fissure (BROWN, 1908), further progress in size and dental characters have occurred, but not until Sangamon and Wisconsin time the full-fledged *Smilodon* known from Rancho La Brea and other asphalt deposits is met with.

It appears highly probable that the migration occurred at a point in time roughly coinciding with the age of the last known *Megantereon*. Thus the migration would be likely to be of Elster date. In this way it may be concluded that the characteristic Elster faunas of the Old World antedate the Yarmouth faunas in North America, and that the Yarmouth is a correlative of the European Holstein.

Smilodon also entered South America, where the earliest form, a relatively small and primitive one, occurs in the Chapadmalal (KRAGLIEVICH, 1948). This appears to give a maximum age for the Chapadmalal: it can hardly antedate the Yarmouth and Holstein.

The outlines of the evolution and migration of the Smilodontini are indicated in the diagram (table 1).

B. Bears of the genus *Ursus*

The bear family, Ursidae, is one of the smallest and most recent carnivore families. Three subfamilies are recognised at present (THENIUS, 1958, 1959), but one, the Agriotheriinae, became extinct at the close of the Pliocene, and does not concern us here. Another, the Tremarctinae, is exclusively American in distribution, and evolved from immigrants dating back to the Pliocene *(Plionarctos)*. The presence of large tremarctine bears *(Arctodus)* may have been a factor in the relatively late spread of *Ursus* into the New World.

The third subfamily, the Ursinae, has a Holarctic and Indian distribution. With a single exception *(Ursus americanus)* all of its species are present in the Old World, and the fossil record shows that it originated in Eurasia. In the Nearctic, ursine bears are not known until the middle Pleistocene, with a single uncertain exception (JOHNSTON & SAVAGE, 1955), a specimen from the Blancan Cita Canyon; it is fragmentary and may, or may not, be ursine.

7 *

Table 1

Evolution and migration of the Smilodontini

South America	North America		Eurasia	
Smilodon neogaeus	Wisconsin	*Smilodon* "*californicus*" (Rancho La Brea etc.)	Würm	
Smilodon ensenadensis	Sangamon	*S. trinitiensis* Texas etc.	Eem	
S. riggii Chapadmalal	Illinoian	*S. troglodytes* Conard Fissure	Riß-Saale	
↑ — — — —	Yarmouth	*S. gracilis* Port Kennedy Cave ↑ ⏐ — — — —	Holstein	
			Mindel- Elster	*Megantereon inexpectatus* Choukoutien
			Cromer & Villa- franchian	*Megantereon* spp. *M. megantereon*

Table 2

Intercontinental migrations of *Ursus*, excluding *U. maritimus*

North America		Eurasia		
Recent	*U. americanus* *U. arctos*	Recent	*U. arctos*	*U. thibetanus*
Wisconsin	*U. americanus* Many locs. *U. arctos* ← Alaska	— — — — Würm	*U. arctos* Many locs.	*U. thibetanus* China
Sangamon	*U. americanus* Trinity River	Eem	*U. arctos* Many locs.	*U. thibetanus* China
Illinoian	*U. americanus* Cumberland Cave, Conard Fissure	Riß-Saale	*U. arctos* Tornewton Cave etc.	*U. thibetanus* Europe, China
Yarmouth	*U. americanus* Port Kennedy Cave ↑ ⏐ ⏐ — — — — —	Holstein	*U. arctos* Grays etc.	*U. thibetanus* China
		Mindel- Elster	*U. arctos* China	
		— — — —	— — — — —	*U. thibetanus* China
		Cromer	*U. thibetanus* China, Europe	
		Villa- franchian	*U. etruscus* Europe	*U. ?thibetanus* China

543

The first certain ursine species to appear in North America is *Ursus americanus*, the black bear. The earliest record appears to be Port Kennedy Cave, as in the case of the smilodonts, and the bears evidently date from the Yarmouth. These early black bears are of moderate size and have several primitive characters, e.g. the large size of the upper carnassial and the small size of the last molar. In these characters, and also in morphology, the Port Kennedy Cave black bear closely resembles the middle Pleistocene *Ursus thibetanus kokeni* (Asiatic black bear) known from Chinese deposits, e. g. Choukoutien. It seems probable that the American form descended directly from the closely allied Asiatic species, and that the migration occurred at the same time as that of the Smilodontini, i. e. during the Elster Glaciation.

Illinoian black bears in North America, from Cumberland Cave and Conard Fissure (GIDLEY & GAZIN, 1938), show definite advance over the stage seen at Port Kennedy. Late Pleistocene forms reached great size; in the Postglacial, the size was secondarily reduced (KURTÉN, in the press).

Two other species of *Ursus* are at present found in North America, the polar bears *(Ursus maritimus)* and the brown and grizzly bears *(Ursus arctos)*. Both are conspecific with Old World species, and the indication is that their migration is of more recent date. Unfortunately the fossil history of the polar bear is practically unknown. As regards *Ursus arctos*, however, the fossil record bears out the contention. Most or all fossils of this species from North America appear to date from Postglacial times or possibly the latest Wisconsin (KURTÉN, 1960). It is possible that the species was present in Alaska at a somewhat earlier date, well back in the Wisconsin. However this may be, the intercontinental migration evidently took place during the Würm = Wisconsin. As a matter of fact, the morphology of the Alaskan and Eastern Siberian *Ursus arctos* indicates two independent migrations; a northern, with narrow-skulled animals from northern Siberia, and a southern, with broad-skulled animals from the Kamchatka population.

C. The glutton, *Gulo gulo*

The glutton or wolverine evolved in the Old World. The ancestral form is the Cromerian *Gulo schlosseri*, known from various localities in Central Europe, and chiefly distinguished by its much smaller size. The first true *Gulo gulo* appear in the Elster (Mosbach; see TOBIEN, 1957) of Europe, and also in China (Choukoutien; see PEI, 1934).

In North America, the earliest record of wolverine appears to come from Cumberland Cave (GIDLEY & GAZIN, 1938), where it is part of the cold elements that appear to date from the Illinoian (HIBBARD, 1958). The Illinoian is thus the minimum age of the migration, but there is also a possibility that the migration dates back to the Elster. It may be noted, however, that the New World population is only very moderately differentiated from that in the Old (KURTÉN & RAUSCH, 1959), and only ranks as a distinct subspecies. This would support a relatively late date, and at present a Saale-Illinoian migration appears the most likely alternative.

Conclusions

The three migration items discussed here support the correlation of glaciations and interglacials in North America and Europe as far back as the Holstein in Europe and Yarmouth in North America, as shown in tables 1 and 2. The exact European correlatives of the Kansan, the Aftonian, and the Nebraskan faunas remain doubtful. Perhaps the definitive clearing up of the migration of the elephants will settle part of the problem. The earliest true elephants in North America appear in faunas assigned to the Kansan, such as the Holloman and Cudahv (TAYLOR & HIBBARD, 1959). This early form is *Elephas haroldcooki*. Detailed analysis of a large material will be necessary to show whether this species may be derived from the primitive mammoths of the Elster, or from advanced *Elephas meridionalis* types of the late Villafranchian.

A correlation of the Kansan with the Elster would make the Aftonian a correlative of the Cromerian in Europe. The latter is a complex stage with two distinct temperate oscillations (WEST, 1961). On the other hand, many American specialists seem to favour a correlation of the Aftonian with the Villafranchian in Europe. Some implications of such a correlation are somewhat disturbing. Correlation of the Aftonian or part of it with the Villafranchian, and of the Yarmouth or part of it with the Holstein, would suggest one of the schemes (1)—(3) below, in contrast with the usual arrangement (4).

(1)

Yarmouth	}	Holstein (Elster) Cromer
Kansan		Günz
Aftonian		Villafranchian
Nebraskan		Donau
Rexroadian		Astian

(2)

Yarmouth		Holstein
Kansan	}	Elster (Cromer) Günz
Aftonian		Villafranchian
Nebraskan		Donau
Rexroadian		Astian

(3)

Yarmouth	Holstein
Kansan	Elster
	Cromer
Aftonian	} (Günz) Villafranchian
Nebraskan	Donau
Rexroadian	Astian

(4)

Yarmouth	Holstein
Kansan	Elster
Aftonian	Cromer
Nebraskan	Günz
Rexroadian	Villafranchian

It would obviously be premature to attempt an evaluation of these and other possible combinations at present, when even the intra-continental correlations (e. g., the dating of the American faunal sequence in terms of the American glacial-interglacial sequence) are somewhat uncertain. One possible source of bias in intercontinental correlation should, however, be mentioned. Strictly homotaxial relations for migrating groups can only be expected at the time of the actual migration, i. e., at the time of a glaciation, as a rule. Apparent homotaxial relations between interglacial faunas may be deceptive. For a hypothetical example, suppose that a typical Holsteinian Old World form migrates to North America in the Saale (Illinoian), and becomes extinct in Eurasia, or perhaps evolves into a more advanced Eemian type. If it is further supposed that the American immigrant happens to be known to us only in strata of Sangamon age, this will be a case of closer homotaxial relationships between the Sangamon and the Holstein, than between the Sangamon and the Eem. This may possibly have some bearing on the apparent relationship between the Aftonian and the Villafranchian. Naturally, migrants in the other direction, from the New World to the Old, will have the opposite effect; but as these migrations have been much less common, their effect is likely to be overshadowed.

Numerous migration items await detailed analysis. Lynxes among felids; wolf and red fox among canids; least weasel, otter and marten among mustelids are obvious cases in the Carnivora, and the only known American hyaenid may be added: *Chasmaporthetes*, which appears in the early Pleistocene Cita Canyon fauna and has a close relationship with the European *Euryboas*, the „sprinter hyena". The importance of the Proboscidea has already been stressed, and numerous examples will be found among the rodents. Horses and camels exemplify migration from the New World to the Old, perhaps a subsequent remigration of modernized *Equus* into North America (McGREW, 1944). Elk

(Alces), reindeer *(Rangifer)*, yak *(Bos)*, bison *(Bison)*, musk ox *(Ovibos)* and sheep *(Ovis)* are important artiodactyl migrants from the Old World to the New, most of them in the late or latest Pleistocene.

References

BROWN, Barnum: The Conard Fissure, a Pleistocene bone deposit in northern Arkansas: With description of two new genera and twenty new species of mammals. - Mem. American Mus. Nat. Hist. 9 (4), 157-208, New York 1908.

GIDLEY, J. W., & GAZIN, C. L.: The Pleistocene vertebrate fauna from Cumberland Cave, Maryland. - Bull. U.S. Nat. Mus. 171, 1-99, Washington 1938.

HIBBARD, Claude W.: Summary of North American Pleistocene mammalian local faunas. - Papers Michigan Acad. Sci. 43, 3-32, Ann Arbor 1958.

JOHNSTON, C. S., & SAVAGE, D. E.: A survey of various late Cenozoic vertebrate faunas of the Panhandle of Texas, part I: Introduction, description of localities, preliminary faunal lists. - Univ. California Publ. Geol. Sci. 31, 27-50, Berkeley 1955.

KRAGLIEVICH,Lucas Jorge: Smilodontidion riggii, n. gen. n. sp. Un nuevo y pequeño esmilodonte en la fauna pliocena de Chapadmalal. - Rev. Mus. Argentino Ci. Nat. 1, 1-44, Buenos Aires 1948.

KURTÉN, Björn: A skull of the grizzly bear *(Ursus arctos* L.) from Pit 10, Rancho La Brea. - Los Angeles County Mus. Contr. Sci. 39, 1-7, Los Angeles 1960.

KURTÉN, Björn & RAUSCH, Robert: Biometric comparisons between North American and European mammals. - Acta Arctica 11, 1-44, Copenhagen 1959.

MCGREW, Paul O.: An early Pleistocene (Blancan) fauna from Nebraska. - Geol. Ser. Field Mus. Nat. Hist. 9, 33-66, Chicago 1944.

MATTHEW, William D.: Critical observations upon Siwalik mammals. - Bull. American Mus. Nat. Hist. 56 (7), 437-500, New York 1929.

MEADE, Grayson E.: The saber-toothed cat *Dinobastis serus*. - Bull. Texas Memor. Mus. 2, 25-60, Austin 1961.

PEI, Wen chung: On the Carnivora from Locality 1 of Choukoutien. - Palaeont. Sinica, ser. C, 8, 1-216, Peking 1934.

SCHAUB, Samuel: Über die Osteologie von *Machaerodus cultridens* Cuvier. - Eclog. geol. Helv. 19, 255-266, Basel 1925.

TAYLOR, Dwight D. & HIBBARD, Claude W.: Summary of Cenozoic geology and paleontology of Meade County area, southwestern Kansas and northwestern Oklahoma. - Pamphlet, 1-157, Ann Arbor 1959.

TEILHARD DE CHARDIN, Pierre: Les formes fossiles. - Les félidés de Chine, 5-35, Peking 1945.

THENIUS, Erich: Über einen Kleinbären aus dem Pleistozän von Slovenien nebst Bemerkungen zur Phylogenese der plio-pleistozänen Kleinbären. - Razprave Slov. Akad. Znan. 4, 633-646, Ljubljana 1958. - - Ursidenphylogenese und Biostratigraphie. - Zeitschr. Säugetierk. 24, 78-84, Berlin 1959.

TOBIEN, Heinz: *Cuon* HODG. und *Gulo* FRISCH (Carnivora, Mammalia) aus den altpleistozänen Sanden von Mosbach bei Wiesbaden. - Acta Zool. Cracoviensia 2, 433-452, Kraków 1958.

WEST, R. G.: The glacial and interglacial deposits of Norfolk. - Trans. Norfolk & Norwich Nat. Soc. 19, 365-375, Norwich 1961.

Manuskr. eingeg. 3. 1. 1963.

Anschrift des Verf.: Dr. B. KURTÉN, Zoologisches Museum der Universität, N. Järnvägsgatan 13, Helsingfors, Finnland.

PLEISTOCENE ZOOGEOGRAPHY OF THE LEMMING, *DICROSTONYX*[1]

John E. Guilday

Carnegie Museum, Pittsburgh, Pa.

Received August 30, 1962

The collared lemmings, genus *Dicrostonyx* Gloger, are currently divided into two subgenera. *Misothermus* Hensel contains a single species, *D. hudsonius* Pallas, isolated on the tundra of northern and coastal Ungava from all other *Dicrostonyx* (see fig. 1). The subgenus *Dicrostonyx* Gloger contains the remaining species of the genus, *D. torquatus* Pallas of the palaearctic, *D. groenlandicus* (Traill) of the nearctic, and *D. exsul* G. M. Allen confined to St. Lawrence Island in the Bering Straits. The *torquatus-groenlandicus-exsul* species group may be conspecific as intimated by Ellerman and Morrison-Scott (1951, p. 653). It is clear that they are more closely related to one another than to the isolated *D. hudsonius*.

In the absence of any fossil record and interpreting on the basis of modern geographical distribution alone, one might argue that the differentiation of *D. hudsonius* dated from the Wisconsin glaciation; that as the ice front and its presumed periglacial tundra belt shrank to the north, the eastern segment of the retreating lemming population was cut off by Hudson Bay. The bay eventually cut the Canadian tundra into an eastern and a western component, each with its distinctive form of collared lemming. This does not appear to be the case, however.

The inaccuracy of this interpretation is shown by the fossil record. The one record from the North American Pleistocene, fragmentary skulls and mandibles of at least four individuals from Sinkhole no. 4, New Paris, Pennsylvania (Guilday and Doutt, 1961), is that of typical *Misothermus* (for characters, see Miller, 1898; Hinton, 1926;

Hall and Kelson, 1959), indistinguishable from the modern *D. hudsonius*. Carbon particles taken from a position five feet higher in the sinkhole matrix were dated at $11,300 \pm 1,000$ years (Yale Univ. lab. no. 727). The age of the lemming remains is somewhat in excess of this. The *Misothermus* dental pattern was fixed prior to the Wisconsin recession and the formation of present Hudson Bay.

There have been two species of collared lemmings described from the palaearctic Pleistocene. *Dicrostonyx gulielmi* Sandford based upon cranial material from Hutton Cave, Somersetshire, England, is a late Pleistocene form of the living *D. torquatus*, and may be conspecific with it (see Kowalski, 1959, p. 229). It is a common Eurasian Pleistocene fossil.

The second Old World fossil form, *D. henseli* Hinton, described from cranial material from a fissure deposit at Ightham, Kent, England, appears to be a typical *Misothermus* (see the description by Hinton, 1926, p. 163). *D. (Misothermus) henseli* has been recorded from the Pleistocene of England, Ireland, Jersey, France, and Germany (Hinton, 1926; Brunner, various papers); *D. (Dicrostonyx) torquatus* (or *gulielmi*), from the Pleistocene of England, Ireland, France, Poland, Czechoslovakia (Hinton, 1926; Kowalski, 1959; Fejfar, 1961). This by no means exhausts the list of Old World Pleistocene *Dicrostonyx* localities. But enough has been cited to indicate a geographical and possibly a chronological overlap between the two species. Both forms were recovered by Hinton from Merlin's Cave, Wye Valley, Herefordshire, England (22 *D. gulielmi* skulls, 4 *D. henseli*) and at Langwith Cave, Derbeyshire (1 *D. gulielmi*, 2 or 3 *D. henseli*). Brunner recorded *D. henseli* from

[1] Research conducted under National Science Foundation Grant no. G-20868.

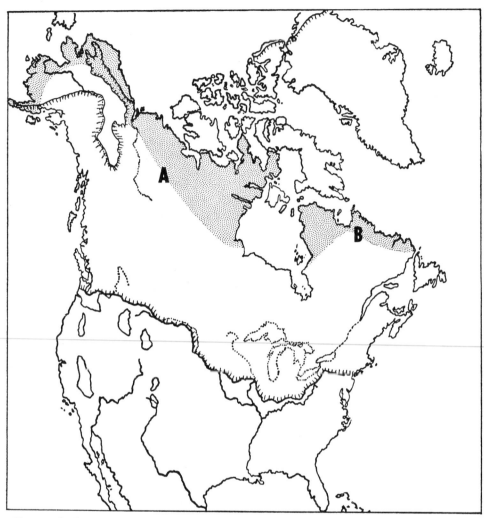

FIG. 1. Outline map of North America, showing approximate limit of continental glaciation. A. Mainland modern distribution of the subgenus *Dicrostonyx* in America. B. Modern distribution of the subgenus *Misothermus*.

thirteen Bavarian cave deposits. In one, the Markgrabenhöhle, Brunner (1952c, p. 465) reported that 25% of the mandibles resembled *D. gulielmi* in possessing a small anteroexternal vestigial angle on M_3, "eine deutliche aüssere Schmeltzfalte." Many uncorrelated fissure, cave, and terrace deposits are involved, however. And while they are all middle to late Pleistocene in age, their sequential position within that time span has not been established with any degree of confidence.

The facts at hand seem to indicate that two forms of the genus *Dicrostonyx* inhabited Eurasia and perhaps North America during the Pleistocene, and that one of them survives as a postglacial relict, isolated in the tundra of Ungava (fig. 2). The pre-Wisconsin origin of the *Misothermus* dental pattern is demonstrated by fossil forms in both continents.

If we assume, as does Hinton (1926), that *Misothermus* is not a true phylogenetic category but that *D. henseli* and *D. hud-*

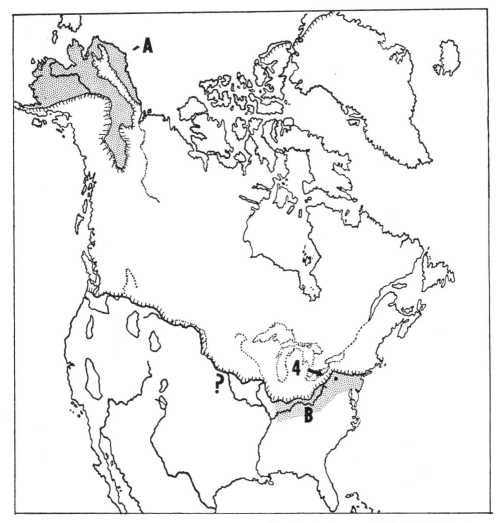

Fig. 2. Postulated distribution of *Dicrostonyx* (A) and *Misothermus* (B) in North America at Wisconsin glacial maximum. Note site of Sinkhole no. 4, New Paris, Bedford County, Pennsylvania. Hachures indicate approximate limit of continental glaciation.

sonius are of independent origin from *Dicrostonyx* proper, we are faced with the apparent coincidence that a form (*henseli*) was replaced by modern *Dicrostonyx* in the Old World while its morphological equivalent, *D. hudsonius*, which ranged as far south as central Pennsylvania during late Wisconsin times, survives today only where it is completely isolated from all contact with the Eurasian-Western Nearctic *Dicrostonyx*.

Both *D. henseli* and *D. hudsonius* appear to have been completely or partially replaced by true *Dicrostonyx*. Is it possible that modern *D. hudsonius* (or some form of *Misothermus*) at one time ranged throughout the holarctic, and that it was replaced during late Pleistocene times by lemmings of the subgenus *Dicrostonyx*, first in the Old World, later in the New; this latter replacement occurring sometime after the post-Wisconsin formation of Hudson Bay and the division of the mainland North

American tundra into an eastern and a western component?

I wish to thank Dr. J. Kenneth Doutt, Curator of Mammals, Miss Caroline A. Heppenstall, Assistant Curator of Mammals, and Dr. Craig C. Black, Gulf Associate Curator of Vertebrate Fossils, Carnegie Museum, for their helpful criticism.

SUMMARY

The modern distribution of *Dicrostonyx hudsonius* Pallas (confined to the tundra of Ungava) is believed, on the basis of the fossil record, to be a relict of a former holarctic pre-Wisconsin distribution.

REFERENCES CITED

BRUNNER, GEORG. 1949. Das Gaisloch bei Münzinghof (Mfr.) mit Faunen aus dem Altdiluvium und aus jüngeren Epochen. Neuen Jahrbuch Mineral., etc. Abhandlungen, **91**(B): 1–34.

———. 1951. Eine Faunenfolge von Würm III Glazial bis in das Spät-Postglazial aus der "Quellkammer" bei Pottenstein (Ofr.). Geol. Bl. NO-Bayern, **1**: 14–28.

———. 1952a. Das Dohlenloch bei Pottenstein (Obfr.). Eine Fundstelle aus dem Würm II Glazial. Abhandlungen Naturhist. Ges. Nürnberg, **27**(3): 49–60.

———. 1952b. Der "Distlerkeller" in Pottenstein Ofr. Eine Faunenfolge des Würm I–III interstadial. Geol. Bl. NO-Bayern, **2**: 95–105.

———. 1952c. Die Markgrabenhöhle bei Pottenstein (Oberfranken). Eine Fauna des Altdiluviums mit *Talpa episcopalis* Kormos u. a. Neues Jahrbuch Geol. Paläontol. Mh., **10**: 457–471.

———. 1953a. Die Heinrichgrotte bei Burggaillenreuth (Oberfranken). Eine Faunenfolge von Würm I-Glazial bis Interstadial. Neues Jahrbuch Geol. Paläontol. Mh., **6**: 251–275.

———. 1953b. Die Abri "Wasserstein" bei Betzenstein (Ofr.). Eine subfossile Fauna mit *Sorex minutissimus* H. de Balsac. Geol. Bl. NO-Bayern, **3**: 94–105.

———. 1955. Die Höhle am Butzmannsacker bei Auerbach (Opf.). Geol. Bl. NO-Bayern, **5**: 109–120.

———. 1956. Nachtrag zur Kleinen Teufelshöhle bei Pottenstein (Oberfranken). Eine Übergang von der letzten Riss-Würm-Warm-fauna zur Würm I-Kaltfauna. Neues Jb. Geol. Paläontol. Mh., **2**: 75–100.

———. 1957a. Die Breitenberghöhle bei Gössweinstein/Ofr. Eine Mindel-Riss-und eine postglaziale Mediterran-Fauna. Neues Jb. Geol. Paläontol. Mb., **7–9**: 352–378, 385–403.

———. 1957b. Die Cäciliengrotte bei Hirschbach (Opf.) und ihre fossile Fauna. Geol. Bl. NO-Bayern, **7**: 155–166.

———. 1958. Das Guckerloch bei Michelfeld (Opf.). Geol. Bl. NO-Bayern, **8**: 158–172.

———. 1959. Das Schmeidberg-Abri bei Hirschbach (Oberpfalz). Paläont. Z., **33**: 152–165.

ELLERMAN, J. R., AND T. C. S. MORRISON-SCOTT. 1951. Checklist of Palaeoarctic and Indian mammals. British Museum (Natural History). London. 810 p.

FEJFAR, OLDRICH. 1961. Review of Quaternary Vertebrata in Czechoslovakia. Instytut Geologiczny. Odbitka z Tomu 34 Prac. Czwartorzed Europy Srodkowej I Wschodniej, p. 109–118.

GUILDAY, J. E., AND J. K. DOUTT. 1961. The Collared Lemming (*Dicrostonyx*) from the Pennsylvania Pleistocene. Proc. Biol. Soc. Washington, **74**: 249–250.

HALL, E. R., AND R. K. KELSON. 1959. The mammals of North America. Vol. II, p. 547–1083.

HINTON, M. A. C. 1926. Monograph of the voles and lemmings (Microtinae) living and extinct. Vol. I. Richard Clay and Sons, Suffolk. 488 p.

KOWALSKI, K. 1959. Katalog Ssakow Plejstocenu Polski. Polska Akademia Nauk. Inst. Zool. Oddzial W Krakowie. 267 p.

MILLER, G. S., JR. 1896. Genera and subgenera of voles and lemmings. North American Fauna no. 12, 80 p., U. S. Dept. Agriculture.

SUBFOSSIL MAMMALS FROM THE GÓMEZ FARÍAS REGION AND THE TROPICAL GRADIENT OF EASTERN MEXICO

By Karl F. Koopman and Paul S. Martin

In the spring of 1953 Byron E. Harrell and P. S. Martin collected superficial animal remains at three cave localities in the Sierra Madre Oriental of southwestern Tamaulipas. Locally, the mountains are sufficiently humid to support small, isolated patches of Cloud Forest and Tropical Evergreen Forest (Martin, 1958). At this latitude, 23° north, these two tropical plant formations appear to reach their limit. In recent years four faunal papers on mammals of southern Tamaulipas have appeared, each reporting certain tropical species unknown at higher latitudes (Baker, 1951; Goodwin, 1954; Hooper, 1953; de la Torre, 1954). Southern Tamaulipas appears to be of primary significance in the Gulf lowlands faunal gradient, a system extending from southern Veracruz to southern Texas. In the lowlands of eastern Mexico many of the dominant tropical American taxa reach their range limits. We seek to define the northeastern section of this gradient with regard to tropical mammal faunas, to relate it to shifts in vegetation and to indicate the relative importance of the Gómez Farías region as a faunal terminus. The Gómez Farías region is defined as the area from 22°48′ to 23°30′ north latitude and 99° to 99°30′ west longitude, or approximately the rectangle enclosed by the towns of Llera, Jaumave, Ocampo and Limon.

In establishing the presence of eight species known in the Gómez Farías region only from the skeletal remains, and in extending the altitudinal or ecological ranges of others, the present collection supplements previous reports. It indicates the value of using owl pellet deposits as a cross check on standard trapping technique.

DESCRIPTION OF DEPOSITS

Cretaceous limestones comprising the precipitous east slope of the Sierra Madre Oriental near the village of Gómez Farías are severely folded. Under torrential summer rains they have eroded into very rough karst terrain with virtually no surface drainage. Caves and sink holes, a characteristic of karst, are a common feature. Most do not appear to be inhabited by bats or owls. Of those that are, the constant high humidity and frequent flushing from

1

percolating rainwater apparently prevent accumulation of deep bone or guano deposits. Large bat colonies numbering thousands of individuals occur in Tropical Deciduous Forest in the Canyon of the Río Boquilla, 8 km. southwest of Chamal, and in a guano cavern at El Abra, 8 km. northeast of Antiguo Morelos. However, no colonies of similar size have been found in the more humid portions of the Gómez Farías region.

Of approximately thirty caves and sink holes explored near Gómez Farías, material from three is represented in our collection. All occur in the mountains west of the village of Gómez Farías, latitude 29°03′ north, longitude 99°09′ west, and within 16 km. of each other (see Table 1). The following information will serve to characterize them:

1. Paraiso. Aserradero del Paraiso is the name of a small sawmill located in Tropical Evergreen Forest 13 km. north-northwest of Chamal. A narrow ravine about 1 km. south of the sawmill harbors several caves, including a deep, wet grotto with permanent water. On the sloping floor of a small

TABLE 1.—*Mammals from cave deposits in the Gómez Farías region (numbers refer to anterior skull parts identified)*

	PARAISO	RANCHO DEL CIELO	INFERNO
Distance from Gómez Farías:	13 km. SW	6 km. NW	7 km. W
Elevation in meters:	420	1050	1320
Vegetation type:	Tropical Evergreen Forest	Cloud Forest	Cloud Forest
Didelphis marsupialis	—	—	1
Marmosa mexicana	—	—	18
Cryptotis pergracilis	3		
Cryptotis mexicana	—	1	10
Chilonycteris parnellii	1		
Enchisthenes harti	—	—	1
Artibeus cinereus	—	3	
Centurio senex	—	—	1
Eptesicus fuscus	1		
Lasiurus cinereus	—	—	1
Antrozous pallidus	1		
Sylvilagus floridanus	1		
Glaucomys volans	1	1	12
Liomys irroratus	1		
Reithrodontomys mexicanus	—	2	5
Reithrodontomys megalotis	—	—	1
Baiomys taylori	1		
Peromyscus boylei	—	1	17
Peromyscus pectoralis	—	—	2
Peromyscus ochraventer	—	—	7
Oryzomys alfaroi	3	—	5
Sigmodon hispidus	29		
Neotoma angustapalata	9	1	17
Total identifications	51	9	98

dry cave on the west wall of this ravine many scattered (apparently water-transported) small bones were found.

2. Rancho del Cielo. A small sink and cave several hundred meters east of this important Cloud Forest collecting locality contained a few bat remains and fresh owl pellets.

3. Inferno (also designated Infernillo). A rather large cave adjacent to an abandoned sawmill of this name is located in upper Cloud Forest roughly 2 km. south of the mill settlement called La Gloria. Abundant pellet remains were found on a dislodged boulder just inside the cave mouth.

All the caves are surrounded by heavy forest, much of it recently lumbered. Jagged, shrub-covered or almost bare karst ridges and pinnacles confine the areas of tall forest to valleys, pockets and gentle slopes. A more complete description of the Gómez Farías region is in preparation (Martin, 1958).

Although the barn owl, *Tyto alba*, occurs in a very large cavern at El Abra north of Antiguo Morelos, the only raptor definitely known to roost in caves in the Gómez Farías region is the wood owl, *Ciccaba virgata*. While it is possible that certain non-volant mammals died from other causes, all skeletal remains of these, and perhaps even some of the bats, can be ascribed to *Ciccaba*.

Cave records do not make ideal locality data, especially when information on ecological or altitudinal distribution is sought. We assume that the forest types and the elevation in the immediate vicinity of each cave represent the habitat in which the owls fed, but this is by no means certain. The hunting range and nocturnal movements of *Ciccaba* are unknown.

There is no stratigraphic evidence for assuming that any material is older than very recent. Limestone deposit on certain bones may represent the concretion of a single rainy season.

SYSTEMATIC TREATMENT OF THE MAMMALS

We emphasize first that only a fraction of the mammalian and none of the bird bones have been identified. Very little attempt has been made to identify any of the post-cranial elements, and in the case of the cricetine rodents only the best of cranial material could be identified with any confidence. Particular difficulty was encountered with lower jaws of cricetines, even when teeth were present. As a result, determinations in this group are considered less reliable than in the others. In particular this applies to the *Peromyscus–Baiomys–Reithrodontomys* group of genera, in which not only specific but also good generic characteristics were hard to find in most of the material. Size, molar form, palatal morphology and shape of the zygomatic plate were found to provide the most useful taxonomic characters in this subfamily.

In many cases a closely adhering limestone drip deposit was present and was sometimes difficult to remove without damaging the underlying bone. No attempt has been made to identify subspecies. In our opinion the theoretical requirements of population sampling and measurement, as employed in subspecific identification, are not met in any but perhaps the best fossil material. For the following species determinations Koopman alone is responsible.

Didelphis marsupialis.—Inferno: one rostral fragment of a young individual. Previously recorded from Gómez Farías (Hooper, 1953).

Marmosa mexicana.—Inferno: three rostral fragments and 15 mandibles. On the basis of dentition, these are clearly opossums, but of the Middle American genera of Didelphidae, all but *Marmosa* are much too large. Of the four species of this genus occurring north of Panama, all but *M. mexicana* differ from the subfossil material either in larger size or greater development of a precondylar crest on the outer side of the mandible. The species has not previously been recorded north of Jalapa, Veracruz.

Cryptotis pergracilis.—Paraiso: one rostrum and two mandibles. Restriction to the family Soricidae and the genus *Cryptotis* may be made on the basis of dentition. In eastern Mexico north of the Isthmus of Tehuantepec (states of Tamaulipas, San Luis Potosí, Hidalgo, Puebla and Veracruz) the following six species of *Cryptotis* are known: *C. parva, C. pergracilis, C. obscura, C. micrura, C. mexicana* and *C. nelsoni.* All except *C. parva* and *C. pergracilis* may be ruled out on the basis of larger size. While clear-cut cranial differences between the latter two species appear to be absent, the eastern race of *pergracilis, C. p. pueblensis,* has a somewhat deeper nasal emargination of the rostrum than the southwestern race of *parva, C. p. berlandieri.* Though slightly broken anteriorly, the subfossil rostrum appears to have a nasal emargination somewhat closer to that of *C. pergracilis pueblensis* than to that of the two Tamaulipan specimens of *C. parva berlandieri* recorded by Goodwin (1954). However, no direct comparison of the Paraiso specimens with *C. pergracilis pueblensis* was made, but only with a sketch. Unfortunately our material seems inadequate to solve the problem of specific status, i.e., are *C. parva* and *C. pergracilis* sympatric in southern Tamaulipas, do they integrade through a narrow hybrid zone, or is there a more complex arrangement?

The northernmost previous record of *C. pergracilis* is Platanito in San Luis Potosí.

Cryptotis mexicana.—Inferno: six rostra, four mandibles, and three humeri; Rancho del Cielo: one rostrum. Specimens were trapped at the latter locality (Goodwin, 1954).

Chilonycteris parnellii.—Paraiso: one mandible. Goodwin (1954) records a series from El Pachon. As Koopman (1955) has pointed out, the mainland *C. rubiginosa* and the West Indian *C. parnellii* are almost certainly conspecific. Since Koopman believed that *C. parnellii* Gray had several months priority over *C. rubiginosa* Wagner, the combined species was called *C. parnellii.* De la Torre (1955) has shown that this is not the case and, finding no way of determining which name was published first, recommended: "In the absence of conclusive evidence, the better known and more widely used name *rubiginosa* should be retained."

Unfortunately, if there is no clear priority, the law of the first reviser would seem to hold, in this case the first to use one of the two names to include both forms, i.e., Koopman (1955). It is not legally possible to withdraw from this position. Therefore it appears that *C. parnellii* must stand as the name assigned to both mainland and West Indian large *Chilonycteris*.

The question of nomenclature should not obscure the more significant taxonomic conclusion that a single species is involved.

Artibeus cinereus.—Rancho del Cielo: two partial skulls, one lower jaw. De la Torre has also recorded this species from Rancho del Cielo.

Enchisthenes harti.—Inferno: one partial skull. The short broad rostrum and characteristic molar pattern rule out all American bats outside the Stenoderminae. Of the Middle American stenodermines, only *Uroderma bilobatum*, *Vampyrops helleri* and *Enchisthenes harti* agree with the Inferno skull in size and dental formula (i^2 o^1 p^2 m^3). Both *Uroderma* and *Vampyrops*, however, have rostra considerably longer than that of the subfossil skull. On the other hand, there is close resemblance to skulls of *Enchisthenes* from Honduras and Ecuador. De la Torre (1955) has recently summarized the known records, specimens from Ciudad Guzman in Jalisco being the closest to Tamaulipas geographically. Four other localities extend the distribution south to Trinidad and Ecuador.

Centurio senex.—Inferno: one rostrum. Of all the North and Middle American bats, only *Centurio senex* agrees with the subfossil skull in dental formula (i^2 c^1 p^2 m^3) and in the palate being more than twice as wide as it is long. The Inferno rostrum resembles a skull of *Centurio senex* in all important respects. De la Torre (1954) recorded a single specimen from Pano Ayuctle.

Eptesicus fuscus.—Paraiso: one mandible. Several characters of this bone immediately narrow the field considerably. These are mandibular length, tooth size, dental formula (i_3 c_1 p_2 m_3), molar pattern and height of the coronoid process. This leaves us with only two North and Middle American species, *Eptesicus fuscus* and *Dasypterus intermedius*. Of these, *Dasypterus* may be ruled out on the basis of its more robust mandibular ramus. Comparison of the Paraiso mandible with *Eptesicus fuscus* reveals no important differences. I have been able to find no other records of this bat in Tamaulipas, the nearest localities being Río Ramos in Nuevo León to the west (Davis, 1944) and Cañada Grande in San Luis Potosí to the southwest (Dalquest, 1953). Good series from Tamaulipas, if they could be obtained, should show integradation between *E. f. fuscus* and *E. f. miradorensis*.

Lasiurus cinereus.—Inferno: one nearly complete skull. All other species of North and Middle American bats may easily be excluded from consideration on the basis of size, rostral shape and dental formula (i^1 c^1 p^2 m^3). It matches *L. cinereus* closely. Since this bat is migratory, it is impossible to say whether this individual belonged to a resident population or was merely a winter visitor. The nearest previous records are Matamoros in northern Tamaulipas (Miller, 1897) and El Salto in eastern San Luis Potosí (Dalquest, 1953).

Antrozous pallidus.—Paraiso: one partial lower jaw. Mandible and tooth size, molar pattern and dental formula (i_2 c_1 p_2 m_3) rule out all North and Middle American bats except *Promops centralis* and *Antrozous*. *Promops* may be excluded by the quite different appearance of the labial surface of the coronoid region. Of the two species of *Antrozous* recognized by Orr (1954), *A. bunkeri* is distinctly larger than the Paraiso mandible. *A. pallidus* resembles it in all respects. The University of Michigan Museum of Zoology has six specimens from Tula, which were mentioned by Orr (1954).

Sylvilagus floridanus.—Paraiso: one maxillary fragment of a young individual. Goodwin (1954) records the species from Gómez Farías, Pano Ayuctle and Chamal.

Glaucomys volans.—Inferno: two palatal fragments, ten mandibles; Rancho del Cielo: one mandible; Paraiso: one mandible. From the dental formula (i_1 c_0^0 p_1^2 m_3^3) these specimens are clearly referable to the Sciuridae, of which all northeastern Mexican species except *Eutamias bulleri*, *E. dorsalis* and *Glaucomys volans* are clearly too large. In *Eutamias*, however, the mandible is much less deep than in the Tamaulipas material. The latter bears a convincing resemblance to *G. volans*. The nearest locality from which the species had previously been obtained is Santa Barbarita in San Luis Potosí (Dalquest, 1953).

Liomys irroratus.—Paraiso: one maxillary fragment. Goodwin (1954) records a series from Pano Ayuctle.

Oryzomys alfaroi.—Inferno: four maxillary fragments, one partial skull; Paraiso: throo mandibles. Coodwin (1954) and Hooper (1953) have each recorded the species from Rancho del Cielo.

R. (Reithrodontomys) megalotis.—Inferno: one partial skull. Identification of this fragment is tentative. The species has already been recorded from Rancho del Cielo by both Goodwin (1954) and Hooper (1953).

R. (Aporodon) mexicanus.—Inferno: five partial skulls; Rancho del Cielo: two partial skulls. The species has been recorded previously from Rancho del Cielo by Goodwin (1954) and Hooper (1953).

Peromyscus boylei.—Inferno: two partial skulls, 15 maxillaries; Rancho del Cielo: one mandible. The species is recorded by Goodwin (1954) from both Rancho del Cielo and Rancho Viejo.

Peromyscus pectoralis.—Inferno: one partial skull, one maxillary. It has been recorded by Goodwin (1954) from both La Joya de Salas and 2 km. west of El Carrizo.

Peromyscus ochraventer.—Inferno: one partial skull, six maxillaries. The species has been recorded from Rancho del Cielo by both Goodwin (1954) and Hooper (1953).

Baiomys taylori.—Paraiso: one mandible. Goodwin (1954) and Hooper (1953) have recorded it from Pano Ayuctle.

Sigmodon hispidus.—Paraiso: one rostral half, ten maxillaries, two premaxillaries, three braincase elements, 16 mandibles. The species has been recorded from Pano Ayuctle by Goodwin (1954) and Hooper (1953).

Neotoma angustapalata.—Inferno: four partial rostra, five maxillaries, eight mandibles; Rancho del Cielo: one maxillary; Paraiso: one maxillary fragment, two premaxillaries, six mandibles. There is also a great deal of additional *Neotoma* material from Inferno that probably belongs here, but which has not been specifically identified. Both Goodwin (1954) and Hooper (1953) list specimens from Rancho del Cielo and El Pachon although, as Hooper points out, the precise status of *N. angustapalata* and its various southern Tamaulipas populations is far from clear. At the present time this name appears to be something of a "catch-all." It is felt, however, that a revision should be based on entire specimens rather than on skeletal fragments.

FAUNAL COMPARISONS AND THE STATUS OF GLAUCOMYS

The identifications summarized in Table 1 represent total number of anterior skull elements and *not* total number of individuals, which may be somewhat less. Although such data do not lend themselves to close quantitative inspection, we feel that the faunas sampled near Inferno and Paraiso reveal important differences. Two quite different habitats, Cloud Forest and Tropical Evergreen Forest, are represented. It would be surprising if the faunas were qualitatively similar. *Peromyscus,* the dominant genus comprising 27 per cent of the Inferno deposits, is unrepresented at Paraiso. In turn, *Sigmodon,* which comprises 57 per cent of the material obtained at Paraiso, is absent from Inferno. Such a discrepancy may reflect an ecological shift in the dominant cricetine form. Other rather common species which appeared only in the Cloud Forest caves include: *Marmosa mexicana* (18%), *Cryptotis mexicana* (10%) and *Reithrodontomys mexicanus* (5%). One genus, *Neotoma,* occurs at both localities with a relatively constant frequency, 17–18 per cent.

Within the Gómez Farías region the following are known only from their skeletal remains: *Marmosa mexicana, Cryptotis pergracilis, Chilonycteris parnellii, Enchisthenes harti, Eptesicus fuscus, Lasiurus cinereus, Antrozous pallidus* and *Glaucomys volans.* Presence of the latter is perhaps of greatest interest. In Middle America flying squirrels are very poorly known, presumably the result of their nocturnal and arboreal habits rather than an inherent scarcity. We are aware of ten other locality records between Chihuahua and Honduras, each represented by one or two specimens. In the Gómez Farías region remains of *Glaucomys* appeared at each cave locality, suggesting a general range through humid forest, both Cloud Forest and Tropical Evergreen Forest, between 420 and 1,320 meters. To our knowledge *Glaucomys* has not previously been collected in lowland tropical forests (below 1,000 meters).

THE LOWLAND TROPICAL GRADIENT IN EASTERN MEXICO

We might imagine a lowland tropical fauna to decline with increasing latitude at a rather regular rate. However, in reality the environment and fauna undergo a series of discrete changes, stepwise (see Fig. 1). In eastern Mexico four major tropical lowland vegetation types are represented. From south to north they terminate in the following sequence: Rainforest, Tropical Evergreen Forest, Tropical Deciduous Forest and Thorn Forest (Leopold, 1950). Each terminus is marked by a steepening of the faunal gradient.

In defining the lowland tropical fauna we have excluded those genera with distributional centers in temperate montane habitats or of limited lowland tropical ranges, e.g., *Idionycteris, Neotoma, Sciurus, Sigmodon* and *Baiomys*. On the other hand, some of the species selected may range into montane habitats or reach temperate latitudes, such as *Marmosa mexicana* and *Didelphis marsupialis*. In general the species listed in Table 2 are of wide distribution in lowland tropical America. Although they occupy tropical environments of the Mexican escarpment and coastal plain, we exclude *Baiomys* and *Sigmodon* because their phylogeny indicates a north temperate origin and in the case of *Baiomys* because it does not range extensively into Central America. Clearly the question of "tropicality" can be vexing. As one might expect the bats have much more extensive ranges than the small terrestrial mammals. The faunas of Trinidad and southern Tamaulipas share 18 species of bats, but none of rodents.

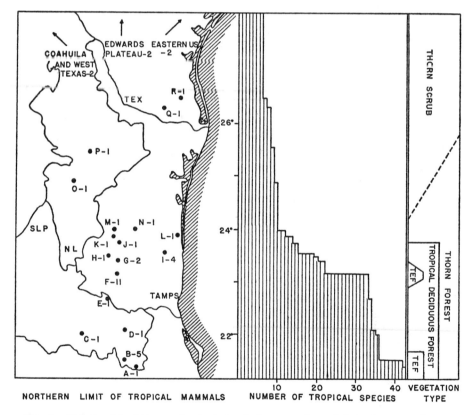

Fig. 1.—Relationship between lowland tropical mammals and latitude in northeastern Mexico. Known range limits for 43 species listed in Table 2 are shown on the map. Major vegetation types are indicated at the right. Tropical Evergreen Forest is abbreviated to TEF.

The general relationship between vegetation, latitude and tropical fauna is shown on Fig. 1. In southern Tamaulipas a rapid "thinning" of tropical forms is evident. Between 23° and 24° north latitude the following 17 genera find their range limits: *Philander, Marmosa, Chilonycteris, Pteronotus, Micronycteris, Macrotus, Glossophaga, Sturnira, Artibeus, Enchisthenes, Centurio, Natalus, Rhogeesa, Molossus, Heterogeomys, Eira* and *Mazama*. At this latitude (24° north) the Tropical Deciduous Forest, well developed and widespread east of the Sierra de Tamaulipas and the Sierra Madre Oriental, disappears. Small, perhaps relict, stands of Cloud Forest and Tropical Evergreen Forest in the Gómez Farías region also enrich the environmental opportunity for tropical mammals. However, these habitats, especially the Tropical Evergreen Forest, are more extensive in southeastern San Luis Potosí and northern Veracruz.

In this region seven genera terminate: *Carollia, Tamandua, Coendou, Cuniculus, Potos, Galictis* and *Ateles*. Although the vegetation near Xilitla, San Luis Potosí, has been designated as Rainforest, it seems preferable to reserve that term for the more luxuriant forests of southern Veracruz with their short dry season. Lowland tropical forests near Xilitla appear taller and richer than those of southern Tamaulipas; however, the area has suffered a long history of intensive Huastecan agriculture. Almost certainly the primeval fauna of southeastern San Luis Potosí included a larger number of tropical genera at their northern limit.

While we have not attempted to represent tropical distributions and plant formations south of San Luis Potosí, the following genera or subgenera approach their northern limit in southern Veracruz: *Caluromys, Vampyrum, Rhynchiscus, Centronycteris, Mimon, Chrotopterus, Hylonycteris, Chiroderma, Alouatta, Tylomys, Dasyprocta, Tayassu, Tapirella* and *Jentinkia*. Is there a relationship between these and the northern limit of Rainforest (see Leopold, 1950)?

Extending from central Tamaulipas northward to Nuevo León and southern Texas is a rather barren Thorn Forest and Thorn Scrub. Gradually these arid habitats lose their tropical character as the Rio Grande Valley is approached. By comparison with plant formations to the south, this environment is poor in tropical fauna. The shift from tropical to temperate thorn scrub involves no sharp faunal boundary among the mammals. *Oryzomys couesi* and *Liomys irroratus* are among the forms reaching southern Texas. At this latitude the herpetological fauna includes such tropical genera as *Coniophanes, Drymobius, Leptodeira, Smilisca* and *Hypopacus*. However, among the reptiles and amphibians, as well as the mammals, the greatest reduction in tropical fauna is found in southern Tamaulipas (Martin, 1958).

DISCUSSION

Although our analysis in confined to eastern Mexico, some interesting comparisons can be made with the tropical biota of the Pacific Coast. Arid tropical vegetation and the genera *Macrotus, Balantiopteryx, Chilonycteris, Pteronotus, Mormoops, Glossophaga, Desmodus, Natalus, Rhogeesa* and *Nasua* extend far-

TABLE 2.—*Northern limits of neotropical mammals*

LOCALITY AND APPROX. ELEV.	REFERENCE	SPECIES PRESENT
San Luis Potosí:		
A. Tamazunchale, 120 m.	Dalquest, 1953	1. *Tamandua tetradactyla*
B. Xilitla and vicinity	Dalquest, 1953	2. *Sturnira ludovici*
630–1350 m.		3. *Coendou mexicanum*
		4. *Cuniculus paca*
		5. *Potos flavus*
		6. *Galictis canaster*
C. Río Verde, 990 m.	Dalquest, 1953	7. *Molossus major*
D. Valles, 75 m.	Koopman, 1956	8. *Balantiopteryx plicata*
E. El Salto, 660 m.	Dalquest, 1953	9. *Carollia perspicillata*
Tamaulipas:		
F. Rancho del Cielo	Goodwin, 1954	10. *Marmosa mexicana*
and vicinity,	Hooper, 1953	11. *Enchisthenes harti*
1000–1320 m.	present report	12. *Oryzomys alfaroi*
		13. *Reithrodontomys*
		(*Aporodon*) *mexicanus*
		14. *Mazama americana*
Pano Ayuctle, 100 m.	Goodwin, 1954	15. *Micronycteris megalotis*
	Hooper, 1953	16. *Sturnira lilium*
	de la Torre, 1954	17. *Artibeus jamaicensis*
		18. *A. lituratus*
		19. *A. cinereus*
		20. *Eira barbara*
G. 2 km. W of El Carrizo,	Baker, 1951	21. *Philander opossum*
800 m.		22. *Heterogeomys hispidus*
H. Jaumave, 730 m.	UMMZ specimens	23. *Macrotus mexicanus*
I. 10–16 mi. WSW	Anderson, 1956	24. *Chilonycteris parnellii*
Piedra, 400 m.		25. *Glossophaga soricina*
		26. *Centurio senex*
		27. *Natalus mexicanus*
J. 2 mi. S Victoria,	Davis, 1951	28. *Molossus rufus*
400 m.		
K. 30 km. NW Victoria,	Málaga-Alba, 1954	29. *Diphylla ecaudata*
1000 m.		
L. La Pesca, 10 m.	Anderson, 1956	30. *Rhogeesa tumida*
M. Rancho Santa Rosa,	Anderson, 1956	31. *Pteronotus davyi*
260 m.		
N. 8 mi. SW Padilla,	Lawrence, 1947	32. *Oryzomys melanotis*
100 m.		
Nuevo León:		
O. 25 km. SW Linares,	Málaga-Alba, 1954	32. *Desmodus rotundus*
700 m.		
P. 20 mi. NW General	Hooper, 1947	33. *Oryzomys fulvescens*
Teran, 300 m.		
Texas:		
Q. Hidalgo Co., 60 m.	Blair, 1952*b*	34. *Oryzomys couesi*
R. Raymondville, 30 m.	Blair, 1952*b*	35. *Liomys irroratus*

TABLE 2.—*Continued*

LOCALITY AND APPROX. ELEV.	REFERENCE	SPECIES PRESENT
Coahuila and West Texas:	Baker, 1956	36. *Choeronycteris mexicana*
	Miller and	37. *Leptonycteris nivalis*
	Kellogg, 1955	
Edwards Plateau, Texas:	Blair, 1952*a*	38. *Mormoops megalophylla*
	Blair, 1952*b*	39. *Nasua narica*
Eastern United States:	Miller and	40. *Dasypus novemcinctus*
	Kellogg, 1955	41. *Didelphis marsupialis*

ther north on the western side. On the other hand the humid tropical fauna and plant formations, e.g., Cloud Forest, Tropical Evergreen Forest and Rainforest, are absent or poorly represented on the Pacific slope north of Chiapas. As a general rule for those species or vicariant species occurring on both sides of Mexico, the arid tropical forms range farther north on the west, and the humid tropical forms farther north on the eastern side. This pattern is evident also in the distribution of lowland tropical birds, reptiles, insects, etc.

Other than noting a rather close "fit," it is beyond our purpose to explore the causal relationship between formation and fauna. In brief our conclusions may be summarized as follows:

1. The decline of the tropical fauna in eastern Mexico corresponds with the vegetation gradient.

2. Where the vegetation gradient steepens and a plant formation is lost, one finds a variety of tropical animals at their range limits.

3. In southern Tamaulipas the northern limit of many tropical mammals corresponds roughly to the boundary of Tropical Deciduous Forest. Contributing to the rich tropical fauna of the Gómez Farías region are relict outposts of Tropical Evergreen Forest and Cloud Forest.

4. The problem of establishing a Nearctic-Neotropical faunal boundary in eastern Mexico can be approached realistically in terms of steps in an environmental gradient.

ACKNOWLEDGMENTS

First we wish to thank Dr. Byron E. Harrell for assistance in the field work. We are indebted to the Mammal Department of the American Museum of Natural History for use of their facilities including collections of comparative material. Mr. Sydney Anderson of the Museum of Natural History, University of Kansas, and Dr. William H. Burt, Museum of Zoology, University of Michigan, kindly advised us concerning specimens in their care.

LITERATURE CITED

ANDERSON, SYDNEY. 1956. Extensions of known ranges of Mexican bats. Univ. Kans. Publ. Mus. Nat. Hist., 9: 349–351.

BAKER, ROLLIN H. 1951. Mammals from Tamaulipas, Mexico. Univ. Kans. Publ. Mus. Nat. Hist., 5: 207–218.

————. 1956. Mammals of Coahuila, Mexico. Univ. Kans. Publ. Mus. Nat. Hist., 9: 125–335.

BLAIR, W. FRANK. 1952a. Bats of the Edwards Plateau in Central Texas. Tex. Jour. Sci., 4: 95–98.

————. 1952b. Mammals of the Tamaulipan Biotic Province in Texas. Tex. Jour. Sci., 4: 230–250.

DALQUEST, WALTER W. 1953. Mammals of the Mexican State of San Luis Potosi. La. State Univ. Studies, No. 1: 1–112.

DAVIS, WILLIAM B. 1944. Notes on Mexican mammals. Jour. Mamm., 25: 370–403.

————. 1951. Bat, *Molossus nigricans*, eaten by the rat snake, *Elaphe laeta*. Jour. Mamm., 32: 219.

GOODWIN, GEORGE G. 1954. Mammals from Mexico collected by Marian Martin for the American Museum of Natural History. Amer. Mus. Novit., No. 1689: 1–16.

HOOPER, EMMET T. 1947. Notes on Mexican mammals. Jour. Mamm., 28: 40–57.

————. 1953. Notes on mammals of Tamaulipas, Mexico. Occ. Pap. Mus. Zool., Univ. Mich., No. 544: 1–12.

KOOPMAN, KARL F. 1955. A new subspecies of *Chilonycteris* from the West Indies and a discussion of the mammals of La Gonave. Jour. Mamm., 36: 109–113.

————. 1956. Bats from San Luis Potosi with a new record for *Balantiopteryx plicata*. Jour. Mamm., 37: 547–548.

LAWRENCE, BARBARA. 1947. A new race of *Oryzomys* from Tamaulipas. Proc. New England Zool. Club, 24: 101–103.

LEOPOLD, A. STARKER. 1950. Vegetation zones of Mexico. Ecology, 31: 507–518.

MALÁGA ALBA, AURELIO. 1954. Vampire bat as carrier of rabies. Amer. Jour. Public Health, 44: 909–918.

MARTIN, PAUL S. 1958. Herpetology and biogeography of the Gómez Farías region, Mexico. Misc. Publ. Mus. Zool., Univ. Mich. In press.

MILLER, GERRIT S., JR. 1897. Revision of the North American bats of the family Vespertilionidae. N. Amer. Fauna 13: 1–155.

———— AND REMINGTON KELLOGG. 1955. List of North American Recent mammals. Bull. U.S. Nat. Mus., 205: 1–954.

ORR, ROBERT T. 1954. Natural history of the pallid bat, *Antrozous pallidus* (Le Conte). Proc. Calif. Acad. Sci., 28: 165–246.

TORRE, LUIS DE LA. 1954. Bats from southern Tamaulipas, Mexico. Jour. Mamm., 35: 113–116.

————. 1955. Bats from Guerrero, Jalisco, and Oaxaca, Mexico. Fieldiana: Zool., 37: 695–703.

Dept. of Biology, Queens College, Flushing, New York and Institut de Biologie, Universite de Montreal, Montreal, Canada. Received August 6, 1957.

Reprinted for private circulation from THE AMERICAN NATURALIST,
Vol. 105, No. 945, September-October 1971, pp. 467-478

MAMMALS ON MOUNTAINTOPS: NONEQUILIBRIUM INSULAR BIOGEOGRAPHY

James H. Brown[*]

Department of Zoology, University of California, Los Angeles, California 90024

INTRODUCTION

MacArthur and Wilson (1963, 1967) have provided a theoretical model to account for variation in the diversity of species on islands. This model, which attributes the number of species on an island to an equilibrium between rates of recurrent extinction and colonization, appears to account for the distribution of most kinds of animals and plants on oceanic islands. However, in addition to these islands, there are many other kinds of analogous habitats. Obvious examples are caves, desert oases, sphagnum bogs, and the boreal habitats of temperate and tropical mountaintops. It is of interest to ask whether the variables which determine the number of species on oceanic islands have similar effects on the biotas of other isolated habitats. For example, it has recently been shown that aquatic arthropods in caves (Culver 1970) and Andean birds in isolated paramo habitats (Vuilleumier 1970) are distributed as predicted by the equilibrium model of MacArthur and Wilson.

The boreal mammals of the Great Basin of North America provide excellent material for testing the generality of the equilibrium model. Almost all of Nevada and adjacent areas of Utah and California are covered by a vast sea of sagebrush desert, interrupted at irregular intervals by isolated mountain ranges. The cool, mesic habitats characteristic of the higher elevations in these ranges contain an assemblage of mammalian species derived from the boreal faunas of the major mountain ranges to the east (Rocky Mountains) and west (Sierra Nevada). Thanks largely to the work of Hall (1946), Durrant (1952), and Grinnell (1933), mammalian distributions within the Great Basin are documented quite thoroughly. I have used the work of these authors and data accumulated during 3 summers of my own field work in the Great Basin to produce the following analysis.

I shall show that the diversity and distribution of small mammals on the montane islands cannot be explained in terms of an equilibrium between colonization and extinction. Boreal mammals reached all of the islands during the Pleistocene; since then there have been extinctions but no colonizations.

* Present address: Department of Biology, University of Utah, Salt Lake City, Utah 84112.

467

METHODS

The Montane Islands

Because mountains do not have discrete boundaries, islands were defined by operational criteria applied to topographic maps (U.S. Geological Survey maps of the states; scale 1 : 500,000). A mountain range was considered an island if it contained at least one peak higher than 10,000 feet and was isolated from all other highland areas by a valley at least 5 miles across below an elevation 7,500 feet. This altitude corresponds approximately to the lower border of montane piñon-juniper woodland. Application of the above criteria defined 17 islands (fig. 1; table 1) which lie in a sea, the Great Basin, between two mainlands, the Sierra Nevada to the west and the central mountains of Utah (a part of the Rocky Mountains) to the east. The sizes (areas) of the islands and the distances between islands and between islands and mainlands have been determined from the topographic maps.

The Boreal Mammals

As with the montane islands, the mammals restricted in range to the higher altitudes must be defined somewhat arbitrarily. I have selected those

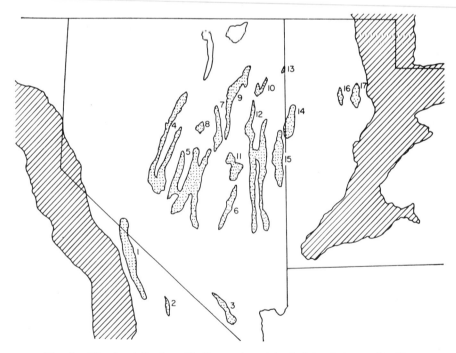

Fig. 1.—The Great Basin, with the montane islands lying between the Sierra Nevada (left) and Rocky Mountains (right). The shaded islands were used for the present analysis and are identified in table 1. The two unshaded islands were not used because they lie on the northern perimeter of the Great Basin and their faunas are poorly known.

TABLE 1
CHARACTERISTICS OF THE MOUNTAIN RANGES CONSIDERED AS ISLANDS

Mountain Range	Area above 7,500 Feet (Sq Miles)	Highest Peak (Ft)	Highest Pass* (Ft)	Nearest Island† (Miles)	Nearest Mainland (Miles)	Boreal Mammal Species (N)
1. White-Inyo	738	14,242	7,000	82	10	9
2. Panamint	47	11,045	5,500	19	52	1
3. Spring	125	11,918	3,500	108	125	3
4. Toiyabe	684	11,353	6,000	...	110	12
5. Toquima-Monitor	1,178	11,949	7,000	9	114	9
6. Grant	150	11,298	7,000	17	138	3
7. Diamond	159	10,614	7,000	7	190	4
8. Roberts Creek ...	52	10,133	7,000	22	216	4
9. Ruby	364	11,387	6,000	...	173	12
10. Spruce	49	10,262	6,500	12	156	3
11. White Pine	262	11,188	7,000	43	150	6
12. Schell Creek-Egan	1,020	11,883	7,000	11	114	7
13. Pilot	12	10,704	5,000	33	114	2
14. Deep Creek	223	12,101	7,000	9	104	6
15. Snake	417	13,063	7,000	76	89	8
16. Stansbury	56	11,031	6,000	4	39	3
17. Oquirrh	82	10,704	5,500	88	19	5

* Elevation of the highest pass separating the island from the mainland or from another island with more species.
† Distance to the nearest island with more species. No distance is given for the Toiyabe and Ruby Mountains because no other islands have more species.

species that occur only at high elevations in the Rockies and Sierra Nevada and are unlikely to be found below 7,500 feet at the latitudes of the Great Basin. I have excluded large carnivores and ungulates from the analysis because their distributions were drastically altered by human activity before accurate records were kept. I have also ignored the bats because their distributions are very poorly known and because their dispersal by flight introduces a completely new variable that could only complicate the present discussion. After these omissions, there remain 15 species of mammals (table 2) which occur in the Sierra Nevada and Rocky Mountains and on at least one of the montane islands of the Great Basin. All of these species occur in piñon-juniper, meadow, or riparian habitats. A striking feature of the mammalian faunas of the isolated peaks is the absence of those species which are restricted to dense forests of yellow pine, spruce, and fir in the Sierra Nevada and Rocky Mountains. None of these species, which include *Martes americana, Aplodontia rufa, Eutamias speciosus, Tamiasciurus hudsonicus, T. douglasi, Glaucomys sabrinus, Clethrionomys gapperi,* and *Lepus americanus,* are found on any of the isolated mountain ranges of the Great Basin, even though some of the large islands have large areas of apparently suitable habitat. However, the well-developed coniferous forests on the large islands have a good sample of the avian species characteristic of these habitats on the mainlands.

Records of occurrence are taken from the literature (Hall 1946; Hall and Kelson 1959; Durrant 1952; Durrant, Lee, and Hansen 1955; Grinnell 1933) and from my own observations which concentrated on the small mountain ranges. Undoubtedly, there are a few errors of omission that will

TABLE 2

CHARACTERISTICS AND DISTRIBUTION IN THE GREAT BASIN OF THE FIFTEEN SPECIES OF BOREAL MAMMALS CONSIDERED

Species	Body Weight (g)	Habitat and Diet	1	2	3	4	5	6	7	8	9	10	11	12	13	14	15	16	17	Total No. Islands Inhabited
*Sorex vagrans**	6.7	Carnivore	…	…	…	x	…	…	…	…	x	…	…	x	…	x	x	…	x	6
Sorex palustris	14	Carnivore	x	…	…	x	x	…	…	x	x	…	…	…	…	x	x	…	…	6
Mustela erminea	58	Carnivore	x	…	…	…	…	…	…	x	x	…	…	…	…	…	…	…	…	3
Marmota flaviventer	3,000	General herbivore	x	…	…	x	x	x	…	…	x	x	x	x	…	x	x	x	…	9
Spermophilus lateralis	147	General herbivore	x	…	x	x	x	x	x	…	x	x	x	x	x	x	x	x	…	13
Spermophilus beldingi	382	Meadow herbivore	…	…	x	x	…	…	…	…	x	…	…	…	…	…	…	…	…	3
Eutamias umbrinus†	57	General herbivore	…	…	…	x	x	x	x	x	x	x	x	x	x	x	x	x	x	14
Thamomys talpoides	102	Deep soil herbivore	x	…	x	x	x	x	x	…	x	x	x	x	…	x	x	…	x	8
Microtus longicaudus	47	General herbivore	x	…	…	x	x	x	…	…	x	x	x	x	x	x	x	x	…	12
Neotoma cinerea	317	General herbivore	x	…	x	x	x	…	x	…	x	x	x	x	x	x	x	x	…	14
Zapus princeps	33	Riparian herbivore	…	x	…	x	…	…	…	x	x	x	…	…	…	…	…	…	…	4
Ochotona princeps	121	Talus herbivore	x	…	…	x	x	…	…	…	x	…	…	…	x	…	…	…	x	4
Lepus townsendi	2,500	Meadow herbivore	…	…	…	…	…	…	…	…	…	…	…	…	…	…	…	…	…	1

* On island 17 the species is *Sorex obscurus*, a Rocky Mountain form, which forms a superspecies with *S. vagrans*.
† On island 3 the species is *Eutamias palmeri*, an endemic, which forms a superspecies with *E. umbrinus*.

be revealed by more intensive collecting, but these should not affect the present analysis significantly.

RESULTS AND DISCUSSION

Island Area and Extinction

The number of mammalian species inhabiting a montane island is closely correlated with the area of the island (fig. 2). When both variables are plotted logarithmically, the data are well described ($r = .82$) by a straight line with a slope (z) of .43. Four comparable areas on the mainland (Sierra Nevada) have more species and the species-area curve has much less slope ($z = .12$). These areas can be considered as "saturated" with suitable species. These relationships are similar to those described for various groups of animals and plants on oceanic islands and continental mainlands (Mac-Arthur and Wilson 1967; Johnson, Mason, and Raven 1968) and for birds on tropical montane islands in South America (Vuilleumier 1970), except that the z value describing the data for mammals on montane islands is

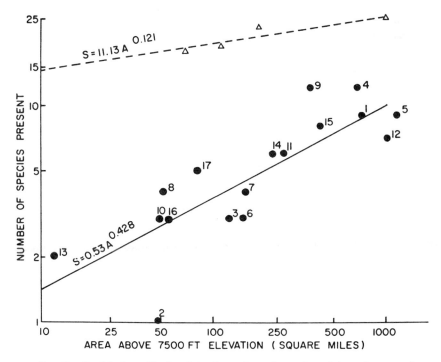

Fig. 2.—Double logarithmic plot of number of species of boreal mammals against area for the 17 montane islands (circles) listed in table 1. The solid line indicates the relationship between number of species (S) and area (A) fitted by least-squares regression. The dashed line represents the "saturation values" based on four areas (triangles) in the Sierra Nevada; data are from Hall (1946) and Grinnell (1933).

567

considerably higher than most of the values (.24–.34) previously obtained
for insular biotas.

Local areas acquire species by immigration and lose them by extinction.
In the absence of historical perturbations and speciation, the number of
species inhabiting an area will represent an equilibrium between the opposing rates of extinction and immigration. The biotas of most continental
areas and oceanic islands (including islands on the continental shelf) approximate such an equilibrium condition (Preston 1962a, 1962b; Mac-
Arthur and Wilson 1963, 1967; Diamond 1969; Simberloff and Wilson
1969). Rates of extinction are dependent largely on population size and,
in the absence of recurrent colonization, should be similar for comparable
areas of islands or mainlands; to the extent that the absence of competitors
results in greater population densities on islands, extinction rates should
be slightly lower for islands than for comparable areas on mainlands. Islands, because of their isolation, have much lower rates of immigration
(colonization) than local areas of comparable size within relatively homogeneous habitat on mainlands. The net result is that the slope of the species-
area curve should be related to the degree of isolation of the islands
concerned—the lower the rate of colonization, the higher the z value. The
fact that the z value obtained in the present study is higher than the range
observed for the biotas of oceanic islands (MacArthur and Wilson 1967)
indicates that boreal mammals have an exceptionally low rate of immigration to isolated mountains.

The population size (and, hence, probability of extinction on an island)
for a species is greatly affected by three variables—body size, trophic level,
and habitat specialization. These variables affect the distribution of mammalian species on montane islands in the manner we would expect from
considerations of their effects on population size: small mammals are found
on more islands than large ones, herbivores are better represented than
carnivores, and herbivores that can live in most montane habitats inhabit
more islands than herbivores which can live only in restricted habitats
(fig. 3). Those species which occur on only a few islands usually are found
only on large islands (see tables 1 and 2). Thus, at least for mammals on
montane islands, insular area not only affects the number of species but also
can be used to predict quite reliably some ecological attributes of those
species which are present.

Distance from the Mainland, Climatic Change, and Colonization

It is frequently observed that the number of species on an island is related to the proximity of the island to the nearest mainland as well as to
the size of the island. This suggests that the probability of colonization of
an island by new species from the mainland is proportional to the distance
involved and that recurrent colonization of islands occurs at a rate sufficient to offset partially the extinction of species. Using this sort of reason-

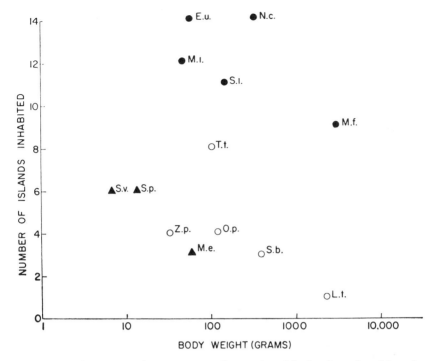

F<small>IG</small>. 3.—Frequency of occurrence on the montane islands of species of boreal mammals plotted against their body weight; shaded circles represent herbivores which are found in most habitats, unshaded circles indicate herbivores with specialized habitat requirements, triangles denote carnivores. The abbreviations are of species names which can be identified by reference to table 2.

ing, MacArthur and Wilson (1963) have proposed a model which represents the number of species on an island as a dynamic equilibrium between rates of colonization and extinction. The effect of recurrent colonization on species diversity is usually assessed by expressing the number of species on an island as a percentage of the number present in an area of the same size on the mainland and plotting this percentage saturation against the distance to the nearest mainland. When this is done for the montane mammals of the Great Basin, no relationship ($r = .005$) is observed (fig. 4). This is in marked contrast to the inverse correlation predicted by the equilibrium model and observed for various groups of organisms on oceanic islands (MacArthur and Wilson 1967) and for birds on montane islands (Vuilleumier 1970).

Before concluding that recurrent colonization has not been a significant factor in determining the diversity of mammals on the montane islands, it is necessary to exclude three alternative explanations: (1) Differences in habitat between islands may be sufficiently great to obscure the distance effect (see Diamond 1969). The montane islands of the Great Basin have similar vegetation, climate, and geomorphology; but they differ somewhat in elevation, which may affect the amount and diversity of boreal habitats

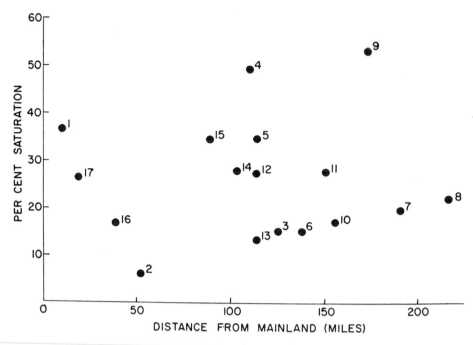

Fig. 4.—Percentage of faunal saturation plotted against distance from the Sierra Nevada or Rocky Mountains, whichever is nearer, for the 17 montane islands listed in table 1. Percentage of saturation is defined in the text.

in a manner different from insular area. (2) The islands may acquire most of their mammalian colonists from other islands (by a stepping-stone process) rather than directly from the nearest mainland, so that distance between islands might be an important variable affecting immigration. If this is so, the nearest island with more species is the most likely source of colonists if it is nearer than the closest mainland. (3) The effectiveness of the desert barriers surrounding the islands rather than the distance between mountains may be the most important variable influencing the rate of colonization. If this is true, the altitude of the highest pass should be the best measure available of the severity of the climatic and habitat barriers which separate the mountain ranges.

The possibility that any of these three explanations may account for some of the variability in insular-species diversity and provide evidence of the effects of immigration can be evaluated by subjecting the data in table 1 to stepwise multiple regression analysis. The results of such analysis using a linear model are shown in table 3. Again it is apparent that the number of mammalian species inhabiting an island is closely related to insular area, although the correlation is not as good as when a logarithmic model is used. Neither elevation of the highest peak nor any of the three variables (distance from nearest mainland, distance from nearest mountain with more species, elevation of highest intervening pass) which might be expected to

TABLE 3

INFLUENCE OF SEVERAL VARIABLES ON THE NUMBER OF SPECIES OF SMALL MAMMALS INHABITING MONTANE ISLANDS IN THE GREAT BASIN ANALYZED BY STEPWISE MULTIPLE REGRESSION USING LINEAR MODEL[a]

Variable	Contribution to R^2	F Value	Order Entered in Equation
Area	.49421	14.65*	1
Highest peak	.00006	0.32	2
Nearest mainland	.00042	0.83	3
Nearest island	.00002	0.16	4
Highest pass	.00000	0.04	5
Total R^2	.49471	...	...

[a] Data are from table 1.
* Significant at 1% level; the other F values are not significant.

influence immigration rate contributes significantly to the reduction of the remaining variability in the number of insular species. Apparently, at the present time the rate of colonization of the islands is effectively zero. The desert valleys of the Great Basin are virtually absolute barriers to dispersal by small boreal mammals.

I conclude that the boreal mammalian faunas of the montane islands of the Great Basin do not represent equilibria between recurrent rates of colonization and extinction. This is somewhat surprising, because conditions of dynamic equilibria with surprisingly high and measurable steady state turnover rates have been documented for some oceanic islands (see Mac-Arthur and Wilson 1967, on Krakatau; Simberloff and Wilson 1969, Diamond 1969). However, my explanation for the distribution of boreal mammals on the montane islands of the Great Basin—that all the islands were inhabited by a common pool of species at some time in the past and subsequent extinctions have reduced the number of species on individual islands to their present levels—is consistent with what we know of the paleoclimatic history of the Southwest and the way small mammals disperse over land.

At intervals during the Pleistocene, colder climates in the Southwest forced piñon-juniper woodland down to elevations approximately 2,000 feet below their present distribution (Wells and Berger 1967). This was sufficient to make piñon-juniper and associated streamside and meadow habitats contiguous across most of the Great Basin perhaps as recently as 8,000 years ago. The climatic and habitat barriers between the isolated peaks and the mainlands were removed for an entire set of species; these paleoclimatic and habitat shifts are sufficient to account for all 15 species of boreal mammals now found on the montane islands. Two lines of evidence indicate that the islands were colonized in this manner. First, the limited fossil evidence indicates that some of the boreal mammals that are now confined to only a few of the islands were more widely distributed in the late Pleistocene; for example, Wells and Jorgensen (1964) found fossils of *Marmota flaviventer* in the Spring Range in southern Nevada. Second, paleobotanical evidence indicates that the climate changes during the Pleistocene were not sufficient to connect the dense coniferous forests of yellow pine, spruce, and fir between the islands and the mainlands. Those mammalian species which are restricted to such forests in the Sierra Nevada and Rocky Mountains are absent from all of the isolated peaks in the Great Basin, even though there are large areas of suitable habitat on several of the larger islands. It is apparent that the islands have been colonized only during those periods in the past when the climatic and habitat barriers which now isolate them were temporarily abolished.

A few thousand feet of elevation, with the associated differences in climate and habitat, constitute a nearly absolute barrier to dispersal by small mammals (with the exception of bats). This is not surprising in view of the fact that small mammals must disperse over land on foot. It may take several days for an individual to travel only a few miles, and during this period it must obtain food and avoid predation in an unfamiliar habitat and survive the stresses imposed by a different climate.

GENERAL DISCUSSION

The mammalian faunas of the isolated mountains in the Great Basin do not represent equilibria between rates of extinction and colonization. This is shown by the unusually steep slope of the species-area curve and by the lack of correlation between the percentage saturation of the insular faunas and any of the variables (distance to mainland, distance between islands, and elevation of surrounding valleys) likely to influence the probability of immigration to an island. The diversity of mammalian species on the mountaintops must be explained by historical events (climatic changes during the Pleistocene) which temporarily abolished the isolation of the mountains and permitted their colonization by a group of mammalian species. Subsequent extinctions, related to island size, have reduced the faunas of each island but not sufficiently to restore equilibrium with the vanishingly small rates of colonization. Approximately 8,000 years after their isolation, the largest islands still have almost all of their original species of boreal mammals, and even on the smallest islands one or more species remain. This suggests that the rate of extinction is very low once the biota of the island has initially adjusted to its isolation; when colonization rates approach zero, extinction of all species is reached only in geological time spans.

Island biotas which do not represent equilibria between rates of extinction and colonization should be fairly common, particularly in the temperate zones where the climatic fluctuations of the Pleistocene have drastically altered the distribution of terrestrial and freshwater habitats. Some organisms are capable of crossing the barriers between isolated habitats, and these may be expected to have distributions predicted by the equilibrium model. Examples are the birds, bats, flying insects, and most kinds of plants on montane islands. However, other groups of organisms are unable to cross the barriers between insular habitats, and these should be distributed as relicts in a nonequilibrium pattern similar to the montane mammals described here. Likely examples are amphibians, reptiles, and large, nonflying arthropods on mountaintops and fish is isolated bodies of fresh water. The fish fauna of the isolated springs and streams of the Great Basin clearly does not represent an equilibrium situation; these habitats were colonized in the Pleistocene when there were aquatic connections between them, and extinctions have subsequently reduced the fauna of each isolate to its present composition (Hubbs and Miller 1948). Certainly, other nonequilibrium patterns of island distributions will be described. In those cases where environmental changes (often the result of human activity) have caused massive extinctions, the number of species on an island will be less than the equilibrium number.

SUMMARY

An analysis of the distribution of the small boreal mammals (excluding bats) on isolated mountaintops in the Great Basin led to the following conclusions:

1. The species-area curve is considerably steeper ($z = .43$) than the curves usually obtained for insular biotas.

2. There is no correlation between number of species of boreal mammals and variables which are likely to affect the probability of colonization, such as distance between island and mainland, distance between islands, and elevation of intervening passes. Apparently the present rate of immigration of boreal mammals to isolated mountains is effectively zero.

3. Paleontological evidence suggests that the mountains were colonized by a group of species during the Pleistocene when the climatic barriers that currently isolate them were abolished.

4. Subsequent to isolation of the mountains, extinctions have reduced the faunal diversity to present levels. Probability of extinction is inversely related to population size and, therefore, is influenced by body size, diet, and habitat. The rate of extinction has been low, and all of the islands still have one or more species of boreal mammals.

5. The mammalian faunas of the mountaintops are true relicts and do not represent equilibria between rates of colonization and extinction.

ACKNOWLEDGMENTS

My wife, Astrid, assisted with all aspects of the study. Dr. F. Vuilleumier made many helpful comments on the manuscript and performed the multiple regression analysis shown in table 3. Drs. G. A. Bartholomew, M. L. Cody, J. M. Diamond, and D. E. Landenberger kindly read and criticized the manuscript. The work was supported in part by a grant (GB 8765) from the National Science Foundation.

LITERATURE CITED

Culver, D. C. 1970. Analysis of simple cave communities. I. Caves as islands. Evolution 29:463–474.

Diamond, J. M. 1969. Avifaunal equilibria and species turnover rates on the Channel Islands of California. Nat. Acad. Sci. (U.S.), Proc. 64:57–63.

Durrant, S. D. 1952. Mammals of Utah, taxonomy and distribution. Univ. Kansas Publ. Mus. Natur. Hist. 6:1–549.

Durrant, S. D., M. R. Lee, and R. M. Hansen. 1955. Additional records and extensions of known ranges of mammals from Utah. Univ. Kansas Publ. Mus. Natur. Hist. 9:69–80.

Grinnell, J. 1933. Review of the recent mammal fauna of California. Univ. California Publ. Zool. 40:71–234.

Hall, E. R. 1946. Mammals of Nevada. Univ. California Press, Berkeley. 710 p.

Hall, E. R., and K. R. Kelson. 1959. The mammals of North America. Ronald, New York. 1083 p.

Hubbs, C. L., and R. R. Miller. 1948. Correlation between fish distribution and hydrographic history in the desert basins of western United States. In The Great Basin, with emphasis on glacial and postglacial times. Bull. Univ. Utah Biol. Ser. 38:20–166.

Johnson, M. P., L. G. Mason, and P. H. Raven. 1968. Ecological parameters and plant species diversity. Amer. Natur. 102:297–306.

MacArthur, R. H., and E. O. Wilson. 1963. An equilibrium theory of insular zoogeography. Evolution 17:373–387.

———. 1967. The theory of island biogeography. Princeton Univ. Press, Princeton, N.J. 203 p.

Preston, F. W. 1962a. The canonical distribution of commonness and rarity. I. Ecology 43:185–215.

———. 1962b. The canonical distribution of commonness and rarity. II. Ecology 43:410–432.

Simberloff, D. S., and E. O. Wilson. 1969. Experimental zoogeography of islands. The colonization of empty islands. Ecology 50:278–296.

Vuilleumier, F. 1970. Insular biogeography in continental regions. I. The northern Andes of South America. Amer. Natur. 104:373–388.

Wells, P. V., and R. Berger. 1967. Late Pleistocene history of coniferous woodland in the Mojave Desert. Science 155:1640–1647.

Wells, P. V., and C. D. Jorgensen. 1964. Pleistocene wood rat middens and climatic change in Mojave Desert: a record of juniper woodlands. Science 143:1171–1173.

ZOOGEOGRAPHY OF THE MONTANE MAMMALS OF COLORADO

By James S. Findley and Sydney Anderson

In Colorado the distribution of montane or boreal habitats is at present closely associated with the local climate produced by the mountains. Peculiarities of this habitat are high precipitation, both in winter and summer, cool temperatures, a continuous water supply and a coniferous forest. Certain mammals are more or less restricted in geographic range, in this part of the continent, to the mountains.

In Pleistocene time the distribution of boreal habitat and hence of boreal mammals has undoubtedly fluctuated widely with the advances and retreats of continental and alpine glaciers. The contemporary pattern of distribution is, at least in part, a result of the most recent major glacial advance and retreat which took place in the Wisconsinan Age. It seems probable to us that most contemporary subspecies have differentiated in late Pleistocene time; otherwise the frequent correspondence of their ranges with current topographical and ecological features, which stem from late Pleistocene events in many cases, seems inexplicable. In the western United States the Boreal Zone is found at higher and higher elevations as one proceeds southward until it is scattered on isolated mountain peaks. The presence of isolated populations of boreal mammals on some of these mountains is evidence of a former displacement southward and downward in altitude of the Boreal Zone in a glacial age, presumably the Wisconsin, and subsequent elevation of the Boreal Zone in altitude and latitude in an ensuing interglacial interval, presumably the Recent. These southern, marginal populations would have been the first to become isolated with the retreat of the ice.

The separation of boreal habitat in the mountains of Colorado from boreal habitat in the Uinta and Wasatch Mountains of Utah and the mountains of northwestern Wyoming is probably of later origin than is the isolation of the southern boreal "islands." We have studied the boreal mammals of Colorado in their relation to those of Utah and northwestern Wyoming. These mammals may be grouped according to the pattern of their variation and distribution as follows:

Group I.—Rare, extinct, or insufficiently known to use in this study: *Alces americana*, *Ovis canadensis*, *Lepus americanus*, *Sylvilagus nuttallii*, *Phenacomys intermedius*, *Mustela erminea*, and *Gulo luscus*.

Group II.—Occurring only north and west of the barrier formed by the Wyoming Basin and the Green River (Fig. 1): *Eutamias amoenus*, *Glaucomys sabrinus*, *Microtus richardsoni*, and *Martes pennanti*.

Group III.—Occurring only southeast of the above mentioned barrier: *Sciurus aberti*.

Group IV.—Occurring in the mountains of northwestern Wyoming and the mountains of Colorado as a single subspecies; this group includes eight of fifteen species that occur on both sides of the barrier shown in Figure 1: *Sorex cinereus*, *Sorex vagrans*, *Sorex palustris*, *Clethrionomys gapperi*, *Microtus montanus*, *Microtus longicaudus*, *Zapus princeps*, and *Erethizon dorsatum*.

Group V.—Occurring in the mountains north of the Wyoming Basin and the mountains southeast of the basin, but as different subspecies: *Martes americana*,

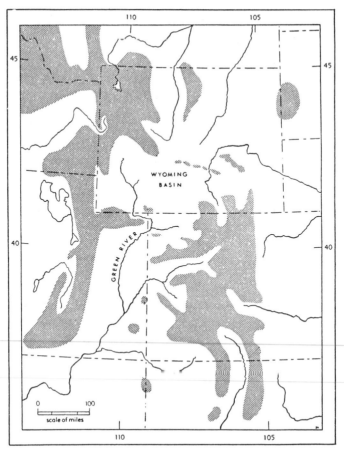

Fig. 1.—The distribution of the Boreal Zone (diagonally lined) in Wyoming, Colorado, and Utah. The major barrier (consisting of the Wyoming Basin and the Green River) separating the boreal habitat in Colorado from the mountains of Utah and northwestern Wyoming is shown.

Marmota flaviventris, Citellus lateralis, Eutamias umbrinus, Tamiasciurus hudsonicus, Microtus pennsylvanicus, and *Ochotona princeps.*

In Figure 1 we have mapped the distribution of boreal habitat and the barrier discussed. We note that the arboreal species, namely, the two tree squirrels, the flying squirrel, the marten, and the fisher, occur either on one side of the barrier only or else have distinct northern and southeastern subspecies. The other species that occur only on one side of the barrier or that have separate subspecies on the north side and on the southeast side of the Wyoming Basin are: *Marmota flaviventris, Ochotona princeps, Microtus richardsoni, Microtus pennsylvanicus, Citellus lateralis,* and two species of *Eutamias.* The species named immediately above and the arboreal species (in comparison to the next group of species to be discussed) seem to be relatively restricted in the range of habitats that they utilize. The pattern in the chipmunks is complicated by other species of less montane chipmunks whose presence may act as a biological barrier. The red squirrel, the marten, and the golden-mantled ground squirrel have populations in Colorado and northern Utah that are alike and differ from corresponding populations on the northern side of the Wyoming Basin.

The species that have a single subspecies occurring on both the north side and the southeast side of the Wyoming Basin are as follows: three species of shrews, three microtines, the jumping mouse, and the porcupine. The porcupine is a ubiquitous creature, prone to wander. The other seven species are small mammals which may migrate by way of narrow avenues found along streamcourses where the water draining from montane areas supports growths of brush, scrub willow, and grasses and sedges. Furthermore, these species do not seem to be dependent upon forests or forest-edge communities. If the montane mammals here dealt with are arranged in order of their decreasing dependence upon montane conditions it is seen that those kinds appearing early in the list are those that occur only on one side of the barrier shown in Figure 1 or those that have distinct northern and southeastern subspecies (that is, belong in Group V). Those that appear later in such a list are in general those that are subspecifically the same north and southeast of the Wyoming Basin (Group IV). It might be concluded that montane meadow and streamside habitats connected the southern and central Rockies across the Wyoming Basin long after they ceased to be connected by continuous forests. The red-backed mouse, *Clethrionomys*, is restricted to the forested areas of the mountains more than the other species. Investigation of the most recent work on *Clethrionomys* (Cockrum and Fitch, Univ. Kansas Publ., Mus. Nat. Hist., 5: 283, 1952), reveals that these authors regarded northwestern Wyoming and the Bighorn Mountains as centers of incipient subspecies. Judging by their comments on *C. gapperi galei* and on *C. g. uintaensis* in Utah we feel that the four populations, (1) in the Uinta Mountains, (2) in northwestern Wyoming, (3) in the Bighorn Mountains, and (4) in northern Colorado, are differentiated from one another and might be regarded equally well as one or as four subspecies. The latter supposition would place them in Group V and would obviate the seeming inconsistency.

On the basis of the information presented above it seems that: (1) The ranges of montane species are correlated with their dependence upon special habitats; the more dependent species are more restricted in range, both locally and regionally. (2) The more dependent, and therefore relatively restricted, species show more differentiation on opposite sides of the Wyoming Basin than the species that are less restricted. (3) The closest affinities of the boreal mammals in the Rocky Mountains of Colorado are with the boreal mammals of the Uinta Mountains across the Green River Canyon rather than with those of the central Rocky Mountains to the north of the Wyoming Basin. (4) The discontinuity in the boreal forest produced by the erosion of the Green River Canyon has become important as a barrier to montane mammals later than the discontinuity caused by the desiccation of the Wyoming Basin.

Department of Zoology, University of New Mexico, Albuquerque, and Museum of Natural History, University of Kansas, Lawrence. Received February 6, 1955.

COMPARISON OF POCKET GOPHERS FROM ALPINE, SUBALPINE AND SHRUB–GRASSLAND HABITATS

We compared the characteristics of pocket gophers, *Thomomys talpoides*, in three contrasting environments within the Cochetopa Creek Drainage in southwestern Colorado. Sampling was done: (1) at Baldy Chato, above timberline in the alpine habitat, at about 12,000 feet elevation; (2) within drainages of Chavez and Pauline creeks, in moist subalpine meadows, at about 10,800 feet elevation; (3) at Cochetopa Park, in arid shrub–bunchgrass valleys, at about 9,300 feet elevation. The distance between the alpine and shrub–grassland areas was about 20 miles, and the subalpine was intermediate.

In mid-August 1961 and 1962 pocket gophers were deadtrapped until about 30 adult females were secured from each area. The number of animals in sex–age-classes depended upon how many were taken with the desired sample of 30 adult females. We do not think the age–sex ratios obtained represented the true ratios as they occur in nature. Traps were set only where gophers were found to be active. Active burrow systems were recognized by the conspicuous mounds of fresh earth gophers push to the surface. Sex and age were determined as reported by Hansen (J. Mamm., 41: 323–335, 1960). Our observations while

TABLE 1.—*Size and reproductive characteristics of male* Thomomys talpoides *from Cochetopa Creek drainage (mean ± SE)*

	Number	Body weight (g)	Total length (mm)	Testes length (mm)	Baculum length (mm)
Adult Males					
Alpine tundra					
1961	14	114 ± 2.9	214 ± 2.1	12.4 ± 0.5	22.1 ± 0.3
1962	4	130 ± 6.8	219 ± 5.1	14.7 ± 0.8	23.2 ± 0.6
Subalpine meadow					
1961	8	134 ± 6.5	219 ± 4.4	12.2 ± 0.4	22.8 ± 1.2
1962	9	129 ± 5.1	214 ± 2.1	12.2 ± 0.4	23.0 ± 0.3
Shrub–bunchgrass					
1961	8	116 ± 4.4	211 ± 3.0	12.2 ± 0.7	21.8 ± 0.5
1962	10	117 ± 4.5	215 ± 1.9	10.6 ± 0.3	22.4 ± 0.5
Young Males					
Alpine tundra					
1961	7	76 ± 7.4	188 ± 5.9	4.3 ± 0.2	10.4 ± 0.9
1962	3	66 —	179 —	4.1 —	9.8 —
Subalpine meadow					
1961	22	73 ± 3.2	188 ± 3.0	4.2 ± 0.6	11.0 ± 0.3
1962	31	92 ± 2.8	202 ± 2.0	4.3 ± 0.4	12.0 ± 0.3
Shrub–bunchgrass					
1961	13	76 ± 3.1	193 ± 2.4	4.1 ± 0.1	11.3 ± 0.4
1962	9	94 ± 5.3	206 ± 3.9	4.5 ± 0.2	12.3 ± 0.6

trapping suggested that gopher population densities were highest in the subalpine meadows, lowest in the shrub–grassland parks and intermediate densities occurred in the alpine area.

The characteristics of gophers from the alpine, subalpine and shrub–grasslands were similar (Tables 1 and 2). The overall differences in size for adults from one area to another were no greater than the year-to-year differences of adults from a given area. Adult males were larger than adult females from all areas and about the same degree of sexual dimorphism was found as was reported by Hansen (*loc. cit.*) for *T. talpoides* from a lower elevation at Livermore, Colorado. In the Great Basin of western North America where one species of pocket gopher occupies both the valley and adjacent mountains there is no sexual dimorphism in gophers at high altitudes and pronounced sexual dimorphism in gophers at low altitudes (Davis, J. Mamm., 19: 338–342, 1938; Hall, Mammals of Nevada, Univ. Calif. Press, Berkeley, 710 pp., 1946). The observed differences in size of adults in this study were not related to altitude but were probably influenced by the age structure of the adult population and nutritional values of gopher foods.

Young gophers were larger in 1962 than 1961 in the subalpine and shrub–grassland samples but were about the same size in both years in the alpine area. Mean size in young probably was related to mean age of young.

There were no pregnant females taken in the shrub–grasslands samples and only one female out of 66 females was pregnant in those from the subalpine. However, in the alpine sample, 4 of 34 adult females had embryos in 1961, and in 1962, 6 of 33 females had embryos. The low frequency of young in the alpine sample also suggests most young

TABLE 2.—*Size and reproductive characteristics of female* Thomomys talpoides *from Cochetopa Creek drainage (mean ± SE)*

	Number	Body weight (g)	Total length (mm)	Ovary length (mm)	Litter size
Adult Females					
Alpine tundra					
1961	34	105 ± 2.1	209 ± 1.6	3.6 ± 0.2	4.8 ± 0.4
1962	33	112 ± 2.3	212 ± 1.7	3.5 ± 0.1	4.3 ± 0.2
Subalpine meadow					
1961	29	108 ± 2.2	210 ± 1.8	3.2 ± 0.1	4.5 ± 0.2
1962	37	127 ± 1.1	217 ± 2.0	3.5 ± 0.1	4.3 ± 0.3
Shrub–bunchgrass					
1961	30	105 ± 2.2	209 ± 1.7	3.4 ± 0.1	5.2 ± 0.3
1962	20	104 ± 2.8	207 ± 2.2	3.2 ± 0.1	4.5 ± 0.3
Young Females					
Alpine tundra					
1961	10	82 ± 3.0	198 ± 2.7	2.3 ± 0.3	
1962	4	84 ± 5.5	199 ± 2.7	2.5 ± 0.4	
Subalpine meadow					
1961	25	71 ± 2.5	188 ± 2.3	2.0 ± 0.1	
1962	29	89 ± 2.4	199 ± 2.2	2.4 ± 0.1	
Shrub–bunchgrass					
1961	17	70 ± 3.7	187 ± 3.8	2.3 ± 0.1	
1962	28	87 ± 1.9	199 ± 1.5	2.2 ± 0.1	

were too small to be readily trapped. The mean number of young per litter was similar in all samples. These data suggest that the reproductive seasons are similar and extensively overlap but the mean date of breeding is probably later for gophers inhabiting the alpine habitat.

This study was aided by National Science Foundation Grants G-8907 and G-6152.— RICHARD M. HANSEN AND GEORGE D. BEAR, *Colorado Agricultural Experiment Station, Fort Collins, Colorado. Received 16 May 1963.*

BAT FAUNAS: A TROPHIC COMPARISON

Don E. Wilson

Abstract

Wilson, Don E. (*Bird and Mammal Laboratories, National Museum of Natural History, Washington, D. C. 20560*) 1973. Bat faunas: a trophic comparison. *Syst. Zool.* 22:14–29.—The bat faunas of the six zoogeographic regions are compared using taxonomic and trophic analyses. Similarity indices are calculated using the percent of total species per area for each genus and the importance of various trophic roles in the areas. Trophic role values were calculated by multiplying generic importance values by indices for the importance of applicable trophic roles for each genus and summing the values for every genus in a given region. Faunal regions are much more similar trophically than taxonomically. Although geographic patterns are discernable from taxonomic approaches, climatic patterns are more important trophically. All of the trophic roles are highly variable from region to region, but foliage gleaning is the least variable. The most important trophic role in all regions is that of the aerial insectivores, followed in order of decreasing importance by frugivory, foliage gleaning, nectarivory, piscivory, carnivory and sanguinivory. [Trophic levels; faunal regimes; bats.]

Lein (1972) recently analyzed the zoogeography of birds utilizing a quantitative approach based on trophic rather than taxonomic groups. His results spurred me to attempt a similar analysis using bats, both for comparative purposes and for the intrinsic value to students of zoogeography in general and of bat biology in particular. Many of the problems involved in such an analysis are magnified when dealing with bats, owing to our lack of knowledge of their basic life histories. On the other hand, I have attempted a finer grained approach which has hopefully alleviated some of these problems by dealing with genera rather than families.

Traditional taxonomic approaches to bat zoogeography were virtually nonexistent until a recent paper by Koopman (1970). Koopman's paper was quite useful in summarizing distribution patterns and laid much of the groundwork for the present analysis. McNab (1971) examined the relationship between food habits and structure of tropical bat faunas. I have attempted to fit all of the genera of bats into seven basic feeding categories and analyze the distribution of these feeding types and the bats which utilize them throughout the world. Five of the feeding categories represent different food sources, i.e., verte-

brates, fish, blood, fruit, and nectar. I have subdivided the food source insects into the feeding categories of aerial insectivores and foliage gleaners. Further subdivision of the categories would be useful, but is impossible at the present time due to the paucity of our knowledge of the food habits of bats.

The underlying concepts for the present type of analysis are those of convergent evolution, ecological equivalence, and functional niches. Modern biology is moving from a descriptive phase to a synthesizing one. Ecologists are anxious to find unifying principles which are applicable over broad areas. In short, we are struggling towards the perfect compromise of maximum generalization with minimum information loss.

The recent blossoming of interest in, and information on, tropical biology has led to a variety of approaches pointing out the similarities and disparities between tropical and temperate regions of the earth. Taxonomic approaches to zoogeography normally emphasize and support the role of past history in current distributional patterns. Ecological approaches tend to point out the importance of present environmental conditions. Both approaches are valuable, and I hope that the present paper will encourage further work in both directions, as well as towards a unified end product.

14

METHODS AND CALCULATIONS

The six faunal zoogeographic regions of the world as they have been accepted by recent workers (Lein, 1972; Koopman, 1970; Darlington, 1957) are as follows:

1. Ethiopian region: Subsaharan Africa, Madagascar, and Southern Arabia.
2. Oriental (Indo-Malayan) region: Tropical Asia and associated continental islands.
3. Palearctic region: Temperate Eurasia and North Africa.
4. Nearctic region: North America excepting tropical Mexico.
5. Neotropical region: South and Central America and tropical Mexico.
6. Australian region: Australia, New Guinea and associated islands.

I have used the 169 genera of bats recognized by Koopman and Jones (1970). Koopman (1970) provides a useful summary of the geographic distribution of these genera, and I have used his compilation as a basis for the calculations.

The following methodology is basically that of Lein (1972). A generic importance index was calculated for each genus in each region by determining the percentage of the total species for that region. For example, *Eidolon* has one species in the Ethiopian region, which has a total of 188 species. Therefore the generic importance index for *Eidolon* in the Ethiopian region is 0.5. Similarity indices for all possible pairs of regions were obtained by summing the common percentages of generic importance for the two regions being compared. For example, *Rousettus* comprises 1.6 percent of the Ethiopian fauna and 1.4 percent of the Palearctic fauna. Since the shared value equals the lower of the two, *Rousettus* yields a contribution of 1.4 to the common generic importance. So the common generic importance is the sum of the shared values for all genera common to both regions. Lein (1972) has pointed out that variations in range and abundance of species are not accounted for by this method. A rare, localized species receives the same weight-

ing as an abundant, widespread species. A possible approach to the distribution problem would be to develop indices based on area covered for each genus. A lack of data on population sizes precludes adjusting for abundance at present.

The seven feeding categories or trophic roles for this study are:

1. Carnivores (feeding on tetrapods)
2. Piscivores (feeding on fish)
3. Sanguinivores (feeding on blood)
4. Foliage gleaners (feeding on insects on foliage or the ground)
5. Aerial insectivores (feeding on insects in the air)
6. Frugivores (feeding on fruit)
7. Nectarivores (feeding on flowers, nectar, and pollen)

The problem of partitioning genera into these several categories was met by assigning indices to each genus for each applicable trophic role. The indices were assigned in tenths with a total of one for each genus. For example, *Myonycteris* rated a 1 for the Frugivore role, and *Pteropus* was assigned 0.9 for Frugivore and 0.1 for the Nectarivore role. These divisions were done on the basis of accounts in the literature (Walker, 1968; Goodwin and Greenhall, 1961; Barbour and Davis, 1969; Allen, 1939; Valdez and LaVal, 1971; Medway, 1967; Lim, 1966, 1970; Jeanne, 1970; Ross, 1961, 1967; Orr, 1954; Storer, 1926; Grinnell, 1918; Hatt, 1923; Borell, 1942; Brosset and Deboutteville, 1966; Hoffmeister and Goodpaster, 1954; Arata, et al., 1967; Carvalho, 1961; Silva and Pine, 1969; Dwyer, 1962; Dunn, 1933; Pine, 1969; Wilson, 1971; Tan, 1965; Tuttle, 1967, 1968; Hooper and Brown, 1968; Greenhall, 1966, 1968; Constantine, 1966; Huey, 1925; Davis, et al., 1964; Poulton, 1929; Ryberg, 1947; Howell, 1919; Sherman, 1935; Fleming, et al., 1972), personal familiarity in the case of many New World species, and brain-picking of colleagues for Old World forms. This experience has impressed upon me just how lacking we are in basic life history data for the order Chiroptera.

TABLE 1. TROPHIC ROLE VALUES ASSIGNED TO EACH GENUS.

	TROPHIC ROLE						
GENERA	Carn	Fish	Blood	F.G.	A.I.	Fruit	Nectar
Eidolon						.9	.1
Rousettus						.5	.5
Myonycteris						1.0	
Boneia						1.0	
Pteropus						.9	.1
Acerodon						1.0	
Neopteryx						1.0	
Pteralopex						1.0	
Styloctenium						1.0	
Dobsonia						.8	.2
Harpyionycteris						1.0	
Plerotes						.7	.3
Hypsignathus						.8	.2
Epomops						.9	.1
Epomophorus						.5	.5
Micropteropus						.8	.2
Nanonycteris							1.0
Scotonycteris						.9	.1
Casinycteris						1.0	
Cynopterus				.1		.7	.2
Megaerops						1.0	
Ptenochirus						1.0	
Dyacopterus						1.0	
Chironax						1.0	
Thoopterus						1.0	
Sphaerias						1.0	
Balionycteris						1.0	
Aethalops						1.0	
Penthetor				.1		.9	
Haplonycteris						1.0	
Nyctimene					.5	.5	
Paranyctimene						1.0	
Eonycteris							1.0
Megaloglossus							1.0
Macroglossus						.5	.5
Syconycteris						.5	.5
Melonycteris						.5	.5
Notopteris					.	.5	.5
Rhinopoma					1.0		
Emballonura					.9	.1	
Coleura					1.0		
Rhynchonycteris					1.0		
Saccopteryx					1.0		
Centronycteris					1.0		
Peropteryx					1.0		
Cormura					1.0		
Balantiopteryx					1.0		
Taphozous					1.0		
Cyttarops					1.0		
Depanycteris					1.0		
Diclidurus					1.0		
Noctilio	.5				.5		
Nycteris				.9	.1		

TABLE 1. (Continued)

GENERA	TROPHIC ROLE						
	Carn	Fish	Blood	F.G.	A.I.	Fruit	Nectar
Megaderma	.8	.1		.1			
Macroderma	.7			.3			
Cardioderma	.5			.5			
Lavia				.6	.4		
Rhinolophus				.5	.5		
Rhinomegalophus				.5	.5		
Hipposideros				.5	.5		
Anthops					1.0		
Asellia					1.0		
Aselliscus					1.0		
Cloeotis					1.0		
Rhinonicteris					1.0		
Triaenops					1.0		
Coelops					1.0		
Paracoelops					1.0		
Pteronotus					1.0		
Mormoops					1.0		
Micronycteris				.5		.5	
Macrotus				.3	.5	.2	
Lonchorhina				1.0			
Macrophyllum				1.0			
Tonatia				.5		.5	
Mimon				1.0			
Phyllostomus	.2			.3		.3	.2
Phylloderma				1.0			
Trachops	1.0						
Chrotopterus	1.0						
Vampyrum	.8					.2	
Glossophaga				.1		.1	.8
Lionycteris							1.0
Lonchophylla				.1		.1	.8
Platalina							1.0
Monophyllus				.1		.1	.8
Anoura						.5	.5
Scleronycteris							1.0
Hylonycteris				.1		.1	.8
Choeroniscus				.1		.1	.8
Choeronycteris				.1		.1	.8
Leptonycteris				.1		.1	.8
Lichonycteris						.1	.9
Carollia				.1		.9	
Rhinophylla						1.0	
Sturnira						1.0	
Brachyphylla				.1		.3	.6
Uroderma						1.0	
Vampyrops				.1		.9	
Vampyrodes						1.0	
Vampyressa						1.0	
Chiroderma						1.0	
Ectophylla						1.0	
Enchisthenes						1.0	
Artibeus				.1		.8	.1
Ardops						1.0	
Phyllops						1.0	
Ariteus				.5		.5	

TABLE 1. (Continued)

GENERA	Carn	Fish	Blood	F.G.	A.I.	Fruit	Nectar
				TROPHIC ROLE			
Stenoderma						1.0	
Pygoderma						1.0	
Ametrida						1.0	
Sphaeronycteris						1.0	
Centurio						1.0	
Erophylla						.1	.9
Phyllonycteris						.1	.9
Diphylla			1.0				
Diaemus			1.0				
Desmodus			.9	.1			
Natalus					1.0		
Furipterus					1.0		
Amorphochilus					1.0		
Thyroptera					1.0		
Myzopoda					1.0		
Myotis		.1		.1	.8		
Lasionycteris					1.0		
Eudiscopus					1.0		
Pipistrellus					1.0		
Nyctalus					1.0		
Glischropus					1.0		
Eptesicus					1.0		
Vespertilio					1.0		
Laephotis					1.0		
Histiotus					1.0		
Philetor					1.0		
Tylonycteris					1.0		
Mimetillus					1.0		
Hesperoptenus					1.0		
Chalinolobus					1.0		
Nycticeius					1.0		
Rhogeessa					1.0		
Baeodon					1.0		
Scotomanes					1.0		
Scotophilus					1.0		
Otonycteris				.5	.5		
Lasiurus				.1	.9		
Barbastella				.2	.8		
Plecotus				.5	.5		
Euderma				.5	.5		
Miniopterus					1.0		
Murina					1.0		
Harpiocephalus					1.0		
Kerivoula					1.0		
Antrozous				.7	.3		
Lamingtona				.5	.5		
Nyctophilus				.5	.5		
Pharotis				.5	.5		
Tomopeas					1.0		
Mystacina				.5	.5		
Tadarida					1.0		
Otomops					1.0		
Neoplatymops					1.0		
Sauromys					1.0		

585

TABLE 1. (Continued)

GENERA	TROPHIC ROLE						
	Carn	Fish	Blood	F.G.	A.I.	Fruit	Nectar
Platymops					1.0		
Myopterus					1.0		
Molossops					1.0		
Eumops					1.0		
Promops					1.0		
Molossus					1.0		
Cheiromeles					1.0		

The next step was to multiply the trophic role indices by the generic importance indices for each genus in each region. The values for each trophic role for all genera in a region were summed to yield a total importance value for that role in that region. Indices of overlap in trophic role importance were then obtained in the manner described above for generic importance.

RESULTS AND DISCUSSION

Table 1 lists the trophic role values assigned for each genus. A genus is considered a specialist in one or two trophic roles if it had a value of .5 or more for those roles. Major contributions are those genera which account for 10 percent or more of the total value for a given region.

Table 2 is a taxonomic comparison of the bat faunas of the six regions. The common generic importance values indicate the degree to which the same genera show the same importance (percent of species) in

the two regions. As with birds (Lein, 1972) the values decrease with distance. Comparison of the Neotropical region to the series Nearctic to Palearctic to Ethiopian yields values of 22.9, 11.2, and 9.7; the series Nearctic to Palearctic to Oriental to Australian yields values of 22.9, 11.2, 9.8 and 7.6.

These findings presumably reflect the dispersal patterns of the various genera through time. It is therefore not surprising to find the distance relationship outlined above. The highest value obtained (47.0) was between the Oriental and Ethiopian regions. The lowest value (7.6) was between the Neotropical and Australian regions. The only genera which these two regions share are the five cosmopolitan genera *Myotis*, *Pipistrellus*, *Eptesicus*, *Nycticeius* and *Tadarida*.

Table 3 lists the total importance values for the seven trophic roles in the six faunal regions, and the variation between regions.

TABLE 2. TAXONOMIC COMPARISON OF THE SIX FAUNAL REGIONS.
The upper figure is the common generic importance, and a value of 100 would indicate complete overlap; the lower figure is the number of shared genera:

	NEOT	NEA	PALE	ETHIO	ORIENT	AUST
Neotropical	—	22.9	11.2	9.7	9.8	7.6
		14	6	6	5	5
Nearctic		—	39.7	19.9	17.9	11.9
			6	6	5	5
Palearctic			—	45.1	46.9	21.4
				16	15	11
Ethiopian				—	47.0	38.3
					18	15
Oriental					—	45.2
						12
Australian						—

TABLE 3. TOTAL IMPORTANCE VALUES FOR THE SEVEN TROPHIC ROLES IN THE SIX FAUNAL REGIONS. Mean values, standard deviations and coefficients of variation are also given.

REGION	TROPHIC ROLE							
	1	2	3	4	5	6	7	TOTAL
Neotropical	1.8	.9	1.4	10.1	43.6	30.2	13.3	101.2
Nearctic	0	3.7	2.4	13.6	75.4	1.0	3.8	99.9
Palearctic	0	2.6	0	12.0	84.0	.7	.7	100.0
Ethiopian	.3	.3	0	14.3	65.8	13.8	5.5	100.0
Oriental	.6	.9	0	15.6	60.5	19.5	3.5	100.6
Australian	1.0	.3	0	9.3	48.4	35.1	6.4	100.5
Mean	.6	1.5	.6	12.5	63.0	16.7	5.5	
SD	.63	1.27	.94	2.25	14.13	13.16	3.91	
CV	105.0	84.7	156.7	18.0	22.4	78.8	71.1	

1. Carnivorous
2. Piscivorous
3. Sanguinivorous
4. Foliage gleaning
5. Aerial insectivore
6. Frugivorous
7. Nectarivorous

Table 4 shows these same values compared on the basis of major climatic regions. A trophic comparison of the six faunal regions is given in Table 5. All regions show a much higher degree of overlap when compared trophically rather than taxonomically. When the same series is examined in the trophic comparison as in the taxonomic comparison above, the values are 60.8, 56.0, 73.6 for the Neotropical region compared to the series Nearctic, Palearctic, Ethiopian. Similarly, for the series Nearctic, Palearctic, Oriental, Australian, the values are 60.8, 56.0, 78.2 and 90.8. The geographic cline shown in the taxonomic comparison disappears entirely in the trophic comparison. Lein (1972) demonstrated a similar relationship for birds.

Trophic similarity is more a reflection of current environmental conditions than of historic or phylogenetic relationship. Table 4 shows that the two temperate areas are more similar to each other than they are to the three tropical areas. The Australian region, which includes both temperate and tropical zones, seems to be trophically more similar to tropical areas. The Neotropical region which is taxonomically most similar

TABLE 4. TOTAL IMPORTANCE VALUES FOR THE SEVEN TROPHIC ROLES COMPARED ON THE BASIS OF MAJOR CLIMATIC REGIONS.

REGION	TROPHIC ROLE						
	1	2	3	4	5	6	7
Tropical							
Neotropical	1.8	.9	1.4	10.1	43.6	30.2	13.3
Ethiopian	.3	.3	0	14.3	65.8	13.8	5.5
Oriental	.6	.9	0	15.6	60.5	19.5	3.5
Intermediate							
Australian	1.0	.3	0	9.3	48.4	35.1	6.4
Temperate							
Nearctic	0	3.7	2.4	13.6	75.4	1.0	3.8
Palearctic	0	2.6	0	12.0	84.0	.7	.7

TABLE 5. TROPHIC COMPARISON OF THE SIX FAUNAL REGIONS.
The figures represent the common trophic role importance. A value of 100 would indicate complete overlap.

	NEOT	NEA	PALE	ETHIO	ORIENT	AUST
Neotropical	—	60.8	56.0	73.6	78.2	90.8
Nearctic		—	91.4	84.5	79.5	62.8
Palearctic			—	79.5	74.8	59.4
Ethiopian				—	92.7	77.6
Oriental					—	81.6
Australian						—

to the Nearctic region, is trophically most similar to the Australian region. Both the Neotropical and Australian regions include substantial amounts of temperate areas. The Oriental and Ethiopian regions are the most similar both taxonomically (47.0) and trophically (92.7). The two least similar areas trophically are the Neotropical and Palearctic (56.0).

All of this represents a quantification of a phenomenon that is quite familiar to anyone working with any group of organisms in different parts of the world, i.e., convergence, or ecological equivalence. Various groups of phylogenetically unrelated bats show a great deal of similarity in feeding habits. This convergence is often accompanied by a corresponding convergence in morphological characters, so that taxonomically distant groups sometimes look alike, as well as act alike.

Table 3 points out the extreme variation in the importance of various trophic roles from region to region. The coefficients of variation are quite high in every case.

The carnivorous trophic role is a minor one for bats. Prey items of carnivorous bats include other bats and rodents, birds, and small lizards. Several of the carnivorous bats are also omnivorous, like *Phyllostomus hastatus*, which probably feeds on insects, fruits, and flowers as well as on vertebrates. Table 6 lists the 7 genera which specialize in carnivory, plus *Phyllostomus*, which is a major contributor to this role in the Neotropical region. *Trachops* and *Chrotopterus* are the only strictly carnivorous genera, although *Megaderma* and *Vampyrum* are probably close. Probably other genera will be found to be partially carnivorous in the future, especially such phyllostomatines as *Tonatia* and *Mimon*. The Phyllostomatinae, a subfamily of the Neotropical family Phyllostomatidae and the Megadermatidae (*Megaderma*, *Macroderma*, *Cardioderma*), are the only groups in which the carnivorous trophic role is found.

There are no carnivorous bats in the temperate zone. Lein (1972), in an at-

TABLE 6. GENERA SPECIALIZING (*) AND/OR CONTRIBUTING MAJOR AMOUNTS TO THE *Carnivorous Trophic Role* (1).

GENERA	FAUNAL REGION					
	NEOT	NEA	PALE	ETHIO	ORIENT	AUST
*Megaderma**					.6	.5
*Macroderma**						.5
*Cardioderma**				.3		
Phyllostomus	.4					
*Trachops**	.5					
*Chrotopterus**	.5					
*Vampyrum**	.4					
Totals for region	1.8	0	0	.3	.6	1.0

TABLE 7. GENERA SPECIALIZING (*) AND/OR CONTRIBUTING MAJOR AMOUNTS TO THE *Piscivorous* Trophic Role (2).

GENERA	FAUNAL REGION					
	NEOT	NEA	PALE	ETHIO	ORIENT	AUST
Noctilio*	.4					
Megaderma					.1	.1
Myotis	.5	3.7	2.6	.3	.8	.2
Total for region	.9	3.7	2.6	.3	.9	.3

tempt to explain a similar, but less well-marked disparity in birds, suggested that there might be a greater abundance of food items in the tropics. Various physiological mechanisms may limit large carnivorous bats to the tropics. The energetic cost of hibernation or migration probably intensifies with body size, and since it is difficult to be carnivorous without being relatively large, it may be physiologically impractical to be a carnivorous temperate zone bat.

Table 7 lists the 3 genera contributing major amounts to the piscivorous trophic role. The only genus which specializes on fish is *Noctilio*, and only one species, *N. leporinus*, is piscivorous. *Megaderma* takes fish occasionally, and some species of *Myotis* are piscivorous. The ubiquity of this otherwise minimally important role is accounted for by the cosmopolitan genus *Myotis*. This role will probably expand also as food habits of other species become known (e.g., *Macrophyllum macrophyllum*).

There are also only 3 genera of sanguinivorous bats (Table 8). *Diphylla*, *Diaemus*, and *Desmodus* are all closely related members of the phyllostomatid subfamily Desmodontinae. They are specialized for feeding on blood of mammals and birds. *Desmodus rotundus* is the most widespread and has probably increased its range into the southern part of the Nearctic region owing to the introduction of cattle raising in recent times. This is the least important trophic role among bats, but the reasons for its being limited to the Neotropical region are not intuitively obvious.

Table 9 lists those genera specializing in or contributing major amounts to the foliage gleaning trophic role. As with birds (Lein, 1972), this is the least variable trophic role in all regions. An important difference is that while this is *the* niche for passerine birds, it is not as important as aerial insectivory and frugivory in bats. The genera *Rhinolophus* and *Hipposideros* are important components of this role in the old world. Foliage gleaning seems to be most important in the Oriental region and least in the Australian. However, the genera *Rhinolophus* and *Hipposideros* are represented by 45 and 26 species respectively in the Oriental region, and only by 5 and 11 species respectively in the Australian region.

Bats are best represented in the aerial

TABLE 8. GENERA SPECIALIZING (*) AND/OR CONTRIBUTING MAJOR AMOUNTS TO THE *Sanguinivorous* Trophic Role (3).

GENERA	FAUNAL REGION					
	NEOT	NEA	PALE	ETHIO	ORIENT	AUST
Diphylla*	.5	2.4				
Diaemus*	.5					
Desmodus*	.4					
Total for region	1.4	2.4				

TABLE 9. GENERA SPECIALIZING (*) AND/OR CONTRIBUTING MAJOR AMOUNTS TO THE *Foliage Gleaning Trophic Role* (4).
Numerical values are given for major contributors. Plus symbols (+) indicate genera present but contributing less than 10 percent to the total value for that region.

GENERA	FAUNAL REGION					
	NEOT	NEA	PALE	ETHIO	ORIENT	AUST
*Nycteris**			1.4	5.3	+	
*Cardioderma**				+		
*Lavia**				+		
*Rhinoloplus**			5.2	4.8	8.6	1.7
*Rhinomegaloplus**					+	
*Hipposideros**			+	3.0	5.0	3.7
*Micronycteris**	2.3					
*Tonatia**	1.2					
*Mimon**	+					
*Phylloderma**	+					
*Ariteus**	+					
Myotis	+	3.7	2.7	+	+	+
*Otonycteris**			+			
*Plecotus**	+	4.9	+	+		
*Euderma**		+				
*Antrozous**	+	1.7				
*Lamingtona**						+
*Nyctophilus**						2.4
*Pharotis**						+
*Mystacina**						+
Total for region	10.1	13.6	12.0	14.3	15.6	9.3

insectivore trophic role (Table 10). Seventy nine out of the total 169 genera of bats are specialized for feeding on insects on the wing. The only genera contributing major amounts to this role are *Rhinolophus, Myotis, Pipistrellus, Eptesicus, Lasiurus, Tadarida,* and *Eumops.* This is an artifact of so many genera sharing the role and is seen in the other roles as well. *Myotis* and *Pipistrellus* are major contributors to this role in 3 of the 6 regions.

This role is most important in the temperate regions. This is in spite of the fact that the seasonal nature of this food supply makes it necessary for these bats to hibernate or migrate during inclement seasons. Aerial insectivores are dominant in every region, although in the Neotropical region, their importance is matched by the frugivorous and nectarivorous roles combined.

Table 11 lists the specialists and major contributors to the frugivorous trophic role. This role shows the greatest temperate-

tropical disparity. The only Nearctic frugivores (*Choeronycteris, Leptonycteris*) are primarily nectarivorous genera which barely penetrate this zone. The Palearctic frugivore is *Rousettus,* which also feeds on insects. As with birds (Lein, 1972), the Neotropical region is particularly rich in frugivores, and the Australian region has an even higher value. Most of the Australian total is accounted for by the genus *Pteropus,* which has 39 species in the Australian region. Many of those species are insular forms with limited distributions, thus skewing the figure ever further. These tropical regions have a much greater diversity of fruit which is available all year than do the temperate zones.

The nectar feeding specialists and major contributors are listed in Table 12. Once again, the temperate zones are poorly represented, undoubtedly due to the climatic restrictions on flower availability. The Neotropical region is the only area with a large diversity of nectarivorous bats. Fif-

TABLE 10. GENERA SPECIALIZING (*) AND/OR CONTRIBUTING MAJOR AMOUNTS TO *Aerial Insectivore Trophic Role* (5).
Plus symbols (+) as in Table 9.

GENERA	NEOT	NEA	PALE	ETHIO	ORIENT	AUST
Nyctimene*					+	+
Rhinopoma*			+	+	+	
Emballonura*				+	+	+
Coleura*				+		
Rhynchonycteris*	+					
Saccopteryx*	+					
Centronycteris*	+					
Peropteryx*	+					
Cormura*	+					
Balantiopteryx*	+					
Taphozous*			+	+	+	+
Cyttarops*	+					
Depanycteris*	+					
Diclidurus*	+					
Noctilio*	+					
Rhinolophus*			+	+	8.6	+
Rhinomegalophus*					+	
Hipposideros*			+	+	+	+
Anthops*						+
Asellia*			+	+		
Aselliscus*					+	+
Cloeotis*				+		
Rhinonicteris*						+
Triaenops*			+	+		
Coelops*					+	
Paracoelops*					+	
Pteronotus*	+					
Mormoops*	+	+				
Macrotus*	+	+				
Natalus*	+					
Furipterus*	+					
Amorphochilus*	+					
Thyroptera*	+					
Myzopoda*				+		
Myotis*	+	29.3	21.2	+	6.7	+
Casinycteris*		+				
Eudiscopus*					+	
Pipistrellus*	+	+	14.7	6.9	10.8	+
Nyctalus*			+		+	
Glischropus*					+	+
Eptesicus*	+	+	8.6	+	+	+
Vespertilio*			+			
Laephotis*				+		
Histiotus*	+					
Philetor*					+	+
Tylonycteris*					+	
Mimetillus*				+		
Hesperoptenus*					+	
Chalinolobus*				+		+
Nycticeius*	+	+	+	+	+	+
Rhogeessa*	+					
Baeodon*	+					
Scotomanes*					+	

591

TABLE 10. (Continued)

GENERA	FAUNAL REGION					
	NEOT	NEA	PALE	ETHIO	ORIENT	AUST
Scotophilus*				+	+	
Otonycteris*			+			
Lasiurus*	+	8.8				
Barbastella*			+	+	+	
Plecotus*	+	+	+	+		
Euderma*		+				
Miniopterus*			+	+	+	+
Murina*			+		+	+
Harpiocephalus*					+	+
Kerivoula*				+	+	+
Lamingtona*						+
Nyctophilus*						+
Pharotis*						+
Tomopeas*	+					
Mystacina*						+
Tadarida*	+	+	+	13.3	+	+
Otomops*				+	+	+
Neoplatymops*	+					
Sauromys*				+		
Platymops*				+		
Myopterus*				+		
Molossops*	+					
Eumops*	4.5	+				
Promops*	+					
Molossus*	+					
Cheiromeles†					+	
Total for region	43.6	75.4	84.0	65.8	60.5	48.4

teen of the 26 genera in Table 11 are Neotropical.

Several groups of bats may be used as illustrations of convergence. The Old World Tropical Megadermatidae seem to be the ecological equivalents of several genera of New World carnivorous Phyllostomatinae.

Fish-eating is practiced by different species in each of the six regions. *Noctilio leporinus* is the Neotropical representative of this role. *Myotis vivesi* occurs in the Nearctic region. Palearctic members include *Myotis daubentoni* and *M. macrodactylus* (Findley, 1972). *Megaderma* is an occasional fish-eater in the Oriental and Australian regions. *Myotis* accounts for the .3 value listed for the Ethiopian region, even though none of the six species are definitely known to fish.

Blood feeding is a highly specialized

role found only in the Desmodontinae, a subfamily of the Neotropical Phyllostomatidae. The absence of this trait in other regions is difficult to explain, since the Neotropical region seemingly has little more to offer in the way of host animals than other regions. It may be that this most specialized role has only appeared once due to the normal selective rigors which attend extreme specialization.

Foliage gleaning is the most evenly distributed food habit in all of the regions. Foliage gleaning bats have a suite of morphological and behavioral characters which go along with this feeding behavior, including large ears, broad wings, and a slow, maneuverable flight. *Tonatia* and some species of *Micronycterus* typify this pattern in the Neotropical region. The Nearctic region has *Antrozous, Plecotus,* and *Myotis evotis, M. auriculus,* and *M.*

TABLE 11.　GENERA SPECIALIZING (*) AND/OR CONTRIBUTING MAJOR AMOUNTS TO *Frugivorous Trophic Role* (6).

Plus symbols (+) as in Table 9.

GENERA	FAUNAL REGION					
	NEOT	NEA	PALE	ETHIO	ORIENT	AUST
Eidolon*				+		
Rousettus*			.7	+	+	+
Myonycteris*				+		
Boneia*					+	
Pteropus*				4.3	8.2	23.8
Acerodon*					+	
Neopteryx*					+	
Pteralopex*						+
Styloctenium*					+	
Dobsonia*					+	+
Harpionycteris*					+	
Plerotes*				+		
Hypsignathus*				+		
Epomops*				1.4		
Epomophorus*				2.2		
Micropteropus				+		
Scotonycteris*				+		
Casinycteris*				+		
Cynopterus*					+	
Megaerops*					+	
Ptenochirus*					+	
Dyacopterus*					+	
Chironax*					+	
Thoopterus*					+	+
Sphaerias*					+	
Balionycteris*					+	
Aethalops*					+	
Penthetor*					+	
Haplonycteris*					+	
Nyctimene*					+	+
Paranyctimene*						+
Macroglossus*					+	+
Syconycteris*						+
Melonycteris*						+
Notopteris*						+
Micronycteris*	+					
Macrotus	+	.4				
Tonatia*	+					
Anoura*	+					
Choeronycteris	+	.3				
Leptonycteris	+	.3				
Carollia*	+					
Rhinophylla*	+					
Sturnira*	4.5					
Uroderma*	+					
Vampyrops*	+					
Vampyrodes*	+					
Vampyressa*	+					
Chiroderma*	+					
Ectophylla*	+					
Enchisthenes*	+					
Artibeus*	+					
Ardops*	+					

TABLE 11. (Continued)

GENERA	FAUNAL REGION					
	NEOT	NEA	PALE	ETHIO	ORIENT	AUST
Phyllops*	+					
Ariteus*	+					
Stenoderma*	+					
Pygoderma*	+					
Ametrida*	+					
Sphaeronycteris*	+					
Centurio*	+					
Total for region	30.2	1.0	.7	13.8	19.5	35.1

keenii. In the Palearctic region, *Hipposideros, Rhinolophus,* and several species of *Myotis* (e.g., *bechsteini, emarginatus*) fill this role.

Aerial insectivores are common in all regions, and it might be better to view this as a basic feeding pattern rather than one of convergence. The family Molossidae is found in all of the tropical regions and is composed entirely of aerial insectivores. The molossid genus *Tadarida* occurs in all six regions. The family Vespertilionidae has four genera (*Myotis, Pipistrellus, Eptesicus,* and *Nyctecieus*) which are cos-

TABLE 12. GENERA SPECIALIZING (*) AND/OR CONTRIBUTING MAJOR AMOUNTS TO *Nectarivorous Trophic Role* (7).
Plus symbols (+) as in Table 9.

GENERA	FAUNAL REGION					
	NEOT	NEA	PALE	ETHIO	ORIENT	AUST
Rousettus*			.7	.8	.6	+
Pteropus				+	.9	2.6
Dobsonia					+	.7
Epomophorus*				2.2		
Nanonycteris*				+		
Eonycteris*					1.1	
Megaloglossus*				+		
Macroglossus*					.4	+
Syconycteris*						1.0
Melonycteris*						1.0
Notopteris*						+
Glossophaga*	+					
Lionycteris*	+					
Lonchophylla*	1.4					
Platalina*	+					
Monophyllus*	+					
Anoura*	+					
Scleronycteris*	+					
Hylonycteris*	+					
Choeroniscus*	1.8					
Choeronycteris*	+	1.9				
Leptonycteris*	+	1.9				
Lichonycteris*	+					
Brachyphylla*	+					
Erophylla*	+					
Phyllonycteris*	+					
Total for region	13.3	3.8	.7	5.5	3.5	6.4

mopolitan and all are primarily aerial insectivores. The monotypic Myzopodidae and Thyropteridae are specialists in this role in the Ethiopian and Neotropical regions, respectively. The Neotropical region also has the families Furipteridae, Natalidae, and Mormoopidae, which are all aerial insectivores. The Emballonuridae, found in every region except the Nearctic, contains 12 genera which are exclusively aerial insectivores. In short, aerial insectivores are the most important component of every region.

Frugivorous types are essentially limited to the tropical areas and are best represented by the Pteropodidae in the Old World and the Phyllostomatidae in the New World.

Nectarivorous bats are also essentially tropical and are mainly members of the above two groups. The subfamilies Glossophaginae and Phyllonycterinae of the Neotropical Phyllostomatidae have specialized for this role, as have several genera of the pteropodid subfamily Macroglossinae in the Oriental, Ethiopian, and Australian regions.

CONCLUSIONS

1. Additional study directed towards elucidation of food habits of bats is a necessary prerequisite to further studies of this sort.
2. The most important trophic role of bats is aerial insectivory, followed by frugivory, foliage gleaning, nectarivory, piscivory, carnivory, and sanguinivory.
3. The tropical faunal regions support a broader base of trophic roles than do the temperate regions.
4. Taxonomic zoogeographic analysis shows a direct correlation between distance and similarity of faunal regions, while a trophic role approach emphasizes ecological similarity between major climatic areas, rather than distance.
5. The lack of correlation between taxonomic and trophic approaches to zoogeography points up the pitfalls facing

a taxonomist dealing with suites of characters which may be influenced by trophic roles.

6. While the vagaries of historical caprice may dictate distributional contiguity, functional niches will be more likely controlled by environmental conditions.

ACKNOWLEDGMENTS

Pat Mehlhop, Ron Pine and Hank Setzer contributed valuable suggestions on drafts of the manuscript. I am grateful to Karl Koopman, both for his painstaking work in summarizing bat distributions and for his helpful criticism of the manuscript, even though he retains grave misgivings about the quantification of poorly known variables.

REFERENCES

ALLEN, G. M. 1939. Bats. Harvard Univ. Press, Cambridge, x + 368 pp.

ARATA, A. A., J. B. VAUGHN, AND M. E. THOMAS. 1967. Food habits of certain Colombian bats. J. Mamm. 48:653–655.

BARBOUR, R. W., AND W. H. DAVIS. 1969. Bats of America. Univ. Press Kentucky, Lexington. 286 pp.

BORELL, A. E. 1942. Feeding habit of the pallid bat. J. Mamm. 23:337.

BROSSET, A., AND C. D. DEBOUTTEVILLE. 1966. Le regime alimentaire du Daubenton, *Myotis daubentoni*. Mammalia 30:247–251.

CARVALHO, C. T. 1961. Sobre los habitos alimentares de phillostomideos (Mammalia, Chiroptera). Rev. Biol. Trop. 9:53–60.

CONSTANTINE, D. G. 1966. New bat locality records from Oaxaca, Arizona and Colorado. J. Mamm. 47:125–126.

DARLINGTON, P. J., JR. 1957. Zoogeography: The geographical distribution of animals. John Wiley and Sons, Ltd. London.

DAVIS, W. B., D. C. CARTER, AND R. H. PINE. 1964. Noteworthy records of Mexican and Central American bats. J. Mamm. 45:375–387.

DUNN, L. 1933. Observations on the carnivorous habits of the spearnosed bat, *Phyllostomus hastatus panamensis* Allen in Panama. J. Mamm. 14:188–189.

DWYER, P. D. 1962. Studies on the two New Zealand bats. Zool. Publ. Victoria Univ., Wellington 28:1–28.

FINDLEY, J. S. 1972. Phenetic relationships among bats of the genus *Myotis*. Syst. Zool. 21: 31–52.

FLEMING, T. H., E. T. HOOPER, AND D. E. WIL-

SON. 1972. Three Central American bat communities: structure, reproductive cycles and movement patterns. Ecology 53:555–569.

GOODWIN, G. G., AND A. M. GREENHALL. 1961. A review of the bats of Trinidad and Tobago. Bull. Amer. Mus. Nat. Hist. 122:191–301.

GREENHALL, A. M. 1966. Oranges eaten by Spear-nosed bats. J. Mamm. 47:125.

GREENHALL, A. M. 1968. Notes on the behavior of the false vampire bat. J. Mamm. 49:337–340.

GRINNELL, H. W. 1918. A synopsis of the bats of California. Univ. Calif. Publ. Zool. 17:223–404.

HATT, R. T. 1923. Food habits of the Pacific pallid bat. J. Mamm. 4:260–261.

HOFFMEISTER, D. F., AND W. W. GOODPASTER. The mammals of the Huachuca Mountains, southeastern Arizona. Illinois Biol. Mongr. 24:1–152.

HOOPER, E. T., AND J. H. BROWN. 1968. Foraging and breeding in two sympatric species of Neotropical bats, genus Noctilio. J. Mamm. 49:310–312.

HOWELL, A. B. 1919. Some Californian experiences with bat roosts. J. Mamm. 1:169–177.

HUEY, L. M. 1925. Food of the California leafnose bat. J. Mamm. 6:196–197.

JEANNE, R. L. 1970. Note on a bat (Phylloderma stenops) preying upon the brood of a social wasp. Science 51:624–625.

KOOPMAN, K. F. 1970. Zoogeography of bats. In About Bats. Southern Methodist University Press. Dallas pp. 29–50.

KOOPMAN, K. F., AND J. K. JONES, JR. 1970. Classification of bats. In About Bats. Southern Methodist University Press. Dallas pp. 22–28.

LEIN, M. R. 1972. A trophic comparison of avifaunas. Syst. Zool. 21:135–150.

LIM, B. L. 1966. Abundance and distribution of Malaysian bats in different ecological habitats. Fed. Mus. J., 11:61–76.

LIM, B. L. 1970. Food habits and breeding cycle of the Malayasian fruit eating bat, Cynopterus brachyotis. J. Mamm. 51:174–177.

McNAB, B. 1971. The structure of tropical bat faunas. Ecology 52:352–358.

MEDWAY, L. 1967. A bat-eating bat Megaderma lyra Geoffroy. Malayan Nature J. 20:107–110.

ORR, R. T. 1954. Natural history of the pallid bat, Antrozous pallidus (Le Conte). Proc. Calif. Acad. Sci. 28:165–246.

POULTON, E. B. 1929. British insectivorous bats and their prey. Proc. Zool. Soc. London 1929:277–303.

PINE, R. H. 1969. Stomach contents of a free-tailed bat, Molossus ater. J. Mamm. 50:162.

ROSS, A. 1961. Notes on food habits of bats. J. Mamm. 42:66–71.

ROSS, A. 1967. Ecological aspects of the food habits of North American insectivorous bats. Proc. West. Found Vert. Zool. 1:205–264.

RYBERG, O. 1947. Bats and bat parasites. Natur. Stockholm xvi + 329 pp.

SHERMAN, H. B. 1935. Food habits of the Seminole bat. J. Mamm. 16:224.

SILVA, T. G., AND R. H. PINE. 1969. Morphological and behavioral evidence for the relationships between the bat genus Brachyphylla and the Phyllonycterinae. Biotropica 1:10–19.

STORER, T. F. 1926. Bats, bat towers and mosquitoes. J. Mamm. 7:85–90.

TAN, K. B. 1965. Stomach contents of some Borneo mammals. Sarawak Mus. J. 12:373–385.

TUTTLE, M. D. 1967. Predation by Chrotopterus auritus on Geckos. J. Mamm. 48:319.

TUTTLE, M. D. 1968. Feeding habits of Artibeus jamaicensis. J. Mamm. 49:787.

VALDEZ, R., AND R. K. LaVAL. 1971. Records of bats from Honduras and Nicaragua. J. Mamm. 52:247–250.

WALKER, E. P. 1968. Mammals of the world (2nd ed.) Vol. I, pp. 182–392. John Hopkins Press, Baltimore.

WILSON, D. E. 1971. Food habits of Micronycteris hirsuta (Chiroptera: Phyllostomatidae). Mammalia 35:107–110.

Manuscript received August, 1972

A Numerical Analysis of the Distributional Patterns of North American Mammals. II. Re-evaluation of the Provinces

EDWIN M. HAGMEIER

Abstract

In an earlier paper, numerical techniques were developed and used to analyze distribution patterns of the native terrestrial mammals of North America. An error in method is here corrected, indicating that 35 provinces, 13 superprovinces, four subregions, and one region may be recognized. The methods used are relatively objective, quantitative, and suited to computerization.

Introduction

In an earlier paper (Hagmeier and Stults, 1964, hereafter referred to as H & S), quantitative and relatively objective methods were used to demonstrate that (1) the range limits of North American terrestrial mammals are grouped, (2) that as a result it was possible to delimit geographic regions of faunistic homogeneity which were termed mammal provinces, and (3) that such provinces could be useful in the analysis of other zoogeographic phenomena.

This paper is concerned with the recalculation of some of the data of the second item above. It was assumed in our earlier paper (H & S) that several of the large provinces of the northern half of the continent required further analysis. On initiation of this analysis, it became apparent that an error in method had been made which required correction. As a consequence of the correction, the number of North American mammal provinces is here increased from 22 to 35, two of which are of uncertain status.

The general philosophy, methodology, and conclusions reached in our earlier paper (H & S) do not differ from those arrived at here, and the earlier paper should be referred to for accounts of these. The ma-

terial given here, since it is essentially revisionary, is presented in as brief a form as possible. Because of the changes reported here, the analysis of mammal areas given in H & S (p. 141–146 and Figs. 6c–8d) needs revision, and this revision will form the subject of a future paper. Since submission of our earlier paper (H & S), Simpson (1964) has considered variation in abundance of species of North American mammals in a superior manner, and his work should be referred to for a treatment of the subject.

Derivation of Provinces

In our earlier paper (H & S), the ranges of all 242 species of native terrestrial North American mammals were converted into a model, first by separately computing the percentage of species and genera whose ranges ended within blocks 50 miles by 50 miles throughout the continent (each such value was called Index of Faunistic Change, or IFC), second by plotting species and genus IFCs on maps of North America, and third by fitting isarithms. The species IFC map resulting was given as H & S, Figure 1. Low IFC values indicated faunistic homogeneity, and regions characterized by such values were termed primary areas.

279

Fig. 1.   Eighty-six primary areas derived through examination of IFC Map (H & S, Fig. 1).

Twenty-four of these were identified through examination of IFC maps of both species and genus and species checklists of each were prepared (H & S, Table 2). The percentage of species common to all combinations of pairs of provinces (Coefficients of Community, or CCs, Jaccard, 1902) were then computed. CCs were then subjected to cluster analysis using the weighted pair-group method, and simple averages (Sokal and Sneath, 1963:180–184, 304–312; H & S: 132, 137), and drawn up in the form of a dendrogram (H & S, Fig. 5). Primary areas pooled at a mean CC level lower than 62.5% were termed mammal provinces, those pool-

ing at below 39% were termed mammal superprovinces, those below 22.5%, mammal subregions, those below 8% as mammal regions.

The basis of our error lay in the fact that we attempted to identify only the 24 North American mammal areas corresponding to those described by Kendeigh (1961) in his modification of Dice's (1943) scheme of biotic provinces. This error became apparent only when subsequent analysis of parts of certain of these provinces showed that the parts in some cases merited full province status.

The correction made and reported here

TABLE 1.

Species	1 Ungavan	2E E. Eskimoan	2W W. Eskimoan	3 Alaskan	4 Aleutian	5 Yukonian	6W W. Hudsonian	6E E. Hudsonian	7E E. Canadian	7W W. Canadian	8 Montanian	9 Coloradan	10 Vancouverian	11 Oregonian	12 Humboldtian	13 Sierran	14 Alleghanian	15 Illinoian	16 Carolinian	17 Kansan	18 Saskatchewanian	19 Balconian	20 Tamaulipan	21 Texan	22 Louisianian	23 Columbian	24 Artemesian	25 Kaibabian	26 Sonoran	27 Yaquinan	28 Mapimian	29 Navahonian	30 Uintan	31 San Matean	32 Mohavian	33 Californian	34 Diablian	35 San Bernardinian
Didelphis marsupialis		X															X	X	X	X		X	X	X	X													
Sorex cinereus		X	X	X	X	X	X	X	X	X	X	X	X	X	X	X	X	X	X	X										X		X	X	X		X	X	X
S. longirostris				X	X	X					X	X	X						X													X	X					
S. vagrans											X	X	X	X	X	X							X			X		X						X				X
S. nanus											X	X	X	X																								
S. palustris									X	X	X	X	X	X	X	X	X				X					X		X				X	X	X		X	X	X
S. bendirei													X	X	X																							
S. fumeus																	X																					
S. arcticus		X	X		X	X	X	X	X	X											X																	
S. dispar																	X																					
S. troubridgei													X	X		X																		X				
S. merriami											X															X	X	X		X		X	X					X
Microsorex hoyi	X					X	X	X	X	X											X																X	X
Blarina brevicauda									X									X	X	X					X													
Cryptotis parva														X	X			X	X	X		X	X	X	X													
Notiosorex crawfordi																						X	X	X				X				X			X			X
Neurotrichus gibbsii													X	X	X																							X
Scapanus orarius													X	X	X																							
S. latimanus																X																						X
S. townsendii													X	X											X									X				
Parascalops breweri									X	X											X																	
Scalopus aquaticus																		X	X	X		X	X	X	X													
Condylura cristata									X	X																												
Dasypus novemcinctus						X																X	X															
Ochotona collaris				X	X	X																				X												
O. princeps											X	X		X	X	X																				X	X	X
Sylvilagus idahoensis																										X	X											
S. bachmani																																						X
S. palustris												X																X		X				X				
S. floridanus																X	X	X	X	X		X	X	X	X	X	X	X			X	X					X	
S. transitionalis																	X																	X				
S. nuttallii		X	X	X	X												X	X		X	X						X	X				X	X		X	X	X	X
S. audubonii												X				X										X	X	X		X		X	X		X			X
S. aquaticus																							X	X	X													

599

Table 1. (*Continued.*)

Species	1 Ungavan	2E E. Eskimoan	2W W. Eskimoan	3 Alaskan	4 Aleutian	5 Yukonian	6W W. Hudsonian	6E E. Hudsonian	7E E. Canadian	7W W. Canadian	8 Montanian	9 Coloradan	10 Vancouverian	11 Oregonian	12 Humboldtian	13 Sierran	14 Alleghanian	15 Illinoian	16 Carolinian	17 Kansan	18 Saskatchewanian	19 Balconian	20 Tamaulipan	21 Texan	22 Louisianian	23 Columbian	24 Artemesian	25 Kaibabian	26 Sonoran	27 Yaquinian	28 Mapimian	29 Navahonian	30 Uintian	31 San Matean	32 Mohavian	33 Californian	34 Diablian	35 San Bernardinian
Lepus americanus	X					X	X	X	X	X	X	X	X	X		X	X																					
L. arcticus	X	X	X	X	X		X	X																														
L. townsendii															X	X				X	X	X	X	X		X	X	X	X	X	X	X	X	X	X	X	X	X
L. californicus												X				X				X						X	X						X					
L. gaillardi																														X								
L. alleni																													X									
Aplodontia rufa														X	X	X																						
Tamias striatus									X								X	X	X																			
Eutamias alpinus												X				X																						
E. minimus										X	X					X					X					X	X						X	X				
E. amoenus															X	X										X												
E. townsendii														X	X	X																						
E. sonomae														X	X																							
E. merriami																X																					X	X
E. dorsalis																											X	X				X						
E. quadrivittatus																																X	X					
E. ruficaudus											X																						X	X				
E. cinereicollis																																X		X				
E. quadrimaculatus																X																						
E. speciosus																X																						
E. panamintinus																X																			X			
E. umbrinus												X														X							X					
Marmota monax																	X	X	X					X														
M. flaviventris								X	X	X	X	X				X										X												
M. caligata						X							X																									
Ammospermophilus harrisii																												X	X									
A. leucurus																											X				X	X			X			
Spermophilus townsendii																										X	X											
S. richardsonii												X									X																	
S. armatus												X																										
S. beldingi																X																						
S. columbianus											X															X												
S. undulatus		X	X	X						X																												
S. tridecemlineatus												X						X		X	X			X										X				

TABLE 1. (*Continued.*)

Province (column) key:
1 Ungavan · 2E E. Eskimoan · 2W W. Eskimoan · 3 Alaskan · 4 Aleutian · 5 Yukonian · 6W W. Hudsonian · 6E E. Hudsonian · 7E E. Canadian · 7W W. Canadian · 8 Montanian · 9 Coloradan · 10 Vancouverian · 11 Oregonian · 12 Humboldtian · 13 Sierran · 14 Alleghanian · 15 Illinoian · 16 Carolinian · 17 Kansan · 18 Saskatchewanian · 19 Balconian · 20 Tamaulipan · 21 Texan · 22 Louisianian · 23 Columbian · 24 Artemesian · 25 Kaibabian · 26 Sonoran · 27 Yaquinian · 28 Mapimian · 29 Navahonian · 30 Uintan · 31 San Matean · 32 Mohavian · 33 Californian · 34 Diablan · 35 San Bernardinian

Species	1	2E	2W	3	4	5	6W	6E	7E	7W	8	9	10	11	12	13	14	15	16	17	18	19	20	21	22	23	24	25	26	27	28	29	30	31	32	33	34	35
S. mexicanus																				×		×	×								×	×	×	×				
S. spilosoma																				×		×	×				×	×	×	×	×	×	×	×				
S. franklinii																		×				×																
S. variegatus																										×	×					×						
S. beecheyi														×	×	×													×							×	×	×
S. tereticaudus																																			×			×
S. mohavensis																																			×	×		
S. lateralis											×	×								×	×	×							×	×				×				×
Cynomys ludovicianus																																						
C. leucurus												×															×							×				
C. parvidens																															×							
C. gunnisoni																												×				×	×					
Sciurus carolinensis																	×	×	×					×	×													
S. griseus														×	×	×																				×	×	
S. aberti																																×						
S. niger												×			×		×	×	×	×			×	×	×					×								
S. apache																													×	×		×						
S. arizonensis																								×														
Tamiasciurus hudsonicus				×		×	×	×	×	×	×																					×	×	×				
T. douglasii											×		×	×		×																						
Glaucomys volans									×								×	×						×	×													
G. sabrinus						×			×	×			×		×	×	×			×																		
Thomomys umbrinus												×										×				×	×	×		×		×	×	×	×	×	×	×
T. talpoides											×		×	×							×					×							×					
T. bulbivorus														×																								
Geomys bursarius																		×													×							
G. arenarius																																		×				
G. personatus																							×															
G. pinetis																			×						×													
Cratogeomys castanops																						×									×							
Perognathus fasciatus																				×	×																	
P. flavescens																						×																
P. merriami																				×		×	×								×			×				
P. flavus																						×								×	×	×						

TABLE 1. (Continued.)

Region columns (number : name): 1 Ungavan; 2E E. Eskimoan; 2W W. Eskimoan; 3 Alaskan; 4 Aleutian; 5 Yukonian; 6W W. Hudsonian; 6E E. Hudsonian; 7E E. Canadian; 7W W. Canadian; 8 Montanian; 9 Coloradan; 10 Vancouverian; 11 Oregonian; 12 Humboldtian; 13 Sierran; 14 Alleghanian; 15 Illinoian; 16 Carolinian; 17 Kansan; 18 Saskatchewanian; 19 Balconian; 20 Tamaulipan; 21 Texan; 22 Louisianian; 23 Columbian; 24 Artemesian; 25 Kaibabian; 26 Sonoran; 27 Yaquinian; 28 Mapimian; 29 Navahonian; 30 Uintian; 31 San Matean; 32 Mohavian; 33 Californian; 34 Diablian; 35 San Bernardinian.

	1	2E	2W	3	4	5	6W	6E	7E	7W	8	9	10	11	12	13	14	15	16	17	18	19	20	21	22	23	24	25	26	27	28	29	30	31	32	33	34	35
P. apache																																×	×	×	×	×		×
P. longimembris																											×											
P. amplus																													×							×		
P. inornatus																												×										
P. parvus																											×								×			
P. formosus																											×											×
P. baileyi																																		×				
P. hispidus																						×	×	×							×							
P. penicillatus																										×	×	×	×	×	×			×				
P. intermedius																										×	×		×	×	×							
P. fallax																																						×
P. californicus														×																					×	×	×
P. spinatus																																						×
Microdipodops megacephalus																											×											
M. pallidus																																			×			
Dipodomys ordii																				×	×	×	×								×	×	×	×				
D. microps																											×								×			
D. panamintinus																																	×				×	
D. elephantinus																																			×			
D. agilis																																					×	×
D. heermanni													×	×																								
D. ingens																																						×
D. spectabilis																										×	×			×	×							
D. elator																							×														
D. merriami																										×	×	×	×	×	×	×	×	×	×	×		×
D. deserti																																		×	×		
Castor canadensis	×					×	×	×	×	×	×	×	×	×						×	×	×	×	×	×	×	×	×	×		×	×	×					×
Oryzomys palustris																	×	×	×						×													
Reithrodontomys montanus											×	×								×	×			×									×	×				
R. humulis																			×						×													
R. megalotis													×	×						×											×	×	×	×	×	×	×	×
R. raviventris																																				×		
R. fulvescens																			×			×	×							×								
Peromyscus crinitus																										×	×						×		×			

TABLE 1. (Continued.)

Faunal zones (column key): 1 Ungavan · 2E E. Eskimoan · 2W W. Eskimoan · 3 Alaskan · 4 Aleutian · 5 Yukonian · 6W W. Hudsonian · 6E E. Hudsonian · 7E E. Canadian · 7W W. Canadian · 8 Montanian · 9 Coloradan · 10 Vancouverian · 11 Oregonian · 12 Humboldtian · 13 Sierran · 14 Alleghanian · 15 Illinoian · 16 Carolinian · 17 Kansan · 18 Saskatchewanian · 19 Balconian · 20 Tamaulipan · 21 Texan · 22 Louisianan · 23 Columbian · 24 Artemesian · 25 Kaibabian · 26 Sonoran · 27 Yaquinian · 28 Mapimian · 29 Navahonian · 30 Uintan · 31 San Matean · 32 Mohavian · 33 Californian · 34 Diablan · 35 San Bernardinian

Species	1	2E	2W	3	4	5	6W	6E	7E	7W	8	9	10	11	12	13	14	15	16	17	18	19	20	21	22	23	24	25	26	27	28	29	30	31	32	33	34	35
P. californicus																×																				×	×	×
P. eremicus																											×	×	×	×	×			×	×		×	×
P. merriami																										×	×	×	×	×	×	×	×	×	×	×	×	×
P. maniculatus				×		×	×	×	×	×	×	×	×	×	×	×	×	×	×	×	×					×	×	×	×	×	×	×	×	×	×	×	×	×
P. polionotus																	×	×	×																			
P. leucopus																	×	×	×	×	×	×	×	×	×													
P. gossypinus																			×					×	×										×	×	×	×
P. boylei																						×	×	×	×			×		×	×			×			×	×
P. pectoralis																						×	×							×	×							
P. truei																												×			×	×	×	×		×		
P. nasutus																															×	×	×					
P. nuttallii																	×	×	×	×																		
P. floridanus																			×																			
Baiomys taylori																						×	×	×	×					×								
Onchomys leucogaster												×								×	×	×	×	×		×	×		×	×	×	×	×	×				
O. torridus																										×			×	×				×	×	×		×
Sigmodon hispidus																						×	×	×	×	×			×	×								
S. minimus																									×					×	×							
S. ochrognathus																														×	×							
Neotoma floridana																			×	×	×	×	×	×	×													
N. micropus																						×	×								×	×	×					
N. albigula																								×				×			×	×	×	×	×	×	×	×
N. lepida																										×	×	×					×	×	×			×
N. stephensi																															×							
N. mexicana																														×	×	×	×	×				
N. fuscipes																×																		×			×	×
N. cinerea									×	×	×	×	×	×	×	×	×				×					×	×					×	×					
Clethrionomys rutilus		×	×	×		×	×																															
C. gapperi							×	×	×	×	×	×		×	×	×	×				×					×												
C. occidentalis									×	×	×	×	×	×	×											×												
Phenacomys intermedius								×	×	×	×	×		×	×											×												
P. albipes														×	×											×												
P. silvicola														×	×																							
P. longicaudus														×	×	×										×												

TABLE 1. (*Continued.*)

	Ungavan 1	E. Eskimoan 2E	W. Eskimoan 2W	Alaskan 3	Aleutian 4	Yukonian 5	W. Hudsonian 6W	E. Hudsonian 6E	E. Canadian 7E	W. Canadian 7W	Montanian 8	Coloradan 9	Vancouverian 10	Oregonian 11	Humboldian 12	Sierran 13	Alleghanian 14	Illinoian 15	Carolinian 16	Kansan 17	Saskatchewanian 18	Balconian 19	Tamaulipan 20	Texan 21	Louisianian 22	Columbian 23	Artemesian 24	Kaibabian 25	Sonoran 26	Yaquinan 27	Mapimian 28	Navahonian 29	Uintan 30	San Matean 31	Mohavian 32	Californian 33	Diablian 34	San Bernardinian 35
Microtus pennsylvanicus	X			X		X	X	X	X	X	X	X	X				X	X	X	X	X																	
M. montanus																X										X	X						X			X	X	X
M. californicus															X	X																						
M. townsendii			X		X									X																								
M. oeconomus				X	X	X																				X	X					X	X	X	X			X
M. longicaudus				X	X	X					X			X	X	X										X	X						X	X				X
M. mexicanus							X			X																				X								
M. xanthognathus							X	X	X		X																											
M. chrotorrhinus								X	X								X																					
M. richardsoni											X	X				X										X	X											
M. oregoni				X		X								X																								
M. miurus																																						
M. ochrogaster																	X	X	X	X				X	X													
M. pinetorum																	X	X	X						X													
M. parvulus																								X	X													
Lagurus curtatus												X														X	X											
Neofiber alleni																						X			X	X												
Ondatra zibethicus	X			X	X	X	X	X	X	X	X	X	X	X	X		X	X	X	X	X					X		X	X	X	X	X	X	X	X			X
Lemmus trimucronatus		X	X	X		X	X										X																					
Synaptomys cooperi							X	X	X	X	X						X			X																		
S. borealis	X	X	X	X	X	X	X	X	X	X	X	X	X	X			X				X																	
Dicrostonyx groenlandicus	X	X	X	X	X		X	X																										X				
Zapus hudsonius	X	X	X	X	X	X	X	X	X	X	X	X	X				X	X	X	X	X																	
Z. princeps											X	X														X	X	X	X		X	X	X	X				
Z. trinotatus													X			X																						
Napaeozapus insignis	X							X	X								X				X																	
Erethizon dorsatum	X	X	X	X	X	X	X	X	X	X	X	X	X	X	X	X	X	X	X	X	X					X	X	X	X	X	X	X	X	X				X
Canis latrans				X		X	X	X	X	X	X	X	X			X	X	X	X	X	X	X	X	X	X	X	X	X	X	X	X	X	X	X	X	X	X	X
C. lupus	X	X	X	X	X	X	X	X	X	X	X	X	X			X	X				X					X	X	X	X	X	X	X	X	X	X	X	X	X
C. niger																			X			X	X	X	X													
Alopex lagopus	X	X	X	X	X	X	X	X																														
Vulpes fulva	X	X	X	X	X	X	X	X	X	X	X	X	X			X	X	X	X	X	X		X	X	X	X	X	X	X	X	X	X	X	X	X	X	X	X
V. velox																										X	X						X	X				
Urocyon cinereoargenteus																	X	X	X	X	X	X	X	X	X	X	X	X	X	X	X	X	X	X	X	X	X	X
Ursus americanus	X			X		X		X	X	X	X	X	X			X	X	X	X	X	X					X	X	X	X	X	X	X	X	X				X

The table records the occurrence (×) of each mammal species across the numbered faunal provinces of North America. Provinces (rows) are numbered 1–35; species (columns) are listed at the foot of the original table.

Province key:
1 Ungavan · 2E E. Eskimoan · 2W W. Eskimoan · 3 Alaskan · 4 Aleutian · 5 Yukonian · 6W W. Hudsonian · 6E E. Hudsonian · 7E E. Canadian · 7W W. Canadian · 8 Montanian · 9 Coloradan · 10 Vancouverian · 11 Oregonian · 12 Humboldtian · 13 Sierran · 14 Alleghanian · 15 Illinoian · 16 Carolinian · 17 Kansan · 18 Saskatchewanian · 19 Balconian · 20 Tamaulipan · 21 Texan · 22 Louisianian · 23 Columbian · 24 Artemesian · 25 Kaibabian · 26 Sonoran · 27 Yaquinian · 28 Mapimian · 29 Navahonian · 30 Uintan · 31 San Matean · 32 Mohavian · 33 Californian · 34 Diablian · 35 San Bernardinian

Species columns (left to right):
Ba = *Bassariscus astutus* · Pl = *Procyon lotor* · Nn = *Nasua narica* · Ma = *Martes americana* · Mpe = *M. pennanti* · Mer = *Mustela erminea* · Mri = *M. rixosa* · Mfr = *M. frenata* · Mni = *M. nigripes* · Mvi = *M. vison* · Gl = *Gulo luscus* · Tt = *Taxidea taxus* · Sp = *Spilogale putorius* · Mme = *Mephitis mephitis* · Mma = *M. macroura* · Cl = *Conepatus leuconotus* · Lc = *Lutra canadensis* · Fo = *Felis onca* · Fco = *F. concolor* · Fpa = *F. pardalis* · Fwi = *F. wiedii* · Fya = *F. yagouaroundi* · Lyc = *Lynx canadensis* · Lru = *L. rufus* · Tj = *Tayassu tajacou* · Cc = *Cervus canadensis* · Dh = *Dama hemionus* · Dv = *D. virginiana* · Aa = *Alces alces* · Rt = *Rangifer tarandus* · Aam = *Antilocapra americana* · Bb = *Bison bison* · Oa = *Oreamnos americanus* · Om = *Ovibos moschatus* · Oc = *Ovis canadensis* · Od = *O. dalli*

Province	Ba	Pl	Nn	Ma	Mpe	Mer	Mri	Mfr	Mni	Mvi	Gl	Tt	Sp	Mme	Mma	Cl	Lc	Fo	Fco	Fpa	Fwi	Fya	Lyc	Lru	Tj	Cc	Dh	Dv	Aa	Rt	Aam	Bb	Oa	Om	Oc	Od
35 San Bernardinian	×	×										×	×	×	×				×					×			×								×	
34 Diablian	×	×										×	×	×	×				×					×			×									
33 Californian		×										×	×	×	×				×					×			×						×			
32 Mohavian	×													×			×		×					×			×							×		
31 San Matean	×	×						×	×				×	×	×	×	×	×	×					×		×	×	×			×		×	×		
30 Uintan	×	×				×		×	×				×	×	×		×		×					×			×	×			×		×			
29 Navahonian	×								×				×	×	×		×	×	×					×			×	×			×					
28 Mapimian	×	×	×					×					×	×	×	×	×	×	×		×				×		×	×			×		×	×		
27 Yaquinian	×	×	×					×					×	×	×	×	×	×	×	×	×		×	×			×	×			×		×	×		
26 Sonoran	×	×	×										×	×	×	×	×	×	×					×			×				×		×	×		
25 Kaibabian	×	×															×		×					×			×				×		×	×		
24 Artemesian	×					×		×					×	×			×		×				×	×			×	×			×				×	
23 Columbian		×				×		×		×			×	×	×		×		×				×	×		×	×	×			×				×	
22 Louisianian		×						×		×				×			×		×				×					×							×	
21 Texan	×	×						×		×			×	×			×		×				×	×			×	×							×	
20 Tamaulipan	×	×	×					×					×	×	×		×	×	×	×	×	×	×	×		×	×	×							×	
19 Balconian	×	×					×	×					×	×	×		×		×				×	×			×	×							×	
18 Saskatchewanian		×				×	×	×	×	×	×	×					×		×				×	×			×	×			×				×	
17 Kansan		×						×	×	×			×	×			×		×				×	×			×	×			×				×	
16 Carolinian		×						×		×			×	×			×		×					×				×							×	
15 Illinoian		×		×				×	×				×	×	×		×		×									×							×	
14 Alleghanian		×		×	×	×	×	×		×			×	×			×		×				×	×		×		×	×	×						
13 Sierran	×	×		×	×	×	×	×					×	×			×		×					×			×			×					×	
12 Humboldtian	×	×		×	×	×		×		×			×	×			×		×				×	×		×	×	×								
11 Oregonian		×		×	×	×		×		×			×	×			×		×				×	×		×	×	×					×			
10 Vancouverian				×	×	×		×		×							×		×				×			×		×					×			
9 Coloradan		×		×	×	×	×	×	×	×		×	×	×			×		×				×	×		×	×	×	×	×	×		×	×		
8 Montanian		×		×	×	×	×	×	×		×	×	×	×	×		×		×				×	×		×	×	×	×	×	×		×		×	
7W W. Canadian				×	×	×	×	×			×						×		×							×	×	×	×	×				×		
7E E. Canadian				×	×	×	×										×									×	×	×	×	×						
6E E. Hudsonian				×		×	×	×									×		×				×	×		×		×	×	×						
6W W. Hudsonian				×		×	×	×									×		×				×	×		×		×	×				×	×		
5 Yukonian			×	×		×	×										×		×				×	×		×	×	×					×			
4 Aleutian				×		×	×										×		×				×			×	×							×		
3 Alaskan			×	×		×	×										×		×				×	×		×	×	×					×			
2W W. Eskimoan		×		×	×		×	×									×		×							×	×			×				×		
2E E. Eskimoan	×	×				×		×									×		×							×	×	×		×				×		
1 Ungavan	×	×				×											×		×							×		×	×	×						×

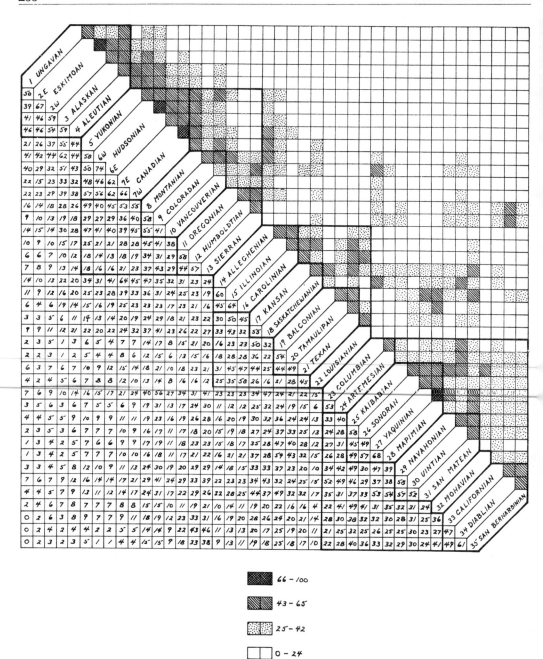

FIG. 2. Final trellis or matrix giving Coefficients of Community (percentage of species common to pairs of provinces). Ordering of provinces is that resulting from cluster analysis. Heavy lines outline superprovinces and subregions. The classes of shading shown in the mirror image are determined by the critical mean CC values used to obtain higher categories of areas.

consisted of laying a transparent overlay over the species IFC map (H & S, Fig. 1) and drawing lines through all regions of high IFC value, delimiting ultimately a total of 86 (rather than the original 24) primary areas. The distribution of these is given as Figure 1. The genus IFC map was not used in the corrected analysis.

Subsequent procedure was that of our earlier paper. Species checklists of each of the primary areas were prepared, and CCs were computed and subjected to cluster analysis. Because of the large number of CCs involved ($86! = 3,741$), calculation of CCs in this and subsequent operations was done by computer.

In our earlier paper (H & S: 137–138), a mammal province was defined as an area with a mean CC of 62.4% or less, when compared to other areas by cluster analysis. This decision was based on the work of Preston (1962), who found that analysis of faunas by means of a "Resemblance Equation" (RE) indicated that values of z (as derived from the RE) of about 0.27 represented the break between faunistic homogeneity and heterogeneity. In our earlier paper we converted z to S (Similarity), where $S = 100 (1 - z)$, and calculated both S and CC for all items in the matrix. These were compared by regression, giving a slope $b = 1.17 \pm 0.02$. Conversion of Preston's critical z value to S gave an S of 73%, and conversion of the critical value of S to CC equaled $73/1.17 = 62.4\%$, and hence our use of this value. We did not, however, allow for the effects of statistical error. If this is incorporated, in the form of plus and minus two standard errors (providing limits at the 95% level of probability), the critical CC value falls within the range 60.30–64.60%. As a result, in this and subsequent papers I propose the use of a CC value of 65% as critical for the determination of mammal provinces. This is a conservative standard, and all values lying between 60 and 65% should be considered suspect, and are reported.

The results of cluster analysis were evaluated according to this new standard. Pairs of primary areas with CCs higher than 65% (as determined by actual calculation or by averaging during cluster analysis) were pooled to create secondary areas. The whole process was then repeated. A new species checklist was prepared for each area, CCs were computed and subjected to cluster analysis, and a new dendrogram prepared. Pooling of areas was again done where CC values were higher than 65%. In all, four such sets of sequential operations were carried out. In the final operation, the total number of primary areas had been reduced from 86 to 38 secondary areas, all but three of which had CCs lower than 65% (2E and 2W, 6E and 6W, and 7E and 7W; see Figs. 2 and 3). These three sets of secondary areas were not pooled because they occur over large geographic areas and because I was concerned with their detailed analysis. The 35 secondary areas with CCs less than 65% constitute mammal provinces by the standards used here and are so treated, although two pairs of these fall within the questionable range 60–65% (15 and 16, 34 and 35, see Fig. 4). Figure 2 is the matrix resulting from cluster analysis, showing ordering of provinces and Coefficients of Community between pairs. Figure 3 is a map showing geographic distribution of the provinces, and Figure 4 is a dendrogram delineating the faunistic relationships existing between provinces, as determined by cluster analysis. A species checklist for each of the provinces, as it was used in the final operation, is given as Table 1.

In our earlier paper (H & S), mammal provinces were grouped into the higher categories of superprovinces, subregions, and regions. The method used in deriving these was to draw lines across the dendrograms at suitable CC levels, the choice of CC level being arbitrary but providing what appeared to be a useful classification (H & S: 139–140, 149). I have tried here to make as little change from the original scheme as possible; however, a small number of minor adjustments have been necessary.

The 0–8% CC range of the dendrogram still stands as a level useful for the category

FIG. 3. Final grouping of primary areas into mammal provinces. Broken lines indicate subdivisions of provinces. The approximate relationships of island faunas are also shown.

of region. At this level, one region, the Nearctic is isolated. Subregions in our earlier paper stood between the 20–25% CC range; this is changed here to the 22–27% CC range, and still encompasses four subregions, following Wallace (1876). The category of superprovince was, in the earlier paper, set at a mean CC level of about 39%. The selection of this value was based on

conclusions reached by Savage (1960), details of which may be obtained from H & S, p. 139–140. The 39% level would, in the case of the dendrogram used here (Fig. 4) give 11 superprovinces. Several cluster at a level very little higher than this, and I have arbitrarily moved the limit up to about 42.5%, so as to encompass these, giving a total of 13 superprovinces. These decisions

→

FIG. 4. Final dendrogram showing relationship between provinces. Ordering of provinces and mean Coefficients of Community (CC) are the results of cluster analysis. Solid vertical lines show mean CC levels at which regions, subregions, superprovinces, and provinces segregate. The three vertical lines for provinces represent the mean critical value plus and minus two standard errors. Per cent similarity is mean Coefficient of Community (CC). The "diamonds" of provinces 2, 6, and 7 represent the mean CCs at which the subdivisions of these provinces pool on cluster analysis.

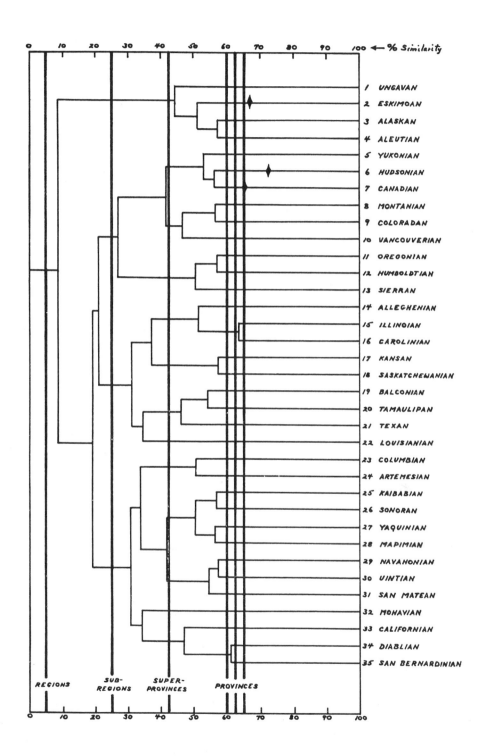

continue to fill the desirable requirements outlined in our earlier paper.

Good single values for each of these hierarchic levels would be: provinces 62.5%, superprovinces 42.5%, subregions 25%, and regions 5%. By this scheme the 35 mammal provinces of North America are grouped into 13 superprovinces, four subregions, and one region. These are named and their distributions mapped on Figures 5a and 5b. They are also blocked out in the matrix (Fig. 2) and marked by lines drawn at appropriate CC levels on the dendrogram (Fig. 4). The problem of nomenclature of mammal areas is discussed subsequently.

The nearest approach to the biotic provinces of Dice (1943), and Kendeigh (1961) through the analysis carried out here, is obtained by fitting a line at about the 54% CC level of the dendrogram. The more intuitive decisions of these workers implies a degree of segregation about 10% lower than the one used here.

Nomenclature and Status of Areas

No changes in the names or status of regions or subregions over those of our earlier paper have resulted from the reanalysis. Because of the increase in numbers of provinces however, minor adjustments in these and in other categories have been obligatory.

As few name changes as possible have been made. Where new names have been applied, an attempt has been made to take them from the literature and to apply them on the basis of priority. Where there has been need to coin names, I have tried to follow the spirit of earlier workers. As an aid to recognition, names of provinces are in the form of adjectives, names of superprovinces in the form of nouns.

Those cases in which provinces segregate within the doubtful 60–65% CC range, or in which segregation occurs at a level only slightly higher than 65%, are here described. More detailed analyses will doubtless result in some changes in status within these groups.

The following provinces are new and have been named by me: no. 1, Ungavan; no. 12, Humboldtian; no. 25, Kaibabian; no. 30, Uintian; no. 31, San Matean; no. 34, Diablian and no. 35, San Bernardinian. Those provinces that are new but that have been given an older name are, together with the source of the name: no. 3, Alaskan (Allen, 1892); no. 5, Yukonian (Cooper, 1859); no. 10, Vancouverian (Van Dyke, 1939); no. 13, Sierran and no. 23, Columbian (Miller, 1951). The names Saskatchewan and Mapimi are here converted to the adjectival forms, Saskatchewanian, and Mapimian.

The Hudsonian (no. 6) is a new province. The name was formerly applied to the province here termed Canadian (no. 7), and the Canadian of earlier workers is here termed Alleghenian (no. 14), following the precedence set by Cooper (1859), and Allen (1892), after Kendeigh (1954). The Carolinian (16) is a new province. The name was applied to what is here largely represented by the Louisianian province (22), in our earlier paper. Its present application is the correct one, however, by the standards of older workers. The Louisianian should properly be called the Austroriparian, following Dice (1943), Kendeigh (1961), and H & S. Because it has superprovince as well as province status, and requires the nounal form of the same as well as an adjectival one, I have applied Allen's (1892) terminology to it.

Not all provinces segregate clearly. The northern limit of no. 3, the Alaskan, was difficult to locate, it being a region of broad transition. Its mapped limit is relatively arbitrary. The Sitkan province of H & S is here included in the Yukonian (no. 5), and the Vancouverian (no. 10), clustering with the former with a CC of 76%. That part of the Yukonian province made up of the Brooks Range very nearly segregates with a CC of 66%.

The Eskimoan, Hudsonian, and Canadian provinces (nos. 2, 6, and 7) have, for reasons given elsewhere, each been split into eastern and western components. Of these,

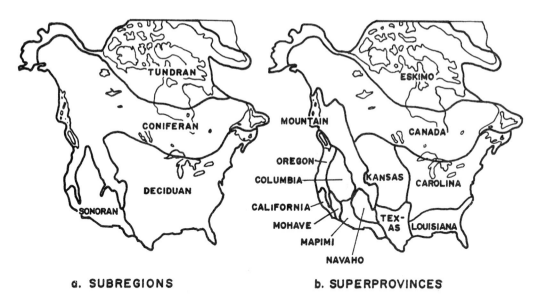

a. SUBREGIONS   b. SUPERPROVINCES

Fig. 5, a and b. The distribution of mammal subregions and mammal superprovinces, as determined from the dendrogram.

the Eskimoan components cluster with a CC of 67%, and the Canadian with a CC of 66%. These are nearly critical values, indicating that the components almost merit province status.

The Oregonian (no. 11) almost segregates into western coastal and eastern Cascadian provinces, pooling with a CC of 66%. The Alleghenian (no. 14) segregates from the eastern component of the Canadian (no. 7) with a CC of only 64%, a critical value. It clusters with the Carolina superprovince on analysis, however, and does so with province rating.

The Carolinian (no. 16) is not a clearly defined province and probably should have been pooled with the Illinoian (no. 15), the two clustering with a CC of 64%. The Carolinian is made up of three distinctive geographic components; that east of the Appalachian Mountains clusters with the rest of the province with a CC of 67%, and that of the Ozark Mountains clusters at 66%. Two of these components are distinct enough from the Illinoian that I have provisionally kept the Carolinian as a full province.

The Balconian (no. 19) was incorrectly identified by H & S as part of what is here called the Tamaulipan (no. 20). The Balconian in its present sense stands as a full province.

The Louisianian (no. 22), under the name Austroriparian, was in part identified as the Carolinian by H & S. Its distribution as determined by reanalysis is the more realistic one.

The larger part of the Columbian province (no. 23) was named Artemesian by H & S. The latter term is here applied to a restricted portion of the Columbian as province no. 24. The Palusian of H & S is here pooled with the Columbian, with a CC of 70%.

The Mohavian province (no. 32) presents something of a puzzle. It was not recognized through examination of the IFC map but appeared through scrutiny of individual species maps. Once recognized, however, cluster analysis caused it to segregate out to the extent of meriting superprovince status, and I have accepted it as this. Its geographic limits, however, have been determined subjectively, and they should be considered as suspect.

The Diablian (no. 34) is also of uncertain status, as it clusters with the San Bernardinian (no. 35) with a CC of 61%, a critical value. Because not all of the latter occurs in the area studied, its analysis is incomplete, and for this reason the distinction is provisionally accepted here.

Of superprovinces, the Texas, Columbia, Mapimi, and Mohave are new, and the names Hudson and Austroriparian of H & S are replaced by the names Canada and Louisiana, for reasons given elsewhere.

Insular Faunas

Sixty-four species (27%) of the total mammal fauna occur on the larger islands adjacent to the continent and on the islands of the Great Lakes. Insular faunas were in each case compared with the faunas of several of the nearest mainland provinces by means of the Coefficient of Community and Simpson's Coefficient (SC). The latter is a measure of the percentage of species occurring on an island that also occur in in the mainland province (Simpson, 1943; H & S: 131). Results are given in Table 2.

Most island faunas give a CC much lower than Preston's critical 65% value when compared with the faunas of adjacent mainland provinces (Table 2). Most islands would therefore merit full province status, if this standard were to be applied. The generally low CC obtained, however, is the result of the small size of insular faunas, a bias being introduced as a consequence of it. Simpson's Coefficient is, in these circumstances, a more reliable measure, and I attribute greater significance to it. No critical value of SC is available, however. Because of this, and because there is more interest in similarities than dissimilarities, island faunas have in all cases been named as part of the fauna of the adjacent province to which they show nearest relationship as determined primarily by Simpson's Coefficient.

TABLE 2.

Island	No. of species	Adjacent provinces	CC	SC
Long Island	29	14 Alleghanian	51	93
		16 Carolinian	52	61
Cape Breton	31	14 Alleghanian	55	94
		7E E. Canadian	64	87
Prince Edward Island	29	7E E. Canadian	68	93
		14 Alleghanian	48	90
Anticosti	5	6E E. Hudsonian	17	100
		7E E. Canadian	13	100
		14 Alleghanian	10	100
Newfoundland	12	6E E. Hudsonian	40	100
		7E E. Canadian	28	92
		14 Alleghanian	19	83
Belcher	3	1 Ungavan	25	100
		6E E. Hudsonian	10	100
		7E E. Canadian	3	33
Manitoulin	25	14 Alleghanian	49	100
		7E E. Canadian	62	96
Isle Royale	9	7E E. Canadian	24	100
		14 Alleghanian	18	100
Arctic Archipelago	10	2E E. Eskimoan	67	100
		1 Ungavan	57	80
Kodiak	12	5 Yukonian	33	100
		4 Aleutian	57	92
		3 Alaskan	39	85
Alexander Archipelago	21	10 Vancouverian	61	95
		5 Yukonian	44	90
Queen Charlottes	12	10 Vancouverian	36	92
		5 Yukonian	24	83
Vancouver	22	11 Oregonian	33	83
		10 Vancouverian	54	79

Long Island shows closest relationship to the Alleghenian province (no. 14), not the Carolinian (no. 16), as might have been expected. Cape Breton is nearest to the Alleghenian by Simpson's Coefficient; I have placed it there though it shows a very high CC with the eastern Canadian (64%). Prince Edward Island lies nearest to the eastern Canadian province (no. 7E), which is surprising in the light of its geographic proximity to the Alleghenian. Anticosti is grouped with the eastern Hudsonian (no. 6E), on the basis of its high CC, as is Newfoundland, the latter on the basis of both coefficients. Belcher Island is, because of its CC only, treated as being most closely related to the Ungavan (no. 1). Manitoulin has highest SC with the Alleghenian, highest CC with the eastern Canadian, but because greater weight is given to Simpson's Coefficient, I have grouped it with the Alleghenian. Isle Royale shows closest affinity with the eastern Canadian province.

Of the islands of the west coast, the Alexander Archipelago and the Queen Charlotte Islands show closest relationship to the Vancouvarian (no. 10). Vancouver Island, on the basis of its SC only, is closest to the Oregonian (no. 11).

Kodiak Island has highest CC with the Aleutian (no. 4), but highest SC with the Yukonian (no. 5), and following the policy set previously is considered most closely related to the latter.

Of the Arctic archipelago and Greenland, the following groups of islands have identical faunas: group 1, Baffin, Southampton, and Coats islands; group 2, Somerset Island; group 3, Banks Island; group 4, Greenland, Sverdrup Islands, Borden and Prince Patrick islands; group 5, Victoria, Prince of Wales, Melville, Bathurst, Cornwallis, Devon, and Ellesmere islands. The faunas of these groups of islands, together with those of the Ungavan and eastern Eskimoan provinces (nos. 1 and 2E) were analyzed by first computing Coefficients of Community between them, then subjecting these to cluster analysis, using the methods outlined previously. All of the island groups cluster at a mean level of 61% or higher, falling within the critical range or better. The groups taken together segregate from the eastern Eskimoan with a mean CC of 54% and from the Ungavan with a mean CC of 44%. The equivalent mean SCs are 100% and 85% respectively. As a result I have treated all of the Arctic archipelago and Greenland as part of the eastern component of the Eskimoan province.

A generalized mapping of these relationships is given in Figure 3. It should be noted that the affinities of Cape Breton, Prince Edward and Long islands are indicated incorrectly here.

Discussion

The general conclusions reached as a result of this re-evaluation differ in no way from those obtained through our earlier analysis (H & S; 147–151), and they are not treated further. It is important that the subjectivity of the methods used here be kept in mind however. The sources and attempted controls of these have been discussed in our earlier paper (H & S: 148–149, 151) and include taxonomic errors, distributional errors, choice of point or block for sample; size of sample block, fitting of isarithms, selection of primary areas, choice of coefficient of association, choice of clustering method, and others.

The methods used here are ideally suited to computer techniques. This reanalysis could not in fact have been completed within reasonable time had such techniques not been available. Miller, Parsons, and Kofsky (1960) have described the use of so-called successive scanning mode microdensitometers, which automatically map the densities of films and other kinds of transparencies. Such devices are sold by Beckman and Whitley of San Carlos, California, under the registered trade name of Isodensitracer. The use of such a device on a transparent map showing the distribution of all North American mammals drawn in inks or paints which gave progressively less translucency as additional layers were added, would be ideal in the development of more refined IFC maps.

The IFC map used in this work (H & S, Fig. 1) was based on the computation of the percentage of species whose ranges ended within blocks 50 miles to a side. The absolute value of an IFC is a function of size of block (H & S: 148). I suggest that any future use of IFCs incorporate as subscript to values given, a statement of the area of the block in kilometers. Converting size of block used here to square kilometers gives an area per block of approximately 6,500 square kilometers, and the IFCs used here are symbolized as $IFC_{6,500}$. Subscripts made up of a statement of length of side of a block rather than area would be less cumbersome. I suspect however that circles may prove more useful than blocks as sampling units, especially if microdensitometers are used, which make the use of area necessary.

Since preparation of our earlier study, a number of similar papers have been drawn to my attention or have been published. Munroe (1956) gave a fine account of the ecologic and zoogeographic features of Canada and an analysis of the insect faunas of the continent. Udvardy (1963) provided an excellent analysis of the bird faunas of North America. His methods differed from ours in that, rather than treating all species simultaneously, he grouped them by type of distribution pattern, and then prepared maps showing numbers of species geographically, by type of pattern. By this method he was able to recognize the presence of 17 primary faunas and 25 secondary ones. The methods used, while different in basic respects from those used here, could easily prove to be more useful.

The following should be added to our earlier summary of coefficients of association (H & S: 131–132). Smith (1960) used the term "Faunistic Relation Factor" (FRF) for the Coefficient of Community, and Huheey (1965) called it a "Divergence Factor" (D), when subtracted from 100. Fager (1965) has devised a new coefficient in the form of $100 \ C/\sqrt{n_1 n_2} - \frac{1}{2}\sqrt{n_2}$, where n_1 is less than n_2. Long (1963) gives a review of coefficients and suggests use of an "average resemblance formula" first used by Kulcznski in 1927, and listed in H & S, p. 132.

In our earlier study (H & S: 128–129, 148) we pointed out that our work has been based on Webb's (1950) analysis of the mammals and herpetofauna of Texas and Oklahoma. The work of Ryan (1963) who improved on Webb's technique in analyzing the mammal faunas of Central America was not known to us at the time. Subsequently, Huheey (1965) published an account of further modifications of the technique in the study of the herpetofauna of Illinois. Since the methods used in all of these are related, and because they are similar in principle, their comparison may be of

interest, and I have attempted to do this briefly in the account following. Webb's (1950) method was to lay a grid of sample points at 100-mile intervals on a map of the area to be studied, to prepare a species checklist for each sample point, to compute Coefficients of Community between sample points and then plot these, to draw lines connecting CCs of equal value, providing a form of "contour map," and to consider "valleys" with CCs of 75% or more as "biogeographic regions." A key point underlying Webb's analysis lies in the fact that he found CCs computed in a north–south plane to differ statistically from those computed in an east–west plane. Since he found that the east–west data gave most significant results, he accepted these in the preparation of his final map and rejected the north–south data. Webb deserves commendation for being first, to my knowledge, to devise a numerical technique for biogeographic analyses in two dimensions.

Ryan's (1963) analysis used a methodology only slightly different from that of Webb. Because of the unusual shape of the area studied, the grid of one portion of it was made up of points 100 kilometers (about 62 miles) to a side, of a second portion of it, 50 kilometers (about 31 miles) to a side. CCs, called "Similarity Values" by both Webb and Ryan, were calculated for both the north–south and east–west planes, and both sets of data were used in preparation of the final contour maps, so far as I can determine. Ryan called the contour lines "isobiots." It is not clear whether Webb's 75% rule for biogeographic regions was used.

Huheey's (1965) method differed to some degree from the preceding. It was to lay a grid of 20 miles to a side onto the area to be studied. Within each block of the grid a species checklist was prepared. For each of the four sides of all blocks, a Divergence Factor (D) was computed, where $D = 100-CC$; thus D is the complement of the Coefficient of Community. Huheey refers to the CC by Smith's (1960) term, "Faunistic Relation Factor," or FRF. The average of the four Ds for each block was computed, this being the mean D for that block. Finally, contour lines called "isometabases" were drawn around mean Ds of equal value. From the contour map herpetofaunal regions were described.

Webb's and Ryan's methods, it will be observed, are essentially the same, differing only in distance between sample points and planes in which CCs are computed. Huheey's method, and the method used in this and in our earlier study differ considerably, though they seek identical ends through development of contour maps depicting faunistic change. I have been led to understand that still other techniques and refinements of those discussed here are in preparation. For example, Valentine (1965) reported a study of the distribution of northeastern Pacific molluscan distributions using methods similar to those employed in this and in our earlier paper. It is apparent that there is need for a comparative testing of the several methods of analysis presently at hand, using the same basic materials in each. I plan to attempt such a study.

Earlier (H & S: 129), we mentioned a partial testing of Webb's original method on the mammal fauna of North America. In view of the preceding, a brief account of the testing follows: Webb's method was followed exactly, except that the grid of sample points was placed on a northeast–southwest plane, giving better coverage of certain coastal areas. A number of variations in the planes in which CCs were computed were attempted. These variations included: (1) computing and plotting CCs in the northeast–southwest plane only; (2) doing the same in the northwest–southeast plane only; (3) averaging adjacent CCs taken in both planes and plotting these; (4) plotting highest CCs only of pairs computed in both planes. We did not try Ryan's device of plotting all CCs taken in both planes. However, of the variants tested, none gave results that appeared anywhere near reasonable in terms of what we knew generally of the distribution of biogeographic and ecologic zones. The variants

used by Ryan and Huheey, however, appear to work well on the basis of their evidence, and my conclusions apply in no way to their results.

No attempt has been made to take into account the effects of altitude on mammal distributions. Dice, in his original study of biotic provinces (1943) described such effects in terms of "life belts" and named a number of these. Kendeigh (1954) on the other hand did not see altitude as a confounding factor in delimitation of biotic provinces. He wrote: "A mountain range may have several life zones represented on it, but only a single biotic province, provided there is a similar tendency for specific or subspecific distinctiveness of the fauna in all the zones. The two concepts therefore have quite different objectives."

I have not been able to decide which of the two views applies in studies of the sort carried out here. It should be realized, however, that the methods employed here are capable of analyzing the effects of altitude on distribution, and segregating altitudinal provinces, if they exist, given distribution maps of sufficient accuracy in the first place. The maps used here failed to show details of vertical distribution, and as a consequence this aspect of the problem has not proven solvable.

An attempt was made to analyze altitudinal distribution in a different way. Each species of mammal was given its life zone distribution, this information being collated from a large number of sources, chiefly certain of the North American Fauna Series. The faunas of each of the life zones within provinces occurring in generally mountainous parts of the continent were treated as primary areas, Coefficients of Community were computed, and the results subjected to cluster analysis. The initial results were unsatisfactory, however, and because of this and because of the circularity of reasoning involved, the method was abandoned.

No attempt has been made to relate the distribution of mammal areas to the distribution of other natural units, either physiographic, climatic, or vegetational, although a comparison of Figure 3 to maps showing the distribution of such features (e.g., Lobeck, 1948; Thornethwaite, 1948; Rowe, 1959; Shantz and Zon, 1923), shows that the relationship is very close.

Summary

1. An earlier study demonstrated that range limits of North American terrestrial mammals were grouped, and that regions of faunistic homogeneity could as a consequence be identified.

2. The method used to identify such regions was to compute percentage of species whose ranges ended in blocks fifty miles on a side (IFCs), and to then fit isarithms. Topographic "valleys" in the map represented regions of faunistic homogeneity, or "primary areas," and for 24 of these, species checklists were prepared. The percentage of species common to pairs of primary areas (CCs) were computed, and the results subjected to cluster analysis, using the method of weighted pair-groups with simple averages. This resulted in a matrix and dendrogram showing relationships and ordering of primary areas. Using a conversion of Preston's Resemblance Equation, a CC of 62.5% was considered critical. Primary areas with a CC lower than this were considered "mammal provinces." By this criterion, 22 mammal provinces grouped into nine superprovinces, four subregions, and one region were recognized.

3. Many more primary areas should have been derived from the IFC map. Starting with 86 primary areas and carrying out four sequential sets of cluster analyses leads to the conclusion that a minimum of from 33 to 35 mammal provinces occur in the continent. These are mapped, named, and briefly described. For statistical reasons, the upper limit of Preston's critical value is raised to a CC of 65%.

4. Higher categories of mammal areas are derived by grouping the provinces on the matrix and dendrogram at appropriate mean CC levels. A mean CC of 42.5% gives 13 superprovinces; a mean CC of 25%, four subregions; a mean CC of 5%, one region (the Nearctic). These are mapped, named, and briefly described. In general, provinces are named as adjectives derived from geographic place-names, superprovinces as nouns. Where possible the names of these and of regions and subregions are taken from the literature on a priority basis.

5. Approximately one-quarter of the mammal species of the continent also occur on nearby continental islands. Island faunas are always smaller than those of the adjacent mainland and always show closest faunistic similarity to nearby provinces.

6. The methods used have the advantage of being relatively objective, repeatable, and well-suited to computer operations. The use of a successive scanning mode microdensitometer may prove useful in the preparation of accurate IFC maps.

7. Accounts of several techniques in biogeographic analysis similar in aim and method to those used here have recently appeared. These are briefly compared. There is need for a critical testing and evaluation of these to determine which provides the best basis for further refinement.

8. While the methods used here are suitable for analyzing the effects of altitudinal zonation on distribution, lack of requisite detail in the distribution maps now available makes such analyses impractical.

9. There appears to be a high degree of correlation between the distribution of mammal areas and other kinds of natural areas.

REFERENCES

ALLEN, J. A. 1892. The geographical distribution of North American mammals. Bull. Amer. Mus. Nat. Hist. 4:199–243.

COOPER, J. G. 1859. On the distribution of the forests and trees of North America . . . . Annual Report Board of Regents, Smithsonian Inst.: 246–280.

DICE, L. R. 1943. The biotic provinces of North America. Univ. Michigan Press, Ann Arbor.

FAGER, E. W. 1963. Communities of Organisms. In Hill, M. N., The sea, vol. 2, p. 415–437.

HAGMEIER, E. M., and C. D. STULTS. 1964. A

numerical analysis of the distributional patterns of North American mammals. Systematic Zool. 13:125–155.

HUHEEY, J. E. 1965. A mathematical method of analyzing biogeographical data. 1. Herpetofauna of Illinois. Amer. Midl. Nat. 73:490–500.

JACCARD, D. 1902. Gezetze der Pflanzenvertheilung in der Alpinen Region. Flora 90:349–377.

KENDEIGH, S. C. 1954. History and evaluation of various concepts of plant and animal communities in North America. Ecology 35:152–171.
1961. Animal ecology. Prentice-Hall, Englewood Cliffs.

KULCZYNSKI, S. 1937. Die pflanzenassoziationen der Pieninen. Bull. Internat. Acad. Polish Sci. Lett. 1927 (2):57–203.

LOBECK, A. K. 1948. Physiographic provinces of North America. Geographical Press, New York.

LONG, C. A. 1963. Mathematical formulas expressing faunal resemblance. Trans. Kansas Acad. Sci. 66:138–140.

MILLER, A. H. 1951. An analysis of the distribution of the birds of California. Univ. Calif. Publ. Zool. 50:531–644.

MILLER, C. S., F. G. PARSONS, I. L. KOFSKY. 1964. Simplified two-dimensional microdensitometry. Nature 202:1196–1200.

MUNROE, E. 1956. Canada as an environment for insect life. Canadian Entomologist 88:372–476.

PRESTON, R. W. 1962. The canonical distribution of commonness and rarity. Ecology 43:185–215, 410–432.

ROWE, J. S. Forest regions of Canada. Bull. Canadian Dept. Northern Affairs and Natural Resources 123, p. 1–71.

RYAN, R. M. 1963. The biotic provinces of Central America as indicated by mammalian distribution. Acta Zoologica Mexicana 6:1–55.

SAVAGE, J. M. 1960. Evolution of a peninsular herpetofauna. Systematic Zool. 9:184–212.

SHANTZ, H. L., and R. ZON. 1924. Atlas of American agriculture. Pt. 1, Sect. E, Natural Vegetation. U.S.D.A., Bureau Agric. Economics, p. 1–29.

SIMPSON, G. G. 1943. Mammals and the nature of continents. Amer. J. Sci. 241:1–31.
1964. Species density of North American Recent mammals. Systematic Zool. 13:57–73.

SMITH, H. M. 1960. An evaluation of the biotic province concept. Systematic Zool. 9:41–44.

SOKAL, R. R., and P. H. E. SNEATH. 1963. Principles of numerical taxonomy. Freeman, San Francisco.

THORNETHWAITE, C. W. 1948. An approach to a rational classification of climate. Geogr. Rev. 38:55–94.

UDVARDY, M. D. F. 1963. Bird faunas of North America. Proc. 13th Internat. Ornithol. Congress:1147–1167.

VALENTINE, J. 1965. Numerical analysis of north-eastern Pacific molluscan distributions. Paper read to the Pacific Division of A.A.A.S., Riverside, California, June 24.

VAN DYKE, E. C. 1939. The origin and distribution of the coleopterous insect fauna of North America. Proc. 6th Pacific Science Congress, 4:255–268.

WALLACE, A. R. 1876. The geographical distribution of animals. 2 vols. Macmillan, London.

WEBB, W. L. 1950. Biogeographic regions of Texas and Oklahoma. Ecology 31:426–433.

EDWIN M. HAGMEIER is a member of the Department of Biology at the University of Victoria, Victoria, B. C., Canada. The work was supported by the University and by the National Research Council. Mr. Peter Darling of the University Computer Center has played an important part in the analyses reported here. The illustrations were prepared by Miss Elizabeth Swemle. The writer extends his thanks to both, and dedicates the study to S.E.H.

ANALYTICAL ZOOGEOGRAPHY OF NORTH AMERICAN MAMMALS[1]

John W. Wilson, III

Department of Biology, George Mason University, Fairfax, Va. 22030

Received January 22, 1973

Latitudinal gradients in species diversity represent, perhaps, one of the longest recognized and most discussed biogeographical phenomena. Simpson (1964) studied the species density pattern of the mammals of North America and found a trend toward increased species density in the tropics; Terent'ev (1963) found similar results in the U.S.S.R. Simpson points out that the increase does not occur in all of the mammals, but in certain groups. The purpose of this paper is to begin the analysis of different mammalian taxa, and to show the significance of topographic relief and climate on which Simpson did not calculate statistics.

METHODS

The material for this paper conforms as much as possible to that used by Simpson (1964). A grid of quadrats 150 miles square (22,500 square miles) centered at 40°N, 100°W was imposed on a map of Lambert's azimuthal equal-area projection (Goode's series by Henry M. Leppard, ed., The University of Chicago Press) of North America to the Panama-Colombia border.

Species lists were generated for each quadrat using the species distribution maps in Hall and Kelson (1959) for mainland species. Whenever Hall and Kelson suspected two species were only subspecies of the same species, I recorded them as one species. In all, species lists were made for 445 quadrats with a total of 671 species of mammals, which were the basis for all analyses. A copy of these species lists is included in Wilson (1972).

The topographic relief (difference in elevation of highest and lowest points within quadrats) in quadrats was determined to the nearest 100 feet (30.5 m) using relief maps in *The World Book Atlas*. Values of the estimated actual evapotranspiration (AE), which is the actual amount of water evaporated from the ground or transpired by plants per year, and which are highly correlated with net annual above ground productivity (Rosenzweig, 1968), were determined from a contour map made by Rosenzweig (pers. comm.), drawn with values of AE calculated from data published by Thornthwaite Associates (1964). Both mean AE and range of AE per quadrat were determined for each quadrat.

To determine the effects of topographic relief and AE on species density of mammals in the temperate region, I used the quadrats between 40° and 50°N, where there is little latitudinal effect on species density, and compared species density in regions of high (5000 feet or more) and low (1000 feet or less) topographic relief and high (600 mm/yr. or more) and low 400 mm/yr. or less) AE. Coastal quadrats which were less than half land were not used.

The distribution of species among higher taxa in faunas, using the Shannon-Weaver information measure, was termed taxonomic hierarchical diversity (THDM). It was designed to be analogous to the hierarchical partitioning of species diversity proposed by Pielou (1967) and is derived in Wilson (1972). The equation used for calculating the measure was:

[1] Portions of this paper were presented at meeting of the Society of Vertebrate Paleontology, 1970; and the Society for the Study of Evolution, 1971.

$$THDM = HD_o + \sum_{i=1}^{o} \frac{S_i}{S} HD_{F,i}$$

$$+ \sum_{i=1}^{o} \sum_{j=1}^{r_i} \frac{S_{ij}}{S} HD_{G,ij}$$

where

$$HD_{G,ij} = - \sum_{k=1}^{g_{ij}} \frac{S_{ijk}}{S_{ij}} \log_e \frac{S_{ijk}}{S_{ij}}$$

where o is the number of orders, F,i is the number of families in the ith order, G,ij is the number of genera in the jth families of the ith order, S is the number of species in the fauna, S_i the number of species in the ith order, S_{ij} the number of species in the jth family of the ith order, HD_O is the diversity of species divided among orders, HD_F is the diversity of species divided among genera. Such a measure of taxonomic hierarchical diversity will have two components: the number of higher taxa and the equitability of the distribution of species among those higher taxa.

The choice of 150 mile square quadrats used by Simpson (1964) is somewhat arbitrary. Large quadrats on the order of 500 miles would probably be inadequate to resolve important zoogeographic changes. On the lower end of the scale the distributional data on the whole are certainly not accurate within 10 miles. Murray (1968), while criticizing conclusions of studies based on data gathered by superimposing quadrats on Hall and Kelson's range maps, points out that Simpson avoids many problems with the basic data by his choice of a relatively large quadrat. Also, supplementary data such as AE are not accurate to much less than 50 miles.

There are limitations with all range maps which draw a smooth line around all peripheral records and indicate the contained area to be a continuous distribution of the species. Species never occur over all of the included range; however they probably are not absent from anything approaching 22,500 square miles. There must be many errors in the distribution maps of individual species. However, it is not likely that there is any systematic bias

of more errors in any one region than others (Simpson, 1964) since the same people reviewed all species. This is a very important point; since there must be errors, a lack of systematic bias makes it likely that the trends I find will not be changed by the constant revision and improvement in our knowledge of the ranges and systematics of individual species of mammals. Unfortunately this lack of systematic bias between areas of taxonomic groups is probably not true in the data for any other part of the world, because no person has reviewed all of the mammals of another entire continent, except Australia, and the mammal faunas are not as completely known in the rest of the world. Hall and Kelson also attempt to produce maps which reflect reasonably natural conditions before gross environmental changes by man have changed drastically many species distributions, especially of larger mammals.

This method overestimates sympatry in quadrats in three ways. (1) When distribution maps overlap, species may never contact each other because of habitat differences. (2) Distribution maps are drawn to include all marginal records of the species. It is doubtful that a species ever inhabits all marginal points at the same time, or in fact, that many marginal points can be considered as places where the species is reproductively successful in even the short term; the presence of the species in these areas may be dependent on continued immigration. (3) The species was included in the species list of a quadrat if its distribution included any part of the quadrat. This will overestimate sympatry most in topographically complex areas.

This study considers only the number of species, genera, families, or orders in the quadrats; it does not take into account relative abundance of individuals or trophic complexity. Such studies would be quite informative, but censuses of mammals which contain no bias in counts of individuals for different taxonomic groups are difficult to compile and do not exist in the literature.

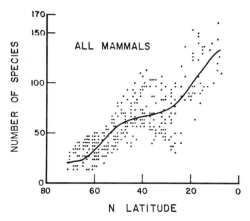

FIG. 1. Relationship of the number of species of all mammals in each quadrat and degrees north latitude shown with best fit polynomial regression line (up to the fifth degree) to indicate the central tendency of the plot.

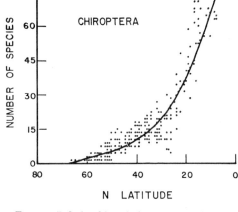

FIG. 3. Relationship of the number of species of Chiroptera in each quadrat and degree north latitude shown with best fit polynomial regression line.

Latitudinal Effects

The general latitudinal increase in species density of the mammals was reported by Simpson (1964) and may be seen in a scatter plot of species against latitude (Fig. 1) and a contour map (Fig. 4). However the increase of species density into the tropics from the temperate regions is due primarily to the order Chiroptera,

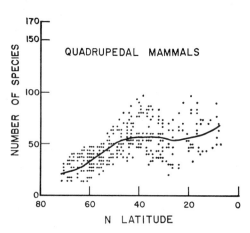

FIG. 2. Relationship of the number of species of quadrupedal mammals (all mammals except the Chiroptera) in each quadrat and degrees north latitude shown with best fit polynomial regression line.

as can be seen in scatter plots of the quadrupedal mammals (the total mammal fauna with the Chiroptera removed) (Fig. 2) and the bats (Fig. 3) as well as in contour maps (Fig. 5, 6). In the graph (Fig. 2) and the contour map (Fig. 5) of quadrupedal mammals, the regions of highest species density in the tropics and temperate region have approximately equal values. Both of these regions have high topographic relief. The bats, unlike the quadrupedal mammals, show a marked latitudinal increase in species density.

While there is no increase in the species density of quadrupedal mammals from the temperate zone to tropics, there is some increase in the genus density (Fig. 7, 8), which is primarily due to the orders Marsupalia, Edentata, and Primates. With the exception of the genus *Marmosa* (Marsupalia) all of the genera of these orders are monotypic in the region studied. In the orders of quadrupedal mammals characteristic of the temperate region (Insectivora, Lagomorpha, Rodentia, Carnivora, Artiodactyla) there is a decline in genus density from temperate to tropical regions (Fig. 9). The pattern of species-per-genus of all mammals and quadrupedal mammals against latitude are essentially the same

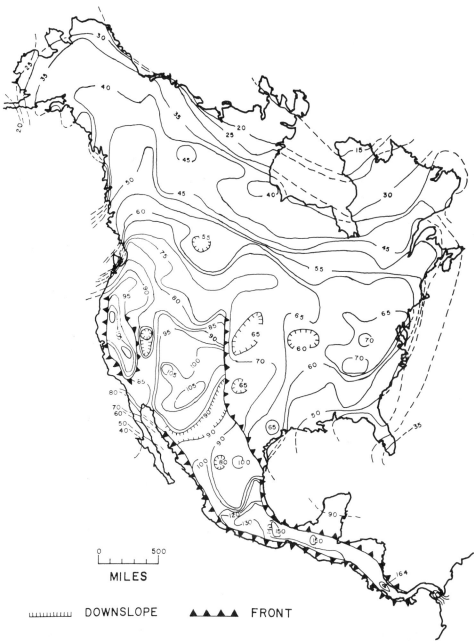

MILES

⊔⊔⊔⊔⊔⊔⊔ DOWNSLOPE ▲▲▲▲▲ FRONT

ALL MAMMALS

FIG. 4. Contour map of the number of species of all mammals in each quadrat, 150 miles square. The contour interval between isograms is 5 species in the part of the map north of the approximate position of the Mexico-United States border south of that the interval is 10. The "fronts" are lines of exceptionally rapid change that are multiples of the contour interval for the region. Indication of the downslope side of a contour is given in the areas where this is not obvious at first sight.

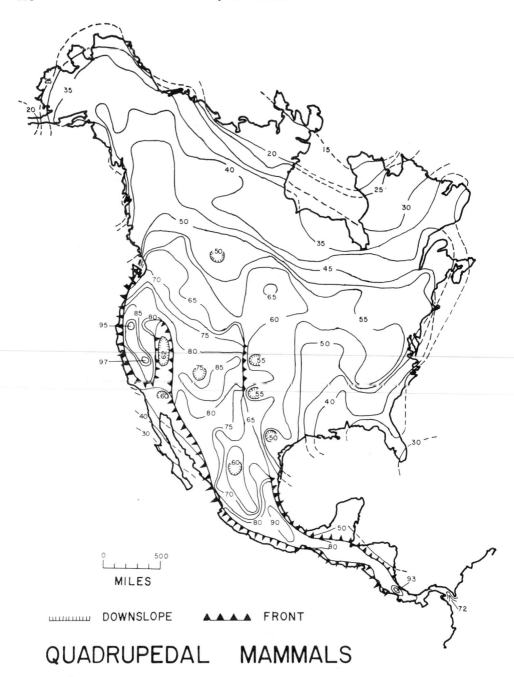

QUADRUPEDAL   MAMMALS

FIG. 5. Contour map of the number of species of quadrupedal mammals (all mammals except the Chiroptera) in each quadrat. The contour interval between isograms is 5 species north of the southern edge of the Mexican Plateau and 10 south of there.

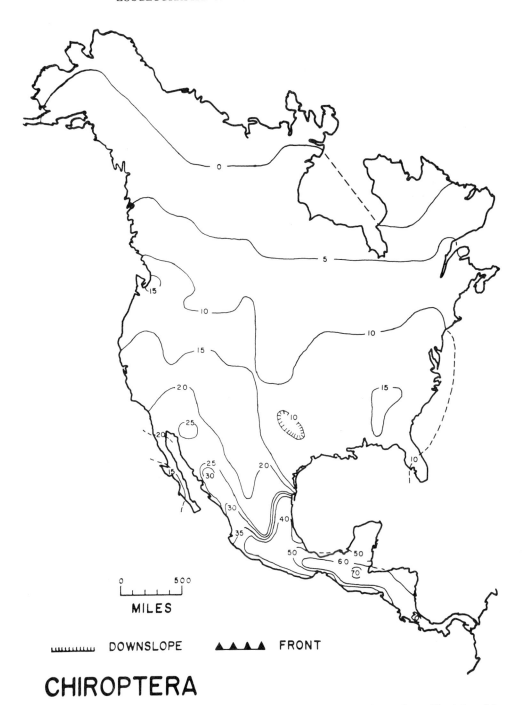

MILES

‿‿‿‿‿‿ DOWNSLOPE ▲▲▲▲ FRONT

CHIROPTERA

FIG. 6. Contour map of the number of species of Chiroptera in each quadrat. The interval between isograms is 5 species north of the southern edge of the Mexican Plateau and 10 south of there.

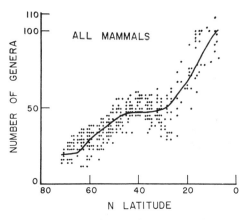

FIG. 7. Relationship of the number of genera of all mammals in each quadrat and degrees north latitude shown with the best fit polynomial regression line.

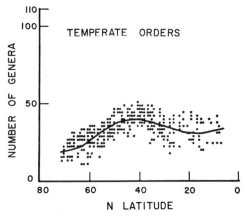

FIG. 9. Relationship of the number of genera in orders of quadrupedal mammals characteristic of the temperature region (Insectivora, Lagomorpha, Rodentia, Carnivora, Artiodactyla) and degrees north latitude shown with the best fit polynomial regression line.

(Fig. 10). The removal of the bats does not significantly change the species-per-genus ratio of mammal faunas. The highest values of the species-per-genus ratio are found in the temperate region.

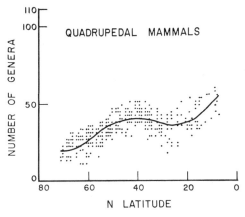

FIG. 8. Relationship of the number of genera of quadrupedal mammals (all mammals except the Chiroptera) in each quadrat and degrees north latitude shown with the best fit polynomial regression line.

→

FIG. 10. Relationship of the ratio of species per genus of (a) all mammals and (b) quadrupedal mammals in each quadrat and degrees north latitude shown with the best fit polynomial regression line.

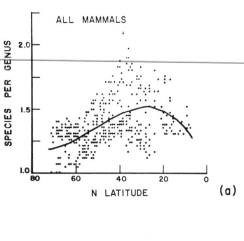

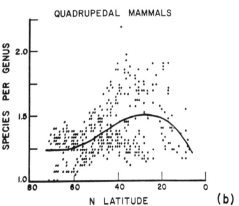

TABLE 1. *Relationship of species density and topographic relief in quadrats between 40°N and 50°N.*

	Region of high topographic relief ($\geqslant$ 5000 feet or 1524 m)	Region of low topographic relief ($\leqslant$ 1000 feet or 304.8 m)	Significance
Number of quadrats	33	27	
Mean Relief	8427 feet (2568 m)	674 feet (205 m)	
Number of all mammal species	82.9	61	$P \ll .001$
Number of quadrupedal mammal species	70.5	52	$P \ll .001$
Number of Chiroptera species	12.5	9	$P < .001$

TABLE 2. *Correlation of species density and topographic relief within regions of high and low relief from quadrats between 40°N and 50°N.*

	Region of High Topographic relief ($\geqslant$ 5000 feet or 1524 m)		Region of low topographic relief ($\leqslant$ 1000 feet or 305 m)	
	Correlation coefficient	Significance	Correlation coefficient	Significance
Number of all mammal species	.58	$P < .001$	.36	$P < .1$
Number of quadrupedal mammal species	.61	$P < .001$	.47	$P < .01$
Number of Chiroptera species	.25	n.s.	−.17	n.s.

TABLE 3. *Relationship of species density and actual evapotranspiration in quadrats between 40°N and 50°N.*

	Region of high actual evapo-transpiration ($\geqslant$ 600 mm/yr.)	Region of low actual evapo-transpiration ($\leqslant$ 400 mm/yr.)	Significance
Number of quadrats	24	34	
Mean actual evapotranspiration	637.5	339.7	
Mean range of actual evapotranspiration per quadrat	27.1	151.5	$P \ll .001$
Number of all mammal species	63.9	79.4	$P \ll .001$
Number of quadrupedal mammal species	54.2	68.1	$P \ll .001$
Number of Chiroptera species	9.8	11.3	$.02 < P < .05$

Effects of Topographic Relief and Actual Evapotranspiration

High species densities are positively associated with regions of high topographic relief (Table 1); this holds for the entire mammal fauna and also for both bats and quadrupeds separately. Within the region of high topographic relief, the species densities of quadrupeds and of all mammals are positively correlated with topographic relief (Table 2). In the region of low relief, only quadrupeds are significantly correlated with relief. In the tropics, the correlation of the species density of quadrupeds and topographic relief in quadrats of high relief is .475 (other correlations are not significant and there are no quadrats of low relief in Central America) which is significantly different from the correlation of quadrupeds with relief in the temperate region at the .001 level by the z transformation.

High species densities of the quadrupeds, bats and all mammals are associated with the region with low values of AE (Table 3). These regions have a much

TABLE 4. *Correlation of species density and actual evapotranspiration within regions of high and low evapotranspiration from quadrats between 40°N and 50°N.*

	Region of High Actual Evapotranspiration ($\geqslant$ 600 mm/yr.)			
	Correlation coefficient with AE	Significance	Correlation coefficient with range of AE/quadrat	Significance
Number of all mammal species	−.08	n.s.	.05	n.s.
Number of quadrupedal mammal species	−.27	n.s.	.24	n.s.
Number of Chiroptera species	.40	$P < .05$	−.37	$P < .1$
	Region of low Actual Evapotranspiration ($\leqslant$ 400 mm/yr.)			
Number of all mammal species	−.37	$P < .05$	.63	$P < .001$
Number of quadrupedal mammal species	−.30	$P < .1$	.58	$P < .001$
Number of Chiroptera species	−.46	$P < .01$	.56	$P < .001$

wider range of values of AE within quadrats due to topographic relief. Within the region of low AE, species densities are negatively correlated with AE, but positively correlated with the range of AE per quadrat (Table 4). Within the region of high AE, bats are positively correlated with AE, and negatively correlated with the range of AE per quadrat. The negative correlation of species density with AE is not significant for quadrupedal and all mammals when the effects of the range of AE are removed by partial correlation, but is significant for the bats.

Hierarchical Diversity

Ordinal level.—The number of orders of mammals increases with latitude (Fig. 11). The ordinal portion of the taxonomic diversity measure for all mammals does not vary greatly with latitude (Fig. 12). In the tropics the value of the measure increases sharply and is higher for the quadrupedal mammals alone than for the total mammal faunas.

Family level.—The number of families increases in the tropics for all mammals and only slightly in the quadruped (Fig. 13). The family portion of the diversity measure increases slightly in the tropics for all mammals and drops sharply in

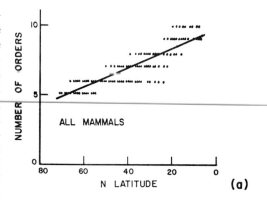

(a)

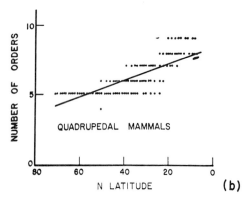

(b)

FIG. 11. Relationship of the number of orders of (a) all mammals and (b) quadrupedal mammals in each quadrat and degrees north latitude shown with a linear regression line.

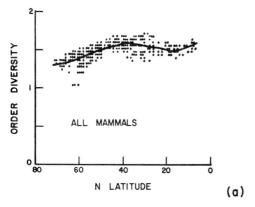

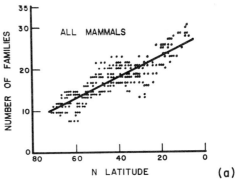

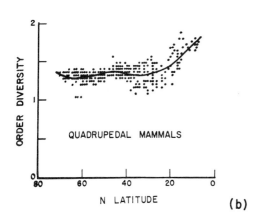

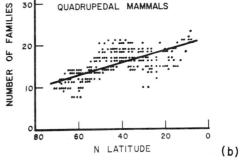

FIG. 13. Relationship of the number of families of (a) all mammals and (b) quadrupedal mammals in each quadrat and degrees north latitude shown with a linear regression line.

FIG. 12. Relationship of the ordinal portion of the taxonomic hierarchical diversity measure for (a) all mammals and (b) quadrupedal mammals in each quadrat and degrees north latitude shown with the best fit polynomial regression line.

quadrupeds (Fig. 14). The values of the family portion of the measure are consistently lower than values of the ordinal portion of the measure.

Generic level.—Only the generic portion of the taxonomic diversity measure shows any marked latitudinal change in value (Fig. 15). The pattern of the values of the generic portion of the measure resembles the pattern of genus density for the total mammal fauna and the quadrupedal mammals alone.

When the three levels are summed to form a taxonomic hierarchical diversity measure and this is plotted against latitude (Fig. 16), the variations of the ordinal and family portions essentially cancel each other except in the far north, and most of the change is due to the generic portion of the diversity measure.

DISCUSSION

It is possible that the lack of increase in species density of quadrupedal mammals is not a latitudinal effect, but is due to the small land area of the isthmus of Central America. To show that this is a general feature of tropical faunas, I generated a species list from Cabrera (1957–1960) of British Guiana (recommended by Hershkovitz as the best known South Amer-

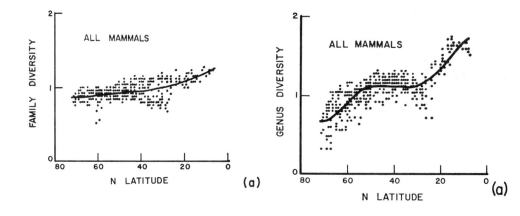

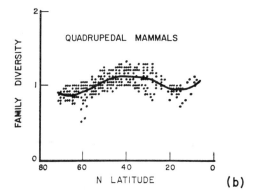

Fig. 14. Relationship of the family portion of the taxonomic hierarchical diversity measure for (a) all mammals and (b) quadrupedal mammals in each quadrat and degrees north latitude shown with the best fit polynomial regression line.

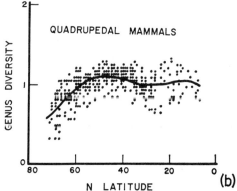

Fig. 15. Relationship of the generic portion of the taxonomic hierarchical diversity measure for (a) all mammals and (b) quadrupedal mammals in each quadrat and degrees north latitude shown with the best fit polynomial regression line.

ican tropical fauna). The fauna of British Guiana is approximately ten per cent smaller than that of Costa Rica for both quadrupeds and bats. On the basis of this evidence, it seems unlikely that any isthmus effect is strong enough to explain the results. Unfortunately, it is not possible to compare directly the faunas of Central America and northern South America because the taxonomy of the South American fauna is taxonomically subdivided to a greater extent (some taxa given species status in Cabrera are listed in subspecies in Hall and Kelson, whose taxonomy I have condensed further) and the South American fauna is not as well known.

Hierarchical Diversity Measure

The measure of taxonomic hierarchical diversity was created to see if species in temperate and tropical faunas were distributed in higher taxonomic categories in a similar manner or if one region contains a fauna where most species are found in a few higher categories while the other has species spread more evenly. Mammal faunas, even ones from large quadrats, do not have many species in the same genus; so the diversity measure at the genus level primarily reflects the number of genera. This is similar to using a species diversity

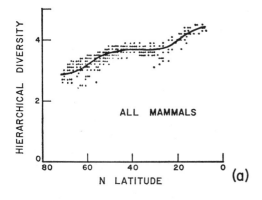

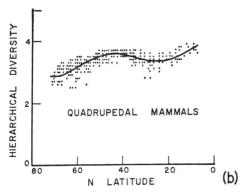

FIG. 16. Relationship of the taxonomic hierarchical diversity measure for (a) all mammals and (b) quadrupedal mammals in each quadrat and degrees north latitude shown with the best fit polynomial regression line.

index; when faunas are equitable, you measure the number of species.

Species in tropical mammal faunas are less equitably distributed among families, due to a larger number of families with few species. Species of the entire mammal fauna are less equitably distributed among orders. The ordinal diversity for quadrupeds increases in the tropics because the most abundant order has been removed which increases the equitability of the rest of the fauna.

The three portions of the diversity measure summed together greatly resemble the pattern of species density. The separate levels of the diversity measure tend to show either no change or a decrease in the equitability of species among higher categories, with the exception of tropical quadrupeds at the ordinal level. This together with the slight general decrease in the species-per-genus ratio in the tropics suggests that the average taxonomic distance between species is higher in the tropics.

Bats

The latitudinal increase in species density of mammals from the temperate zone to the tropics is due primarily to the bats, which are more successful in exploiting new food resources available in the tropics than are the quadrupedal mammals. These new resources are the year-round presence of active insects, fruit, and flowers. Bats utilize additional food resources in the tropics; there are six species of carnivorous bats, two allopatric fishing bats, and two vampires. However, these account for less than one-sixth of the increase in species density from the temperate region to the tropics. Therefore, in order to explain the increased diversity of bats in the tropics, we must examine what aspects of both the physical and biotic environment lead to the year round diversity of this set of food resources.

Tropical Fruit and Flowers

The year-round presence of active insects, fruit and flowers is allowed by the absence of frost in the tropics. Frost would kill or harm insects and plants such that immediate or future supplies of these resources would be interrupted. However, the abundance and diversity of fruit and flowers is increased in the tropics as a direct result of adaptations of plants to increased distance between plants of the same species and a greater number of plant species in an area (spatial heterogeneity of plants). The spatial heterogeneity of plants is a result of the interaction of tropical plants and their insect seed and seedling predators which has been developed in a theoretical paper by Janzen (1970)

and subsequent papers of examples (Janzen, 1971a, 1971b, 1971c, 1972).

Because seeds or seedlings are unlikely
to survive in the close proximity of a parent or adult of the same species, a number
of strategies have been developed by plants
to insure the dispersal of seeds away from
the vicinity of the parent. One of these
strategies is fruit (by this I include any
sweet succulent tissues associated with the
seed regardless of their embryologic origin)
which achieve active organism mediated
seed dispersal with a relatively low expenditure of energy by the plants (Snow,
1971). Fruits are composed primarily of
water and carbohydrate which are much
easier for the plant to produce than protein or lipid.

With an increased distance between
plants of the same species which are reproductive at any one time, there is a
greater dependence on organism mediated
pollination. Nectar (primarily water and
sugar) decreases the cost of pollination to
the plant. Pollinators consume both nectar and pollen. It is more expensive energetically to support vertebrate pollinators
than insect ones, but they have the advantages of being able to fly greater distances
and a greater memory enabling them to
visit daily the same locations over a long
period when plants are widely spaced.
Both fruit and flowers are adaptations of
plants to cope with the problems of spatial heterogeneity which is forced upon
them, by the activity of their insect predators. That both fruit and flowers are food
resources utilized by bats does not mean
that their distribution in the habitat will
be the same as any of their food resources
(Fleming, 1973). They depend nonetheless on the underlying spatial heterogeneity
of the plants which produce their food
resources.

Resources Antecede Bats

During the latter part of the early Cretaceous, the angiosperms (Doyle and
Hickey, 1972) and the insects underwent
an adaptive co-radiation when many of

their important mutual interactions (her-
bivory and defense against it, pollination)
were initiated. The early establishment of
these interactions with subsequent coevolution is evidenced by the fact that many
families of both angiosperms and insects
go back to the Cretaceous. During the
Cretaceous, birds radiated and at this time
very likely some birds initiated their relationship of seed dispersal, fruit eating, and
pollination. Thus the biotic environment
into which mammals radiated contained
pollinating insects and fruit eating birds,
so that the food resources of fruit and nectar were already present when the first
bats evolved sometime in Paleocene or
early Eocene time. The prior presence of
food resources would increase the likelihood that the radiation of bats which en-
abled them to utilize insects, fruit, and
flowers occurred rapidly once the ability
to fly had been evolved.

Bat Predation

Bats are certainly preyed upon, but do
not form the major food sources of predators in general (in contrast to rodents and
passerine birds, Gillette and Kimbrough,
1970). Low predation rates are difficult
to prove empirically, but they may be inferred from the bat's relatively long life-
span (Barbour and Davis, 1969; Paradiso
and Greenhall, 1967; Wilson, 1970, 1971)
and low reproductive rates (Walker, 1964;
Barbour and Davis, 1969). Thus, the
bat's habitus (flying and primarily nocturnal) is a relatively safe one.

Increased Bat Diversity

Nocturnal flight gives bats greater ac-
cessability to the food resources of insects,
fruit, and flowers; increases their mobility
between scattered resources; and increases
their safety. It is no wonder that bats are
so dominant in tropical mammal faunas.

Perhaps the harder question is: Why
are there so many different species? First,
as the seasonal differences in important
environmental variables (temperature and
day length everywhere and moisture in

many regions) decreases, important episodes of environmental fluctuation are
more likely to be randomly distributed in
time (Lloyd et al., 1968) and in a fauna
of several species with different strategies
or temporal differences, species abundances
are more likely to fluctuate with respect
to each other and dominance may change.
It is possible that under these non-equilib-
rium conditions more species may coexist
than under constant conditions which is
a frequent picture of tropical conditions.

Second, all else being equal, it would be
advantageous for plants which depend
upon other organisms for pollination and
seed dispersal to receive as exclusive attention of these organisms as possible during
the time of flowering and fruit ripening.
This would lead to disruptive selection on
bats and their food resources which would
tend to increase the number of co-existing
bat species in an area.

Why No Temperate Fruit Bats?

Bats are primarily a tropical order with
only one family (Vespertilionidae) really
successful in the temperate region. The
vampire bats are unable to survive in regions with mean winter temperatures lower
than 10C (McNab, 1973) because their
food source is not rich in either carbohy-
drates or lipids. Carnivorous bats require
a diverse fauna of bats, small mammals,
birds, lizards, and large insects to maintain
them (McNab, 1971). The one carnivorous bat, *Antrozous pallidus*, which extends
into the temperate zone is primarily an
insectivore in this part of its range (Barbour and Davis, 1969).

In order to be successful in the temperate region a mammal must either remain active, hibernate during the winter,
or migrate out of the region before winter.
Any of these activities requires the developement of a good fat body. The limiting
factor is probably the success of the juvenile which must both mature and de-
velope the fat body. Fruit ripens at the
wrong time, not until at least a third to
half of the way through the growing season. Even then it consists primarily of
water and sugar which is not sufficient for
rapid developement and fattening of the
juveniles. The further north a bat moves
during the summer, the shorter its safe
foraging period (night) becomes. The increased day length which makes migration
to higher latitudes advantageous for diurnal birds is not in the bat's favor. Lastly,
bats are active at night which is cooler
than the day in the temperate region.
Warm seasons at night will be shorter than
those available to diurnal organisms. Bats
are not well insulated compared with birds
whose feathers are both a flight mechanism
and insulation. The food of the insectivore is richer in protein and lipids and is
available for a greater part of the year,
which I believe allows only insectivorous
bats in the temperate region.

Quadrupedal Mammals

While the species density of bats is primarily effected by the shift from tropics
to the temperate region with the presence
of frosts, quadrupeds do not show a reduced species density until 50°N. Their
major adaptive success is the ability to
withstand winters of the length and severity found in the United States without
reduction in the number of species. The
pattern of species density in quadrupeds
appears to be more one of success in the
temperate region due to endothermy than
one of failure in the tropics.

There seem to be a greater number of
adaptive types of quadrupeds in the tropics but no more species. The lower species-
per-genus ratio in the tropics, together
with the greater number of orders and
families which are on the average as old
as temperate families and orders (Van
Valen, 1969) suggest that quadrupeds in
the tropics are more separated taxonomically from each other than in the temperate region. Since mammal taxonomists use,
as characters in the classification of mammals, features important to mammals in
exploiting their food resources, it would
appear that quadrupeds are more widely

spread through an adaptive space than are their temperate counterparts.

Topographic Relief

High species density is associated with high topographic relief because more habitats occur due to altitudinal zonation of the habitat (Simpson, 1964). Janzen (1967) suggests mountain passes of the same altitude should fracture the habitat of organisms to a greater extent in the climatic regime of the tropics. This effect is not seen in quadrupedal mammals where the correlation of species density and topographic relief is significantly stronger in the temperate region than in the tropics. In the temperate region cooler temperatures are usually associated with shorter seasons, whereas higher altitudes in the tropics are cooler, but do not have seasons of different lengths than lower altitudes. Also, tropical mountains may never get warm, even for a short time which might explain the absence of any relationship of bat species density with relief in the tropics. If temperatures are lower, but no change in season length, endothermic mammals might easily be able to inhabit wider altitudinal ranges. My data are not sufficient to explore these relationships further.

It is possible that the interaction of the multiple north-south shifts of the climatic zones in the temperate region during the Pleistocene with the topographically complex western Cordillera led to a greater amount of speciation due to repeated isolation of populations on habitat "islands". Bats show a lower correlation with topographic relief than other mammals. This may be due to lowered density at high altitudes due to low air temperatures at night, which would affect both bats and their flying insect food adversely. Bats would also be less affected by the influence of the Pleistocene since they can fly over less habitable areas.

Climatic Factors

High species density in the region of low actual evapotranspiration (AE) is due to two factors. A greater range of AE in quadrats allows a greater number of habitats and is analogous to having more topographic relief. However, a quadrat with a low mean value of AE and the same range of AE as another with a high mean value of AE will contain a greater number of habitats. This occurs because small differences in the amount of water available become more significant to organisms when water is scarce. This feature can be seen in Holdridge (1967, Fig. 7) where isopleths of AE are plotted on his classification of world plant formations and life zones.

In addition to the number of habitats, there is an important change in food sources available to herbivores. The only grazers (eating the leaves and stems of grass) in all of the vertebrates are mammals. Three orders presently alive in North America contain grazers: artiodactyls, rodents, and lagomorphs; there were other orders in the recent past: perissodactyls, probosidea, and possibly edentates. The leaves and stems of the grasses contain silica (see Gould, 1968) which require the special adaptation of high crowned teeth for grazers to successfully consume grasses. In regions with low values of AE, grasses compose a much greater portion of the herbivorous biomass than in moister areas. This is an important new food resource which only mammals among the vertebrates have been able to utilize and represents a new adaptive zone. There are a wide range of different strategies utilizing this new resource which leads to niche separation of different grazers (Bell, 1970). If there is a diversity of mammalian herbivores, there will be a consequent increase in their carnivores as well.

Kiester (1971) observed a zoogeographic complementarity of mammal and reptile faunas in the United States, suggesting a competition on a large evolutionary scale and the relative exclusion of mammals from regions and habitats where reptiles were successful. I rather think that the reptiles, which are primarily insectivorous and carnivorous, are responding to a density of their prey items which are more abundant in areas of high AE. Mammals in a totally separate feature, contain the only vertebrate grazers and therefore expand their importance in drier regions by becoming far more dominant as the herbivores.

Bats, in regions of high AE, show positive correlation with AE but negative correlation with the range of AE per quadrat. I suspect that this is due to greater insect abundance and diversity in regions with high values of AE; but regions with high ranges of AE per quadrat are regions with higher topographic relief and bats are less successful at high altitudes due to lowered nocturnal temperatures.

SUMMARY

The latitudinal increase of mammal species per unit area from the temperate zone to the tropics is not characteristic of the class Mammalia as a whole, but is primarily due to the order Chiroptera. Bats have evolved a way of life that is relatively predation-free and have filled most of the new niches which are associated with the canopy and terminal branches of trees, where productivity is highest. North of the tropics, where year round fruit, flowers, and active insects constitute typically tropical resources, mammals do not suffer any decrease in density due to increasingly severe winters until they reach the severity found in Canada (north of the 50th parallel). Mammals have evolved homeothermy to avoid consequences of being poikilothermic in an environment with temperature fluctuations.

The ratio of species per genus is generally lower in tropical faunas when compared with temperate ones. Temperate mammals are relative generalists because often they cannot depend on the same year round food source. Tropical mammals are relative specialists, living in a region with year round fruit, flowers, and insects. These resources are more predictable and allow greater specialization but, except for those utilized by bats, appear to be no more abundant. This leads to greater "taxonomic distance" between species in the tropics. Part of this may be due to a greater tendency of mammal taxonomists to split tropical taxa, but at least some portion of the drop in species per genus in the tropics is a real reflection of the difference in selective environment.

The latitudinal trends in present terrestrial communities may be due to the co-evolution of the angiosperms and insects in the absence of a cold season in lower latitudes. The birds and mammals have radiated into this more diverse biotic environment and are often used by the plants for pollination of flowers and dispersal of fruit and seeds.

A measure of hierarchical taxonomic diversity was devised to study the distribution of species among higher taxa in faunas. As faunas increase in size and higher taxa become more numerous, the equitability of the distribution of species among those higher taxa decreases.

ACKNOWLEDGMENTS

This research was supported by a traineeship from the NDEA, Title IV, and NSF predoctoral fellowship, a Ford Foundation Population Biology Grant to the Biology Dept., of the Univ. of Chicago, the L. B. Block Fund, and Henry H. Hinds Fund at the Univ. of Chicago.

I thank Leigh Van Valen for valuable advice. Daniel H. Janzen, Monte Lloyd, and many graduate students answered my questions about the tropics. George Gaylord Simpson allowed me to use his data and sent me materials to begin my work.

LITERATURE CITED

BARBOUR, R. W., AND W. H. DAVIS. 1969. Bats of America. Univ. Kentucky Press, Lexington. 286 p.

BELL, R. H. V. 1970. The use of the Herb Layer by Grazing Ungulates in the Serengeti. *In: Animal Populations in Relation to Their*

Food Resources. Brit. Ecol. Soc. Symp. #10, Adam Watson ed., Blackwell Scientific Publ., Oxford; 1970 p. 111–124.

CABRERA, A. 1957–1960. Catalogo de los mamiferos de America del Sur. Mus. Argentino Cienc. Nat., Revista, Cienc. Zool. 4:1–732.

DOYLE, J. A., AND L. J. HICKEY. 1972. Coordinated Evolution in Potomac Group Angiosperm Pollen and Leaves (Abst.). Amer. J. Bot. 59:660.

FLEMING, T. H. 1973. Numbers of mammal species in North and Central American forest communities. Ecology 54:555–563.

GILLETTE, D. D., AND J. D. KIMBROUGH. 1970. Chiropteran mortality, p. 262–283. *In* B. H. Slaughter and D. W. Walton (eds.) About bats, a chiropteran biology symposium. Southern Methodist Univ. Press, Dallas.

GOULD, F. W. 1968. Grass systematics. McGraw-Hill, New York. 382 p.

HALL, E. R., AND K. R. KELSON. 1959. The mammals of North America. 2 vols. Ronald Press, New York.

HOLDRIDGE, L. E. 1967. Life Zone Ecology. rev. ed. Tropical Science Center, San Jose, Costa Rica.

JANZEN, D. H. 1967. Why mountain passes are higher in the tropics. Amer. Natur. 101:233–249.

———. 1970. Herbivores and the number of tree species in tropical forests. Amer. Natur. 104:501–528.

———. 1971a. Seed predation by animals. Ann. Rev. Ecol. Syst. 2:465–492.

———. 1971b. Escape of juvenile *Dioclea megacarpa* (Leguminosae) vines from predators in a deciduous tropical forest. Amer. Natur. 105:97–112.

———. 1971c. The fate of *Sheelea rostrata* fruits beneath the parent tree: predispersal attach by bruchids. Principes 15:89–101.

———. 1972. Association of a rainforest palm and seed-eating beetles in Puerto Rico. Ecology 53:258–261.

KIESTER, A. R. 1971. Species density of North American amphibians and reptiles. Syst. Zool. 20:127–137.

LLOYD, M., R. F. INGER, AND F. W. KING. 1968. On the Diversity of Reptile and Amphibian species in a Bornean Rain Forest. Amer. Natur. 102:497–515.

McNAB, B. D. 1971. The structure of tropical bat faunas. Ecology 52:352–358.

———. 1973. Energetics and the distribution of vampires. J. Mammal. 54:131–144.

MURRAY, K. F. 1968. Distribution of North American mammals. Syst. Zool. 17:99–102.

PARADISO, J. L., AND A. M. GREENHALL. 1967. Longevity records of American bats. Amer. Midl. Natur. 78:251–252.

PIELOU, E. C. 1967. The use of information theory in the study of the diversity of biological populations. Fifth Berkeley Symp. Math. Stat. Probability, Proc. 4:167–177.

ROSENZWEIG, M. L. 1968. The strategy of body size in mammalian carnivores. Amer. Midl. Natur. 80:299–315.

SIMPSON, G. G. 1964. Species density of North American recent mammals. Syst. Zool. 13:57–73.

SNOW, D. W. 1971. Evolutionary aspects of fruit-eating by birds. Ibis 113:194–202.

TERENT'EV, P. V. 1963. Attempt at application analysis of variation to the qualitative richness of the fauna of terrestrial vertebrates of the U.S.S.R. Vestnik Leningradsk Univ. Ser. Biol. 18:19–26. Translation published by Smithsonian Herpetological Information Services, 1968.

THORNTHWAITE (C.W.) ASSOCIATES. 1964. Average climatic water balance data of the continents, North America. Lab of Climatol., Publ., in Climatology 17:231–615.

VAN VALEN, L. 1969. Climate and evolutionary rate. Science 166:1656–1658.

WALKER, E. P. 1964. Mammals of the World. 3 vols. Johns Hopkins Press, Baltimore.

WILSON, D. E. 1970. Longevity Records for *Artibeus jamaicensis* and *Myotis nigricans*. J. Mammal. 51:203.

———. 1971. Ecology of *Myotis nigricans* (Mammalia: Chiroptera) on Barro Colorado Island, Panama Canal Zone. J. Zool. Lond. 163:1–13.

WILSON, J. W. 1972. Analytical zoogeography of North American mammals. Ph.D. Thesis, Univ. of Chicago. 397 p.

THE WORLD BOOK ATLAS. 1965. Field Enterprises Educational Corp., Chicago. 391 p.

LITERATURE CITED

ADAMS, L., and S. G. WATKINS
    1967. Annuli in tooth cementum indicate age in California ground squirrels. Jour. Wildlife Mgt., 31:836-839.

ALCOCK, J.
    1975. Animal behavior: an evolutionary approach. Sinauer Assoc., Sunderland, Mass., xi+547 pp.

    1939. A checklist of African mammals. Bull. Mus. Comp. Zool., 83:1-763.
    1938, 1940. The mammals of China and Mongolia. Amer. Mus. Nat. Hist., New York, xxv+1-620; xxvi+621-1350 pp.

ALLEN, H.
    1864. Monograph of the bats of North America. Smithsonian Misc. Coll., 7(165): xxiii+1-85.

ANDERSON, S.
    1961. Mammals of Mesa Verde National Park, Colorado. Univ. Kansas Publ., Mus. Nat. Hist., 14:29-67.
    1972. Mammals of Chihuahua, taxonomy and distribution. Bull. Amer. Mus. Nat. Hist., 148:149-410.

ANDERSON, S., and J. K. JONES, JR. (eds.)
    1967. Recent mammals of the world—a synopsis of families. Ronald Press, viii+453 pp.

ANTHONY, H. E.
    1928. Field book of North American mammals. G. P. Putnam's Sons, xxvi+674 pp.

ARMSTRONG, D. M.
    1972. Distribution of mammals in Colorado. Monogr. Mus. Nat. Hist., Univ. Kansas, 3:x+1-415.

ASDELL, S. A.
    1964. Patterns of mammalian reproduction. Comstock Publishing Associates, 2nd ed., viii+670 pp.

BAKER, R. H., and J. K. GREER
    1962. Mammals of the Mexican state of Durango. Publ. Mus. Michigan State Univ., Biol. Ser., 2:25-154.

BANFIELD, A. W. F.
    1974. The mammals of Canada. Univ. Toronto Press, xxv+438 pp.

BEDDARD, F. E.
    1902. Mammalia. Macmillan, xiii+605 pp.

BIRNEY, E. C.
    1973. Systematics of three species of woodrats (genus Neotoma) in central North America. Misc. Publ. Mus. Nat. Hist., Univ. Kansas, 58:1-173.

BLACK, C. C.
    1963. A review of the North American Tertiary Sciuridae. Bull. Mus. Comp. Zool., 130:109-248.

BLACKWELDER, R. E.
    1967. Taxonomy—a text and reference book. John Wiley and Sons, xiv+698 pp.

BLAIR, W. F.
    1968. Introduction. Pp. 1-20, in Vertebrates of the United States (W. F. Blair et al.), McGraw-Hill, ix+616 pp.

BOCK, W. J.
    1969. Discussion: The concept of homology. Ann. New York Acad. Sci., 167:71-73.

BOURLIÈRE, F.
    1954. The natural history of mammals. Alfred A. Knopf, xxi+363+xi pp.

BOWLES, J. B.
    1975. Distribution and biogeography of mammals of Iowa. Spec. Publ. Mus., Texas Tech Univ., 9:1-184.

BROWN, J. H.
    1968. Adaptation to environmental temperature in two species of woodrats, Neotoma cinerea and N. albigula. Misc. Publ. Mus. Zool., Univ. Michigan, 135:1-48.

BURT, W. H., and R. P. GROSSENHEIDER
    1964 A field guide to the mammals. Houghton-Mifflin, xxiii+284 pp.

CABRERA, A.
    1958. Catalogo de los mamíferos de America del Sur [I. Metatheria, Unguiculata, Carnivora]. Revista Mus. Argentino Cien. Nat. "Bernardino Rivadavia," Cien. Zool., 4(1):iv+1-307.
    1961. Catalogo de los mamíferos de America del Sur [II. Sirenia, Perissodactyla, Artiodactyla, Lagomorpha, Rodentia, Cetacea]. Revista Mus. Argentino Cien. Nat. "Bernardino Rivadavia," Cien. Zool., 4(2):xxii+309-732.

CARLQUIST, S.
    1965. Island Life. . . . Natural History Press, viii+451 pp.

CHOATE, J. R.
  1970. Systematics and zoogeography of Middle American shrews of the genus Cryp-
        totis. Univ. Kansas Publ., Mus. Nat. Hist., 19:195-317.
COCKRUM, E. L.
  1962. Introduction to mammalogy. Ronald Press, viii+455 pp.
CRAIGHEAD, J. J., F. C. CRAIGHEAD, JR., J. R. VARNEY, and C. E. COTE.
  1971. Satellite monitoring of black bear. BioScience, 21:1206-1212.
CRANDALL, L. S.
  1964. The management of wild mammals in captivity. Univ. Chicago Press, xv+761
        pp.
CROMPTON, A. W., and A. SITA-LUMSDEN.
  1970. Functional significance of the therian molar pattern. Nature, 227:197-199.
DARLINGTON, P. J., JR.
  1957. Zoogeography. John Wiley and Sons, xi+675 pp.
DAVIS, D. D.
  1964. The giant panda—a morphological study of evolutionary mechanisms. Fieldiana-
        Zool., Mem., 3:1-339.
DAVIS, D. E., and F. B. GOLLEY
  1963. Principles in mammalogy. Reinhold Publishing Co., xiii+335 pp.
DAWSON, M. R.
  1958. Later Tertiary Leporidae of North America. Univ. Kansas Publ., Paleo. Contrib.
        (Vertebrata), 6:1-75.
DE BLASE, A. F., and R. E. MARTIN.
  1974. A manual of mammalogy, with keys to families of the world. Wm. C. Brown
        Co., Dubuque, Iowa, xv+329 pp.
DICE, L. R.
  1952. Natural communities. Univ. Michigan Press, x+547 pp.
DORST, J., and P. DANDELOT.
  1970. A field guide to the larger mammals of Africa. Houghton Mifflin, Boston, 287 pp.
ELLERMAN, J. R.
  1940. The families and genera of living rodents. Volume 1. Rodents other than
        Muridae. British Mus., xxvi+689 pp.
  1941. The families and genera of living rodents. Volume 2. Family Muridae. British
        Mus., xii+690 pp.
  1948. The families and genera of living rodents. Volume 3 [additions and corrections].
        British Mus., v | 210 pp.
ELLERMAN, J. R., and T. C. S. MORRISON-SCOTT
  1951. Checklist of Palaearctic and Indian mammals 1758 to 1946. British Mus., 810 pp.
EMLEN, J. M.
  1973. Ecology: an evolutionary approach. Addison-Wesley, xiv+493 pp.
ENDERS, A. C. (ed.)
  1963. Delayed implantation. Rice Univ. and Univ. Chicago Press, x+318 pp.
EWER, R. F.
  1968. Ethology of mammals. Plenum Press, xiv+418 pp.
FINDLEY, J. S., A. H. HARRIS, D. E. WILSON, and C. JONES.
  1975. Mammals of New Mexico. Univ. New Mexico Press, Albuquerque, xxii+360 pp.
FLINT, V. E., YU. D. CHUGUNOV, and C. M. SMIRIN.
  1965. Mlekopitayushchie SSSR (Mammals of the USSR). Mysl, Moscow, 435 pp.
        (in Russian).
FLOWER, W. H., and R. LYDEKKER
  1891. An introduction to the study of mammals living and extinct. Adam and Charles
        Black, xvi+763 pp.
FOSTER, J. B.
  1965. The evolution of the mammals of the Queen Charlotte Islands, British Columbia.
        Occas. Papers British Columbia Prov. Mus., 14:1-130.
GENOWAYS, H. H.
  1973. Systematics and evolutionary relationships of spiny pocket mice, genus Liomys.
        Spec. Publ. Mus., Texas Tech Univ., 5:1-368.
GILL, T., and E. COUES
  1877. Material for a bibliography of North American mammals. Pp. 951-1081, in Coues
        and Allen, U.S. Geol. Surv. Territories, 11:x+1-1091.
GLASS, B. P., and R. J. BAKER
  1968. The status of the name *Myotis subulatus* Say. Proc. Biol. Soc. Washington, 81:
        257-260.
GRASSÉ,P.-P. (ed.)
  1955. Mammifères: les ordres, anatomie, éthologie, systématique. Traité de zoologie,
        17:1-2300.
GREGORY, J. T., J. A. BACSAI, B. BRAJNIKOV, and K. MUNTHE
  1973. Bibliographies of fossil vertebrates, 1969-1972. Mem. Geol. Soc. Amer., 141:
        1-733.

GRODZINSKI, W., B. BOBEK, A. DROZDZ, and A. GORECKI
  1970. Energy flow through small rodent populations in a beech forest. Pp. 291-298, *in* Energy flow through small mammal populations (K. Petrusewicz and L. Ryszkowski, eds.), Warsaw.
GRZIMEK, B. (ed.)
  1972. Grzimek's animal life encyclopedia. Vol. 10. Mammals I. Van Nostrand Reinhold, New York, 627 pp.
HALL, E. R.
  1926. Changes during growth in the skull of the rodent Otospermophilus grammurus beecheyi. Univ. California Publ. Zool., 21:355-404.
  1946. Mammals of Nevada. Univ. California Press, xi+710 pp.
HALL, E. R., and K. R. KELSON
  1959. The mammals of North America. Ronald Press, 1:xxx+1-546+79 and 2:viii +547-1083+79.
HAMILTON, W. J., III
  1962. Reproductive adaptations of the red tree mouse. Jour. Mamm., 43:486-504.
HARPER, F.
  1927. The mammals of the Okefinokee Swamp region of Georgia. Proc. Boston Soc. Nat. Hist., 38:191-396.
HARRISON, D. L.
  1964. The mammals of Arabia. Ernest Benn Ltd., London, 1:xx+1-192.
HENNIG, W.
  1966. Phylogenetic systematics. Univ. Illinois Press, Urbana, 263 pp.
HESSE, R., W. C. ALLEE, and K. P. SCHMIDT
  1937. Ecological animal geography. John Wiley and Sons, xiv+597 pp.
HIBBARD, C. W.
  1950. Mammals of the Rexroad Formation from Fox Canyon, Kansas. Contrib. Mus. Paleo., Univ. Michigan, 8:113-192.
HILL, W. C. O.
  1953. Primates—comparative anatomy and taxonomy. I—Strepsirhini. Univ. Press, Edinburgh, xxiii+798 pp.
HINDE, R. A.
  1970. Animal behavior. . . . 2nd ed. McGraw-Hill, xvi+876 pp.
HOOPER, E. T.
  1952. A systematic review of the harvest mice (genus *Reithrodontomys*) of Latin America. Misc. Publ. Mus. Zool., Univ. Michigan, 77:1-255.
HOWELL, A. B.
  1926. Anatomy of the woodrat. Monogr. Amer. Soc. Mamm., 1:x+1-225.
  1930. Aquatic mammals. . . . Charles C. Thomas, xii+338 pp.
HUXLEY, J.
  1932. Problems of relative growth. Dial Press, xix+276 pp.
  1943. Evolution—the modern synthesis. Harper and Brothers, 645 pp.
JACKSON, H. H. T.
  1928. A taxonomic revision of the American long-tailed shrews (genera *Sorex* and *Microsorex*). N. Amer. Fauna, 51:vi+1-238.
  1961. Mammals of Wisconsin. Univ. Wisconsin Press, xii+504 pp.
JOHNSON, D. H., M. D. BRYANT, and A. H. MILLER
  1948. Vertebrate animals of the Providence Mountains area of California. Univ. California Publ. Zool., 48:221-376.
JONES, J. K., JR.
  1964. Distribution and taxonomy of mammals of Nebraska. Univ. Kansas Publ., Mus. Nat. Hist., 16:1-356.
JONES, J. K., JR., and S. ANDERSON (eds.)
  1970. Readings in mammalogy. Monogr. Mus. Nat. Hist., Univ. Kansas, 2:ix+ 586 pp.
KAUFMAN, D. W.
  1974. Adaptive coloration in *Peromyscus polionotus*: experimental selection by owls. Jour. Mamm., 55:271-283.
KAYSER, C.
  1961. The physiology of natural hibernation. Pergamon Press, vii+325 pp.
KINGDON, J.
  1971. East African mammals. . . . Academic Press, London and New York, 1:x+1-446.
KLEVEZAL, G. Z., and S. E. KLEINENBERG
  1969. Age determination of mammals from annual layers in teeth and bones (translated from Russian edition, 1967). TT 69-55033, U.S. Dept. Interior and National Science Foundation, Washington, D.C., iii+128 pp.
KLINGENER, D.
  1964. The comparative myology of four dipodoid rodents (genera *Zapus, Napaeozapus, Sicista,* and *Jaculus*). Misc. Publ. Mus. Zool., Univ. Michigan, 124:1-100.

637

Krebs, C. J.
  1972.   Ecology. . . . Harper and Row, x+694 pp.
Layne, J. H.
  1960.   The growth and development of young golden mice, *Ochrotomys nuttalli*. Quart. Jour. Florida Acad. Sci., 23:36-58.
  1966.   Postnatal development and growth of *Peromyscus floridanus*. Growth, 30:23-45.
Lidicker, W. Z., Jr.
  1960.   An analysis of intraspecific variation in the kangaroo rat *Dipodomys merriami*. Univ. California Publ. Zool., 67:125-218.
Lowery, G. H., Jr.
  1974.   The mammals of Louisiana and its adjacent waters. Louisiana State Univ. Press, xxiii+565 pp.
Lyne, A. G., and A. M. W. Verhagen
  1957.   Growth of the marsupial *Trichosurus vulpecula*, and a comparison with some higher mammals. Growth, 21:167-195.
MacArthur, R. H., and E. O. Wilson
  1967.   The theory of island biogeography. Monogr. Pop. Biol., Princeton Univ. Press, 1:xi+1-203.
McKenna, M. C.
  1973.   Sweepstakes, filters, corridors, Noah's arks, and beached Viking funeral ships in paleogeography. Pp. 295-308, *in* Implications of continental drift to earth sciences (D. H. Tarling and S. K. Runcorn, eds.), vol. 1, Nato Study Inst., Academic Press.
Maglio, V. J.
  1973.   Evolution of mastication in the Elephantidae. Evolution, 26:638-658.
  1973.   Origin and evolution of the Elephantidae. Trans. American Phil. Soc. new series, 63(3):1-149.
Marler, P. and W. J. Hamilton III
  1966.   Mechanisms of animal behavior. John Wiley and Sons, xi + 771 pp.
Matthew, W. D.
  1030.   Climate and evolution. Spec. Publ. New York Acad. Sci., 1:xii+1-223.
Mayr, E.
  1963.   Animal species and evolution. Harvard Univ. Press, xiv+797 pp.
  1969.   Principles of systematic zoology. McGraw-Hill, xi+428 pp.
Mayr, E., E. G. Linsley, and R. L. Usinger
  1953.   Methods and principles of systematic zoology. McGraw-Hill, ix+336 pp.
Mearns, E. A.
  1907.   Mammals of the Mexican boundary of the United States. Bull. U.S. Nat. Mus., 56:xv+1-530.
Miller, G. S., Jr., and R. Kellogg
  1955.   List of North American Recent mammals. Bull. U.S. Nat. Mus., 205:xii+1-954.
Mörzer Bruyns, W. F. J.
  1971.   Field guide of whales and dolphins. Uitgeverij tor/n.v. Uitgeverij v.h. C. A. Mees, Amsterdam, 258 pp.
Musser, G. G.
  1968.   A systematic study of the Mexican and Guatemalan gray squirrel, *Sciurus aureogaster* F. Cuvier (Rodentia: Sciuridae). Misc. Publ. Mus. Zool., Univ. Michigan, 137:1-112.
Ognev, S. I.
  1962.   Mammals of eastern Europe and northern Asia (translated from Russian edition, 1928). Available from Nat. Tech. Information Service, Springfield, Va. 22161, xii+487+XV pp.
Osgood, W. H.
  1909.   Revision of the mice of the American genus Peromyscus. N. Amer. Fauna, 28:1-285.
Packard, R. L.
  1960.   Speciation and evolution of the pygmy mice, genus Baiomys. Univ. Kansas Publ., Mus. Nat. Hist., 9:579-670.
Pallas, P. S.
  1778.   Novae species quadrupedum e glirium ordine. . . . Erlangae, 388 pp.
Palmer, R. S.
  1954.   The mammal guide. Doubleday and Co., 384 pp.
Pearson, O. P.
  1958.   A taxonomic revision of the rodent genus Phyllotis. Univ. California Publ. Zool., 56:391-496.
Peterson, R. L.
  1966.   The mammals of eastern Canada. Oxford Univ. Press, xxxii+465 pp.
Pocock, R. I.
  1924.   The external characters of the pangolins (Manidae). Proc. Zool. Soc. London, pp. 707-723.

PRATER, S. H.
    1965. The book of Indian mammals. Bombay Natural History Soc. and Prince of Wales Mus. of Western India., 2nd ed., revised, xxii+323 pp.
RHEINGOLD, H. L. (ed.)
    1963. Maternal behavior in mammals. John Wiley and Sons, viii+349 pp.
RICKLEFS, R. E.
    1973. Ecology. Chiron Press, x+861 pp.
RIDE, W. D. L.
    1970. A guide to the native mammals of Australia. Oxford Univ. Press, Melbourne, xiv+249 pp.
RINKER, G. C.
    1954. The comparative myology of the mammalian genera *Sigmodon, Oryzomys, Neotoma,* and *Peromyscus* (Cricetinae), with remarks on their intergeneric relationships. Misc. Publ. Mus. Zool., Univ. Michigan, 83:1-124.
ROMER, A. S.
    1966. Vertebrate paleontology. Univ. Chicago Press, 3rd ed., ix+468 pp.
ROSEVEAR, D. R.
    1965. The bats of West Africa. British Mus. (Nat. Hist.), London, xvii+418 pp.
    1969. The rodents of West Africa. British Mus. (Nat. Hist.), London, xii+604 pp.
ROSS, H. H.
    1974. Biological systematics. Addison-Wesley Publ. Co., Reading, Massachusetts, 345 pp.
ROWLANDS, I. W. (ed.)
    1966. Comparative biology of reproduction in mammals. Symp. Zool. Soc. London, Academic Press, 15:1-559.
ROWLANDS, I. W., and B. J. WEIR
    1974. The biology of hystricomorph rodents. Symp. Zool. Soc. London, Academic Press, 34:1-482.
SADLIER, R. M. F. S.
    1969. The ecology of reproduction in wild and domestic mammals. Methuen, London, 321 pp.
SEARLE, A. G.
    1968. Comparative genetics of coat colour in mammals. Logos Press, xii+308 pp.
SEVERINGHAUS, C. W.
    1949. Tooth development and wear as criteria of age in white-tailed deer. Jour. Wildlife Mgt., 13:195-216.
SIMPSON, G. G.
    1945. The principles of classification and a classification of the mammals. Bull. Amer. Mus. Nat. Hist., 85:xvi+1-350.
    1961. Principles of animal taxonomy. Columbia Univ. Press, xii+247 pp.
SMITH, J. D.
    1972. Systematics of the chiropteran family Mormoopidae. Misc. Publ. Mus. Nat. Hist., Univ. Kansas, 56:1-132.
SNEATH, P. H. A., and R. R. SOKAL
    1973. Numerical taxonomy. W. H. Freeman and Co., San Francisco, xv+573 pp.
SMITHERS, R. H. N.
    1966. The mammals of Rhodesia, Zambia and Malawi. Collins, London, 159 pp.
    1971. The mammals of Botswana. Mus. Mem., Nat. Mus. Rhodesia, 4:xi+340.
STOLL, N. R. (ed.)
    1964. International code of zoological nomenclature. . . . Internat. Trust Zool. Nomenclature, London, xx+176 pp.
THOMPSON, D'A. W.
    1942. On growth and form. Cambridge Univ. Press, 2nd ed., 1116 pp.
TROUGHTON, E.
    1965. Furred animals of Australia. Halstead Press, 8th ed., xxxii+376 pp.
TURNER, R. W.
    1974. Mammals of the Black Hills of South Dakota and Wyoming. Misc. Publ. Mus. Nat. Hist., Univ. Kansas, 60:1-178.
UDVARDY, M. D. F.
    1969. Dynamic zoogeography. . . . Van Nostrand Reinhold Co., xviii+445 pp.
VAN DEN BRINK, F. H.
    1967. A field guide to the mammals of Britain and Europe. Collins, 221 pp.
VAUGHAN, T. A.
    1966. Morphology and flight characteristics of molossid bats. Jour. Mamm., 47:249-260.
    1972. Mammalogy. W. B. Saunders Co., Philadelphia, viii+463 pp.
WALKER, E. P., and associates
    1964. Mammals of the world. Johns Hopkins Press, 3 vols.
WEBER, M.
    1927-28. Die Säugetiere. . . . Fischer, 2nd ed., 2 vols. (1:xv+1-444 and 2:xxiv+1-898).

WIGHT, H. M., and C. H. CONAWAY
    1962.   A comparison of methods for determining age of cottontails. Jour. Wildlife Mgt., 26:160-163.
WILSON, E. O.
    1975.   Sociobiology. The new synthesis. Harvard Univ. Press, Cambridge, Massachusetts, ix+697 pp.
WILSON, R. W.
    1960.   Early Miocene rodents and insectivores from northeastern Colorado. Univ. Kansas Publ., Paleo. Contrib. (Vertebrata), 7:1-92.
    1972.   Evolution and extinction in early Tertiary rodents. 24th International Geological Congress, Sec. 7, pp. 217-224.
WOOD, A. E.
    1959.   Are there rodent suborders? Syst. Zool., 7:169-173.
YABLOKOV, A. V.
    1974.   Variability of mammals (translated from Russian edition, 1966). Available from Nat. Tech. Information Service, Springfield, Va. 22161, xvi+350 pp.
YOUNG, J. Z.
    1957.   The life of mammals. Oxford Univ. Press, xv+820 pp.
ZITTEL, K. A.
    1891-93.   Handbuch der Palaeontologie. Band 4, Vertebrata (Mammalia). R. Oldenbourg, xi+799 pp.